The Dictionary of Science for Gardeners

The Dictionary of Science for Gardeners

6000 Scientific Terms Explored and Explained

Michael Allaby

TIMBER PRESS
Portland, Oregon

Published in 2015 by Timber Press, Inc.

The Haseltine Building
133 S.W. Second Avenue, Suite 450
Portland, Oregon 97204-3527
timberpress.com

Printed in China
Text design by Patrick Barber
Cover design by Anna Eshelman

Library of Congress Cataloging-in-Publication Data

Allaby, Michael, author.
The dictionary of science for gardeners : 6000 scientific terms explored and explained / Michael Allaby. -- First edition.
pages cm
Includes bibliographical references and index.
ISBN 978-1-60469-483-3
1. Science--Dictionaries. 2. Gardening--Dictionaries. I. Title.
Q123.A498 2015
503--dc23 2015006909

A catalog record for this book is also available from the British Library.

Contents

Introduction

A garden is a place of delight. Conceived in the imagination of the gardener and realised through hours of toil in rain and shine, its produce feeds the body and its shapes, colours, and structures feed the soul. It is a paradise in miniature.

It is also a haven for wildlife. Flower petals are coloured, after all, in order to attract pollinating insects. And where there are insects there will be animals that feed on insects, and on surplus seeds, and on each other. The well-tended soil that garden plants require teems with living organisms, a world of its own inhabited by beings, most too small to be visible, that feed on detritus, and the terrifying monsters that hunt them.

The modern garden has evolved, at least in part, from gardens that were not intentionally ornamental. Monastic gardens grew plants with medicinal properties and these became physic gardens, and the gardeners were often apothecaries, the ancestors of our pharmacists. The Chelsea Flower Show is held in what was once known as the Apothecaries' Garden. It was where trainee apothecaries learned their botany. Physicians, administers of physic, were also trained in botany. Carolus Linnaeus, perhaps the world's most famous botanist, was a physician with a highly successful practice in Stockholm that gave him the time and money to further his studies of plants. He followed Olof Rudbeck (*Rudbeckia* bears his name) as director of the Botanic Garden at the University of Uppsala, which he rearranged and where he developed the system of biological classification that we use to this day. His garden is still there, and open to visitors. Botanic gardens exist, and have always existed, for purposes of scientific research and education, and nowadays for plant conservation. The fact that members of the public enjoy them is a bonus.

Gardens, then, are real. But not so many decades ago many scientists thought the modern private garden wholly artificial, merely a contrived collection of cultivated varieties of plants that had been selectively bred purely for their appearance, resistance to disease, and ability to thrive locally. Garden plants were far removed genetically from their wild ancestors, the 'proper' plants. A garden was aesthetically pleasing perhaps, but it was of no botanical or ecological interest. Scientists conducted their fieldwork in ancient woodlands, natural grasslands, and alpine meadows, and scorned orchards, plantations, lawns, and rockeries.

Gardeners, of course, were not in the least troubled by this and probably were unaware of it. They gardened for pleasure and their skills were entirely practical. It was enough to know that a particular place was a frost hollow without knowing why it was, no knowledge of biogeography was needed to understand which plants were hardy and which were not, and gardeners knew the invertebrate animals that would attack their plants and the best ways of dealing with them. Seed catalogues and garden centres labelled plants with names that were familiar and no one but a specialist needed to know about the evolutionary relatives of any particular plant, or even how it came by its name.

Scientists began to take gardens more seriously when ecologists became concerned about the threats to wildlife that arose mainly from the intensification of arable farming. Meadows, a riot of flowers in summer, were being ploughed up and sown to much more productive temporary pastures containing a small number of grass species, where wild flowers were regarded as weeds. Hedges were being grubbed out to make larger fields more suited to modern farm machines, fields where combine harvesters could work three or more abreast. Everywhere, it seemed, natural habitats were disappearing or, more commonly, being degraded.

That was when scientists realised that ordinary gardens are not isolated entities. They offer privacy to their owners, and may be enclosed by hedges or fences, but far from being barriers to small animals, hedges provide cover, fences are perches and singing posts, and both can be tunnelled beneath or flown over. What is more, the gardens are adjacent to each other on either side of these imaginary boundaries, together occupying truly vast areas in urban settings. Scientists also noted that these entirely artificial gardens provide food and nesting sites to a wide range of species, including those being driven from farmland. Gardens are viable ecosystems growing many different plants, exotic it is true but edible to wildlife nonetheless, and they are already contributing to the conservation of species. *Buddleja*, native to tropical and warm climes and entirely alien to northern gardens, is now known as the butterfly bush. Clearly, gardens had to be taken seriously.

Excellent conservation areas though they are, gardens could be improved ecologically and gardeners soon found themselves being exhorted to provide small areas of enhanced habitat to attract hedgehogs, for example, and other small animals. They were advised about the preferred food plants for caterpillars, how to extend the growing season for flowers that supplied pollen for bees and other insects, and the desirability of wet areas and ponds to attract frogs and newts. They were warned not to provide so many nesting boxes that the surrounding area would provide too little food for the birds using them, so the young would starve. In TV programmes and magazine articles, gardeners were urged not to buy peat-based composts and soil conditioners because the industrial-scale mining of peat was destroying valuable habitat.

All of this advice and information drew on the findings of years of scientific research. Increasingly, and whether they were aware of it or not, gardeners were being invited to modify their methods and objectives, to introduce elements of applied science. Gardening books and magazines began to include scientific information. The material was treated lightly, but it encouraged readers to penetrate to the real science behind the simplification. Gardeners were being exposed to scientific writing. This is a trend that is likely to continue and accelerate, and as it does so amateur gardeners will encounter more and more scientific terms.

Many of these terms are likely to be unfamiliar, possibly intimidating, and some are misleading because they also have an everyday meaning that is subtly different. So what do you do when confronted with a strange vocabulary? You reach for a dictionary. The problem is that until now there has been no appropriate up-to-date dictionary of a convenient size available at an affordable price. This book aims to fill that gap.

How do you set about compiling a dictionary? I began by listing the branches of science gardeners might find relevant, and I found 16. They include plant classification, the science of how and why plants are grouped into genera and families, plant geography or how the world breaks down floristically, plant evolution, with the genetic code as an appendix, plant structure and function, or how plants work, fungi, insects, other invertebrate animals, vertebrate animals, bacteria and viruses, the way major nutrients move through cycles, pesticides, soil science including the way soils are classified, ecology, conservation, and weather and climate. I prioritized these, allotting a proportion of the dictionary to each, and then I commenced the writing and compilation.

Illustrations appear only where I judged that they helped explain the meaning of a word or phrase. In many definitions it is impossible to explain the meaning without using other technical terms that also demand explanation. I have written such terms **like this** to indicate that they are defined in their own right. Many of the entries include this icon ↗. It directs you to a list of web addresses at the end of the dictionary where you can find more detailed explanations.

This is not a dictionary of gardening. It does not explain terms familiar to every gardener. I have not defined *spade* or *lawn*, *bedding plant*, *trellis*, or *raised bed*. Nor is it a textbook. It will not tell you how to cultivate your garden. But it will help you with articles and books that describe the scientific background to modern gardening. Will it make you a better gardener? How am I to know? But it may throw an interesting light on what is happening above and below that surface of soil.

MICHAEL ALLABY
Tighnabruaich, Argyll, Scotland
michaelallaby.com

A

abaca (*Musa textilis*) *See* Musaceae.

abaptation The ability of a species to prosper in a particular **environment** because of characteristics it has inherited.

abaxial Directed away from the **axis**.

ABC soil A soil in which the upper three (A, B, and C) **soil horizons** are clearly defined.

Abelmoschus esculentus (okra) *See* Malvaceae.

abiotic Non-living.

ablation The removal from the ground of ice and snow by melting and **sublimation**.

abrasion (corrasion) The erosive effect of dragging rock particles across, or impacting them against, a solid surface.

abscisic acid (abscisin II, dormin) A **terpenoid** that is one of the five major **plant hormones**. It is synthesized primarily in the **chloroplasts** and occurs throughout the plant body, especially in the leaves, fruits, and seeds. It inhibits growth, promotes leaf **abscission** and **senescence**, and induces seed and **bud dormancy** and the closing of **stomata**.

abscisin II *See* abscisic acid.

abscission The rejection by a plant of one of its organs, e.g. leaves.

absolute humidity The mass of water vapour present in a specified volume of air, expressed in grams per cubic meter. *See* humidity.

absolute porosity The proportion of the volume of a medium, e.g. rock or soil, that consists of **pore** spaces. The spaces are not necessarily connected and not all may be capable of holding or transmitting fluids. *Compare* effective porosity, porosity.

absorption The uptake of substances by cells or tissues.

abstriction The release of a **spore** by the constriction of the tissue to which it is attached.

Abutilon (family **Malvaceae**) A genus of of **annual** or **perennial herbs**, or **evergreen** or **deciduous shrubs** or small **trees**. The plants are **pubescent**, their leaves **simple**, **alternate**, and **palmate**. The flowers are pendulous, resembling Chinese lanterns, and have five petals. The **stigmas** are usually apical. The fruit is a **schizocarp**. There are about 150 species native to warm regions of the Western Hemisphere, especially South America. Many species are cultivated for ornament. Some are food for butterfly and moth **caterpillars**.

Abutilon* yellows** A viral disease, possibly caused by a species of ***Crinivirus that affects ***Abutilon*** species and other members of the **Malvaceae**. It is transmitted by the banded-wing whitefly (*Trialeurodes abutilonea*).

Abyssinian banana (*Ensete ventricosa*) *See* Musaceae.

Ac *See* altocumulus.

Acacia (family **Fabaceae**, subfamily **Mimosoideae**) A genus mainly of **trees** and **shrubs**, with a few climbers, native to Australia, where there are nearly 1000 species (wattles); the Australian national flower is the golden wattle (*A. pycnantha*). Until 2003, about 350 species of closely related plants occurring in dry tropics and subtropics of Africa and America (thorn trees) were included in the genus. Scientists then recognized that the Australian and African-American species should be separated. The name *Acacia* has been retained for 948 Australian species and 7 species occurring in the Pacific Islands, with 1 or 2 in Madagascar and 10 in Asia, are now called *Vachellia*. Names have not yet been agreed for the African-American species. Leaves are **bipinnate** or phyllodic (*see* phyllode), many (though fewer of the Australian species) bearing thorns, and in some species leaves are swollen and inhabited by ants. Many species are important commercially for timber, perfumes, and other products.

acacia ant (*Pseudomyrex ferruginea*) *See* co-adaptation.

Acacia hindsii *See* co-adaptation.

Acalymma vittatum (striped cucumber beetle) *See* bacterial wilt.

Acalypha (family **Euphorbiaceae**) A genus of **dioecious shrubs**, **trees**, and **annual herbs** with **apetalous** flowers. Female *A. hispida* (cat's tail, red-hot cattail, chenille plant) bears long, purple to bright red **catkins** and is widely cultivated outdoors or as a houseplant. Other species are cultivated for their attractive foliage. Cats find the roots of *A. indica* irresistible. There are 450–500 species native to the tropics and subtropics, with a few temperate.

Acanthaceae (order **Lamiales**) The acanthus family, comprising mostly **annual** or **perennial herbs**, **shrubs**, and climbers, but also some large trees, including several mangroves. Leaves are **opposite** and usually **entire**, without **stipules**, and spiny in some species. In the tribe Acantheae and in the subfamilies Nelsonioideae and Thunbergioideae the leaves have **cystoliths** that are visible with a magnifying glass. The **bracts** are often showy. Flowers are **gamopetalous** and **zygomorphic** with 4 or 5 fused **sepals** and **petals**; most have a 2-lipped **corolla**, sometimes with the upper lip lacking, and 2 or 4 **epipetalous** stamens. The **ovary** has 2 **carpels**, developing into a **capsule** with 2–16 seeds. There are about 220 genera with about 3000 species, occurring throughout the tropics and subtropics. Many are cultivated as ornamentals (in conservatories or as houseplants in temperate regions). Black-eyed Susan (*Thunbergia alata*) belongs to this family.

Acanthopanax senticosus (Siberian ginseng) *See* Araliaceae.

Acanthus (family **Acanthaceae**) A genus mainly of thorny **xerophytes** with flowers that lack the upper lip. **Nectaries** are outside the flower. There are 30 species, native to the tropics and subtropics. *Acanthus spinosus* (oyster plant) is believed to be the pattern for the decoration of Corinthian capitals. *Acanthus mollis* is bear's breech.

acaricide A pesticide that kills ticks (Acarina).

acaulescent Lacking a stem, or having a stem that is very short or hidden below ground.

accelerated erosion A rate of **erosion** which exceeds the rate that would occur in the absence of land use by humans.

accessory bud A **bud** that forms above, below, or to the side of an **axillary** bud.

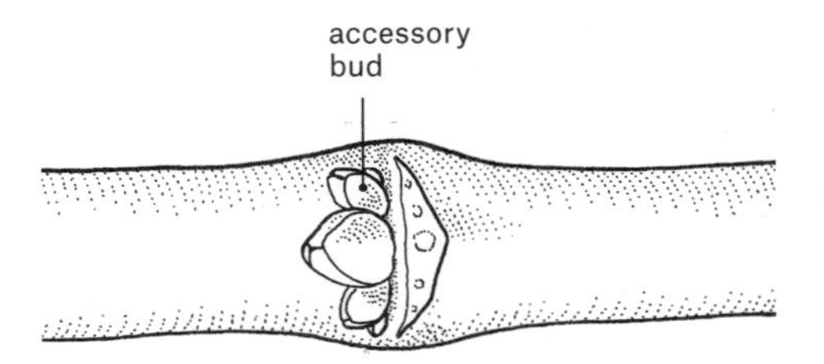

accessory cell *See* subsidiary cell.

accessory cloud A small cloud that is close to or attached to a larger cloud belonging to an identifiable cloud genus (*see* cloud classification).

accessory fruit *See* pseudocarp.

accessory pigments In **photosynthesis**, pigments that absorb light energy and pass it to **primary pigments**.

accidental drought *See* contingent drought.

accidental species In the phytosociological (*see* phytosociology) scheme devised by the school led by Josias **Braun-Blanquet**, one of the five classes of fidelity (*see* faithful species) that describe and classify plant communities. Accidental species occur rarely, as relicts from a previous community or invaders from another community. *Compare* exclusive species, indifferent species, preferential species, selective species.

acclimation 1. A response that allows a plant or animal to tolerate a change in one factor in its environment. **2.** In a plant, hardening. *See* acclimatization.

acclimatization 1. A reversible, physiological response, sometimes involving a behavioural change, that allows a plant or animal to tolerate an environmental change. **2.** (acclimation, hardening) The physiological changes that protect a plant against cold.

accumulated temperature The sum of the extent to which the mean air temperature exceeds or falls below a specified datum level over a prolonged period. If the temperature remains m° above or below (in which case it has a negative value) the datum for n hours (i.e. $n/24$ days) the accumulated temperature for that period is $mn/24$ days. Accumulated temperatures can be calculated for a week, month, or year.

-aceae The suffix that indicates a plant family (e.g. Rosaceae, the rose family).

acellular slime moulds *See* Myxogastria.

Acer (maples; family **Sapindaceae**) A genus of **monopodial** and **sympodial trees** and **shrubs**, most with **entire**, **palmate** leaves, but a few **pinnate** (e.g. box elder, *A. negundo*). Flowers are regular and **pentamerous**, borne in **racemes**, **corymbs**, or **umbels**; the fruit is a **samara**. There are 111 species native to the mountains of Northern Hemisphere tropical and temperate regions. Many are cultivated for timber or ornament (e.g. Japanese maple, *A. palmatum*).

acervulus An asexual structure formed by certain plant-parasitic **Fungi**, comprising a mat of tissue bearing **conidiophores** that erupts through the **epidermis** of the host to release conidia (*see* conidium).

Acetabularia *See* mermaid's cup.

acetamiprid *See* chloronicotinyls.

acetic acid bacterium A **bacterium** that obtains energy by oxidizing **ethanol** to acetic acid (CH_3COOH). Acetic acid **Bacteria** are rod-shaped, aerobic, and Gram-negative (*see* Gram reaction).

acetylglycerols *See* glycerides.

Achariaceae (order **Malpighiales**) A family of five genera and 24–25 species of neotropical **trees** and **shrubs** with **simple**, **alternate**, **entire**, **serrate**, or **dentate** leaves with simple **trichomes**, and flowers that are **pentamerous**, solitary or **inflorescences** of **fascicles**, **racemes**, or **panicles**. The fruit is a **berry** or **capsule**.

Achatocarpaceae (order **Caryophyllales**) A family of small **trees** and **lianas** (snake eyes, limonacho) with fruits that are one-seeded berries. There are two genera. *Achatocarpus* comprises five to ten species occurring in Central and northern South America; *Phaulothamnus* is **monotypic** and native to North America.

Achatocarpus *See* Achatocarpaceae.

achene A small, dry, **indehiscent** fruit, often small.

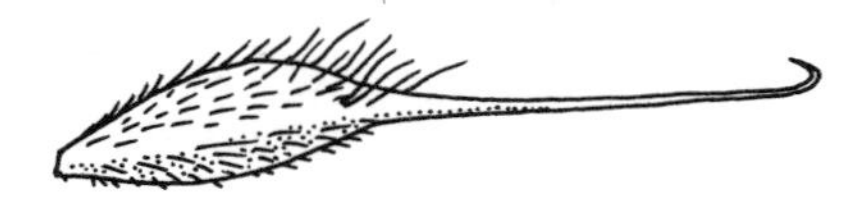

Achene fruit (*Geum urbanum*).

Achras The former generic name for plants now placed in the genus ***Manilkara***.

A-chromosome A member of the normal set of **chromosomes**.

Acicula fusca (point snail) A minute **snail** with an **operculum** and a conical shell up to 1 mm long that inhabits litter in undisturbed, **broad-leaved deciduous** woodland, sea cliffs, and among mosses (**Bryophyta**) on roadside banks. It has a scattered distribution in Europe and is endangered through loss of **habitat**.

acicular Pointed, like a needle.

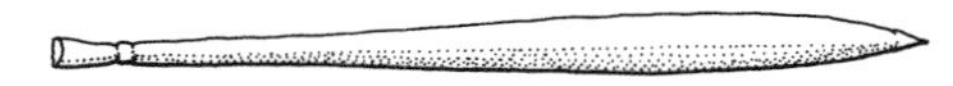

An acicular leaf.

acid A substance that in solution releases hydrogen **ions** (protons), and that acts as an electron acceptor. It has a **pH** of less than 7.0 and reacts with a **base** to yield a **salt** and water.

acidic rock **Igneous** rock that contains more than about 60 percent **silica** by weight.

acid mist A form of **acid precipitation** that coats the leaves of plants and is more harmful than **acid rain**.

acidophile *See* extremophile.

acid precipitation **Precipitation** that has a **pH** lower than about 5.0, which is the average pH of water droplets in clouds. The increase in acidity may be natural, e.g. from volcanic gases, or due to emissions most commonly resulting from the burning of **fossil fuels**. *See* acid mist, acid rain, dry deposition.

acid rain A form of **acid precipitation** that evaporates or runs off plant surfaces, limiting its harmful effects, but that may reach the soil.

acid soil Soil in which the **pH** is less than 7.0. If the pH is less than 5.0 the soil is considered to be very **acid**.

Acleris comariana (strawberry tortrix moth) A species of tortrix moths (**Tortricidae**) that vary in colour from dark brown to orange-brown, often with pale markings, and with a wingspan of 13–18 mm. **Caterpillars** are green and up to 15 mm long. The moth overwinters as eggs mainly on the underside of strawberry leaves, hatching in spring when the caterpillars feed on the opening leaves, spinning webs that sometimes extend to blossoms and developing fruits, also feeding in the flowers. They pupate between spun leaves or in folded leaves. There are two generations a year. Infestations can cause severe **defoliation**. The moth occurs throughout Europe and in China, Japan, and North America.

Acoela A class of soft-bodied, unsegmented worms that lack a **coelom**, hindgut, and anus. Most are free-living and move by means of cilia (*see* cilium) but some live as commensals (*see* commensalisms) or symbionts (*see* symbiosis) of **green algae (Chlorophyta)**.

acoelomate Lacking a **coelom**.

Acoraceae (order **Acorales**) A family of sweet-smelling **monocotyledons** (calamus, sweet flag), with two-ranked, **ensiform** leaves that are **equitant** and oriented edge-on to the stem (isobifacial). The **peduncle** has two separate vascular systems. There are no **bracts** or **bracteoles**. Flowers are small, **perfect**, usually **trimerous** in a densely **spicate inflorescence**

A

overtopped by a leaf-like **spathe**. The **superior** ovary has 2–3 **carpels**. The fruit is a **capsule** or **berry**. There is one genus (***Acorus***) with two to four species found from eastern North America to east and south Asia, and possibly **naturalized** elsewhere in North America and in Europe.

Acorales An order of **monocotyledons** in which the **inflorescence** is a **spadix** with a large **spathe** and sessile flowers. The order contains one family (**Acoraceae**).

Acorus (calamus, sweet flag; family **Acoraceae**) A genus of two to four species of sweet-smelling **monotocyledons** found in marshy **habitats**. **Rhizomes** of *A. calamus* have long been used medicinally and the oil (calamus oil) is used in perfumery.

acrasin The chemical substance that induces **chemotaxis** in the **Myxamoebae** of **Acrasiomycetes**.

Acrasiomycetes (cellular slime moulds) A class of slime moulds (**Myxogastria**) in which the feeding stage usually consists of **myxamoebae**, each with a single nucleus, which usually aggregate into a **pseudoplasmodium** prior to producing **spores**. They occur in soil and decaying plant material.

acre A unit of land area equal to 0.4047 hectares (1 ha = 2.471 acres).

acre-foot The volume of water that would submerge 1 **acre** of a level surface to a depth of 1 foot. It amounts to 43,560 ft^3 = approximately 1233.48 m^3 = approximately 271,328 imperial gallons = approximately 325,853 U.S. gallons.

Acrididae (locusts, short-horned grasshoppers) A family of medium to large grasshoppers (**Orthoptera**), which are active by day. Wings are well developed in most species and the insects move short distances by jumping. They inhabit open country. There are about 10,000 species with a **cosmopolitan distribution**. All are **herbivores** and locusts are serious pests.

Acris crepitans (northern cricket frog) A **diurnal**, grey or pale brown frog with a lumpy skin, dark bands on the legs, and a triangular mark on the head, 20–40 mm long, that is a terrestrial member of the tree frog family (Hylidae). It lives on the banks of ponds and streams and feeds on insects. It occurs throughout the central and eastern United States.

Acris gryllus (southern cricket frog) A frog very similar to ***Acris crepitans***, but smaller, 15–33 mm long, and more slender. It lives in shallow ponds adjacent to meadow, and in marshes and **bogs**, and feeds on insects, especially mosquitoes. It occurs in the temperate southeastern United States.

acrisols **Acid soils** with an argic B horizon (*see* argic horizon) and a **cation exchange capacity** of less than 24 $cmol_c$/kg. Acrisols comprise a reference soil group in the **World Reference Base for Soil Resources**.

acrocarpous Describes a moss (**Bryophyta**) that has an erect stem bearing the archegonia (*see* archegonium) contained in a capsule at the tip.

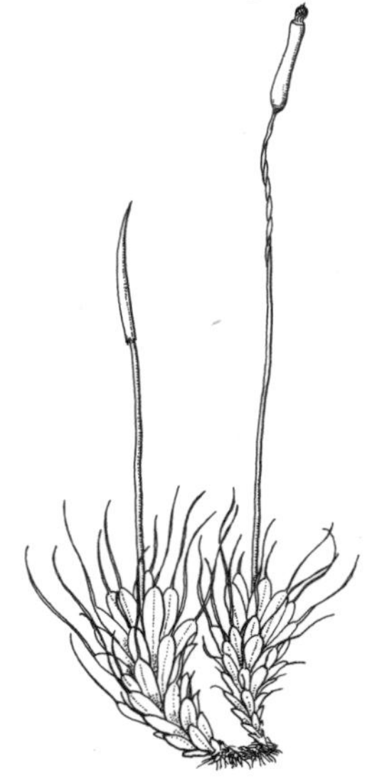

An acrocarpous moss. Note the archegonia borne in capsules at the tips of the stems.

acrocentric Describes a **chromosome** in which the **centromere** is closer to one end than to the other. *See also* holocentric, metacentric, telocentric.

acropetal Growing upward from the point of attachment or base. *Compare* basipetal.

acrotelm The upper layer of a **peat bog**, where organic material decomposes aerobically and more rapidly than in the lower **catotelm**.

AC soil A soil that possesses only an A and C **soil horizon**.

Actinidiaceae (order **Ericales**) A family of **trees**, **shrubs**, and woody **lianas** (Kiwi vine, Chinese gooseberry). Young growth is frequently hairy. Leaves are **exstipulate**, **alternate**, **simple**, the margins **entire** or **serrate**. The **axillary inflorescences** are **cymose** with a few up to 500 unisexual flowers that are **monoecious**, **dioecious**, or **hermaphrodite** depending on genus. Flowers are usually **pentamerous** and **actinomorphic**. There are usually five **sepals**, free or **adnate** to the petals. The petals are usually white, but sometimes red or brown-yellow. There are 10–15 or many **stamens**, the **ovary** is **superior** with 3–5 **carpels**, usually with 5 **locules** but sometimes 3 or up to 20. The fruit is a **berry** or **capsule**. The plants are native to Central and northern South America, temperate central and southeastern Asia. There are 3 genera with 355 species. *Actinidia chinensis* is widely cultivated as the Chinese gooseberry or Kiwi fruit (so named because it was first commercialized in New Zealand).

Actinobacteria A **phylum** of Gram-positive (*see* Gram reaction) **Bacteria** found in freshwater and marine environments and in soils, where they are among the most common organisms, playing an important part in the decomposition of organic material. Many produce branching **filaments** resembling fungal mycelia (*see* mycelium) and were formerly mistaken for **Fungi** and known as actinomycetes.

actinomorphic (regular) Radially symmetrical.

Actinomycetales An order of **Actinobacteria** that are either rod-shaped or filamentous, resembling mycelia (*see* mycelium). They are aerobic, **pleomorphic**, non-**motile**, and Gram-positive (*see* Gram reaction). Some are free-living, others commensals (*see* commensalisms), and some are pathogens.

actinomycetes *See* Actinobacteria.

actinorrhiza *See* root nodule.

actinorrhizal plants *See Frankia.*

actinostele A **monostele** type of **protostele** in which the **xylem** is lobed or star-shaped in cross-section.

Actinotus (family **Apiaceae**) A genus of comprising about 20 species of plants, most native to Australia with 2–3 native to New Zealand. *Actinotus helianthi* is the flannel flower, a **herb** or **shrub** which is common around Sydney and widely cultivated for ornament. The leaves are **alternate**, deeply lobed, grey, and velvety in texture (hence the common name). The flowers are 12–20 mm across and in **umbels** 2.5–8 cm across. *Actinotus schwarzii* (desert flannel flower) occurs only in parts of the Northern Territory and is classed as vulnerable.

activation energy (energy of activation) The energy that must be supplied to a system in order to initiate a reaction.

active absorption The **absorption** of water by plant roots with the help of energy derived from metabolic processes in the root cells.

active acidity The concentration of hydrogen **ions** present in the soil water solution. *Compare* reserve acidity.

A

active chamaephyte A chamaephyte that produces **buds** on horizontal stems.

active front A weather **front** that produces much cloud and precipitation.

active layer The surface layer above **permafrost** that thaws during summer. It is up to 3 m deep.

active pool That part of a **biogeochemical cycle** in which an **essential element** is moving rapidly between living organisms and non-living components, e.g. water, air, or mineral. *Compare* reservoir pool.

active site The part of an **enzyme** molecule that binds to the **substrate**.

active transport The movement of substances across a membrane against a concentration gradient.

actual evapotranspiration The amount of water that evaporates from surfaces and by **transpiration** if the supply of water is restricted.

aculeate Pointed or with prickles.

Aculus schlechtendali (apple rust mite) An orange brown mite (order Acarina) up to 0.18 mm long as an adult that overwinters beneath loose **bark** or **bud** scales on apple trees, emerging in spring to feed on opening flower buds, flowers, and leaves. The mites lay eggs on flower and leaf buds leading to a second generation that emerges in May. Generations overlap through summer, with eggs growing to adults in one to two weeks.

acuminate Tapering to a point.

Acyrthosiphon malvae (geranium aphid, pelargonium aphid) An aphid (**Aphididae**) that feeds on members of at least 25 plant families, but is especially harmful on *Pelargonium*, *Geranium*, and ***Fragaria*** (strawberries), transmitting viral infections as well as causing feeding damage. It is distributed worldwide.

Acyrthosiphon pisum (pea aphid) A large aphid (**Aphididae**) up to 4 mm long that is a major pest of **legumes**, feeding on the underside of leaves, **buds**, and pods. Females lay eggs in autumn that hatch the following spring as a generation of females that moult four times, then reproduce, each individual producing 4–12 female **nymphs** a day, which mature in 7–10 days. Adults live for about 30 days. They occur in temperate regions worldwide. The species is a model organism for biologists and its **genome** has been sequenced.

Adanson, Michel (1727–1806) A French naturalist and plant collector who worked at a trading mission in Senegal, where he collected many previously unknown plants. He was the first European to describe the baobab, and its genus, *Adansonia*, is named for him. Adanson returned to Paris in the 1750s with a large collection of plants and seeds, which he described in his *Histoire Naturelle du Sénégal*, publishing the first volume in 1757.

adaptable enzyme *See* inducible enzyme.

adaptation Any characteristic that fits an organism, generally or specifically, to survive in the **environment** it inhabits.

adaptedness The state of being adapted to an environment. *See* adaptation.

adaptive breakthrough The evolutionary acquisition of an **adaptation** that permits a **taxon** to another **adaptive zone**.

adaptive radiation The rapid evolution of taxa (*see* taxon) descended from a single ancestor, each new **taxon** exploiting resources in a different way.

adaptive value *See* fitness.

adaptive zone The abstract concept of a region a **taxon** inhabits by virtue of its **adaptations** together with the environment, **habitat**, or **niche** that it occupies.

adaxial Directed toward the **axis**.

adductor muscle A muscle that draws the limb of an animal toward the centre of the body.

adecticous In **Arthropoda**, possessing **mandibles** that are often reduced, are not articulated, and in most species are not used to escape from the **coccoon**.

adelgids (woolly conifer aphids) A family (Adelgidae) of bugs (**Homoptera**) that are closely related to aphids (**Aphididae**), from which they can be distinguished by the absence of **cornicles** and the fact that they feed on coniferous trees. They reproduce by laying eggs, never asexually. Their **nymphs** are called sistentes. They are covered with a thick layer of wax, resembling wool. There are about 50 species found throughout the Northern Hemisphere but introduced to the Southern Hemisphere.

adelphoparasite A parasite that has a host closely related to itself.

adelphous Describes an **androecium** with fused **filaments**.

adenine *See* DNA.

adenosine A **nucleoside** formed when adenine (*see* DNA) is linked to a **ribose** sugar.

adenosine diphosphate (ADP) A **coenzyme** consisting of **adenosine** linked to a **pentose sugar** backbone and two phosphate groups. With expenditure of energy (in plants through **photosynthesis**), ADP can acquire an additional phosphate group, becoming **adenosine triphosphate**.

adenosine monophosphate (AMP) An **ester** of phosphoric acid and **adenosine**, consisting of a **ribose** backbone linked to adenine and a phosphate group. It can be produced by **hydrolysis** of **adenosine diphosphate**.

adenosine triphosphate (ATP) A **coenzyme** containing three phosphate groups that is the principal energy-transporting compound in all living cells. Processes requiring energy convert ATP to adenosine diphosphate with the release of energy (30.6 kJ/mol at **pH** 7.0). The ATP–ADP reaction is the means by which energy is stored, transported, and supplied.

adhesion The attachment that holds two surfaces together. Water molecules cling to surfaces, wetting them, by adhesion. *Compare* cohesion.

adiabatic cooling and warming The decrease in temperature in a body of air that is rising or increase in a body of air that is subsiding that involves no exchange of energy with the surrounding air. It is due to the molecular effects of decompression as air rises and compression as it subsides.

Adiantum (maidenhair fern; family Pteridaceae) A genus of ferns in which all the fronds are alike, usually with fan-shaped **leaflets** and black **petioles**. The **sori** are borne on the reflexed tips of the leaflets. There are about 200 species occurring widely throughout the moist tropics and warm temperate regions, as far north as southern Britain. Many species are cultivated for ornament.

adnate 1. Describes the close attachment of two organs. **2.** Describes the **gill** of an **agaric** that is attached to the **stipe** by all or most of its width.

adnexed 1. Describes the loose attachment of two organs. **2.** Describes the **gill** of an **agaric** that is attached to the **stipe** by only part of its width.

Adoketophyton A genus, now extinct, of vascular plants (*see* Tracheophyta) that lived in the Early Devonian epoch (416–397.5 million years ago). It had leafless stems, 1–2.5 mm in diameter, with a central cylinder of **xylem** that divided **dichotomously** in three dimensions, and leaf-like appendages.

Adoxaceae (order **Dipsacales**) A family of **herbs**, **shrubs**, and **trees** in which the leaves are **opposite**, the flowers less than 5 mm across, **pentamerous** or (rarely) 4-merous, the **style** short, one **carpel** with one **ovule**, borne in **cymose inflorescences**. The single seed is a **drupe**. There are 4 genera (*Adoxa*, ***Sambucus*** [elder],

A

Sinadoxa, and *Viburnum*) with about 200 species found throughout the temperate regions of the Northern Hemisphere, tropical mountains, and southeastern Australia, but rare in Africa. *Adoxaceae moschatellina* is moschatel.

ADP *See* adenosine diphosphate.

adpressed Pressed close.

adrenal gland An organ in vertebrates that secretes **hormones**.

adrenalin (epinephrine) The **hormone** principally involved in the 'fight-or-flight' response in mammals. Its effects include stimulating **glycogen** breakdown, thereby raising the blood-sugar level, and mobilizing free **fatty acids**.

adret An inclined ground surface that faces toward the equator and, therefore, the noonday sun.

adsorption The attachment of an atom or molecule bearing an electrical charge to a surface bearing an opposite charge.

adsorption complex The soil ingredients, principally **clay** and **humus**, that are capable of adsorbing **ions** and **cations**.

advection The transport of heat by the movement of air or water, usually horizontally.

advection fog **Fog** that forms when warm, moist air is carried across a markedly cooler surface by a wind blowing at 10–32 km/h.

adventitious Arising from an unusual position, e.g. roots that grow from the stem.

adventive polyembryony *See* polyembryony.

aecidiosorus *See* aecium.

aecidiospore *See* aeciospore.

aeciospore (aecidiospore) A fungal **spore** produced by a **rust** fungus in an **aecium**.

aecium (aecidiosorus) A structure, often cup-shaped, formed in plant tissues by certain **rust** fungi. Chains of **aeciospores** are produced at the base of the aecium.

Aegithalos caudatus (long-tailed tit, long-tailed bushtit) A black and white tit (**Paridae**) that is 140 mm long including the long tail, with a wingspan of 160–190 mm. They inhabit forests, hedges, and gardens, and feed on insects. They occur throughout Eurasia.

aeolian Pertaining to the action of wind.

aeration (aerification) Exposure to oxygen, e.g. by bubbling air through a liquid, dissolving oxygen in a liquid, circulating air through soil.

aerenchyma Plant tissue that contains large air spaces.

aerial mycelium That part of a fungal **mycelium** that is held clear of the substrate.

aerification *See* aeration.

aerobe An organism that requires oxygen and cannot survive without air.

aerochory *See* anemochory.

aerogenic Gas-producing, e.g. **Bacteria** that release gas as a by-product of their metabolism of certain substrates.

aerole An area surrounded by **veins** in a leaf with **reticulate vernation**.

aerophore A root that grows upward out of water or waterlogged soil.

aerosol A mixture of liquid and solid particles, e.g. soil, **dust**, and smoke particles, salt crystals, **spores**, etc., that are suspended in the air, falling about 10 cm a day.

aerotaxis A change in the direction of locomotion in response to a change in the concentration of oxygen.

aerotolerant Describes an organism, e.g. **bacterium**, that is normally an **anaerobe**, but that survives exposure to air.

Aesculus (family **Sapindaceae**) A genus of **trees** and **shrubs** with **opposite**, **palmate**

leaves that are often large. Flowers are 4-merous or **pentamerous**, the petals fused into a **corolla** tube, and the **inflorescence** borne as a **panicle**, often showy. The fruit is a large, usually **globose capsule** containing 1–3 seeds (often incorrectly called **nuts**), the **hilum** showing as a large pale scar. There are 13–19 species found in throughout the temperate Northern Hemisphere. The genus includes the North American buckeyes and *A. hippocastanum*, the European horse chestnut. Some species are known as red chestnut and white chestnut.

aestivation (estivation) The arrangement of **sepals** and **petals** in an unopened **flower bud**.

aethalium A large, stalked or **sessile fruiting body** containing masses of **spores** held within a **cortex** that is formed by some members of the **Myxogastria**.

aetherolea *See* essential oil.

Aextoxicaceae (family **Berberidopsidales**) A family of **dioecious trees**, up to 25 m tall, found only in the rain forests of Chile. The leaves are **opposite** and covered in scales, **entire** and without **stipules**. The flower is **actinomorphic**, unisexual, and born in **racemes**. The fruit is a single-seeded **drupe** resembling a small olive, hence the common name, olivillo. There is one genus and one species, *Aextoxicon punctatum*.

aflatoxins A group of at least 14 toxins produced by ***Aspergillus** flavus* and *A. parasiticus*, parasitic **Fungi** that infest a range of crops including cereals. They are among the most poisonous natural substances known.

African-Indian desert floral region North Africa (Sahara) eastward to the Rajasthan Desert, India, part of the **Palaeotropical region**, containing about 50 **endemic** species. *Phoenix dactylifera* (date palm) occurs here.

African violet (*Saintpaulia*) *See* Gesneriaceae; East African steppe floral region.

Afro-alpine vegetation Shrubland and grassland that occur above the **elfin woodland** on African mountains.

agamospermy *See* apomixis.

agaric 1. A fungus belonging to the class **Agaricomycetes**. **2.** Any fungus with a **fruiting body** resembling a mushroom.

Agaricomycetes (Homobasidiomycetes) A class of filamentous **Fungi** that produce **basidiocarps**, multicellular **fruiting bodies** that include mushrooms, toadstools, puffballs, and **bracket fungi** but range in size from those a few millimetres across to those of *Rigidoporus ulmarius*, which weigh more than 300 kg. Honey fungus (***Armillaria***) belongs to this class. There are about 16,000 species found worldwide.

Agaricomycotina One of the three major **clades** of **Basidiomycota**, comprising about 20,000 species of **Fungi** that are either **saprotrophs** or live in close association with plants, animals, and other fungi. Many live in mycorrhizal associations (*see* mycorrhiza) with plants. The group includes most of the edible fungi and some that cause illness.

Agaricostilbomycetes A class of **Fungi** in the subphylum **Pucciniomycotina**, comprising organisms that go through a **yeast** stage in their life cycle. There are 10 genera with 47 species.

Agaricus arvensis (horse mushroom) A species of **basidiomycete fungi** in which the **fruiting body** has a yellowish white **pileus** up to 250 mm across and white **gills** that later turn grey and finally brown. It occurs in grasslands, frequently close to stables (hence the common name), and may form **fairy rings**. It occurs widely throughout Europe, western Asia, and North America. Horse mushrooms are edible.

Agaricus campestris (field mushroom, meadow mushroom) A species of

A

basidiomycete fungi in which the **fruiting body** has a creamy-white **pileus** 30–100 mm across, deep pink **gills** that darken as they mature, and a **stipe** 30–100 mm tall. The fungi are **saprotrophs** found in fields, especially if grazed by livestock, worldwide. They are edible and wholesome when cooked, provided they are not eaten too frequently.

agarophyte A seaweed from which sugar (agar) is obtained.

Agave (family **Asparagaceae**) A genus of **perennial, succulent monocotyledons** that produce **rosettes** of thick, fleshy leaves with spiny margins and sharp tips, arising from a very short stem. After many years the rosette produced a single terminal **inflorescence**, after which the plant dies (*see* semelparity). There are 208 species, found mainly in Mexico but also in central and tropical South America, and the southern and western United States. *Agave americana* is the century plant, used to make the beverage pulque. Other species are cultivated for ornament.

age polyethism *See* polyethism.

aggregate A group of soil particles that are adhering together.

aggregate fruit (etaerio) A fruit that is formed from a **flower** in which there are several free **carpels**, so each carpel forms a separate **fruitlet**. Raspberries and blackberries are aggregate **drupelets**.

aggregation 1. The process by which soil particles coalesce to form **aggregates**. **2.** The accumulation of water droplets or ice crystals around a nucleus. **3.** A group of animals drawn to a shared resource, e.g. food.

aggregative response The preference **consumers** have for spending most of their feeding time in small areas with the highest concentration of food.

aggression Behaviour, not linked to predation, with which one animal seeks to intimidate or injure another.

Agkistrodon contortix (copperhead) A species of thick-bodied pit vipers (**Crotalinae**), about 76 mm long, with reddish brown bodies and heads with darker markings. They inhabit semi-aquatic **habitats** and feed on rodents, other small vertebrates, and insects, occasionally climbing trees to pursue prey. They occur in much of the eastern and southern United States. They are not aggressive to humans but are very well camouflaged and many bites result from treading on them. The bites are seldom fatal.

Agkistrodon piscivorus (cottonmouth) A species of black, olive, and brown, semi-aquatic pit vipers (**Crotalinae**), about 1.8 m long, with black bands along their bodies and large heads. The inside of the mouth is brilliant white, and the snake will **display** it if threatened. Cottonmouths inhabit wetlands and on land close to water. They feed mainly on fish and mammals, but also eat turtles, frogs, and other snakes. They are not aggressive toward humans, but their bite can be fatal. They occur throughout the southern and central United States.

agonistic behaviour The behaviour of two rivals of the same species that often occurs when there is conflict between fear and **aggression**. It may involve threat, actual aggression, appeasement, or **avoidance**, and it may be ritualized.

agouti Hair pigmentation in mammals that consists of alternating light and dark bands of **melanin**.

agric horizon A soil **diagnostic horizon** consisting of **clay**, **silt**, and **humus** that has moved downward from an upper, cultivated horizon. The horizon results from **cultivation**.

agricultural drought A **drought** that reduces agricultural crop yields.

Agrimonia (family **Rosaceae**) A genus of **perennial herbs** with **pinnate** leaves that are interrupted, small **leaflets** alternating with larger ones. **Inflorescences** are **racemes** of flowers with 5 **petals** and **sepals**, many **stamens**, and 2 **carpels**. The fruits have burs that attach them to passing animals. *Agrimonia eupatoria* (common agrimony, also known as church steeples and sticklewort) has many traditional medicinal uses. *Agrimonia* flowers attract hoverflies, bees, and flies, and the foliage is food for **caterpillars** including the grizzled skipper (*Pyrgus malvae*) and large grizzled skipper (*P. alveus*).

agrimony *See Agrimonia.*

Agrobacterium A genus of Gram-negative (*see* Gram reaction) **Alphaproteobacteria** that uses **horizontal gene transfer** to insert its genes into plants, causing crown gall disease. Because of their ability to move genes to other cells *Agrobacterium* species are widely used in **genetic engineering.** ↗

Agrobacterium tumefaciens A species of **Alphaproteobacteria** that is a pathogen causing crown gall disease in more than 140 species of plants. It occurs widely in soils and invades plants by inserting a tumour-inducing **plasmid** into the host **genome.**

agroclimatology The study of the influence of climate on agriculture.

agro-ecosystem An agricultural **ecosystem**, i.e. a field crop together with the wild plants and animals present in the field.

agroforestry A farming system in which areas growing **annual** crops are interspersed with areas of tree crops. If the tree crops are in rows the system is often called alley cropping or corridor farming. ↗

agrometeorology The study of the effects of day-to-day weather on agriculture and horticulture, and the preparation of specialized weather forecasts for growers.

Agropyron spicatum (blue-bunch wheatgrass) *See* Palouse prairie.

Agrotis exclamationis (heart and dart moth) A noctuid moth (**Noctuidae**) with pale to dark brown forewings with prominent stigmata (*see* stigma), very pale hind wings, and a wingspan of 30–40 mm. **Caterpillars** are brown and grey and feed on a wide range of plants, overwinter in the soil as larvae and pupate in spring. *Agrotis exclamationis* larvae are **cutworms** that damage plants at the base. The moth is common throughout Eurasia.

Ailanthus altissima (tree of heaven) *See* Simaroubaceae.

air frost The condition in which the air temperature is below freezing.

air mass A body of air extending from the surface to the **tropopause** and covering a large surface area, e.g. most of a continent or ocean, throughout which the physical characteristics of temperature, density, **humidity**, and lapse rate are approximately constant at every height.

air pressure *See* atmospheric pressure.

air sac A thin-walled extension to a bird's lung that penetrates the body cavity and extends into the bones. It increases the amount of oxygen available to the bird and reduces the weight of the bones.

air temperature *See* shelter temperature.

Aix sponsa (wood duck, Carolina duck) A duck, 470–540 mm long with a 700–730 mm wingspan, that exhibits marked **dimorphism.** Males are slightly larger than females and have an iridescent green, blue, and purple head with a white line on either side, red eyes, rust breasts, bronze flanks, and black backs and tails. Females are brown or grey with white rings round the eyes, white throats, and grey breasts. They inhabit woodlands and wetland areas of all kinds, and feed on nuts, fruit, aquatic plants, and invertebrates. They occur throughout eastern North America. ↗

A

Aizoaceae (order **Caryophyllales**) A family of **succulent** plants, ranging from **herbs** to **perennial** woody **shrubs**, that have **entire**, **opposite** or **alternate** leaves, usually small and sometimes tiny and **imbricate.** Flowers are **actinomorphic**, **bisexual**, 4- or 5-merous, with **tepals** forming a **perianth** that is green on the outside and green, white, yellow, or pink inside, with 4 to many **stamens**, 2–20 **carpels** fused into an **ovary** with 1–10 **locules**, each with 1 **ovule.** There is a **hypanthium.** The fruit is a **capsule** with one to many seeds. There are 123 genera with about 2035 species, found throughout the tropics and subtropics; they account for 50 percent of the species and 90 percent of the **biomass** in the Succulent Karoo of southwestern Africa. Some use the **C4 pathway**, others the **CAM pathway.** *Carpobrotus edulis* (ice plant) has been widely introduced and is invasive in some places; its leaves contain so much water they will not burn and people plant it around their homes as a firewall. It has edible leaves, as has *Tetragonia tetragonioides* (New Zealand spinach), grown as a leaf vegetable.

Ajuga (family **Lamiaceae**) A genus of **annual** and **perennial herbs** with **radical simple** or lobed, **opposite** leaves, sometimes with toothed margins, without **stipules**, that form a **rosette.** Whorls of 2 to many flowers often form a terminal **inflorescence.** The flowers are **bisexual**, with 5 fused **sepals** that form a toothed bell shape, the **corolla** has a short upper and larger, 3-lobed lower lip, with a ring of hairs in the corolla tube. The **stamens** are attached to the corolla tube. The **ovary** is **superior**, with 2 **carpels**, each with 2 **locules.** The fruit is an oval **nutlet.** There are 40–50 species found throughout temperate regions of the Old World. *Ajuga reptans* is common bugle; other species are known as bugleweed, ground pine, or carpet bugle. Some are cultivated for ornament.

Akaniaceae (order **Brassicales**) A family of small **deciduous** or **evergreen trees** that have **pinnate** leaves with **entire** or toothed **leaflets.** The plants are **hermaphrodite**, the **inflorescences** borne as **panicles.** The fruit is a **capsule.** There are two **monotypic** genera. *Akania bidwilli* occurs in eastern Australia, *Bretschneidera sinensis* in southwestern China, adjacent Vietnam, and Taiwan.

akinete A thick-walled resting cell formed by certain **cyanobacteria.**

Alabama jumper *See Amynthas gracilis.*

alabaster *See* gypsum.

Alar *See* daminozide.

alar cell In mosses (**Bryophyta**), a cell found where the **phyllode** joins the stem.

Alaria esculenta *See* dabberlocks.

alarm response A signal an animal emits when it has perceived a hazard. Others observing the signal take it as a warning. The signal may be visual, e.g. the white tail of a rabbit that is exposed when running, vocal, e.g. when a monkey sees a snake, or olfactory, as in certain insects.

alarm substance A substance released from the skin of an injured fish that causes nearby fish to disperse.

alate Having wings or structures resembling wings.

Albany pitcher plant (*Cephalotus follicularis*) *See* Cephalotaceae.

albedo A measure of the extent to which a surface reflects light, expressed as a percentage or decimal fraction of the radiation falling on the surface that is reflected. Radiation that is not reflected is absorbed and converted to heat, so the albedo of a surface indicates how strongly sunlight warms the surface.

albeluvisols Soils that have an **agric horizon** with an irregular upper boundary as the B **soil horizon.** Albeluvisols comprise a reference soil group in the **World Reference Base for Soil Resources**.

albic horizon An almost white **soil horizon** that is composed of **sand** or **silt** particles with very little **clay** coating them. It may lie at or below the surface.

albinism In plants, a condition resulting from a deficiency of the **chromoplasts** that impart colour to flowers and fruit.

Albugo A genus of water moulds (**Oomycota**) that includes several species of plant pathogens, causing white blister and white rust disease in a variety of cultivated plants.

Albugo candida A species of water moulds (**Oomycota**) that is a parasite of brassicas (**Brassicaceae**), causing white rust, a disease characterized by pustules, 1–2 mm across, filled with sporangia (*see* sporangium) on leaves. It occurs worldwide.

albuminous Describes seeds that contain **endosperm** when they are mature.

Alchemilla (family **Rosaceae**) A genus of **perennial herbs**, all with the common name lady's mantle. Most form clumps or mounds with fan-shaped basal leaves, covered with soft hairs, arising from woody **rhizomes**. The leaves are **alternate** and **palmate** or lobed. The small, green flowers have no **petals**, but 4 **sepals** borne in **cymes** and an **epicalyx** with 4 lobes. There are 4 or 5 **stamens**. The fruit is dry. Many species are apomictic (*see* apomixis). There are about 300 species found in cool to temperate and subarctic regions or Eurasia, with a few in mountainous regions of Africa and America.

alcohol An organic compound in which a hydroxyl (OH) group is attached to a **carbon** atom that has bonds to other atoms; the general formula is $C_nH_{2n+1}OH$. Alcohols are designated primary, secondary, and tertiary according whether the carbon is attached to one, two, or three other atoms.

alcoholic fermentation The chemical process by which **glucose** breaks down to **ethanol** and carbon dioxide with the release of energy to convert ADP to ATP. It is a form of anaerobic **respiration**.

alcrete A **duricrust** dominated by sesquioxides of aluminium.

alder (*Alnus*) *See* Betulaceae; root nodule.

aldrin (aldrine) An **organochlorine** seed dressing that is harmless to insects but reacts in insect bodies to form **dieldrin**, which kills them. Aldrin bioaccumulates (*see* bioaccumulation) and its use led to declines in the populations of some birds of prey. It was withdrawn from use in the 1960s.

aldrine *See* aldrin.

-ales The suffix that indicates a plant order (e.g. **Brassicales**).

aletophyte A plant that grows on road verges.

Aleuria aurantia (orange-peel fungus) A species of **ascomycete fungi** in which the **fruiting body** is cup-shaped, up to 100 mm across but often smaller, and bright orange, resembling pieces of discarded orange peel. It is a **saprotroph**, growing on disturbed soil and **clay** and occurs throughout Europe and North America.

aleuriospore A single fungal (*see* Fungi) **spore** which develops on the tip of an aerial hypha (*see* aerial mycelium); the end of the **conidiophore** swells and becomes separated by a **septum**.

aleurone grain A seed that stores **protein** in its cells.

aleurone layer In the **testa** of some seeds, cells containing **enzymes** that aid digestion of the material in the **endosperm**.

aleuroplast A **leucoplast** involved in the storage of **proteins**.

Aleutian low A large area of low **atmospheric pressure** centred over the Aleutian

Islands, at about 50° N, which generates many storms that travel eastward along the **polar front** and tend to merge. It is present most of the winter, with pressure lowest in January.

Aleyrodidae (whitefly) A family of small bugs (**Homoptera**) with bodies and wings covered in powdery, white wax that feed by sucking **sap**, typically feeding on the underside of leaves. They cause damage to **phloem** and are also vectors for a number of diseases. There are more than 1550 species with a **cosmopolitan distribution**.

alfisols (grey-brown podzolics) **Alkaline** to intermediate, mineral soils in which the B **soil horizon** is argillic (*see* argic horizon) or enriched in **clay** and has a **base saturtion** of more than 35 percent.

alga (pl. algae) A protoctist (*see* Protoctista) that contains **chlorophyll** *a*, performs **photosynthesis**, and resembles a plant, but that is not differentiated into root, leaf, or stem, and has no true **vascular tissue** or protective layer of cells surrounding the reproductive organs. The algae comprise three groups: **red algae** (Rhodophyta), **brown algae** (Phaeophyta), and several groups of **green algae** (**Chlorophyta**). Algae range from single-celled organisms to seaweeds that can be several metres long. Algae occur in all **habitats**, but most are found in marine or freshwater environments.

algal bloom A sudden and rapid proliferation of aquatic algae (*see* alga), often in spring or early summer when rising temperature and abundant nutrients cause the algae to reproduce faster than **herbivores** can consume them. Nutrient enrichment can also cause such blooms, which are then an indication of **eutrophication**.

algal layer The tissue in a **lichen** that contains the **phycobiont**.

algal mat A covering of **cyanobacteria** that forms on the surface of sediments in shallow water.

algin A **salt** of alginic acid present in the **cell walls** and intercellular spaces in brown seaweeds (Phaeophyta). When mixed with water many aligns form viscous solutions used commercially as stabilizers in ice cream as well as in pharmaceutical products and paints.

algology *See* phycology.

Alismataceae (order **Alismatales**) The water plantain family, comprising aquatic **monocotyledons**, mostly **perennials** but some **annual** or **perennial**. The stems resemble **corms** or **stolons**. Most have leaves that are **simple** and **entire** often with a **sagittate** or **hastate blade** and distinct **petiole**, with an expanded, sheathing base. Some species have two leaf forms: submerged leaves are narrow, emerged or floating leaves are broader. The plants produce **latex**. The **inflorescence** is a **panicle** or **raceme**, with whorls of flowers, single flowers, or groups resembling **umbels**. The flowers are regular, **monoecious** or **dioecious**, with 3 **deciduous** petals (but petals absent in some *Burnatia* spp.) and 3 **sepals** that often persist in the fruit. There are 3, 6, 9, or many **stamens**, the **ovary** is **superior** with 3 to many free **carpels**, each with 1 (rarely 2) **locule**. The fruit is a head of **nutlets** except in *Damasonium* spp., where it is 6–10 **dehiscent** or semidehiscent **follicles** in whorls united at the base or **adnate** to the **receptacle** and spread in a star shape. There are 15 genera with 88 species, with a **cosmopolitan distribution** except for deserts and polar regions. A few *Sagittaria* spp. are cultivated for ornament, *S. sagittifolia* is grown in Asia for its edible corms, and *S. latifolia* roots were formerly used as food by Native Americans.

Alismatales An order comprising 14 families, 166 genera, and 4560 species of aquatic **monocotyledons**, including the only fully submerged, marine **angiosperms** (sea grasses), as well as plants of freshwater and marsh **habitats**. *See*

Alismataceae, Aponogetonaceae, Araceae, Butomaceae, Cymodoceacea, Hydrocharitaceae, Juncaginaceae, Maundiaceae, Posidoniaceae, Ruppiaceae, Potamogetonaceae, Scheuchzeriaceae, Tofieldiaceae, and Zosteraceae.

alisols Soils that have an argic B horizon (*see* argic horizon) with a **cation exchange capacity** of more than 24 $cmol_c$/kg of **clay** and a **base saturation** of less than 50 percent within 100 cm of the soil surface. Alisols contain large amounts of aluminium and are a reference soil group in the **World Reference Base for Soil Resources.**

alisphenoid One of a pair of structures that extend to either side of the **sphenoid** bone in the skull of vertebrates.

alkaline Having a **pH** greater than 7.0.

alkaline soil Soil with a **pH** greater than 7.0. If the **base saturation** is 100 percent the pH is 7.0 or higher.

alkaliphile *See* extremophile.

alkaloid A complex nitrogen compound, usually **heterocyclic**, belonging to a group of more than 1000 such substances found combined with organic acids in many plants. They are secondary compounds that may confer a degree of protection against **herbivores.** Many are poisonous to animals including humans.

alkanet *See Anchusa.*

alkyl mercury The mercury compounds ethyl mercury and methyl mercury, which were formerly used as **fungicides** and seed dressings. They are highly toxic and banned or severely restricted in most countries.

Allamanda (family **Apocynaceae**) A genus of tropical **shrubs** or vines native to Central and South America and known as yellow bell, golden trumpet, or buttercup flower. The leaves are leathery, **lanceolate**, and **opposite** or in whorls of 3 or 4. The flowers are trumpet-shaped and bright yellow, although cultivated forms may be white, orange, purple, or pink, with a delicate scent. They grow naturally along river banks and in the open, where the ground is permanently moist. They are cultivated widely and are naturalized throughout the tropics. *Allamanda cathartica* is invasive in Queensland, Australia.

Allee effect A positive association between population density and individual **fitness** over a specified period of time, such that individuals are healthier and reproduce more at high rather than low population densities, and it is under-crowding that inhibits growth. The effect occurs in small or sparsely distributed populations and was first described in 1931 by the American zoologist Warder Clyde Allee (1885–1955).

allele One of two or more forms of a **gene** occupying the same **locus** on **homologous chromosomes** and having different effects on the **phenotype.**

allelopathy The release by an organism of a chemical substance that inhibits the **germination** or **growth** of other organisms, e.g. root secretions by barley inhibit competing weeds.

Allen's rule The observation that **endotherms** inhabiting a cold climate typically have smaller protruding body parts—i.e. ears, snout, legs, tail—than members of the same species inhabiting a warm climate, as an **adaptation** which minimizes the loss of body warmth. The rule was first proposed by the American zoologist Joel Asaph Allen (1838–1921).

alley cropping *See* agroforestry.

alliance A group of closely related **plant associations.**

Allium (family **Amaryllidaceae**) The genus of **monocotyledons** that includes the onions and all their relatives. They are **perennial herbs** with undivided, parallel-veined, **radical** leaves. The flowers are borne in **umbels** enclosed in a **spathe**

A

which later splits. The **perianth** has 6 free, equal segments and 6 **stamens**; **bulbils** often replace the flowers. The flowers are erect (pendent in some species) with 2 whorls each of 6 petal-like **tepals**. The fruits are **capsules** that open longitudinally. The **bulbs** contain oils with a pungent onion or garlic smell. There are about 750 species, found throughout the temperate regions of the Northern Hemisphere as far south as Mexico. Garlic is *A. sativum*, onion and shallot *A. cepa*, leek *A. ampeloprasum*, and chive *A. schoenoprasum*.

Allium root rot *See Sclerotium cepivorum*.

allochthonous Acquired or transported from its original location.

allodapic Describes material that has been deposited on land by **mass flow** or on the seafloor by a turbidity current.

allogamy **Fertilization** of a **flower** by **pollen** from a different flower.

allogenic Describes a change in a **succession** that is due to a change in an **abiotic** feature of the environment.

allograft (homograft) A **graft** of tissue from one plant on to another plant of the same **species**.

Allolobophora chlorotica (green worm) An earthworm (**Annelida**), 30–80 mm long and 3–7 mm wide, that occurs in two forms, green and pink, and has three pairs of discs resembling suckers on the underside of the **clitellum**. It occurs in woodland, farmland, and gardens, in Europe, North and South America, Africa, and New Zealand.

allometric macroecology A branch of **ecology** that studies the relationship between species in terms of differences in their body sizes.

allometry The growth of one part of an organism faster or slower than the growth of the whole organism.

alloparenting Care of the young by an individual that is not related to them.

allopatric speciation The appearance of new **species** due to the geographic separation or fragmentation of the ancestral population.

allopatry The occurrence of **species** in different geographic locations.

allopolyploidy **Polyploidy** that results from the union of genetically distinct sets of **chromosomes**, usually from different **species**. Bread wheat (*Triticum aestivum*) is an allopolyploid with 42 chromosomes (42*n*), resulting from interbreeding between emmer (*T. turgidum*, 28*n*) and a wild wheat (*T. tauschii*, 14*n*).

allosteric effect The binding of a **effector molecule** to an **allosteric site** on a **protein** molecule, thereby regulating the action of an **enzyme** or other protein.

allosteric site A site on a **protein** molecule other than the **active site**.

allothetic Describes information about its orientation in the **environment** that an animal obtains from external clues.

Allothrombium fuliginosum (red velvet mite) A large, red mite (**Arachnida**), 2.5 mm long and with a velvety appearance, that is active in spring among the foliage of fruit trees. It is a predator of small animals including aphids (**Aphididae**) and caterpillars (**Lepidoptera**). There is one generation a year and the mites lay their eggs in the soil or among dead leaves. Larvae begin their lives as ectoparasites before detaching themselves and climbing trees in search of prey.

allspice (*Pimenta dioica*) *See* Myrtaceae.

alluvial Transported by a river.

alluvial soil Soil composed of material that has been transported by rivers.

alluvium A deposit of material transported by a river.

almond *See Amygdalus.*

Alnus (alder; family **Betulaceae**) A family of **monoecious**, mainly **deciduous trees** that have **alternate**, **simple** leaves with **serrated** margins and unisexual flowers borne in short male and longer female **catkins**. The female flowers develop into a persistent, woody structure resembling a **cone**. The flowers are mainly wind-pollinated, but also visited by bees. The roots have nitrogen-fixing nodules (*see* root nodule). There are 20–30 species found on river banks and in other wet **habitats** throughout the Northern Hemisphere temperate zone and in the Andes. The **bark** has been used medicinally as an anti-inflammatory agent, the timber is commercially valuable, and some species are grown for ornament.

alpha-amino acid An **amino acid** that has the **amino group** adjacent to the **carboxyl** group, attached to the 'alpha' **carbon** and to side chains of amino acids. Alpha-amino acids are the building blocks of **peptides**.

alpha diversity (local diversity) The mean of the number of species present in a subunit of a set of data. *See* beta diversity, gamma diversity.

alpha helix The helical configuration of a **polypeptide** chain, which is the secondary structure of some **protein** molecules. It is maintained by **hydrogen bonds** between CO and NH **peptide** bonds and is usually right-handed, occasionally left-handed.

Alphaproteobacteria A class comprising ten orders of **Proteobacteria** that include most of the **phototrophic** genera. Some are symbionts (*see* symbiosis) of plants and others are pathogens.

alpine zone The region that lies between the **tree line** and **snow line**.

Alseuosmiaceae (order **Asterales**) A family of **shrubs**, **subshrubs**, and small **trees** with **simple** leaves with **serrate** margins and **axillary** tufts of hairs, borne in a **pseudowhorl** of up to 6 leaves. Flowers are **actinomorphic**, **bisexual**, **perigynous**, and mostly **pentamerous**, with an **inferior ovary** with 2–3 **locules** each with 1 or several seeds. **inflorescences** are borne in axillary or terminal **cymes**. There are four genera and ten species occurring in Australasia, New Guinea, and New Caledonia. *Aleusomia* is cultivated for its colourful, sweet-scented flowers.

Alstroemeria (Inca lily) *See* Alstroemeriaceae.

Alstroemeriaceae (order **Liliales**) A family of **perennial monocotyledon herbs** or vines that have **rhizomes** and often swollen roots. Leaves are **resupinate**, **simple**, **alternate** along the stem or in a basal **rosette**. Flowers are **bisexual**,

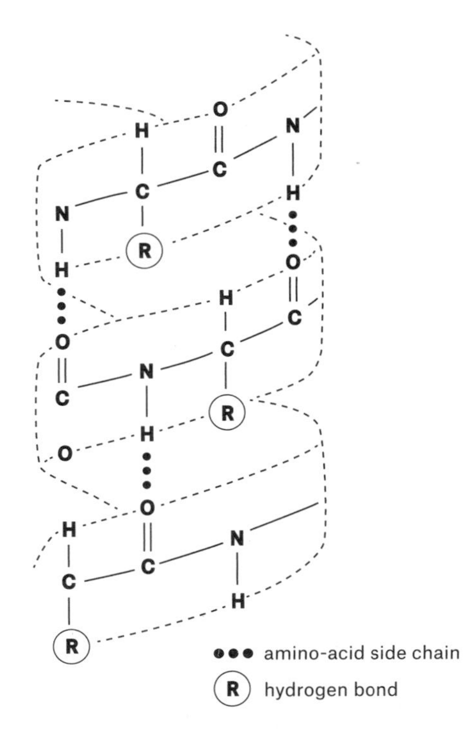

Alpha helix. The spiral (helical) shape of the secondary structure of certain protein molecules.

actinomorphic or **zygomorphic** with 2 whorls of 3 petaloid **tepals** and 2 whorls of 3 **stamens**, **inferior ovary** with 3 **carpels** each with 1–3 **locules**. The fruit is a **dehiscent loculicidal capsule**, seeds are **globose** often with an orange-red **aril**. There are 5 genera and 170 species found in Central and South America and Australasia. Several species are cultivated for ornament; *Alstroemeria* is Inca lily. *Bomarea* species are grown for their edible, starchy roots.

Alternaria A genus of **ascomycetes** comprising about 300 species, many of which are plant pathogens. They occur worldwide.

alternaria black rot of carrot *See Alternaria radicina.*

Alternaria radicina The species of **ascomycetes** that causes the disease black rot of carrot, also called black crown of carrot or alternaria black rot of carrot. Infection causes the plant to rot, starting at the crown, where the leaves are attached, and eventually may kill all the leaves, making the carrots difficult to harvest.

alternate 1. Describes leaves that arise one at each **node**. **2.** Describes **stamens** that are located between the **petals**.

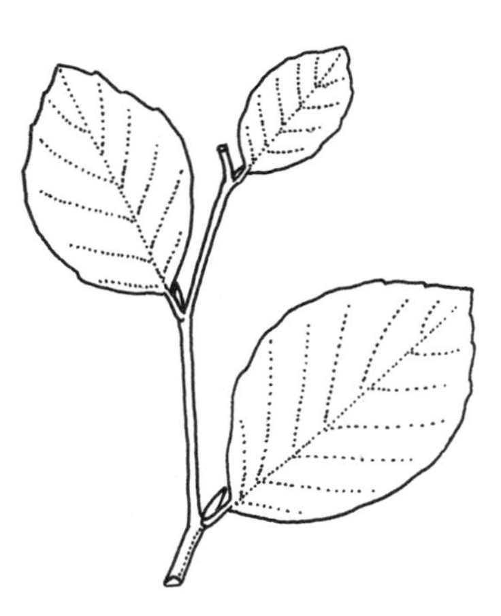

Alternate leaves.

alternation of generations The alternate development of two different forms during the life cycle of all vascular plants (**Tracheophyta**), mosses (**Bryophyta**), many algae (*see* alga), and some **Fungi**, in which a **haploid gametophyte** generation alternates with a **diploid sporophyte** generation. The generations may be **isomorphic** or **heteromorphic**.

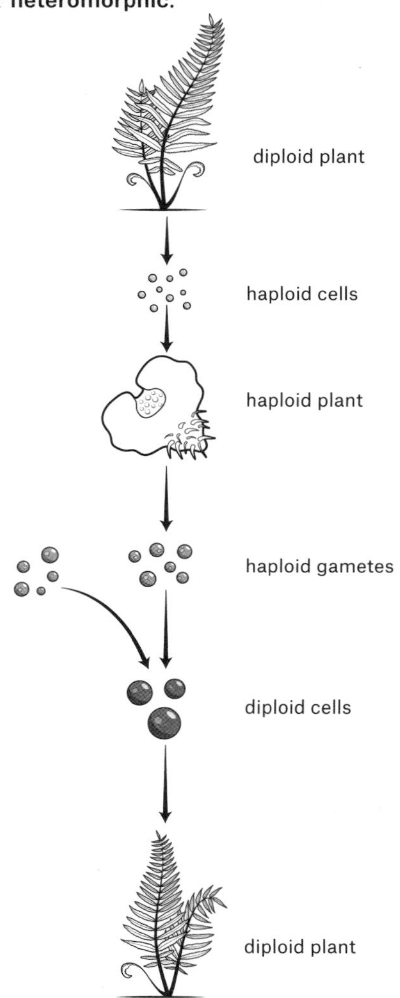

The life cycles of all vascular plants, mosses, many algae, and some fungi involve an asexually reproducing haploid generation alternating with a sexually reproducing diploid generation.

Althaea (family **Malvaceae**) A genus of **herbs** with **palmate** leaves with 3–7 lobes and flowers that are **actinomorphic** with 5 petals and an **epicalyx** with 6–9 segments that form an **involucre** hiding the **calyx**. There are 6–12 species occurring in temperate Eurasia, several of which are cultivated. *Altaea rosea* is hollyhock, *A. officinalis* is marshmallow, *A. cannabina* is hemp-leaved marshmallow, and *A. hirsute* is hairy marshmallow.

Altingiaceae (order **Saxifragales**) A family of **monoecious**, **deciduous trees** with spiral leaves with **stipules** on the bases. The very small flowers are **staminate** with 4–10 **anthers**, borne in **capitate inflorescences**, the **buds** are scaly, the fruit **dehiscent** with many small seeds. There is 1 genus and 13 species, found in Central America, eastern North America, the eastern Mediterranean, and eastern Asia to Malesia. They produce valuable timber.

altitudinal vegetation zones Horizontal belts of vegetation found on mountainsides, each reflecting the climatic conditions, of temperature and precipitation, at that elevation.

altocumulus (Ac) A genus (*see* cloud classification) of middle-level cloud, composed of water droplets, that is very variable in appearance, but that consists of white, grey, or mixed white and grey elements, about the thickness of three fingers held at arm's length, arranged in lines or waves. The edges of the elements may merge, however, to produce a sheet of cloud.

altostratus (As) A genus (*see* cloud classification) of middle-level cloud, composed of water droplets, that forms a fibrous veil or uniform grey or blue-grey sheet. Sometimes the Sun or Moon can be seen through it. It usually indicates approaching precipitation.

altricial Describes young mammals that are helpless at birth. Their eyes and ears are sealed and they are unable to walk, maintain a constant body temperature, or excrete without assistance.

altruism The performance by an individual of an act that involves a cost to itself and from which another individual derives a benefit.

alula (bastard wing) A small projection on the leading edge of the wing in some birds, or a lobe on the trailing edge of the wing of a fly (**Diptera**).

aluminium soil An **acid soil** that is enriched in aluminium. Aluminium is present in all soils but remains bound harmlessly to particles, especially of **clay** and organic material, until the **pH** falls, when it becomes soluble. A concentration of 2–5 parts per million (ppm) damages the roots of sensitive plants and more than 5 ppm is toxic to most plants.

alveolus **1.** A thin-walled sac in the lung, surrounded by blood vessels, through which gases are exchanged. **2.** A sac at the end of a glandular duct. **3.** The socket holding a tooth.

Alzateaceae (order **Malvales**) A family of **trees** or **shrubs** with **opposite**, **obovate** to **elliptical** or **oblong** leaves, and **actinomorphic**, **bisexual**, bell-shaped flowers borne in **cymes**. There are 5–6 persistent **calyx** lobes, no **corolla**, 5 **stamens** and **sepals**, **superior ovary** with 2 **carpels** each with 2 **locules**. The fruit is a **capsule**. There is one genus (*Alzatea*) with one or two species found from Costa Rica south to Peru.

Amanita A genus of **agaric fungi** comprising about 600 species that include some of the most toxic of all fungi, e.g. ***A. muscaria***, ***A. phalloides***, *A. virosa* (**destroying angel**), containing **amatoxins**. A few species are edible, e.g. **blusher**, ***A. vaginata***. The **fruit bodies** resemble mushrooms and have white **gills**. Typically there is a **volva** and remnants of the **universal veil** may cling to the **pileus**.

A

Amanita bisporigera *See* destroying angel.

Amanita caesarea (Caesar's mushroom) A species of **agaric fungi** found in woodlands with oak (***Quercus***) and conifers, in southern Europe and North Africa. The **fruiting body** is edible and was highly esteemed by the aristocracy of the Roman Empire. The fruiting body is shaped like a mushroom with an orange **pileus** and yellow **gills** and stem.

Amanita citrina (false death cap) A species of **agaric fungi** found in woodlands in Europe and North America. Its **fruiting body** has a white or pale yellow **pileus**, 40–100 mm across, a **stipe** about 80 mm tall, with a large **volva**. It is edible, but mildly toxic, but best avoided because of its close resemblance to the death cap (***A. phalloides***).

Amanita fulva (tawny grisette) *See Amanita vaginata.*

Amanita muscaria (fly agaric) A species of **agaric fungi** found associated with trees, especially ***Betula*** and ***Pinus*** species, usually in woodland, throughout the temperate Northern Hemisphere. Its **fruiting body** has white **gills** and is initially covered by a white **universal veil**, fragments of which remain on the **pileus**, which is scarlet or more rarely orange. The fruiting body is toxic; ingestion causes hallucinations, but fatalities are rare.

Amanita novinupta (blusher) *See Amanita rubescens.*

Amanita ocreata *See* destroying angel.

Amanita pantherina (panther cap, false blusher) A species of **agaric fungi** found growing beneath **broad-leaved** trees in Europe and western Asia. It has an ectomycorrhizal (*see* ectomycorrhiza) association with the adjacent trees. The **fruiting body** is shaped like a mushroom with a brown or olive **pileus**, 40–110 mm across, with many small, white patches, white or greyish **gills**, and a **stipe** 50–140 mm tall. The *A. pantherina* found in western North America may be a different species, although it is very similar. Both are toxic and hallucinogenic.

Amanita phalloides (death cap, death cup) A species of **agaric fungi** found close to **broad-leaved** trees, with which it has a **mycorrhizal** association, throughout Europe, North Africa, and parts of Asia, and that has been introduced along with imported timber in the United States, South America, and Australia. The **fruiting body** has a white **pileus** 50–150 mm across that turns yellow, bronze, or olive, white **gills** that turn cream, and a **stipe** 70–150 mm tall with a swollen base. It is deadly poisonous.

Amanita vaginata (grisette) A species of **agaric fungi** found throughout Europe and North America in ectomycorrhizal (*see* ectomycorrhiza) association with **broad-leaved** trees. The **fruiting body** has a grey **pileus** 50–100 mm across, white **adnexed gills**, and a **stape** 120–200 mm tall with a large **volva**. *Amanita fulva* (tawny grisette) is very similar. Grisettes are edible, but not highly recommended.

Amanita verna *See* destroying angel.

Amanita virosa *See* destroying angel.

Amaranthaceae (order **Caryophyllales**) A large family of **annual** or **perennial herbs** and **shrubs**, with some **trees** and climbers, in which the leaves are **entire**, **opposite** or **alternate**, and lack **stipules**. Flowers are **actinomorphic**, **bisexual** or **unisexual** (and **dioecious**), solitary or borne in **axillary** dichasial (*see* dichasium) **cymes** often with prominent **bracts**. There are 3–5 **perianth** segments, sometimes united, usually as many **stamens** as perianths, **superior ovary** of 2–3 fused **carpels** each with 1 **locule** containing 1 to many **ovules**. The fruit is a **berry**, **pyxidium**, or **nut**. There are 174 genera with 2050–2500 species with worldwide distribution, especially in dry or saline **habitats**. The family now includes the former family

Chenopodiaceae, comprising goosefoot and other weeds, but also spinach, beets, sugar beet, chard, and quinoa. Seeds of some amaranth species were formerly used widely as cereals in Central and South America and some species are grown as ornamentals.

amargo (*Quassia amara*) *See* quassia.

Amaryllidaceae (order **Asparagales**) A family of **perennial monocotyledon herbs**, with some aquatic species and some **epiphytes**, most of which produce **bulbs**, with **distichous** or spirally arranged, **simple**, **linear** or **elliptic** leaves, occasionally with a **pseudostem** formed from a sheathing base. The **inflorescence** is initially enclosed in a **scarious spathe**. Flowers are **actinomorphic** or **zygomorphic** with 6 similar **perianth** segments in 2 whorls, and sometimes a trumpet-like **corona**; there are 6 **stamens**. There are 1 to many flowers, which are often large and showy. There are 73 genera with 1605 species, with worldwide distribution. Many are cultivated as ornamentals; 30–60 *Narcissus* species are known as daffodils, narcissi, and jonquils.

amatoxins A group of at least eight toxins found in **Fungi** of the genera ***Amanita***, ***Conocybe***, and ***Lepiota***. They cause severe gastrointestinal symptoms, followed by rapid degeneration of the **liver** and **kidneys**. Ingestion of even a small amount may be fatal.

Amazon floral region The Amazon basin, from the Andean foothills to the Atlantic coast, part of the **Neotropical region**. The **flora** is one of the world's richest, and comprises the areas (igapo) above and below (ete) the seasonal floods.

Amborellaceae (order **Amborellales**) A **monotypic** family (*Amborella trichopoda*), which is a **dioecious shrub**, small **tree** (up to 8 m), or **liana** that have **distichous**, **exstipulate** leaves with coarsely **serrate** or **undulate** margins and a long **petiole**. The small flowers have an undifferentiated **perianth**, and are **staminate** with 11–14 **sessile anthers** arranged in a spiral on the **receptacle**; the 5–8 **carpels** are free and the fruits are **drupes**. The plants are **endemic** to New Caledonia, where they are locally abundant.

Amborellales An order of plants that contains only one family, **Amoborellaceae**.

ambrosia beetles *See* Scolytidae.

ambrosia fungi **Fungi** that grow in the tunnels excavated in wood by ambrosia beetles (**Scolytidae**). There are several species.

ambush bugs *See* Reduviidae.

Ambystoma maculatum (spotted salamander) A species of mole salamanders (those that spend most of their time in burrows) that are 150–250 mm long. They are stout with a broad snout, and dark with 2 irregular rows of 24–45 round, yellow or orange spots from the head to tail. The salamanders have poison glands in their skin. They inhabit forests close to rivers and other moist places with access to ponds for breeding, and feed on small animals. They occur in eastern North America.

Ambystoma opacum (marbled salamander) A species of mole salamanders (those that spend most of their time in burrows) that are 90–110 mm long, stocky, and their dark bodies marked with strong grey (female) or white (male) bands. They inhabit moist woodland close to water, breeding in dried out pools and ditches, the eggs hatching after the pool or ditch has refilled. They feed on invertebrates. The salamanders occur throughout most of the eastern United States.

ameba *See* amoeba.

ameboid *See* amoeboid.

ameiosis A form of **meiosis** in which the **nucleus** divides only once, so the number of **chromosomes** is not reduced.

A

Amelanchier (family **Rosaceae**) A genus of **deciduous shrubs** and small **trees** that have **alternate**, **simple**, **lanceolate**, **elliptic**, or **orbiculate** leaves, and flowers that are **pentamerous**, with 2–5 white (rarely yellow, pink, or with red streaks) **carpels** united to the **receptacle** below but free above, each carpel divided into 2 cells by a false **septum**. The **inflorescences** are terminal with groups of 1–20 flowers, or borne in terminal **racemes** of 4–20 flowers. The fruit is a **pome** that resembles a **berry**. There are about 20 species, native to temperate regions of the Northern Hemisphere but most diverse in eastern North America. Several species are cultivated for ornament, often becoming naturalized and known as shadbush, June berry, sugar plum, wild plum, saskatoon, and chuckley pear.

amenity horticulture (hobby farming) Cultivating plants or livestock for recreational or aesthetic, as opposed to commercial, reasons.

amenity planting The planting of **shrubs** or **trees** for aesthetic reasons or to provide a public amenity.

amenity resource Any natural resource that is valued primarily for its non-commercial qualities.

amensal *See* amensalism.

amensalism An interaction between two populations in which one is inhibited but the other, known as the amensal, remains unaffected.

ament *See* catkin.

American badger *See Taxidea taxus.*

American bullfrog *See Rana catesbiana.*

American cobra *See Micrurus fulvius.*

American goldfinch *See Carduelis tristis.*

American mourning dove *See Zenaida macroura.*

American robin *See Turdus migratorius.*

American toad *See Bufo americanus.*

American yellow warbler *See Setophaga petachia.*

amictic Describes a **meromictic** lake in which the water does not become thermally stratified and does not overturn.

amide A compound formed by replacing one or more of the hydrogen (H) atoms in ammonia (NH_3) by an organic acid group. Amides are designated primary, secondary, and tertiary according to the number of hydrogen atoms replaced.

amide herbicide A group of about 30 products used as **herbicides** against **broad-leaved** weeds and **annual** grasses. They prevent leaf production but have no effect on plants that have emerged above ground, and break down rapidly and disappear from the soil.

amine A **base** formed by replacing one (primary amine), two (secondary), or three (tertiary) of the hydrogen (H) atoms in ammonia (NH_3) with an organic radical group, e.g. methyl (CH_3). The simplest amine is methylamine (CH_3NH_2, also called methanamine or aminomethane).

amino acid A compound formed from a **carboxyl** group linked to an **amino group**. Depending on the configuration of the molecule, amino acids are classed as neutral, acidic or basic, or as non-polar, polar, or charged. They are the building blocks of **peptides** and **proteins**.

amino group The radical chemical group $-NH_2$.

aminopeptidase An **enzyme** that catalyzes the **hydrolysis** of **amino acids** in a **polypeptide** chain.

aminotransfer *See* transamination.

aminotransferase *See* transaminase.

aminozide *See* daminozide.

amitosis A form of nuclear division in which the **chromosomes** remain invisible, the **nuclear envelope** remains intact, and no **mitotic spindle** forms.

ammonification The conversion by soil **Bacteria** or **Fungi** of organic compounds of nitrogen present in decomposing organic material into ammonium (NH_4). This is part of the **nitrogen cycle**.

ammonium fixation The **adsorption** of ammonium **ions** (NH_4^+) on to sites in the layers of **clay minerals**, thereby rendering them unavailable to plants.

amnion A layer of tissue that lines the inner surface of the fluid-filled **amniotic sac** in reptiles, birds, and mammals. *See also* chorion.

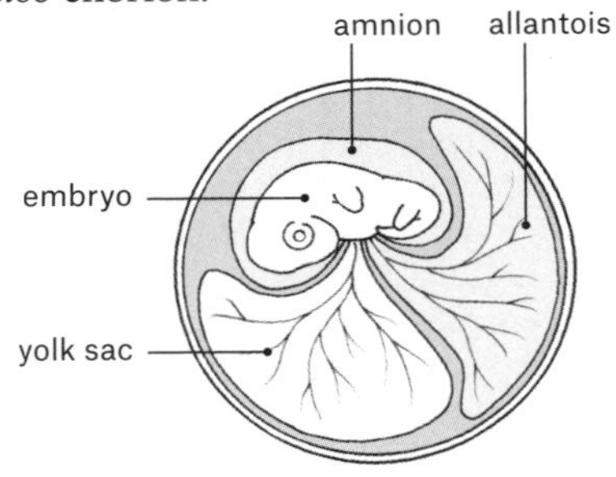

A lining that surrounds the embryo sac.

amniote A type of development found in reptiles, birds, and mammals in which the **embryo** is contained in an **amniotic sac** filled with fluid.

amniotic sac In **amniotes**, the sac in which the foetus develops. It is filled with fluid that cushions the foetus while allowing it to move and grow.

amoeba (ameba) A genus of Protozoa comprising single-celled **eukaryotes** that have no definite shape and that move by means of pseudopodia (*see* pseudopodium).

amoeboid (ameboid) Resembling or moving in the manner of an **amoeba**.

Amoebozoa A group of **amoeboid Protozoa** that have blunt, finger-like pseudopodia (*see* pseudopodium). Most are unicellular and move by circulating cell **cytoplasm**. They are common in soils and in water. Some are symbionts (*see* symbiosis), others are pathogens. The group includes the **slime moulds**.

amorphous clay **Clay** that is not composed of crystals. It occurs principally in volcanic ash and forms during the early stages of **weathering** in soils derived from volcanic material.

AMP *See* adenosine monophosphate.

Ampelovirus A genus of **Closteroviridae viruses** that are long, **flexuous**, and have no envelope. They penetrate plant cells where they produce more virus particles, causing yellowing and necrosis. They occur worldwide.

amphi-Atlantic species Plant species that occur along both the eastern coast of North America and the western coast of Europe. Their seeds may have been carried by wind or birds toward the end of the most recent (Weichselian) ice age and as the ice sheets retreated.

Amphibia (amphibians) A class of vertebrates that includes three **extant** groups, the caecilians (Apoda), frogs and toads (**Anura**), and salamanders and newts (**Urodela**). Most amphibians live on land but must return to water to breed. Eggs are fertilized externally and hatch into aquatic larvae (tadpoles, pollywogs). Amphibians are **poikilotherms** with soft, smooth skin with many mucus and poison glands, and they engage in **cutaneous respiration**. There are about 7000 species of which more than 6000 are frogs. They occur on all continents except Antarctica.

amphibians *See* Amphibia.

amphibiotic **1.** Describes an organism that is able to live parasitically or mutualistically (*see* mutualism) with a host organism. **2.** Describes an aquatic organism that lives in water as a larva and on land as an adult.

A

amphicribral Describes a **vascular bundle** in which the **phloem** surrounds the **xylem**.

amphid An anterior sense organ in **Nematoda** that probably detects odours.

amphidromous Describes the behaviour of fish that migrate between fresh water and the oceans.

amphigastrium A leaf that develops on the underside of the stem in leafy liverworts (**Marchantiophyta**).

amphimixis Sexual reproduction.

amphiphloic Describes a **vascular bundle** in which the **phloem** occurs as concentric cylinders both inside and outside the **xylem**.

amphiphloic siphonostele A **monostele siphonostele** in which one layer of **phloem** lies outside the **xylem** and another layer lies inside the xylem surrounding the **pith**.

amphithecium In mosses (**Bryophyta**), an outer layer of cells in a young **sporophyte**.

amphitrophic Describes an organism that is able to perform **photosynthesis** in the presence of light, and also able to grow in darkness.

amphitropical species Plant species that occur in separate ranges on either side of the equator. Their seeds may have been carried across the equator during the ice ages, when climatic belts were telescoped.

amphitropous Describes the position of an **ovule** when it lies with its long **axis** parallel to the **placenta**, attached at its centre.

amphivasal Describes a **vascular bundle** in which the **xylem** surrounds the **phloem**.

Amphophora idaei *See* raspberry aphid.

Amphophora rubi (blackberry aphid) An aphid (**Aphididae**) that is widely distributed and that feeds on blackberry (*Rubus fruticosus*) and dewberry (*R. caesius*). The wingless form is 2.6–4.1 mm long, pale green or yellowish green, with long legs and antennae (*see* antenna). Eggs hatch in spring and the aphids feed at the tips of leaf **buds**, moving later to the undersides of leaves. After two generations of wingless forms, winged forms appear in summer and migrate to other plants or other parts of the same plant. Eggs are laid in late autumn, usually not higher than 30 cm from ground level. The aphids act as vectors of a range of viral diseases.

amplexicaul Describes a **stipule** that surrounds the twig.

ampulliform Bottle- or flask-shaped.

amygdaliform Almond-shaped.

Amygdalus (family **Rosaceae**) A genus of **deciduous shrubs** and small **trees** adapted to arid and semiarid environments. They have **axillary buds** produced in threes, flower before the leaves open, and the fruit is a fleshy **drupe** often with a downy skin in which the wrinkled or pitted **pericarp** (hull) dries out at maturity splits along the central suture, and separates from the hard endocarp (shell), a feature that separates *Amygdalus* from the otherwise similar ***Prunus***, in which the pericarp remains attached to the stone. There are 40 species, many of which are grown for their fruits or stones. The genus includes almonds, peaches, apricots, and nectarines.

amylase A member of a group of **enzymes** that catalyze the **hydrolysis** of **starch** and **glycogen**.

amyloid **Starch**-like.

amylopectin One of the two **polysaccharides** that make up **starch**.

amyloplast A **plastid** that synthesizes and stores **starch**.

amylose One of the two **polysaccharides** that make up **starch**.

amylum *See* starch.

Amynthas gracilis (Alabama jumper, Georgia jumper) A species of earthworms that thrash about violently when disturbed (hence their common name). They burrow deeply into the soil, moving organic surface material deeper and faster than other worms and can be used to improve soils and to make **compost**. They should not be introduced where they do not occur naturally, because they are invasive. Jumpers occur widely in the tropics, subtropics, and warm temperate regions.

anabatic wind A wind that blows up the side of a hill.

anabolism The metabolic reactions that require energy and synthesize necessary compounds.

Anacampserotaceae (order **Caryophyllales**) A family of small, fleshy **subshrubs** with **opposite** leaves and flowers in which 2 **sepals** surround 5 **petals**. The fruit has 2 layers of **pericarp** that separate **periclinally** and seed in which 2 coats of the **integument** also separate periclinally. There are 3 genera with 32 species with a very scattered distribution in subtropical and warm temperate regions.

Anacardiaceae (order **Sapindales**) A family of **trees** and **shrubs**, with some climbers and **lianas**, that produce a milky sap that often turns black on exposure and that is often toxic, causing severe contact dermatitis. The leaves are **exstipulate**, often **pinnate** but sometimes **simple**. The small, inconspicuous flowers are **actinomorphic**, **bisexual** or sometimes **unisexual** (then the plants **dioecious**), usually with 5 **sepals** and free **petals**. Female flowers have **staminodes**. The **ovary** is usually **superior** with 1–3 or 5 usually united **carpels** with 1–5 **locules** each with 1 **ovule**. There are 5–10 or more **stamens**. The fruit is usually **indehiscent** and **drupe**-like, but in some species a **samara**, **syncarp**, or **achene**. There are 81 genera with 873 species found throughout tropical and temperate regions. Many species of ***Cotinus***, *Pistacia*, *Schinopsis*, and *Rhus* are cultivated for their **tannins**, used to tan leather, resin from *Rhus verniciflua* is the basis of Chinese lacquer, *Anacardium occidentale* yields cashew nuts and cashew apples, *Pistacia vera* yields pistachio nuts, and the fruit of *Mangifera indica* is the mango. Others are grown for ornament (e.g. *Schinus molle*, the pepper tree).

Anacardium occidentale (cashew) *See* Anacardiaceae.

anachoresis Living in holes or crevices to avoid predators.

anadromous Describes the migratory behaviour of fish that spend most of their lives at sea but return to fresh water to spawn.

anaerobe An organism that is able to grow only in the absence of oxygen.

anaerobic respiration *See* fermentation.

anaflexistyly A morphological (*see* morphology) change in which a **flower** functions first as a male and later as a female.

anagenesis Evolutionary change along a single, unbranching lineage.

Anagrus atomus A tiny wasp (**Hymenoptera**), about 2 mm long, that is a **parasitoid** of **leafhoppers**; since 1994 it has been used in **biological control** of this pest. The female wasp lays eggs inside leafhopper eggs and the wasp larvae eat the egg contents.

anal glands (anal sacs) In all carnivorous mammals apart from bears, a pair of sacs, one on each side of the anus, containing glands that secrete a liquid with a strong smell by which individuals recognize each other.

anal sacs *See* anal glands.

analogous variation Features with similar function that have developed independently in unrelated taxa (*see* taxon), e.g. **phyllodes** found in ***Acacia*** species are analogous to leaves.

anamniote A type of development in fish and **Amphibia** in which the egg lacks a protective shell or membranes, and must, therefore, be laid in water.

anamorph The asexual reproductive stage in **Ascomycota** and **Basidiomycota**. *Compare* teleomorph.

Ananas comosus (pineapple) *See* Bromeliaceae.

anaphase A stage in **cell division** that occurs once in **mitosis** and twice in **meiosis**. During anaphase the **chromosomes** separate into two sets under the control of the **mitotic spindle**. In anaphase II of meiosis the **centromere** doubles and daughter chromosomes separate from the equator and move toward the poles of the spindle.

Anarthriaceae (order **Poales**) A small family of **evergreen**, **caespitose**, **dioecious**, **monocotyledon herbs** in which the leaves are either **linear** and folded longitudinally with a split, sheathing base, or rudimentary with only the sheaths well developed. The **inflorescence** is a **raceme** with few or many flowers consisting of 6 membranaceous or tough **tepals** in 2 whorls. Male flowers have 3+3 **stamens** opposite the inner tepals; in female flowers the **ovary** is **superior** with 3 **locules**. The fruit is a **capsule**. There are 3 genera and 11 species **endemic** to West Australia.

anastomosis The growing together of **branches** or roots in woody plants.

anatomy Details of the physical structure of an organism.

anatropous Describes the position of an **ovule** when it lies close to the **funicle**, with the **micropyle** facing the **placenta**.

Anaxyrus americanus *See Bufo americanus*.

Anaxyrus terrestris (*Bufo terrestris*, southern toad) A toad that is 44–92 mm long, with a dark red to black mottled back, pronounced knobs on its head, a warty skin, many of the warts tipped with spines, and **parotoid** glands. It inhabits sandy areas including cultivated fields and feeds on invertebrates. It occurs in the southeastern United States.

Anchusa (family **Boraginaceae**) A genus of **annual**, **biennial**, or **perennial** plants, mainly **herbs**, most of which are covered in bristly hairs. Leaves are **oblong**, **simple** or **undulate**, and covered with stiff hairs. Flowers are small, blue or purple, **pentamerous** with scales closing the throat of the **corolla** tube. There are about 40 species distributed widely in Europe and Asia, and introduced in North America. Several species, known as alkanets, are cultivated for ornament.

ancient countryside In Britain, an area of countryside in which the field boundaries, paths, roads, and so on were in their present locations prior to 1700.

ancient woodland Primary or secondary woodland that has occupied its present site continuously since a particular date, in Britain about 1600. *See* primary woodland, secondary woodland, old-growth forest.

Ancistrocladaceae (order **Caryophyllales**) A family of **sympodial** woody climbers in which each **branch** ends in a coiled hook that acts as a grapnel. Leaves are **entire**, **alternate** with **petioles** and **stipules** that soon drop, and have glands in pits on the **abaxial** surface. The flowers are small, have no scent, **actinomorphic**, **bisexual**, with articulated **pedicels**, and borne in **cymes**. Flowers have 5 overlapping **sepals**, 5 **petals**, 10 (occasionally 5) **stamens**, **anthers** with 2 **pollen sacs**. The semi-inferior **ovary** consists of 3 fused **carpels** with 1 **locule** containing 1 **ovule**. The fruit is a **nut**. There is 1 genus with 12 species found from tropical Africa to western Borneo and Taiwan. *Ancistrocladus korupensis* produces naphthylisoquinoline **alkaloids** that show strong anti-HIV activity.

Andean floral region The area of the Andes, extending from Colombia to central Chile, south of Santiago, and including the Galápagos Islands, part of the

Neotropical region. The region is rich in **endemic** species.

andic horizon A **soil horizon** that contains weathered (*see* weathering) volcanic rock.

andisols In the U.S. Department of Agriculture **soil taxonomy**, an order of soils that are derived from volcanic ash and contain glass and **amorphous clay**.

andosols A group of soils derived from volcanic materials, with a **vitric horizon** more than 30 cm below the surface or an **andic horizon** within 25 cm of the surface. Andosols are a reference soil group in the **World Reference Base for Soil Resources**.

Andreaea (rock moss) *See* protonema.

Andricus kollari *See* oak-marble gall.

androchory (anthropochory, brotochory) Dispersal of **seeds** or **spores** by humans.

androchronial scale *See* androconial scale.

androconial scale (androchronial scale, androconium) A modified scale on the wing or body of a butterfly or moth (**Lepidoptera**) that disperses a **pheromone**.

androconium *See* androconial scale.

androdioecious Describes a **dioecious** plant in which **hermaphrodite** and male flowers occur on different plants.

androecious Bearing only male flowers.

androecium The collection of **stamens** that constitute the male reproductive organ of an **angiosperm**.

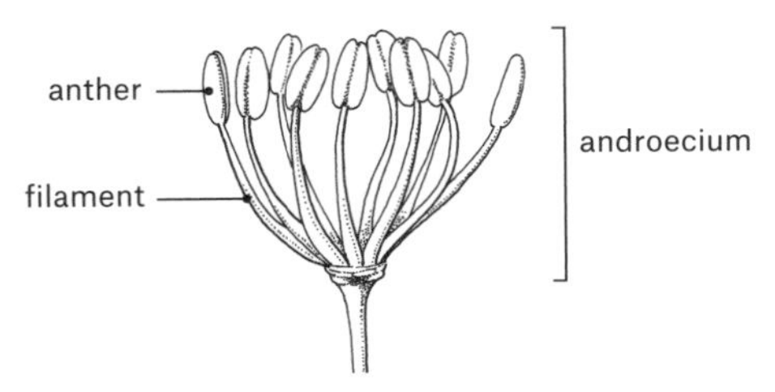

The male reproductive organs of a flowering plant (stamens), comprising the filaments and anthers.

androgens **Hormones** that regulate the development of male secondary sexual characters. They are secreted mainly by the testes.

androgynophore A stalk bearing both the **androecium** and **gynoecium**.

andromonoecious Describes a **monoecious** species that bears **hermaphrodite** and male flowers separately on the same plant.

anecic Describes an earthworm that burrows vertically into the soil.

anemo- Associated with wind (from the Greek *anemos*, wind).

anemochory (aerochory) Dispersal of **seeds** or **spores** by wind.

anemometer An instrument that measures surface wind speed.

Anemone (family **Ranunculaceae**) A genus of **perennial herbs** or rarely small **shrubs** with **rhizomes**, leaves that are **palmate** and lobed or **simple** with **serrate** or **entire** margins. Some leaves are **radical**. Stem leaves occur in a whorl of 3 below the flowers. Flowers are single or with 2–3 growing together, borne in **cymes** of 2–9 or **umbels**. The flower is a **perianth** of 1 **petal**-like whorl, with many free **stamens** and **carpels**. The fruit is an **achene**. There are about 120 species found throughout northern temperate regions. The plants are poisonous. Many are cultivated for ornament, sometimes called wind flowers. *Anemone nemorosa* is wood anemone.

anemophily **Pollination** by wind.

anestrus *See* anoestrus.

aneucentric Describes a **chromosome** with more than one **centromere**.

aneuploid Describes an organism or cell **nucleus** with more **chromosomes** than is normal for that species, but by only a small number.

Angelica (family **Apiaceae**) A genus of **perennial herbs** with stout, hollow stems and 2–3 **bipinnate** leaves with wide, oval leaflets. The white or greenish flowers are borne in **umbels** and are pollinated by a wide variety of insects. Fruits are flattened dorsally and ovoid. *Angelica angelica* is cultivated for its stems, which are candied.

angel's trumpets *See Datura.*

angiosperm A flowering plant that produces seeds which are completely enclosed by the fruits. Angiosperms are the most highly evolved of plants and the most diverse, with at least 260,000 species in 453 families found in every **habitat** except for mountaintops, the polar regions, and the oceans below the limit of light penetration. The earliest angiosperm **fossils** date from the early **Cretaceous**, about 132 million years ago.

Angiosperm Phylogeny Group (APG) A group of plant taxonomists who have established a classification of **angiosperms** based on **phylogeny**, such that all groups within the classification are **monophyletic**. They published their first classification in 1998 and have updated it at intervals since. They added **gymnosperms** in 2005. Names and descriptions of plant families and orders in this dictionary are taken from the 12th version of the APG system.

Anguis fragilis (slow worm, blind worm) A legless lizard that burrows and shelters beneath stones and logs. It resembles a snake, and has a forked tongue, but it possesses eyelids. It can detach its tail if threatened (*see* autotomy). A slow worm is up to 500 mm long and feeds on invertebrates, including many garden pests. It is common in gardens, especially in **compost** heaps.

anholocyclic Describes an insect that produces only asexual females. A wingless **fundatrix** produces both winged and wingless viviparous (*see* vivipary) females, which in turn produce viviparous females that overwinter as wingless insects or as **nymphs**. *Compare* holocyclic.

anhydrobiosis A type of **cryptobiosis** in which organisms tolerate extreme **desiccation**, in the case of certain invertebrate animals by shrinking in size and producing a sugar that allows them to survive by substituting a sugar solution for water. Plants achieve the same end using a different sugar. *See* resurrection plant.

Anigozanthos manglesii (kangaroo paw) *See* Haemodoraceae.

Animalia (Metazoa) The taxonomic kingdom which includes all multicellular organisms that develop from **embryos** derived from **gametes** formed within multicellular sex organs and never within unicellular structures. The kingdom excludes protozoons (*see* Protoctista), and sponges (Porifera) are sometimes excluded because their structure is markedly different from that of other animals. Animalia is one of the three kingdoms of multicellular life (the others being **Fungi** and **Plantae**).

anion *See* ion.

anion exchange capacity The total quantity of **anions**, measured in moles per gram, that a soil is able to adsorb.

aniso- Unequal (Greek *an* not, and *iso* equal).

anisodactylous Describes the foot of a bird in which three toes face forward and one is opposable to them and faces to the rear.

anisogamy Fusion of **gametes** of different sizes.

Anisophylleaceae (order **Cucurbitales**) A family of **shrubs** or **trees** in which the leaves are **alternate** in two rows or in four rows with the upper rows reduced in size, **simple**, somewhat **coriaceous**, often asymmetrical at the base, **exstipulate**, the margins **entire**. The small, inconspicuous flowers are **actinomorphic**, **unisexual**

(the plants **dioecious**), with 3–5 persistent **sepals**, 3–5 **petals** twice as many free **stamens** as petals, **ovary inferior**, **syncarpous**, with 3–4 **carpels**. Flowers are borne in **axillary racemes** or **panicles**. The fruit is a **drupe**. There are 4 genera and 34 species distributed throughout the tropics. Some of the trees produce valuable timber; *Poga oleosa* also yields edible oil.

anisophylly Having leaves of different sizes on the **dorsal** and **ventral** sides of the shoot.

Anna's hummingbird *See Calypte anna.*

annatto (*Bixa orellana*) *See* Bixaceae.

Annelida A **phylum** of **coelomate**, segmented worms that have a well-developed head as well as respiratory, vascular, and nervous systems. The phylum includes earthworms, bristleworms, and leeches. Annelid worms occur throughout the world in terrestrial, freshwater, and marine environments.

Annona (family **Annonaceae**) A genus of **evergreen** or semideciduous **shrubs** and **trees** in which flowers arise from **pedicels** growing from **axils** or from **axillary buds**. The 3–4 **sepals** are smaller than the outer **petals**; the 6–8 petals forming 2 whorls. There are many **stamens** and many **pistils**. Flowers are pollinated by beetles. Each flower produces one large, ovate or spherical fruit comprising an aggregate of **berries** enclosed in the fleshy **receptacle**. There are about 110 species occurring in the American and African tropics. Many of the fruits are edible and known as custard apples. That of *A. reticulata* is also known as bullock's heart, *A. cherimola* produces cherimoya, and *A. squamosa* produces sweetsop or sugar apple.

Annonaceae (order **Magnoliales**) A family of **evergreen** or **deciduous trees** and **shrubs** with **simple**, **alternate** leaves without **stipules**, usually in 2 ranks. The fragrant flowers are **actinomorphic**, **hermaphrodite**, occasionally **unisexual**, with 3 partially free **sepals**, 6 **petals**, sometimes in 2 whorls of 3, many **stamens** in spirals, **ovary superior**, **apocarpous**, usually with many **carpels**. The fruit is usually apocarpous and consists of an aggregate of 2 to many **berries**. There are 129 genera with 2220 species distributed throughout most of the tropics. Several species are cultivated for their fruits (*see Annona*).

annosum foot rot *See Heterobasidion annosum.*

annual Describes a plant that completes its life cycle, from **germination** to the production of **seed**, within a single growing season.

annual meadow grass (*Poa annua*) *See Poa.*

annual ring *See* tree ring.

annular Ring-shaped.

annulate lamellae **Cisternae** that have regularly spaced pores.

annulus 1. In the fruit body (*see* fruiting body) of certain **agarics**, a ring of tissue around the **stipe**; it is a remnant of the **partial veil**. **2.** In some ferns (**Pteridophyta**), a ring of cells around the **sporangium**.

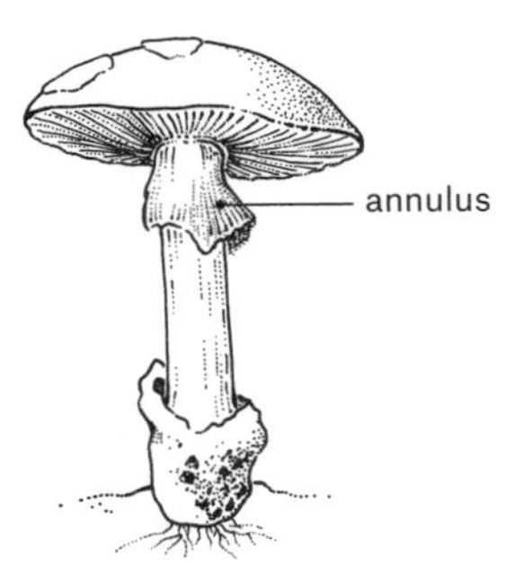

A ring of tissue around the stipe of a mushroom or toadstool.

anoestrus (anestrus) The stage in the **oestrus cycle** when the female reproductive organs are inactive.

Anolis carolinensis (Carolina anole, green anole) A species of green, grey,

A

or brown lizards, 40–80 mm long with males larger than females, that feed on insects and other invertebrates, and will also take seeds and other items when these are available. They are popular as pets. The lizards occur throughout the southeastern United States.

anomocytic Describes a **stoma** which lacks morphologically distinct (*see* morphology) **subsidiary cells**.

anoxic Depleted of oxygen, or lacking oxygen entirely.

anoxybiosis A type of **cryptobiosis** in which organisms tolerate a lack of oxygen, for years or decades in the case of some invertebrate animals.

antagonistic resources Resources that can substitute one for another unless they are exploited together, when some partly offset the effects of others, so **consumers** taking them together need more than consumers taking each separately.

Antarctic region The floristic region comprising the **New Zealand**, **Patagonian**, and the **south temperate oceanic-island regions**.

antenna (pl. antennae) One of a pair of sensory structures borne on the head of an invertebrate animal.

antennate To communicate by touching antennae (*see* antenna).

anteriad Pointing forward.

anther In an **angiosperm flower**, the male reproductive organ carrying **pollen sacs** containing **pollen** and borne at the tip of a **filament**.

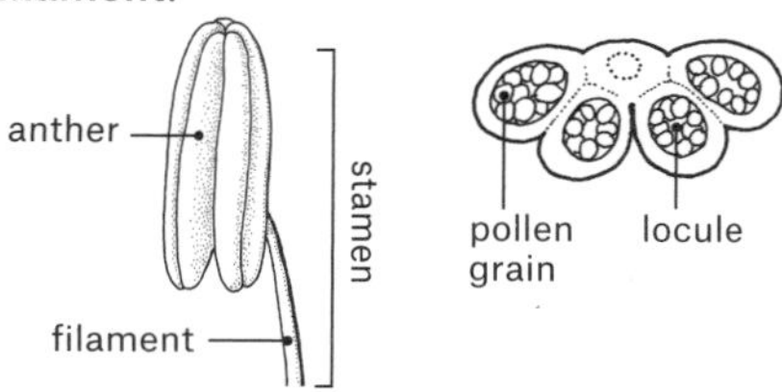

The anthers are each carried on a filament, together forming the stamen. The transverse section through an anther shows the locules (pollen sacs) containing pollen grains.

antheridiogen A chemical substance released by the archegonia (*see* archegonium) of **embryophytes** that stimulates **gametophytes** to produce antheridia (*see* antheridium).

antheridiophore In some **thallose** liverworts (**Marchantiophyta**) the structure bearing the anteridia (*see* antheridium).

antheridium In algae (*see* alga), **Bryophyta**, **Marchantiophyta**, **Pteridophyta**, and **Fungi**, the male sex organ (**gametangium**).

antherocyte (spermatocyte) A cell that develops into an **antherozoid** without further **cell division**.

antherozoid (spermatozoid) In algae (*see* alga), **Bryophyta**, **Marchantiophyta**, **Pteridophyta**, and **Fungi**, the male **gamete**, produced in the antheridia (*see* antheridium).

anthesis **1.** The opening of a **flower bud**. **2.** The time of flowering of a plant.

Anthocerotophyta (hornworts) A **phylum** of small **cryptogams** found on disturbed ground and along streamsides, in which the **sporophyte** remains attached to the **gametophyte** throughout its life, but continues to grow upward, resembling a horn, hence the name. As it grows, the sporophyte splits open, releasing **spores** as they mature. There are about 100 species.

anthocorid bugs (flower bugs, minute pirate bugs) A family (Anthocoridae) of bugs (**Hemiptera**), 1.5–5 mm long, that feed on other insects, insect eggs, and **spider mites**, cutting a hole into the prey, injecting saliva, and drinking the dissolved body contents, making them useful in **biological control**, although they can also deliver a painful bite to humans. There are 400–600 species, found world-wide on trees, flowers, under **bark**, or in the nests of birds or mammals. *See* *Orius*.

anthocyanins Water-soluble pigments that are red, purple, or blue depending on the **pH**, that contribute to the autumn colours of many **deciduous** plants.

Anthonomus pomorum (apple blossom weevil, apple weevil, pear flower bud weevil) A dark brown or black weevil (**Curculionidae**), 4.5–5.0 mm long, the body covered in fine, grey hairs and with a grey V-shaped mark on the rear of the **elytra**. The **rostrum** is about 1/3 the length of the body. Its larvae are legless, yellowish white, and 6–8 mm long. These insects produce one generation a year, adults emerging in May or June, and feeding on leaves of apple and pear trees. Larvae feed on **ovaries**, **stamens**, and **petals**.

Anthonomus rubi (strawberry blossom weevil) A dull black weevil (**Curculionidae**) covered with fine hairs, 2–4 mm long with striped **elytra** and a slightly curved **rostrum**, and white larvae with brown heads 3.0–3.5 mm long. It produces one generation a year and feeds on the flowers and flower stalks mainly of strawberry, but also of raspberry, bramble, and roses.

anthophyte A member of an extinct group of plants that includes the **angiosperms**, **Cycadales**, **gnetophytes**, and ***Pentoxylon***.

anthracnose Any one of several fungal diseases of woody plants that affect growing shoots and leaves. *See Colletotrichum.*

anthraquic horizon An **anthropedogenic horizon** that is flooded for part of the year and has a puddle layer (*see* puddling) and a plough **pan**.

anthrax *See Bacillus.*

Anthriscus (family **Apiaceae**) A genus of **herbs** that have fern-like **bipinnate** or tripinnate leaves, hollow, upright, branched stems, bearing at their tips **compound umbels** of small, white flowers. Fruits are oblong to ovoid, smooth or with spines. There are 15 species found throughout Europe and temperate Asia, growing in meadows and road verges. The plants provide food for several lepidopteran (*see* Lepidoptera) species. Some species are noxious weeds. *Anthriscus sylvestris* is cow parsley; *A. cereifolium* is chervil.

anthropedogenic horizon A **soil horizon** that is the result of prolonged **cultivation**. Its qualities vary depending on the type and intensity of management. *See* anthraquic horizon, hortic horizon, hydragric horizon, irragic horizon, plaggic horizon, terric horizon.

anthropic horizon A surface **soil horizon** that consists of unconsolidated material resulting from landfill, mining spoil, garbage dumps, dredgings, and similar material deposited by people and that has not been present long enough to have been subjected to soil-forming processes.

anthropochory *See* androchory.

anthropogenic Of human origin or caused by human activity. Strictly, the term pertains to human origins rather than human activities.

anthropomorphism The attribution of human characteristics, especially emotions, to non-humans.

anthrosols Soils that have been strongly affected by **cultivation**. Anthrosols are a reference soil group in the **World Reference Base for Soil Resources**.

anticlinal At right angles to a surface, e.g. describes the **cell wall** that is perpendicular to the surface of the plant. *Compare* periclinal.

anticlinal division Cell division in which the wall between **daughter cells** is **anticlinal** to the plant surface, thus increasing the circumference of the organ. *Compare* periclinal division.

anticyclone An area in which the surface air pressure is higher than in the surrounding air, and pressure decreases with distance from the centre.

anticyclonic Describes the direction of the wind around an **anticyclone** or **ridge**. This is clockwise in the Northern Hemisphere and anticlockwise in the Southern Hemisphere.

antigibberellin A chemical compound, e.g. **maleic hydrazide**, that has an effect on plants opposite to that of **giberellins**, or causes stunting.

antipetalous Describes a plant structure that occurs in front of or opposite to the **petals**, rather than alternating with them.

antipodal The location opposite the **micropyle** end of an **ovule**, usually applied to the **haploid nuclei** formed during **sporogenesis**.

antipredator behaviour Behaviour by which animals seek to deter predators, e.g. by forming compact groups such as schooling in fish, or by mobbing, e.g. by small birds against a raptor.

Antirrhinum (family **Scrophulariaceae**) A genus of **herbs** in which the upper leaves are **alternate** and the lower leaves **opposite**; all leaves are **simple**. Flowers, borne in **racemes**, have strongly 2-lipped **corollas**, the lower lip with 3 lobes projecting forward. There are about 20 species, native to southern Europe, all of which are able to form fertile **hybrids** with each other and with *A. majus*, the snapdragon, which is cultivated widely.

antisepalous Opposite or in front of the **sepals**.

antlers Outgrowths that are borne on the heads of deer (**Cervidae**). They are made from bone and are shed each year at the end of the mating season (rut) and regrown the following year, each year producing more branches. As they grow they are covered with skin richly supplied with blood vessels (velvet) that is shed when growth is complete. Except for reindeer, also called caribou (*Rangifer tarandus*), only male deer have antlers.

ant lions *See* Neuroptera.

ants *See* Apocrita, Hymenoptera.

Anura (frogs, toads) An order of **Amphibia** in which adults lack tails, the backbone is short, the forelegs are stout and the hind limbs adapted for jumping and swimming and with webbed digits, and the eyes are large and prominent high on the head. Anurans lay eggs that are fertilized externally. Males are often smaller than females. Most species are described as either frogs or toads, but there is no scientific difference between them. There are more than 4000 species, found in all continents except Antarctica.

Anystidae A family of reddish, soft-bodied mites (**Arachnida**), 0.5–1.5 mm long, that are fast-running, generalist predators, attacking any arthropod they chance to encounter and are able to overcome, their prey including other mites, springtails (**Collembola**), bugs (**Hemiptera**), leafhoppers (**Cicadellidae**), aphids (**Aphididae**), and small caterpillars (**Lepidoptera**), making them important agents of **biological control**. There are about 20 genera with 100 species. They live in the soil and on plants and have a worldwide distribution.

AONB *See* Area of Outstanding Natural Beauty.

aorta The main **artery** in mammals, conveying blood to all parts of the body.

aperturate Having openings (apertures).

aperture The area of a mollusc (**Mollusca**) shell in which the animal lives and from which it emerges.

apetalous Lacking **petals**.

apex 1. The outermost (distal) part of a stem, leaf, shoot, or root. **2.** The pointed tip of a conical gastropod (**Gastropoda**) shell.

APG *See* Angiosperm Phylogeny Group.

aphanic species *See* sibling species.

Aphanopetalaceae (order **Saxifragales**) A family of **shrubs**, stragglers, and climbers, most with **opposite**, **linear** leaves with **entire** margins. Flowers are **actinomorphic**, **bisexual**, and 4-merous, **apetalous** but with 4 large, white **sepals**, 8 **stamens** in 2 whorls, **ovary** semi-**inferior** with 4 **locules** each with a single **ovule**. The fruit is a single-seeded **nut**. There is one genus and two species found in western and eastern Australia. They are cultivated for ornament on a small scale.

Aphelinidae A family of small (up to 1.5 mm long), yellow or brown wasps (**Hymenoptera**), most of which are **parasitoids** of **Hemiptera**, especially **Aphididae**, **Aleyrodidae**, and **Coccidae**, and some feeding on the eggs of **Lepidoptera**, **Orthoptera**, and **Diptera**. They are the most widely used of all **biological control** agents. There are 33 genera and 1168 species with a worldwide distribution.

Aphelocoma californica (western scrub-jay) A blue, white, and grey corvid (**Corvidae**) bird, 270–310 mm long with a 400 mm wingspan, that inhabits low scrub in western North America. It feeds on small vertebrates, seeds, and berries, and caches food for later use.

aphicide An **insecticide** that kills aphids (**Aphididae**).

aphid *See* Aphididae.

Aphididae (aphids) A family of soft-bodied insects of the suborder **Homoptera** that includes greenfly, blackfly, and plantlice. Aphids feed on plant sap by means of a **rostrum** between the front pair of legs, and some are serious pests, damaging plants directly or as disease **vectors**. Many species are tended by ants (**Formicidae**), which feed on the **honeydew** excreted by the aphids. Most aphids reproduce sexually once a year, laying eggs that overwinter, commonly on woody plants. The following year several generations reproduce asexually and viviparously (*see* vivipary). There are at least 4000 species found mainly in temperate regions of the Northern Hemisphere.

Aphis fabae (black bean aphid) An aphid (**Aphididae**) that is black or olive green, 1.3–2.6 mm long in the winged form and 1.5–3.1 mm long in the wingless form. Eggs spend the winter usually on spindle (*Euonymus europaeus*), hatching in early spring. Winged forms appear in early summer and populations peak in late summer. Ants often attend the aphids, which feed on many plant species. They are serious pests of field beans, broad beans, most garden beans, and sugar beet.

Aphis grossulariae (gooseberry aphid) A species of grey-green aphids (**Aphididae**) that infest the shoot tips of gooseberry **bushes**. Infestations can cause severe deformation of plant shoots. **Alate** individuals 1.6–2.4 mm long, dark green to yellow, **apterous** individuals 1.2–1.8 mm with black head and thorax and green abdomen with dark stripes. The aphids are usually attended by ants.

Aphloiaceae (order **Crossosomatales**) A **monotypic** family (*Aphloia theiformis*) of **evergreen shrubs** or **trees** with **distichous**, **serrate** leaves and often persistent **stipules**. The sweet-scented, **actinomorphic**, **bisexual**, **axillary** flowers have a **hypanthium**, undifferentiated **perianth**, many **stamens**, and a single **carpel**. The fruit is a white, fleshy **berry**. The plants occur only in East Africa, Madagascar, the Mascarenes, and Seychelles.

aphyllous Lacking leaves.

Apiaceae (order **Apiales**) A family (formerly known as Umbelliferae) of **annual**, **biennial**, and **perennial herbs**, some with **stolons**, others **rosette** or **cushion plants**. Leaves are **alternate**, without **stipules**, and usually dissected. The **inflorescence** is usually a **compound umbel**, but sometimes a **raceme** or **panicle**. In the umbels each **pedicel** rises from the same point on the **peduncle** and they are of different lengths so the flowers are all at the same height with the small umbels (umbellets or umbellules) arranged in umbels. There are about 434 genera and 3780 species with worldwide distribution but especially in temperate regions. The plants have many uses. ***Daucus** carota* is the carrot, *Pastinaca sativa* the parsnip, and the family also includes culinary herbs and spices.

Apiales An order of **dioecious** plants with articulated **pedicels**, bearing terminal, branched **inflorescences** of small flowers. It contains 7 families, 494 genera, and 5489 species. *See* Apiaceae, Araliaceae, Griseliniaceae, Myodocarpaceae, Pennantiaceae, Pittosporaceae, and Torricelliaceae.

apical dominance A condition, controlled by **auxins** produced in the apical **bud**, in which the stem **apex** prevents the growth of **branches** close to the apex. Removing the apical bud allows **branches** to develop.

apical meristem **Meristem** cells at the growing tip of a shoot or root.

Apidae (bumblebees, carpenter bees, cuckoo bees, honeybees, orchid bees, stingless bees) A family of hymenopterans (**Apocrita**) that transport **pollen** in corbiculae (*see* corbiculum) on the outside of their hind legs. Stingless bees (*Melipona* and *Trigona* spp.) and **honeybees** are among the most highly social of all insects. There are more than 200 genera and about 5750 species with a worldwide distribution.

Apis A genus of bees (**Apidae**) all of which are **eusocial**. There are at least four species: *A. florea* (dwarf bee) of southern Asia; *A. dorsata* (giant bee, rock bee) of Asia; *A. cerana* (Asian bee) of Asia; and *A. mellifera* (common bee, honeybee) of Eurasia and Americas. Only *A. cerana* and *A. mellifera* have been domesticated.

Apium (family **Apiaceae**) A genus of **glabrous herbs**, most with **pinnate** leaves and flowers forming **umbels** opposite the leaves. Fruits are oblong or broadly ovoid. There are about 20 species with **cosmopolitan distribution**, some commercially important. *Apium graveolens* var. *dulce* is celery and *A. graveolens* var. *rapaceum* is celeriac.

aplanate Arranged in a plane.

aplanetism The condition in which no **motile** stage occurs.

aplanogamete A non-**motile gamete**.

aplanospore An asexual, non-**motile spore**.

apocarpous The condition in which the **carpels** are free.

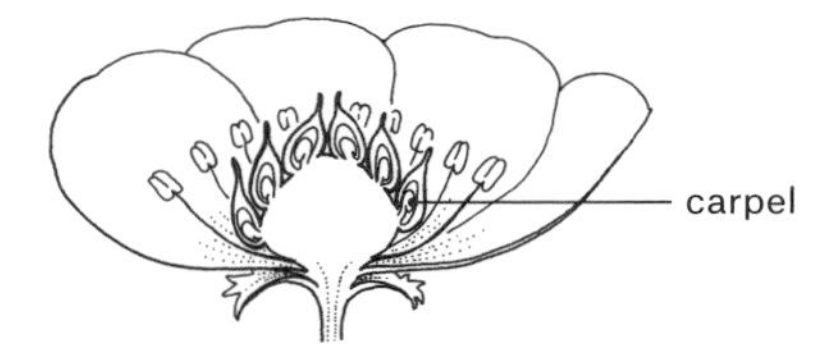

The carpels are free, rather than being fused together.

apocrine gland A gland in which the **apex** breaks down after it has released its secretion. It opens near the skin surface close to a hair follicle and bacterial action imparts an odour to the odourless secretion.

Apocrita (ants, bees, wasps) The larger of the two suborders of **Hymenoptera** and distinguishable from the **sawflies** by the tight constriction (petiole) between the first two segments of the abdomen. There are many families and about 105,000 species distributed throughout the world.

Apocynaceae (order **Gentianales**) A large family of **trees**, **shrubs**, and **lianas** that climb by twining, and **herbs**, mostly **perennial** but some **annual** or **ephemeral**

and some **succulent**, producing a white (occasionally yellow or red) **latex**. Leaves are **simple** and usually **entire**, **opposite**, **alternate**, or whorled. Flowers are **actinomorphic**, **bisexual**, most **pentamerous** (but not the **carpels**), **petals** fused at least partly, **corolla rotate**, **campanulate**, tubular, or funnel-shaped. **Stamens** are inserted on the corolla tube or at its base, **ovary superior** to subinferior, **syncarpous**, with 2 **carpels**. Fruits are **dehiscent follicles**, but **drupes**, **berries**, and **capsules** in the subfamily Rauvolfioideae. There are 415 genera with 4555 species forming 5 subfamilies, distributed worldwide apart from polar regions. Many species are used as sources of arrow poisons, medicines, and India rubber, or are cultivated as ornamentals.

Apodanthaceae (order **Cucurbitales**) A family of parasitic **herbs** with no **chlorophyll** that resemble a fungal **mycelium**, consisting entirely of tissues residing inside the host with only the flowers visible outside the host. There is no stem and the leaves are reduced to scale-like **bracts** subtending the flowers, which are small, **unisexual** with 3 or 4 **whorls** of 2–6 bracts. There are 3 genera with at least 23 species found from California southward through South America, East Africa, southwestern Asia, and southwestern Australia. Different species parasitize genera in the **Brassicaceae** and **Flacourtiaceae**.

Apodemus sylvaticus (field mouse, long-tailed field mouse, wood mouse) A species of mice (**Muridae**) that are 60–150 mm long with a tail 70–145 mm long. They are grey-brown with a pale underside, white feet, and large ears and eyes, and have an extremely keen sense of smell that allows them to locate buried seeds without digging randomly. They jump and swim well. They live in grassland and cultivated land, sometimes moving into houses in winter, but usually wintering in nests they construct in deep burrows. They feed on seeds, berries, roots, and insects, and can cause considerable damage to cultivated plants by digging up seeds before they germinate and damaging roots. They occur throughout most of Europe and in parts of central and southwestern Asia.

apodous Lacking legs.

apoenzyme The **protein** portion of an **enzyme** that has a non-protein component.

apogamy Reproduction without **fertilization**, i.e. the formation of **sporophytes** by **parthenogenesis** of **gametocytes**. It occurs most commonly in ferns (**Pteridophyta**).

apogeotropism **Tropism** in which plant organs grow in the opposite direction to the force of gravity.

apomict A plant produced by **apomixis**.

apomixis **Asexual reproduction** without **fertilization**, i.e. excluding **vegetative reproduction**. Since the progeny are genetically identical, they constitute **microspecies**. In **angiosperms** apomixis commonly involves the production of seeds; it is then known as agamospermy.

apomorph An evolutionarily advanced (i.e. derived) **character** state.

apomorphy species concept A definition of **species** as all the descendants of a single ancestral population, recognizable by their possession of **apomorph character** states.

Aponogetonaceae (order **Alismatales**) A **monocotyledon** family of **perennial** aquatic plants with **corms** or short **rhizomes**. Leaves are arranged spirally around the base of the stem; floating leaves are **ovate** or **lanceolate**, submerged leaves are **linear** to **oblong** or **elliptic**. The **inflorescence** is **spicate**, borne on long stalks and held above the water. Flowers are usually small and **zygomorphic**, usually with 2 white, persistent or **caducous tepals**, 6 **stamens** in **whorls** of 3, 2–9 free, **sessile**, **superior ovaries**, with usually 3 **carpels**. Fruits are

follicles. There is 1 genus with 43 species, found throughout the Old World, mainly in tropical and warm temperate regions. Some have edible tubers and several are popular aquarium plants.

apopetalous With the **petals** separate.

apophysis 1. A swollen region between the **seta** and **capsule** in the **sporophyte** of mosses (**Bryophyta**). **2.** A projection from a bone, usually for muscle attachment.

apoplast The parts of a plant that lie outside the **cell membrane**, e.g. **cell walls** and **xylem**. *Compare* symplast.

apoptosis The controlled, regulated, death of a cell; programmed cell death.

Aporrectodea caliginosa (grey worm) A species of **anecic** earthworms (**Annelida**), 50–150 mm long and 2–4 mm wide, with a body **anterior** to the **clitellum** in three shades of grey, off-white, and brown. It lives in topsoil and feeds on soil. It is one of the most common of British earthworms, found in woodland, pasture, and gardens. It occurs throughout the **Palearctic** and has been introduced to North America.

Aporrectodea icterica (mottled worm) A species of **anecic** earthworms (**Annelida**) that are variable in colour, commonly grey, brown, or yellow, 50–140 mm long and 3–6 mm wide, with a long **clitellum** located farther to the posterior than in most worms. The worm inhabits gardens, orchards, and meadows, and occurs throughout western Europe and eastern North America.

Aporrectodea longa (black-headed worm) A species of **anecic**, grey or brown earthworms (**Annelida**) with a dark-coloured head, 80–120 mm long, that live in permanent burrows and feed on soil, producing worm casts up to 50 mm high around the entrances. They live in **alkaline soils** and are common in gardens. It is common in Europe and North America and has been introduced in Australasia.

Aporrectodea rosea (rosy-tipped worm) A species of **anecic** earthworms (**Annelida**), 20–110 mm long and 2–4 mm wide, with a body that is pale from the anterior end to the orange **clitellum**, then pale or rose-pink. It occurs in woodland, pastures, and gardens worldwide.

aposematic coloration *See* aposematism.

aposematism (aposematic coloration) Bright skin colours or prominent patterns born by a potential prey animal to warn predators that they are dangerous, e.g. venomous or possessing a formidable sting, or unpalatable.

aposepalous With the **sepals** separate.

apospory The development of a **diploid embryo sac** by the division of a **nucellus** or **integument** cell without undergoing **meiosis**. It is a form of agamospory (*see* apomixis).

apotepalous With the **tepals** separate.

apothecium A disc- or cup-shaped **ascocarp** in which the asci (*see* ascus) line the inner surface, so they are exposed to the atmosphere.

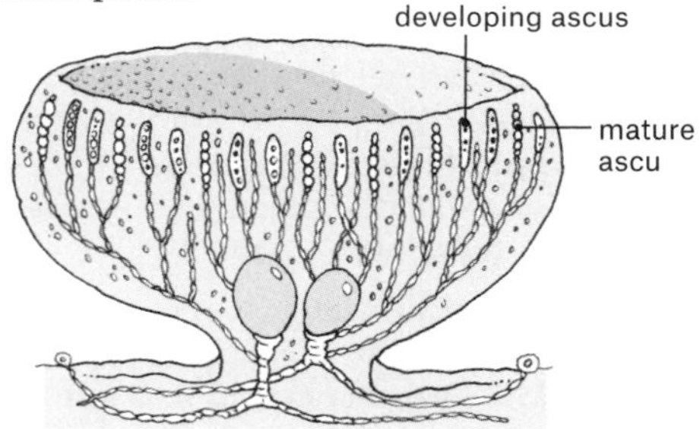

Apothecium. The asci are exposed to the atmosphere inside the open, dish-shaped structure.

appeasement Behaviour with which one animal seeks to prevent an attack by another without trying to escape or avoid the aggressor.

appendicular skeleton The part of the vertebrate skeleton consisting of the fins or limbs and shoulder and pelvic girdles.

appetitive Describes the behaviour of an animal that appears to be in pursuit of a goal.

apple *See Malus.*

apple bark louse *See Lepidosaphes ulmi.*

apple blossom weevil *See Anthonomus pomorum.*

apple canker *See Nectria galligena.*

apple capsid *See Plesiocorus rugicollis.*

apple comma scale *See Lepidosaphes ulmi.*

apple leaf curling midge *See Dasineura mali.*

apple leaf midge *See Dasineura mali.*

apple maggot *See Cydia pomonella.*

apple mussel scale *See Lepidosaphes ulmi.*

apple rust mite *See Aculus schlechtendali.*

apple sawfly *See Hoplocampa testudinea.*

apple scab *See Venturia inaequalis.*

apple sucker *See Psylla mali.*

apple weevil *See Anthonomus pomorum.*

apple worm *See Cydia pomonella.*

appressorium A flattened **hypha** that presses close tissue of the host plant at an early stage of infection by certain parasitic **Fungi**.

apricot *See Amygdalus.*

Aptandraceae (order **Santalales**) A family of **trees** and **shrubs** that probably are root parasites. Leaves are **simple**, **alternate**, and **exstipulate**. Flowers are **actinomorphic**, **bisexual** or plants **dioecious**, **tetramerous** or **pentamerous**, **apopetalous** or **sympetalous**, 4–5 free or fused **stamens**, **ovary superior** with 1 **locule**. The fruit is a **drupe** or **nut**. There are 8 genera with 34 species found throughout the tropics.

apterous Wingless.

Apterygota The smaller of the two subclasses of **Insecta** (*see* Pterygota) comprising insects that are primarily wingless, i.e. they are descended from wingless ancestors. They are terrestrial and have biting mouthparts. Most are free-living, 3–15 mm long, develop directly from larva to adult and continue to moult throughout life, and feed on **Fungi**, **pollen**, **lichens**, and small amounts of **protein**. There are about 2800 species with a worldwide distribution. ↗

aquiclude (aquifuge) A rock that is impermeable to **groundwater**, so it will confine an **aquifer** horizontally or form a vertical boundary blocking its flow.

aquifer A body of permeable rock, e.g. gravel or sand, through which **groundwater** flows, and that lies above a layer of impermeable material. If the permeable rock continues to the surface with no overlying layer of impermeable material, the aquifer is said to be unconfined.

Aquifoliaceae (order **Aquifoliales**) A family of **shrubs** and small **trees** with some **epiphytes**, usually **evergreen** but a few **deciduous**, with leaves that are **alternate** (occasionally **opposite**), **simple**, often leathery, with spiky **dentate** margins, but with dentate and **entire** leaves on the same or separate plants in some species. Flowers are small, inconspicuous, white or cream to pale green or yellow, **actinomorphic**, **unisexual** (plants are often **dioecious**), with 4–5 or 6–8 **sepals**, 4–5 or occasionally 6–9 or more **petals**, usually as many **stamens** as petals. **Staminodes** occur in **pistillate** flowers and **pistillodes** in **staminate** flowers. **Ovary superior** with 2–3, 4–5 or more united **carpels**, each with 1 **locule**. Flowers are borne in **axillary cymose**, **racemose**, or sub-umbellate **inflorescences**. Fruit is a **drupe**. There is 1 genus (*Ilex*) with 405 species, with worldwide distribution. Many species are grown for ornament as hollies and some have medicinal uses; *I. paraguarensis* is

A

used in South America to prepare yerba maté.

Aquifoliales An order of plants with **arcuate petiole** bundles, **valvate corolla**, and fruits that are 1-seed **drupes**. There are 5 families, with 21 genera and 536 species. *See* Aquifoliaceae, Cardiopteridaceae, Helwingiaceae, Phyllonomaceae, and Stemonuraceae.

aquifuge *See* aquiclude.

Aquilegia (family **Ranunculaceae**) A genus of poisonous **perennial herbs** with **alternate**, **compound** leaves divided into 2 or 3 lobes. Flowers are showy, **actinomorphic**, **pentamerous** with **petaloid sepals**, **petals** with long spurs, many free **stamens** and **carpels**, and fruits that are **follicles**. There are about 70 species found throughout the temperate region of the Northern Hemisphere. Some are cultivated as columbines.

aquitard A rock that is almost, but not quite, impermeable, so **groundwater** flows through it more slowly than through more open **aquifers** nearby.

Arabidopsis thaliana (family **Brassicaceae**) A small plant (thale cress) that is widely used as a model in the study of plant **genetics** because of its small **genome**, comprising 25,498 genes that encode **proteins** from 11,000 protein families.

Arabis (family **Brassicaceae**) A genus of **annual** or **perennial calcicole herbs** with **simple**, **entire** to lobed, **oblong**, **sessile** leaves that are usually densely covered in hairs, and small, white, pink, purple, or pale yellow flowers with 4 **petals**, borne in **racemes**. The fruit is a long **capsule**. There are about 100 species found throughout the temperate region of the Northern Hemisphere. Some species are considered weeds, others are cultivated for ornament as rock cress.

Araceae (order **Alismatales**) A family of **monocotyledon perennial herbs** with a variety of forms, occurring as **geophytes**, **epiphytes**, climbers, and a few as free-floating aquatics. The leaves usually have net veins and vary in shape (those of *Monstera deliciosa*, the Swiss cheese plant, develop **fenestrated** leaves). Roots are **adventitious** and often thick, growing from **rhizomes** or **tubers**; the aquatics have simple roots or lack roots entirely. Flowers are **sessile**, **bisexual** or **unisexual**, usually with male flowers borne above female flowers in a bisexual **spadix inflorescence**. **Tepals** are often more or less **connate** in bisexual flowers and absent in unisexual flowers. Usually there are 2 **whorls** each of 2–3 **stamens**. The **ovary** is **superior** or embedded in the inflorescence, with 1–3 **carpels** and **locules**. The fruit is a **berry**, **drupe**, or occasionally a **utricle**. There are 117 genera with 4095 species (possibly more) found mainly in the tropics. Several species have edible tubers (*see Colocasia*). Others are grown as ornamental garden (e.g. *Arum*) or house plants.

Arachnida (harvestmen, mites, palpigrades, pseudoscorpions, scorpions, spiders, whipscorpions) A class of **Arthropoda** that have **book lungs** or tracheae evolved from gills, indicating their descent from aquatic ancestors. Most now live on dry land. Except in mites, the body is in two portions, the anterior **prosoma** bearing 4 pairs of legs, up to 12 eyes, **chelicerae**, and **pedipalps**, the posterior **opisthosoma** holding most of the body organs. Many mites are parasites, but most other arachnids are predators or scavengers. There are more than 60,000 species, with a worldwide distribution.

aragonite A mineral ($CaCO_3$), chemically identical to **calcite** but different in its physical properties and less stable, from which the shells of living or recent molluscs (**Mollusca**) are made; in **fossil** shells the aragonite is converted to calcite or replace by another mineral. It is named after Aragon, Spain.

A

Araliaceae (order **Apiales**) A family of **evergreen shrubs**, **trees**, and some climbers or rhizomatous (*see* rhizome) **herbs** with **alternate**, often large, **compound pinnate** or **palmate** leaves (but sometimes **simple**) and sometimes crowded at the ends of shoots. **Stipules** are small or conspicuous and leaf bases are sometimes sheathing. The small, greenish flowers are **actinomorphic**, **bisexual** or **unisexual** (when plants are **dioecious**; rarely there are bisexual, male, and female flowers on the same plant). There are 4–5 reduced **sepals**, 5–10 free or partially fused **petals**, 5–10 free **stamens**, and a **disc** above the **ovary**. The ovary is **inferior** with 5–10 fused **carpels** each with 1 pendulous **ovule**. The fruit is a **drupe** or sometimes **baccate**, with 2–5 seeds. There are 43 genera with 1450 species with a cosmopolitan but mainly tropical distribution. Three *Panax* species yield ginseng, Siberian ginseng is obtained from *Acanthopanax senticosus*, and Chinese rice paper is made from *Tetrapanax papyrifera*. Varieties of *Hedera helix* (ivy) are widely cultivated.

Araneae (spiders) A large order of **Arachnida**, all of which are predators feeding on other **Arthropoda**, except for some large species that catch nestling birds, tadpoles, and small fish. The **prosoma** has an upper **carapace** bearing two to eight eyes, the **chelicerae** are large and bear fangs that inject venom, and the **pedipalps** resemble legs. Males use their pedipalps to transfer sperm during mating. There are more than 50,000 species found throughout the world except for Antarctica.

Araucaria (family **Araucariaceae**) A genus of tall, **monopodial**, coniferous **trees** with radial limbs. Leaves are **lanceolate**, in crowded **whorls**, tough, and with sharp points. In some species the leaves are narrow, in others broad and overlapping. Most species are **dioecious** with male and female **cones** on separate trees, but occasionally individual trees are **monoecious** or change gender. Female cones are **globose**, contain 80–200 edible seeds, and are borne near the top of the tree. Male cones are smaller and cylindrical. There are 19 species with a **disjoint distribution** in Chile, Argentina, and southern Brazil, and Australia, New Guinea, New Caledonia, and Norfolk Island. *Araucaria angustifolia* yields paraná pine, *A. bidwillii* is bunya bunya of Queensland, *A. araucana* is widely cultivated as the monkey puzzle tree.

Araucariaceae (order **Pinales**) A family of large coniferous **trees** that have **branches** in whorls and leaves that are broad to **lanceolate**. Branches are periodically shed. The male **cone** is a **catkin**, the female cone is **globose** and often large, disintegrating on maturing. Winged seeds develop from a **bract** scale. There are 3 genera with 33 species found in southern South America and from Malesia to eastern Australia and New Zealand.

arbuscule A tuft of branching **hyphae** produced by certain **mycorrhizae**.

Arbutus (family **Ericaceae**) A genus of **shrubs** and small **trees** that have red, flaking **bark**. Flowers are small, bell-shaped, and white or pink. The fruit is a red **berry** somewhat resembling a strawberry, hence the common name strawberry tree. Fruit develops about five months after pollination so flowers open while the previous season's fruit is still ripening. There are about 14 species, found in North America, parts of western Europe, and the Mediterranean region. U.S. (but not Canadian) species are known as madroñes. North American *A. menziesii*, madroña laurel, produces useful timber and bark rich in tannins, used in tanning. *Arbutus unedo* is widely cultivated.

Archaea One of the three **domains** into which organisms are divided in the three-domain system of classification, and in the older **five-kingdom system** a subkingdom (also called Archaebacteria) in the kingdom **Bacteria**. Archaea are single-celled organisms, including **methanogens**, **extremophiles**, and sulphur-reducing organisms, that are phylogenetically (*see* phylogeny) distinct from Bacteria and **Eukarya**.

Archaefructus A genus, now extinct, of **seed plants** (Spermatophyta) that lived about 125 million years ago, during the Early Cretaceous epoch. It was an aquatic **herb** that lacked **sepals** and **petals**, and with **carpels** and **stamens** borne on a long stem, rather than in a **flower**. It is one of the earliest genera of **angiosperms**, with three known **species**.

arch cloud A stationary cloud, usually **altostratus**, that forms a long arch above a mountain range. A **foehn wind** blows down the mountainside beneath the arch and the appearance of an arch cloud indicates the approach of the wind.

Archaeocalamites scrobiculatus (formerly *A. radiatus*) One of the earliest recorded horsetails (Equisetaceae), which lived during the early Carboniferous (Mississippian, 359.2–318.1 million years ago). Like all equisetaleans it had a jointed stem with a ring of short **branches** at each joint.

Archaeopteris A genus, now extinct, of ancestors of the **gymnosperms** that lived during the Late Devonian epoch (385.3–359.2 million years ago). It was the first modern **tree**, growing up to 30 m tall, with a trunk more than 1 m in diameter, and it formed extensive forests in many parts of the world.

Archaeosporales An order of mycorrhizal (*see* mycorrhiza) **Fungi** that form either **arbuscules** or **endocytosymbioses** with **cyanobacteria**.

archegoniophore In some liverworts (**Marchantiophyta**) the structure bearing the archegonia (*see* archegonium).

archegonium In **Marchantiophyta**, **Bryophyta**, **Pteridophyta**, and most **gymnosperms**, the female sex organ.

archesporium The tissue from which **spore** mother cells develop.

Archilochus colubris (ruby-throated hummingbird) A hummingbird, 75–90 mm long with an 80–110 mm wingspan, in which adults are iridescent green on the back and head, and pale grey on the underside. Males have a bright red throat and forked tail, females and juveniles a grey throat and square tail. They feed on nectar, especially from red flowers, and small insects. During the breeding season they inhabit forests and forest edges, and gardens. They winter in southern Mexico, Central America, and the West Indies. The birds breed throughout the eastern United States and in southern Canada.

Archips podana (fruit tree tortrix moth, large fruit-tree tortrix moth) A light brown tortrix moth (**Tortricidae**) with a wingspan of 18–26 mm that is active in midsummer. Females lay large batches of eggs the colour of leaves. **Caterpillars** are yellow with black heads and become grey-green with brown heads as they mature. They feed on fruit **buds** and blossoms, then join young leaves together to form shelters in which they pupate. The moth occurs through out Europe and parts of Asia, and has become established in North America.

arctic-alpine species Species that occur both in high latitudes and high elevations, e.g. ***Salix** herbacea* (least willow).

arctic and subarctic floral region The area covering the whole of northern Canada, Alaska, and Eurasia, including Iceland, Greenland, and the islands of the Arctic Basin, part of the **boreal region**. The region contains few **endemic** species if any.

arctic heath **Heathland** that occurs in the low and middle arctic **tundra**, usually in well drained, sheltered sites that are covered by snow in winter. Heaths (**Ericaceae**) commonly dominate the vegetation.

arctic scrub Vegetation of the **low arctic tundra** that occurs in moist hollows and beside open water and is dominated by plants averaging 60 cm in height, e.g. dwarf willows (***Salix***) and birches (***Betula***).

arcuate Arched or curved.

Ardis brunniventris (rose shoot sawfly, rose tip infesting sawfly) A black sawfly (**Symphyta**), 5.5–6.5 mm long with brown-white larvae up to 12 mm long that feeds on rose plants (***Rosa***). Adults are active from early summer. They lay eggs in leaf tissue in young terminal **buds**. The larvae feed on the tissue, then bore into the tip of the shoot, leaving through an exit hole when fully grown. Larvae overwinter in the soil and pupate in spring. The damage can be severe. The sawfly occurs throughout Europe and is present in North America.

area cladistics A technique biologists use to reconstruct the past distribution of organisms and the positions of continents independently of geological data. By comparing the **genomes** and **morphology** of several groups it is possible to determine patterns of relationships that allow their former geographic distributions to be inferred. Plotting the resulting patterns reveals the location of former barriers (e.g. mountain chains, oceans) to migration that led to reproductive isolation and speciation.

Area of Outstanding Natural Beauty (AONB) In England, Wales, and Northern Ireland, a landscape designated under the National Parks and Access to the Countryside Act 1949 as possessing a distinctive character and natural beauty that merits their protection in the national interest. There are 33 AONBs in England, 4 in Wales, and 1 straddling the England–Wales border. ↗

Arecaceae (order **Arecales**) A **monocotyledon** family (formerly known as Palmae) of **trees**, **shrubs**, and climbers, all of which have woody stems terminating in a crown of leaves. The stems are solitary or clustered, seldom branched (but **dichotomous** when it occurs), and often covered in leaf scars. The stem diameter increases only through **primary growth**, so it remains constant once the plant has matured. Leaves are usually in a spiral, rarely **distichous** or tristichous, with an initial sheath, sometimes with spines or prickles, that may split open. Leaves are **entire**, **palmate**, **pinnate**, **bifid**, or occasionally **bipinnate**, and usually **plicate**, the folds on one side of a long central **rachis** (feather palms) or radiate from a short rachis or **costa** (fan palms). Flowers occur singly or in small, **cymose** groups or large cincinnate (*see* cincinnus) clusters. They are **bisexual** or **unisexual** (plants **monoecious**, **andromonoecious**, or **dioecious**), **trimerous**, with 3 **sepals**, 3 **petals** (male flowers may have 4 of each), often more than 6 **stamens** (*Ammandra* spp. have more than 1000), **ovary superior** of 3 (or up to 10) **carpels** each with 1 **ovule**. **Inflorescences** are **axillary**, simple or as large **panicles** (with more than 10 million flowers in *Corypha* spp.) The fruit is a one-seeded **berry** or a **drupe** with one seed (but occasionally up to 10). Fruits vary in size; the double coconut (from *Lodoicea maldivica*), weighing up to 18 kg, is the world's largest fruit and seed. There are 183 genera with 2361 species, found throughout the tropics, subtropics, and warm temperate regions. Palms supply construction materials, fibres, and food, and many species are grown as ornamentals.

Arecales The order that contains only one family (the **Arecaceae**), of 183 genera with 2361 species.

Arenaria (family **Caryophyllaceae**) A genus of **herbs** and small **shrubs** that have small, **opposite** leaves and small, usually white flowers with 5 **petals**, 10 **stamens**, and 3 (or rarely more) **styles**. The fruit is a **capsule**. There are about 150 species with a **cosmopolitan distribution**, but especially common in northern temperate regions. Several mat-forming species are cultivated rock garden plants (sandworts).

arenite *See* sandstone.

arenosols Weakly developed soils with a coarse texture. Arenosols are a reference soil group in the **World Reference Base for Soil Resources**.

areola (pl. areolae) A small area bounded by lines or crevices, e.g. the **veins** of a leaf.

areolate Divided into small areas (areolae), e.g. by the **veins** of a leaf.

areole In cacti (**Cactaceae**), a small, rounded protrusion, sometimes lighter or darker in colour than the adjacent tissue, that bears spines or **glochids**.

Argasidae A family of soft ticks in which **nymphs** pass through up to eight stages before becoming adult and adult females lay eggs after each meal of blood. Many feed on mammals and birds, some transmitting viral diseases. There are 183 species distributed worldwide.

argic horizon (argillic horizon) A subsurface **soil horizon** that contains significantly more **clay** than the overlying horizon, derived by **illuviation**.

argillans *See* clay skins.

argillic horizon *See* argic horizon.

Argophyllaceae (order **Asterales**) A family of **shrubs** and small **trees** with **alternate**, **simple**, **exstipulate** leaves with **entire** or **dentate** margins. Flowers are small, yellow, green, or white, **actinomorphic**, **hermaphrodite**, 4- to 5-merous (or 8), **petals** usually **valvate**, **ovary inferior** with 1–2 (6) **locules**. The fruit is a single-seeded **drupe**. There are 2 genera with 21 species found in eastern Australia and New Zealand. Several are grown as ornamental shrubs.

aridisols An order of soils consisting mainly of mineral particles with very little organic matter near the surface, but often rich in calcium carbonate or **gypsum** ($CaSO_4\ 2H_2O$), with or without accumulated soluble salt. These are desert soils, prone to **erosion** and supporting only sparse vegetation.

aril A third **integument**, forming an outgrowth from a seed **hilum** or **funicle**; it is often fleshy. The spice mace is the dried aril surrounding the nutmeg, which is the seed, the edible parts of the pomegranate and lychee are arils, and the red aril of the yew (*Taxus baccata*) is the only part of that plant which is not poisonous.

arillate Possessing an **aril**.

Arion ater (black slug, black arion) A species of round back slug (**Arionidae**) that is usually black but can be brown, red, or white. It reaches maturity when it is 25 mm long, but eventually grows to 100–150 mm. It is covered in a mucus that tastes foul and deters predators. The slug is mainly nocturnal, avoiding bright light, and feeds on plant material, carrion, and fungi. It occurs throughout western Europe and northwestern North America.

Arion hortensis (garden slug, black field slug, small striped slug, yellow-soled slug) A round back slug (**Arionidae**), 25–35 mm long, with a black back, paler sides, and yellow foot. It inhabits woodland, cultivated ground, and gardens, and feeds on plants of all kinds. It is a serious pest of strawberries, lettuces, and seedlings. It belongs to a species complex that also includes the very similar *A. owenii* and *A. distinctus*. These are native to western and southern Europe, and have been introduced in North America, and Australasia.

Arionidae (round back slugs) A family of terrestrial, air-breathing **slugs** (**Gastropoda**) that have no internal shell, the respiratory pore at the centre of the **mantle**, and that grow up to 250 mm long. If threatened they retract their heads, becoming hemispherical, and fix themselves firmly to the substrate. Most feed on plant material but some are scavengers and a few are predators of other gastropods and small invertebrates. There are 54 species occurring in North America and Eurasia.

Arion rufus (chocolate arion, European red slug, large red slug, red slug) A species of round back slugs (**Arionidae**), 70–140 mm long, that are orange, red, brown, or black in colour. They have no **keel** and the respiratory pore is near the front of the **mantle**. They are very similar to ***Arion ater*** and occur in woodland, near coasts, and in gardens. It probably originated in Europe but has been introduced in many other parts of the world.

Aristolochiaceae (order **Piperales**) A family of rhizomatous (*see* rhizome) **herbs**, **subshrubs**, and **lianas** with **alternate**, **simple** leaves. Flowers, solitary or in terminal or **axillary cymes** or **racemes**, are **actinomorphic** to strongly **zygomorphic**, **bisexual**, the strongly developed **calyx** fused into a tube with 3 equal lobes or a longer tube forming an S-shaped pitcher with a swollen base and 1 lobe much expanded. The **corolla** is absent or consists of 3 usually vestigial **petals**. There are 6–12 (or up to 40) **stamens** in 1–4 **whorls**, **ovary** semi-**inferior** with 4–6 **carpels**, usually fused, 4–6 **locules** each usually with many **ovules**. The flowers are pollinated by flies and smell strongly of carrion. The fruit is either **indehiscent** or a **capsule**. There are 5–8 genera with 480 species with worldwide distribution. They have long been used in traditional medicine, hence their common name birthwort.

Armadillidum nasatum (blunt snout pill bug, southern pill woodlouse) A terrestrial pill bug (Armadillidae) that rolls itself into an imperfect ball, with the antennae (*see* antenna) protruding, if disturbed. It is up to 12 mm long with a narrow, protruding 'snout' and a dark grey body with pale, longitudinal stripes. It produces two or more generations a year. Native to southern Europe it occurs as far north as southern England and has been introduced to North America. It also occurs in greenhouses. It feeds on plant roots and causes damage to many plants, especially seedlings.

Armeria (sea pink, thrift) *See* Plumbaginaceae.

Armillaria (bootlace fungus, honey fungus) A genus comprising about 40 species of **agaric fungi**, found worldwide, that are parasites of trees and shrubs, and occasionally of **perennial herbs**. The **fruiting body** is mushroom-like with a yellow-brown, sticky **pileus** and the **stipe** has no **volva**. The fungus spreads by **rhizomorphs** that resemble bootlaces, but infection can also spread through contact between roots of adjacent plants. The rhizomorphs kill the roots and spread to the stem, forming strands beneath the **bark**, eventually **girdling** and killing the entire plant. The fruiting body is edible when cooked, and highly prized.

armoured scale insects *See* Diaspididae.

army worms *See* Noctuidae.

arrowroot (*Tacca leontopetaloides*) *See* Taccaceae; (*Curcuma angustifolia*) *See* Zingiberaceae.

arsenic (As) A metalloid element that occurs widely in soils, at average concentrations of 1–10 parts per million. It is toxic and carcinogenic in humans, and bioaccumulates (*see* bioaccumulation). Arsenic compounds have been used as **insecticides**, **fungicides**, and wood preservatives, but have been or are being phased out.

arsenical pesticides **Insecticides** made from **arsenic** compounds. They are

A

effective, but are being replaced by less toxic products.

Artemisia (family **Asteraceae**) A genus of **perennial** (with some **annual**) **herbs** and low **shrubs** with **alternate**, **pinnate** leaves divided into narrow segments. Flowers are very small, tubular, and overlapping, with a flat, naked **receptacle** and **bracts** having **scarious** edges. The **inflorescence** is a **cyme** or **raceme**. There are 200–400 species, most occurring in semi-arid grasslands of the Northern Hemisphere and arctic. The plants have many uses and cultural associations. *Artemisia vulgaris* is common mugwort, *A. tridentata* is big sagebrush, *A. absinthum* is wormwood, *A. abrotanum*is southernwood, and *A. dracunculus* is tarragon.

artery A blood vessel that conveys oxygenated blood away from the heart.

Arthoniomycetes A class of **Ascomycota** that comprises **Fungi** that have **bitunicate** apothecia (*see* apothecium). Most Arthoniomycetes are **mycobionts** in tropical or subtropical **lichens**.

Arthrobacter A genus of **Actinobacteria** that are **obligate aerobes**, common in soil. They are Gram-positive (*see* Gram reaction) and are rods while growing and cocci in their resting stage. They are able to degrade a number of toxic substances, including pesticides.

Arthropoda A **phylum** of animals with jointed limbs and a body enclosed in an **exoskeleton** made from **chitin**. The phylum includes insects, arachnids, crustaceans, centipedes, millipedes, and other groups, and accounts for more than 75 percent of all animal species.

arthrospore A fungal **spore** that forms by the fragmentation of a **hypha**.

Arthurdendyus triangulatus (New Zealand flatworm) A dark brown flatworm (**Platyhelminthes**) with a buff **ventral** surface, a smooth, unsegmented body 5–170 mm long, pointed at both ends, that rolls itself into a tight ball when at rest. It is covered with a sticky mucus. Native to New Zealand, it entered the United Kingdom in the 1960s. It feeds almost exclusively on earthworms and poses a threat to animals that feed on earthworms. It is an offence to release this animal into the wild or allow it to escape, and sightings of it should be reported. *See also Australoplana sanguinea.*

articular bone In all vertebrates other than mammals, the bone of the lower jaw that articulates with that of the upper jaw. In mammals the articular bone has become the malleus of the inner ear.

Artocarpus (breadfruit, jackfruit) *See* Moraceae.

arum lily (*Zantedeschia aethiopica*) *See Zantedeschia.*

Arundo (family **Poaceae**) A **monocotyledon** genus of strong bamboo-like **perennial**, **evergreen** or **deciduous**, rhizomatous (*see* rhizome) grasses with broad, **linear** leaves up to 60 cm long and small flowers borne in a feathery terminal **panicle**. They can grow to 5 m. There are two or three species found in the Mediterranean region and from India to China and Japan. *Arundo donax* is used in basketry, making musical instruments (the original pan pipes), reeds for clarinets and organ pipes, walking sticks, and fishing rods, and several varieties are cultivated for ornament.

arylphenalenones Orange, red, or purple pigments that occur only in the **rhizomes**, **bulbs**, **corms**, and roots of members of the **Haemodoraceae**.

As **1.** *See* altostratus. **2.** *See* arsenic.

Ascencion and St Helena floral region The region of the **Palaeotropical region** that contains these two islands. There are five **endemic** species (three being members of the **Asteraceae**). Endemism was formerly higher, but many species have been destroyed by introduced domestic animals.

ascidium Any bottle- or pitcher-shaped plant organ, e.g. the pitcher of a pitcher plant (*see* Nepenthaceae).

Asclepias (family **Apocynaceae**) A genus of **perennial herbs** with **linear**, **alternate** or less commonly **opposite** leaves and greenish or orange flowers, in some species tinged with purple and borne in hemispherical or spherical **umbels**. **Pollen** is produced in pollinia (*see* pollinium) that attach themselves to visiting insects, which detach a pair of pollinia each time one flies away. Fruits are **follicles**. There are more than 140 species, almost all confined to North and Central America. The plants produce a milky juice, hence their common name, milkweeds. Some are poisonous but their nectar is an important food source for many insects and milkweeds are the only larval food for monarch butterflies (*Danaus plexippus*), making the plants popular with gardeners.

ascocarp The fruit body (*see* fruiting body) of an **ascomycete**.

ascogonium A female reproductive organ in certain **ascomycetes**.

ascomycete A fungus belonging to the **Ascomycota**.

Ascomycota (sac fungi) A **phylum** of **Fungi** belonging to the subkingdom **Dikarya** that reproduce by **ascospores**. The phylum includes most of the **mycobionts** present in **cyanobacteria** as well as **yeasts**, e.g. *Saccharomyces cerevisiae* used in breadmaking and brewing, and some species that yield antibiotics, e.g. *Penicillium chrysogenum*, the source of penicillin. Approximately 75 percent of all fungi are **ascomycetes**. There are more than 64,000 species, occurring worldwide.

Ascophyllum nodosum *See* egg wrack.

ascorbic acid Vitamin C, a water-soluble compound found in many fruits and leaf vegetables.

ascospore A **haploid spore** that forms within an **ascus**.

ascostroma A fungal **fruiting body** in which asci (*see* ascus) form within a **stroma**.

ascus In **ascomycete fungi**, a minute sac in which **spores (ascospores)** develop. Most asci contain four or eight ascospores and often discharge them explosively.

asepalous Lacking **sepals**.

asexual reproduction Reproduction without the formation of **gametes**, e.g. **apogamy**, **apomixis**, **apospory**, **budding**, and **vegetative reproduction**.

ash *See Fraxinus*, Oleaceae.

ash dieback (Chalara dieback) A fungal disease of ash trees (***Fraxinus** excelsior* and *F. angustifolia*) that causes the loss of leaves, the death of **branches** and their tips (crown dieback), and that often results in the death of the infected tree. It is caused by *Chalara fraxinea* and its **teleomorph**, *Hymenoscyphus pseudoalbidus*.

Asian bee (*Apis cerana*) *See Apis*.

Asian lady beetle *See Harmonia axyridis*.

Asparagaceae (order **Asparagales**) A **monocotyledon** family of erect or **scandent**, **perennial herbs** or small **shrubs** with **rhizomes** (a few producing **tubers**) that lack true leaves but possess leaf-like, sometimes spiny, **phylloclades** in which they perform **photosynthesis**. The plants are **monoecioous** or **dioecious**, with 6 **tepals** joined in pairs (3+3), 3+3 **stamens**, often free, **ovary superior** with 3 **carpels** and 3 **locules**. Flowers are borne in **umbels** or **racemes**. The fruit is a **berry** or less commonly a **capsule**. There are 153 genera with 2480 species distributed worldwide. Some are cultivated as vegetables.

Asparagales A large order that

A

comprises 14 families, 1122 genera, and 26,070 species, making it one of the most diverse **monocotyledon** orders. It first appeared about 122 million years ago. *See* Amaryllidaceae, Asparagaceae, Asteliaceae, Blandfordiaceae, Boryaceae, Doryanthaceae, Hypoxidaceae, Iridaceae, Ixioliriaceae, Lanariaceae, Orchidaceae, Tecophilaeaceae, Xanthorrhoeaceae, and Xeronemataceae.

Asparagus (family **Asparagaceae**) A **monocotyledon** genus of small **shrubs**, **lianas**, and **perennial herbs**, most with **rhizomes** and green, photosynthesizing **phylloclades** replacing the leaves, which are reduced to papery scales. There are up to 300 species found throughout the Old World except Australia. *Asparagus officinale*, asparagus, is grown as a vegetable and several species for ornament.

asparagus beetle Two species of beetles, *Crioceris asparagi* (common asparagus beetle) and *C. duodecimpunctata* (spotted asparagus beetle). Both are oval and 6–8 mm long, with prominent antennae (*see* antenna). *Crioceris asparagi* has a red thorax and black **elytra** with 6 cream-coloured blotches, and pale grey larvae with a black head; *C. duodecimpunctata* has reddish orange elytra with 12 black spots, and orange larvae. The larvae of both resemble slugs. Both overwinter as eggs in sheltered places and hatch as the asparagus spears are emerging above ground, and move to the foliage to feed. After about two weeks they fall to the ground and pupate in the soil, adults appearing after about one week to start a second generation. Both larvae and adults feed on the foliage and **bark** of asparagus. *See* Chrysomelidae.

aspect The direction that sloping ground faces.

aspen (*Populus tremula*) *See Populus.*

aspergillosis A group of diseases of air-breathing vertebrates caused by **spores** of some species of ***Aspergillus***. In humans the most common form is a lung infection that can develop to pneumonia, especially in those with a weakened immune system.

Aspergillus A genus comprising several hundred species of saprotrophic (*see* saprotroph) **Fungi** that grow as moulds on substrates containing carbon. They release into the air **spores** that are inhaled by air-breathing animals. Most are harmless but those of some species can cause **aspergillosis**, and *A. flavus* and *A. parasiticus* produce **aflatoxins**.

asporogenous Describes an organism which does not produce **spores**.

assassin bugs *See* Reduviidae.

assemblage A collection of plants or animals that are characteristic of a particular type of **environment** and can be used to identify that environment.

assimilate 1. The portion of the nutrient energy absorbed or consumed by an organism that is metabolized. **2.** To perform **assimilation**.

assimilation The incorporation into the tissues of an organism of substances acquired through **photosynthesis** or by ingestion or **absorption**.

assisted migration 1. A technique used to establish a population of migratory animals in an area the species has not previously occupied by training them to move between that area and one with which they are familiar, thereby establishing a seasonal migration route. **2.** The establishment of a population beyond the edge of the historic range for that species.

associes In **phytosociology**, a sub-**climax** plant community.

assortative mating Non-random sexual reproduction, in which males and females of particular types tend to breed with each other, e.g. wind-pollinated plants of similar height.

Asteliaceae (order **Asparagales**) A family of **monocotyledon perennial herbs** with **rhizomes** in which the leaves are **linear** to **lanceolate**, arranged in often dense spirals, with a sheathing base. Flowers are small, **actinomorphic**, **unisexual** (plants usually **dioecious** or **gynodioecious**) or **bisexual** with 3–7 **tepals**, free or basally fused arranged in 2 **whorls**, 6 often-free **stamens**, **ovary superior** with 3 (or up to 7) **carpels** with 1 or 3 **locules**. The fruit is a **berry** or **capsule**. There are 4 genera and 36 species found in Chile, New Zealand to New Guinea, and in the Pacific Islands as far east as Hawaii.

Aster (family **Asteraceae**) A genus of mainly **perennial herbs** in which the stem is erect and branched, the leaves **simple**. The white, blue, red, or lilac (but never yellow) flowers have **imbricate bracts**, and the **receptacles** are naked and flat. There are about 180 species found throughout Eurasia.

Asteraceae (order **Asterales**) A family, formerly known as Compositae, of **annual** or **perennial herbs**, **shrubs**, or climbers, with some **lianas**, **trees**, and **epiphytes**, most of which possess **resin** canals or **latificers** and produce **latex**. Leaves are **alternate** or **opposite**, occasionally in **whorls**, **exstipulate**, usually **simple**, often lobed or toothed. Flowers are very small but are gathered into a **capitulum** resembling a single flower that is the family's most characteristic feature, although it is sometimes highly modified. The **florets** are surrounded by an **involucre** of protective **bracts**. The capitula form **cymose inflorescences**, commonly as terminal or terminal and upper **cymes** or **panicles**. Florets often lack a **calyx** but in some a scaly or hairy **pappus** develops in the fruit. The **corolla** is tubular or strap-like with 5 lobes, the 5 **stamens** are joined in a tube, **ovary inferior** with 1 **carpel**. In some all the florets are similar and either tubular or strap-like and female or neuter; in others the outer florets are strap-like and female or neuter and the inner ones tubular and usually **hermaphrodite**. The fruit is an **achene**. This is the largest **angiosperm** family with 1620 genera and 23,600 species, with worldwide distribution. They are very important to **biodiversity**, in some regions accounting for more than 10 percent of the **flora**. Many are cultivated for food or grown as garden flowers.

Asterales A plant order of 11 families, with 1743 genera and 26,870 species. *See* Alseuosmiaceae, Argophyllaceae, Asteraceae, Calyceraceae, Campanulaceae, Goodeniaceae, Menyanthaceae, Pentaphragmataceae, Phellinaceae, Rousseaceae, and Stylidiaceae.

Asteropeiaceae (order **Caryophyllales**) A **monogeneric** family of **trees** and scrambling **shrubs** with **alternate**, **exstipulate**, **simple**, **entire** leaves and small flowers with 5 free **sepals** and **petals**, 9–15 **stamens**, **ovary superior** with 2–3 **carpels**, and **indehiscent** fruit. Flowers are borne in **axillary** or terminal **panicles**, with the terminal unit **cymose**. There are eight species found only in Madagascar.

Asteroxylon A genus, now extinct, of vascular plants (**Tracheophyta**) that lived during the Middle Devonian epoch (397.5–385.3 million years ago), known from **fossils** found in chert at Rhynie, Aberdeenshire, Scotland. The plant was erect, up to 40 cm tall with **dichotomously** branching stems up to 12 mm in diameter bearing leaf-like protrusions, growing from a **rhizome** from which **rhizoids** extended about 20 cm below the surface. ↗

aster yellows A disease caused by the aster yellows **phytoplasma** that affects about 300 species of plants, most in the **Asteraceae**. Infected plants suffer a variety of symptoms including vein-clearing leading to **chlorosis**, and the production of multiple adventitious roots resembling witches' broom. The infection is transmitted mainly by leafhoppers.

astragalus The ankle bone.

asulam A **systemic pesticide** of the **carbamate** group that is used to control docks and bracken.

atactostele A type of **dictyostele**, found in **monocotyledons**, in which the **vascular bundles** are scattered randomly in the stem tissue.

atavism The reappearance of a **character** after it has been absent for several generations, due to the expression of a **recessive gene** or of **complementary genes**.

Atgard *See* dichlorvos.

Atherospermataceae (order **Laurales**) A small family of **trees** and **shrubs** with **opposite**, **petiolate**, **simple**, **serrate**, **exstipulate** leaves. Flowers are **actinomorphic** or slightly **zygomorphic**, with 2 **sepals**, 7–20 **petals**, 4–6 (or numerous) **stamens**, **ovary superior** or **inferior**, 3 to many **carpels**. The fruit is an aggregate of **achenes**. There are 6–7 genera with 16 species found in Chile, New Zealand, Australia, New Guinea, and New Caledonia.

Atlantic coast slimy salamander *See Plethodon chlorobryonis.*

Atlantic North American floral region The area of North America covering southern Alaska and Canada south of Hudson Bay to the Gulf of Mexico and northern Florida, and from the Atlantic coast to the Rocky Mountains, part of the **boreal region**. It is divided into two parts approximately along the Canadian–U.S. border but passing south of the Great Lakes. The region is floristically rich, with 100–200 **endemic** species.

atmospheric circulation *See* general circulation.

atmospheric composition Air is a mixture of gases. The table shows the composition of the present atmosphere.

Atmospheric Composition

GAS	SYMBOL	ABUNDANCE
Major		
Nitrogen	N_2	78.08 percent
Oxygen	O_2	20.95 percent
Argon	Ar	0.93 percent
Water vapour	H_2O	variable
Minor		
Carbon dioxide	CO_2	390 ppmv
Neon	Ne	18 ppmv
Helium	He	5 ppmv
Methane	CH_4	2 ppmv
Krypton	Kr	1 ppmv
Hydrogen	H_2	0.5 ppmv
Nitrous oxide	N_2O	0.3 ppmv
Carbon monoxide	CO	0.05–0.2 ppmv
Xenon	Xe	0.08 ppmv
Ozone	O_3	variable
Trace		
Ammonia	NH_3	4 ppbv
Nitrogen dioxide	NO_2	1 ppbv
Sulphur dioxide	SO_2	1 ppbv
Hydrogen sulphide	H_2S	0.05 ppbv

ppmv = parts per million by volume;
ppbv = parts per billion by volume

atmospheric pressure (air pressure) The force exerted over a unit area by the weight of a column of air of similar cross-sectional area extending upward to the top of the atmosphere. At sea level this averages 100 kilopascals (kPa) = 1 **bar** = 1 kg/m^2 = 14.7 lb/in^2.

atmospheric structure The atmosphere forms concentric layers (spheres around the spherical Earth) defined by changes of temperature with increasing altitude. Except for the uppermost layer, each has an upper boundary where the

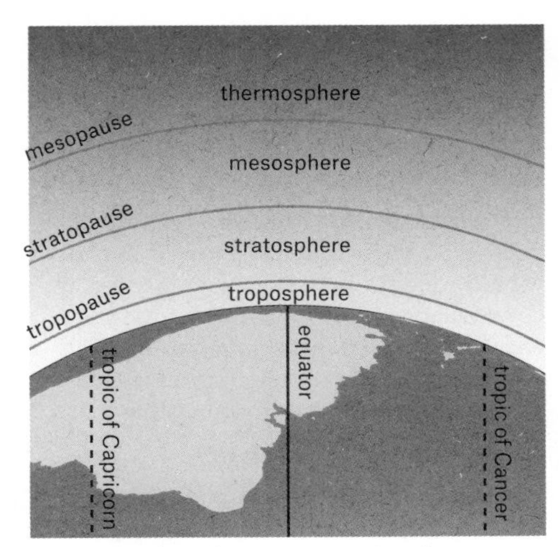

The atmosphere forms concentric spheres around the (spherical) Earth, bounded by layers in which the air temperature, and therefore density, does not change with altitude.

temperature, and therefore density, of the air remains constant with increasing altitude.

ATP *See* adenosine triphosphate.

Atractiellomycetes A class of **Pucciniomycotina** that are **mycobionts** of orchids (**Orchidaceae**). There are 34 species.

Atrichum undulatum *See* Catherine's moss.

atropous *See* orthotropous.

Atterberg limits A sequence of three thresholds, devised in 1911 by the Swedish soil physicist A. M. Atterberg, that are observed when the water content of a sample of fine-grained soil changes. The shrinkage or contraction limit occurs when sufficient water is added to cause cracks in a dry sample to close. With the addition of more water the sample can be shaped and rolled into a thread, at the plastic limit. The liquid limit occurs when sufficient water is added to cause the soil to behave like a liquid.

aubergine (*Solanum melongena*) *See Solanum.*

Aucuba (family **Garryaceae**) A genus of **evergreen**, **dioecious shrubs** in which the glossy leaves are **opposite**, **lanceolate** and similar to those of laurel. Flowers are small, with 4–8 **petals**, borne in a loose **cyme**. The fruit is a red **berry**. There are three to ten species, native to eastern Asia. They are widely cultivated for ornament as spotted laurel.

auger A tool used to sample soil and other sediments, the simplest type consisting of a screw head at the end of a rod.

Aulacapsis rosae (rose scale) An armoured scale insect (**Coccidae**) that is usually found on the stems and **branches** of roses and ***Rubus*** species, but that may spread to **petioles** and leaf stalks, where they lay up to 80 eggs, producing several overlapping generations that result in a dense infestation, individuals living for up to one year. Eggs overwinter beneath the **test** of the female. They occur throughout the world.

Aulacorthum solani (glasshouse-potato aphid) An aphid (**Aphididae**), 1.5–3.0 mm long, with a pale green, or yellow with a bright green or rust-coloured spot, shiny, pear-shaped body. It is able to overwinter as eggs on a range of plant species, but most individuals overwinter as adults on sprouting potatoes and glasshouse plants. It feeds on more than 200 species of plants, but not grasses. It seldom causes serious damage but it does transmit about 40 viral diseases, albeit inefficiently.

auricle A small projection, often resembling an ear, from the base of a leaf.

Auricularia (jelly fungi) A genus of **Basidiomycota** in which the **fruiting bodies** are gelatinous, some being flabby and others firm, and shaped like ears or shells, or forming narrow, **imbricate** brackets. The **hymenium** is borne on one side of the fruiting body. Many are edible and some are grown commercially. There are about eight species with a widespread distribution.

A

Australian bluebell *See Wahlenbergia.*

Australian flatworm *See Australoplana sanguinea.*

Australian laurel *See Pittosporum.*

Australian region The floristic region covering the whole of Australia, divided into three parts: **North and East**; **Central**; and **Southwest**. It is rich in **endemic** species.

Australoplana sanguinea (Australian flatworm) A pink or orange flatworm, (**Platyhelminthes**) 20–80 mm long and 3–8 mm wide, with a smooth, flattened, unsegmented body, pointed at both ends, and many very small eyes that does not coil its body when at rest. Native to Australia, it entered Britain in 1980. It preys on earthworms and poses a threat to animals that feed on earthworms. It is an offence to release this animal into the wild or allow it to escape, and sightings of it should be reported. *See also Arthurdendyus triangulatus.*

Austrobaileyaceae (order **Austrobaileyales**) A **monogeneric** family of woody **evergreen lianas** with **opposite**, **entire** leaves with small **stipules**, large, pendulous flowers with about 12 free, pale green **petals**, 6–11 **stamens**, **ovary superior**, and fruits that are small **berries**. There are two species occurring in northeastern Australia.

Austrobaileyales An order of plants that produce seeds with starchy **endosperm**, comprising 3 families with 5 genera and 100 species. *See* Austrobaileyaceae Schisandraceae, and Trimenuiaceae.

autecology The study of the **ecology** of individual organisms and populations.

auticidal control The introduction of sterile male or genetically engineered (*see* genetic engineering) insects into a pest population in order to control an infestation.

autochory Dispersal of **seeds** or **spores** by the parent.

autochthonous Describes material that originated in its present location, i.e. it has not been transported from elsewhere.

autochthonous landrace A variety of cultivated plant that has evolved in the region where it is traditionally grown, in consequence of which it has a high tolerance for **biotic** and **abiotic** stress and, therefore, produces reliable yields and grows well under low-input regimes.

autogamy Self-**fertilization**.

autogenic Describes a change in a **succession** that occurs because the vegetation has modified the environment.

autolysis The destruction of all or part of a cell through the action of its own **enzymes**.

automimicry Having a **polymorphism** for palatability in which members of a species that a predator would find palatable are coloured or patterned in a way that makes them indistinguishable from unpalatable members of the same species.

autophagic vacuole A **vacuole** that contains material to be digested.

autopolyploidy The condition in which a single **genome** multiplies, so the autopolyploid individual carries **chromosomes** all derived from a single **species**. This has been applied commercially in breeding certain crop plants, e.g. sugar beet, tomatoes, to improve their vigour, although there is some loss in **fertility**.

autosome Any **chromosome** other than a **sex chromosome**.

autotetraploidy The condition in which an autopolyploid (*see* autopolyploidy) organism carried four similar **genomes**. Many crop plants are autotetraploids.

autotomy The voluntary severance by an animal of a body part, usually the tail, that later grows back. It is a strategy used to escape from a predator.

autotroph 1. An organism that is able to synthesize complex organic compounds from simple inorganic precursors, e.g. plants that synthesize **carbohydrates** by **photosynthesis**. 2. An organism that uses carbon dioxide as its only source of carbon.

autumn crocus (*Colchicum autumnale*) *See* Colchicaceae.

auxin A member of a class of **hormones** that are produced at the growing tips of stems and roots and that increase the rate at which cells grow longer (they do not promote **cell division**). This causes longitudinal growth and curvature as cells on one side of a stem grow more than those on the opposite side. Together with **cytokinins**, auxins also initiate activity in **cambium** tissue and they may be involved in the growth of fruit and leaf fall.

auxotroph An organism that has lost, usually as a result of a **mutation**, the ability to synthesize a substance necessary for its survival; it must, therefore, acquire that substance from its environment.

avalanche *See* flow.

avermectin A group of toxins produced by the actinomycete (*see* Actinomycetales) **bacterium** *Streptomyces avermitilis* that are used as **pesticides** against mites and **leaf miners**.

Aves (birds) A class of vertebrates that are descended from theropod dinosaurs and that most taxonomists consider to be dinosaurs. Birds are **endotherms**. They are bipedal, have **feathers**, and possess anatomical and physiological modifications equipping them for flight; some, descended from flying ancestors, have abandoned flight and lost some of these features. Birds have a beak made from horn, no teeth, a four-chambered heart, a large, muscular stomach, and lay **amniote** eggs with hard shells. There are about 9000 species found worldwide.

avicide A chemical compound that is used to kill birds.

avocado pear (*Persea americana*) *See* Lauraceae, *Persea*.

avoidance Behaviour by which an organism minimizes its exposure to a hazard. It may be learned, e.g. when an animal refuses to eat an item that previously made it sick, or innate, e.g. when a young bird utters distress calls and tries to hide when presented with a shadow reminiscent of a hawk, despite having no experience of hawks.

awn In a grass (**Poaceae**), a bristle-like continuation of the central **nerve** of a **lemma** or **glume** that projects from the tip of a **spikelet**.

axial Pertaining to the **axis**.

axial skeleton The part of the vertebrate skeleton consisting of the cranium, **notochord**, vertebrae (*see* vertebra), and **visceral skeleton**.

axil The angle between a **petiole** and stem or between a small stem and a larger one.

axile Attached to the central **axis**.

axile placentation **Placentation** in which the **ovules** are attached to the **axis** of the **ovary**. *See* basal placentation, free-central placentation, parietal placentation.

axillary Borne in an **axil**.

axis 1. The main or central stem. 2. The skull and spine of a vertebrate animal.

axoneme The cytoskeletal structure of a **cilium** or **flagellum**, consisting of an outer ring of nine **microtubules** and, in a cilium, no inner microtubules (9+0 configuration), or in a **motile** flagellum two inner microtubules (9+2 configuration).

Azadirachta indica (neem tree) *See* Meliaceae.

Azalea (family **Ericaceae**) A genus of **perennial shrubs** closely related to ***Rhododendron*** and often included in that genus, when **deciduous** azaleas are placed in the subgenus ***Pentathera*** and **evergreen** azaleas in the subgenus *Tsutsusi*. They have **elliptic** leaves, varying greatly in size depending on species, and colourful, showy **inflorescences.** There are about 800 species, originally from Asia but now widely cultivated ornamentals falling into three types: azaleas, alpine rhododendrons, and tropical rhododendrons.

Azomonas A genus of **motile**, Gram-negative (*see* Gram reaction) **Bacteria** that are usually oval or spherical and produce copious amounts of slime. Most are aquatic, but some occur in soil. They are able to fix nitrogen (*see* nitrogen fixation).

azonal soils An **immature soil** formed from particles that either move away, e.g. mineral particles produced by **weathering** on steep slopes that soon slide or are washed downhill, or that are regularly buried beneath fresh deposits, e.g. **alluvial soil** on a **flood plain**. The material does not remain in place long enough for soil to develop.

Azores high (Bermuda high) An **anticyclone** permanently centred over the Azores and often extending westward as far as Bermuda, when it is known in North America as the Bermuda high. The difference in **atmospheric pressure** between the Azores high and **Icelandic low** drives weather systems in an easterly direction across the North Atlantic. *See* North Atlantic oscillation.

Azotobacter A genus of aerobic (*see* aerobe), **motile**, oval or spherical **Proteobacteria** that form thick-walled **cysts** and produce large amounts of slime. They are free-living in soil and are able to fix nitrogen (*see* nitrogen fixation), releasing it into the soil as ammonium (NH_4). They are found worldwide.

azygospore *See* parthenospore.

Azotobacter A genus of aerobic (*see* aerobe), **motile**, oval or spherical **Proteobacteria** that form thick-walled **cysts** and produce large amounts of slime. They are free-living in soil and are able to fix nitrogen (*see* nitrogen fixation), releasing it into the soil as ammonium (NH_4). They are found worldwide.

B

B 1. *See* bubnoff unit. 2. *See* boron.

B-995 *See* daminozide.

BAC *See* bacterial artificial chromosome.

baccate Resembling a **berry**.

bachelor's buttons *See Bulgaria inquinans*.

Bacillariophyta *See* diatom.

Bacillus A genus of rod-shaped, Gram-positive (*see* Gram reaction) **Bacteria** that can be **obligate aerobes** or **facultative anaerobes**. Some species are free-living, others pathogens. Under stress but in the presence of air, the bacteria can form **endospores** that can remain dormant for long periods. *Bacillus* species occur worldwide in a wide variety of **habitats**. *Bacillus anthracis* causes anthrax; ***Bacillus thuringiensis*** is an insect pathogen.

bacillus Any rod-shaped bacterial cell.

Bacillus thuringiensis (Bt) A species of Gram-positive (*see* Gram reaction) **Bacteria** that are common in soil and that synthesize compounds which are toxic to flies (**Diptera**), butterflies and moths (**Lepidoptera**), bees, wasps, ants, and sawflies (**Hymenoptera**), beetles (**Coleoptera**), and nematodes (**Nematoda**). Bt toxins are used as **insecticides** and certain crops have been genetically engineered (*see* genetic engineering) to produce the toxins.

backcross A cross between an **F_1 hybrid** or **heterozygote** and an individual genetically identical to one or other parent.

backing An anticlockwise change in the wind direction. *Compare* veering.

back mutation A **reverse mutation** in which a mutant **gene** reverts to the **wild type** form.

backswamp An area on a **flood plain**, some distance from the river channel, that is low-lying and poorly drained, where **silts** and **clays** accumulate and the vegetation is typical of a marsh.

Bacteria In the **three-domain system** of taxonomic classification, one of the three domains; in the older **five-kingdom system**, one of the kingdoms. Most are single-celled and have a rigid **cell wall**. They usually reproduce by **binary fission** and **mitosis** never occurs. Bacteria comprise 11 groups of **prokaryotes**: purple (photosynthesizing); Gram positive; **cyanobacteria**; green non-sulphur; spirochaetes; flavobacteria; green sulphur; Planctomyces; Chalmydiales; Deinococci; and Thermatogales. Bacteria occur in almost every environment. They contribute to the decay of organic material and recycling of nutrients and **biogeochemical cycles**, aid the **absorption** of nutrients as part of the gut flora in animals, and are used in the production of many dairy products, antibiotics, and fermented products. They are also agents of spoilage of food and other substances.

bacterial artificial chromosome (BAC) A section of **DNA**, typically 100–300 kilobases long, assembled in a laboratory and containing DNA of interest taken from one organism, that is inserted into the DNA of a **bacterium**. Cloning the recipient amplifies the BAC and the inserted genes can then be used in sequencing.

B

bacterial rot of apples and pears *See Gluconobacter oxydans.*

bacterial wilt Any wilting disease, in which leaves wilt and die, leading to the death of the plant, that is caused by a **bacterium** that multiplies in the vascular system, eventually blocking the **xylem**. In cucurbits (**Cucubitaceae**) other than watermelons, which are immune, the disease is caused by ***Erwinia*** *tracheiphila*, transmitted by the striped cucumber beetle (*Acalymma vittatum*) and spotted cucumber beetle (*Diabrotica undecimpunctata*). In other families the disease is caused by ***Ralstonia solanacearum***.

bactericide A chemical compound that kills **Bacteria**.

bacteriochlorophylls Pigments involved in **photosynthesis** that are found contained in small bodies continuous with or attached to the **cell membrane** in certain **anaerobe**, photosynthetic **Bacteria**. There are small structural differences between bacteriochlorophylls and the **chlorophylls** found in **cyanobacteria** and plants.

Bacteriodetes *See Cytophaga-Flavobacterium* group.

bacteriophage (phage) A **virus** that infects **Bacteria**.

bacteriorhodopsin A pigment found in certain **Archaea** that captures light energy and uses it to expel protons from the cell, thereby creating a proton gradient that is converted to chemical energy.

bacteriostatic Describes a compound that inhibits bacterial growth but without killing the **bacteria**.

bacterium A single bacterial cell (*see* Bacteria).

bacteroid A structurally modified **bacterium**, e.g. ***Rhizobium*** found in the **root nodules** of legumes (**Fabaceae**).

badderlocks *See* dabberlocks.

badger *See Meles meles, Taxidea taxus,* Mustelidae.

badlands An area with sparse vegetation that is eroded (*see* erosion) into an intricate pattern of steep-sided channels.

baeocyte A type of reproductive cell found in certain **cyanobacteria**.

Baeolophus bicolor (*Parus bicolor*, tufted titmouse) A species of birds, 150–170 mm long with a 230–280 mm wingspan, that have grey backs, rust-coloured sides, white undersides, and prominent crests. They inhabit **deciduous** woodlands and are common in city parks and gardens, and feed on invertebrates. They occur throughout eastern North America.

balanced polymorphism (overdominance) A **polymorphism** in a population that is maintained by **natural selection** because **heterozygotes** for particular **alleles** are fitter (*see* fitness) than either **homozygote**.

Balanites aegyptiaca (desert date) *See* Zygophyllaceae.

Balanopaceae (order **Malpighiales**) A **monogeneric** family (*Balanops*) of tall, **evergreen trees** that have **alternate**, pseudoverticillate (*see* verticillate), **coriaceous**, **exstipulate**, **dentate** leaves. Flowers **unisexual** (plants **dioecious**), male flowers as **catkins** usually with 5–6 **stamens**, female flowers with 2–3 **carpels**, **ovary superior**. The fruit is a **drupe** with 1 or 2 seeds. There are nine species found in northeastern Australia, New Caledonia, and islands in the southwestern Pacific.

Balanophoraceae (order **Santales**) A family of **obligate** root parasites that contain no **chlorophyll**. They have an underground structure resembling a **tuber** that in some species consists entirely of parasite tissue and in others is part parasite and part host, and that is connected to the host by **haustoria**. The 'tuber' ruptures to release the **inflorescence**, which is the only part of the plant to appear above ground, leaving the remains of the tuber as a collar-like structure. The yellow, red, or brown inflorescence is terminal, **racemose**, **spicate**, **globose**, or club-shaped, sometimes branched, its stalk either naked or bearing scales or **bracts**. Individual flowers are small or minute and numerous, **actinomorphic**, **unisexual**, and structurally very varied. There are 17 genera with 50 species found throughout the tropics with some species in the subtropics.

Balea biplicata (common door snail, two-lipped door snail) A species of terrestrial **snails** that has an approximately spoon-shaped cover (clausilium) it can slide across to seal the **aperture** (hence the name door snail). The shell is brown, 15–18 mm high and 3.8–4.5 mm wide, and the snail is found in leaf litter or between stones in forests and under herbs. It occurs in central and southeastern Europe and in southern England.

ballistospore A fungal **spore** that is discharged violently.

ballooning A method of dispersal in young spiders in which a spiderling climbs to the top of a plant and pays out a length of silk until the wind catches it, lifting the spiderling up to 1 km above the ground and transporting it often for a long distance before it settles.

Balsaminaceae (order **Ericales**) A family of **annual** and **perennial herbs** that have translucent, watery stems and usually spiral, toothed leaves. Flowers are **bisexual** and **zygomorphic** usually with 3 free **sepals**, the **adaxial** sepal with a spur that appears to be on the **abaxial** part of the flower because the flower is upside down. There are 5 unequal **petals**, 5 **stamens**, **ovary superior** of 5 fused **carpels**. The fruit is usually an explosive **capsule**. There are 2 genera with 1001 species widely distributed in temperate regions. ***Impatiens*** is widely cultivated for ornament.

Baltimore oriole *See Icterus glabula.*

banana *See Musa*, Musaceae.

banded rattlesnake *See Crotalus horridus.*

banded sage hopper *See Eupteryx melissae.*

banded-wing whitefly (*Trialeurodes abutilonea*) *See Abutilon* yellows, *Diodia* vein chlorosis, tomato chlorosis virus.

Banks, Sir Joseph (1743–1820) An English explorer who was an important patron of science and sent botanical expeditions to many parts of the world. While sailing with James Cook he named Botany Bay, Australia, for its rich **flora**. He became honorary director of the Royal Botanic Gardens at Kew, London. His **herbarium** is held by the Natural History Museum, London, and his books and manuscripts, of major scientific importance, are shared between the Natural History Museum and the British Library.

Banksia (family **Proteaceae**) A genus of **trees** and woody **shrubs** with very variable leaves, most with **serrate** margins and borne in irregular spirals. **Inflorescences** are dense, showy **spikes** of up to 1000 or more (6000 alleged in *B. grandis*) yellow or sometimes orange, red, pink, or violet flowers. As they fade and dry, inflorescences become **cone**-shaped. The fruit is a **follicle**. There are 173 species, all but one native to Australia, but widely cultivated. The genus is named for Sir Joseph **Banks**.

bank worm *See Dendrodrilus rubidus.*

banner cloud A narrow cloud that extends downwind from the summit of a mountain, resembling a banner flying in the wind.

banyan (*Ficus benghalensis*) *See* strangler fig.

bar 1. A low ridge of sand or gravel formed in shallow water by the action of waves and tides. **2.** A unit of pressure equal to 10^5 pascals (10^5 newtons per square metre).

B

barachory (clitochory) Dispersal of **seeds** or **spores** by gravity, i.e. they fall to the ground.

Baragwanathia longifolia A species, now extinct, of vascular plants (**Tracheophyta**) that was one of the first plants to grow on land after ***Cooksonia hemispherica***. It lived during the Late Silurian (about 420 million years ago) or Early Devonian (about 410 million years ago). It had stems up to 1 m long and 2–4 mm thick, with sporangia (*see* sporangium) either along them or on the bases of the 4-cm-long leaves.

barb One of the branches arising on either side of the shaft of a **feather**.

barberry (*Berberis vulgaris*) *See Berberis*.

Barbeuiaceae (order **Caryophyllales**) A **monotypic** family (*Barbeuia madagascariensis*), which is a **liana** with **alternate**, **petiolate**, **simple**, **exstipulate** leaves. Flowers are **actinomorphic** with 5 **imbricate sepals**, no **petals**, 30–100 free **stamens** in 2–4 **whorls**, **ovary superior** of 2 fused **carpels** with 2 **locules**. The fruit is a **capsule**. The family occurs only in Madagascar and is close to extinction.

Barbeyaceae (order **Rosales**) A **monotypic** family (*Barbeya oleioides*), which is a small, **evergreen tree** with **simple**, **opposite**, **lanceolate**, **entire**, **exstipulate** leaves. The small flowers are **actinomorphic**, **unisexual** (the plants **dioecious**), with 3–4 **tepals**, 6–12 **stamens**, **ovary superior** of 1–2 **carpels** with 2–3 **locules**. The fruit is a **nutlet**. The species occurs in northeastern Africa and southwestern Arabia.

bark The outer layers of the stem, **branches**, and roots of a woody plant, lying outside the **vascular cambium** and comprising an outermost layer of dead tissue, often sculpted, beneath that a layer of **cork**, and beneath that the living **phloem** tissues.

bark beetles *See* Scolytidae.

barley-root nematode *See Meloidogyne naasi*.

barley yellow dwarf A disease of cereals and other grasses (Poaceae) caused by the barley yellow dwarf **virus**, also called cereal yellow dwarf virus (*Luteovirus* species), transmitted by many species of aphids (**Aphididae**), that infects **phloem** cells. Leaves of infected plants turn yellow, orange, red, or purple, tillering and flowering are reduced, plants are stunted, and yields are reduced.

barn swallow *See Hirundo rustica*.

baroclinic Describes the common atmospheric condition in which surfaces of constant pressure and constant air density intersect, so that the air density changes along each **isobar**.

bar of Sanio *See* crassula.

barophile *See* extremophile.

barotropic Describes the atmospheric condition in which surfaces of constant **atmospheric pressure** and constant air density lie approximately parallel at all heights. There is little change in temperature or wind direction over horizontal distances, or of wind direction with height, and atmospheric conditions are fairly uniform over a large area.

barren Describes land where vegetation covers less than half of the available area.

basal body A structure in **cell cytoplasm**, usually composed of nine **microtubules**, from which flagella (*see* flagellum) project.

basal bristle A small **feather** with a small or no shaft at the base of the beak of a bird.

basal placentation **Placentation** in which the **ovules** are attached to the base of the **ovary**. *See* axile placentation, free-central placentation, parietal placentation.

basalt A fine-grained, dark-coloured, **igneous** rock formed by the partial melting of peridotite in the mantle. Basalt flows cover about 70 percent of the Earth's surface.

base 1. A substance that in solution can bind hydrogen **ions** (protons) and that acts as an electron-pair donor. It has a **pH** greater than 7.0 and reacts with an **acid** to yield a **salt** and water. **2.** One member of a **base pair**.

base analogue A base (*see* base pair) with a structure that differs slightly from that of the normal base but that is similar enough to act as a **mutagen** when inserted into **DNA**.

base number (basic number) The number of **haploid chromosomes** in a **genome**.

base pair 1. Two **nucleotides** on separate strands of **DNA** that are linked by **hydrogen bonds**. **2.** A unit of measurement applied to a length of double-stranded DNA.

Basellaceae (order **Caryophyllales**) A family of **perennial** climbing vines or spreading **herbs**, often producing **tubers**, with mostly **alternate**, **simple**, **ovate**, **exstipulate** leaves. The small flowers are **actinomprhic**, **bisexual** (occasionally **unisexual** with plants **monoecious**), **perigynous**, the **perianth** with 5 segments, 5 **stamens**, **ovary superior** of 3 fused **carpels**, with 1 **locule**. The fruit is a **drupe**. There are 4 genera with 19 species, occurring in tropical and subtropical America and Africa. Tubers of *Ullucus tuberosus* are an important food in the Andes. Leaves of *Basella rubra* and *B. alba* (Malabar spinach) are widely eaten in the tropics.

base saturation The extent to which sites on soil particles are occupied, i.e. saturated, by exchangeable **cations** with a **pH** greater than 7.0, or by cations other than hydrogen or aluminium, expressed as a percentage of the total **cation exchange capacity**.

basic grassland Vegetation dominated by grasses (**Poaceae**) growing on **basic soil**. These are often rich in herbs.

basic number *See* base number.

basic rock A rock containing 45–53 percent **silica** by weight and a high concentration of iron, magnesium, and calcium.

basic soil A soil with a **pH** greater than 7.0.

basidiocarp The **fruiting body** of a member of the **Basidiomycota**.

basidiomycete A member of the **phylum Basidiomycota**.

Basidiomycota (club fungi) A **phylum** of **Fungi**, belonging to the subkingdom **Dikarya**, comprising **yeasts** and asexual species, but also those with **fruiting bodies** in the form of mushrooms, toadstools, brackets, etc. Most are **saprotrophs**, many feeding on wood in buildings, and others are plant parasites including those that cause **rust** and **smut** diseases. There are about 30,000 species found worldwide on land and also in freshwater and marine **habitats**.

basidiospore A **spore** that is produced sexually on a **basidium**.

basidium A structure in **Basidiomycota**, visible only with a microscope, on which sexually produced **spores** form.

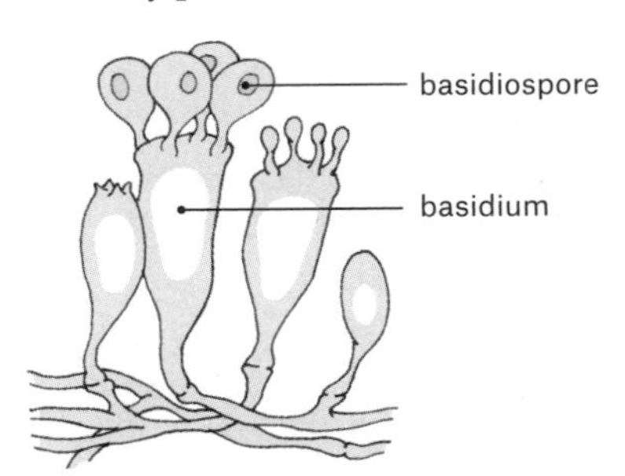

A basidium is the microscopic structure on which spores (basidiospores) form in the Basidiomycota (club fungi).

basifixed Describes an **anther** that is attached to its **filament** by its base.

B

basifugal movement Growth or movement away from the base.

basil (*Ocimum*) *See* Lamiaceae.

basipetal Growing from the **apex** toward the base so the oldest part is at the apex. *Compare* acropetal.

basipetal movement Movement toward the base.

basket fern Any fern (**Pteridophyta**) with an erect **habit**.

basophilic Describes a cell or cell product that can be stained by a basic dye.

bass *See* bast.

Bassia (family **Amaranthaceae**) A genus of **annual** and **perennial**, **halophyte herbs** and small **shrubs** with deeply penetrating roots. The leaves are highly variable. The small flowers are borne in a **cymose inflorescence** and are mostly wind-pollinated (some bee-pollinated). The fruit is an **achene**. There are 25 species occurring mainly in temperate Eursasia, but introduced elsewhere. Some are cultivated for ornament (e.g. *B. scoparia*, summer cypress, or *B. scoparia* f. *trichophylla*, firebush or burning bush). Others (e.g. *B. hyssipifolia*, fivehorn smotherweed) are invasive and troublesome weeds. *Bassia saxicola*, **endemic** in the Mediterranean islands of Capri and Sicily, is endangered.

basswood *See Tilia*.

bast (bass) Fibre obtained from the **phloem** tissue of a non-woody plant.

bastard wing *See* alula.

Bataceae (order **Brassicales**) Saltwort, a **monogeneric** (*Batis*) family of **halophyte subshrubs** with **opposite**, **decussate**, **obovate** or **linear**, **simple**, **entire**, fleshy leaves with minute **stipules**. Flowers are **staminate** or **carpellate**, male flowers 4-merous with joined **sepals** unequal in size, the large one forming one lobe and the three others a second. Female flowers are greatly reduced with 2 **carpels** each with 2 **locules**. **Inflorescence** is a **spike**. The fruit is a **drupelet**. There are two species found along the coasts of warm temperate and tropical America, and tropical Australia and New Guinea. The leaves are sometimes eaten in salads.

Bates, Henry Walter (1825–92) An English naturalist who befriended A. R. **Wallace** and in 1848 accompanied him on an expedition to the Amazon, where Bates collected nearly 15,000 species of insects, of which 8000 were new to science. His studies of them led him to propose the form of mimicry now named after him.

Batesian mimicry The possession by palatable species of bright colours or distinctive markings very similar to those carried by unpalatable species. Predators that learn to avoid the unpalatable species also avoid the palatable species. This form of mimicry was first observed by Henry Walter **Bates**.

batology The study of brambles (***Rubus***).

batrachotoxins A group of substances that are the most poisonous known. They are secreted by *Dendrobates* and *Philobate* frogs in South America, where the poison is used to tip the points of spears and arrowheads, and in the skin and **feathers** of *Pitohui* and *Ifrita* birds in New Guinea. These animals are believe to acquire the poison by feeding on *Choresine* beetles, which contain it.

Bauhin, Gaspard (1560–1624, also known as Caspar Bauhin) A Swiss botanist and anatomist who wrote *Pinax theatri botanici* (published 1623). This was a concordance to an earlier system of botanical nomenclature and an attempt at a formal system of plant classification. He also competed three of the planned twelve parts of *Theatrum botanicum*, of which only one part was published in his lifetime (in 1658). *Bauhinia* is named for the Bauhin family.

Bauhin, Jean (1541–1613) A Swiss physician and botanist, the elder brother of Gaspard **Bauhin**, who, with help from his son-in-law Jean Henri Cherier (*c.* 1570–*c.* 1610), wrote *Historia plantarum universalis*, the first international **flora**, describing 5266 plants. It was unfinished at Bauhin's death, but was published in 1650–51 and became a standard reference work for the next century.

bay laurel (*Laurus nobilis*) *See* Lauraceae, *Laurus*.

bay-winged cowbird (*Molothrus badius*) *See* Icteridae.

B-chromosome A **chromosome** that is additional to the set of **A-chromosomes** and shows no **homology** to it. It contains no functional **genes**. Such chromosomes occur widely in **angiosperms**.

BC soil Soil in which only the B and C **soil horizons** are present.

bdelloplast The two-celled structure formed by a ***Bdellovibrio*** **bacterium** in the **periplasmic space** of its host.

Bdellovibrio A genus of Gram-negative (*see* Gram reaction) **Bacteria** that are **obligate aerobes**. They are shaped like curved rods with a single, small **flagellum** covered by a sheath, and swim very fast. They parasitize other Gram-negative bacteria by attaching to the outer membrane, making a hole in it, then entering the **periplasmic space**. The parasite then seals the hole, forms a **bdelloplast**, and releases **enzymes** that break down the contents of the host, forming new *Bdellovibrio* cells. When the host is exhausted the parasite lyses (*see* lysis) the host and its new cells leave.

beach nourishment *See* beach replenishment.

beach replenishment (beach nourishment) The addition to a beach of sand or other beach material in order to stabilize a beach that is eroding or to restore a severely eroded beach.

beak 1. In an orchid (**Orchidaceae**) flower, a projection separating an **anther** from the surface of the **stigma** below it. **2.** A rigid projection from the tip of a fruit.

bean seed flies Two species of flies (**Diptera**), *Delia platura* and *D. florilega*, about 5 mm long, that live mainly as scavengers, but as larvae (about 7 mm long) feed on the developing shoots of more than 40 plant species, including onions, beans, cucurbits, spinach, brassicas, radish, beet, asparagus, maize (corn), cereals, and clover. They occur worldwide.

bean weevil *See* Bruchidae.

beard lichens *See Usnea*.

bear's breech (*Acanthus mollis*) *See Acanthus*.

Beaufort wind scale A classification of wind strengths that was devised in 1805 by Francis Beaufort, a British naval officer, originally to allow naval commanders to record wind strengths. *See* appendix.

Beauveria bassiana A species of **Fungi** that occurs in soils worldwide and that is a **parasite** of insects, causing white muscardine disease, in which fungal **spores** germinate on contact with an insect body, producing **hyphae** that penetrate the **cuticle**, rapidly killing the host and then producing a white mould on the exterior of the body that produces more spores. The fungus is used in **biological control** of a wide range of insect pests.

bedrock The solid rock that lies beneath the unconsolidated surface materials, e.g. soil.

bee bread The food of bee larvae, consisting of a mixture of **pollen**, honey, sometimes plant oils and gland exudates.

beech *See* Fagaceae, *Fagus*.

bee dance *See* dance language.

beefsteak fungus *See Fistulina hepatica*.

beefsteak polypore *See Fistulina hepatica*.

bee mites *See* Laelapidae.

bees *See* Apocrita, Hymenoptera.

beeswax Wax that is secreted by glands of bees (**Apidae**) and used in constructing nests.

beet *See* Amaranthaceae.

beet cyst eelworm *See Heterodera schachtii.*

beet leaf miner *See Pegomyia betae.*

beetles *See* Coleoptera.

beet mosaic virus A **virus** belonging to the **Potyviridae** that is transmitted by aphids (**Aphididae**) and causes a mosaic disease in ***Beta** vulgaris* (beet) and in members of the **Solanaceae**, **Fabaceae**, and **Amaranthaceae**, e.g. spinach (*Spinacia oleracea*). It occurs worldwide, especially in temperate regions. Infected beet leaves show vein-clearing, in other species yellow flecks appear on leaves; plants are usually stunted.

beet pseudo-yellows virus (cucumber yellow virus, muskmelon yellows virus) A *Closterovirus* (**Closteroviridae**) that causes yellowing, thickening, and sometimes curling of leaves or leaves becoming brittle in a wide range of plants, especially ***Beta** vulgaris* (beet), ***Lactuca** sativa* (lettuce), ***Cichorium** endiva* (endive), *Capsella bursa-pastoris* (shepherd's purse), *Cucumis sativa* (cucumber), *Taraxacum officinale* (dandelion), and *Conium maculatum* (hemlock). It is transmitted by the greenhouse whitefly *Trialeurodes vaporariorum.*

beet yellow stunt A disease caused by a *Closterovirus* (**Closteroviridae**) that causes twisting and stunted growth in ***Beta** vulgaris* (beet) and **chlorosis** and collapse in ***Lactuca** sativa* (lettuce). It is transmitted by aphids (**Aphididae**) and sowthistle (*Sonchus* spp.) is the principal reservoir. It occurs in Britain and parts of the United States.

beet yellows A disease caused by a **virus** of the **Closteroviridae** and transmitted by aphids (**Aphididae**) that causes yellowing and thickening of leaves that then become brittle in ***Beta** vulgaris* (beet) and *Spinacia oleracea* (spinach). It occurs worldwide.

Begoniaceae (order **Cucurbitales**) A family mainly of fleshy **herbs**, with some members that are **acaulescent** and some woody and up to 10 m tall with aerial roots. Leaves are **alternate**, sometimes lobed, with **serrate** margins. Flowers **actinomorphic** or **zygomorphic**, **unisexual** (plants **monoecious**), male flowers with usually 4 **tepals** and 4 to many **stamens**, female flowers with 4–5 (or 9 or 10) tepals, **ovary inferior** with usually 3–8 **locules**. The fruit is a **berry** or **capsule**. There are 2 genera and 1501 species, occurring mainly in the tropics. *Begonia* is widely cultivated as an ornamental with an estimated 10,000 **cultivars**.

bellflower *See Campanula.*

bell moths *See* Tortricidae.

bell pepper (*Capsicum annuum*) *See Capsicum.*

Beltian bodies In certain trees of the African **savanna** (*see Acacia*), sausage-shaped organs at the tips of leaves that secrete **proteins** and oils used as food by ants living in nests they have made in the leaf bases. The ants also feed on **nectar**. In return they cut away adjacent plant tissue that threatens to shade the leaves and defend the plant from attack by herbivores.

belt transect A strip, commonly 1 m wide, that is marked out through an area of **habitat** and within which an investigator records all the species present in order to determine the distribution of species within the habitat.

Bemisia tabaci (sweetpotato whitefly, silverleaf whitefly) A bug (suborder **Homoptera**) that feeds by piercing **phloem** tissue, causing physical damage and facilitating infection by more than 100 **viruses.** It excretes **honeydew**, providing a **substrate** for fungal infection, and inhibiting **photosynthesis** by blocking stomata (*see* stoma). It occurs in all continents except Antarctica and is most likely to be encountered on poinsettia (***Euphorbia** pulcherrima*). It is treated mainly with **neonicotinoid insecticides.** *See* lettuce infectious yellows virus, sweet potato chlorotic stunt, tomato chlorosis virus.

Bennettitales A group of seed plants (**Spermatophyta**), now extinct, that resembled cycads (**Cycadaceae**) for which they were mistaken, but are now classed as **anthophytes.** They had woody stems with persistent leaf bases, and **simple** or **pinnate** leaves, and flower-like reproductive structures. They lived from the Triassic (251–199.6 million years ago) to the Late Cretaceous (99.6–65.5 million years ago) epochs.

Bentham, George (1800–84) An English botanist who assembled a large **herbarium** and wrote several important botanical works. These included the first of the colonial floras published by the Royal Botanic Garden at Kew: *Flora Hongkongensis* appeared in 1861 and seven volumes of *Flora Australiensis* between 1863 and 1878. His other major work was *Genera Plantarum* written in collaboration with Sir Joseph Dalton **Hooker** and published in three volumes between 1862 and 1883. His *Handbook of the British Flora* was first published in 1858, its seventh edition in 1924; the fifth and sixth editions were prepared by Hooker, and botanists came to know the work as 'Bentham and Hooker'.

benthic Describes organisms that live on or near the sea bed.

bentonite An absorbent **clay** rich in **montmorillonite** formed from volcanic deposits.

benzene An organic **hydrocarbon**, C_6H_6, with a molecule consisting of six carbon atoms joined in a ring with one hydrogen atom attached to each. It is colourless, has a sweet smell, and is highly flammable.

Berberidaceae (order **Ranunculales**) A family of **herbs** and woody **shrubs** that have **rhizomes** or **tubers**, and usually **alternate**, **simple**, **pinnate**, or **ternate** and **exstipulate** leaves. In ***Berberis*** leaves on long shoots are modified into spines while those on short shoots are simple leaves. Flowers **actinomorphic**, **bisexual**, the **perianth** of several **whorls** each with 6 or 4 segments, 6 **stamens**, **ovary superior** of 1 **carpel** and 1 **locule**. The fruit is **succulent** in some genera, in others a **capsule** with 2 **valves**, or a papery **bladder** containing seeds resembling **berries.** There are 14 genera with 701 species found in temperate regions of the Northern Hemisphere, in western South America, and in North Africa, East Africa, and eastern Asia. Many species are cultivated for ornament (*see Berberis*) and a few have medicinal uses.

Berberidopsidaceae (order **Berberidopsidales**) A family of **evergreen perennial** woody scramblers with **alternate**, **entire** or with spiny-toothed, **ovate**, **petiolate**, **exstipulate** leaves that are shiny on the upper side and **glaucous** on the underside. Flowers **actinomorphic**, **bisexual**, the **perianth** of 9–12 (or 15) **petaloid** segments, usually red, 8–15 (or more) **stamens**, **ovary** of 3 or 5 fused **carpels** each with 1 **locule** with 2 to many **ovules**. The fruit is a **berry.** There are two genera with three species found in Chile and eastern Australia.

Berberidopsidales An order comprising two families, three genera, and four species. *See* Aextoxicaceae and Berberidopsidaceae.

Berberis (family **Berberidaceae**) A genus of **shrubs** in which the leaves on long shoots are modified into 3-spined thorns and the **bud** in the **axil** of each of these leaves develops a short shoot bearing a normal leaf that is **simple** and **entire** or with a spiny margin. The yellow or orange flowers are **trimerous** with 6 **sepals** and **petals** in alternating **whorls** of 3. Sepals and petals are usually the same colour. The **inflorescences** are borne in **racemes** of up to 20 flowers. The fruit is a red or dark blue **berry**, often waxy with a violet or pink bloom and edible in some species (e.g. *B. vulgaris*, barberry). There are 450–500 species found throughout temperate regions apart from Australia, and in the mountains of tropical Africa. Several are cultivated as ornamentals.

Bergeron-Findeisen-Wegener mechanism A theory of how cloud droplets grow into raindrops in clouds containing both liquid droplets and ice crystals. It holds that water evaporates from the droplets and is deposited on the crystals. These grow, collide, and form snowflakes that collide with more supercooled (*see* supercooling) droplets as they fall and continue growing. If the temperature in the lower part of the cloud or below the cloud base is above freezing the snowflakes will melt and fall as rain.

Bergmann, Karl Georg Lucas Christian (1814–65) A German biologist, who in 1847 proposed the relationship between climate and body size known as **Bergmann's rule**.

Bergmann's rule In closely related species of **homeotherms**, body size increases along a gradient from warm to cool climates, i.e. animals living in a cold climate are larger than closely related species living in a warm climate. The idea was proposed by Karl **Bergmann**.

Beringia The area, now lying beneath the sea, between Siberia and Alaska. At various times during the late Mesozoic and Cenozoic it has lain above sea level, allowing plants and animals to migrate between the Palaearctic and Nearctic biogeographical regions..

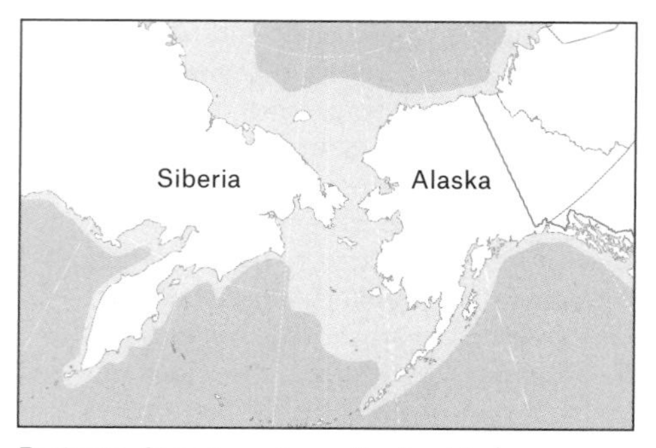

Beringia. At various times the land between Siberia and Alaska has lain above sea level, allowing plants and animals to migrate between the continents.

Bering land bridge A strip of land that linked Siberia and Alaska several times during the Cenozoic era.

Bermuda high *See* Azores high.

berry A fleshy, **indehiscent** fruit containing many seeds and with no hard parts other than the seeds. A banana, grape, tomato, gooseberry, and date are berries.

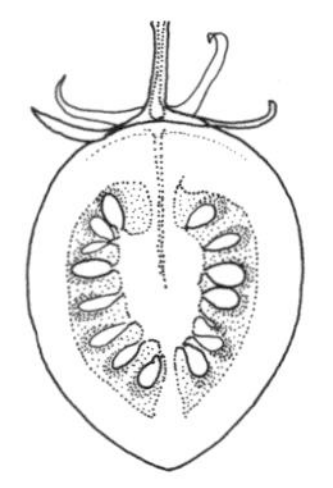

A tomato is a berry.

Bertholletia excelsa (Brazil nut) *See* Lecythidaceae.

Bessey, Charles Edwin (1845–1915) An American botanist who did much to establish botany in the United States and who compiled a survey of the world's main plant groups, arranging them in trees to show their relationships and the way ancestral forms had diverged. His work contributed greatly to the development of modern classifications based on evolutionary relationships.

Beta (family **Amaranthaceae**) A genus of **biennial** or **perennial herbs** with heart-shaped leaves and small, **pentamerous** flowers borne in dense **spikes**. The fruit is a cluster of **nutlets**. There are 11–13 species found from Europe through the Near East to India. *Beta vulgaris* subsp. *maritima* (sea beet) has been cultivated since Assyrian times and its derivatives are classified as *B. vulgaris* subsp. *vulgaris*, all of which store sugar in their roots, in the case of sugar beet accounting for up to 20 percent of the plant by weight. The varieties are also grown as beetroot, chard, spinach beet, and mangelwurzel.

beta diversity (species turnover) A measure of **biodiversity** that is interpreted in different ways. The simplest definition is that it is equal to **gamma diversity** divided by **alpha diversity** ($\beta = \gamma/\alpha$), i.e. the number of species that are unique to each of the subunits in a set of data compiled from an inventory of the species in a **habitat**.

betalains Red and yellow pigments found in members of the **Caryophyllales**, where they replace **anthocyanins**.

Betaproteobacteria A class of **Proteobacteria** that are Gram-negative (*see* Gram reaction) **aerobes**, comprising 390 species. Some are **phototrophs**. Betaproteobacteria contribute to **nitrogen fixation** in a number of plants. They occur worldwide.

beta sheet *See* pleated sheet.

bet-hedging The behaviour of animals living in an **environment** subject to irregular fluctuations in conditions that release their young into several different environments. This increases the chance that in the event of change at least some young will survive. *See K*-selection.

Betula (family **Betulaceae**) A genus of fast-growing, **deciduous trees** (birch) and **shrubs** with **simple**, **alternate**, singly or doubly **serrate** or lobed, **petiolate**, **stipulate** leaves. Flowers are **monoecious**, male flowers borne as **catkins** that disintegrate when mature (unlike ***Alnus***), female flowers in clusters; there is no **perianth**. The fruit is a **samara**. There are 35–60 species occurring in northern temperate and arctic regions. Birch timber is used to make furniture and plywood. *See* arctic scrub.

Betulaceae (order **Fagales**) A family of small **deciduous trees** and **shrubs** (alders, birches, hazels, and hornbeams), many of which have smooth **bark** that peels off in thin layers. Leaves are **simple**, **alternate**, **dentate** or almost **entire**, with **stipules**. Flowers are **unisexual** (plants **dioecious**), male flowers in long, pendulous **catkins** comprising clusters of 1–3 flowers, female flowers in erect or pendulous clusters of 2 or 3 on a stiff **axis** subtending a leafy **involucre**. When present, the **perianth** is in a variable number of scaly segments. There are usually 4–6 **stamens**, 0–6 **petals**, **ovary inferior** of 2 fused **carpels**. The fruit is a **nut** or 2-winged **samara** There are 6 genera with 145 species occurring throughout the northern temperate region, in the Andes, and in Sumatra. Birches and some alders provide valuable timber; *Ostrya* (hophornbeam) has very hard wood, used to make tools.

Betula lenta (black birch, cherry birch) *See* methyl salicylate.

bicarpellate Derived from two **carpels**.

bicentric distribution The natural occurrence of a **taxon** in two widely separated locations but nowhere in between, e.g. ***Liriodendron***, found in eastern North America and China, and *Nothofagus* (*see* Nothofagaceae), found in New Guinea and Australasia and in Chile.

bicollateral bundle A **vascular bundle** with **phloem** tissue on both sides of the **xylem**.

Biebersteiniaceae (order **Sapindales**) A **monogeneric** family (*Biebersteinia*) of woody **perennial herbs** with **rhizomes** or **tubers.** Leaves are **alternate, pinnately compound** with lobed leaflets, toothed margins, and **stipules.** Flowers are **actinomorphic, hermaphrodite,** with 5 free **sepals** and **petals,** 10 **stamens, ovary superior** of 5 **carpels,** each **locule** with a single **ovule.** The fruit is a **schizocarp** of 5 one-seeded **mericarps.** There are five species occurring from Greece to Central Asia.

biennial Describes a plant that lives for two years, producing flowers and seeds in its second year.

bifid Divided in two.

bifurcate Forked, with two **branches.**

big-bang reproduction *See* semelparity.

Bignoniaceae (order **Lamiales**) A family of **trees, shrubs,** and woody climbers (often with tendrils). **Compound** leaves are usually **opposite,** but sometimes in **whorls** or **alternate, palmate** or **pinnate;** in climbers the terminal leaflet often modified into a simple or branched tendril. Flowers have 5 fused **sepals** and **petals,** usually 5 **stamens, ovary** of 2 **capsules** and 1, 2, or 4 **locules.** Flowers are solitary or in **axillary** or terminal **racemes.** The fruit is a **capsule.** There are 110 genera with 800 species occurring throughout the tropics, but mainly in South America. Many are cultivated as ornamentals. Calabashes, used to carry water or as maracas, are from *Crescentia* fruits.

big sagebrush (*Artemisia tridentata*) *See Artemisia.*

big tree (*Sequoiadendron giganteum*) *See* Pacific coast forest.

bilabiate Two-lipped.

bilateral symmetry An arrangement of the parts of a body such that an imaginary central plane divides the body into halves that are approximate mirror images of each other.

bilberry (*Vaccinium myrtillus*) *See Vaccinium.*

bilins **Tetrapyrroles** used to capture light by **cryptophyte** marine algae (*see* alga).

billing Behaviour in which two courting birds clasp or touch each other's beaks.

billow cloud Parallel rolls of cloud that form bars separated by clear sky.

bilocular With two **locules.**

binary fission Division into two identical or very similar parts, e.g. of a cell.

binding hyphae **Filaments** that lie across the **prosenchyma** tissue in the **medulla** of brown seaweeds.

bindweed *See Calystegia,* Convolvulaceae.

binomial nomenclature The international standard system for naming plants and animals. The name has two parts, both in Latin and conventionally written in italic. The first name is that of the genus (the generic name), written with an intial capital letter, the second that of the species (known as the specific or trivial name), written all in lower case. The carrot belongs to the species *carota* in the genus ***Daucus***, so its botanical name is *Daucus carota.* A subspecies or variety name may be added as third Latin name (in italic); a varietal name is preceded by 'var.' in roman letters.

bioaccumulation (bioamplification, bioconcentration, biological magnification) The increase in the amount and concentration of a substance in the bodies of animals along a **food chain** that occurs because the animals ingest the substance faster than their bodies can break it down and excrete its products. In a food chain A → B → C, the substance accumulates because animal B ingests all the molecules ingested by A and ingests more itself; animal C ingests all those from B (A + B), plus those it ingests, and so on.

bioamplification *See* bioaccumulation.

biochemical oxygen demand *See* biological oxygen demand.

biochore An area supporting a characteristic plant and animal community. It was one of several terms biologists used prior to the introduction of the term **ecosystem**.

biocoenosis The biological component of a **biogeocoenosis**.

bioconcentration *See* bioaccumulation.

biocontrol *See* biological control.

biodegradable Capable of being broken down (decomposed) by living organisms.

biodiversity A contraction of biological diversity that is applied to the number of species in a specified area, the amount of genetic variation within a specified area, or the complexity of an **ecosystem**.

biogeochemical cycle The movement of a chemical element from the physical environment, through living organisms, and back to the physical environment.

biogeochemistry The scientific study of the distribution and movement of chemical elements in the **biosphere**, and the effect living organisms have on the rocks below the ground surface.

biogeocoenosis A term equivalent to **ecosystem** that is often found in Russian and East European scientific literature.

bioherm A mound or reef formed from the accumulation of living organisms. *Compare* biostrome.

biological amplification *See* bioaccumulation.

biological conservation Management that aims to encourage the maximum number of species and the maximum genetic variation within each species as a means of ensuring the long-term maintenance of the resource.

biological control (biocontrol) The use of naturally occurring predators, parasites, or competitors to control the size of a pest population.

biological magnification *See* bioaccumulation.

biological oxygen demand (biochemical oxygen demand, BOD) The weight of oxygen utilized by the organisms inhabiting 1 litre of effluent that has been stored in darkness for five days at a constant 20°C. It is a measure of the polluting capacity of the effluent.

biological pesticide (biopesticide) Compounds derived from living organisms that are used as **pesticides**.

biological species concept (isolation species concept) A definition of a **species** as a group of organisms that are able to interbreed, but are either unable to breed with members of other groups or produce infertile offspring when they do. It is difficult to apply this definition to plants.

biomass (standing crop) The total mass of all the living organisms, or of a particular set of them, present in an **ecosystem** or at a specified **trophic level** in the ecosystem. It is expressed as the dry weight or, more precisely, as the carbon, nitrogen, or calorific content per unit area.

biome The largest **biotic** community that is recognized, broadly corresponding to a major climatic region and defined in terms of all the living organisms together with their interactions with their physical environment. The number of recognized biomes varies somewhat with different authorities, but most biogeographers accept polar regions, deserts, wetlands, mountains, temperate forests, tropical forests, temperate grasslands, and tropical grasslands. Some also regard the oceans as a biome.

biopesticide *See* biological pesticide.

bioreclamation *See* bioremediation.

bioremediation (bioreclamation) A contraction of biological remediation (reclamation), which is the use of living organisms to concentrate and thereby remove pollutants from soil, water, or air.

Biorhiza pallida *See* oak-apple gall.

biospecies A group of interbreeding organisms that is reproductively isolated from all other groups.

biosphere (ecosphere) The part of the Earth in which living organisms occur. It extends from below the floor of the ocean and deep inside continental rocks to the **stratosphere**.

biosphere reserve A **conservation** site designated by the U.N. Educational, Scientific, and Cultural Organization (UNESCO) under its Man and the Biosphere Programme, as part of an international network of protected areas representing all of the world's major vegetation types, in order to promote sustainable development by local communities based on sound scientific principles.

biostrome A layered accumulation of living organisms, forming a sheet. *Compare* bioherm.

biotechnology The use of microorganisms for human purposes, especially when this involves the genetic manipulation of those organisms (e.g. **genetic engineering**).

biotic Describes the living component of an environment.

biotope A region with specified environmental conditions that supports a characteristic community of organisms.

biotopographic unit **1.** A small **habitat** with a distinctive topography that is made by a living organism or colony of organisms, e.g. a termite mound. **2.** A topographic unit, e.g. a hill, that comprises a distinctive micro-environment.

biotroph A parasite that feeds on the living tissues of its host.

Bipalium kewense (greenhouse planarian) A predatory flatworm (**Platyhelminthes**) with a half-moon-shaped head that feeds on earthworms, slugs, and insect larvae; they are also cannibalistic. They secrete a mucus that is distasteful to other animals, so they have few predators. They are native to southeastern Asia but are widespread in the southern United States.

bipectinate Resembling a comb with teeth on both sides of the central stem. Most often used to describe insect antennae (*see* antenna).

bipinnate Of leaves, having a central **axis** arising from the **rachis**, with **leaflets** (*see* pinna) on either side (i.e. doubly **pinnate**)

A bipinnate leaf.

biradial symmetry An arrangement of the parts of a body such that they are situated around a central **axis** so that each quarter of the body mirrors the quarter opposite but differs from the two adjacent quarters.

biramous With two **branches**.

birch *See Beta*, Betulaceae.

birch bracket *See Piptoporus betulinus.*

birch polypore *See Piptoporus betulinus.*

bird-of-paradise flower *See Strelitzia.*

birds *See* Aves.

bird's nest fungi *See* Nidulariaceae.

birthwort *See* Aristolochiaceae.

bisaccate With two **bladders** or **air sacs**.

biseriate In two rows.

bisexual **1.** A **species** with both male and female individuals. **2.** A **hermaphrodite**

possessing both male and female reproductive organs, e.g. a flower with both **stamens** and **pistils**.

Biston betularia (peppered moth) A nocturnal moth (**Lepidoptera**) that exists in two forms, one pale (*typica*) and the other dark (*carbonaria*). The moths rest on tree trunks, against which they are very well camouflaged, and prior to 1848 the *typica* form was the more common, but by 1895 in the Manchester area 98 percent were *carbonaria*. In the 1950s H. B. D. Kettlewell found experimentally that against **bark** coated in soot and that had lost its **lichens** because of air pollution, the *carbonaria* form was better hidden and escaped predation by birds; as air quality improved and the lichens returned, the *typica* form thrived. This selection of the dark form was called industrial melanism and is clear evidence of **natural selection** in operation.

bitegmic With two **integuments**.

bitter ash (*Quassia amara*) *See* quassia.

bitter wood (*Quassia amara*) *See* quassia.

bitunicate Describes an **ascus** in which the inner and outer layers of the wall separate as the **ascospores** are released.

bivalent Describes the condition of two **homologous chromosomes** when these are paired during **prophase I** of **meiosis**.

Bixaceae (order **Malvales**) A **monogeneric** family of **evergreen shrubs** and **trees** that have yellow, orange, or red sap. Leaves are **alternate**, **simple**, **entire**, **stipulate**, and covered with **peltate** scales on the underside. Flowers are **actinomorphic**, **hermaphrodite**, with 5 free, **caducous sepals**, 5 free white or pink **petals**, many **stamens**, **ovary superior** of 2 **carpels**. The fruit is a **loculicidal capsule**. There are 4 genera with 21 species occurring throughout the tropics. *Bixa orellana* is grown for the orange to red colouring annatto obtained from its seeds.

Bixa orellana (annatto) *See* Bixaceae.

black arion *See Arion ater.*

black bean aphid *See Aphis fabae.*

blackberry *See Rubus.*

blackberry aphid *See Amphorophora rubi.*

black-billed magpie *See Pica pica.*

black birch (*Betula lenta*) *See* methyl salicylate.

blackbird *See Turdus merula*, Turdidae.

black bulgar *See Bulgaria inquinans.*

black-capped chickadee *See Poecile atricapillus.*

black crown of carrot *See Alternaria radicina.*

blackcurrant (*Ribes nigrum*) *See Ribes.*

blackcurrant aphid *See Cryptomyzus galeopsidis.*

blackcurrant gall mite *See Cecidophyopsis ribis.*

black-eyed Susan (*Thunbergia alata*) *See* Acanthaceae.

black field slug *See Arion hortensis.*

blackfly *See* Aphididae.

black frost (hard frost) **Frost** that occurs when the air is dry. It leaves plants blackened as water freezes in and between cells, but with no ice crystals on their surfaces.

black-headed grosbeak *See Pheucticus melanocephalus.*

black-headed worm *See Aporrectodea longa.*

black ice A layer of ice that forms when rain close to freezing falls on surfaces below freezing. Raindrops spread on impact and freeze.

black jelly drops *See Bulgaria inquinans.*

black-kneed capsid *See Blepharidopterus angulatus.*

B

blackleg disease *See Leptosphaeria maculans.*

black pepper (*Piper nigrum*) *See* Piperaceae.

black rot of carrot *See Alternaria radicina.*

black scab *See Synchytrium endobioticum.*

black scurf *See Rhizoctonia solani.*

black slug *See Arion ater.*

black snake *See Elaphe obsoleta.*

black-staining polypore *See Meripilus giganteus.*

black stem rust *See Puccinia graminis.*

black tang *See* bladder wrack.

black tany *See* bladder wrack.

black vine weevil *See Otiorhynchus sulcatus.*

black walnut (*Juglans nigra*) *See Juglans.*

bladder A sac in which air or metabolic products may be stored. It opens and closes by means of a valve that responds to stimuli detected by trigger hairs.

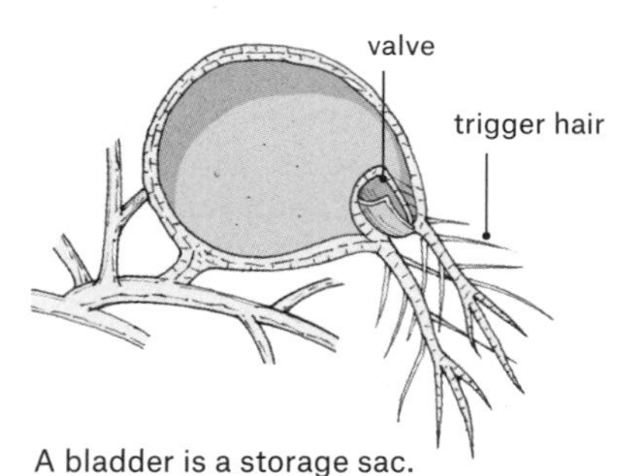

A bladder is a storage sac.

bladder fucus *See* bladder wrack.

bladderwort (*Utricularia*) *See* Lentibulariaceae.

bladder wrack (black tang, black tany, bladder fucus, cut weed, dyers fucus, red fucus, rockweed, rock wrack, sea oak) The brown seaweed *Fucus vesiculosus* in which the **thallus** is flattened and strap-like and has **bladders** containing air that give it **buoyancy**. It has been used in herbal medicine as a source of iodine.

blade Either an entire leaf apart from the **petiole**, or the **lamina**.

Blandfordiaceae (order **Asparagales**) A **monocotyledon**, **monogeneric** family (*Blandfordia*) of **perennial**, **caespitose herbs**. Leaves are **linear**, **alternate**, **distichous**, with a sheathing base. Flowers are **actinomorphic**, **bisexual**, with 6 **petaloid tepals** fused into a **campanulate corolla**, with 2 **trimerous whorls** of **stamens** fused to the corolla tube along half its length, **ovary superior** with 3 **carpels** and 3 **locules**. The fruit is **capsule**. There are four species **endemic** to eastern Australia.

blanket bog An **ombrogenous bog** with a deep surface layer of **peat** that forms on level ground or very shallow gradients in wet, **maritime climates**.

Blastobasis decolorella (straw-coloured apple moth) A moth (**Lepidoptera**) that is native to Portugal and Madeira but that has been present in Britain since the late 1970s. There is a main flight in midsummer and a smaller second flight in autumn and early winter. The purple-brown **caterpillars** ordinarily feed on **detritus**, and on apples they occur on short-stalked varieties, especially where two fruits are touching. They feed on the base of the fruit, removing skin and flesh and causing wounds that are sometimes covered with sticky **frass**. They also spin silk to tie leaves together. The moth also feeds on almonds and peaches, and on hawthorn berries and rose hips.

blastochory Plant dispersal by means of offshoots or **runners**.

blastospore A fungal **spore** produced by **budding**.

bleeding In plants, exuding the contents of the **xylem** at a cut surface of a root or of the **phloem** from a stem, **petiole**, or fruit.

bleeding heart (*Dicentra*) *See* Papaveraceae.

Blepharidopterus angulatus (black-kneed capsid) A blue-green, omnivorous capsid bug (**Miridae**) that has black patches on its antennae (*see* antenna), variable yellow markings on the **scutellum** and forewings, and very long antennae in the male. The bug is common through temperate regions of the Old World, mainly in **broad-leaved** trees. It feeds on **red spider mites**, aphids (**Aphididae**), and **whitefly**.

blepharoplast A spherical structure in the **cytoplasm** of a developing **antherozoid** that breaks down to produce the **basal body** of a **flagellum**.

blewit *See Clitocybe.*

blight Any plant disease that causes extensive **chlorosis**, **wilting**, browning, and rotting of tissues.

blind worm *See Anguis fragilis.*

blizzard A wind accompanied by heavy snow and low temperature. The U.S. National Weather Service defines a blizzard as a wind of at least 56 km/h, temperature no higher than −7°C, and sufficient snow falling to produce a layer at least 25 cm thick or snow blown from the surface that reduces visibility to below 400 m. The temperature requirement has been dropped in some places.

blocking The situation in which a particular weather pattern, commonly an **anticyclone**, remains stationary for a prolonged period and moving weather systems are deflected around it.

blossom blight *See Sclerotinia sclerotiorum.*

blue-bunch wheatgrass (*Agropyron spicatum*) *See* Palouse prairie.

blue-green algae *See* Cyanobacteria.

blue-grey worm *See Octolasion cyaneum.*

blue jay *See Cyanocitta cristata.*

blue-leg *See Clitocybe.*

blue mould The mat of **mycelium** formed by certain *Penicillium* species, bearing blue conidia (*see* conidium).

Blue Ridge two-lined salamander *See Eurycea wilderae.*

bluethroats *See* Turdidae.

blue tit *See Cyanistes caeruleus.*

blunt snout pillbug *See Armadillidum nasatum.*

blusher The edible **fruiting body** of the fungus ***Amanita*** *rubescens*, found in Europe and eastern North America and the closely related *A. novinupta*, found in western North America. The fruiting body resembles a mushroom with a reddish brown **pileus** retaining whitish fragments of the **universal veil**. The name blusher refers to the tendency of the flesh to turn pink on bruising or exposure to air, a feature that distinguishes it from the toxic ***Amanita pantherina***.

B-Nine *See* daminozide.

BOD *See* biological oxygen demand.

bog A plant community that develops in wet areas where acid conditions reduce the rate of decomposition, allowing plant material to accumulate and encouraging the formation of **peat**. Bogs occur in some tropical forests, e.g. near the coasts of Sarawak, but are more common in high northern latitudes, where there are two principal types: **blanket bog** and **raised bog**. **Valley bog** is similar, but more correctly described as a **mire** or **fen**.

bog forest A forest that develops as water drains from a **bog** and the ground dries.

bog myrtle (*Myrica gale*) *See* Myricaceae; root nodule.

bog soil A poorly drained **peat soil** or **muck soil** that lies above a grey mineral soil.

Boletales An order of **agaric fungi** that produce a variety of **fruiting bodies**,

B

boletes being the most familiar but also including earthballs, puffballs, and false truffles. Most Boletales are ectomycorrhizal (*see* ectomycorrhiza) and occur close to trees, but some are **saprotrophs** or parasitic. Some are edible, e.g. ***Boletus edulis***, but others are toxic. There are more than 1300 species distributed worldwide.

bolete A fungal **fruiting body**, typical of the **Boletales**, that have a **pileus** resembling a mushroom, but tubes on the underside rather than **gills**, the ends of the tubes giving the appearance of pores or the surface of a sponge.

Boletus edulis (cep, king bolete, penny bun, porcino) A species of **Boletales** with a **fruiting body** that is an edible and much sought-after **bolete**. It is ectomycorrhizal (*see* ectomycorrhiza), so is found close to trees, and occurs throughout Eurasia and North America (although the American king bolete may be a different species), and it has been introduced in southern Africa, Australia, and New Zealand.

bolochory Dispersal of **seeds** or **spores** by expelling them, so they are spread by an action of the plant itself.

bolting Flowering and producing seed prematurely.

Bombus (bumblebees) A genus (**Apidae**) of social insects, most with black and yellow, long, soft, body hairs but some with black, orange, or red markings. The hind legs have prominent corbiculae (*see* corbiculum). Queens (reproductive females) overwinter, emerging in spring and immediately seeking a suitable nest site where they prepare wax cells in which they lay eggs that hatch into larvae. Bumblebee workers are able to lay **haploid** eggs that develop into males; only the queen lays **diploid** eggs. Bumblebees feed on **nectar** and **pollen** and are important pollinators. There are more than 250 species found throughout the Northern Hemisphere and South America, and they have been introduced to New Zealand and Tasmania. Bumblebees are threatened in many parts of the world due to loss of **habitat** and damage from insecticides.

Bombycilla cedrorum (cedar waxwing) A species of gregarious, nomadic, grey-brown birds with a pale yellow breast and underside, a black face mask, a crest, red tips to the **secondary feathers**, and a yellow band at the tip of the square tail. They are about 155 mm long. They inhabit areas with many small **trees** and **shrubs**, and visit suburban gardens, feeding on fruit in winter and insects in summer They occur only in North America and are partly migratory.

bone 1. A unit of the vertebrate skeleton, e.g. rib, **femur**, etc. **2.** The material from which the vertebrate skeleton is made. It is about 70 percent calcium salts and 30 percent **collagen**.

bone marrow The tissue found inside bones. There are two types. Red marrow produces red and white blood cells and platelets (involved in clotting and preventing blood loss); yellow marrow consists mainly of fat cells.

Bonnetiaceae (order **Malpighiales**) A family of **evergreen trees** and **shrubs** that excrete resinous sap. Leaves are **alternate**, **simple**, **exstipulate**, **entire** or **dentate**. Flowers are **actinomorphic**, **hermaphrodite**, with 5 unequal, persistent **sepals**, 5 red or pink free, **contorted petals**, **ovary superior** of 3 or 5 **carpels**. The fruit is a **capsule**. There are 3 genera with 35 species found in Cambodia, Malesia, Cuba, and South America.

Bonpland, Aimé (1773–1858) A French botanist and physician, born Aimé-Jacques-Alexandre Goujaud, who accompanied Alexander von **Humboldt** on his expedition to South America and who classified most of the 3600 previously unknown plants that they collected. In 1804 Bonpland was made director of

Empress Josephine's private botanical garden at Malmaison and in 1808 Josephine appointed him her official botanist. In 1816 he returned to South America as professor of natural sciences in Buenos Aires, also practising medicine.

book lungs Respiratory organs found in spiders (**Araneae**). They are situated in the abdomen and consist of many fine leaves, providing a large surface area over which blood passes and absorbs oxygen. Openings to the book lungs are located on the **ventral** surface of the abdomen and can be closed to prevent water entering.

bootlace fungus *See Armillaria.*

bor Open woodland dominated by Scots pine (*Pinus sylvestris*) found on dry, sandy soils in the **boreal forest** of Russia and Canada.

borage (*Borago*) *See* Boraginaceae.

Boraginaceae (not placed in an order) A family of plants ranging from large, tropical **trees** to **shrubs**, woody climbers, and **annual** and **perennial herbs**, all of them bearing conspicuous hairs with bulbous bases. Leaves are **alternate**, sometimes with the lower leaves **opposite**, **simple**, **entire** or occasionally **serrate** or serrate-**spinulose**, a few **succulent**. Flowers **gamopetalous**, hypogynous (*see* hypogyny), **actinomorphic** or slightly **zygomorphic**, sometimes **unisexual** (plants **dioecious**), usually **pentamerous**, often heterostylous (*see* heterostyly), **corolla** tubular to trumpet-shaped, rarely **campanulate**, **stamens** attached to the corolla tube, **ovary** of 2 **carpels** with up to 4 **locules**. **Inflorescence** typically a **cyme**. The fruit is usually a **schizocarp** that splits into 4 **nutlets**, less commonly a **drupe** or **dehiscent capsule**. There are 148 genera with 2755 species found throughout the tropics and temperate regions. Some trees provide timber and edible fruits, some species are cultivated as ornamental shrubs, some have **rhizomes** that provide dyes, *Borago* spp. leaves are a culinary herb (borage), and many species are grown as ornamental herbs.

Bordeaux mixture A **fungicide** made from copper sulphate ($CuSO_4$) and slaked lime ($Ca[OH]_2$) that is mixed with water and used as a spray against **downy mildew**, **powdery mildew**, and other fungal diseases.

bordered pit A **pit** in a **cell wall** between **tracheids** or **vessel elements** in which the cavity is partly covered by an extension of the cell wall. *Compare* simple pit.

boreal Pertaining to the north.

boreal forest The subarctic belt of forest, dominated by conifers, that extends across Eurasia and Canada. It is bounded by **tundra** in the north and by **temperate deciduous forest**, **steppe**, or semi-**desert** to the south. *See* taiga.

boreal region The floristic region that includes all of the Northern Hemisphere as far south as southern Japan, the Himalayas, the North African coast, and the Gulf of Mexico.

boron (B) An element essential for healthy plant growth that occurs in the soil solution as boric acid.

Borthwickiaceae (order **Brassicales**) A **monotypic** family (*Borthwickia trifoliata*), which is a small **tree** with **opposite**, **trifoliate** leaves and large **pentamerous** to 8-merous flowers with a large **calyx**, small **petals**, and many **stamens**. The fruit is a **capsule** that hangs open, revealing the red seeds. The species occurs in southwestern Yunnan, China, and adjacent Myanmar.

boscus (subboscus) Undergrowth or wood that is produced by coppicing (*see* coppice).

bostryx A **cymose inflorescence** in which **branches** arise on only one side of the **rachis**.

B

Boswellia sacra (frankincense) *See* Burseraceae.

botanical pesticide A chemical compound derived from a plant that is used as a **pesticide**, e.g. **pyrethrum** and **rotenone**.

botanic garden An area in which a variety of plants are cultivated for educational, research, and **conservation** purposes. Most botanic gardens are open to the public.

Botrytis cinerea A species of **ascomycete fungi**, found worldwide, that lives as a **saprophyte** but is a **facultative parasite** causing grey mould in a wide variety of crops. It infects flowers and developing fruits, but requires a wound to provide access to a green plant. The fungus survives periods when hosts are scarce as sclerotia (*see* sclerotium) on dead plant tissue. Patches of grey mould appear on infected plants and may spread rapidly under humid conditions. Soft brown rot affects some fruits, especially strawberries and grapes. If the **humidity** falls after grapes have been infected, the fungus draws water from the fruits, increasing their sugar content and sweetening the wines. This is called noble rot; it reduces the volume but improves the quality of the wine.

Bougainvillea (family **Nyctaginaceae**) A genus of **evergreen** or **deciduous**, **sarmentose shrubs** and woody vines, bearing thorns tipped with a black, waxy substance. Leaves are **alternate**, **simple**, **ovate acuminate**. Flowers are small, usually white, borne in clusters of 3 surrounded by 3 or 6 papery, brightly coloured pink, magenta, purple, red, orange, white, or yellow **bracts**. The fruit is a 5-lobed **achene**. There are 4–18 species, native to South America but widely cultivated in warm climates, with many varieties.

boundary layer A layer of air adjacent to a surface within which conditions are strongly influenced by the proximity of the surface. *See* planetary boundary layer.

bower vine (*Pandorea jasminoides*) *See Pandorea.*

bowstring hemp (*Sansevieria zeylanica*) *See Sansevieria.*

box elder (*Acer negundo*) *See Acer.*

boxthorn *See Lycium.*

brachium The upper forelimb.

Brachycaudus helichrysi (leaf-curling plum aphid) A greenish yellow **plum aphid** (**Aphididae**), 1.5–2.0 mm long, that lays eggs in autumn at the base of shoot **buds** on fruit trees. These hatch in spring and form large colonies on the underside of leaves, feeding on **sap** from leaf veins; they also feed on flowers. In late summer the aphids migrate. The primary hosts are plum, peach, apricot, and almond trees, the secondary hosts are plants in the **Asteraceae** and the aphid causes significant damage to asters, chrysanthemums, and sunflowers. It occurs worldwide.

brachypterous Of an insect, having both sets of wings reduced.

bracket fungi Fungal **fruiting bodies** that grow out as **annual** or **perenial** shelves, up to 600 mm across, from the main stems and **branches** of trees. Many species produce brackets that are specific to particular hosts. Brackets cause the decay of **heartwood**, weakening trees so eventually they may fall. *See* polypore.

brackish Describes water that is salty, but less so than sea water.

Braconidae A family of **parasitoid** wasps (**Hymenoptera**) that feed mainly on the larvae of beetles (**Coleoptera**), flies (**Diptera**), butterflies and moths (**Lepidoptera**), aphids (**Aphididae**), and **Heteroptera**. Some are used in **biological control**. Braconids are 1–40 mm long and most are dark brown or black. The females lay eggs on the bodies of their hosts, the larvae then feeding internally. Most kill the host, but some cause the host to become sterile and less active. There are probably

40,000–50,000 species with a worldwide distribution.

bract A modified leaf that subtends a **flower** or **inflorescence**.

bracteate Having **bracts**.

bracteole A small **bract** on the **pedicel** below the **calyx** and above the bract.

Bradyrhizobium A genus of **Alphaproteobacteria** that are rod-shaped, Gram-negative (*see* Gram reaction) and more than 55 species of which fix nitrogen (*see* nitrogen fixation). They occur widely in soil and form symbiotic relationships (*see* symbiosis) with legumes (**Fabaceae**), forming **root nodules**.

brain fungus *See Sparassis crispa.*

brake An area covered by scrub, underwood, or bracken.

bramble leafhopper *See Ribautiana tenerrima.*

bramble shoot moth *See Epiblema uddmanniana.*

branch A lateral stem that arises from the main stem or from another branch.

brandling worm *See Eisenia fetida.*

Brassica (family **Brassicaceae**) A genus of **annual** and **perennial herbs** with **taproots** and erect, branched stems (some used as walking sticks!). The flowers are 4-merous with yellow **petals** borne in a compact **raceme** or **corymb**. The fruit is a 2-valved **silique** that shatters to release its seeds. There are 35 species occurring in Eurasia but with a distribution centred on the Mediterranean. Many species are cultivated for food, oil, or ornament (e.g. cabbage, cauliflower, turnip, canola, etc.).

Brassicaceae (order **Brassicales**) A family, formerly known as Cruciferae, mainly of **annual** and **perennial herbs** that have flowers with 4 petals arranged in a cross (crucifer, hence the former name), with a few **shrubs**, climbers, and aquatics. Leaves are **alternate**, **simple** and **entire** or **pinnate**, **exstipulate**, sometimes heterophyllous (*see* heterophylly). Flowers usually **actinomorphic**, **bisexual**, hypogynous (*see* hypogyny) with 4 **sepals** sometimes swollen at the base, 4 cruciform **petals**, 6 **stamens** (2 short and 4 long), **ovary superior** of 2 **carpels**, **syncarpous**, usually with 2 **locules**. The fruit is a 2-locular **capsule**, called a **silique** if it is more than three times longer than it is wide and a **silicle** if it is less than three times as long as it is wide. There are 338 genera with 3710 species with a worldwide distribution, but especially common in northern temperate regions and **dry climates**. Many species are cultivated as vegetables or for edible oil or seeds.

brassica cyst nematode *See Heterodera cruciferae.*

Brassicales An order of plants with a **racemose inflorescence**, leaves often spiral, and small **stipules**. There are 18 families with 398 genera and 4765 species. *See* Akaniaceae, Bataceae, Borthwickiaceae, Brassicaceae, Capparaceae, Caricaceae, Cleomaceae, Emblingiaceae, Gyrostemonaceae, Koeberliniaceae, Limnanthaceae, Moringaceae, Pentadiplandraceae, Resedaceae, Salvadoraceae, Setchellanthaceae, Tovariaceae, and Tropaeolaceae.

brassinosteroids A class of plant **hormones** that, with **auxin**, promote cell elongation and expansion, **differentiation** of **vascular tissue**, and are thought to influence cell division and the regeneration of **cell walls**.

Braun-Blanquet, Josias (1884–1980) A Swiss botanist who developed the most widely used scheme for describing vegetation communities (*see* phytosociology), which he first described in his book *Pflanzensoziologie,* first published in Berlin in 1924, the third edition appearing in 1964. He worked at the University of Zürich and later was director of the Station Internationale de Géobotanique Méditerranéenne et Alpine, at Montpellier, France; the team he led became known as the Zürich-Montpellier School of Phytosociology.

braunerde *See* brown earth.

Brazil nut (*Bertholletia excelsa*) *See* Lecythidaceae.

breadfruit (*Artocarpus*) *See* Moraceae.

breakage and reunion The breaking and cross-wide rejoining of **chromatids** (i.e. **crossing over**) during the pairing of **homologous chromosomes** at **prophase I** of **meiosis**.

breckland A **habitat** consisting of grass **heathland** with gorse (*Ulex europaeus*), bracken (*Pteridium aquilinum*), and heather (***Calluna*** *vulgaris*) found in northern Suffolk and southern Norfolk, England, on land that was once forest but cleared in Neolithic times.

breeding true Producing offspring possessing particular **characters** identical to those of the parents. Homozygous (*see* homozygosity) organisms invariably breed true; heterozygous (*see* heterozygosity) organisms seldom do so.

breeze A light wind; on the **Beaufort Wind Scale** a wind blowing at 2–14 m/s.

Brevicoryne brassicae (cabbage aphid, mealy cabbage aphid) A species of grey-green aphids (**Aphidae**) covered with a white-grey wax resembling meal. Eggs hatch in spring and from midsummer the aphids can form dense colonies on brassicas that can inhibit growth and kill young plants. Winged aphids migrate during summer to establish colonies on other plants. These aphids absorb **glucosinolates** from the plants, producing mustard oils that deter predators. The species is native to Europe but is now found in many other parts of the world.

brigalow scrub Semi-arid scrub vegetation, dominated by ***Acacia*** species, found in parts of Australia.

brimstones *See* Pieridae.

broad bean (*Vicia faba*) *See Vicia.*

broadhead skink *See Eumeces laticeps.*

broad-leaved Describes the leaves of most **angiosperms**, which are broader than the scale- or needle-like leaves of **gymnosperms**.

broad-leaved evergreen forest A forest dominated by **angiosperm** trees that remain in leaf throughout the year. Such forests occur where rainfall is abundant and distributed evenly through the year. They are found in parts of the tropics, and in temperate regions of the Northern Hemisphere on the coastal plain of the Gulf of Mexico, central China, and southern Japan; in the Southern Hemisphere most temperate forests are broad-leaved, except in southern South America.

broad mite *See Polyphagotarsonemus latus*, Tarsonemidae.

Bromeliaceae (order **Poales**) A family of **monocotyledon epiphytes** and **rosette trees**. The rosette trees have fully developed roots, the roots of epiphytes serve only to anchor the plant; some species lack roots except when seedlings. Leaves are arranged in spirals, in some the overlapping leaf bases forming reservoirs that hold water and **humus** absorbed by **adventitious** roots that grow up from the leaf bases. Flowers are usually **actinomorphic** but sometimes **zygomorphic**, usually **bisexual**, **trimerous** with a **petalloid corolla**, 6 free **stamens**, **ovary superior** or **inferior** of 3 fused **carpels** each with 3 **locules**. Flowers are borne in a terminal **inflorescence** as a **spike**, **raceme**, or **panicle**. The fruit is a **berry** or **capsule**. There are 57 genera with 1770 species occurring in subtropical America, with 1 species (*Pitcairnia feliciana*) in West Africa, and a few on islands in the Pacific. Several species produce edible fruit, the most famous being *Ananas comosus* (pineapple), and some are grown as ornamentals.

bromomethane *See* methyl bromide.

Brongniart, Adolphe Théodore (1801–76) A French botanist who had a special interest in the classification and distribution of **fossil** plants, and their relationships to modern plants. He is sometimes called the father of palaeobotany.

bronze frog (*Rana clamitans melanota*) *See Rana clamitans clamitans.*

brood parasitism (nest parasitism) A form of **parasitism** in which the parasite lays its eggs in the nest of the host, and the host incubates them and raises the young. In some species, e.g. cuckoo (*Cuculus canorus*) the parasitic young kill the young of the host or eject them from the nest.

broomrape *See* Orobranchaceae.

brotochory *See* androchory.

Brown, Robert (1773–1858) A Scottish botanist who did much to develop a natural system of plant classification and who was the first person to distinguish between **angiosperms** and **gymnosperms**. He made a special study of plant reproductive processes and was the first to study **fossil** plants microscopically. He was also the first person to observe the random motion of microscopic particles now known as Brownian motion.

brown algae Mostly marine seaweeds (Phaeophyta), many of which are olive-brown in colour and almost black when dry. The **blades** can reach more than 30 m in length. Their cells lack plasmodesmata (*see* plasmodesma) and do not produce **starch.**

brown earth (brown forest soil, brown soil, braunerde) A well-drained soil that is brown below the surface with weakly defined **soil horizons**. The soil is well-weathered (*see* weathering) and slightly leached (*see* leaching). In humid, temperate regions brown earths usually form beneath **deciduous** forest, but their high fertility makes them valuable agricultural soils. They fall within the **inceptisols** of the U.S. Department of Agriculture **soil taxonomy.**

brown forest soil *See* brown earth.

brown frog *See Rana temporaria.*

brown garden snail *See Cornu aspersum.*

brown-lipped snail *See Cepaea nemoralis.*

brown podzolic soil A free-draining soil that is leached (*see* leaching) to an early stage of **podzolization**, has a **mor** surface horizon and a B **soil horizon** enriched in iron oxide.

brown roll-rim *See Paxillus involutus.*

brown rot *See Monilinia fructicola.*

brown soft scale *See Coccus hesperidium.*

brown soil *See* brown earth.

Bruchidae (bean weevil, seed weevils) A family of small beetles (**Coleoptera**), 2–5 mm long, usually mottled, with a pronounced neck and the head curved under the body, and well-developed wings. Adults feed on plants, especially **Fabaceae**, and lay eggs on developing seeds. The larvae are less than 3 mm long and C-shaped, resembling those of **Scarabeidae.** They feed on seeds and on stored plant material. There are about 1300 species with a worldwide distribution.

Brunelliaceae (order **Oxalidales**) A **monogeneric** family (*Brunellia*) of **evergreen trees** in which the leaves are **opposite** or in **whorls**, **simple** or **compound ternate** or **pinnate**, with **stipules**. Flowers are **actinomorphic**, **hermaphrodite** or **unisexual** (plants **dioecious** or **gynodioecious**), 4- to 8-merous, with a single whorl of valvate **sepals**, twice as many **stamens** as sepals, usually in 2 whorls, **ovary superior** of 2–8 free **carpels**. Flowers are borne in **axillary inflorescences** on new shoots. The fruit is a **follicle**. There are 55 species found in Central and South America and the Antilles.

B

Bruniaceae (order **Bruniales**) A family of **shrubs** with leaves that are small, tough, sometimes **imbricate**, spiral, **entire**, with **stipules**. Flowers are small and usually white, **actinomorphic**, **bisexual**, **pentamerous**, usually **perigynous**, with 5 **sepals**, **petals** and **stamens**, **ovary inferior** or occasionally **superior** of 1–3 fused **carpels** with 1–5 **locules** each with 1–16 **ovules**. Flowers are borne in a terminal **spike** or spherical **capitulum** with up to 400 flowers. The fruit is **indehiscent** or a **capsule** with 2–4 **valves**. There are 12 genera with 75 species found only in South Africa, almost all in Cape Province.

Bruniales An order of **evergreen**, woody plants found in South Africa and Chile. There are 2 families with 14 genera and 79 species. *See* Bruniaceae and Columelliaceae.

brunizem (prairie soil) A dark-coloured **prairie** soil developed from **loess**.

bryology The study of mosses (**Bryophyta**).

Bryophyta A division of plants that includes only the mosses, of which there are 110–120 families with 700 genera and approximately 10,000 species. All mosses are small with green, leafy shoots that are the **gametophytes** (**haploid**) on which the **diploid embryo** develops into the mature **sporophyte**. The sporophyte contains **chlorophyll** and performs **photosynthesis** only during its early development and the rest of the time it is dependent on the gametophyte, which is the prominent stage in the life cycle.

Bryopsida A class that includes those mosses (**Bryophyta**) in which the **peristome** surrounding the **sporophyte** capsule is formed from articulated remnants of a **cell wall**. The Bryopsida accounts for more than 95 percent of all mosses, with 90–110 families.

Bt *See Bacillus thuringiensis.*

bubnoff unit (B) A unit used in reporting **erosion**; 1 B = 1 μm/yr = 1 mm/10^3 years. The unit is named after the Russian-born German geologist Serge von Bubnoff.

buccal cavity Mouth.

buccal force pump In **Amphibia**, a respiratory mechanism in which the floor of the mouth is raised while the nostrils are closed, forcing air into the lungs.

buckeye *See Aesculus.*

buckthorn (*Rhamnus catharticus*) *See* Rhamnaceae.

buckwheat (*Fagopyrum esculentum*) *See* Pologonaceae.

bud An immature shoot, protected by scale leaves, from which a leaf, flower, or new stem growth may emerge.

Budapest slug *See Tandonia budapestensis.*

budding 1. (gemmation) A form of **asexual reproduction** in which an offspring develops in the **cell wall** or body wall of a mature individual, grows into a swelling, then detaches itself to live independently. **2.** The formation of a **bud** as a result of **cell division** stimulated by **cytokinins**. **3.** Grafting (*see* graft) a bud on to a plant.

Buddleja (family **Scrophulariaceae**) A genus of **shrubs** and **trees** usually with **lanceolate** leaves in **opposite** pairs or **whorls** (**alternate** in one species), and **simple**. Flowers are **unisexual** (plants **dioecious**), with 4 fused often unequal **sepals**, **corolla** tubular with 4 lobes, 4 **stamens**, **ovary superior** of 2 **carpels** each with 2 **locules**. The fruit is a dry **capsule**. There are about 100 species occurring in America, Africa, and Asia. Many buddlejas are cultivated for ornament. The flowers are attractive to many butterflies and the plant is sometimes called the butterfly bush.

bud fission In certain **Fungi**, a type of **budding** in which a **septum** separates the daughter cell from the parent.

bud scale *See* cataphyll.

buffer A solution that consists of a weak acid and its **conjugate base**, or a weak base and its **conjugate acid**. Its **pH** changes only slightly when a strong acid or base is added to it, so it prevents large changes in pH. Many living organisms can survive only within narrow pH limits, so buffer solutions are common biologically.

buffer strip (buffer zone, filter strip) An area of undisturbed vegetation adjacent to a protected **habitat**.

buffer zone *See* buffer strip.

Bufo americanus (*Anaxyrus americanus*, American toad) A brown, grey, olive, or red toad (**Bufonidae**), 50–107 mm long, with prominent eyes and large warts on the **dorsal** side of its hind legs, that feeds at night on invertebrates. It occurs in a variety of **habitats** including cities in northeastern and eastern North America.

Bufo bufo (common toad, European toad) A toad (**Bufonidae**) that is about 150 mm long, brown or grey-brown, with bulging eyes, a wide mouth, a warty skin, and short forelegs with the feet turned inward. It possesses **paratoid glands** that secrete a toxin. It is nocturnal, spending the day concealed and sometimes travelling a considerable distance while hunting, moving by walking or in short hops using all four legs. It feeds on invertebrates and small vertebrates. The toad hibernates during winter. It occurs throughout Eurasia and parts of North Africa.

Bufo fowleri (*Anaxyrus fowleri*, Fowler's toad) A brown, grey, or olive toad (**Bufonidae**) with dark spots on the back and a pale **dorsal** stripe, 50–100 mm long. It inhabits woodland and sandy soil, moving into burrows during dry weather and in winter. It feeds on insects and other small invertebrates, and occurs along the eastern coastal plain of the United States.

Bufonidae (toads) A family of **Anura**, 20–250 mm long, with squat bodies and short, powerful legs that move by crawling. They lack teeth. The skin is usually dry with many wart-like protrusions and many species secrete toxins from parotoid glands behind the head. There are more than 350 species found worldwide except for Antarctica, Australasia, and Madagascar.

Bufo terretris (southern toad) *See Anaxyrus terrestris.*

bugleweed *See Ajuga.*

bulb An underground storage organ growing from a short, flat stem with roots beneath, and comprising fleshy leaves or leaf bases protected by a surrounding cover of scale leaves. It allows the plant to survive from one season to the next and may divide, thereby allowing **vegetative reproduction**.

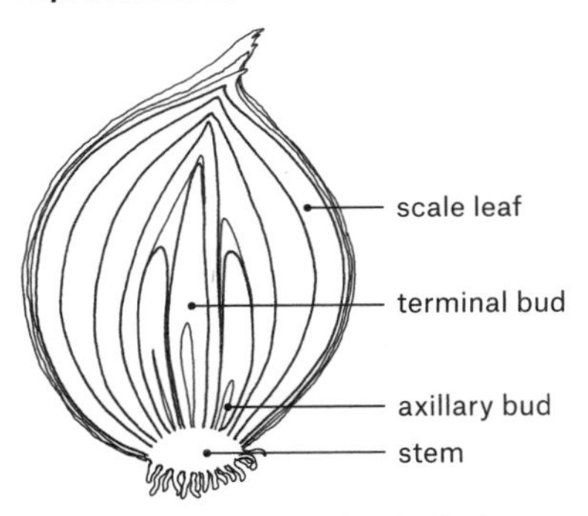

A cross-section through a bulb showing the internal structure, with scale leaves enclosing a terminal bud and axillary bud, and the stem at the base.

bulbil Any small, **bulb**-like structure that detaches from the plant and develops into a new plant.

Bulgaria inquinans (black bulgar, black jelly drops, bachelor's buttons, poor man's licorice, rubber buttons) A species of **ascomycete fungi** in which the **fruiting body** is round, flat becoming cup-shaped, shiny black with dark brown sides and flesh, 5–40 mm across and about 10 mm tall. The fungus is a **saprotroph** growing in dense masses on fallen trees. It has a **cosmopolitan distribution**.

B

bulk density The mass per unit volume of a soil that has been dried to a constant weight at 105°C. It ranges from 1.1–1.4 g/cm^3 in **clay** soils to 1.3–1.7 g/cm^3 in sandy soils.

bulk volume The volume of a dried soil sample divided by its weight.

bulla A thin-walled projection of the skull that encloses the middle ear in most mammals.

bullate Blister-like or blistered.

bulliform cell Large, bubble-like, epidermal (*see* epidermis) cells found in groups near the midrib on the upper surface of the leaves of many grasses (**Poaceae**). When full of water the cells push the sides of the leaf, opening it out; lack of water empties the cells so they contract, pulling the sides of the leaf together, thereby shielding the **stomata** and reducing water loss through **transpiration**.

bullock's heart (*Annona reticulata*) *See Annona.*

bulrush (*Scirpus lacustris*) *See Scirpus; see also Typha.*

bumblebees *See* Apidae, *Bombus.*

bunch grass prairie *See* Palouse prairie.

bundle cap **Sclerenchyma** or thickened **parenchyma** tissue that forms a layer over the tip of a **vascular bundle**.

bundle sheath **Parenchyma** or sometimes **sclerenchyma** cells that are densely packed to form a sheath around leaf **veins** in plants that use the **C4 pathway** of **photosynthesis**.

bunt (stinking smut) A disease of wheat caused by *Tilletia tritici* and *T. laevis*, which are **Fungi** belonging to the **Basidiomycota**. In infected plants the seed kernels become brown sori (*see* sorus), called bunt balls, filled with dark brown or black masses of **teliospores**. The **spores** are released at harvest and can persist on other kernels or in the soil.

bunya bunya (*Araucaria bidwillii*) *See Araucaria.*

buoyancy The upward force exerted on a body that is immersed in a fluid of lower density. If buoyancy is positive the body will rise, if it is negative the body will descend, and if it is neutral the body will remain at the same level.

buried soil Soil that has been covered by a deposit of **alluvium** or **colluvium**, or by an **aeolian**, glacial, organic, or other deposit.

Burmanniaceae (order **Dioscoreales**) A **monocotyledon** family of **annual** and **perennial herbs**, most of which are small saprophytes (*see* saprotroph) or hemisaprophytes with slender, upright stems that grow from **rhizomes** and roots that are fleshy or produce **tubers**. The saprophytes lack **chlorophyll** and are often white, yellow, or red, with **alternate**, **sessile** scale-like leaves. Hemisaprophytes are green, with well-developed alternate leaves. Flowers are **actinomorphic**, **bisexual**, with 6 **tepals** in 2 **whorls**, fused at the base into a tube, 3 or 6 **stamens**, **ovary inferior** of 3 **carpels** and 1 or 3 **locules**. **Inflorescence** usually a terminal **cyme**. The fruit is a **capsule**. There are 9 genera with 95 species, found mostly in the tropics, especially America.

burning bush (*Bassia scoparia* f. *trichophylla*) *See Bassia.*

burn-off The morning clearance of **fog**, **mist**, or low cloud as sunshine intensifies, the air temperature increases, and the water droplets evaporate.

Burseraceae (order **Sapindales**) A family of **trees** and **shrubs** with flaky **bark** that sometimes peels away in large sheets. Leaves are **alternate**, in spirals, **imparipinnate**, usually **exstipulate** or with pseudostipules. Flowers are small, **actinomorphic**, **unisexual** (plants **dioecious** or **polygamodioecious**, usually hypogynous;

see hypogyny), with 3–5 **petals**, as many or twice as many **stamens** as petals, **ovary superior** of 2–5 **carpels** with 2–5 **locules**. **Inflorescence** is an **axillary** or terminal **panicle** or **raceme**. The fruit is a **drupe**. There are 19 genera with 755 species found throughout the tropics. Many species produce aromatic resins. Frankincense is obtained from several, especially *Boswellia sacra*; myrrh is obtained mainly from *Commiphora myrrha*.

bush **1.** A shrub. **2.** In Australia, forest. **3.** Uncultivated land.

busy lizzie (*Impatiens walleriana*) *See Impatiens*.

Butomaceae (order **Alismatales**) A **monotypic** (*Butomus umbellatus*), **monocotyledon** family of aquatic **perennial** plants with **rhizomes** and **linear**, three-angled leaves up to 1 m long by which they are easily recognized. Flowers are **actinomorphic**, **perianth** in 2 **whorls** of 3 **petal**-like **sepals** and 3 **petals**, 6 **stamens** in 2 **whorls**, **ovary superior** usually of 6 **carpels**. The **axillary inflorescence** resembles an **umbel**, terminating in a **scape** without leaves and consisting of a single terminal flower surrounded by 3 **cymes**. Fruit is a **follicle**. The plants are found throughout temperate Eurasia and are naturalized in northeastern North America. The rhizomes are eaten in some places; in others the plant is a weed.

butt The base, e.g. of a **tree** trunk.

buttercup (*Ranunculus*) *See* Ranunculaceae.

buttercup flower *See Allamanda*.

butterflies *See* Lepidoptera.

butterfly bush *See Buddleja*.

butterfly effect A metaphor illustrating the fact that weather systems develop in ways that are extremely sensitive to their initial conditions, so differences too small to be detected cause apparently identical systems to diverge dramatically over time. This is encapsulated in the saying that the flapping of a butterfly's wings in Brazil might trigger a tornado in Texas.

butternut (*Juglans cinerea*) *See Juglans*.

buttress root A root similar to a **stilt root**, but with a continuous connection with the trunk, usually for the whole of its length; if it partly separates from the trunk it is called a flying buttress.

butyrous Of the consistency of butter.

Buxaceae (order **Buxales**) A family of **evergreen shrubs** and **trees** in which leaves are **alternate** or **opposite** and **decussate**, **simple**, **exstipulate**, **entire** or **dentate**, often **coriaceous**. **Staminate** flowers are **actinomorphic**, **unisexual** (plants **monoecious** or (in most *Styloceras* spp.) **dioecious**, 4 **tepals** (none in *Styloceras*), 4 **stamens** usually with **pistillode** or 6–8 without pistillode or up to 45 in *Styloceras*; **pistillate** flowers have 5–6 or up to 20 stamens. **Ovary superior** of 2 or 3 fused **carpels**. Fruit is a **loculicidal capsule** or resembles a **drupe**. There are 6 genera with 70 species, with a worldwide but scattered distribution. Several species are cultivated for ornament, hedging, topiary, or, as with *Buxus macowani* (Cape box), for hard, dense wood. *Buxus sempervirens* is common box.

Buxales An order of **trees** and **shrubs** comprising 2 families, 5 genera, and 72 species. *See* Buxaceae and Didymelaceae.

Buys Ballot's law The rule enunciated by the Dutch meteorologist C. H. D. Buys Ballot (1817–90) that in the Northern Hemisphere if you stand with your back to the wind there is an area of low **atmospheric pressure** to your left; the directions are reversed in the Southern Hemisphere.

buzz pollination Shedding **pollen** when the vibration of bee wings is in sympathy with the natural frequency of protruding **anthers**, e.g. in some members of the **Solanaceae**.

B

Byblidaceae (order **Lamiales**) A **monogeneric** family (*Byblis*) of **shrubs**, **subshrubs**, and **ephemeral herbs**. Mirid bugs often stick to a substance secreted by glands on the leaves; there is no evidence that the plant absorbs nutrients directly from the insects (i.e. is carnivorous) but it may absorb nutrients from their excreta. Leaves are **alternate** in spirals, **simple**, **linear**, **entire**. Flowers are **hermaphrodite**, slightly **zygomorphic**, with 5 **sepals**, 5 **petals**, 5 **stamens**, **ovary superior** of 2 fused **carpels**. Flowers are solitary, borne in **axils**. The fruit is a **loculicidal capsule**. There are six species found in western and northern Australia and New Guinea.

Byssochlamys A genus comprising four species of **ascomycete fungi** which produce **ascospores** that can grow at temperatures of 98–100°C, allowing the fungi to cause spoilage of canned and pasteurized fruit. The fungi occur in the soil and harvested fruit may already be contaminated. They also occur in environments where fires occur and may have potential as **biological control** agents for pathogens in such forests. The fungi have a widespread distribution.

Byturus tomentosus (raspberry beetle) A beetle (**Coleoptera**), 3.2–4.0 mm long with light brown **elytra**, that lays its eggs on the flowers of ***Rubus*** plants. The pale brown larvae feed on the developing fruit. It occurs throughout Eurasia and has been recorded in North America.

C

C *See* carbon.

C2 cycle *See* glycolate cycle.

C3 pathway The sequence of reactions in **photosynthesis** that occurs in most plants (including all **trees**), in which the first product in the **Calvin cycle** is 3-phosphoglycerate, a compound with three carbon atoms in its molecule.

C4 pathway A sequence of reactions in **photosynthesis** that occurs in many plants of low latitudes, which are adapted to high temperatures and light intensities, e.g. *Zea mays* (maize, corn), *Saccharum officinarum* (sugar cane). The first product in the **Calvin cycle** is oxaloacetate, with four carbon atoms in its molecule.

Ca *See* calcium.

caatinga A semi-arid region of northeastern Brazil and the vegetation it supports, and patches of similar vegetation elsewhere in the Amazon basin. It resembles **savanna**, but differs in having little or no grass, owing to the more arid climate. **Deciduous** thorn **trees** and **shrubs** dominate the vegetation, with cacti (**Cactaceae**) and **annual** herbs.

cabbage aphid *See Brevicoryne brassicae.*

cabbage butterfly *See Pieris brassicae.*

cabbage fly *See Delia radicum.*

cabbage gall weevil *See Ceutorhynchus assimilis.*

cabbage moth *See Mamestra brassicae, Pieris rapae.*

cabbage palm (cabbage tree) *See Cordyline.*

cabbage root fly *See Delia radicum.*

cabbage white *See Pieris brassicae, P. rapae.*

Cabombaceae (order **Nymphaeales**) A family of water lilies, which are aquatic, **perennial herbs** with **rhizomes**. Submerged leaves **opposite**, floating leaves **alternate**; floating leaves **peltate**, **elliptical** to oval-elliptical, **entire**, covered in mucilage on the underside; submerged leaves, also coated with mucilage, are feathery and fanlike. Flowers **actinomorphic**, **hermaphrodite**, usually **trimerous**, 3 (occasionally 2) **petaloid sepals**, 3 **petals**, 3–6 (or 18–36 or more) **stamens**, **ovary superior** of 1–4 (or 4–18) free **carpels**. Fruits are **achene**-like or **follicles**. There are two genera with six species with worldwide but scattered distribution. Several species grown as aquarium plants; *Brasenia* shoots are eaten in Asia.

cacomistle *See* Procyonidae.

Cactaceae (order **Caryophyllales**) A family of **perennial**, **succulent**, **trees**, **shrubs**, and climbers, most bearing spines, in which the spines, **branches**, and flowers arise from **areoles** arranged singly on small prominences or serially along ridges. Areoles may also have **glochids**. **Photosynthesis** takes place in young, green shoots. These shoots become corky with age and in tree species develop into a woody trunk without spines. Roots are typically close to the ground surface and in larger species spread widely. Flowers are usually **actinomorphic**, **bisexual**, with numerous **petals**, **sepals**, **stamens**, and **bracts**, arranged spirally, **ovary inferior** on an areole and often covered in hairs or spines, with 2 to many **carpels**. Fruit is **baccate** There are 131 genera with 1866 species, most confined to arid regions of America, but one species (*Rhipsalis bassifera*) in Africa, Madagascar, and Sri Lanka. *Opuntia* spp. (prickly pear) widely naturalized. A few (e.g. *Opuntia*) are grown for their fruit, others for interest.

caducous Soon dropping off.

caecilians *See* Amphibia.

caecotrophy Passing food through the digestive system twice.

Caesalpinoideae (famiy **Fabaceae**) One of the three subfamilies of the Fabaceae, comprising tropical and subtropical **trees** and **shrubs**, most with **pinnate**, sometimes **bipinnate** leaves with **stipules**. Flowers more or less **zygomorphic**, with free **sepals** and **petals**, 10 **stamens** or fewer. There are 160 genera with 1930 species.

Caesar's mushroom *See Amanita caesarea.*

caespitose In dense tufts.

cal *See* calorie.

calabash (*Crescentia* spp.) *See* Bignoniaceae.

Calamites cistiiformes One of the earliest species of plants with jointed stems (**Sphenopsida**) that were an important component of the **flora** of Carboniferous swamps (359.2–299 million years ago) throughout the Northern Hemisphere. Some grew up to 18 m tall.

calamus *See* Acoraceae and *Acorus.*

Calandrinia (family **Portulacaceae**) A genus of low-growing **annual**, **succulent herbs** with **alternate**, basal leaves and **ephemeral** red, white, or purple flowers in **racemes** or **panicles**. There are about 150 species found in arid or semi-arid regions of America and Australia. Some are cultivated for ornament.

calcaneum The heel bone.

calcareous soil Soil that contains so much calcium carbonate that it effervesces visibly when treated with dilute hydrochloric acid.

Calceolariaceae (order **Lamiales**) A family of **herbs** and **shrubs** with **opposite**, **serrate** leaves that are sometimes joined at the base. Flowers are **tetramerous**, usually with 2 **stamens**. The fruit is a **loculicidal capsule**. There are 2 genera and 260 species found in upland tropical and western temperate regions of South America and New Zealand.

calcic horizon A mineral **soil horizon** more than 150 mm thick in which enough calcium carbonate has been deposited to bring the content to more than 15 percent of the soil by weight and to more than 5 percent greater than the content of the lower horizons or the **parent material**.

calcicolous Describes an organism that occurs mainly, or only, where the soil is rich in **calcium**.

calcification The deposition of calcium carbonate from other parts of the **soil profile**.

calcifuge Describes a plant that rarely occurs in soils containing free calcium carbonate.

calcisols Soils that have a **calcic horizon** within 125 cm of the surface. Calcisols are a reference soil group in the **World Reference Base for Soil Resources**.

calcite A widely distributed carbonate mineral, $CaCO_3$, formed by reactions between dissolved carbon dioxide and **calcium** compounds dissolved from rocks. It is precipitated from water to form **limestone** and is a common ingredient of the shells of invertebrate animals, from which **chalk** forms. *See* aragonite.

calcium (Ca) An element essential for healthy plant growth, in **eukaryotes** found mainly in the **apoplast**. It strengthens **cell walls** and membranes, and protects roots from the effects of low **pH** and **ion** imbalances and toxicity. Impaired root development and die back are symptoms of calcium deficiency.

calcrete *See* caliche.

caliche (calcrete) A carbonate **soil horizon** (*see* duricrust) that forms by the precipitation from solution of calcium carbonate in regions where the mean annual temperature is about 18°C and the rainfall is 20–60 mm. A caliche forms over several thousand years, at first with the development of nodules, called glaebules, and later of massive layers that may become **cemented** on exposure.

California lilac (*Ceanothus*) *See* root nodule.

Californian poppy (*Eschscholzia californica*) *See* Papaveraceae.

Caliroa cerasi (cherry slug, pear slug, pear sawfly) A black and yellow, stout-bodied sawfly (**Tenthredinidae**), 5–8 mm long, that emerges in early summer. Each female lays a single egg on the underside of a leaf, which hatches into a slug-like larva that grows to 12 mm long. The larva feeds for about a month before dropping to the soil to pupate; a second generation emerges in late summer. The larvae feed on the leaves of pear, plum, cherry, and other woody plants. They occur throughout the temperate Northern Hemisphere.

Callistemon (family **Myrtaceae**) A genus of bird-pollinated **shrubs** and small **trees** with **alternate**, **terrate**, **lanceolate**, or obovate-lanceolate, **entire**, **sessile**, sometimes pungent leaves. Flowers **pentamerous**, **sepals** and **petals** circular with more petals than sepals, more **stamens** than petals, **ovary half-inferior** with 3 or 4 **locules**. Fruit a woody **loculicidal capsule**. There are about 26 species found in Australia and 4 in New Caledonia. Several are cultivated for ornament.

Callistophytales An order, now extinct, of pteridosperms (seed ferns) that lived during the Pennsylvanian epoch (318.1–299 million years ago) and possibly Permian period (299–251 million years ago). They were scrambling, **shrub**-like plants with many **branches**, **adventitious** roots, and **pinnate compound**, fern-like leaves.

callose An insoluble **glucan** that is produced in the **cell wall** and laid down at **plasmodesmata**, **cell plate**, during **cytokinesis**, and during the production of **pollen**. It is also produced in response to wounding, infection, and chemical damage.

Calluna (family **Ericaceae**) A **monotypic** genus (*C. vulgaris*, ling, heather) of low-growing, **evergreen shrubs** with tiny, **adpressed**, **decussate**, **opposite** leaves. Flowers are **tetramerous**, the **calyx** larger and more deeply lobed than the **corolla** but the same colour, mauve but white in some cultivated varieties. *Calluna vulgaris* occurs throughout Europe and western Asia. In western Europe it is the dominant species on large areas of acid **heathland**.

callus Protective tissue that forms over a wound.

Calocoris norvegicus (potato capsid) A green or brown bug (**Miridae**), about 6 mm long with long legs and antennae (*see* antenna), that is a minor pest of **Asteraceae**.

Calophyllaceae (order **Malpighiales**) A family of **evergreen trees** and **shrubs** with spiral to **opposite**, **exstipulate**, **entire** leaves. Flowers are 4- or 5-merous with free **sepals** and usually white or pink **petals**, many **stamens**. The fruit is a **berry**

or **drupe**. There are 13 genera with 460 species found throughout the tropics.

calorie (cal) A unit of energy, being the energy need to raise the temperature of 1 gram of water by 1°C at standard **atmospheric pressure** (101.325 kPa); 1 cal = 4.128 joules. The Calorie or kilocalorie, often used in reporting the energy value of foods, is 1000 calories.

calorific value The energy released when a unit weight of a substance is burned in oxygen.

Calvatia gigantea (giant puffball) A species of **basidiomycete fungi** in which the **fruiting body** is white when young, approximately spherical, 100–700 mm in diameter and occasionally up to 1.5 m, and weighs up to 20 kg. It occurs in grassland and forests throughout Europe and North America, and is edible until the **spores** have formed.

Calvin cycle A sequence of chemical reactions in the **stroma** of **chloroplasts** during which carbon dioxide is fixed chemically and reduced to **glucose** using energy from ATP (*see* adenosine triphosphate) and **nicotinamide adenine dinucleotide phosphate** plus hydrogen (NADPH) formed during the **light-dependent stage** of **photosynthesis**.

calybium The fruit of **Fagaceae**; it is a **nut** that develops from an **inferior ovary** and is held in a **cupule**.

Calycanthaceae (order **Laurales**) A family of small, **evergreen trees** and aromatic, **deciduous shrubs** with **opposite**, **simple**, **entire**, **exstipulate** leaves. Flowers are **actinomorphic**, **hermaphrodite**, with 10–40 often fleshy **tepals** inserted on the outside of the **hypanthium**, 5–30 **stamens**, **ovary superior** of 5–35 or 1–2 **carpels** with 1 **locule** each with 1–2 **ovules**; carpels are inside the **receptacle**. Flowers are solitary and often fragrant. Fruits are **achenes**. There are 5 genera with 11 species found in North America, eastern Asia, and northeastern Australia. Some shrubs are cultivated for their fragrant flowers.

Calyceraceae (order **Asterales**) A family of **annual** and **perennial herbs** with **entire** to **pinnatisect** leaves, often in a basal rosette. Flowers are small, **actinomorphic** to slightly **zygomorphic**, **bisexual**, the **sepals** are spines or thick and aerenchymatous (*see* aerenchyme), the outer tube of the **corolla** is photosynthetic, **stamens** are free, **ovary inferior** with 2 fused **carpels**. **Inflorescence** is **capitate**. Fruits are **achenes**. There are 4 genera with 60 species found in southern South America.

calyces *See* calyx.

calyciform Cup-shaped; shaped like a **calyx**.

calyculus A group of **bracts** that resemble a **calyx**, or a small, rim-like calyx.

Calypte anna (Anna's hummingbird) A hummingbird, 100–110 mm long, with green flanks, a bronze-green back, pale front and underside, and males have a bright red crown and throat and dark tail. They feed on nectar and small invertebrates, and occur along the west coast of North America. They inhabit woodlands, parks, and gardens.

calyptra In a moss (**Bryophyta**) or liverwort (**Marchantiophyta**), a protective cap on the **capsule**.

Calystegia (family **Convolvulaceae**) A genus of **annual** and **perennial** twining vines with long, trailing, twining stems. Leaves are **alternate**, **simple**, **exstipulate**. Flowers are **actinomorphic**, **bisexual**, with an **involucre** of large **bracteoles** below the **calyx**, 5 **sepals**, 5 fused **petals** forming a bell or funnel shape, **ovary superior** of 2 fused **carpels**, 2 **locules**, each with 2 **locules**. Fruit a **dehiscent capsule**. There are about 25 species found throughout temperate and subtropical regions, concentrated in California. Leaves are food for larvae of several lepidopteran species. Some species are troublesome weeds, others cultivated for their flowers and known as bindweed, false bindweed, and morning glory.

calyx (pl. calyces) All the **sepals** of a **flower.**

cambic horizon A mineral **soil horizon** that is weakly developed and occurs in the middle of the B horizon of **brown earths** and **gleys.** It shows signs of **weathering** and **gleying.**

cambisols Soils that have a **cambic horizon** or **mollic horizon** above a B horizon that has a **base saturation** in the upper 100 cm of less than 50 percent, or an **andic horizon, vertic horizon,** or **vitric horizon** with an upper boundary 25–100 cm below the surface. Cambisols are a reference soil group in the **World Reference Base for Soil Resources.**

cambium In the stems, **branches,** and roots of vascular plants (**Tracheophyta**), a layer of tissue lying between the **phloem** and **xylem.** Cambium cells continue to divide throughout the life of the plant, producing **secondary phloem** and xylem cells as well as more cambium cells. *See* secondary growth.

Camellia (family **Theaceae**) A genus of **evergreen shrubs** and small **trees** with thick, usually glossy **alternate, simple, serrate** leaves. Flowers, usually large and conspicuous, with 5–9 white, pink, red, or yellow **petals, stamens** prominent and often of a contrasting colour. Fruit is a dry **capsule.** There are 248 species found in eastern Asia, and about 30,000 **cultivars.** Camellia leaves provide food for larvae of several lepidopteran species. Many camellias are cultivated for ornament, for tea (*C. sinensis, C. assamica*), tea oil (*C. oleifera*), oilseed (*C. sasanqua*), etc.

camellia yellow mottle virus A virus, transmitted by propagating and root **grafts** from diseased stock, that causes patches of discoloration, often yellow, on the leaves of camellias.

c-AMP *See* cyclic AMP.

Campanula (family **Campanulaceae**) A genus of **annual, biennial,** and **perennial herbs** with **alternate, entire** or **serrate** leaves. Flowers with 5-lobed **corolla,** 5 **sepals,** and borne in **panicles.** Fruit is a **capsule.** They are food plants for larvae of several lepidopteran species. There are more than 500 species found throughout the temperate region of the Northern Hemisphere. Many are cultivated for ornament, as bellflowers.

Campanulaceae (order **Asterales**) A family of **annual, biennial,** and **perennial herbs, shrubs,** climbers, and **pachycaul rosette plants** with **alternate** (sometimes **opposite** or in **whorls**), **simple** or **pinnatisect** leaves. Flowers **actinomorphic** to strongly **zygomorphic, bisexual** (rarely **unisexual**), 5-, 6-, or up to 9-merous, with long, **acute sepals, corolla** regular, 2-lipped, or cleft down one side, tubular or cup-shaped, **ovary** semi-**inferior** or **superior** of 2, 3, or 5–9 **carpels** usually with 1 **locule. Inflorescence** a **raceme, cyme,** or **capitulum.** Fruit is a **capsule.** There are 84 genera with 2380 species with worldwide distribution. Many are cultivated as ornamentals.

campanulate Bell-shaped.

CAM pathway (crassulacean acid metabolism) A sequence of reactions in **photosynthesis** that occurs in certain plants of arid climates, including cacti (**Cactaceae**) and pineapple (*Ananas comosus*). It was first observed in the **Crassulaceae,** hence the name. CAM plants absorb carbon dioxide at night, store it in the form of organic acids until daylight, then close their **stomata,** use the energy of sunlight to produce **adenosine triphosphate** (ATP) and nicotinamide adenine dinucleotide phosphate plus hydrogen (NADPH), and release the stored carbon dioxide into the **Calvin cycle.**

camphor (*Cinnamomum*) *See* Lauraceae.

C

campo Savanna grasslands with scattered **broad-leaved** trees, that occur in Brazil. They develop on soils poor in nutrients. Campo cerrado is relatively rich in trees; campo sujo has few trees.

campo cerrado *See* campo.

campodeiform larva A predatory insect larva that has a flattened body with **sclerites** and six legs.

campo sujo *See* campo.

campylotropous Describes the orientation of an **ovule** that lies horizontally with the **funicle** attached midway between the **chalaza** and **micropyle**.

Campynemataceae (order **Liliales**) A family of **monocotyledon herbs** with short **rhizomes**. There is a single **linear** to **elliptical**, basal leaf or several basal clusters of leaves. Flowers have green **tepals** in **pentamerous** or **trimerous whorls**, **ovary inferior** with 1 or 3 **carpels** and **locules**. Fruit is a **capsule**. There are two genera with four species found in New Caledonia and Tasmania.

canalization (canalizing selection) The containment of a developmental process within narrow bounds. Over succeeding generations a **phenotype** is affected by genetic variations and environmental perturbations, but **natural selection** eliminates those **genotypes** that allow deviation from the phenotype that is optimal in the most common environment. ↗

canalizing selection *See* canalization.

Candolle, Alphonse Louis Pierre Pyramus de (1806–93) A Swiss botanist, son of Augustin Pyramus de **Candolle**, who completed his father's work by publishing the final three volumes of the *Prodromus* (one in collaboration with his own son, Anne Casimir de Candolle). In his own published works he speculated on why particular plant species occur in certain places but not others, thus helping to establish the basis of **biogeography**.

Candolle, Augustin Pyramus de (1778–1841) A Swiss botanist who coined the term *taxonomy* (in *Théorie élémentaire de la botanique*, 1813) and sought to place plant classification on a more natural basis than the system promoted by Carolus **Linnaeus**. He conducted a botanical survey of the whole of France on behalf of the government. His most important work was *Prodromus systematis naturalis regni vegetabilis* (Introduction to natural classification of the vegetable kingdom), covering the taxonomy, **ecology**, and geography of all known seed plants. He completed seven of the planned ten volumes by 1839, when ill health forced him to abandon the project, which was completed by his son Alphonse de **Candolle**.

cane blight *See Leptosphaeria coniothyrium.*

canebrake rattlesnake *See Crotalus horridus.*

Canellaceae (order **Canellales**) A family of aromatic **evergreen trees** and **shrubs** with **alternate**, **simple**, **exstipulate**, **entire**, often **coriaceous** leaves. Flowers **actinomorphic**, **hermaphrodite**, with 3 persistent, **imbricate sepals**, 5–12 **petals** in 1–2 or more **whorls**, 6–12 **stamens**, **ovary superior**, **syncarpous**, of 2–6 **carpels** with 1 **locule** of 2 to many **ovules**. Flowers solitary or in terminal or **axillary cymes**. Fruit is a **berry**. There are 5 genera with 13 species scattered in southern Florida, Caribbean, eastern South America, East Africa, and Madagascar. **Bark** of *Canella winterana* is white cinnamon, a condiment and stimulant. Other species used medicinally.

Canellales An order of plants with 2 families, 9–13 genera, and 75–105 species. *See* Canellaceae and Winteraceae.

canescent Grey and covered densely with hairs.

canker Any of several bacterial, fungal, or viral diseases of woody plants, in which **bark** formation is prevented locally and a small area of dead tissue forms and grows slowly over a number of years. Canker may be unimportant, but some can spread, eventually threatening the life of the plant.

Cannabaceae (order **Rosales**) A family of **annual herbs** and **climbers** with **opposite**, **serrate**, **stipulate**, heart-shaped leaves. Flowers are small, green, **actinomorphic**, **unisexual** (plants usually **dioecious**, sometimes **monoecious**). Male flowers have 5 free **tepals**, female flowers have tepals fused into a tube **adnate** to the **ovary**, 5 **stamens**, ovary **superior** of 2 **carpels** united into 1 **locule**. **Inflorescence** a **cyme** in leaf **axils**. Male cymes loose and in panicles with small **bracts**; female either (*Cannabis*) crowded along bracts with each flower enveloped by a bract or (*Humulus*) **cone**-like with many broad, papery **imbricate**, persistent bracts. Fruit a **nut** or small **achene**. There are 11 genera with 170 species with worldwide distribution. *Cannabis sativa* (**monotypic** genus) cultivated for its fibre (hemp) and for use as a drug. Female inflorescences of *Humulus lupulus* are hops, used to flavour beer.

Cannabis sativa (hemp) *See* Cannabaceae.

Cannaceae (order **Zingiberales**) A **monogeneric** family (*Canna*) of large, erect **monocotyledon herbs** with starchy **rhizomes**. Leaves **distichous** or spiral with sheathing bases. Flowers are large, short-lived, strongly asymmetric, **bisexual**, **perianth** of 3 **imbricate sepals**, usually green or purple, 3 basally fused, yellow or white **petals** much smaller than the sepals and with one smaller than the others, 6 brightly coloured **stamens** resembling petals, 5 of them **staminodes**, **ovary inferior** of 3 fused **carpels** with 3 **locules** each with 2 **ovules**. **Inflorescence** terminal with single flowers or **cymes** of a few flowers. Fruit is a **capsule**. There are ten species found in the American subtropics. *Canna edulis* is widely cultivated in the tropics for its rhizomes, the source of Queensland arrowroot.

canopy The uppermost layer of a forest or woodland, comprising the crowns of trees and all their animal, **epiphyte**, and other inhabitants.

Cantharellus (chanterelle) A genus of **agaric fungi** that form mycorrhizal (*see* mycorrhiza) associations with several tree species and consequently occur in forests. The **fruiting body** is vase- or trumpet-shaped with a well-defined **pileus** and **stipe**. They are edible and highly prized. There are at least 100 species, distributed worldwide.

Cantharidae (leather-winged beetles, soldier beetles, sailor beetles) A family of beetles (**Coleoptera**) with soft bodies covered with fine hair, 5–15 mm long, and parallel-sided **elytra** which are sometimes bright red, but usually brown or black and trimmed with yellow, orange, or red, reminiscent of a soldier's uniform. They have long, thread-like antennae (*see* antenna). Larvae are flattened with long legs and a well-defined head. They are active in winter and sometimes called snow worms. Adults are often seen on flowers. There are about 3500 species with a worldwide distribution. Some species feed on **nectar** and **pollen** but most are predators, especially of aphids (**Aphididae**).

canyon wind *See* mountain-gap wind.

cap 1. A thin, impermeable layer of particles covering the surface of soil. It is produced by the impact of rain on exposed soil. **2.** *See* pileus.

Cape box (*Buxus macowani*) *See* Buxaceae.

Cape floral region The area covering Cape Province, and the only region within the South African region. It is one of the world's richest floristic regions, with many **endemic** species. It is a UNESCO-designated protected area.

Cape primrose *See Streptocarpus.*

capers (*Capparis spinosa*) *See* Capparaceae.

capillarity (capillary movement) The process by which a liquid moves in any direction through a very narrow passage. Molecules of the liquid are attracted electromagnetically to molecules in the side of the passage, so the edges of the liquid move upward. This is not a stable configuration and surface tension establishes a convex surface. The edges of the liquid then move upward and the process repeats.

capillary The narrowest type of blood vessel, with walls only one cell thick through which oxygen and nutrient molecules pass into cells and cell waste products are absorbed.

capillary fringe (capillary zone) A region above the **water table** in which water is being drawn upward by **capillarity.** The depth of the fringe depends on the particle size of the soil. In **clay** it might average 3 m, in sand less than 10 cm.

capillary moisture (capillary water) Water that remains in the soil once **gravitational water** has drained away. Capillary moisture exists as a film on the surface of soil particles, held there by **surface tension.**

capillary movement *See* capillarity.

capillary water *See* capillary moisture.

capillary zone *See* capillary fringe.

capillitium In certain **slime moulds** and **Fungi**, sterile threads among the **spores** in the **fruiting body.** They are involved in spore dispersal and are important in identifying the species.

capitate Having an enlarged, rounded tip resembling a head.

capitulum An **inflorescence** comprising flowers or **florets** packed closely and without **pedicels**, all arising at the same level on a flattened **axis.** It is subtended by an **involucre** of **bracts**, so it resembles a single flower.

Capitulum.

Capparaceae (order **Brassicales**) A family of woody **trees**, **shrubs**, **herbs**, and twining **lianas** with leaves that are **alternate**, often **simple**, **exstipulate** or with small **stipules.** Flowers usually **hermaphrodite**, 4-merous, **actinomorphic** to **zygomorphic**, sometimes with the **receptacle** as a **disc**, **cone**, or tube that may have 4 **nectaries** or other appendages, 3–5 or 7 free **sepals**, sometimes **petaloid**, 0–4 **petals**, 3–8 or many **stamens**, some with 2–7 **staminodes**, sometimes **androgynophore**, **ovary** usually **gynophore**, **superior**, usually with 1 **locule.** **Inflorescence** usually a terminal or **axillary raceme**, single flowers in leaf **axils**, rarely as a **corymb** or **fascicle.** Fruit **indehiscent.** There are 16 genera with 480 species, most tropical. Flower **buds** of *Capparis spinosa* are pickled and eaten as capers.

capping inversion A **temperature inversion** that develops when dry air advances against moist air more slowly at ground level than it does above the **planetary boundary layer**, the dry air overrunning the moist air and preventing the development of **convective clouds.** Capping inversions are often associated with **dry lines.**

Capreolus capreolus (roe deer) *See* Cervidae.

Caprifoliaceae (order **Dipsacales**) A family of **trees**, **shrubs**, and woody climbers with **opposite**, sometimes lobed, **entire** to **serrate** leaves. Flowers with 5 (or 2–4) **imbricate sepals** and **petals**, tubular or funnel-shaped **corolla**, 4 or 5 **stamens**, sometimes 2 long and 2 short, **ovary inferior** of 2–5 or up to 8 **carpels** with 1–5 **locules. Inflorescence** is a compact or diffuse terminal or **axillary cyme** of 2 or a few flowers, or a broad **panicle** of small capitula (*see* capitulum). Fruit is a **capsule**, **berry**, or fleshy **drupe**. There are 42 genera with 890 species, most occurring in northern temperate and warm temperate regions, a few on tropical mountains. Several species grown for ornament as honeysuckles, *Symphoricarpus* grown for its fruit (snowberries).

capsaicin *See Capsicum.*

Capsicum (family **Solanaceae**) A genus of **shrubs** and **annual herbs** with **alternate** or **opposite**, **simple**, **entire** leaves. Flowers **pentamerous**, pendulous, fruit a many-seeded **berry** with 2 or 3 chambers. There are 20–27 species, native to tropical America, of which 5 are widely cultivated for the fruits, which contain capsaicin (methyl vanillyl nonenamide), giving them a pungent flavour of varying strength. The most widely grown is *C. annuum* with fruits known as bell pepper, sweet pepper, pimento, cayenne pepper, and chilli pepper.

capsid The **protein** envelope enclosing a **virus**.

capsid bugs *See* Miridae.

capsule 1. A dry, usually **dehiscent**, fruit. **2.** In a moss (**Bryophyta**) or liverwort (**Marchantiophyta**) the structure bearing the **spores**. **3.** (sheath) In **prokaryotes**, the gelatinous outer layer of the cell surface.

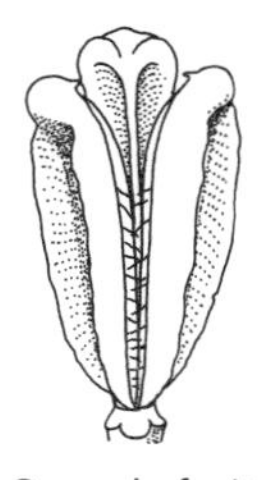

Capsule fruit (*Lilium*).

Carabidae (ground beetles) A family of beetles (**Coleoptera**), most of which are black, brown, or metallic in colour, and 2–35 mm long. Many are flightless, their **elytra** fused. They have long, strong legs adapted for running and in some species for digging. Some species are herbivorous but most are predators of other invertebrates, including many plant pests. Many ground beetles are nocturnal and spend the day concealed. There are more than 40,000 species with a worldwide distribution.

Carabus violaceus (rain beetle, violet ground beetle) A species of nocturnal ground beetles (**Carabidae**) that are shiny black with violet or indigo edges to its oval **elytra** and thorax. Adults do not fly, are 20–30 mm long, and are active from March to October. The beetles occur throughout Europe and in Japan, in forests, parks, and gardens, where both adults and larvae feed on other insects possibly including weevils (**Curculionidae**), worms, slugs, and snails.

carapace The upper, usually domed shell of a turtle or tortoise (**Chelonia**). *See also* plastron.

carbamate herbicides A group of **herbicides**, including **asulam**, carbetamide, **chlorpropham**, and propham, that inhibit cell division.

carbamate insecticide A class of **insecticides** that are derived from carbamic acid (NH_2COOH). They are less toxic and more easily biodegraded than most **organophosphate** compounds.

carbaryl A **carbamate**, **contact insecticide** that inhibits the **enzyme** cholinesterase. It is used against mosquitoes, earwigs, winter moth (***Operophtera brumata***), and earthworms in turf. It is toxic to humans and harmful to other insects, crustaceans, and fish. Its use is banned in the United Kingdom, Angola, Austria, Denmark, Germany, and Sweden.

carbendazim A **carbamate**, systemic **fungicide** that is used against black spot, **mildew**, and **blight**. It is of low toxicity to birds and mammals.

carbetamide *See* carbamate herbicides.

carbohydrate Any compound consisting only of carbon, hydrogen, and oxygen, with the general formula $C_x(H_2O)_y$, i.e. any carbon hydrate. They are **sugars** (saccharides) and classed as mono-, oligo-, or polysaccharides depending on the size of the molecule.

carbolic acid *See* phenol.

C

carbon (C) An element that is able to form chains or rings, leading to large, highly complex molecules in which carbon atoms bond to atoms of other elements. It forms the basis of organic chemistry (the chemistry of carbon compounds), and it is the basic ingredient of all living organisms. It enters plants through **photosynthesis** and **heterotrophs** (e.g. animals) by consuming plant material.

carbonates A group of minerals containing carbon that are found mainly in **dolomite** and **limestone** rocks. **Calcite** is the most common carbonate mineral. *See also* aragonite.

carbonation A process in **chemical weathering** in which dilute **carbonic acid** reacts with a mineral, e.g. the reaction with **limestone** that releases **ions** of calcium and bicarbonate into the solution: $CaCO_3 + H^+ + HCO_3^- \Rightarrow Ca^+ + 2HCO_3^-$.

carbon cycle The flow of carbon in a cycle that takes it through the atmosphere, water, living organisms, soils, and sedimentary rocks. **Photosynthesis** utilizes carbon taken from atmospheric carbon dioxide. The carbon passes through **heterotrophs** and is returned to the atmosphere by **respiration**. In certain places the decomposition of organic material is arrested and the carbon it contains is stored below ground or below the sea bed, eventually to become **fossil fuels**, a process that was much more extensive in the distant past. Carbon dioxide also dissolves in water to form **carbonic acid** (H_2CO_3), which reacts with calcium silicate ($CaSiO_3$) in rocks to form calcium carbonate ($CaCO_3$) and silica (SiO_2). The $CaCO_3$ is carried to the sea and precipitates to form **carbonate** sediments.

carbonic acid An acid (H_2CO_3) that forms naturally when carbon dioxide (CO_2) dissolves in water (H_2O).

carbon-nitrogen ratio The ratio of the mass of carbon a substance contains to the mass of nitrogen. This is important in composting, which depends on the activity of organisms that utilize carbon as a source of energy and nitrogen for building cell structure.

carboxyl An organic functional group that consists of a carbon atom double-bonded to an oxygen atom and single-bonded to a hydroxyl group. It is written as –COOH or –C(=O)OH, or $-CO_2H$.

carboxylase An **enzyme** that catalyzes reactions in which carbon dioxide is incorporated in an organic compound.

carboxysome An **organelle**, made entirely of **protein**, that is found in **cyanobacteria** and chemoautotrophs (*see* autotroph) and that concentrates and fixes carbon dioxide.

cardenolides A group of **steroids** ($C_{23}H_{34}O_2$), found in a number of plant families, most of which are very poisonous; they cause heart failure.

Cardinalis cardinalis (northern cardinal, common cardinal, redbird) A bird that is 210–235 mm long with a distinctive crest, long tail, and thick, orange or red beak. Males are bright red with a black mask, females are pale brown or greenish with traces of red. They inhabit woodland edges, hedgerows, and gardens, often visiting feeders, and occur throughout eastern and central North America and parts of Central America.

Cardiopteridaceae (order **Aquifoliales**) A family of twining herbs with spiral, **alternate**, **entire**, **exstipulate** leaves and small, actinomorphic, hermaphrodite flowers with 4–5 sepals and petals, 4–5 **epipetalous** stamens, ovary superior with 1 locule, and fruit that is a 2-winged **samara**.

The family also includes **trees** and **shrubs** with fruits that are **drupes**. There are 5 genera with 43 species scattered throughout the tropics; *Citronella* accounts for 21 of the species. *Citronella mucronata* is cultivated for its foliage and flowers but *Citronella* is not the source of the insect-repellant citronella oil.

Carduelis carduelis (goldfinch, European goldfinch) A bird 120–130 mm long with a wingspan of 210–250 mm, that has a black and white head, red face, brown upper parts, white underside, buff sides and breast, and black wings with a vivid yellow stripe. They are gregarious and inhabit lowland woods, often visiting garden feeders, and feeding mainly on small seeds. They occur throughout Europe to Central Asia, and in North Africa.

Carduelis chloris (European greenfinch, greenfinch) A bird that is about 150 mm long with a wingspan of 245–275 mm, and mainly green with some yellow in its tail and wings. It inhabits woodland edges, hedgerows, and gardens, sometimes forming flocks, and feeds on seeds and berries. It occurs throughout Europe, southwest Asia, and North Africa, and has been introduced to Australia and New Zealand.

Carduelis pinus (pine siskin) A bird 110–140 mm long with a wingspan of 180–220 mm, that has a brown back and wings and pale underside. They inhabit conifer forests but are also found in parks, roadsides, grasslands, and gardens, and often visit feeders offering small seeds. They feed on seeds and insects, and occur throughout Alaska, Canada, and the northern United States.

Carduelis spinus (siskin, Eurasian siskin) A bird 110–125 mm long with a 200–230 mm wingspan, in which the male has a grey-green back, black wings with a yellow stripe, yellow rump, yellow stripes on the tail, pale yellow underside, black cap, and black chin patch. The female is more olive. They inhabit woodlands and visit garden feeders. Siskins occur throughout most of Eurasia and are migratory over part of their range.

Carduelis tristis (American goldfinch, eastern goldfinch) A bird 110–140 mm long with a wingspan of 190–220 mm, a conical beak, pink for most of the year but orange when moulting, and bright yellow **feathers** with a black cap and white rump. It breeds across North America from southern Canada to northern California and North Carolina and winters from southern Canada to northern Mexico, often congregating in large flocks. It feeds mainly on seeds, but also on insects, **buds**, and berries. It commonly enters gardens and takes food from feeders.

Carex (family **Cyperaceae**) A **monocotyledon** genus of **perennial herbs** with **rhizomes**, **stolons**, or short rootstocks, and stems that are solid, leafy, and triangular in cross-section. Leaves are **linear**, often inrolled with **keels**, with a sheathing base and **ligule**. Flowers **unisexual** (plants usually **monoecious**) with no **perianth**, **ovary superior** of 3 fused **carpels** and 1 **locule**. Male flowers usually have 2 or 3 **stamens**. **Inflorescence** varies from a single **spike**, with female flowers at the bottom and male flowers at the top, to a branched **panicle**. Fruit is a single-seeded **indehiscent achene** or **nut**. There are more than 1500 species, known as sedges, with worldwide distribution, especially in marshes and other wet soils.

Caribbean floral region The area covering Central America and the islands of the Caribbean, part of the **Neotropical region**. There are about 13,000 plant species of which about half are single-island **endemics**.

caribou *See* antlers.

caribou moss (*Cladonia rangiferina*) *See* reindeer moss.

Caricaceae (order **Brassicales**) A family of small **trees** and a few vines with articulated **latificers** containing milky **latex** and many with spiny stems. Leaves **alternate**, lobed or foliate, **exstipulate** or with **stipules** resembling spines. Flowers **unisexual** (plants usually **dioecious**), rarely **bisexual**, **actinomorphic**, **pentamerous**, **sepals** free, **corolla connate**, valvate, or **contorted**, the corolla tube long in male flowers, short in female flowers, **ovary superior** of 5 fused **carpels** with 1–5 **locules**. **Inflorescence axillary** of 1 flower or a many-flowered **thyrse**. Fruit is a **berry**. There are 4 genera with 34 species found mainly in tropical America, with 1 genus (*Cylicomorpha*) in Africa. *Carica papaya* is widely grown throughout the tropics for its fruit, the papaya or pawpaw.

carina *See* keel.

carinal canal A longitudinal canal inside the **metaxylem** and produced by the disintegration of the **protoxylem**, found in ***Equisetum*** and some related **fossil** plants.

Carlemanniaceae (order **Lamiales**) A family of **perennial herbs** and **shrubs** with leaves that are **opposite**, **serrate**, **exstipulate**, with more or less swollen **nodes**. Flowers 4- or 5-merous, **calyx** lobes often markedly unequal, **corolla** funnel-shaped to **campanulate**, **zygomorphic**, with 2 **stamens** inserted in the corolla tube, **ovary inferior** of 2 **carpels** and 2 **locules**. **Inflorescence** a terminal or **axillary cyme**. Fruit is a fleshy, **dehiscent capsule**. There are two genera with five species found in southern Asia and Sumatra.

carmine spider mite *See Tetranychus cinnibarinus.*

carnassial A modified premolar or molar tooth, found in many mammal **carnivores**, that allow teeth of the upper and lower jaw to move against each other with a scissor-like shearing action. It usually involves the last upper premolar and first lower molar.

carnation (*Dianthus caryophyllus*) *See* Caryophyllaceae, *Dianthus.*

carnation necrotic fleck virus (carnation streak virus) A **virus** of the **Closteroviridae**, transmitted by aphids (**Aphididae**), that causes streaks and mottling in tissue that becomes necrotic. The disease affects ***Dianthus*** species (carnation) and occurs worldwide.

carnivore An animal that feeds on flesh.

carnivorous plants *See* insectivorous plants.

Carolina anole *See Anolis carolinensis.*

Carolina duck *See Aix sponsa.*

Carolina wren *See Thryothorus ludovicianus.*

carotene (carotin) An orange photosynthetic pigment (*see* photosynthesis) that is a **hydrocarbon carotenoid**. It occurs in carrots (hence the name), sweet potatoes, cantaloupe melons, etc.

carotenoid A group of about 600 pigments that form 2 groups, **carotenes** and **xanthophylls**. They absorb blue light and act as accessory photosynthetic pigments (*see* photosynthesis) in some plants and photosynthesizing **bacteria**, protect **chlorophyll** from damage by light, and can be converted to vitamin A in animals.

carotin *See* carotene.

carpal One of the bones of the wrist.

carpal spur A sharp projection, covered in horn, from the **carpus** of some birds; it is used in combat.

carpel One of the female reproductive organs in a **flower**, i.e. part of the **gynoecium**. It consists of an **ovary** containing one or more **ovules** borne on a **placenta**, usually with a **style** ending in a **stigma**.

carpellate Possessing a **gynoecium** but no **androecium**.

carpellode A sterile **carpel**.

carpenter bees *See* Apidae.

carpenter moths *See* Cossidae.

carpet bugle *See Ajuga.*

Carphophis amoenus (worm snake) A colubrid snake (**Colubridae**), 190–280 mm long, that has a brown back and pink belly. It burrows and spends much of its time below ground, so it is seldom seen. It inhabits moist areas and feeds on invertebrates, especially earthworms. If handled it may release a foul-smelling liquid, but it will not bite. The snake occurs throughout the eastern United States.

Carpinus (family **Betulaceae**) A genus of smooth-barked (*see* bark) **deciduous trees** (hornbeams) with **alternate**, **simple**, **serrate** leaves. Trees **monoecious** with male and female flowers on separate **catkins**. Male flowers without **bracteoles**, with about 10 forked **stamens**. Fruit is a small **nut** with a large leafy or **bract**-like **involucre** on one side, making the fruit spin as it falls. There are 30–40 species found throughout the northern temperate zone. Many are grown for their timber (ironwood). Hornbeams are food plants for the larvae of several species of Lepidoptera.

Carpobrotus edulis (ice plant) *See* Aizoaceae.

Carpodacus purpureus (*Haemorhous purpureus*, purple finch) A finch, 150 mm long with a wingspan of 250 mm, in which the male is mainly red and females brown. It inhabits forest edges, hedgerows, pastures, and urban parks and gardens. It feeds on seeds, augmented with insects in spring and fruits in summer. It occurs throughout North America.

carpogonium A **gametangium** (female sex organ) in **red algae** (Rhodophyta). It usually consists of a thin, hair-like cell with a swollen base and a long, twisted, gelatinous upper part, which receives the male **gamete**.

carpospore A spore produced by a **carpogonium** in **red algae** (Rhodophyta).

carposporophyte The **sporophyte** produced by the fusion of **gametes** in **red algae** (Rhodophyta); it is **diploid** and lives inside the **haploid** female **gametophyte**. Cells at the tips of individual **filaments** of the carposporophyte become **carpospores** that are released into the water.

carpus The wrist joint; in birds (**Aves**) it is the outermost wing joint.

carrion crow *See Corvus corone.*

carrot (*Daucus carota*) *See* Apiaceae, *Daucus.*

carrot cyst nematode *See Heterodera carotae.*

carrot fly *See Psila rosae.*

carrot root aphid *See Dysaphis foeniculus.*

carrot root nematode *See Heterodera carotae.*

carrot yellow leaf virus A **virus** of the **Closteroviridae**, transmitted by aphids (**Aphididae**), that causes discoloration, often yellow, in the leaves of ***Daucus*** *carota* (carrot). It occurs in Japan, Britain, and parts of mainland Europe.

carrying capacity The largest population of a particular species that a specified **environment** can sustain without causing environmental damage. It is the saturation value for a population showing an **S-shaped growth curve**, but can be modified, e.g. by applying **fertilizers** to boost crop yields.

Cartap *See* nereistoxin analogue insecticides.

cartilage Flexible skeletal tissue in vertebrates that forms most of the skeleton of **embryos** and is found in adults at the ends of bones, between **vertebrae**, and in the ear **pinna**.

caruncle 1. A reduced **aril**, often brightly coloured. 2. A fleshy appendage or protuberance, e.g. the wattle of a bird.

Carya (family **Juglandaceae**) A genus of **deciduous trees** (hickory) with **pinnate compound** leaves. Flowers are small **catkins**, the fruit is a **nut**. There are 17–19 species, most from North America and Mexico, 5–6 from Asia. The nuts are edible in some species (*C. illinoinensis* yields pecan, *C. glabra* yields hognut). Hickory wood is tough and springy, traditionally used to make drumsticks, lacrosse sticks, baseball bats, shafts of golf clubs, etc.

Caryocaraceae (order **Malpighiales**) A family of **evergreen trees**, some with large **buttress roots**, and a few **shrubs**. Leaves **opposite** or **alternate**, **palmate**, **trifoliate**, **serrate**, usually with **caducous stipules**. The large, showy, nocturnally bat-pollinated flowers are 6-merous, **actinomorphic**, **bisexual**, **petals** free, 55–750 **stamens** the inner ones **staminodes**, **ovary superior** of 4–20 fused **carpels** with 4–6 or 8–20 **locules** each with 1 **ovule**. The fruit is a **drupe**. There are 2 genera with 21 species found in tropical America. Some species are cultivated for their edible nuts or timber.

caryogamy *See* karyogamy.

Caryophyllaceae (order **Caryophyllales**) A family of **annual** and **perennial herbs** that die back to the crown, with a few **shrubs** and small **trees**, with **opposite** (occasionally **alternate**), **simple**, **entire**, a few **succulent**, usually **exstipulate** leaves. Flowers **actinomorphic**, **bisexual** (a few **unisexual**, plants then **dioecious** or **monoecious**), 4–5 free **sepals** or united sepals with a 4- or 5-lobed **apex**, 4–5 free **petals** (sometimes 0), usually twice as many **stamens** as petals, **ovary superior** of 2–5 united **carpels**, usually with 1 **locule**. **Inflorescences cymose**, or a dichasial (*see* dichasium) **panicle**, or **monochasia**, or single flowers. Fruit an **achene** or **utricle**. There are 86 genera with 2200 species, most found in temperate regions, especially Eurasia. Many are cultivated as ornamentals, e.g. ***Dianthus*** spp. as carnation (*D. caryophyllus*), pinks, and sweet William.

Caryophyllales An order of plants with 35 families comprising 811 genera with 11,510 species, including 2 species (*Hypertelis* and *Macarthuria*) not placed in any family. *See* Achatocarpaceae, Aizoaceae, Amaranthaceae, Anacampserotaceae, Ancistrocladaceae, Asteropeiaceae, Barbeuiaceae, Basellaceae, Cactaceae, Caryophyllaceae, Didiereaceae, Dioncophyllaceae, Droseraceae, Drosophyllaceae, Frankeniaceae, Gisekiaceae, Halophytaceae, Limeaceae, Lophiocarpaceae, Microteaceae, Molluginaceae, Montiaceae, Nepenthaceae, Nyctaginaceae, Physenaceae, Phytolaccaceae, Plumbaginaceae, Polygonaceae, Portulacaceae, Rhabdodendraceae, Sarcobataceae, Simmonsiaceae, Stegnospermataceae, Talinaceae, and Tamaricaceae.

caryopsis An **achene** in which the **ovary** wall is united with the seed coat. A cereal grain is a caryopsis.

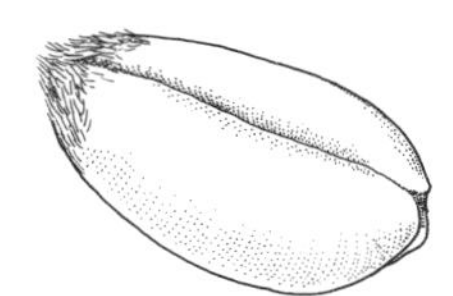

Caryopsis fruit (wheat).

cascade effect Any sequence of events in which each one establishes the conditions necessary for the next. In **ecology**, a cascade occurs when the organisms present at each stage in a **succession** provide resources that are exploited by organisms that form the following stage.

cashew (*Anacardium occidentale*) *See* Anacardiaceae.

Casparian strip A band of specialized **cell wall** tissue on the sides and walls of the **endodermis** of the roots of vascular plants (**Tracheophyta**) that prevents water and solutes from entering the **pericycle** except by passing through the **cytoplasm** of endodermal cells, producing **root pressure** that forces water to enter the **xylem** by **osmosis**.

cassava (*Manihot esculenta*) *See* Euphorbiaceae.

Castanea (family **Fagaceae**) A genus of **deciduous trees** and **shrubs** with **ovate** or **lanceolate**, **simple** leaves. **Inflorescences catkins**, some of only male flowers, others with small clusters of female flowers forming a **calybium**. Flowers have 8 or 10–12 **stamens**. Fruits are **nuts** enclosed in a **cupule** covered in spines. There are eight to nine species found in temperate regions of the Northern Hemisphere. Several species yield valuable timber and some produce edible nuts (chestnuts).

caste In social insects, a group of functionally specialized individuals distinguished morphologically (*see* morphology), by age, or both, e.g. honeybee colonies, which contain three castes: a reproductive female (queen), non-reproductive workers that perform a range of functions depending on their age, and reproductive males (**drones**).

caste polyethism *See* polyethism.

castor oil plant (*Ricinus communis*) *See* Euphorbiaceae.

Casuarinaceae (order **Fagales**) A family of tall **evergreen trees** and some **shrubs** with slender, jointed **branches** that give the trees a weeping **habit**. Scale-like leaves are **whorls** of 4–16 many-toothed sheaths around the joints in the branches. Flowers greatly reduced, usually **unisexual** (plants **monoecious** or **dioecious**). **Staminate** flowers with 1 **stamen** and **perianth** of 2 lobes each subtended by 2 or more scale-like **bracteoles**. **Pistillate** flowers usually on lower branches, with **ovary** of 2 fused **carpels**. Fruit a 1-seeded **nut** enclosed in hard bracteoles that open to release the seeds, giving the fruit a **cone**-like appearance. There are 4 genera with 95 species occurring from southeastern Asia and Malesia through the southwestern Pacific, especially in Australia. Many species yield hard timber, some are grown for ornament. Introduced species have become invasive in some places.

catabolism Metabolic reactions that release energy from nutrient molecules.

catadromous Describes the behaviour of fish that spend most of their lives in fresh water but breed at sea, e.g. the common eel (*Anguilla anguilla*).

cataflexistyly A morphological (*see* morphology) change in which a **flower** functions first as a female and later as a male.

catalysis The acceleration of a chemical reaction through the action of a **catalyst**.

catalyst A substance that participates in certain chemical reactions, facilitating or accelerating them, without undergoing alteration itself.

cataphyll (bud scale) A scale-like leaf that covers a dormant **bud** on a **deciduous** plant; cataphylls often contain **resin**.

catchment The area from which a river system or **groundwater** gathers its water. In U.S. usage a catchment is known as a watershed.

catena A sequence of related soils that are all of similar age and usually derived from the same **parent material** that repeats down a hill slope.

caterpillar A butterfly or moth (**Lepidoptera**) larva; the term is sometimes extended to **sawfly** larvae.

cathemeral Describes an organism that is equally active by day and by night.

Catherine's moss (common smoothcap) The moss *Atrichum undulatum* (**Bryophyta**), with erect stems up to 7 cm tall, long, narrow, dark green leaves with transverse undulations, and cylindrical **capsules** with a very long **beak**, borne on setae (*see* seta) 2–4 cm long. The moss is widespread in woodland and also occurs on heaths and waste ground.

cation *See* ion.

cation exchange A process in which **cations** in the soil solution change places with cations held at exchange sites on soil particles.

cation exchange capacity (CEC) The total amount of exchangeable **cations** that a soil is able to adsorb at a given **acidity**. Exchangeable cations occur mainly on the surface of **clay** and **humus colloids**. CEC is measured in units of centimol positive per kilogram of soil ($cmol_c/kg$) or milli-equivalents per 100 grams of soil (meq/100 g); these units are equal (1 $cmol_c/kg$ = 1 meq/100 g).

catkin (ament) An **inflorescence** that is a pendulous **spike**, usually comprising **simple**, **unisexual flowers**.

catkin curl *See Taphrina.*

catmint (*Nepeta*) *See* Lamiaceae.

catotelm The lower, anaerobic part of a **peat bog**, where organic material decomposes much more slowly than in the **acrotelm**.

cat's tail (*Acalypha hispida*) *See Acalypha.*

cattail *See Typha.*

caudal Relating to the tail.

Caudata (Urodela, salamanders, newts) An order of amphibians (**Amphibia**) that have tails at all stages in their lives, limbs set at right angles to the body, and usually with all four limbs of similar size. Their life cycle involves an aquatic larval form with gills and a terrestrial adult form with lungs, but some species retain their larval appearance throughout their lives, and others hatch on land and lack the aquatic larval stage. Most salamanders are less than 150 mm long. There are about 550 species occurring in almost all temperate regions of the Northern Hemisphere.

caudex (pl. caudices) **1.** A thick, swollen, short, persistent stem of a **perennial herb** that occurs underground or close to ground level. **2.** The main stem of a palm or tree fern.

caulescent In the process of growing a stalk.

caulid The main stem of a moss (**Bryophyta**).

cauliflory The production of flowers or fruit on the main trunk of a woody plant.

cauliflower fungus *See Sparassis crispa.*

cauline Pertaining to the stem.

Cavariella pastinaceae (parsnip aphid) A **holocyclic** aphid (**Aphididae**) that feeds on parsnip, celery, carrot, parsley, and fennel, and transmits a number of viral diseases. It overwinters on ***Salix*** species and occurs throughout Europe and North America.

cay A small, flat, offshore island formed from sand or coral.

cayenne pepper (*Capsicum annuum*) *See Capsicum.*

Cb *See* cumulonimbus.

Cc *See* cirrocumulus.

CCN *See* cloud condensation nuclei.

Ceanothus (California lilac) *See* root nodule.

CEC *See* cation exchange capacity.

cecidium A plant **gall**.

cecidization The formation of a plant **gall**, especially by gall midges of the family **Cecidomyidae**.

Cecidomyiidae (gall midges, gall gnats) A large family of small flies (**Diptera**), most 1–5 mm long, some up to 8 mm, with long legs, long antennae (*see* antenna), and simple wings (i.e. with few veins). In several genera larvae reproduce (**paedomorphosis**). Most are **gall** makers, some feed on plants and are serious pests, a few are **detritivores**, predators, parasites, or live as inquilines (*see* inquilism) with ants or termites. There are more than 6000 species distributed worldwide.

Cecidophyopsis ribis (blackcurrant gall mite) A mite (**Arachnida**) up to 0.3 mm long, that causes big bud (a swelling of the **buds**) and is a vector for blackcurrant reversion virus. A plant may have up to 100 galled buds and up to 35,000 mites emerge from each bud as it begins to open. They may then migrate a short distance to other buds, but most are transported passively by animals, wind, or rain. They feed on **sap**.

cedar *See Cedrus*, Pinaceae.

cedar-apple rust *See Gymnosporangium juniperi-virginianae.*

cedar of Lebanon (*Cedrus libani*) *See* diageotropism.

cedar waxwing *See Bombycilla cedrorum.*

Cedrus (family **Pinaceae**) A genus of tall **evergreen** coniferous **trees** (cedars) in which the needle-like leaves grow in dense spiral clusters on short shoots and in more open spirals on long shoots. **Cones** are barrel-shaped. There are four species occurring on high ground in the Mediterranean region and in the western Himalayas. Cedars yield valuable timber and are often grown for ornament.

Cedrus libani (cedar of Lebanon) *See* diageotropism.

Celastraceae (order **Celastrales**) A family of **trees**, **shrubs**, and climbers with **opposite**, **alternate**, or sometimes whorled, **simple**, **entire** to **serrate** leaves, usually with **petioles**, **exstipulate** or with very small **stipules**. Flowers **actinomorphic** or rarely **zygomorphic**, **bisexual** or **unisexual** (plants **dioecious**), 2 fused, 4- to 5-merous, free **sepals** and **petals**, 3–5 **stamens**, sometimes alternating with **staminodes**, **ovary superior**. Fruit a **capsule**, **drupe**, **berry**, or **schizocarp**. There are 94 genera with 1400 species, with worldwide distribution. Many provide medicial or insecticidal compounds, edible fruits and seeds, timber, and other products, and some are grown for ornament.

Celastrales An order of mainly tropical and subtropical plants comprising at least 2 families (possibly more) with 101 genera and about 1403 species. *See* Celastraceae and Lepidobotryaceae.

celeriac (*Apium graveolens* var. *rapaceum*) *See Apium.*

celery (*Apium graveolens* var. *dulce*) *See Apium.*

celery fly *See Euleia heraclei.*

celery-leaf fly *See Euleia heraclei.*

celery-leaf miner *See Euleia heraclei.*

cell The fundamental unit of all living organisms. It consists of a **cell membrane** containing **cytoplasm** and genetic material (**DNA**). **Prokaryotes** are simpler and smaller than **eukaryotes**; many have cilia (*see* cilium) or flagella (*see* flagellum). Eukaryote cells possess a **nucleus**, **ribosomes**, mitochondria (*see* mitochondrion), **Golgi bodies**, and **vacuoles**.

cell culture A mass of cells that are maintained *in vitro*.

cell cycle The sequence of events in the life of a cell between two cell divisions.

cell differentiation The processes by which a cell becomes specialized for a particular function, in which it continues for the rest of its life.

cell fusion The merging of nuclei (*see* nucleus) and **cytoplasm** from different **somatic cells** to produce a **hybrid** cell.

cell growth An irreversible increase in the size of a cell.

cell line A group of cells related to each other by cell division, i.e. they are all descended asexually from a single ancestor cell.

C

cell membrane (plasmalemma, plasma membrane) The **selectively permeable** membrane, 7.5–10 nm thick, that encloses the contents of a cell. It consists of a double layer of **lipid** and **protein** molecules.

cell plate During **cytokinesis**, the partition that forms between **daughter cells** and later provides the framework for the new **cell walls**.

cell sap The liquid content of a **vacuole** in a plant cell.

cellular slime moulds *See* Acrasiomycetes.

cellulase An **enzyme** that catalyzes the **hydrolysis** of **cellulose** to **glucose**.

cellulolytic Able to break down **cellulose**.

cellulose An insoluble **polysaccharide** that is the principal structural material in plants; it is believed to be the most abundant organic compound in the world.

cellulytic Able to break down cells.

cell wall The outer layer of a plant, fungal, or **prokaryote** cell (animal and protozoan cells do not have walls). Plant cell walls form the interface between adjacent cells and collectively provide a structure to the whole plant. They control the movement of molecules into and out of the cell, maintaining the internal composition, and protect against **pathogens**. They are made from **polysaccharides** especially **cellulose**, **hemicelluloses**, **pectins**, and variable amounts of **protein**, **lipid**, **lignin**, **tannin**, and mineral salts.

cementation The process by which substances deposited from the water, rich in minerals, that flows through the **pore** spaces of a rock cement sedimentary particles together.

cemented Describes a mineral soil that is massive and indurated (*see* induration), giving it a hard and often brittle consistency.

centipedes *See* Chilopoda.

Central Australian floral region The area covering all of central Australia, much of it desert but with extensive ***Acacia*** thorn forest.

centre of diversity (gene centre) A geographical area where a particular **taxon** exhibits greater genetic variation than it does anywhere else. A centre of diversity may also be a **centre of origin**.

centre of origin A geographical area where a **taxon** is believed to have originated. Many authorities believe centres of origin are also **centres of diversity**, because as the taxon spreads from its centre of origin, some of its variants are more successful than others, so the amount of genetic variation decreases with distance from the origin.

centric diatom A **diatom** that has radial symmetry.

centrifugal Developing outward from the centre.

centriole In most **eukaryote** cells, but not in vascular plants (**Tracheophyta**) or most **Fungi**, an **organelle**, usually comprising nine groups each of three **microtubules** forming a cylinder with a central cavity. Centrioles occur in pairs at right angles to each other. They form part of the **centrosome** and are involved in the formation of the **mitotic spindle**. Centrioles are also present in the basal region of cilia (*see* cilium) and flagella (*see* flagellum).

centripetal Developing inward from the exterior.

Centrolepidaceae (order **Poales**) A **monocotyledon** family of small, **caespitose**, **annual** or **perennial herbs** with **linear** leaves resembling bristles, that are basal in annuals and **imbricate** and crowded along the stem in perennials. Flowers are very small, **unisexual**, the male with 1 **stamen** the female with 1 **carpel**, **inflorescence** unisexual or **bisexual pseudanthia**. Fruit is a **follicle**. There are 3 genera with 35 species occurring in southern South

America, Southeast Asia, Australia, and New Zealand.

centromere (spindle attachment) The region containing the **kinetochore** where the two halves of a **chromosome** are joined to the **spindle** during **mitosis** and **meiosis**.

Centroplacaceae (order **Malpighiales**) A family of **evergreen trees** with unbranching **inflorescences** and fruits that are **capsules**. There are two genera, both formerly placed in other families: *Centroplacus* (formerly in **Euphorbiaceae**) with five species found in West Africa and *Bhesa* (formerly in **Celastraceae**), with one species found in Indochina and Malesia.

centrosome The **organelle**, containing the **centrioles**, where **microtubules** are assembled and disassembled. The cells of vascular plants (**Tracheophyta**) and most **Fungi** lack centrosomes.

cep *See Boletus edulis.*

Cepaea hortensis (white-lipped snail) A terrestrial **snail** with a shell 15–16 mm high and 16–22 mm wide that is usually yellow, sometimes pink, red, or brown, with variable amounts of brown banding and a white lip at the **aperture**. It occurs throughout Europe in woodland, coastal dunes, and grassland, feeding on nettles, hogweed, and ragwort.

Cepaea nemoralis (brown-lipped snail, grove snail) A terrestrial **snail** with a spiral shell 12–22 mm high and 18–25 mm wide marked by 4.5–5.5 whorls of brown, yellow, or white bands of varying relative thickness, so the appearance is highly variable and the subject of genetic studies that have made this snail a model organism. The **aperture** has a dark brown, occasionally white lip. It occurs throughout Europe and has been introduced to North America. It inhabits a wide range of **habitats** and feeds mainly on dead or dying plants. It is not a crop pest and is prey to birds and rodents.

cephalic Relating to the head.

cephalodium In the **thallus** of a **lichen** belonging to the **Chlorophyta**, a region where there are **cyanobacteria** which carry out **nitrogen fixation**.

Cephalotaceae (order **Oxalidales**) A **monotypic** family (*Cephalotus follicularis*) of carnivorous **evergreen**, **herbs** with strongly dimorphic (*see* dimorphism) leaves. These are either spiral, **simple**, **entire**, and **exstipulate**, or form a pitcher 2–6 cm tall and half-filled with water, the opening surrounded with ridges bearing teeth pointing inward overhung by a lid. Both types usually present but varying with the season. Flowers are small, white, **actinomorphic**, **hermaphrodite**, 6-merous, with no **petals**, 6 long and 6 short **stamens** in 2 **whorls**, and 6 **carpels**. **Inflorescence** an erect **scape** bearing **cymose** clusters of flowers. Fruit is a **follicle**. The plant is found in southwestern Australia and is widely grown for ornament as the Albany pitcher plant.

Cephalotus follicularis (Albany pitcher plant) *See* Cephalotaceae.

Ceratiomyxa fruticulosa (coral slime) A species of slime mould (**Myxogastria**) that grows on dead wood. The **fruiting bodies** are white and translucent, erect, 1–10 mm tall and 0.5–1.0 mm wide, and occur in large numbers covering substantial areas of the substrate. They appear fuzzy because they bear their **spores** on the exterior surface. The species occurs worldwide.

Ceratiomyxomycetes A class of **Myxogastria** that produces a true **plasmodium** but differs from other **slime moulds** in bearing its **spores** on the surface of columnar **fruiting bodies**.

Ceratocystis fagacearum A species of **ascomycete fungi** that causes the disease oak wilt, affecting all oak species (*Quercus*). Oaks with pointed leaves are most susceptible and usually die within months of being infected; oaks with rounded leaves often live for several years after infection. The fungus spreads through the distribution of **spores** by insect vectors, especially beetles attracted to mats of perithecia (*see* perithecium); **root grafts** can also spread the fungus. Leaves of infected trees become discoloured, wilt, and die, and the tree is defoliated. The fungus occurs only in North America.

Ceratophyllaceae (order **Ceratophyllales**) A **monogeneric** family (*Ceratophyllum*) of free-floating submerged aquatics with **whorls** of 3–10 finely divided leaves, often branching **dichotomously** 1–4 times, and slender, branched stems. Flowers are **actinomorphic**, **unisexual** (plants **monoecious**), **sessile**, **axillary**, with 1 female or 1–4 male flowers at each **node**, **perianth** with 9–12 free or 8–15 fused lobes, male flowers with up to 45 **stamens** in whorls around a **pistillode**, females with **ovary superior**. Fruit is an **achene**. There are about six species with worldwide distribution. Some are grown as aquarium plants.

Ceratophyllales An order that contains only the family **Ceratophyllaceae**.

cerci A pair of conical or cylindrical appendages, often with a sensory function, that protrude from the abdomen of some insects, e.g. mayfly (**Ephemeroptera**). *See also* forceps.

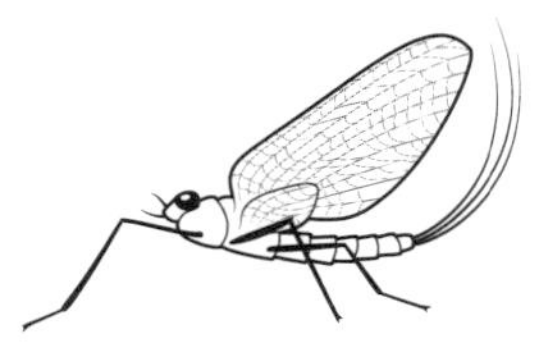

The three 'tails' protruding from the abdomen of this mayfly are cerci, and probably serve a sensory function.

Cercidiphyllaceae (order **Saxifragales**) A **monogeneric** family (*Cercidiphyllum*) comprising two species of tall **deciduous trees** that produce long shoots in the first year of growth followed by long-lived short shoots in subsequent years. Leaves are **ovate**, **opposite**, and **stipulate**. The plants are **dioecious**. The **unisexual** flowers lack a **perianth**. **Staminate** flowers have up to 40 **stamens**, female flowers have 2–7 **carpels**. **Inflorescence** is a **raceme**. Fruit is a **follicle**. There are two species, *C. japonicum* occurring throughout Japan, southern Korea, and in China as far west as Sichuan, and *C. magnificum* only on mountains in Honshu. The trees are cultivated for ornament and their wood is used in cabinet-making.

Cercopsis vulnerata (red and black leafhopper, red and black froghopper) A large bug (**Cicadellidae**), 9–11 mm long with a wingspan of 11 mm, that is brightly marked with red or orange and black. Adults are active in spring and summer and feed on a wide range of plants, including fruit trees, causing discoloration of leaves (angular leaf spot). The **nymphs** feed on stems and roots of fruit trees and hops, sheltered beneath a protective covering of foam (**cuckoo-spit**). There is one generation a year. It occurs throughout Europe.

cereal-root nematode *See Meloidogyne naasi.*

cerebriform Convoluted, like the surface of a brain.

Cervidae (deer) A family of ruminant mammals that feed by browsing or grazing. Males of all species except *Hydropotes* (Chinese water deer) possess **antlers**; some, e.g. *Hydropotes*, have canine teeth modified to form tusks. Deer vary greatly in size. Most, but not all, are gregarious, their herds having a complex social organization. They live in a wide variety of **habitats** and some species, e.g. *Capreolus capreolus* (roe deer) enter suburban gardens, where they feed on cultivated plants. There are 23 genera with 47 species, occurring throughout Eurasia and the Americas, but Africa has only 1 species and deer are absent from Australasia and Antarctica.

Cetraria islandica *See* Iceland moss.

Ceutorhynchus assimilis (cabbage gall weevil, turnip gall weevil) A grey-black weevil (**Curculionidae**), 2.0–2.5 mm long, with white, curved larvae with brown heads 3–4 mm long, that feeds on the shoots, flower **buds**, or flowers of brassicas (**Brassicaceae**), each larva living inside a **gall** until it matures, then falling to the ground to pupate. There is one generation a year. The weevil occurs throughout Europe.

chaetae Bristles found on most worms belonging to the **Oligochaeta** and the predominantly marine Polychaeta.

Chaetosiphon fragaefolli (strawberry aphid) A species of yellowish green or almost white aphids (**Aphididae**), 1.3–1.5 mm long with green or yellow **nymphs** 0.8–1.1 mm long, that feeds on new shoots and on the underside of **leaflets** of strawberry (***Fragaria*** spp.). They inhibit plant growth, cause leaves to curl, and excrete **honeydew** that attracts **sooty mould**. They also transmit viral diseases. The aphids occur in Europe, North America, South Africa, and Australasia.

Chaetothyriomycetidae *See* Eurotiomycetes.

chafers *See* Scarabeidae.

chaffinch *See Fringilla coelebs.*

chain kingsnake *See Lampropeltis getula.*

chain response A behavioural sequence in which each behaviour provokes the next, e.g. in courtship rituals.

Chalara fraxinea **Anamorph** of *Hymenoscyphus pseudoalbidus.*

chalaza The base of an **ovule**, comprising an **embryo sac** enclosed by **integuments**.

chalcid seed flies *See* Torymidae.

chalcid wasps *See* Torymidae.

chalk A porous, fine-grained rock composed principally of the crushed and compressed shells and skeletons of marine organisms, made from calcium carbonate ($CaCO_3$).

Chamaecyparis (family **Cupressaceae**) A genus of medium to tall **evergreen**, coniferous **trees** with **adpressed**, scale-like leaves in **opposite** pairs, the foliage forming sprays flattened dorsiventrally (*see* dorsiventral). **Cones** are small, **globose** to oval, in opposite, **decussate** pairs. There are about six species, occurring in eastern Asia and eastern and western North America. Known as cypress or false cypress, they are widely grown for hedging and as ornament, with many **cultivars**. *Chamaecyparis lawsoniana* is Lawson cypress.

chamaephyte A plant in which the **perennating bud** or shoot **apex** is borne very close to the ground. It is one of the life form categories described by Christen **Raunkiær**, and there are four types. *See* active chamaephyte, cushion chamaephyte, passive chamaephyte, suffruticose chamaephyte.

channelled wrack The brown seaweed *Pelvetia canaliculata*, found on rocky shores throughout Europe. Its **thallus** is flattened, with **branches** up to 16 cm long that have inrolled margins, forming channel-like gutters that retain water, allowing the plant to grow higher on the shore than any other seaweed, even above the high-tide mark. It is sometimes fed to sheep and cattle.

chanterelle *See Cantharellus.*

chaparral The **sclerophyllous vegetation** of western California and adjacent areas that has developed mainly on land where fire destroyed a former forest. The climate is dry and with average temperatures ranging from about 10°C in winter to 40°C in summer. Fires are common.

Similar types of vegetation occur in other regions with a Mediterranean climate. ↗

character (trait) Any recognizable feature of the **phenotype** of an organism.

Charadrius vociferus (killdeer) A species of plovers whose common name refers to their call. They are 230–270 mm long with a wingspan of 175 mm, and have a brown back and wings, tawny rump, white breast with two black bands, a brown face with a white forehead, and a red or orange eye ring. They inhabit open grassland, forest, sand **bars**, and mudflats, and feed on invertebrates and berries. If disturbed while incubating they perform a 'broken wing' behaviour, running away from the nest pretending to be injured and thereby diverting the predator. They occur throughout most of North America and parts of South America. ↗

chard *See* Amaranthaceae.

Charophyceae A group of freshwater **green algae** known as stoneworts, in which the **thallus** comprises a main stem from which emerge **whorls** of lateral **branches** and **rhizoids**. Thalli are **monoecious** or **dioecious**. Oogonia (*see* oogonium) and antheridia (*see* antheridium) are borne at **nodes** of the lateral branches. In some ways the charophytes resemble bryophytes (**Bryophyta**). ↗

charophytes *See* Charophyceae.

chats *See* Turdidae.

check dam (jack dam) A small dam that is built across a minor channel or **gully** in order to slow the water flowing down a slope and thereby reduce **erosion**. ↗

cheese plant (*Monstera deliciosa*) *See Monstera*.

chelation A reaction between a metal **ion** and an organic molecule in which the components are linked by more than one bond. The metal ion is called the complexing agent and the organic molecule the ligand.

chelicera One of the first pairs of legs on the **prosoma** of **Arachnida**. In most species the chelicerae are held forward horizontally. They are used in defence, for digging, and spiders, harvestmen, and windscorpions use them to seize and kill prey.

Chelonia (Testudines, turtles, terrapins, tortoises) An order of **Reptilia** in which the body is enclosed between an upper **carapace** and lower **plastron** made from plates of bone covered in scales of horn. In many species the carapace is fused to the ribs and vertebrae. There are more than 290 species inhabiting marine, freshwater, and terrestrial **habitats**.

Chelydra serpentina (common snapping turtle) A species of dark brown or black turtles (**Chelonia**), 200–215 mm long, with **tubercles** on the neck and legs. They inhabit fresh or **brackish** water with a muddy bottom and are omnivorous, feeding on carrion, invertebrates, small vertebrates, and plant material. They occur throughout North America south of southern Canada. ↗

chemical control The application of chemical compounds to control invertebrate pests, weeds, fungi, etc.

chemical oxygen demand (COD) A measure of the amount of oxygen chemicals in effluent absorb. Potassium dichromate is added to a sample as the oxidizing agent and the reaction takes two hours, which is much quicker than the test for **biological oxygen demand**; since the ratio of BOD to COD is fairly constant, the COD test is the more widely used.

chemical weathering Sequences of chemical reactions involving the dissolving, **hydration** and **hydrolysis**, and **redox** reactions.

chemotaxis A change in the direction of locomotion in response to a gradient of concentration of a chemical substance.

chemotroph (chemotrophic organism) An organism that derives its energy from the **oxidation** of organic or inorganic chemical compounds.

chemotrophic organism *See* chemotroph.

chenille plant (*Acalypha hispida*) *See Acalypha.*

Chenopodiaceae The goosefoot family, now included in the **Amaranthaceae**.

cherimoya (*Annona cherimola*) *See Annona.*

chernic horizon A type of **mollic horizon** that is deep, well-structured, black (*chern* is Russian for black) with a high **base saturation** and a high content of organic matter.

chernozems (black earth) A free-draining, black soil (*chern* is Russian for black, *zemla* for soil), associated with temperate grassland. It is rich in **humus**, plant nutrients, and exchangeable **cations**. These are among the world's most productive agricultural soils. Chernozems are a reference soil group in the **World Reference Base for Soil Resources**.

cherry birch (*Betula lenta*) *See* methyl salicylate.

cherry blackfly *See Myzus cerasi.*

cherry slug *See Caliroa cerasi.*

chervil (*Anthriscus cereifolium*) *See Anthriscus.*

chestnut *See Castanea.*

chestnut blight *See Cryphonectria parasitica.*

chestnut canker *See Cryphonectria parasitica.*

chestnut soils A **zonal soil** that develops in warm temperate, semi-arid climates under grassland with some shrubs. It is dark brown at the surface with reddish tints, underlain by red soil with accumulations of lime. These are valuable agricultural soils.

chestnut worm *See Lumbricus castaneus.*

chiasma (pl. chiasmata) The cross-shaped point of contact between the four-strand bundle of non-sister **chromatids** of **homologous chromosomes**, where genetic material is exchanged, that appears at the diplotene stage of **prophase I** of **meiosis**.

chiasma interference The frequency of non-random distribution of chiasmata (*see* chiasma) during **meiosis**; if the frequency is higher than expected by chance, the chiasma interference is said to be negative, if the frequency is lower it is said to be positive.

chiasmata *See* chiasma.

chickadees *See* Paridae.

chicken fungus *See* sulphur fungus.

chicken mushroom *See* sulphur fungus.

chicken of the woods *See* sulphur fungus.

chicory (*Cichorium intybus*) *See Cichorium.*

Chilean firebush (*Embothrium coccineum*) *See Embothrium.*

Chilean holly *See Desfontainea.*

chilling injury *See* chilling-sensitive plant.

chilling-sensitive plant A plant that experiences a marked reduction in its rate of growth at temperatures between 0°C and 12°C, with a subsequent expression of symptoms of stress, called chilling injury.

chilli pepper (*Capsicum annuum*) *See Capsicum.*

Chilopoda (centipedes) A class of **Arthropoda** that have segmented bodies with one pair of legs to each segment, long antennae (*see* antenna), and large claws beneath the mouthparts that inject venom. Despite the name suggesting 100 legs, different species have 15–191 pairs, always an odd number. Most centipedes run on the ground surface but some burrow in the soil. All are predators, most of small arthropods but some large tropical species will attack bats, small mammals, snakes, birds, and other animals. There are about 8000 species with a worldwide distribution.

chimera Tissue containing two or more genetically distinct types of cell, or an individual composed of such tissues.

Chinese gooseberry (*Actinidia chinensis*) *See* Actinidiaceae.

Chinese rice paper (*Tetrapanax papyrifera*) *See* Araliaceae.

Chinese water chestnut (*Eleocharis dulcis*) *See Eleocharis.*

Chinese water deer (*Hydropotes*) *See* Cervidae.

chinook A warm, dry, **foehn wind** that occurs on the eastern side of the Rocky Mountains in North America, most commonly in late winter and spring.

chiropterophily Pollination by bats (Chiroptera).

chitin A long-chain polymer $(C_8H_{13}O_5N)_n$ of N-acetyl-D-glucosamine that is the principal component of fungal **cell walls**, the **exoskeleton** of **Arthropoda**, the **radula** of **Mollusca**, and the beaks and internal shells of squids and octopuses.

chitin inhibitor An **insecticide** that inhibits the formation of **chitin**; when an exposed insect **moults** it is unable to form a new **exoskeleton** and dies.

chive (*Allium schoenoprasum*) *See Allium.*

chlamydospore A thick-walled resting **spore** produced by some **Fungi** and **Oomycota**.

Chloranthaceae (order **Chloranthales**) A family of tropical **trees**, many with **prop roots**, also **shrubs** and **herbs**, with **opposite**, **decussate**, **simple** leaves with **serrate** margins. Leaves usually aromatic when crushed. Flowers are small, green, **unisexual** (plants **dioecious**, rarely **monoecious**) or **bisexual**, **staminate** flowers without **sepals**, 1 **stamen** or 3 fused, borne in large numbers on **racemose inflorescences**; **pistillate** flowers with **calyx** fused to **inferior ovary** with 1 **locule**, borne in small numbers on a **raceme** or **spike**. Fruit is a fleshy **drupe**. There are 4 genera with 75 species found throughout much of the tropics and subtropics (but not Africa). Some have medicinal uses.

Chloranthales An order of tropical plants containing only the family **Chloranthaceae**.

chlordane A **cyclodiene insecticide** that was formerly used to kill termites and earthworms in turf. It is now banned in most countries.

chlorenchyma Tissue that contains **chloroplasts**.

chlorine (Cl) An element needed for normal plant growth. It controls **turgor** and may be involved in the **light-dependent stage** of **photosynthesis**. Plants lacking chlorine suffer **wilting** and young leaves become shiny and **glaucous**, later turning bronze-coloured and becoming chlorotic (*see* chlorosis).

chlorite **1.** A member of an important group of soft, green, **clay minerals**. **2.** Chlorine dioxide (ClO_2^-).

chlormequat A plant **growth regulator** that inhibits cell elongation and is used to strengthen stems in cereals, thereby reducing **lodging**.

chloronicotinyls A class of **systemic pesticides** that are used in greenhouses against insect pests. The group includes marathon, acetamiprid (trade name Tristar), and imidacloprid (trade name Flagship). They act on the central nervous system to disrupt nerve transmission.

chlorophyll The green pigment, a magnesium-porphyrin derivative $(C_{55}H_{72}O_5N_4Mg)$, that contributes to **photosynthesis** by absorbing light, predominantly in the blue (435–438 nm) and red (670–680 nm) regions of the spectrum. There are two principal types of chlorophyll in land organisms, designated *a* and *b* and differing only in the composition of a side-chain ($-CH_3$ in *a* and $-CHO$ in *b*); marine algae (*see* alga) have chlorophylls *c* and *d*.

Chlorophyta A **phylum** of **green algae** that contain **chlorophyll** *a* and *b*; their storage product (**starch**) is formed in the **chloroplasts** and not in the **cytoplasm**. Some are single-celled, others multicellular. They are the closest relatives to land plants and probably closely resemble the first **eukaryotes**. They occur in both fresh and salt water, in forms ranging from pond scum to leaf-like seaweeds. Some form symbioses with **Fungi** to form **lichens**. There are about 7000 species found worldwide.

chloroplast The **plastid** in the cells of green plants in which **photosynthesis** occurs; it contains some of its own **DNA**. **Parenchyma** cells each contain between 10 and 100 chloroplasts. A chloroplast is biconvex or plano-convex, 5–10 µm long and 2–3 µm wide. It is enclosed in a double membrane and contains stacks (grana, sing. **granum**) of flattened membranous discs (**thykaloids**) linked by lamellae (*see* lamella) and embedded in a matrix (**stroma**). The thykaloids contain photosynthetic pigments including **chlorophyll** and photosynthesis takes place on the thykaloid outer membrane. The stroma also contains **ribosomes** and **lipid** and **starch** granules. Chloroplasts are believed to have originated from **cyanobacteria** and become plant-cell **organelles** by **endosymbiosis**.

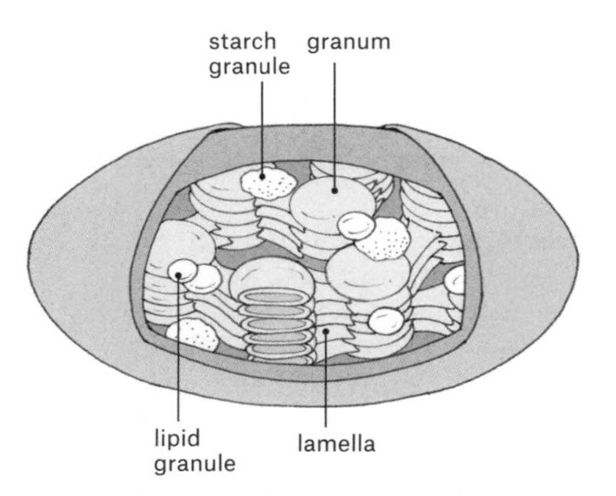

Chloroplast. The organelle within which photosynthesis occurs.

chlorosis A reduction in green coloration that is a sign of disease or nutrient deficiency that inhibits the formation of **chlorophyll**. Chlorotic plants appear pale green or yellow.

chlorpropham A **carbamate herbicide** that is used to control germinating weeds and to prevent sprouting in stored potatoes.

chlorthiamid A **herbicide** that is applied to the soil or slow-flowing water for total control of weeds. It acts by inhibiting the formation of **cell walls** and it poses very little harm to animals.

chocolate arion *See Arion rufus.*

Choisya (family **Rutaceae**) A genus of **evergreen shrubs** with **opposite**, palmately **compound**, glossy, leathery leaves. Flowers have 4–7 white **petals** and 8–15 **stamens**. Fruit is a **capsule**. There are seven to nine species occurring in southern North America and Mexico. Several are cultivated as mock orange or Mexican orange.

choke of grasses *See Epichloë typhina.*

Chondrodendron tomentosum (curare) *See* Menispermaceae.

Chondrostereum purpureum A species of **basidiomycete fungi** that causes the disease silver leaf in trees, attacking most members of the **Rosaceae**, especially ***Prunus***. The fungus enters through wounds, e.g. from pruning, and patches of crust appear on the **bark**. These grow into **fruiting bodies** that are undulating, intergrowing brackets about 30 mm across. The maturing brackets are violet-coloured on the underside and grey on the upper side covered with pale hairs, later drying out and becoming brown and rubbery. The leaves develop a silver sheen and affected **branches**, and eventually the entire plant, die. A paste made from **spores** is smeared on the bark of unwanted trees to kill them.

chorda dorsalis *See* notochord.

Chordata A **phylum** of animals that includes all those possessing a **notochord**. the phylum includes the tunicates (Urochordata), amphioxus, fish, **Amphibia**, **Reptilia**, birds (**Aves**), and **Mammalia**.

chorion The outer of two layers of tissue surrounding the fluid-filled **amniotic sac** in reptiles, birds, and mammals. *See also* amnion.

choughs *See* Corvidae.

Christmas berry *See Lycium.*

Christmas box *See Sarcococca.*

Christmas rose *See Helleborus.*

chromatid One of the two daughter strands of a **chromosome** that has undergone division during **interphase**.

chromatin The substance from which **chromosomes** are made, comprising **DNA**, chromosomal **proteins**, and chromosomal **RNA**.

chromatophore 1. In a plant cell, a **plastid** containing pigment. **2.** In a **prokaryote**, a **vesicle**, bound by a membrane and containing photosynthetic pigment. **3.** In some animals (squid, octopus, lizards, etc.), an organ or cell containing pigment.

chromomere (idiomere) During **prophase** of **mitosis** and **meiosis** in **eukaryotes**, a bead-like or granular structure visible in the **chromosome**. In many organisms, chromomeres on **homologous chromosomes** pair during meiosis.

chromophore A molecule that absorbs light energy at certain wavelengths; those it does not absorb it reflects, and the reflected wavelengths define its colour.

chromoplast A **plastid** containing pigments other than **chlorophyll**.

chromosome A structure found in the cell **nucleus** of all **eukaryotes** that consists of DNA wound tightly around a thread of **histone**. Each chromosome has a constriction (the **centromere**) dividing it into two arms; the shorter arm is known as the p arm, the longer as the q arm. Chromosomes occur in **homologous** pairs (the **diploid** state) except in **gametes** and **gametophytes**, in which they occur singly (the **haploid** state). The DNA carries the **genes**. Each species has a characteristic number of chromosomes (e.g. ***Arabidopsis thaliana*** 10, maize 20, **einkorn wheat** 14, **bread wheat** 42).

chromosome map A diagram that shows the positions of **genes** on a **chromosome**.

chromosome polymorphism The presence within an interbreeding population of two or more alternative forms of one or more **chromosomes**.

chromosome substitution The replacement of one or more **chromosomes** by totally or partially **homologous chromosomes** from a different **species** or strain of the same species.

chrysalis *See* pupa.

chrysanthemum leafhopper *See Eupteryx melissae.*

chrysanthemum white rust *See Puccinia horiana.*

Chrysemys picta (painted turtle) A turtle (**Chelonia**) that is 100–250 mm long, with a smooth, oval, flat-bottomed shell. The **carapace** is olive or black, the **plastron** yellow or red, sometimes with dark marks near the centre, olive or black skin with red and yellow stripes on the neck, legs, and tail, and yellow stripes on the face with a yellow spot and streak behind each eye and yellow stripes on the chin.

They live in shallow, fresh water with a muddy bottom and feed on plants and small animals. They occur throughout North America from southern Canada to Mexico.

Chrysobalanaceae (order **Malpighiales**) A family of **trees** and **shrubs** with **alternate, simple, entire** leaves with **stipules** that are often **caducous**. Flowers **actinomorphic** or **zygomorphic, bisexual** (plants **monoecious** or **polygamous**), with 5 **sepals** and usually 5 **petals**, 2–100 or more **stamens, ovary superior** of 3 **carpels**. Fruit is a fleshy **drupe**. There are 17 genera with 460 species found throughout the tropics, especially America. Several species have fruit that is eaten locally.

Chrysomelidae (leaf beetles) A family of robust beetles (**Coleoptera**), 1.5–2.2 mm long, that have smooth, metallic or brightly coloured **elytra**. Larvae are grub-like, with short legs and often coloured. All leaf beetles are **herbivores** and many are pests of cultivated plants, e.g. Colorado beetle (***Leptinotarsa decemlineata***), asparagus beetle (*Crioceris asparagi*), and **flea beetles**. There are more than 35,000 species, distributed worldwide.

Chrysophyta A **phylum** of golden algae (*see* alga), most single-celled and free-swimming in fresh water, but some filamentous and colonial. Almost all are photosynthetic. They are a primary food for zooplankton, but they feed on **diatoms** and **Bacteria** when light levels are low. There are more than 1000 species.

chuckley pear *See Amelanchier.*

Chytridiomycetes *See* Chytridiomycota.

Chytridiomycota A **phylum** of mainly aquatic **Fungi** with **gametes** that swim by means of a **flagellum**, a characteristic possessed by no other fungi. Many chytrids have an affinity for and grow on **pollen grains**. Some are unicellular, others produce mycelia (*see* mycelium), some are **saprotrophs**, others parasites on animals or plants. There are 127 genera with about 1000 species, found worldwide. Chytridiomycetes is the most important class, with members that parasitize aquatic species.

Ci *See* cirrus.

Cicadellidae (leafhoppers, sharpshooters) A family of bugs (**Homoptera**), most less than 13 mm long, with hind legs adapted for jumping. Many feed on **sap** or cell contents and transmit diseases, making them serious pests. There are at least 22,000 species and possibly more than 100,000, with a worldwide distribution.

Cichorium (family **Asteraceae**) A genus of **perennial herbs** with **alternate** leaves and blue (occasionally white or pink) **florets**. There are six to eight species occurring in Europe and naturalized in North America. *Cichorium intybus* (chicory) and *C. endiva* (endive) are cultivated for their leaves. Roots of *C. intybus* are blended with coffee or used as a coffee substitute.

cider sickness Spoilage of cider that is caused by *Zymomonas anaeroba*, a member of the **Alphaproteobacteria** that ferment sugars to **ethanol** but that also produce acetaldehyde, which spoils the appearance and flavour.

cilium A small (2–10 µm long, 0.5 µm wide), hair-like structure on the outside of a cell, usually present in large numbers on those cells that possess them at all, which function in locomotion and/or feeding, their coordinated beating generating currents in the fluid around the cell.

cincinnus A **monochasium** with the **branches** on alternate sides of the stem and the **inflorescence** often bent to the side.

Cinnamomum (cinnamon, camphor) *See* Lauraceae.

cinnamon (*Cinnamomum*) *See* Lauraceae.

cinnamon fungus *See Phytophthora cinnamomi.*

C

Cionus scrophulariae (figwort weevil) A weevil (**Curculionidae**) with a brown or black body with two black spots on the **elytra** that feeds on figwort (*Scrophularia* spp.), mulleins (*Verbascum* spp.), and orange-ball buddleja (***Buddleja*** *globosa*). When disturbed it drops from the plant and is well camouflaged against the soil. Its larvae are covered in a sticky, shiny substance that is distasteful to predators. The weevil is native to Europe but has become established in New York State and is spreading.

circadian rhythm A pattern of metabolic processes with a period of approximately 24 hours (Latin *circa* about, *diem* day).

Circaeasteraceae (order **Ranunculales**) A family of **annual herbs** with a persistent **hypocotyl** terminating in a rosette of leaves that surround the **inflorescence**. Leaves **simple**, occasionally bilobed or approximately **orbicular**, with **dichotomous venation**. Flowers **actinomorphic**, **bisexual**, with a **uniseriate perianth**, 2–3 inconspicuous or 5–7 **petaloid tepals**, 1–3 **stamens** or 8–12 **staminodes**, **ovary superior** of 1–3 or 5–9 free **carpels**. **Inflorescence** either a single flower or compact, compound **fascicle**. Fruit is an **achene**. There are two genera with two species occurring from northern India to western and southwestern China.

circinate *See* vernation.

circle of vegetation In **phytosociology**, a geographic region with a distinctive **flora**. It is the highest classification used by the Zürich-Montpellier School of Phytosociology. *See* Braun-Blanquet, Josias.

circumaustral distribution The distribution pattern of organisms that occur all around the high latitudes of the Southern Hemisphere.

circumboreal distribution The distribution pattern of organisms that occur all around the high latitudes of the Northern Hemisphere.

circumnutation The spiralling of a growing shoot **apex**.

circumpolar distribution The distribution pattern of organisms that occur all around the North or South Pole, their ranges extending into the polar regions.

circumscissile Opening or splitting along a circumference, e.g. in a seed **capsule**.

cirriform In long, fine **filaments**, resembling **cirrus**.

cirrocumulus (Cc) A genus (*see* cloud classification) of high-level clouds composed entirely of ice crystals that forms small patches or sheets arranged in approximately regular patterns resembling ripples in the sand on a seashore or, less commonly, in lines or groups.

cirrostratus (Cs) A genus (*see* cloud classification) of high-level clouds composed entirely of ice crystals that appears as a thin veil. It often gives rise to haloes, but the Sun and Moon are clearly visible through it.

cirrus 1. In palms (**Arecaceae**), a long, protruding, spiny or whip-like leaf tip. **2.** (Ci) A genus (*see* cloud classification) of high-level clouds composed entirely of ice crystals that forms long, wispy **filaments**, narrow bands, or white patches, always with a fibrous appearance.

Cistaceae (order **Malvales**) A family mainly of aromatic **shrubs** but with some **annual** or **perennial herbs**, with **alternate** or **opposite**, **simple**, **entire** leaves, with or without **stipules**. Flowers, often showy, are **actinomorphic**, with 5 **sepals**, the 2 outer ones small and resembling **bracteoles**, 5, 3, or no **petals**, 3–10 or more free **stamens**, **ovary superior**, usually of 3 **carpels**. Fruit a **loculicidal capsule**. There are 8 genera with 175 species occurring in North America, southern South America, Eurasia especially around the Mediterranean, and North Africa. Several species of *Cistus*, *Halimium*, and ***Helianthemum*** cultivated for their flowers.

cisternae Long, flattened, membrane discs found in certain **organelles**, e.g. in the **Golgi body** and **endoplasmic reticulum**.

cistrans test *See* complementation test.

cistron A section of a **DNA** molecule that encodes for the formation of a particular **polypeptide** chain.

citric acid cycle (Krebs' cycle, tricarboxylic acid cycle) A series of chemical reactions in most living cells that provide hydrogen and electrons for the production of ATP (**adenosine triphosphate**).

Citrus (family **Rutaceae**) A genus of spiny, usually **evergreen shrubs** and small **trees**, many of which are cultivated for their fruit, which is a **hesperidium**. Leaves shiny, **coriaceous**, **alternate**, **unifoliate**, and **entire**. Flowers, often strongly scented, with 5 (rarely 4) **petals**, many **stamens**. *Citrus* species hybridize readily, making it difficult to determine the number of true species, but probably there are about 20, originating from southern and southeastern Asia to eastern Australia.

citrus stubborn disease A disease caused by ***Spiroplasma** citri*, a **bacterium** belonging to the **Mollicutes**, and transmitted by leafhoppers. It blocks **phloem** tissue, causing reduced size and upright position in the leaves of ***Citrus*** plants and deformation of fruits. A number of other plant families host the bacterium, especially brassicas (**Brassicaceae**). The disease occurs in the Mediterranean region, Middle East, North Africa, and the United States.

Cl *See* chlorine.

clade In **cladistics**, the organisms on a branch that results from a split in an ancestral lineage. Each such split produces two new taxa (*see* taxon), each represented as a branch (clade) in a diagram depicting **phylogenies**. In the **Angiosperm Phylogeny Group** classification, **orders** are grouped into clades and clade is the highest-ranking taxon below kingdom.

cladistics A classification system that arranges organisms according to their evolutionary relationships. In the branching diagrams depicting those relationships, a dividing lineage always produces two equal daughter taxa, so each daughter is a **monophyletic** group sharing a stem **taxon** (ancestor) with its sister group.

cladode (phylloclade, phyllocladium) A flattened green stem, resembling a leaf so closely it is often difficult to distinguish, that is a plant's principal organ of **photosynthesis**.

Cladonia *See* cup lichen, reindeer moss.

Cladosporium A genus of **ascomycete fungi** that occur as some of the most widespread black, brown, and olive-green moulds, often growing on living or dead plant material. It occurs both indoors, on walls and carpets, and outdoors.

Cladoxylopsida A class of plants closely related to ferns and **Sphenopsida** (horsetails) that flourished in the Middle Devonian epoch (397.5–385.3 million years ago) and disappeared in the Mississippian (359.2–318.1 million years ago) and that included the earliest **trees**, having a central stem with smaller lateral **branches** growing from it.

clamp connection A bulge that appears on one side of the site of a **septum** in a **dikaryotic** fungal **hypha**. It forms during cell division and helps maintain the dikaryotic state.

clavate Club-shaped.

Claviceps purpurea (ergot) A species of **ascomycete fungi** that grows on the ears of cereal grasses (**Poaceae**). **Spores** infect flowers, inducing cells to expand and divide to form sclerotia (*see* sclerotium); these fall to the ground and remain dormant until the following spring, when they produce **stroma** with spores that infect the next crop. The infection reduces yield but otherwise causes little harm to the plant, but the sclerotia contain **alkaloids** that cause ergotism if ingested.

clavicle The collar bone, a bone on the **ventral** side of the shoulder, found in many vertebrates.

clay **1.** Soil that contains at least 20 percent clay particles. **2.** A soil particle smaller than 2 µm in size, or a mineral particle smaller than 4 µm in size.

clay films *See* clay skins.

clay loam Soil that consists of 27–40 percent **clay**, 20–45 percent **sand**, and the remainder **silt**.

clay minerals A group of hydrous aluminium silicate minerals that occur as small, plate-like or fibrous crystals arranged in layers. They absorb and lose water readily.

clay pan A soil **pan** that has a very high **clay** content.

clay skins (cutan, clay films, argillans, tonhäutschens) A coating of clay particles that have moved downward through the soil and been deposited on stones and **peds**.

clear ice *See* glaze.

cleavage polyembryony *See* polyembryony.

cleidoic egg An egg which is enclosed by a shell that isolates it from the surrounding environment.

cleistocarp *See* cleistothecium.

cleistogamy Self-pollination in an unopened flower.

cleistothecium (cleistocarp) An **ascocarp** that completely encloses the asci (*see* ascus). When the wall of the cleistothecium ruptures the **ascospores** are released.

Clematis (family **Ranunculaceae**) A genus mostly of woody **lianas**, but with some **shrubs** and **perennial herbs**. Those originating in cool climates are **deciduous**, those from warmer climates **evergreen**. It is the only genus of large, woody plants in the family. Leaves **opposite**, **compound**, with some **petioles** forming tendrils that twine around supporting structures. Flowers with 4 **petaloid perianth** segments and long, plumed **styles**. The fruit is an **achene**. There are about 300 species, most occurring in northern temperate regions, with a few in tropical Africa and Oceania. Many are cultivated for ornament.

Clements, Frederic Edward (1874–1945) An American botanist who proposed that plant communities develop by means of constant adjustments of the relationships between species, i.e. a **succession**, and finally reach an optimal **climax community**. Clements believed that in some ways a plant community behaves as though it were a superorganism.

Cleomaceae (order **Brassicales**) A family of **annual** and **perennial herbs** and **shrubs** with **alternate**, **simple** or palmately **compound**, **petiolate** leaves with **stipules** sometimes forming thorns. Flowers with usually 4 **sepals**, 4 clawed **petals**, usually 6 **stamens**, **ovary superior** of 2 fused **carpels** with 1 **locule**. Fruit is a **silique** or **silicula**. There are 10 genera with 300 species found throughout tropical and warm temperate regions, especially America. Some grown for ornament.

Clethraceae (order **Ericales**) A family of large **shrubs** or **trees** with **alternate**, **entire** or **serrate** leaves without **petioles** and **stipules**. Flowers **actinomorphic** or with irregular **sepals**, **bisexual**, 5 free or fused sepals and **petals**, 10 **stamens** in 2 **whorls**, **ovary superior** of 3 or 3–5 fused **carpels** and 3 or 3–5 **locules**. Fruit is a **loculicidal capsule**. There are 2 genera with 75 species found in southeastern United States,

Central and northeastern South America, Madeira, and southeastern Asia. Several species cultivated for ornament.

Clianthus (family **Fabaceae**) A genus of **shrubs**, 1–2 m tall, and climbers, with **alternate**, **pinnate** leaves comprising 1–24 short **leaflets**. Flowers, 7.5–10 cm long or longer, are pea-like, turning downward with one upturned **petal**. There are 2 species, both **endemic** to New Zealand and widely cultivated for ornament. *Clianthus puniceus* has matt, grey-green, narrow leaves and salmon-pink flowers; *C. maximus* has broad, glossy, green leaves and bright red or orange flowers. Both species are endangered in the wild.

click beetles *See* Elateridae.

Clifford, George (1685–1760) A wealthy Anglo-Dutch financier who lived on a large estate, Hartekamp, near Haarlem, Netherlands, where he expanded the garden and added a menagerie, aviary, orangery, and four tropical houses, and where he assembled a **herbarium** comprising 3461 sheets of mounted specimens. In 1735, Carolus **Linnaeus** visited Clifford, who then became his patron. Linnaeus wrote a detailed description of the estate and herbarium, *Hortus Cliffortianus*, published 1738.

climacteric An increase in **respiration** and **ethylene** production that occurs as fruit is ripening and approaching **senescence**.

climate The weather conditions experienced in a particular location averaged over a long period. *See* climatic normal.

climate change Variations in the **climate** of a location or the world that occur over time, e.g. the transitions into and out of ice ages.

climate classification The ordering of **climates** according to their most important characteristics, e.g. summer and winter temperatures, amount and distribution of precipitation, so each type can be assigned a name or code by which it can be identified unambiguously. *See* Köppen climate classification, Thornthwaite classification.

climate diagram A graph that shows the average monthly temperature and precipitation amount for a particular place, with

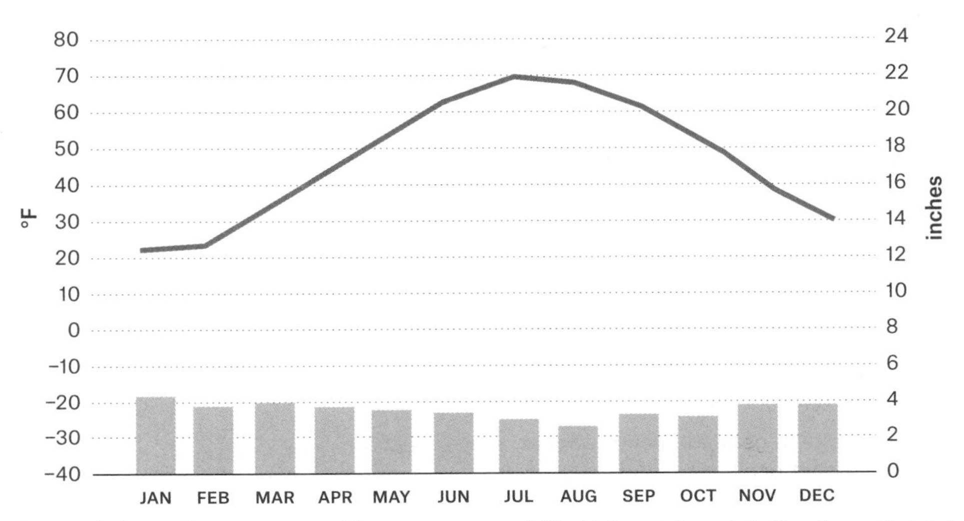

The graph shows the average monthly temperature and (the histogram) precipitation, beneath details of the location. The resulting diagram describes the local climate, in this case in Portland, Maine.

the latitude, longitude, and elevation of the location. It allows the typical climate of the location to be seen at a glance.

climate types The **climates** that are listed in systems of **climate classification**. *See* Köppen climate classification, Thornthwaite classification.

climatic normal The mean values for temperature, **humidity**, and precipitation at a particular location over a fixed period that is used to define the **climate** of that location. In the United States and many other countries the fixed period is 30 years and it changes every 10 years.

climatology The scientific study of **climates**. *Compare* meteorology.

climax The ultimate stage of a plant **succession**, in which the community attains an equilibrium with its **abiotic environment** and in the absence of perturbation becomes self-perpetuating.

climax community (climax vegetation) The plant community of a **climax**.

climax theory The ecological theory which proposes that a community of organisms develops through a **succession** until it reaches a **climax** where it is in equilibrium with its physical environment. There are two versions: **monoclimax** and **polyclimax**.

climax vegetation *See* climax community.

climosequence A sequence of **soil profiles**, usually derived from the same **parent material**, with differences that reflect variations in local climatic conditions.

clinal speciation A form of **allopatric speciation** that occurs when a **species** is divided into two segments by a geographic barrier that falls across a **cline**.

cline A change in **gene frequency** or **character** states that occurs in a **species** gradually across the area it occupies.

clinosequence A sequence of soils on an inclined surface in which the **soil profiles** reflect changes in the gradient.

clitellum (saddle) In mature terrestrial earthworms (**Oligochaeta**) and leeches (**Hirudinea**), a swollen region of the body near the head that secretes a mucus which holds two worms together during copulation and later secretes a cocoon that envelopes the worm and into which eggs are laid.

clitochory *See* barachory.

Clitocybe (*Lepista*) A genus of agaric **fungi**, comprising about 50 species, including the wood blewit (*C. nuda*) and field blewit or blue-leg (*C. saeva*). The wood blewit **fruiting body** is mushroom-like with a smooth, purple **pileus** that turns brown with age, bright purple **gills**, and lilac **stipe**. The field blewit cap is grey-brown and the stipe blue. Both are edible and occur in grassland and woodland throughout Europe and North America.

clod A compact block of soil that retains its form when the soil is dug or ploughed.

clonal dispersal Plant dispersal by means of **stolons** or **rhizomes** from which new plants develop that are genetically identical to one another and to the parent (i.e. they are **clones**). Bracken (*Pteridium aquilinum*) and trembling aspen (*Populus tremuloides*) are among the plants that disperse in this way.

clone A group of genetically identical cells or organisms derived from an ancestor asexually.

closed drainage basin (endorheic basin) A **drainage** basin in which water remains within the basin, there being no streams carrying water away.

Closteroviridae A family of viruses that consist of long, **flexuous** particles containing **RNA**. There are three genera:

Ampelovirus, transmitted by mealybugs, *Closterovirus*, transmitted by aphids (**Aphididae**), and *Crinivirus*, transmitted by whiteflies. Many cause serious plant diseases.

clothesline effect The effect of **advection** when warm, dry air enters and flows through vegetation, e.g. a field crop or forest. Near where the air enters it raises the temperature and rate of **evaporation**, tending to dry the soil. Farther in, the moving air cools, increasing the **relative humidity**.

clothianidin *See* neonicotinoid.

cloudberry *See Rubus.*

cloudburst A sudden, brief, very heavy rain shower that occurs when the downdrafts in a **cumulonimbus** cloud suppress the updrafts, causing the mechanism sustaining the cloud to fail. As the cloud dissipates it releases all of its moisture.

cloud classification The naming of cloud types in order to identify them unambiguously. Clouds are grouped first by the usual height of the cloud base, as high, middle, and low, then by their appearance into 10 genera. The genera are further divided into 14 species with 9 varieties. There are also **accessory clouds**. The table shows the conventional classification.

Cloud Base (metres)

POLAR REGIONS	TEMPERATE REGIONS	TROPICS
High		
3000–8000	5000–13,000	5000–18,000
Medium		
2000–4000	2000–7000	2000–8000
Low		
0–2000	0–2000	0–2000

Cloud genera:
High: cirrocumulus, cirrostratus, cirrus
Medium: altocumulus, altostratus, nimbostratus
Low: stratus, stratocumulus, cumulus, cumulonimbus

cloud condensation nuclei (CCN) Microscopic particles drifting in the air onto which water vapour condenses to form cloud droplets.

cloud forest A tropical **montane forest**, usually more than 1000 m above sea level, that is shrouded in mist for most of the time. The moisture encourages the growth of **epiphytes**.

cloud formation **1.** A pattern of clouds with particular shapes. **2.** The processes by which clouds form.

cloud seeding The injection of material, e.g. dry ice (solid carbon dioxide), silver iodide, salt (sodium chloride), or calcium chloride, into air containing supercooled (*see* supercooling) water droplets in order to modify the characteristics of the cloud and thereby cause rain to fall where otherwise it might not have fallen, or inhibit the formation of **hail**.

clove The dried young flower **bud** of *Syzygium aromaticum* (**Myrtaceae**), used as a spice.

clover *See Trifolium.*

club fungi *See* Basidiomycota, coral fungi.

clubroot *See Plasmodiophora brassicae.*

club rush *See Scirpus.*

Clusiaceae (order **Malpighiales**) A family, formerly called Guttiferae, of **evergreen** or **deciduous trees** and **shrubs**, and **annual** and **perennial herbs**, with **opposite**, rarely **alternate** or as **whorls**, **entire**, usually **simple** leaves. Flowers **actinomorphic**, **hermaphrodite** or **unisexual** (plants **dioecious**), 4–5 (can be 2–20) free or fused **sepals**, 4–5 (or 3 or 4–8 or no) **petals**, sometimes with **epipetalous ligules**, 5 **stamens** in 2 whorls, **ovary superior** usually of 2–5 (or 2–20) united **carpels**. Fruit is a **capsule**, **berry**, or **drupe**. There are 14 genera with 595 species found throughout the tropics. Several species yield valuable timber, chemicals are extracted from the

bark, leaves, and flowers, some provide fats and oils, and some have edible fruits (e.g. mangosteen, *Garcinia mangostana*).

cluster cup The cup-shaped **aecium** formed on leaves infected by certain **rust fungi**.

CMV *See* cucumber mosaic virus.

Cneoraceae A family now included in the **Rutaceae**.

Co *See* cobalt.

co-adaptation The development and maintenance in two or more groups of organisms of genetic traits that benefit each of the groups and thus allow the relationship to continue, e.g. between **angiosperms** and pollinating insects, and between the **acacia ant** *Pseudomyrex ferruginea* and ***Acacia*** *hindsii*, in which the ant remains active 24 hours a day, providing constant protection for the plant, and the plant bears leaves throughout the year, providing food for the ant.

coalescence 1. (fusion) The union of parts of a **flower**. **2.** The possession by different **species** of **genes** descended from a common ancestral gene. **3.** The merging of two or more cloud droplets into a larger droplet.

coal tit *See Periparus ater.*

coarctate Describes a **puparium** formed from the **cuticle** of the final larval **instar**.

coastal redwood (*Sequoia sempervirens*) *See* Pacific coast forest.

coatis *See* Procyonidae.

cobalt (Co) An element that is an essential trace nutrient for all animals; **Bacteria** and **Archaea** convert cobalt salts to cobalamin (vitamin B_{12}). Cobalt is also essential for **cyanobacteria** and for the **root nodule** bacteria in legumes (**Fabaceae**), where it is involved in **nitrogen fixation**, and it occurs in many species of algae (*see* alga).

cobble A stone 64–256 mm or 60–200 mm in size, depending on the classification used.

cobnut *See Corylus.*

cobra lily *See* Sarracenciaceae.

cobras *See* Elapidae.

cocaine *See* Erythroxylaceae.

Coccidae (mealybugs, scale insects, soft scales, tortoise scales, wax scales) A family of very small bugs (**Homoptera**) in which the females are wingless, usually sedentary, with long, oval bodies covered with wax beneath which they feed on the host plant; in some genera they lack legs and antennae (*see* antenna) are short, sometimes absent. Males have a single pair of wings or are wingless and do not feed. Newly hatched **nymphs** are dispersed by the wind, most older nymphs are sedentary. Coccid females and nymphs feed on plant **sap** and some are serious pests. There are about 4000 species with a worldwide but mainly tropical distribution.

Coccinellidae (ladybirds, ladybugs, lady cows, lady beetles) A family of rounded, shiny beetles (**Coleoptera**), 1–10 mm long, with red, black, orange, or yellow **elytra** usually marked with spots, bands, or geometric shapes in a contrasting colour. Most of the head is covered by the **pronotum**, the antennae (*see* antenna)are short and clubbed. Larvae are highly active and vary in appearance according to species. Both adults and larvae will bleed reflexively from their leg joints, exuding a sticky, foul-smelling, and in some species irritant fluid. Adults and larvae of most species feed on plant-eating insects, especially aphids (**Aphididae**) and some are used in **biological control**, but a few are herbivorous and pests. There are nearly 6000 species with a worldwide distribution.

coccolithophorids Unicellular, marine protists (**Protista**) that, at some stage in their life cycle, are covered in calcareous plates (coccoliths) embedded in a gelatinous sheath. Coccoliths are a major component of deep-sea calcareous oozes.

coccoliths *See* coccolithophorids.

Coccothraustes vespertinus (evening grosbeak) A species of bulky finches, 160–220 mm long with a wingspan of 300–360 mm, that have large beaks, short tails, black wings with a white patch in males, males with a brown head with a bright yellow forehead and body, females olive-brown with a grey underside. They inhabit forests and feed on berries, seeds, and insects. They occur across Canada and in mountainous areas of the western United States and Mexico.

coccus A spherical or oval **bacterium**, resembling a **berry**.

Coccus hesperidium (brown soft scale) A species of oval, flat, reddish brown, scale insects (**Coccidae**), about 6 mm long, that feed on the sap of a wide range of plants, mainly indoors. Females lay eggs that hatch almost immediately into nymphs called crawlers that find a feeding site within a few days, then moult into passive nymphs that continue feeding. Winged adult males and females appear after about one month; they can produce six or seven generations a year. Their damage to **phloem** tissue stresses the plant and they produce copious amounts of **honeydew** that coats surfaces and provides as substrate for **sooty mould**. The insect has a worldwide distribution.

Cochlicopa lubrica (glossy pillar, slippery-moss snail) A **snail** with a brown or yellowish, translucent, glossy shell, 5–7 mm high and 3 mm wide, found in woodlands, river banks, and among rocks. It occurs in Europe, North America, Australasia, and Sri Lanka.

cock's-foot *See Dactylis.*

cock-tail beetle *See Ocypus olens.*

cocoon *See* pupa.

COD *See* chemical oxygen demand.

codling moth *See Cydia pomonella.*

codon A sequence of three **nucleotides** in **messenger-RNA** that codes for an **amino acid** during **protein** synthesis.

coefficient of consolidation The factor controlling the rate at which a particular soil can be compressed. This depends on **permeablity**, which determines the rate at which water leaves soil **pores**.

coefficient of inbreeding (*F*) A measure of the probability that two **genes** at any **locus** in an individual are identical to genes in the common ancestor of both parents, i.e. the degree to which two **alleles** are more likely to be homozygous (*see* homozygosity) than heterozygous (*see* heterozygosity).

coelom The fluid-filled body cavity found in most animals that separates the muscles of the body wall from the gut and provides space for the growth of the internal organs.

coelomate Possessing a **coelom**.

coelomoduct In **coelomate** invertebrate animals, a duct that links the lining of the **coelom** to the exterior of the body.

coenobium A colony of algae (*see* alga) comprising a fixed number of unspecialized cells that behaves as a single organism.

coenocline A sequence of plant communities that can be traced across an environmental gradient.

coenocyte A cell with many nuclei that are not separated by **cell walls**. Coenocytes occur through the repeated division of the **nucleus** of the original cell, but not of the **cytoplasm**. *Compare* syncytium.

coenozygote A **zygote** with many nuclei.

coenzyme An organic compound that is not a **protein** and that acts as a **cofactor** for an **enzyme**.

C

coenzyme Q *See* ubiquinone.

co-evolution The complementary evolution of **species** with close associations, e.g. between certain **flowers** and the insects that pollinate them.

cofactor A non-**protein** compound required for the functioning of an **enzyme**, to which the enzyme is bound.

Coffea arabica *See* West African rain forest floral region.

coffee (*Coffea*) *See* Rubiaceae.

cohesion A force, due to **hydrogen bonds**, that draws water molecules together and causes them to attach to molecules on the surface of other substances. Water is the most highly cohesive of all non-metallic liquids, readily forming drops.

cohort 1. A group of individuals all of the same age. **2.** A group of related plant families. **3.** A group of animal orders.

col The region between two centres of high or low **atmospheric pressure**, where the **pressure gradient** is low.

Colaptes auratus (northern flicker) A species of woodpeckers, 300–350 mm long with a wingspan of 540 mm, which has grey-brown bars on its back and a white rump. It inhabits woodlands with stands of dead trees, farmland, and suburban areas, and feeds on insects, mainly ants, other arthropods, and berries. It occurs throughout most of North America.

Colchicacae (order **Liliales**) A **monocotyledon** family of **perennial herbs** with **rhizomes** or **corms** and occasionally **tubers**, erect sometimes **scandent** stems, and **distichous**, occasionally **verticillate**, often sheathing, **ovate**, **lanceolate**, or **linear** leaves. Flowers **actinomorphic**, usually **bisexual**, **sessile** or with a **pedicel**, 6 **tepals**, 6 **stamens**, **ovary superior** of 3 partly or completely **syncarpous carpels** with 3 **locules**. Fruit is a **capsule**. There are 15 genera with 245 species found in temperate and tropical regions, but absent from South America. About 90 species of *Colchicum* cultivated as naked ladies, *C. autumnale* is meadow saffron or autumn crocus, other species have medicinal uses.

cold cloud Cloud in which the temperature is below freezing throughout.

cold front A **front** that advances with the air behind it cooler than the air ahead of it.

cold-front thunderstorm A **thunderstorm** generated on a **cold front**, as advancing cold air pushes beneath warmer, moist air, raising it and rendering it unstable (*see* instability), causing **cumulonimbus** clouds to form.

cold lightning **Lightning** that does not ignite forest fires because the **lightning stroke** is not sustained for long enough. *Compare* hot lightning.

cold low (cold pool) A **cyclone** in which cool air at a low **atmospheric pressure** is surrounded by warmer air at a higher pressure. Such lows often develop in winter in the middle **troposphere** over northeastern North America and northeastern Siberia. They are often persistent.

cold pole One of the places that experiences the lowest mean temperatures; these do not coincide with the geographic or magnetic poles. The Southern Hemisphere cold pole is at Vostok Station, Antarctica, 78.46° S, 106.87° E (mean temperature −55.1°C). There are two cold poles in the Northern Hemisphere, at Verkhoyansk, Siberia, 67.57° N, 133.85° E (mean temperature −17.2°C) and Snag, Yukon, 62.37° N, 140.40° W (mean temperature −5.8°C).

cold pool *See* cold low.

cold sector During the development of a **frontal system**, the cold air that partly encloses the **warm sector**.

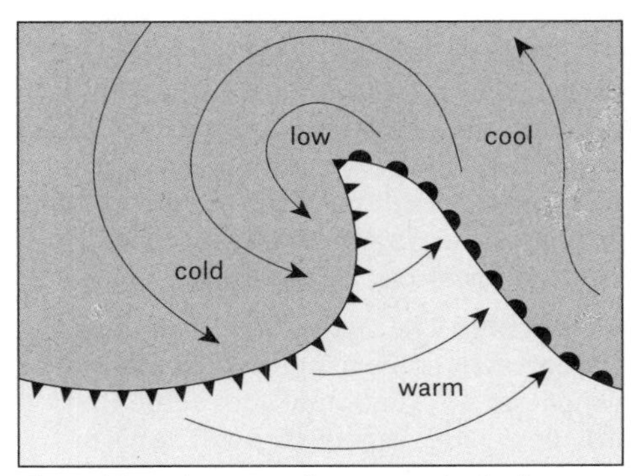

The cold air that partially surrounds the low pressure at the crest of a frontal wave.

cold tongue A long, narrow protrusion from a cold **air mass** that extends toward the equator.

cold wave The arrival of cold air bringing a large and sudden drop in temperature. Across most of the United States, a cold wave is defined as a decrease in temperature of at least 11°C, continuing for no longer than 24 hours, and reducing the temperature to −18°C or lower. In California, Florida, and the states of the Gulf Coast the temperature decrease must be at least 9°C to 0°C or lower.

Coleoptera (beetles) An order of insects whose name means 'sheath wings' (Greek *koleos* and *ptera*) because their forewings are hardened to form **elytra** that cover the membranous hind wings and abdomen. The head is well-developed with biting mouthparts. Larvae possess antennae (*see* antenna) and **mandibles**; **pupae** are **adecticous** and **exarate**. There are probably 3,000,000 species found worldwide except for Antarctica. Many are crop pests, but others are scavengers or predators. Beetles range in size from the fringed ant beetle (*Nanosella fungi*), 0.25 mm long, to the goliath beetle (*Goliathus giganteus*), up to 100 mm long.

coleoptile A sheath that surrounds the **apical meristem** of **monocotyledon** seeds, protecting the growing tip of the shoot and embryonic leaves as they push through the soil toward the surface.

coleorhiza A protective sheath that surrounds the **radicle** of a **monocotyledon** seed.

collagen A fibrous **protein** that has a high tensile strength and is fairly inelastic. It is a major constituent of **connective tissue**.

collared dove *See Streptopelia decauocto.*

collateral bundle A **vascular bundle** in which the **phloem** occurs on only one side of the **xylem**.

Collembola (springtails) An order of eyeless, wingless **Arthropoda**, most of which are less than 6 mm long. They have a springing organ, the furcula, which is held by a catch against the underside of the six-segmented abdomen; when the catch is released the furcula moves downward against the substrate, propelling the animal into the air. Immature springtails resemble adults. A few species feed on plants and are pests, but most live in the soil, under **bark**, in decaying wood, and among **Fungi**, and contribute to the decomposition of organic material. There are more than 6000 species found worldwide, often in very large numbers; 60,000 per square metre have been counted.

collenchyma Tissue that supports and strengthens young shoots and leaves. It consists of elongated cells with thickened **cell walls**, containing **protoplasm** and sometimes **chloroplasts**.

collet The point where the stem and root of a vascular plant (**Tracheophbyta**) meet.

Colletotrichum A genus of about 600 species of **ascomycete fungi** that are **obligate endophytes**, many of them pathogens attacking more than 3200 species of plants. They form dome-shaped appressoria (*see* appressorium) that puncture the host, then **hyphae** develop first within epidermal cells and later throughout the plant. The diseases they cause include **anthracnose**, crown rot, and smudge.

C

collision theory An explanation of the way raindrops form in **warm clouds** containing droplets of varying sizes. Large droplets fall faster than smaller ones and as the large droplets fall they collide and coalesce with smaller droplets along their path.

colloid **1.** A substance that consists of two homogeneous substances, one dispersed evenly throughout the other. **2.** In soil, a component consisting of very small mineral, e.g. **clay**, or organic, e.g. **humus**, particles that have a very large total surface area in relation to their volume. Soil colloids usually have a high **cation exchange capacity**.

colluvium Weathered (*see* weathering) rock **debris** that has moved down a slope.

colonization The establishment of a viable population by a species arriving in a new **habitat**.

Colorado beetle *See Leptinotarsa decemlineata.*

Colorado potato beetle *See Leptinotarsa decemlineata.*

colpate Describes a **pollen grain** that has one or more colpi (*see* colpus).

Colpomenia peregrina *See* oyster thief.

colporate Describes a **pollen grain** that is both **colpate** and **porate**.

colpus (germinal furrow) An elliptical or approximately rectangular aperture or groove, at least twice as long as it is wide, on the surface of a **pollen grain**. Pollen grains are identified largely through the shape and arrangement of their colpi.

Coluber constrictor (black racer) A colubrid snake (**Colubridae**) that is 900–1900 mm long and black, dark blue, or olive-brown with yellow underside as an adult, and brighter red, brown, and grey as a juvenile. It inhabits a wide range of **habitats**. Juveniles feed on invertebrates, frogs, and other small vertebrate animals, adults on larger animals including other snakes. They are not aggressive to humans and non-venomous, but will bite if handled carelessly. They help control insect and rodent pests. There are 11 subspecies occurring throughout most of the United States and Central America.

Colubridae The largest family of snakes (**Squamata**), comprising snakes most of which are not venomous, but some that are, delivering venom through fangs at the rear of the jaw, and a few species can deliver bites fatal to humans. There are 304 genera and 1938 species, found worldwide, but the family is polyphyletic (*see* polyphyletism).

Columba livia (domestic pigeon, feral pigeon, rock dove) A species of pigeons (**Columbidae**) with blue-grey heads and bodies, with red or green iridescent patches on the neck and wings, usually two dark bars across the wings, and a pale bar on the tail. They nest on sea cliffs, in farm buildings, and on city buildings. They feed on the ground mainly on seeds. They are native to Europe, southwestern Asia, and North Africa, but also occur throughout North America. They have been domesticated and bred as racing and fancy pigeons and for food, and have become **feral**, occurring in many urban areas, often in large numbers.

Columba palumbus (wood pigeon) A pigeon (**Columbidae**), 380–450 mm long with a 680–800 mm wingspan, that is grey with a pinkish breast and white patches on its neck and wings. It inhabits woodland, parks, and gardens, and feeds on plant material. It occurs throughout Eurasia.

Columbidae (pigeons, doves) A family of stocky birds with small heads, short beaks and legs, ranging in size from 150 mm to 750 mm, the smaller species usually known as doves and the larger as pigeons. Most are arboreal, some ground-dwelling, and they fly strongly; many are migratory. They are gregarious, often forming large flocks. They feed on seeds or fruit, the seed-eaters being generally grey, brown, or pink, the fruit-eaters being more brightly coloured. There are 42 genera with 308 species, occurring in most terrestrial **habitats** in all continents except Antarctica.

columbine *See Aquilegia.*

columella The central spiral in the shell of a gastropod (**Gastropoda**) that forms as the growing shell coils tightly around an **axis**.

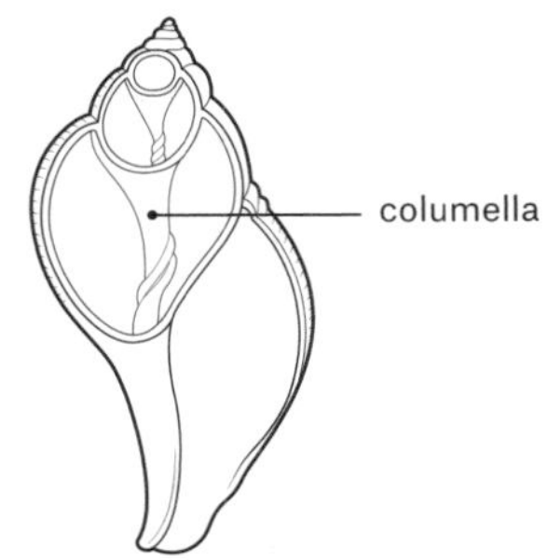

The central section of a gastropod shell.

Columelliaceae (order **Bruniales**) A family of **evergreen shrubs** and **trees** with **opposite**, **simple**, **entire** or **dentate**, **exstipulate** leaves that are conspicuously asymmetrical. Flowers somewhat **zygomorphic**, 8–16 fused or 10 free **sepals** and **petals** in **whorls**, **gamopetalous**, 2 **stamens**, **ovary inferior** of 2 **carpels** with 1 **locule**. Flowers solitary or in **cymes**. Fruit a **capsule**. There are two genera with five species found from southern Colombia to Bolivia.

column The **style** and **stigma** considered together.

columnella In some **Fungi** and **gymnosperms**, a central column of sterile tissue in the **spore**-bearing structure.

Combretaceae (order **Myrtales**) A family of **evergreen** and **deciduous trees** up to 50 m tall, **shrubs**, and **lianas** with **opposite** or **alternate**, **simple**, **entire** leaves. Flowers **actinomorphic** or slightly **zygomorphic**, **hermaphrodite**, sometimes **unisexual** (plants **dioecious** or **andromonoecious**), 4–5 sometimes 5–8 free, **valvate sepals**, 4–5 sometimes 5–8 valvate or **imbricate petals** (sometimes petals absent), 4–5 or 8–10 **stamens** in 1 or 2 **whorls**, **ovary inferior** of 2–5 **carpels** with 1 **locule**. Fruit is like a **drupe** or **nut**, or a false **samara**. *Laguncularia racemosa* has **pneumatophores**. There are 14 genera with 500 species occurring throughout the tropics. Many species yield valuable timber, some are cultivated for ornament, and many have local medicinal uses.

comfort zone The temperature range within which people feel comfortable, for most people 18–24°C, although **wind chill** and **relative humidity** make the air feel colder or warmer. An apparent temperature of 27°C presents no risk to most people; at 27–32°C people should exercise caution; at 32–41°C extreme caution is advisable; 41–54°C is dangerous; and temperatures higher than 54°C are extremely dangerous.

Commelinaceae (order **Commelinales**) A **monocotyledon** family mainly of **perennial herbs** that have **stolons** and **rhizomes**, with some **annual** or **epiphytes**. Leaves **distichous** or in spirals, **entire**. Flowers usually **actinomorphic** but sometimes **zygomorphic**, **bisexual** or **unisexual**, 3 **sepals**, 3 **petals**, 6 **stamens** in 2 **whorls**, often with some reduced to **staminodes**, **ovary superior** of 3 fused **carpels** with 3 (occasionally 2) **locules**. Flowers typically last only one day. **Inflorescence** usually a **cyme** with parts a **cincinnus**. Fruit usually a **dehiscent capsule**, occasionally a **berry**. There are 40 genera with 652 species found in tropical and temperate regions. Several are widely cultivated as ornamentals, e.g. *Tradescantia*.

C

Commelinales An order of **monocotyledons** that includes about 5 families with 68 genera and 812 species. *See* Commelinaceae, Haemodoraceae, Hanguanaceae, Philydraceae, and Pontederiaceae.

commensalism A relationship between two species in which one, the commensal, benefits from its association with the other, but the other, sometimes called the host, suffers no disadvantage.

Commiphora myrrha (myrrh) *See* Burseraceae.

common asparagus beetle (*Crioceris asparagi*) *See* asparagus beetle.

common bee (*Apis mellifera*) *See Apis.*

common bladder moss (*Physcomitrium pyriforme*) *See* pear moss.

common box (*Buxus sempervirens*) *See* Buxaceae.

common box turtle *See Terrapene carolina.*

common bugle (*Ajuga reptans*) *See Ajuga.*

common cardinal *See Cardinalis cardinalis.*

common coral snake *See Micrurus fulvius.*

common door snail *See Balea biplicata.*

common earthworm *See Lumbricus terrestris.*

common frog *See Rana temporaria.*

common garter snake *See Thamnophis sirtalis.*

common green capsid *See Lygocoris pabulinus.*

common kingsnake *See Lampropeltis getula.*

common mugwort (*Artemisia vulgaris*) *See Artemesia.*

common myrtle (*Myrtus communis*) *See* Myrtaceae.

common newt *See Triturus vulgaris.*

common reed (*Phragmites australis*) *See Phragmites.*

common roll-rim *See Paxillus involutus.*

common shrew (*Sorex araneus*) *See Sorex.*

common smoothcap *See* Catherine's moss.

common snail *See Cornu aspersum.*

common snapping turtle *See Chelydra serpentina.*

common toad *See Bufo bufo.*

common vetch (*Vicia sativa*) *See Vicia.*

community A group of organisms that live together in the same environment, i.e. the **biotic** component of an **ecosystem.**

companion planting The cultivation of a variety of crops in close proximity, the species being selected to control pests, improve pollination, optimize the use of space, etc.

competition An interaction between individuals of the same species (intraspecific competition) or different species at the same **trophic level** (interspecific competition) from which some individuals derive benefit and others suffer a disadvantage. Continued, competition leads either to the species with a competitive advantage replacing the other, or to a process of selective **adaptation**, whereby the competitors minimize the competition, e.g. by adopting different feeding habits.

competitive exclusion principle (exclusion principle, Gause principle) The principle, demonstrated experimentally in 1934 by G. F. Gause, that two or more species with identical patterns of resource use and both limited by the availability of resources cannot exist together in the same environment. Inevitably, one species will be better adapted and will eliminate the others.

competitive release The expansion of its **range** by a species that follows the disappearance of an environmental competitor.

complementary genes **Alleles** at different loci (*see* locus) that have mutated in such way as to complement each other and restore the **wild type phenotype**.

complementary resources Resources that can substitute for one another and that augment one another when both are available.

complementation map A diagram of **genes** in which a line or bar indicates each **mutation** and overlaps those of other mutations it does not complement.

complementation test (cistrans test) A test that determines whether two **mutant** sites on a **gene** occur on the same **cistron**.

complete flower A **flower** that possesses **sepals**, **petals**, **stamens**, and **carpels**.

complete penetrance The genetic condition in which a particular **genotype** invariably produces the same **phenotype**. It happens because a **dominant gene** and a **recessive gene** in the homozygous (*see* homozygosity) state always produce a phenotypic effect.

complexing agent *See* chelation.

complex low An area of low **atmospheric pressure** that has two or more centres.

Compositae *See* Asteraceae.

compost Organic material that has fully decomposed and that is used to improve soil structure and as a source of plant nutrients.

compost worm *See Eisenia veneta.*

compound Describes a flower or leaf that has two or more parts.

conceptacle A hollow, **urceolate** chamber in the **receptacle** (swollen tip of the **thallus**) of certain **brown algae** (Phaeophyta) and **Fungi** in which **gametes** are formed. It opens through a small **ostiole**.

concerted evolution The homogenization of **genes** that occurs in long, repetitive sequences in the **RNA** or **ribosomes**.

conchocelis stage The **diploid sporophyte** stage (once thought to be a different species) in the life cycle of the edible seaweed *Porphyra*. It produces rows of fertile cells (conchosporangia) by **meiosis**, which release **haploid** conchospores that develop into the **gametophyte**.

conchosporangia *See* conchocelis stage.

conchospores *See* conchocelis stage.

concrescent Growing together of adjacent tissues.

concrete minimum temperature The lowest temperature registered by a thermometer that remains in contact with a concrete surface for a specified period.

concretion A local concentration of a substance in the form of a nodule.

condensation The change in phase from gas to liquid.

conditional instability The situation that arises when the **environmental lapse rate** (ELR) is greater than the **saturated adiabatic lapse rate** (SALR) but lower than the **dry adiabatic lapse rate** (DALR). If stable air (*see* instability) is forced to rise to the **lifting condensation level**, **condensation** releases **latent heat**, warming the air and slowing its **lapse rate** from the DALR to the SALR. The SALR is lower than the ELR, so the rising air remains warmer than the surrounding air and continues to rise. The air is then unstable, the condition being that it must first be forced to rise.

conduplicate *See* vernation.

condyle A knob of bone that fits into a socket to form a ball-and-socket joint.

cone In coniferous plants (Pinopsida), the structure that in males carries **pollen** and in females bears **ovules** and seeds on sterile scales. The scales are attached to a **rachis** and a **bract** subtends each scale. *See also* strobilus.

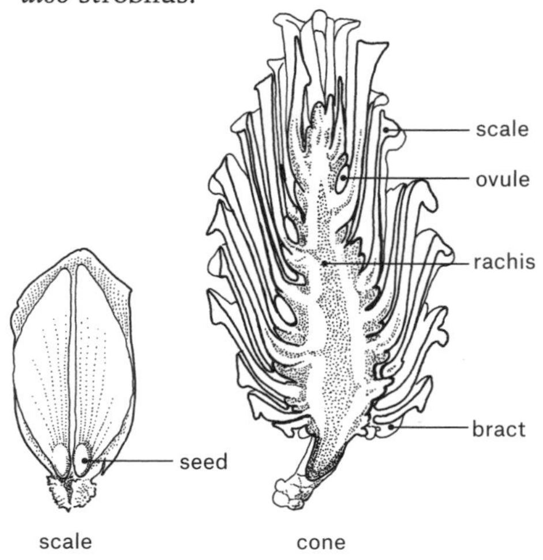

cone heads *See Conocybe.*

confined aquifer An **aquifer** that lies beneath a layer of impermeable rock.

conflict The situation in which an animal is motivated to perform two or more different activities at the same time.

confluence A flow of air in which several **streamlines** approach each other and the air accelerates.

congelifluction *See* gelifluction.

congelifraction *See* frost wedging.

congeliturbation *See* geliturbation.

congenital Describes an inherited **character** or condition that, therefore, is present at birth or **germination** or becomes evident during growth.

conglomerate (puddingstone) A coarse-grained rock containing rounded fragments greater than 2 mm in size.

conidiophore A **hypha** bearing conidia (*see* conidium).

conidiospore *See* conidium.

conidium (conidiospore) In many **Fungi**, a thin-walled **spore** that is produced asexually.

conjugate acid A base, i.e. a solution with a **pH** greater than 7.0, to which a hydrogen **ion** (H^+) has been added.

conjugate base An acid, i.e. a solution with a **pH** less than 7.0, from which a hydrogen **ion** (H^+) has been removed.

conjugated protein A **protein** that is bonded to a component which is not composed of **amino acids**.

conjugation The process in which two **Bacteria** exchange **DNA** through direct contact. The donor carries a DNA sequence called a fertility factor (F-factor) that allows it to extend a tube-like structure called a pilus that makes contact with the recipient and draws the cells together. The donor then transfers DNA, usually in the form of a **plasmid**, to the recipient. The recipient also receives the F-factor, allowing it to become a donor, so both cells may act as donors and recipients. Conjugation occurs only in unicellular organisms, but the term is often applied more widely to the union of **gametes**, especially in **isogamy**.

conk A **fruiting body** of a wood-rotting fungus, especially a **polypore**.

Connaraceae (order **Oxalidales**) A family of mainly **evergreen** but some **deciduous trees** and climbing **shrubs** with **alternate**, usually **imparipinnate**, **exstipulate** leaves. Flowers **actinomorphic**, usually **hermaphrodite**, with 5 **imbricate** sometimes **valvate sepals**, 5 imbricate **petals**, 10 **stamens** in 2 **whorls**, **ovary superior** of 1–5 free **carpels**. **Inflorescence** an **axillary panicle**. Fruits are **follicles**. There are 12 genera with 180 species occurring throughout the tropics.

Some species provide valuable timber, others have medicinal uses.

connate Describes the condition in which similar organs (e.g. **sepals**, **petals**) are joined together.

connecting band (girdle band) In **diatoms**, a band connecting the two halves of the **testa** (**epitheca** and **hypotheca**).

connective The tissue that connects the **pollen sacs** of an **anther**.

connective tissue Supporting or packing, fibrous tissue found in vertebrates. It consists mainly of **collagen** with some more elastic material.

Conocybe A genus of **ascomycete fungi** in which the **fruiting body** is conical and the **stipe** slender and delicate. They are known as dunce caps or cone heads and grow in grassland, sand dunes, and on decayed wood, and are very common in pastures and lawns. Some contain **amatoxins**. There are more than 240 species, occurring worldwide.

consensus tree A **phylogenetic tree** for a particular **taxon** that contains information shared by other trees derived by different methods.

conservation The maintenance of the resources and environmental quality within an area, and the sustainable production of goods and services, through appropriate management within economic and societal constraints. By allowing for natural change and commercial exploitation, the concept contrasts with preservation, which aims to prevent change or human utilization of the resource.

conservation tillage Preparation of the soil with the least disturbance of the surface. Plant residues are left and the ground is neither ploughed nor dug deeply. The aim is to reduce **erosion** and retain moisture.

conserved DNA A **DNA** sequence that is almost identical across many taxa that are only distantly related.

consistence (consistency) The extent to which a soil resists physical operations, e.g. digging or ploughing. This is determined by the strength of **adhesion** between soil particles. The consistence of dry soil may be described as hard, soft, or loose, and of wet soil as plastic or sticky.

consistency *See* consistence.

consociation In **phytosociology**, a plant community with a single dominant species.

constitutive enzyme An **enzyme** that is produced whether or not a suitable substrate is present, i.e. whether or not the enzyme is immediately useful. *Compare* inducible enzyme.

consumer Any **heterotroph**, i.e. an organism that feeds on other organisms, living or dead, and that is unable to synthesize organic compounds from simple, inorganic precursors. *See* macroconsumer, microconsumer.

consummatory Behaviour performed in direct pursuit of a goal, e.g. consuming food to relieve hunger, as opposed to **appetitive** behaviour.

consumptive use A use of a natural resource that reduces the supply, i.e. the resource is not recycled.

contact herbicide A **herbicide** that kills a plant on contact by scorching its leaves. It is not selective but leaves no residue. It can be used to kill **annual** weeds or **perennial** weed seedlings as they emerge between crop plants, first covering the crop plants to protect them.

contact inhibition The cessation of movement, growth, and division that occurs when a cell being grown in a **cell culture** makes physical contact with another cell.

contact insecticide An **insecticide** that kills insects that come into contact with it, i.e. they need not ingest it. It leaves very little residue. Products administered as **aerosols** or fogs are of this type.

contagious distribution *See* overdispersion.

contest competition Unequal **competition** for a resource, in which some competitors obtain all they need while others obtain less than they need. *Compare* scramble competition.

continental air Air that has lost most of its moisture crossing a continent and forms a very dry **air mass**.

continental climate A **climate** typical of the deep interior of a continent in **continental air**. The climate is dry and hot in summer and cold in winter.

continental Southeast Asia floral region The area that includes southwestern China, Myanmar, Thailand, Vietnam, Cambodia, Malaysia, and Indonesia, part of the **Palaeotropical region**. There are probably about 250 **endemic** species.

contingent drought (accidental drought) A **drought** that can occur anywhere, without warning, and that ends as abruptly as it began; it is unpredictable.

contorted Describes **sepals** and **petals** that are twisted while in the **bud**, so each one overlaps its neighbour on one side and is overlapped by its neighbour on the opposite side.

contour A line drawn on a map or an imaginary line on a surface that remains at a constant distance from a reference, e.g. at a constant height above sea level and, therefore, at right angles to the gradient.

contour cultivation **Cultivation** of a hillslope working parallel to the **contours**, i.e. at right angles to the gradient.

contour feather *See* feather.

contractile root A specialized root that contracts as soon as its tip contact soil particles below the surface. Its contraction pulls downward the structure from which it grows.

contractile vacuole A **vacuole**, found in many freshwater unicellular organisms, that expands and contracts with a pulsating motion, gathering water from inside the cell and expelling it from the cell, thereby regulating the water content of the cell.

contraction limit *See* Atterberg limits.

Contrarinia pisi (pea midge, pea gall midge) A black or yellow fly (**Diptera**) about 2–3 mm long with a very long **ovipositor**, that overwinters in a **puparium** in the soil. Adults become active as **legumes** are forming **buds** and starting to flower, mating the same day, and each female lays 20–80 eggs in about 20 flower buds. Larvae, 1–3 mm long, of the first generation feed on the flowers, those of the second generation inside pods, with 70–80 or more larvae in a single pod. Damaged flowers are deformed and fall, damaged pods contain small, wrinkled seeds. Infestation can seriously reduce yield. The midge occurs in most temperate parts of the world but is a serious pest in Europe.

Contrarinia pyrivora (pear midge, pear leaf curling midge) A dark-coloured fly (**Diptera**), 2.5–4.0 mm long, that overwinters as pupae in the soil and flies in spring. Females lay eggs in pear flowers. These hatch in 4–6 days and up to 100 larvae feed in each developing fruit, forming a black cavity filled with maggots. The damaged fruits turn black and fall. The larvae eat their way out of the fruit and burrow into the soil to pupate. The midge attacks only pears and occurs throughout the temperate Northern Hemisphere.

controlled pollination In plant hybridization (*see* hybrid), a technique in which **pistillate** flowers of one species are sealed in bags to prevent **pollen** reaching them, and when mature dusted with pollen collected from **staminate** flowers of the other species.

controlling gene A **gene** that turns the **transcription** of a **structural gene** on or off.

convection The transport of heat by the vertical movement of molecules within a fluid.

convection cell A vertical air circulation in which warm air rises by **convection**, cools and becomes denser, and subsides by gravity to be warmed again.

convective cloud A cloud that forms by **condensation** in air that is rising by **convection. Cumuliform** clouds are of this type.

convective condensation level The height at which **condensation** commences in a body of air that becomes saturated as it rises by **convection** through air in which temperature decreases at the **dry adiabatic lapse rate** and becomes unstable (*see* conditional instability) above the height at which it becomes saturated.

convergence A flow of air in which **streamlines** approach from different directions, producing an increase in pressure where they meet and causing air to rise.

convergent evolution The evolution of similar features in organisms that are only distantly related as they adapt to similar environmental conditions.

convolute *See* vernation.

Convolvulaceae (order **Solanales**) A family of **annual** and **perennial herbs** and **shrubs** with many herbaceous or woody climbers, and a few **trees**. Climbers have no tendrils and twine to the left. Leaves **alternate**, usually **simple**, **entire**; some have no leaves. Flowers usually **actinomorphic**, **bisexual**, occasionally functionally **unisexual** (plants **dioecious**), 4- or 5-merous, **sepals** free and overlapping, **corolla campanulate** to funnel-shaped or **urceolate**, **ovary superior** of 2 or 3–5 **carpels. Inflorescence** an **axillary cyme** or single flower, or sometimes a **raceme** or terminal **thyrse.** Fruit usually a **dehiscent capsule**, sometimes a **berry** or **utricle.** There are 57 genera with 1625 species, with worldwide distribution. Tubers of *Ipomoea batatas* are sweet potatoes, others are cultivated as ornamentals or have medicinal uses, ***Cuscuta*** (dodder) is a plant parasite, *Convolvulus arvensis* and ***Calystegia*** spp. (bindweeds) are aggressive weeds.

Convolvulus arvensis (bindweed) *See* Convolulaceae.

Cooksonia hemispherica A very early vascular plant (*see* Tracheophyta) that possessed an **epidermis** and **stomata**, as well as a **rhizome**, and that branched **dichotomously.** It lived from the Late Silurian to Early Devonian epochs (422.9–397.5 million years ago).

co-operation Behaviour among several animals that is mutually beneficial and may involve **altruism**, e.g. collaboration in hunting, care of young, etc.

Cope's gray tree frog *See Hyla chrysoscelis.*

copper (Cu) An element that plants require in trace amounts. It occurs bound to **proteins** and is involved in **redox reactions.** Plants with copper deficiency may be chlorotic (*see* chlorosis) or have leaves that are dark green, **bark** of woody plants may blister, and **shrubs** may be abnormally bushy.

copperhead *See Agkistrodon contortix.*

coppice 1. A method of **woodland** management in which **broad-leaved** trees are cut almost at ground level. This allows a number of poles to grow from the stump. These are harvested at intervals of 12–15 years, so the trees in an area can be cut in rotation to provide an annual crop of wood for fuel, fencing, etc. **2.** To cut trees in order to produce poles. **3.** Trees that regenerate from stumps to produce poles. **4.** An area of land managed in this way.

Coprinopsis *See* ink cap.

coprophagy The ingestion of faecal pellets.

coprophilous Growing on or in animal dung.

coral fungi (club fungi) **Fungi** belonging to the **Basidiomycota** in which the **fruiting bodies** (**basidiocarps**) are erect and either simple, resembling clubs, or branching, resembling corals.

coralloid Branching repeatedly; resembling a coral.

coral slime *See Ceratiomyxa fruticulosa.*

coral snakes *See* Elapidae.

coral spot *See Nectria cinnarbarina.*

corbiculum (pollen basket) In female bees (**Apidae**), an area on the outer side of the middle section (tibia) of each hind leg comprising a smooth depression lined by stiff bristles into which the bee packs mud, resin, dung, or **pollen** for transport to the nest.

corbie *See Corvus cornix.*

Cordaitales An order, now extinct, of woody **gymnosperms** that may have been early conifers or ancestors of the orders **Pinales**, **Cycadales**, and **Ginkgoales**. They were trees up to 30 m tall with strap-like leaves and reproductive structures resembling **cones**. They appeared during the Pennsylvanian and died out during the early Permian epochs (318.1–27.6 million years ago).

cordate Of a leaf, heart-shaped.

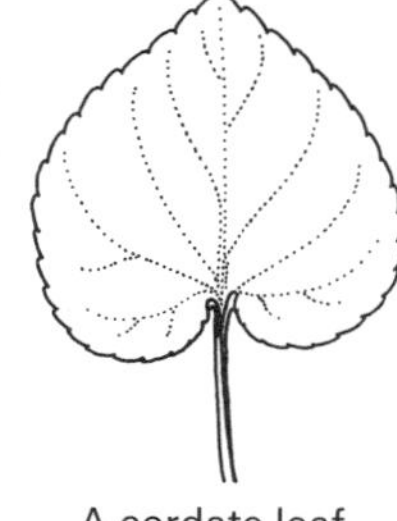
A cordate leaf.

Cordyline (family **Asparagaceae**) A genus of **evergreen shrubs** the larger of which resemble **trees**. They are unusual among **monocotyledons** in having woody stems. These grow from enlarged underground **rhizomes**. The long, narrow, palm-like leaves grow in tufts or rosettes. Flowers are sweet-scented, sometimes in **panicles**. Fruit is a **berry**. There are 15 species occurring in Southeast Asia, the Pacific Islands, Australia, and New Zealand. Many are cultivated, mainly for hedging and ornament, as cabbage tree or cabbage palm.

core area That part of a **range** where an animal or group of animals can rest securely and tend their young, and to which some species carry food they have obtained elsewhere. Animals will often defend their core area against intruders.

coremium A bunch of fungal **conidiophores** or **hyphae**.

CorF *See* Coriolis effect.

coriaceous Leathery.

Coriara (sumach) *See* root nodule.

Coriariaceae (order **Cucurbitales**) A **monogeneric** family (*Coraria*), which is a **perennial herb** with **rhizomes**, **shrub**, or small **tree** with **opposite**, **sessile**, **entire**, **lanceolate** or broadly **ovate** leaves. Flowers **actinomorphic**, hypogynous (*see* hypogyny), **pentamerous**, 10 long **stamens**, **ovary superior** usually of 5 free **carpels**. **Inflorescence** a **raceme**. Fruit is an **achene**. There are five species, interesting because of their **disjunct distribution**, in Central and western South America, the western Mediterranean region, around the Pacific, and in China and the Himalayas. Several are cultivated for ornament.

Coriolis effect (CorF) The apparent deflection, to the right in the Northern Hemisphere and to the left in the Southern Hemisphere, a body experiences when it moves across the surface due to the rotation of the Earth. Its magnitude is proportional to the latitude and the speed of the moving body, and is zero at the equator and at a maximum at each pole. It was first described in 1835 by the French physicist Gustave de Coriolis and the abbreviation CorF reflects the fact that it was initially thought to be a physical force rather than the consequence of moving across the surface of a rotating sphere.

cork (phellem) A protective layer of dead cells found below the **epidermis** of woody plants. It is derived from the **phellogen** and coated with **suberin**.

cork cambium *See* phellogen.

corm An underground storage organ comprising a swollen stem base protected by scale leaves and bearing **adventitious** roots.

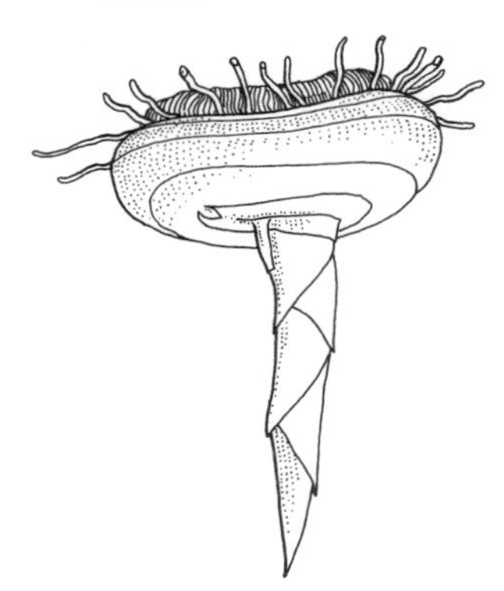

Gladiolus corm.

corn (*Zea mays*) *See Zea.*

Cornaceae (order **Cornales**) A family of **shrubs**, **trees**, and a few **perennial herbs** with woody **stolons** or **rhizomes**. Leaves **alternate** or **opposite**, usually **entire** occasionally **serrate**. Flowers **actinomorphic**, **epigynous**, sometimes **unisexual**, small **calyx**, 4–10 free and **valvate petals**, usually as many or twice as many **stamens** as petals, **ovary inferior** of 1–2 **carpels** and **locules**. **Inflorescence** a terminal or occasionally **axillary thyrse** occasionally subtended by large, showy, white **bracts**. Fruit resembles a **drupe**. There are 2 genera with 85 species scattered throughout temperate regions, but not in South America. Many species cultivated, mainly as ornamentals (*Cornus* is dogwood). *Davidia involucrata* is the handkerchief tree, named for its large bracts.

Cornales An order comprising 6 families with 51 genera and 590 species. *See* Cornaceae, Curtisiaceae, Grubbiaceae, Hydrangeaceae, Hydrostachyaceae, and Loasaceae.

cornicle (siphuncle) One of a pair of tubes that point to the rear on the underside of the last segment of the abdomen of **Aphididae**. They are sometimes mistaken for **cerci**.

corn salad *See Valerianella.*

corn smut *See Ustilago maydis.*

corn snake *See Pantherophis guttatus.*

corn stunt disease A disease of maize (*Zea mays*) that is caused by ***Spiroplasma*** *kunkelii*. Affected plants are stunted, chlorotic (*see* chlorosis) stripes appear on leaves, and **internodes** are shortened, producing a proliferation of secondary shoots. Ears are small.

Cornu aspersum (brown garden snail, common snail, garden snail) A **snail** with a spiral shell that is yellow to brown, often with darker bands, 20–35 mm high and 25–40 mm wide. The aperture is wide and has a white lip. The snail has a patchy worldwide distribution. It is widely regarded as a garden pest, but it is also edible. It was formerly known as *Helix aspersa*.

corolla All the **petals** of a **flower**.

corona A series of **petal**-like structures that are either outgrowths from the petals or modified **stamens**.

corpus 1. In **angiosperms**, a layer of cells below the **tunica** of the **apical meristem**. **2.** The body of a **pollen grain** that has air-filled sacs or **bladders**.

corrasion *See* abrasion.

corridor farming *See* agroforestry.

Corsiaceae (order **Liliales**) A **monocotyledon** family of small, **perennial herbs** that lack **chlorophyll** and are **saprophytes**. Leaves more or less **distichous**, **simple**, **ovate**. Flowers **zygomorphic**, **bisexual**, with 6 **tepals**, 6 **stamens**, **ovary inferior syncarpous** of 3 **carpels**. Flowers terminal and solitary. Fruit is a **capsule**. There are 3 genera with 30 species occurring mainly in humid forests in southern China, southern South America, New Guinea, Solomon Islands, and northern Australia.

cortex An outer layer of tissue, in plants between the **epidermis** and **vascular bundles**.

cortical canal *See* vallecular canal.

corticated Containing **cortex**.

corticolous Growing on or in **bark**.

Corvidae (crows, choughs, jackdaws, jays, magpies, nutcrackers, ravens, rooks) A family of small to large, black, black and white, and brightly coloured birds with strong legs, large, heavy beaks, and **rictal bristles**. Some have crests and some have long tails. They are the most intelligent of all birds and comparable to primates. They are gregarious and playful. They are omnivorous and distributed worldwide. There are more than 120 species.

Corvus cornix (hooded crow, corbie) A crow (**Corvidae**) that is very similar to the carrion crow (***Corvus corone***), but has a pale grey body, resembling an academic hood. It is omnivorous and occurs in Ireland, Scotland, and throughout central and northern Europe.

Corvus corone (crow, carrion crow) A black bird (**Corvidae**) with a purple or green sheen, black legs and feet, and **feathers** covering the nostrils. They are gregarious and feed on carrion, seeds, invertebrates, and small mammals, and commonly occur in towns. They are distributed throughout Europe.

Corvus monedula (jackdaw, Eurasian jackdaw, western jackdaw) A member of the **Corvidae** that is 340–400 mm long and black with a grey nape and distinctive pale grey irises. It is gregarious, living in groups with a complex social structure, and feeds on urban food waste, invertebrates, and plant material. It inhabits a variety of **habitats**, preferring those with some tall trees, buildings, and open spaces, and occurs throughout much of Eurasia and parts of North Africa.

Corydalis (Dutchman's breeches) *See* Papaveraceae.

Corylus (family **Betulaceae**) A genus of **deciduous shrubs** and small **trees** with **simple**, **serrate** leaves. Flowers **monoecious catkins**, male catkins long and yellow, females very small and concealed in the **bud** with only the red **styles** protruding. Seeds are **nuts** (hazelnut, filbert, cobnut) surrounded by a leafy **involucre**. There are about 15 species occurring in northern temperate regions.

corymb An **inflorescence** that is **racemose**, but with the lower **pedicels** longer than the upper ones, so the inflorescence has a flat or slightly domed top.

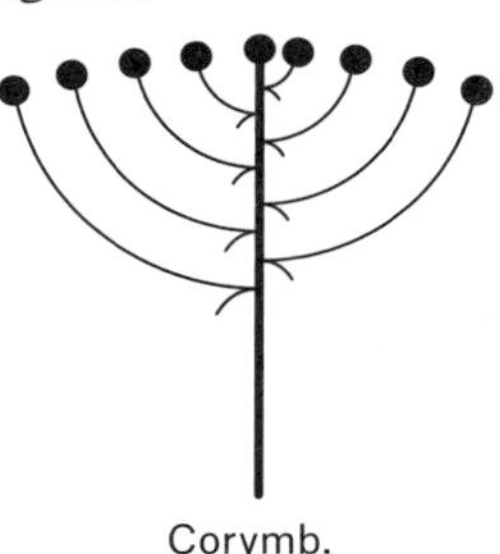

Corymb.

corymbose Resembling a **corymb**.

Corynebacterium A genus of **Actinobacteria** that are Gram-positive (*see* Gram reaction), and in which the cells are straight or slightly curved rods, and **pleomorphic**. They are found in a wide range of **habitats**. Most are harmless, but *C. diphtheriae* causes diphtheria.

coryneform Describes bacterial cells that are club-shaped and **pleomorphic**, like those of ***Corynebacterium***.

Corynocarpaceae (order **Cucurbitales**) A **monogeneric** family (*Corynocarpus*) of tall **trees** and **shrubs** with thick, leathery, **alternate**, **petiolate**, **entire** leaves. Flowers **actinomorphic**, **bisexual**, hypogynous (*see* hypogyny), **pentamerous**, **sepals** and **petals** free and **imbricate**, 5 **stamens** alternating with 5 **petaloid staminodes**, **ovary superior** of 2 **carpels** but 1 **locule**. **Inflorescence** a terminal **panicle** of **cymes**. Fruit is a **drupe**. There are six species occurring from New Guinea to New Zealand, and introduced in Hawaii.

cosmopolitan distribution A worldwide distribution.

Cossidae (carpenter moths, goat moths) A family of nocturnal moths (**Lepidoptera**) with much-reduced mouthparts and **bipectinate** antennae (*see* antenna). Most adults are grey, camouflaged to mimic leaves or **bark**, with wingspans of up to 240 mm and some with narrow wings. Most **caterpillars** bore into wood and pupate inside their tunnels. The larvae of *Cossus cossus* (goat moth) are about 70 cm long and aggressive; if handled they can bite and exude a foul smelling liquid, hence the common name. It is a pest of orchard trees. There are at least 700 species of Cossidae, with a worldwide distribution.

Cossus cossus (goat moth) *See* Cossidae.

costa A ridge or vein.

Costaceae (order **Zingibales**) A family of **perennial herbs** with **rhizomes**, a few **epiphytes** with approximately **elliptical**, **ligulate** leaves in spirals. Flowers **zygomorphic**, 2- to 3-lobed **calyx**, 3-lobed **corolla**, 1 **stamen** often **petaloid** and 5 **staminodes**, **ovary** usually **inferior**, of 3 **carpels** with 3 **locules**. **Inflorescence** a dense, **globose** or **cone**-like **spike**, with persistent, **imbricate bracts**, each supporting 1 or 2 flowers. Fruit usually a **loculicidal capsule**. There are 6 genera with 110 species occurring throughout the tropics.

Cotinus (family **Anacardiaceae**) A genus of **deciduous shrubs** and small **trees** with **alternate**, **simple**, **oval** leaves. Grey-buff flowers borne in terminal **panicles** resembling a cloud of smoke, hence the common name smokebush or smoke tree. There are two species found in the warm temperate Northern Hemisphere. They are widely cultivated for ornament.

Cotoneaster (family **Rosaceae**) A genus of **shrubs**, some with prostrate **habit**, and small **trees** with **dimorphic** shoots, long shoots producing structural growth, short shoots bearing the flowers. Leaves **alternate**, **simple**, **ovate** to **lanceolate**, **entire**. Flowers with 5 **petals**, 10–20 **stamens**, 5 **styles**, solitary or borne in **corymbs**. Fruit is a **pome**. There are about 260 species found throughout temperate regions of the Northern Hemisphere. They are close to hawthorn (*Crataegus*), onto which some species can be grafted (*see* graft), but without thorns. Several are cultivated for ornament.

cotton (*Gossypium*) *See* Malvaceae.

cotton-belt climate A **climate** with dry winters and warm, wet summers, typical of cotton-growing regions in the United States and China.

cottonmouth *See Agkistrodon piscivorus.*

cottontails *See* Leporidae.

cottony grape scale *See Pulvinaria vitis.*

cottony maple scale *See Pulvinaria vitis.*

cottony rot *See Sclerotinia sclerotiorum.*

cottony vine scale *See Pulvinaria vitis.*

cotyl The junction of the **epicotyl** and **hypocotyl**, where **cotyledons** arise.

cotyledon The seed leaf that emerges from a plant **embryo**.

Coulaceae (order **Santalales**) A family of tall, **evergreen trees** with **alternate**, **simple**, **entire**, **petiolate**, **ovate** to **elliptic** leaves. Flowers **actinomorphic**, hypogynous (*see* hypogyny), 3 free or 4–5 or 6–7 fused

sepals and **petals**, **apopetalous** or **sympetalous** at base, 4–5 or 12–20 **stamens**, **ovary superior** of 3 **capsules** and **locules**. **Inflorescence axillary racemes** or **panicles**. Fruit is a **drupe**. There are three genera with three species scattered throughout the tropics.

country park In Britain, an area of countryside within easy reach of an urban population that is set aside for public recreation.

covalent bond A chemical bond in which two atoms share one or more pairs of electrons.

cover The proportion of the ground, usually expressed as a percentage, that is directly beneath the above-ground parts of a plant, i.e. the area covered by a perpendicular projection downward from the extremities of the plant.

cover crop Plants that are grown between crops to avoid leaving the ground bare, thereby reducing **erosion**.

cowbirds *See* Icteridae.

cow parsley (*Anthriscus sylvestris*) *See Anthriscus.*

crab apple *See Malus.*

cranberry *See Vaccinium.*

crane flies *See* Tipulidae.

crane flower *See Strelitzia.*

cranesbill *See Geranium.*

crassula (bar of Sanio) A thickening in the **cell wall** and intercellular material found in pairs associated with the **bordered pits** in the **tracheids** of **gymnosperms**.

Crassulaceae (order **Saxifragales**) A family of mainly **perennial**, **succulent herbs** and small **shrubs**, with one species (*Crassula helmsii*) an **emergent** aquatic. Leaves mainly **entire**, **exstipulate**, in rosettes. Leaf surfaces often waxy and covered with hairs or bristles. Flowers with 3–30 fused or 5 free **sepals**, **petals**, and **carpels**, **ovary superior**. Fruit is a group of **follicles**. There are 34 genera with 1400 species with **cosmopolitan distribution**. Several genera cultivated for ornament (e.g. *Sedum*, stonecrops, *Sempervivum*, houseleeks).

crassulacean acid metabolism *See* CAM pathway.

crawler *See Coccus hesperidium.*

creep The slow movement downhill of soil or other surface material.

crenate Of a leaf margin, scalloped or with rounded teeth.

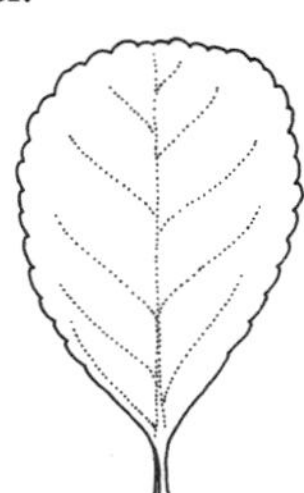

A leaf with a crenate margin.

crenulated With fine notches.

creosote bush (*Larrea tridentata*) *See* Zygophyllaceae.

crepuscular Active at twilight.

crepuscular rays Bands of light that radiate from the position of the Sun when the Sun is low in a sky partially covered by cloud. The rays consist of columns of sunlit air separated by columns of air darkened by cloud shadows. The rays are almost parallel, but appear to diverge because of perspective.

Crescentia (calabash) *See* Bignoniaceae.

Cricetidae (hamsters, lemmings, muskrats, New World rats and mice, voles) A family of rodents (**Rodentia**) that have long bodies, prominent ears and **vibrissae**, and most have long tails. Most are terrestrial, some living in burrows, and others are semi-aquatic. The family includes **carnivores**, **herbivores**, and omnivores. Some species cache food. There are 130 genera and more than 680 species. They occur worldwide except for Antarctica, Australasia, and Malaysia.

Crinivirus A genus of **Closteroviridae** that are **RNA viruses**, which cause a number of plant diseases.

Crinum (family **Amaryllidaceae**) A genus of **perennial, monocotyledon herbs** that grow from **bulbs**, with a **pseudostem** formed from sheathing bases of old leaves. Typically long, strap-like leaves grow from the base. Large, showy flowers with a long **perianth** tube forming a trumpet-shape and borne as **umbels**. There are 60–100 species occurring in America, Africa, southern Asia, and Australia. Many are cultivated for ornament.

Crioceris asparagi (asparagus beetle) *See* Chrysomelidae.

criss-cross inheritance The transmission of a **gene** from father to daughter or mother to son.

crista An infolding of the inner **membrane** of a **mitochondrion** that bears structures involved in the synthesis of ATP (**adenosine triphosphate**).

critical habitat A **habitat** on which an endangered or threatened species depends for its survival.

Crocosmia (family **Iridaceae**) A genus of **evergreen** and **deciduous perennial herbs** that grow from **corms** that form vertical chains, the oldest at the bottom with **contractile roots** that pull the chain deeper into the soil. Leaves **lanceolate, entire**. Red or orange flowers are **hermaphrodite, sessile**, forming a **spike**; terminal flowers may form a **cyme** or **raceme**. Fruit is a **dehiscent capsule**. There are about 13 species, native to South Africa. They are widely cultivated ornamentals, with more than 400 **cultivars**, but their chains of corms allow them to be invasive in some places.

Crocus (family **Iridaceae**) A genus of **herbs** with tufts of narrow, **ensiform, entire** leaves, usually with as white central stripe, that grow from **corms**. Flowers are solitary, goblet-shaped, **actinomorphic** with 6 similar **perianth** segments and a long perianth tube resembling a **pedicel**, typically 3 **stamens, ovary inferior** of 3 **carpels**. Fruit is a **capsule**. There are 90 species occurring in temperate regions of the Old World, especially around the Mediterranean. They are widely cultivated.

Croesus septentrionalis (hazel sawfly, nut sawfly) A sawfly (**Symphyta**) species in which adults are 8–10 mm long with a black head and thorax and black and reddish brown abdomen, and yellow to blue-green larvae with a black head and orange markings, up to 22 mm long. Females lay eggs in leaf veins in early summer and larvae feed along the edges of leaves; if disturbed they thrash violently. They pupate in the soil, spinning a brown **cocoon**. A second generation emerges in late summer and autumn. The larvae can cause serious **defoliation**.

Cronquist, Arthur John (1919–92) An American botanist who developed a scheme for classifying **angiosperms** by dividing them into two classes: Magnoliopsida comprising the **dicotyledons**, and Liliopsida comprising the **monocotyledons**. Within these classes he grouped orders into subclasses.

crop An extension of the oesophagus in birds and insects, used to store food items.

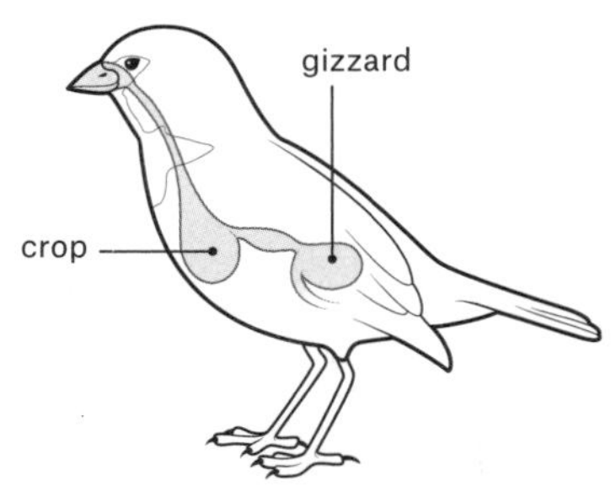

The crop is an extension of the oesophagus used to store food. Food to be digested is broken into small fragments in the gizzard.

crop rotation The growth of a series of different crops in a regular sequence, so that each crop is repeated at regular intervals. This helps prevent the accumulation of species-specific pathogens.

cross-breeding Reproduction by parents of different **genotypes**. In plants, this involves the transfer of **pollen** from the **anthers** of one plant to the **stigma** of another.

crossing over The exchange of **genes** between **homologous chromosomes** during **breakage and reunion**.

Crossomatales An order comprising 7 families with 12 genera and 66 species. *See* Aphloiaceae, Crossosomataceae, Geissolomataceae, Guamatelaceae, Stachyuraceae, Staphyleaceae, and Strasburgeriaceae.

Crossosomataceae (order **Crossomatales**) A family of **deciduous shrubs** with small leaves usually **alternate** or as **fascicles**, usually **entire**, **stipules** small and **caducous**. Flowers **actinomorphic**, **bisexual**, **perigynous**, **hypanthium** short and cup-shaped, 4 or 5 free **sepals**, 4 or 5 **petals**, as many or 2–10 times as many **stamens** as petals, in **whorls**, **ovary superior** of 1–5 (or as much as 9) free **carpels**. **Inflorescence** a solitary flower. Fruit is a **follicle**. There are 4 genera with 12 species found in western North America.

cross-over region The part of a **chromosome** that lies between two **genes** used as markers and where **crossing over** occurs during **recombination**.

cross-pollination The transfer of **pollen** between **flowers** of different **genotypes**, but usually of the same **species**.

Crotalinae (pit vipers) A subfamily of venomous snakes (**Viperidae**) that possess two heat-sensing organs in depressions (pits), between the eyes and nostrils on both sides of the head, which help them in hunting prey. There are 7 genera and 54 species of pit vipers in the Americas and 11 genera and 97 species in the Old World.

Crotalus adamanteus (eastern diamondback rattlesnake) A pit viper (**Crotalinae**) 800 mm–1.8 m long but sometimes larger, with a thick, heavy body and large head. A row of brown diamonds with cream edges run along its back, the body is olive, brown, or very dark, with a paler, banded tail and a well-developed rattle. The snake inhabits dry regions along the coastal lowlands of the southeastern United States. It is **crepuscular** and feeds on small mammals and birds; many of the mammals it eats are pests, so the snake is very beneficial. It will bite only if severely provoked.

Crotalus horridus (timber rattlesnake, banded rattlesnake, canebrake rattlesnake) A species of pit vipers (**Crotalinae**) that grow to 900 mm–1.5 m long, with the males being larger than females. They vary in colour, but all have transverse bands of contrasting colour. They inhabit forests on rocky hillsides and swamps, and up to 60 snakes hibernate together among boulders or in south-facing crevices in cliffs. They hibernate for up to seven months, always returning to the same den to do so. They feed on small mammals. They are not aggressive toward people and bite only in self-defence. They occur throughout the eastern United States, but their distribution is patchy.

crow *See Corvus corone.*

crown rot *See Colletotrichum*; *Erwinia rhapontici*; *Sclerotinia sclerotiorum.*

crows *See* Corvidae.

crozier 1. (fiddlehead) The tightly coiled leaf of a fern, prior to its opening. **2.** The curled tip of a developing **ascus** on the ascogenous (*see* **ascogonium**) **hypha** of the fruit body (*see* fruiting body) of an **ascomycete** fungus.

cruciate (cruciform) Cross-shaped.

Cruciferae *See* Brassicaceae.

cruciform *See* cruciate.

crumb structure A soil structure in which individual **peds** are approximately spherical, or crumb-like.

crust A surface soil layer that is harder and more compact than the soil beneath, and that may be enriched in calcium carbonate, iron oxide, or **silica**.

crustaceous *See* crustose.

crustose (crustaceous) Crust-like. *See also* lichen.

cryergic Describes surface processes controlled by ice, e.g. **frost heaving**, **frost wedging**.

cryic horizon A **soil horizon** that is permanently frozen. The soil temperature has been below 0°C for two or more years in succession.

cryobiosis A type of **cryptobiosis** in which organisms tolerate temperatures below freezing, usually by allowing water to freeze in certain locations and by preventing the growth of large ice crystals.

cryogenic Describes materials or features produced by the action of ice.

cryosols A group of soils that have a **cryic horizon** within 1 m of the surface. Cryosols are a reference soil group in the **World Reference Base for Soil Resources**.

cryosphere That part of the Earth's surface that is permanently frozen. It comprises ice sheets, glaciers, **permafrost**, and areas covered by sea ice for part of the year.

cryoturbation *See* geliturbation.

Cryphonectria parasitica A species of **ascomycete fungi**, formerly known as *Endothia parasitica*, that causes chestnut blight (chestnut canker), a disease that destroyed most American chestnut trees (**Castanea** *dentata*) during the first half of the 20th century. The fungus causes a **canker** on the surface from which its **hyphae** spread in and beneath the **bark**, eventually killing the tree. The fungus originated in Asia and is now widespread.

crypsis (cryptic coloration) Coloration or markings that make an animal difficult to see against its background.

Crypteroniaceae (order **Myrtales**) A family of tall **trees** and **shrubs** with **opposite**, **entire** leaves up to 30 × 18 cm, **stipules caducous**. Flowers **actinomorphic**, **bisexual** or **unisexual** (plants **polygamodioecious**), 4- or 5-merous, 4 or 5 **sepals** inserted on rim of **hypanthium**, **petals** hooded over the **stamens**, as many stamens as sepals or twice as many, **ovary superior** of 2–5 fused **carpels**. Fruit a **capsule**. There are three genera with ten species occurring in Southeast Asia, Malesia, and Sri Lanka.

cryptic coloration *See* crypsis.

cryptic species *See* sibling species.

cryptobiosis A state of **dormancy** into which an organism enters in order to survive a period of adverse environmental conditions, e.g. **anhydrobiosis**, **anoxybiosis**, **cryobiosis**, **osmobiosis**.

cryptogam An **alga**, bryophyte (**Bryophyta**), or pteridophyte (**Pteridophyta**), i.e. a plant that reproduces by **spores**, producing no flowers and no seeds.

cryptomonads (cryptophytes) Unicellular, asymmetric, aquatic, flagellated (*see* flagellum) **eukaryotes**, most of which are photosynthetic and contain **plastids** varying in pigmentation. They acquired **photosynthesis** by **endosymbiosis**, having absorbed an algal **symbiont** at some point in their evolutionary past.

Cryptomycocolacomycetes A class of **Fungi** of the **Pucciniomycotina** that comprises two **monotypic** genera. They do not go through a **yeast** stage in their life cycle and are parasites of **ascomycetes**. They are known only from Central America.

Cryptomyzus galeopsidis (blackcurrant aphid) A pale green aphid (**Aphididae**) that lays eggs along the shoots of blackcurrant and gooseberry **bushes**. These hatch as the leaf **buds** begin to open and before blossom appears. The aphids live on the underside of leaves and although they cause little serious damage they excrete much **honeydew** and affected leaves then become covered in black mould. The first two generations are wingless, subsequent generations are winged. One variety of *C. galeopsidis* remains on blackcurrant all summer, a second variety moves to hemp-nettle (*Galeopsis* spp.) in early summer, and a third variety lives its whole life on redcurrant and whitecurrant plants.

Cryptomyzus ribis (currant aphid, currant blister aphid, red currant blister aphid) A species of pale yellow aphids (**Aphididae**) that feeds on red, white, and blackcurrants. Colonies form in spring and early summer on the underside of leaves. They cause obvious distortion and discoloration of leaves, but otherwise the host seems little affected. In midsummer the aphids migrate to wild plants (in Britain, hedge woundwort [*Stachys sylvatica*]), returning to the currant plant in late summer to lay eggs.

cryptophyte A plant that produces its **perennating buds** below the ground or water surface. It is one of the life form categories described by Christen **Raunkiær**.

cryptozoa Invertebrate animals that live in dark places such as leaf litter and the upper soil, and that are large enough to be visible to the naked eye.

crystallochory Dispersal of seeds or **spores** by glaciers.

Cs *See* cirrostratus.

Ctenolophonaceae (order **Malpighiales**) A **monogeneric** family (*Ctenolophon*) of **evergreen trees** up to 40 m tall with **opposite**, **simple**, **ovate** to **elliptic**, **acuminate**, **entire** leaves and **stipules** between the **petioles**. Flowers **actinomorphic**, **bisexual**, **pentamerous**, 5 **imbricate sepals**, 5 narrow, **caducous petals**, 10 unequal **stamens**, **ovary superior** of 2 **carpels** with 2 **locules**. **Inflorescence** a terminal, sometimes **axillary**, **cymose panicle**. Fruit a woody, ribbed **capsule**. There are three species found in West Africa and Malesia.

Cu **1.** *See* copper. **2.** *See* cumulus.

cuckoo bees *See* Apidae.

cuckoo-spit A protective covering of foam produced from the anus of leafhopper (**Cicadellidae**) **nymphs**.

cucumber mosaic virus (CMV) A **virus** belonging to the genus *Cucumovirus* that is a pathogen of possibly more plant species than any other virus. It occurs worldwide and can be transmitted directly between plants through **sap** and occasionally seed, and indirectly by aphids (**Aphididae**). Infected plants have yellow mottling on the leaves, distorted leaves, and stunted growth.

cucumber yellow virus *See* beet pseudo-yellows virus.

Cucurbitaceae (order **Cucurbitales**) A family of climbing **perennial**, sometimes **annual herbs**, woody **lianas**, a few **shrubs** and **trees**, with **alternate**, **simple** sometimes **ternate** or palmately **compound** leaves with three or more **leaflets**. Usually a single, simple or branched **tendril** arises on each side of the **petiole** base, coils around any suitable support, then coils like a spring, drawing the plant upward. Flowers **actinomorphic**, usually **unisexual** (plants **monoecious** or **dioecious**), usually 5 **sepals** and **petals** at the top of an expanded **hypanthium**, 1–5 **stamens** usually 3 with 2 double. In **pistillate** flowers **ovary inferior** of 3 fused **carpels** with 1 or 2–5 **locules**. **Inflorescence axillary**, solitary or **cyme**, **raceme**, or **panicle**. Fruit a **berry**. There are 97 genera with 960 species of warm temperate and tropical distribution. These are major food plants, yielding cucumbers, squashes, pumpkins, marrows, gourds, melons, courgettes, etc.

Cucurbitales An order comprising 7 families with 129 genera and 2295 species. *See* Anisophylleaceae, Apodanthaceae, Begoniaceae, Coriariaceae, Corynocarpaceae, Cucurbitaceae, Datiscaceae, and Tetramelaceae.

cucurbit yellow stunting disorder virus (CYSDV) A species of ***Crinivirus*** that is transmitted by the whitefly ***Bemisia tabaci*** and causes **chlorosis** in which leaf veins remain green but the remaining leaf tissue turns yellow, and leaves often roll upward and become brittle. The disease occurs predominantly in **Cucurbitaceae.** The virus occurs in Europe, the Middle East, North Africa, and North America.

cull 1. To kill selected individuals of an animal species in a specified are in order to reduce the size of its population. **2.** To kill individual animals that are judged inferior. **3.** The operation of culling animals.

culm A jointed stem found in grasses (**Poaceae**) and sedges (**Cyperaceae**).

cultivar A strain or variety of plant that has been selectively bred for its desirable properties, can be maintained by propagation, and does not exist in the wild.

cultivation Operations that prepare the ground for crop-growing, including ploughing, digging, harrowing, draining, etc.

cultural landscape A landscape that has been modified by people.

cultural services *See* ecosystem services.

culture A population of unicellular organisms maintained experimentally in a nutrient medium.

Cumberland turtle *See Trachemys scripta.*

cumuliform Resembling **cumulus** or **cumulonimbus** cloud.

cumulonimbus (Cb) A genus of low **convective cloud** (*see* cloud classification) that often extends to a great height, with a smooth top marking the limit above which air is unable to rise by **convection.** Seen from below the cloud is dark and menacing due to its depth, which provides ample space for light to be scattered by ice crystals and cloud droplets. Cumulonimbus brings precipitation, often heavy, and is associated with **thunderstorms**, **tropical cyclones**, and **tornadoes.**

cumulus (Cu) A genus of low **convective cloud** (*see* cloud classification) that develops vertically and is billowing and fleecy, often with blue sky visible between individual clouds. Small, scattered, cumulus clouds that form on summer afternoons from moisture evaporated by the warm sunshine are known as fair-weather cumulus.

cuneate Wedge-shaped.

Cunoniaceae (order **Oxalidales**) A family of **evergreen trees**, **shrubs**, and **stranglers** with **opposite**, rarely in **whorls**, leathery, usually **pinnately compound**, **serrate** or **entire** leaves. Flowers **actinomorphic**, **hermaphrodite** or rarely **unisexual** (plants **dioecious**), 3–6 free or up to 10 fused **imbricate** or **valvate sepals** alternating with **petals** (petals absent in some species), 4, 5, 8, 10, or many **stamens**, **ovary superior** of 1 or 2–5 free or fused **carpels** usually with 2 **locules. Inflorescence** a **panicle**, **raceme**, **thyrse**, **capitate**, or occasionally solitary, terminal, **axillary**, occasionally cauliflorous (*see* cauliflory). Fruit is a **capsule.** There are 27 genera with 280 species occurring mainly in temperate and tropical Southern Hemisphere, a few in Africa.

cup fungi *See* Pezizaceae.

cup lichen A **lichen** (*Cladonia* spp.) in which the podetia (*see* podetium) are cup-shaped.

cupola *See* raised bog.

Cupressaceae (order **Pinales**) A family of mostly **evergreen** but three genera

deciduous (*Glyptostrobus, Metasequoia, Taxodium*) coniferous **trees** (cypresses) with leaves that are needle-like on young plants becoming scale-like, arranged spirally or in **decussate** pairs or **whorls**. Plants are **monoecious**, or rarely **dioecious. Cones** are woody or leathery or (***Juniperus***) resembling **berries.** There are 30 genera with 133 species occurring throughout temperate regions, especially in the Northern Hemisphere. Many are important timber trees, others grown for ornament or hedging.

Cupressus (family **Cupressaceae**) A genus of **evergreen trees** (cypress) and large **shrubs** with scale-like leaves in **opposite decussate** pairs on plants more than two years old, needle-like on younger plants. **Cones** are long, **globose** or **ovate**. There are 16–25 species found throughout warm temperate regions. Many grown for ornament, some for timber.

cupulate Cup-shaped.

cupule A sheath that holds and protects the developing fruit (**calybium**) in **Fagaceae**. If the **nut** is single the cupule only partly encloses it (e.g. the acorn cup of oaks). In some species the cupule is scaly and in some, e.g. chestnut (***Castanea***), the scales are modified to form spines that deter **herbivores.**

curare A toxic compound containing **alkaloids**, used as a muscle relaxant and in South America as an arrow poison, that is obtained principally from the **bark** of *Strychnos toxifera* and *Chondrodendron tomentosum. See* Loganiaceae, Menispermaceae.

Curculionidae (weevils, snout beetles) A large family of stout beetles (**Coleoptera**) with toughened and often elaborately sculptured **elytra** and the head produced into a long, narrow **rostrum** with **mandibles** at the tip and a pair of short, jointed, clubbed antennae (*see* antenna) part way along. Weevils range in size from 1 to 50 mm and most are brown or grey, although some are red, green, or black and shiny. Some are flightless, their elytra fused. Larvae are grub-like, most living inside plants or close to their roots. Adults sometimes feed on **nectar** or **pollen**, but adults and larvae feed mainly on plants and some are serious crop pests. There are more than 40,000 species, distributed worldwide.

currant aphid *See Cryptomyzus ribis.*

currant blister aphid *See Cryptomyzus ribis.*

currant clearwing *See Synanthedon tipuliformis.*

currant gall *See* oak-spangle gall.

currant stem aphid *See Rhopalosiphoninus ribesinus.*

current competition Competition that restricts the competitors to smaller **niches** than they would occupy were the other competitors absent.

curry leaf (*Murraya koenigii*) *See Murraya.*

Curtisaceae (order **Cornales**) A **monotypic** family (*Curtisia dentate*) comprising an **evergreen tree** with **opposite**, **ovate** to **elliptical**, **dentate** leaves. Flowers **actinomorphic**, subsessile, 4-merous, **epigynous**, **sepals** and **petals** inconspicuous, **ovary inferior**with 4 **locules.** Fruit is a **drupe.** The tree occurs in southwestern Africa. Its timber is valuable.

Cuscuta (family **Convolvulaceae**) A genus of parasitic climbers (dodder) that have very little or no **chlorophyll.** The thin stems bear leaves reduced to tiny scales and haustoria (*see* haustorium). Small white, cream, yellow, or pink flowers, somewhat **campanulate**, often borne in dense clusters. Fruit is a **berry.** There are 100–170 species with worldwide distribution in temperate and tropical regions.

cushion chamaephyte A compacted **suffruticose chamaephyte**.

cushion plant A plant with leaves held close to the ground, often forming a hummock. It is an **adaptation** to cold, windy conditions.

cuspidate With a sharp, pointed tip.

custard apple *See Annona.*

cutan **1.** *See* clay skins. **2.** *See* cutin.

cutaneous respiration The exchange of gases through pores in the skin, e.g. in **Amphibia**, which secrete a mucus that maintains a moist body surface as an aid to respiration.

cuticle A thin, waxy layer that protects leaves and stems of plants or the **epidermis** of invertebrate animals.

cutin One of the two waxy polymers from which **cuticle** is made; the other polymer is cutan.

cutinization The deposition of **cutin** on leaves or stems.

cutoff high An **anticyclone** that forms in middle latitudes and moves into a higher latitude where it becomes detached from the westerly air flow. It can then cause **blocking**.

cutoff low A **cyclone** that forms in middle latitudes and moves into a lower latitude where it becomes detached from the westerly air flow. It can then cause **blocking**.

cutting-off The process by which a **cyclone** or **anticyclone** that forms in middle latitudes becomes detached from the prevailing westerly air flow. This usually happens in the upper **troposphere** and produces slow-moving **cutoff highs** and **cutoff lows**.

cut weed *See* bladder wrack.

cutworm A moth caterpillar (**Lepidoptera**) that feeds by biting through plant stems, usually at the base or below ground level, and some species also climb plants and feed on foliage. Most are nocturnal. A number of species feed in this way.

cutworms *See Agrotis exclamationis*, Noctuidae.

Cyanistes caeruleus (blue tit) A tit (**Paridae**), about 120 mm long long with a wingspan of 175–200 mm, with bright blue crown, wings, and tail, yellow underside, greenish back, white cheeks, and black eyestripe, bib, and collar. They feed mainly on insects, especially **caterpillars**, and also eat seeds. It inhabits **deciduous** and **mixed woodland** and gardens and is resident throughout temperate and subarctic Europe and western Asia.

cyanobacteria A large and diverse group of **bacteria**, formerly known as blue-green algae, that contain **chlorophyll** *a* and perform **photosynthesis**. They also contain the bluish photosynthetic pigment **phycocyanin**, which gives them their name. **Chloroplasts** are descended from cyanobacteria. Some cyanobacteria are single-celled, others form **filaments**, some form colonies, and some are capable of gliding across a solid surface. Some colonial filamentous forms are able to fix nitrogen (*see* nitrogen fixation). All cyanobacteria are aquatic and found in freshwater and marine environments, in soil, on rocks, and on plants as **epiphytes** or symbionts (*see* symbiosis). Some release toxins. ↗

Cyanocitta cristata (blue jay) A corvid (**Corvidae**) that is 220–300 mm long with a bright blue upper side and pale grey underside and a black collar. They are aggressive to other birds and omnivorous, feeding on fruit, seeds, and nuts, insects, small vertebrates, other small birds, and eggs they take from nests. Like other jays, they cache food for consumption later; because they often fail to return to their food stash, the birds help disperse seeds. They inhabit **mixed woodland**, suburban areas, and city parks, and occur throughout the **Nearctic**. ↗

cyanogenesis An enzymatic reaction (*see* enzyme) that releases hydrogen cyanide (HCN). Such reactions occur widely in **angiosperms**.

cyanogenic Describes a plant that releases hydrogen cyanide (HCN) when cut or bruised.

cyanophyte A cyanobacterium (**cyanobacteria**).

cyathia *See* Euphorbiaceae.

Cycadaceae (order **Cycadales**) A **monogeneric** family (*Cycas*) of woody, unbranched or sparsely branching **trees** and **shrubs** with thick stems and **alternate**, **pinnately compound**, frond-like leaves in spiral clusters at the top of the stem, giving cycads a somewhat palm-like appearance. The plants are **dioecious**, females producing **ovules** with two to eight seeds naked on the **petioles** of reduced leaves. Males **cones** have many scales. There are 91 species, found in equatorial regions of Africa, Asia, Australia, and Polynesia. *Cycas revoluta* is cultivated as the sago palm or king sago palm.

Cycadales An order that comprises the cycads, with 3 families of 289 species. *See* Cycadaceae, Stangeriaceae, and Zamiaceae.

Cyclamen (family **Primulaceae**) A genus of **perennial herbs** with leaves and flowers that grow as rosettes from large **tubers**. Leaves are shiny, **simple**, each growing from the tuber on its own **petiole**, but varying in shape according to the species. Leaves also vary in colour with many variegated. Flowers have 5 **sepals** and upswept **petals**, with large, **reflexed corolla** lobes, and grow singly on long, leafless stalks. There are 23 species occurring in Europe, around the Mediterranean, and to Iran, with one species in Somalia.

Cyclanthaceae (order **Pandanales**) A **monocotyledon** family of **perennial epiphytes**, **lianas**, and **herbs** with **rhizomes**. Leaves **distichous** or in spirals, with sheathing bases, usually with **petioles**, rarely palm- or fan-like. Flowers **unisexual** (plants **monoecious**). **Inflorescence** a terminal or **axillary spadix**, unbranched and subtended by 3–4 (or 2–11) **bracts** or **spathes**, the crowded flowers in a spiral along the spadix, or in groups with 1 female flower surrounded by 4 males, or individual flowers not discernable. There are 12 genera with 225 species occurring in Central and tropical South America. *Carludovica palmata* is the panama hat plant with petiole and leaf fibres from which the hats are woven.

cyclic AMP (c-AMP) **Adenosine** 3′:5′-cyclic monophosphate, formed from **adenosine triphosphate** (ATP), acts as a mediator (second messenger) in the activity of some **hormones**, in **gene** regulation, and as an **enzyme** activator in animals, **slime moulds**, and **Bacteria**. It has been detected in vascular plants (**Tracheophyta**) but its function remains unknown.

cyclic photophosphorylation In **photosynthesis**, **photophosphorylation** in which electrons move along an **electron-transport chain** from P_{700}, then return to P_{700}, which is the specialized form of **chlorophyll** *a* involved in **photosystem I**, ready to re-enter the electron-transport chain, i.e. the electrons are recycled.

cyclodiene insecticides A group of **organochlorine insecticides**, e.g. **aldrin**, **dieldrin**, **endosulfan**, **endrin**, and **heptachlor**, all of which were banned in the 1970s due to their toxicity, persistence, and tendency to bioaccumulate (*see* bioaccumulation).

cyclogenesis The sequence of events on the **polar front** that lead to the formation of a **cyclone**.

cyclolysis The weakening and eventual disappearance of the **cyclonic** circulation of air as a family of **cyclones** dissipate and high pressure comes to dominate.

cyclone 1. (depression) An area of low **atmospheric pressure** that forms at the crest of a **frontal wave. 2.** A **tropical cyclone** that develops in the northern Indian Ocean or Bay of Bengal.

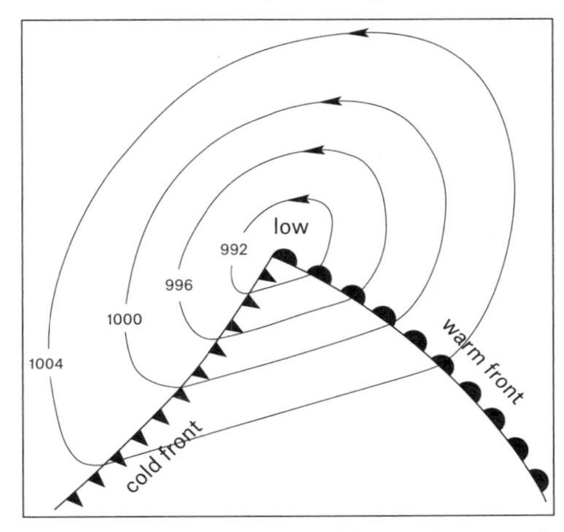

The cyclone (low, depression) lies at the crest of the frontal wave. The arrows show the direction of the geostrophic wind. The lines are isobars, labelled with the pressures in millibars.

cyclone family (depression family) A series of **cyclones** that form beneath waves in the **jet stream** and travel from west to east, carried by the prevailing westerly air flow in middle latitudes. They bring prolonged periods with grey skies and precipitation, interrupted by the **ridges** between one cyclone and the next.

cyclonic Describes the direction in which air moves around a **cyclone** or **trough.** This is anticlockwise in the Northern Hemisphere and clockwise in the Southern Hemisphere.

cyclonic rain Precipitation that is associated with a **cyclone.**

cyclosis The circulation of **protoplasm** within a cell.

Cydia nigricana (pea moth) A small, drab, brown tortrix moth **(Tortricidae)** with black and white bars on its forewings and a wingspan up to 16 mm. It pupates in the soil, emerging in early summer. Females lay eggs on the underside of leaves and on **petioles**, stems, and flowers. The creamy-white **caterpillars** with dark spots, up to 14 mm long, emerge after seven to ten days and feed inside pea pods, eating their way out of the pods in late summer. It occurs throughout Europe.

Cydia pomonella (codling moth) A small, grey tortrix moth **(Tortricidae)** with copper-coloured wing stripes and a wingspan of about 17 mm. The moths overwinter as **pupae** emerging in spring to mate and lay eggs on the leaves and early fruit of pear, walnut, but especially apple trees. Larvae commence feeding immediately on hatching, tunnelling into the fruit to eat the seeds, leaving the fruit after about three weeks to pupate. In North America there are usually two, sometimes three generations a year. The larva is often called an apple worm, apple maggot, or worm in the apple, although it is not related to true worms. The moth is native to Europe but now occurs throughout the world.

cyme (determinate thyrse) A **sympodial inflorescence** in which each lateral **branch** bears a terminal flower. The oldest flower is at the tip of the main **axis.** In a compound cyme **(dichasium)** the branches bear cymes. In a scorpioid cyme **(monochasium)** there are branches on only one side of the main axis, each branch bearing a compound cyme, so the inflorescence forms a coil. *See* dichasium, monochasium.

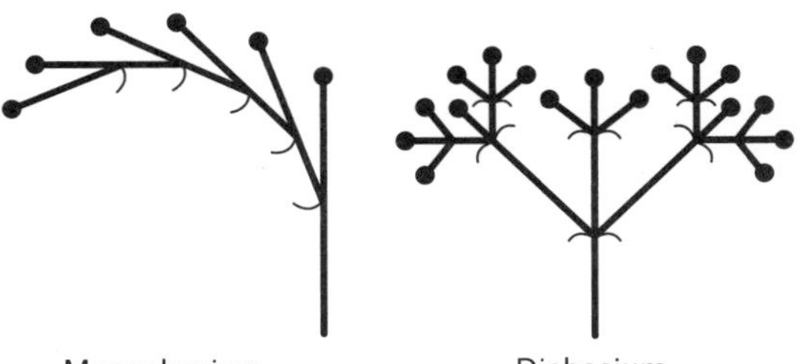

Monochasium. Dichasium.

Cymodoceaceae (order **Alismatales**) A **monocotyledon** family of sea grasses, with **rhizomes**. Leaves **distichous**, **linear**, or **terete** with a persistent sheathing base. Flowers **unisexual** (plants **monoecious** or **dioecious**), **apetalous**, male **pedunculate** or nearly **sessile**, 2 **stamens** fused laterally, female sessile with 2 free **carpels**. **Inflorescence** in **cymose** groups enclosed by **bracts**. Fruit an **achene** or **drupelet**. There are 5 genera with 16 species found in tropical and warm temperate seas.

cymose Resembling a **cyme**.

Cynipidae (cynipids) A family of wasps (**Apocrita**) most of which belong to the subfamily Cynipinae. These form **galls** on a variety of plants but especially oak (***Quercus***). Most adults are hump-backed, with the rear segments of the abdomen tucked beneath the body. Galls are fully enclosed and may hold one or several larvae. Spring galls on leaves, flowers, and **buds** produce short-lived adults that lay eggs which hatch into the larvae that produce the harder autumn galls where the larvae may spend one or two winters before emerging. There are about 1300 species with a worldwide distribution. Other cynipids are parasites of flies (**Diptera**), hyperparasites of **Braconidae**, or inquilines (*see* inquilism) of other gall wasps.

cynipids *See* Cynipidae.

Cynomoriaceae (order **Saxifragales**) A **monogeneric** family (*Cynomorium*) of **obligate** root parasites. They are reddish brown to purple, with an underground **rhizome** with many haustoria (*see* haustorium) with which they penetrate the roots of their many hosts. Only the **clavate**, **spike**-like **inflorescence** of tightly packed flowers appears above ground. Flowers **unisexual** and **bisexual** (plants **polygamomonoecious**) with 1–8 fused or 4–5 free **sepal**-like segments, **staminate** flowers with 1 **stamen**, **pistillate** flowers with **inferior ovary** of 1 **carpel**. Fruit is a **nut**. There are two species occurring from the Mediterranean region to Central Asia. The plants have traditional medical uses.

Cyperaceae (order **Poales**) A **monocotyledon** family mainly of **caespitose**, **perennial herbs** (sedges) resembling grasses, with solid, **trigonous** stems and **rhizomes**, with some **annual** herbs, **shrubs**, and **lianas**. Some have **corms** or **tubers**. Leaves usually in 3, rarely 2 ranks. **Inflorescence** terminal **spikelets** consisting of 1 to many **glumes** in spirals or **distichous**, supporting small **bisexual** or **unisexual** flowers (plants usually **monoecious**), **perianth** of 3–6 scales, hairs, or bristles, or absent, 3 **stamens**, **ovary superior** of 2 or 3 **carpels** with 1 **locule**. Fruit is a **utricle**. There are 98 genera with 5430 species with worldwide distribution. Corms of ***Eleocharis*** *dulcis* are water chestnuts. *Cyperus papyrus* (papyrus sedge, paper reed, Indian matting plant, Nile grass) was used in ancient times to make papyrus paper and its stems to make boats; it is now cultivated for ornament but is almost extinct in the wild.

cyphella 1. A small pore surrounded by a white ring on the lower surface of the **thallus** of some **lichens**. **2.** a genus of **Fungi** (Cyphellaceae).

cypress *See Chamaecyparis*, Cupressaceae, *Cupressus*.

cypsela A single-seeded fruit that is derived from a **unilocular**, **inferior ovary**.

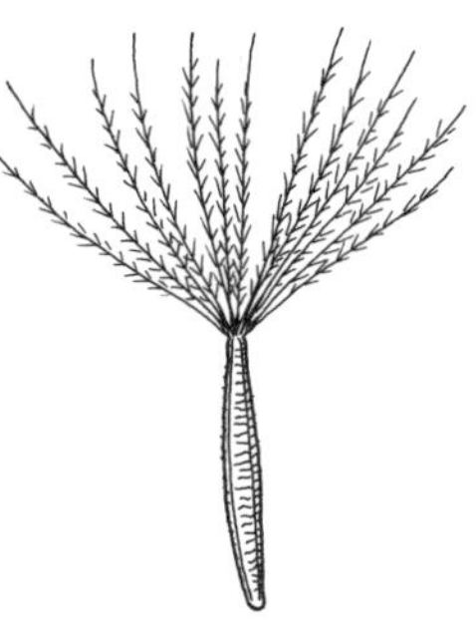

Cypsela. A single-seeded fruit.

Cyrillaceae (order **Ericales**) A family of **evergreen** with some **deciduous trees** and **shrubs**, often with prominent, black **collets**, and spirally arranged, **simple**, **entire**, **exstipulate** leaves with

petioles. Flowers **actinomorphic**, **hermaphrodite**, with 5 free or partly fused persistent, **imbricate sepals**, 5 imbricate **petals**, 5 **stamens** or 10 in 2 **whorls**, **ovary superior** of 2–4 or 3–5 fused **carpels**, each with 1 **locule**. **Inflorescence** a terminal or **axillary raceme**. Fruit is a **capsule**. There are two genera with two species, occurring from the southern United States to northern South America. Some grown for ornament.

CYSDV *See* cucurbit yellow stunting disorder virus.

cyst 1. A resting cell in some **Bacteria** and other unicellular organisms. **2.** A closed sac.

cyst-forming nematodes Sedentary **endoparasite** nematodes (**Nematoda**) that attack plant roots and in which the female retains most of her eggs inside her body until she breaks through the root surface, when she dies and her body hardens to form a cyst, protecting the eggs. *Compare* root knot nematodes.

cyst nematodes *See* Heteroderidae.

cystolith A stalk-like deposit of amorphous calcium carbonate ($CaCO_3$) and silica (SiO_2) that forms in specialized leaf epidermal cells of certain plants.

cytidine The **nucleotide** formed by **cytosine** linked to **ribose sugar**.

Cytinaceae (order **Malvales**) A family of endoparasites (*see* parasitism) that lack **chlorophyll**; the plant is an **endophyte** resembling a fungal **mycelium**. There are no stems. Leaves reduced to scales around the flowers, which are borne in **racemes**, sometimes **capitate**. Flowers **unisexual** (plants **dioecious** or **monoecious**), **perianth** tubular with 4–9 **imbricate** lobes, often brightly coloured, **staminate** flowers with 8–20 **stamens**, **pistillate** flowers with **inferior ovary** with 1 **locule**. Fruit a **berry**. There are two genera with two species occurring in Mexico, the Mediterranean region, South Africa, and Madagascar.

cytochrome A member of a class of **proteins** containing iron (haem proteins) that function as **electron carriers** in many **redox reactions**, including cell **respiration** and **photosynthesis**.

cytochrome oxidase An **enzyme** containing **copper** and iron that reduces oxygen to water and catalyzes the final reaction in **oxidative phosphorylation**.

cytogenetics The combined study of **cytology** and **genetics**, i.e. the study of heredity at the cellular level and especially the study of **chromosomes**.

cytokinesis The separating of the constituents of the **cytoplasm** during cell division. In plants it usually begins in early **telophase** with the formation of a **cell plate** that fuses with the **cell wall**, dividing the cell in two.

cytokinin A class of plant **hormones** that promote **cytokinesis** and lateral growth in shoots and roots.

cytology The study of cells, including their structure and function.

***Cytophaga-Flavobacterium* group** (*Cytophaga-Flexibacter-Bacteriodes*, *Bacteriodetes*) Two genera of heterotrophic (*see* heterotroph), Gram-negative (*see* Gram reaction), rod-shaped **Bacteria** that possess gliding motility. Most are **aerobes**, but some are **facultative anaerobes**. They occur worldwide in soils and are important in decomposing organic matter; they can digest **cellulose** and a wide range of other substances.

Cytophaga-Flexibacter-Bacteriodes *See Cytophaga-Flavobacterium* group.

Cytophagales An order of Gram-negative (*see* Gram reaction), rod-shaped and often **pleomorphic**, non-**motile** or motile by gliding **Bacteria** that are often yellow, orange, or red. Some are **aerobes** others

obligate or **facultative anaerobes**. They are free-living in soil and aquatic **habitats** where they decompose organic matter; a few are pathogens if animals.

C

cytoplasm The gel-like content of a cell enclosed by the **plasma membrane** but excluding the **nucleus**.

cytoplasmic-genetic male sterility *See* male sterility.

cytoplasmic inheritance The inheritance of **characters** through **genes** present in **organelles** outside the **nucleus** of the cell, e.g. viruses, mitochondria (*see* mitochondrion), and **plastids**.

cytoplasmic male sterility *See* male sterility.

cytoplasmic streaming (streaming) The continuous movement of **cytoplasm** and **organelles** within the cell and between connected cells.

cytosine A **pyrimidine base** present in both **DNA** and **RNA**.

cytosis The movement of substances into or out of a cell. *See* endocytosis, exocytosis, phagocytosis, pinocytosis, transcytosis.

cytoskeleton The framework or scaffolding found throughout the **cytoplasm** of all cells that is involved in cell motility, **cytokinesis**, and the arrangement of **organelles**.

cytosol The liquid component of **cytoplasm**.

D

2,4-D (2,4-dichlorophenoxyacetic acid) An **auxin**-type **herbicide** that is used against **broad-leaved** weeds. It acts by causing uncontrolled growth. It is of low toxicity to mammals.

2,4-dichlorophenoxyacetic acid *See* 2,4-D.

dabberlocks (badderlocks, winged kelp) The edible seaweed *Alaria esculenta* that has long fronds which can be eaten raw or cooked.

Dactylis (family **Poaceae**) A **monotypic** genus (*D. glomerata*, cock's-foot) of **perennial** pasture grasses with an erect **culm** 15–140 cm tall, forming dense tussocks, leaves 20–50 cm long. **Inflorescence** a contracted **panicle** of solitary **spikelets**, each of 2–5 fertile **florets**. **Glumes** shorter than spikelets and persistent. Florets compressed laterally, keeled (*see* keel), fertile **lemma obovate**, **ovary glabrous**. The grass occurs in Europe, Asia, and northern Africa.

daddy-long-legs *See* Tipulidae.

daffodil (*Narcissus*) *See* Amaryllidaceae.

Dahlia (order **Asteraceae**) A genus of **perennial herbs** with tuberous roots (*see* tuber) and flowers borne singly, one on each leafy stem, growing 30 cm–2.4 m tall, and up to 8 m in the wild. The brightly coloured but unscented flowers are very variable due to *Dahlia* being octoploid (*see* polyploidy). There are at least 36 species, native to Mexico, Central America, and Colombia, with many **hybrids** that are cultivated for ornament.

daisy-bush *See Olearia.*

Daktulosphaira vitifoliae (grape phylloxera) *See* Phylloxeridae.

dalapon A selective **herbicide** and plant **growth regulator** that is used to control **annual** and **perennial** grasses, especially in potatoes, carrots, asparagus, and fruits. It is moderately toxic when inhaled and can cause skin irritation.

DALR *See* dry adiabatic lapse rate.

daminozide (Alar, aminozide, B-Nine, B-995, Dazide, Kylar) A plant **growth regulator** (Alar is the trade name) that was formerly applied to regulate the growth and improve the colour of apples, and make them easier to harvest. The manufacturer withdrew the product for use on food crops in November 1989 following public concern over studies suggesting it could be harmful to children if ingested in large amounts. It may now be used only on crops not intended for human consumption.

damping-off *See Rhizoctonia solani.*

damsel bugs *See* Nabidae.

damson-hop aphid *See Phorodon humuli.*

dance language (bee dance) Ritualized behaviour by which a honeybee forager returning to the hive having located a previously unknown source of food communicates the direction and distance of the source to other worker bees, thereby recruiting them to forage in that location. When the bees reach the general location odours guide them to the individual plants. There are two dances, both performed inside the hive on the vertical wax comb. The round dance indicates the existence and distance to a food source within 50 m of the hive and after completing its dance the successful forager gives food (**pollen**) to the audience, to demonstrate its quality. If the source is 50–150 m from the hive the dance is modified to indicate direction. The waggle or wagtail dance communicates the distance and direction to the source. The bee runs in a figure-eight circuit while wagging its abdomen from side to side and buzzing its wings.

dancing devil *See* dust whirl.

dandelion (*Taraxacum officinale*) *See Taraxacum.*

Daphniphyllaceae (order **Saxifragales**) A **monogeneric** family (*Daphniphyllum*) of **evergreen trees** and **shrubs** with **alternate** leaves that appear **verticillate**. Leaves **simple**, **entire**, **exstipulate**, often **glaucous** on underside. Flowers **actinomorphic**, **unisexual** (plants **dioecious**), 2 free or 3–6 fused **sepals**, **petals** absent, **staminate** flowers with 5–12 free or up to 24 fused **stamens**, sometimes with **staminodes**, **pistillate** flowers with **superior ovary** of 2 occasionally 4 **carpels** of 2(–4) **locules**. **Inflorescence** an **axillary raceme**. Fruit is a **drupe**. There are 10 species occurring in eastern Asia and Malesia.

Darcy, Henry Philibert Gaspard (1803–58) A French engineer who studied the flow of fluids. He invented the pitot tube and discovered **Darcy's law**. The unit of intrinsic **permeability** is named for him: 1 darcy = $0.987 \times 10^{-12} m^2$.

Darcy's law A description of the relationships among the factors affecting the flow of **groundwater**, formulated in 1856 by Henry **Darcy**. The law states that $Q = kIA$, where Q is the rate of groundwater flow, k is the **permeability** of the rock or soil, I is the gradient down which the groundwater is flowing, and A is the cross-sectional area through which the groundwater is flowing.

dark-eyed junco *See Junco hyemalis.*

dark reactions *See* light-independent stage.

dart leader A small **lightning stroke** that travels along the **lightning channel**, ionizing it, and that is followed by a major lightning flash.

Darwin, Charles Robert (1809–92) The English naturalist who proposed **natural selection** as the mechanism by which **species** evolve, first in 1858 in a joint presentation with Alfred Russel **Wallace** to the Linnean Society, 'On the tendency of species to form varieties: and on the perpetuation of varieties and species by natural means of selection' (*Journal of the Linnean Society*, vol. 3), and in 1859 in his book *On the Origin of Species by Means of Natural Selection.*

Darwin, Erasmus (1731–1802) An English physician, naturalist, philosopher, and botanist, who, in his two-volume *Zoonomia; or the Laws of Organic Life* (1794) proposed a theory of evolution. He translated the works of Carolus **Linnaeus** into English, devising many of the plant names still used today, and popularized Linnaeus's work in his long poem *The Loves of the Plants*; he also wrote *The Economy of Vegetation* and these were published together as *The Botanic Garden* (1791). He was the grandfather of Charles **Darwin**.

Dasineura mali A small fly (**Diptera**) that lays eggs in folds in immature apple leaves. These hatch into very small pink or orange larvae which feed on the leaves, causing the margins to curl tightly, eventually becoming discoloured and falling from the tree. This also stunts terminal shoots. *Dasineura mali* (apple leaf midge, apple leaf curling midge) attacks apple trees, *D. pyri* (pear leaf midge, pear leaf curling midge) attacks pear trees.

dasycladacean algae A family (Dasycladaceae) of large unicellular **green algae** (phylum **Chlorophyta**), most with some **calcification**, in which the **thallus** is radially symmetrical, with an erect, branching **axis**. They are known as **fossils** from the lower Palaeozoic era (about 500 million years ago). Living dasycladaceans occur in tropical and subtropical seas, with a few in warm temperate waters from the surface to depths of about 30 m.

Dasypogonaceae A **monocotyledon** family of **perennial shrubs** and **trees**, some of the trees with **stilt roots**, that has not been placed in any order. Leaves in spirals. Flowers **bisexual** of 2 **whorls** of **tepals**, 6 **stamens**, 1 **carpel** and **locule**. Fruit **indehiscent** and enclosed in the **perianth**. There are 16 species found in southwestern and southern Australia.

date palm (*Phoenix dactylifera*) *See* African–Indian desert floral region.

Datiscaceae (order **Cucurbitales**) A **monogeneric** family (*Datisca*) of robust, **perennial herbs**, with **imparipinnate** or deeply **pinnatifid** leaves. Flowers **actinomorphic**, **unisexual** (plants **dioecious** or **androdioecious**), **staminate** flowers with short **calyx** tube, no **petals**, 6–15 free or up to 25 fused **stamens. Pistillate** flowers with **inferior ovary** of 3–5 fused **carpels** with 1 **locule. Inflorescence** a **spike.** Fruit a **capsule.** There are two species found in western North America and from Crete to India.

Datura (family **Solanaceae**) A genus of **annual** and short-lived **perennial herbs** that can grow 2 m tall, with **alternate**, but often **opposite** near the top of the stem, lobed or **serrate**, with **petioles**, usually **glabrous** leaves. Flowers solitary, trumpet-shaped, the **calyx** sharply toothed, 5 fused **petals**, 5 **stamens**, **ovary** with 4 **locules.** Flowers held in the **axils.** Fruit is a spiny **capsule.** There are nine species occurring in America. Several are cultivated for ornament, known as angel's trumpets, moonflowers, and thorn-apples, and some have medical uses. All are poisonous. *Datura stramonium* is jimsonweed.

Daucus (family **Apiaceae**) A genus of mostly **biennial** but also **annual** and **perennial herbs** with bristly stems and **alternate**, 2–3 **pinnatisect** leaves with sheathing bases, and tiny flowers forming **umbels.** There are 25 species with worldwide distribution. **Cultivars** of *D. carota*, wild carrot, are grown for their edible **taproots.**

daughter cells The two cells resulting from the division of a single cell by **mitosis.**

daughter nuclei The two nuclei that result from the division of a single **cell nucleus** by **mitosis.**

Davidia involucrata (handkerchief tree) *See* Cornaceae.

day degrees The amount by which the average daily temperature departs from a specified datum level, e.g. the minimum temperature required to grow a particular crop. It is calculated as the number of days on which the temperature is above or below the datum multiplied by the number of degrees (plus or minus) by which it deviates.

day-neutral plant A plant in which the time of flowering is not determined by **photoperiod**, e.g. dandelion (***Taraxacum*** *officinale*).

Dazide *See* daminozide.

DCMU *See* diuron.

DDT (dichlorodiphenyltrichloroethane) An **organochlorine insecticide** that is persistent and bioaccumulates (*see* bioaccumulation). It was used very extensively in the 1940s and 1950s in agriculture and to control malarial mosquitoes, but was implicated in reproductive failures in several birds of prey and was banned in the United States in 1973 and soon after that in most other countries.

D

DDVP *See* dichlorvos.

deamination The removal of an **amino group** from an organic compound.

dear enemy recognition The different response that an animal defending its territory makes to an intruder it recognizes, compared to its response to a stranger. The difference develops as a new arrival establishes its own territory and frequently encounters its neighbours in the border areas. At first the threats are vigorous, but with time the neighbours become familiar to each other and the border threats are shorter and less intimidating.

death cap *See Amanita phalloides*, phallatoxins.

death cup *See Amanita phalloides.*

debris Rubble, consisting of rocks of various sizes mixed with other material.

debris dam (landslide dam) Material that blocks the flow of a river as a result of **mass wasting**.

debris flow Large rocks mixed with mud and water that move slowly down a gradient by the force of gravity.

debris slide A shallow landslide involving rock **debris**.

decarboxylase An **enzyme** that facilitates the removal of carbon dioxide from the **carboxyl** group of an organic compound.

deceiver *See Laccaria laccata.*

deciduous Describes parts of an organism (e.g. leaves, deer **antlers**) that are shed seasonally.

deciduous summer forest The most extensive type of forest in temperate regions of the Northern Hemisphere, but absent from the Southern Hemisphere, dominated by **broad-leaved** trees that shed their leaves in winter.

declinate Curving downward.

decollate Tapering to a blunt end.

decollate snail *See Rumina decollata.*

decomposer *See* microconsumer.

decorticate To remove **bark**; describes the **branch** or stem of a woody plant from which the bark has been stripped.

decumbent Growing along the ground with the tip curving upward.

decurrent Describes fungal **gills** with edges that are attached to the **stipe** and extend down it.

decussate Having a pair of leaves arising at each **node**, with each pair at right angles to the pairs above and below.

deepening A decrease in the **atmospheric pressure** at the centre of a **cyclone**.

deep percolation The downward movement of water through the **soil profile** to a level beyond the reach of plant roots.

deep soil *See* effective soil depth.

deer *See* Cervidae.

deflation The removal of surface material by the action of wind.

deflation hollow An enclosed depression caused by wind **erosion**.

deflected climax A **climax** that is maintained by browsing, grazing, mowing, or other interventions.

deflocculation The breaking down of **aggregates** into individual soil particles.

defoliation The removal of the leaves from a plant.

deforestation (disafforestation) The permanent clearing of an area of forest or woodland.

degenerate code A term sometimes applied to the **genetic code** because more than one **codon** codes for most **amino acids**.

Degeneriaceae (order **Magnoliales**) A **monogeneric** family (*Degenaria*) of large **trees** with **alternate**, **simple**, **entire**, **petiolate**, **exstipulate** leaves. The solitary flowers are **actinomorphic**, **hermaphrodite**, with 3 free **sepals**, 12–13(–18) free **petals** in 3–5 **whorls**, many **stamens** and **staminodes**, **ovary superior** of 1 **carpel**. Fruit is large and **indehiscent follicle**. There are two species **endemic** to Fiji.

dehiscent Splitting or bursting open when mature.

dehydrogenase An **enzyme** that facilitates the removal of hydrogen.

deimatic behaviour A threat by which an animal aims to deter predators. The threat may precede an attack, as when a skunk stands on its front legs in preparation for spraying, or bluff, as when a toad inflates itself.

delayed flow The movement into a river of water that has flowed through underground channels or as **groundwater**. *Compare* surface flow.

deletion The loss of a section, of any length from a single **nucleotide** to an entire **gene**, from a **chromosome**. If the loss occurs at the end of the chromosome it is called a terminal deletion, if it occurs elsewhere it is an intercalary deletion.

Delia antiqua (onion fly) A small fly (**Diptera**) resembling a housefly that lays its eggs on the shoots, leaves, and **bulbs** and in the adjacent soil of its host plant. The maggot larvae feed on bulbous plants, especially onions, shallots, leeks, salad onions, and garlic, as well as ornamental ***Allium*** plants. The flies pupate in the soil and overwinter as pupae. It occurs throughout the temperate Northern Hemisphere.

Delia echinata (spinach stem fly) A fly (**Diptera**) that lays eggs on the upper surface of leaves. The maggots mine the leaves on which they hatch then excavate a tunnel through the stem to reach another leaf. They feed on a wide range of plants and are widespread throughout Europe, also occurring in Japan and North America.

Delia radicum (cabbage fly, cabbage root fly, root fly, turnip fly) A species of flies (Anthomyiidae) that are grey, about 25 mm long, and resemble a housefly. They overwinter in the soil as pupae and emerge in spring to feed on **nectar**. They mate and lay eggs close to brassica plants. The eggs hatch after about six days and the larvae (cabbage maggots, root maggots) feed on the roots of the plant, with up to 300 larvae on a single plant, inhibiting plant growth and causing leaves to acquire a bluish colour before withering. The flies occur throughout Europe. They produce three generations a year, but it is the first that causes most damage.

Delichon urbicum (house martin) A migratory bird which breeds throughout temperate Eurasia and spends the winter in sub-Saharan Africa. It is about 130 mm long with a wingspan of 260–290 mm and has a steel-blue back, white rump, and white underside. It inhabits open country with low vegetation and is often seen around dwellings, and feeds on insects that it catches in flight.

Delphinium (family **Ranunculaceae**) A genus of **annual** and **perennial herbs** with **palmate** leaves with 3–7 **serrate**, pointed lobes. Flowers with usually 5 **petaloid sepals**, the posterior one with a spur giving the plant its name (larkspur), forming a hollow socket inside which are 4 inconspicuous **petals**. Fruit is a **follicle**. There are about 300 species occurring throughout the Northern Hemisphere. Many are cultivated for ornament.

Deltaproteobacteria A class of **Proteobacteria**, most of which are **aerobes**, but that contains a number of **anaerobes**, including most of the species that reduce sulphur and sulphate. All are Gram-negative (*see* Gram reaction).

D

dematiaceous Dark-coloured, especially of mould **Fungi**.

deme A group of interbreeding organisms in a particular place and possessing distinct **genetic** or cytological (*see* cytology) **characters**.

denaturation Altering the biological activity of a **protein** or **nucleic acid** by changing its structure, e.g. by changing the temperature, solvent, or **pH**, but without breaking the bonds between **amino acids** or **nucleotides**. The effect may be permanent or reversible.

dendritic Branched, like a tree.

dendritic drainage A **drainage pattern** resembling a tree, with branches feeding into a main channel.

Dendrobates *See* batrachotoxin.

Dendrobium (family **Orchidaceae**) A genus of **sympodial** orchids, most of which are **epiphytes**, with a few **lithophytes**. They produce pseudobulbs that in some species grow several metres long, from the bases of which shoots emerge, in spring or less commonly in autumn, followed by new roots. Leaves are **ovate**, usually **alternate**, **inflorescences axillary**, insignificant in some species but in others up to 1 m long. There are about 1200 species occurring throughout much of southern, southeast, and eastern Asia, and the Pacific Islands to New Zealand. Many are cultivated and highly prized.

Dendrocopos major (great spotted woodpecker) A species of woodpeckers (Picidae) that are 230–260 mm long with a wingspan of 380–440 mm, glossy black and white, white on the sides of the face and neck, a large white patch on the shoulders, crimson on the underside of the tail, and males with a crimson patch on the nape. It inhabits woodlands and parklands, sometimes visits garden feeders, and feeds on insects and seeds. It occurs throughout Europe and northern Asia.

Dendrodrilus rubidus (bank worm, jumbo red worm, jumper, jumping red wiggler, pink worm, red trout worm, red wiggler worm, red wiggler, trout worm wiggler) A species of brightly coloured, **epigeic** earthworms (**Annelida**) less than 100 mm long that inhabits the leaf litter and upper soil in coniferous forests. It is native to Europe, where it contributes to the breakdown and recycling of organic material, but has spread to parts of North and South America, Australia, Russia, and some sub-Antarctic islands. It is invasive and harmful in some North American soils that lack native earthworms.

dendroid Shaped like a tree.

denitrification The conversion of nitrate (NO_3) or nitrite (NO_2) to gaseous nitrogen (N_2) or nitrous oxide (N_2O) by **denitrifying bacteria**.

denitrifying bacteria Bacteria that carry out **denitrification**. They occur worldwide in soils and aquatic **habitats**, and include such species as *Thiobacillus denitrificans*, *Micrococcus denitrificans*, *Paracoccus denitrificans*, and several *Pseudomonas* species.

density dependence The limit to the size of a population that is due to the size the population has attained. It occurs because of environmental resistance as **competition** for resources and predation increase. It may manifest as increasing mortality or decreasing **fecundity**. *Compare* density independence.

density independence The limit to the size of a population that is due to factors unrelated to the size the population has attained. The population continues to grow until an environmental factor, e.g. the onset of winter, removes resources or renders them unavailable and the population crashes. *See* J-shaped growth curve; *compare* density dependence.

dentate Bearing teeth or serrations.

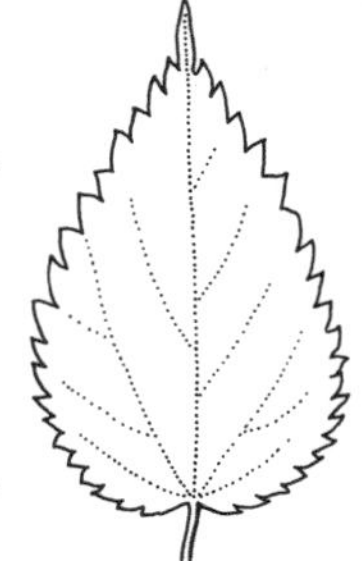

A leaf with a dentate margin.

denticulate Bearing very small teeth or serrations.

denudation The removal of surface material as a result of **erosion** or **weathering** and transport, exposing bare rock.

deoxyribonucleic acid *See* DNA.

depensation An increase in **parasitism** and **predation** that occurs in certain species when its population density falls below a certain threshold.

deplasmolysis The entry of water into a plasmolyzed (*see* plasmolysis) cell, so the **cell membrane** returns to the **cell wall**.

deposit feeder An animal that lives on or below the surface of seabed sediment and ingests the sediment, from which it extracts nutrients.

deposition The formation of ice on a solid surface by the direct change of water vapour to ice without passing through the liquid phase.

depression *See* cyclone.

depression family *See* cyclone family.

depth hoar (sugar snow) A layer of **frost** that forms by **deposition** beneath the surface of a layer of snow.

dermal Pertaining to the skin (**epidermis**).

Dermaptera (earwigs) A family of long-bodied insects with **cerci** modified to form **forceps**, short, leathery forewings, and in many species large, membranous hind wings, although some species are wingless. **Nymphs** resemble adults but are wingless. Some females guard their eggs until they hatch and exhibit parental care. Earwigs are nocturnal scavengers or **herbivores**, some being crop pests. A few species are predators. There are about 1800 species distributed worldwide.

D

dermatophyte A parasitic fungus (*see* Fungi) that lives on skin.

Deroceras reticulatum (field slug, grey field slug, grey garden slug) A species of **slug**, 35–50 mm long, usually grey or cream, although highly variable in colour, but with dark spots behind the **mantle**, dark tentacles, and a short **keel** at the posterior end of the body. It occurs in cultivated areas, sheltering by day beneath stones and in leaf litter and feeding at night on leaves and fruit. It is a serious pest, but several ground beetles (**Carabidae**) prey on it. The slug is native to Europe but has been introduced to North and South America, New Zealand, Tasmania, and parts of Asia.

derris An organic **insecticide** derived from the root of several species of *Derris*, especially *D. elliptica*, leguminous climbing plants (**Fabaceae**) of Southeast Asia. When crushed the root releases **rotenone**. It was applied as a powder, but owing to its high toxicity it is now banned.

desalination (desalinization) Processes that remove sufficient salt to render salt or **brackish** water potable.

desalinization *See* desalination.

desert date (*Balanites aegyptiaca*) *See* Zygophallaceae.

desert devil *See* dust whirl.

desert flannel flower (*Actinotus schwarzii*) *See Actinotus*.

desert pavement *See* yermic horizon.

desert thorn *See Lycium*.

Desfontainea (family **Columelliaceae**) A **monotypic** genus (*D. spinosa*, Chilean holly) of **shrubs** with small, holly-like leaves and scarlet, tubular flowers with yellow tips, pollinated by hummingbirds. It occurs in rain forests and on mountainsides from Costa Rica to Cape Horn, and is the national flower of Bolivia. It is cultivated for ornament.

D

desiccation **1.** Drying out. **2.** The long-term drying out of the land as a consequence of climatic change.

desiccation cracks (mud cracks, shrinkage cracks, sun cracks, syneresis cracks) Cracks that develop in mud that has dried out, marking the surface with a pattern of polygons.

desilication The removal of **silica** from a **magma** or rock, e.g. by reaction between magma and **limestone** that deposits silicate minerals on the rock wall adjacent to the magma.

desmids Unicellular **green algae** (**Chlorophyta**) in which the cells are in two halves. Cells are usually solitary but form colonies in some species. They occur mainly in fresh water and their presence indicates that the water is unpolluted.

Desmognathus fuscus (dusky salamander, northern dusky salamander) A species of brown, gray, olive, or reddish brown salamanders (**Salamandridae**) with a pale underside, in which males are about 95 mm long and females 85 mm. They are lungless salamanders (Plethodontidae), relying entirely on **cutaneous respiration**. They inhabit woodland areas close to flowing water and feed on a wide variety of invertebrates. In some places they are collected and sold for fishing bait as 'spring salamanders' or 'spring lizards'. They occur throughout eastern and central North America. ↗

destroying angel The **fruiting body** of *Amanita virosa*, a species of **agaric fungi** found throughout European forests, especially in upland areas, and of the very similar *A. verna* (also known as spring amanita), *A. bisporigera* found in eastern North America, and *A. ocreata* found in western North America, and found in western Europe. The **pileus**, **gills**, and **stipe** are pure white, the pileus 50–100 mm across, the stipe 90–150 mm tall. The fungus is deadly poisonous; ingestion is usually fatal. ↗

Desulfovibrio A genus of **Deltaproteobacteria** that are Gram-negative (*see* Gram reaction), rod-shaped, and **motile** by means of flagella (*see* flagellum). They occur in water rich in organic matter and waterlogged soils, and reduce sulphate, releasing hydrogen sulphide (H_2S) which has the smell of rotten eggs and reacts with metals to form sulphides that blacken mud. They are **anaerobes** but **aerotolerant**.

detachment **1.** The separation of surface material on a hillslope from the underlying rock. **2.** A rock fault with considerable horizontal movement, caused by the instability of a raised block. **3.** A zone of deformed, ductile rock extending all the way through the Earth's **crust** and caused by extension, as two rock masses move apart.

determinate inflorescence (monotelic inflorescence) Describes a **cyme**, i.e. an **inflorescence** with a flower at the tip of the **axis**, preventing further growth of the **peduncle**.

determinate thyrse *See* cyme.

detorsion In certain gastropods (**Gastropoda**), the untwisting of the viscera during development.

detrital Describes material resulting from the breakdown of rock by **weathering** or **erosion**.

detrital pathway (detritus food chain) A **food chain** based on organic **detritus** that is consumed by **detritivores**, which in turn provide food for predators.

detritivore (detritus feeder) A **heterotroph** which feeds on organic **detritus**.

detritus 1. Small fragments of organic material. **2.** Loose fragments of rock that has been transported away from the site where they were produced by **weathering**, **erosion**, or **abrasion**.

detritus feeder *See* detritivore.

detritus food chain *See* detrital pathway.

Deuteromycota *See* Fungi Imperfecti.

Deutzia (family **Hydrangeaceae**) A genus of **deciduous**, and some **evergreen**, **shrubs** with **opposite**, **simple** leaves with a **serrate** margin. Flowers usually white, sometimes green or red, borne in **panicles** or **corymbs**. Fruit is a **capsule**. There are about 60 species found in Central and eastern Asia, Central America, and Europe. Many are cultivated.

devil's coach-horse beetle *See Ocypus olens.*

dew Moisture that condenses on to surfaces, e.g. plant leaves, on cool nights when there is little wind. Surfaces radiate away the warmth they absorbed during the day until they cool to the **dewpoint temperature**.

dewatering The deliberate removal of **groundwater**, e.g. by **drainage** or abstraction from wells, in order to reduce pressure or the rate of flow.

dewberry *See Rubus.*

dew gauge (surface wetness gauge) An instrument that measures and records the amount of **dew** that forms overnight.

dewpoint front *See* dry line.

dewpoint temperature The temperature at which a body of air would become saturated if it cooled with no change in the amount of moisture it contained or in the **atmospheric pressure**.

dew worm *See Lumbricus terrestris.*

dextral coil A **snail** shell that coils clockwise when viewed from above.

dextrorse Developing in a clockwise spiral when viewed from above.

dextrose *See* glucose.

D-fructose *See* fructose.

D-glucose *See* glucose.

Diabrotica undecimpunctata (spotted cucumber beetle) *See* bacterial wilt.

diadelphous Describes **stamens** with **filaments** that are fused, forming two groups.

Diadophis punctatus (ringneck snake) A species of olive, brown, blue-grey, or black colubrid snakes (**Colubridae**) with a distinctive neck band that is red, yellow, or orange. Most are 250–380 mm long. They are nocturnal and secretive, but inhabit a variety of **habitats** where the soil is warm. They feed small amphibians, reptiles, and invertebrates and occur widely throughout eastern and central North America. ↗

diadromous Describes fish that regularly migrate between fresh water and the sea.

diageotropism A **tropism** in which a plant organ grows at right angles to gravity (horizontally), e.g. the **branches** of *Cedrus libani* (cedar of Lebanon), strawberry (***Fragaria*** spp.).

diagnostic horizon A **soil horizon** that contains materials typical of that type of soil.

dialect Vocal communications among a population of a particular species that differ in certain respects from the communications of other populations of the same species. Many songbirds have local dialects.

diallelic Describes a polyploid (*see* polyploidy) individual possessing more than two sets of **chromosomes** with two different **alleles** at a particular **locus**.

Dianthus (family **Caryophyllaceae**) A genus of **perennial** (a few **annual** or

biennial) herbs with **opposite**, **simple**, usually **linear** leaves that are often **glaucous**. The **calyx** is tubular, tightly enclosed at the base by an **epicalyx** of 1–3 scales. There are 5 **petals**, typically with a frilled margin. Fruit is a 4-toothed **capsule**. There are about 300 species occurring mainly in Eurasia, with a few in South Africa and one in northern North America. Several are cultivated for ornament, e.g. *D. caryophyllus* (carnation), *D. plumarius* (pink), and *D. barbatus* (sweet William).

diapause A temporary cessation of growth and development, often associated with a period of unfavourable environmental conditions.

Diapensiaceae (order **Ericales**) A family of **perennial herbs** and **subshrubs** with **alternate**, **simple** leaves with **entire** to **serrate** margins. Flowers are **actinomorphic**, **bisexual**, **pentamerous**, with usually free **sepals** and **petals** free or fused into a 5-lobed **corolla** tube, 5 **stamens** sometimes alternating with 5 **staminodes**. **Ovary superior** of 3 fused **carpels** and 3 **locules**. **Inflorescence** a long, many-flowered **spike** or a **raceme** of 1–15 flowers on a long **scape**, or flowers solitary and terminal. Fruit is a **loculicidal capsule**. There are 6 genera with 18 species occurring through the arctic and scattered in northern temperate regions, especially eastern Asia and eastern North America. Some are cultivated as rock-garden plants.

diaphototropism A **tropism** in which a plant organ grows at right angles to the direction of light. It commonly occurs in **petioles** of **broad-leaved** plants, so the leaves are fully exposed to the light.

diaphragm A partition, made from muscle and tendon, that separates the thoracic and abdominal cavities in vertebrates. Flexure of the diaphragm alters the volume of the thorax, assisting breathing.

diaphysis The main part of a bone.

Diaporthales An order of **Fungi** belonging to the **Pezizomycotina** that are pathogens of many plants. The order includes ***Cryphonectria parasitica***, the cause of chestnut blight as well as fungi causing several stem rot, **canker**, and other diseases. Most species, however, are **saprotrophs** living on rotting wood. There are 11 genera with 14 species, found in Europe and North America.

diarch A root that has two strands of **xylem**.

Diaspididae (armoured scale insects) A family of scale insects (**Hemiptera**) that protect themselves beneath more substantial coverings than other coccids, comprising wax from two **nymph instars** as well as from the adult, mixed with faeces and fragments of material from the host plant. It is the largest family of scale insects, with more than 2650 species with a worldwide distribution.

diaspore A seed or **spore**, together with any attached tissues, that functions in the dispersal of the plant, i.e. as a **propagule**. In some plants (tumbleweeds) the whole plant (in a few species just the **inflorescence**) detaches from the roots and is blown along, scattering seeds as it goes.

diastema A natural gap in a row of teeth.

diatom A **phylum** (Bacillariophyta) of mostly unicellular, occasionally colonial or filamentous, algae (*see* alga) in which the **frustule** is made from **silica** and is in two halves, one of which overlaps the other. Frustules are often delicately ornamented by tiny holes, some covered by a membrane. Most diatoms are photosynthetic (*see* photosynthesis), others living as **heterotrophs** in decaying organic material. Diatoms account for about 20 percent of the world's total **fixation** of carbon by **photosynthesis**. Bilaterally symmetrical diatoms are said to be pinnate; radially symmetrical diatoms are centric. Pennate diatoms occur in both freshwater and marine environments; centric diatoms are predominantly marine and form part of the **plankton**. There are more than 200 genera with about 100,000 species.

diatomite A sediment rich in the siliceous **cell walls** of **diatoms**.

diatropism A **tropism** in which a plant or plant organ grows at right angles to a stimulus.

diazotroph An organism capable of fixing atmospheric nitrogen (*see* nitrogen fixation).

dicaryotic *See* dikaryotic.

Dicentra (family **Papaveraceae**) A genus of about eight species of clump-forming, **perennial herbs** with **rhizomes** or **tubers**, occurring in North America and eastern Asia. The popular garden plant old-fashioned bleeding heart, formerly *D. spectabilis*, is now classed as ***Lamprocapnos spectabilis***.

dicentric Describes a **chromosome** or **chromatid** that has two **centromeres**.

Dichapetalaceae (order **Malpighiales**) A family of **trees**, **shrubs**, and **lianas** with **alternate**, **simple**, **entire** leaves with **stipules**. Flowers are **actinomorphic** or somewhat **zygomorphic**, **hermaphrodite** or **unisexual** (plants **monoecious**), with 5 **imbricate**, often unequal, free or **connate** **sepals**, 5 imbricate **petals** that are free or connate in a tube, 3 or 5 **stamens**, **ovary** **superior**, **syncarpous**, of 2 or 3 **carpels** and **locules**. **Inflorescence** an **axillary** **cyme**. Fruit is a **drupe**. There are 3 genera with 165 species occurring throughout the tropics. Leaves and seeds of some species contain fluoroacetic acid, used to kill vertebrates.

dichasium A **cymose inflorescence** in which two **branches** arise from each branch.

dichlobenil A **herbicide** that is applied to the soil to control **broad-leaved** weeds and grasses in agricultural, residential, and industrial areas and to kill tree roots. It also kills weeds in still or slow-flowing water. It is of low toxicity.

dichlorodiphenyltrichloroethane *See* DDT.

dichlorphos *See* dichlorvos.

dichlorprop A **systemic herbicide** that is used to control **broad-leaved** weeds in grass, on roadsides, and in forests. It can cause eye irritation but otherwise is of low toxicity.

dichlorvos (Atgard, DDVP, dichlorphos, Divipan, Equigard, Vapona) An **organophosphorus insecticide** that is used indoors against crawling and flying insects, and also in agriculture and veterinary products. It inhibits the **enzyme** anticholinesterase. It was banned in the European Union in 2012, but is still used in many countries.

dichogamy The condition in which the male and female parts of a flower mature at different times.

dichotomous branching Repeated division into two parts.

Dichotomous branching. Growth proceeds as each branch divides into two equal parts.

dichthadiiform A permanently wingless hymenopteran (*see* Hymenoptera) with a much enlarged **gaster**.

diclinous 1. Having **stamens** and **pistils** in separate flowers, either on different plants or on the same plant. **2.** Unisexual (*see* unisexual flower).

dicotyledon (dicot) An **angiosperm** in which the **embryo** contains at least two **cotyledons** (some dicots have more than two).

D

Dicroidium A genus, now extinct, of **seed ferns** that had leaves similar to those of modern ferns, but forked so they looked like two leaves united at the base. There were about nine species occurring throughout **Gondwana** during the Triassic period (251–199.6 million years ago).

dictyosome The **Golgi body** in a plant cell.

dictyostele A **stele** with several strands (meristeles); these may be **protosteles** or **siphonosteles**.

Dictyosteliomycetes A class of cellular slime moulds (**Acrasiomycetes**) in which the feeding stage consists of independent **myxamoebae**, which feed mainly on bacteria and remain independent so long as food is available. When the food supplies dwindles they aggregate into a slug-like **pseudoplasmodium** that moves about until it finds suitable conditions, when it forms a **fruiting body**. Some of the cells then become **spores** while the remainder form stalks that raise the spores above the substrate, thus increasing the likelihood that they will be carried away by air currents. There are about 70 species.

Dictyostelium A genus of **Dictyosteliomycetes** in which the slug-like **pseudoplasmodium** is 0.5–2.0 mm long and the **fruiting body** is a mass of **spores** borne at the tip of a tapering, branched stalk. There are many species found worldwide in decaying plant material, dung, and soil.

Didiereaceae (order **Caryophyllales**) A family of **succulent** or woody **shrubs** and **trees** that are **xerophytes** superficially resembling cacti. Stems are **dimorphic**, starting succulent and growing woody with age. Leaves are **alternate**, **simple**, **entire**, and **exstipulate**. Flowers **unisexual** (plants **dioecious** or **gynodioecious**), with 2 opposite **sepals** resembling **petals** and 4 overlapping petals, 8–10 **stamens**, **ovary superior** of 3 fused **carpels** of 3 **locules**. **Inflorescence** a **thyrse**. Fruit resembling an **achene**. There are 7 genera with 16 species occurring in South Africa, East Africa, and Madagascar. All species are rare; some cultivated as succulents.

Didymelaceae (order **Buxales**) A **monogeneric** family (*Didymeles*) of **evergreen trees** with **alternate**, **simple**, **entire**, **exstipulate**, **glabrous** leaves. Flowers are small, **actinomorphic**, **unisexual** (plants **dioecious**), **staminate** flowers with 2 **stamens**, **pistillate** flowers with **ovary superior** of 12 **carpel**. Fruit a **drupelet**. There are two species occurring in eastern Madagascar.

didymous In pairs.

didynamous With two **stamens** that are longer than the others.

die-back *See Phytophthora cinnamomi.*

dieldrin A **cyclodiene insecticide** that was developed as an alternative to **DDT** and that is the active ingredient in **aldrin**. It is persistent and bioaccumulative (*see* bioaccumulation) and it was banned in the 1970s.

diestrus *See* dioestrus.

dietary fibre *See* fibre.

differential resource utilization *See* resource partitioning.

differentially permeable membrane A **membrane** that allows small molecules to cross but prevents the passage of larger ones. *See also* partially permeable membrane, selectively permeable membrane.

differentiation Changes in the structure and function of cells in a developing organism as tissues become increasingly specialized.

diffusion The random thermal movement of molecules that takes them from a region of higher solute concentration to one of lower concentration.

Digitalis (family **Plantaginaceae**) A genus of **perennial** or **biennial herbs** and **shrubs** with **alternate** leaves and tall **spikes** of drooping, 2-lipped, bell-like, purple, pink, white, or yellow flowers, some with spots on the lower lip. There are about 20 species occurring in western and southwestern Europe and in western and central Asia. They are cultivated for ornament (foxgloves) and for digitalin, a group of drugs used for heart stimulation.

digitate *See* palmate.

digitigrade Describes a gait in which only the digits make contact with the ground, as in cats and dogs.

Diglossa (flowerpeckers) *See* nectar robber.

dihyodonty The condition of having two sets of teeth in the course of a lifetime. *See* diphyodonty.

Dikarya A subkingdom of **Fungi** that comprises the **Ascomycota** and **Basidiomycota**, both of which produce dikarya (*see* dikaryon) and lack flagella (*see* flagellum). The subkingdom includes the great majority of all fungi.

dikaryon A fungal **mycelium** of **hypha** composed of cells each of which contains two **haploid** nuclei.

dikaryotic (diacaryotic) Describes a cell containing two **haploid nuclei**.

Dilleniaceae (order **Dilleniales**) A family of **trees**, **shrubs**, **lianas**, and **perennial herbs** with **alternate**, **simple**, persistent or **caducous**, **exstipulate** leaves. Flowers **actinomorphic**, **bisxual** or **unisexual** (plants **monoecious** or functionally **dioecious**), with 3–20 fused or 5 free persistent, **imbricate sepals**, 3–5 imbricate, **deciduous petals**, crumpled in the **bud**, many **stamens**, **ovary superor** of 1–5 or more **carpels** with 1–5 **locules**. **Inflorescences axillary** or terminal, resembling a **cyme**, **panicle**, or **fascicle**. Fruits are **follicles**, **nuts**, or **berries**. There are 10 genera with 300 species occurring in tropical and warm temperate regions. Some cultivated for ornament.

Dilleniales An order of plants comprising one family of ten genera and 300 species. *See* Dilleniaceae.

dilution effect The effect of behaviour in which animals crowd together in the proximity of predators. Crowding reduces the probability that any individual will be taken. If the predator seeks a prey animal in a group of ten, there is a 10 percent chance that a particular individual will be taken, but if the prey animals are in a group of 100, the chance is 1 percent.

dimer A **protein** that consists of a pair of **polypeptide** chains or units; if the two are identical the protein is homomeric, if they differ it is heteromeric.

dimerous Of a **flower**, having parts in twos.

dimethoate An **organophosphate insecticide** and **acaricide** that is used to control insects and mites on fruit, vegetable, and ornamental crops. It kills insects on contact and also acts systemically. It breaks down quickly after application, but is moderately toxic.

dimethyl sulphide (DMS, methylthiomethane) An insoluble chemical compound ($[CH_3]_2S$) that is produced by many species of marine algae (*see* alga) from the precursor dimethylsulphonioproprionate (DMSP), probably to control the amount of salt in their cells and as protection against other chemical stresses. A proportion of DMS is released into the air, where it is oxidized to sulphate particles, which act as **cloud condensation nuclei**, encouraging **cloud formation** over the ocean. DMS is also the principal compound involved in the transfer of sulphur from sea to land.

dimictic Describes a lake in which there are two seasonal periods during which the water circulates freely. Such lakes are typical of temperate **climates**, in which the two seasons are spring and autumn;

in summer the water becomes thermally stratified and in winter water close to freezing expands, becoming less dense than the warmer water below, and causing a reversed stratification.

D

dimidiate 1. Divided in two. **2.** Of a fungal fruit body (*see* fruiting body), semicircular in outline.

dimorphic enantiostyly *See* enantiostyly.

dimorphic fungi **Fungi** that are able to exist either as **yeast**-like single cells or as mycelia (*see* mycelium), depending on the environmental conditions.

dimorphism The existence of morphological (*see* morphology) differences that divide a **species** into two distinct groups.

dinitro-*ortho*-cresol (DNOC) A **dinitro herbicide**, **insectide**, and **acaricide** that kills on contact. It was formerly used to control **broad-leaved** weeds and the overwintering stages of many insect and mite pests. It is highly toxic and is now banned.

dinitro pesticides Compounds that contain a dinitro group, i.e. two nitrogen atoms (N_2).

dinitrophenol (DNP) A **dinitro** compound that was formerly used as a dieting aid, but was banned in 1938 owing to its toxicity. It is used as a wood preservative, **insecticide**, **acaricide**, and **fungicide**.

dinocap (DNOPC) A **dinitro** compound that is used as an **acaricide** and **fungicide** on vegetable, fruit, and ornamental crops. It is highly poisonous to fish, but only slightly to moderately toxic to humans.

dinoflagellate *See* Pyrrophyta.

Dinophyta *See* Pyrrophyta.

dinoseb (DNBP, DNSBP, DNOSPB) A **herbicide** formerly used to control grasses and **broad-leaved** weeds. It is highly toxic and has been banned in most countries including the European Union, United States, and Canada.

***Diodia* vein chlorosis** A disease producing **chlorosis** and vein-clearing in *Diodia virginiana* (Virginian buttonweed) that is caused by a crinivirus and transmitted by the banded-winged whitefly (*Trialeurodes abutilonea*). It occurs in the United States.

dioecious With male and female reproductive organs on separate individuals (e.g. plants).

dioestrus (diestrus) The period between two **oestrus cycles**.

dioicious Of a moss (**Bryophyta**), having a **gametophyte** stage that bears antheridia (*see* antheridium) or archegonia (*see* archegonium) but not both.

Dioncophyllaceae (order **Caryophyllales**) A family of woody **lianas** and **shrubs** with leathery, **alternate**, **exstipulate** leaves that are **simple** and **linear** or with pairs of **recurved** hooks by which the plant climbs. Flowers are **actinomorphic**, **hermaphrodite**, with 5 **sepals** that are free, fused into a tube, or **valvate**, 5 free **petals** that alternate with the sepals, 10 or 25–30 **stamens**, **ovary superior** of 2 or 5 **carpels** with 1 **locule**. **Inflorescence cymose**. Fruit a dry, **dehiscent capsule**. There are three genera of three species occurring in tropical West Africa. Some yield medicines.

Dioscoreaceae (order **Dioscoriales**) A **monocotyledon** family of vines and **lianas** that are **perennial geophytes**, most with **rhizomes** or **tubers**. Leaves are **alternate**, sometimes **opposite** or in **whorls**. Flowers **actinomorphic**, **unisexual** (plants **dioecious**) or **bisexual**, with 2 whorls of similar **tepals** or 2 whorls of 3 **stamens**, **ovary inferior** of 3 fused **carpels** and 3 **locules**. **Inflorescence axillary**, and a **spike**, **raceme**, or **panicle**. Fruit is a **berry** or **samara**. There are 4 genera or 870 species found mainly in the tropics, but some subtropical. *Dioscorea* spp. (yams) are a staple food in parts of the tropics.

Dioscoreales An order comprising about 5 families of 21 genera and 1037 species.

See Burmanniaceae, Dioscoreaceae, Nartheciaceae, Taccaceae, and Thismiaceae.

Dipentodontaceae A family of uncertain affinity, comprising a small, **deciduous tree** or **shrub** with **alternate**, **simple** leaves with **petioles**. Flowers are small, **actinomorphic**, **bisexual**, with 5–7 **sepals** and **petals**, 5–7 **stamens**, **ovary superior** of 3 fused **carpels** with 1 **locule**. **Inflorescence** an **umbel**. Fruit a **capsule**. There are 2 genera and 16 species occurring in southeastern Asia to Malesia, and from Mexico to Peru.

dipeptide *See* peptide.

diphyodonty A type of **dihyodonty** in which a first set of **deciduous** teeth are shed and followed by the eruption of a second, permanent set.

diplococcus A round **bacterium** that occurs as two cells (cocci, *see* coccus) are joined together.

diplohaploplontic Describes a life cycle involving an **alternation of generations** in which **sporophytes** produce **spores** by **meiosis** and **gametophytes** produce **gametes** by **mitosis**.

diploid Having two sets of **chromosomes**.

Diplolepis rosae *See* robin's pincushion gall.

diploplontic Describes a life cycle with no **alternation of generations**, in which a **diploid** organism produces **gametes** by **meiosis**.

Diplopoda (millipedes) A class of **Arthropoda** in which the first four body segments bear one pair of legs each and all other segments bear two pairs; despite the name, no millipede has 1000 legs. If threatened, most curl into a ball. Females lay 10–300 eggs, some species constructing a nest for them. The young have three pairs of legs and four legless segments; as they grow they add more segments and more legs. Most millipedes feed on decomposing plant material but a few species are omnivorous or carnivorous. There are more than 10,000 species with a worldwide distribution.

dipole Describes a molecule with an unevenly distributed charge, such that one pole has a net negative charge and the other a net positive charge.

dipole moment The difference in charge at opposite ends of a **polar molecule**.

Dipsacales An order comprising 2 families of 46 genera and 1090 species. *See* Adoxaceae and Caprifoliaceae.

Dipsacus (family **Caprifoliaceae**) A genus of **biennial herbs** (teasels) with **lanceolate** leaves **connate** at the base of the stem, forming a cup that collects water, possibly preventing insects from climbing to the **inflorescence**, and **opposite** higher up the stem. The purple, lavender, or pink inflorescence is a **capitulum**. The leaves and stem are very prickly, and the seed heads bear **recurved** spines. Adding dead insects to the leaf cups increases the number of seeds that set, suggesting *Dipsacus* is partly carnivorous. There are about 15 species occurring in Eurasia, tropical Africa, and Sri Lanka. Dried flower heads of *D. fullonum* (fuller's teasel) were formerly used industrially to raise the nap on woven fabrics and are still preferred to their metal substitutes by some craft weavers. Some species are grown for ornament. Teasel seeds are an important food for some birds.

Diptera (flies, true flies, two-winged flies) An order of insects (**Insecta**) in which adults have a single pair of membranous wings, the ancestral hind wings having been modified into **halteres**. Mouthparts are usually adapted for sucking and form a **proboscis**, often used for piercing; **mandibles** are absent in many families. Larvae are **eruciform** with up to 12 abdominal segments. There are about 120,000 species, found worldwide.

Dipterocarpaceae (order **Malvales**) A family of mainly **evergreen trees**, some up to 70 m tall and often with **buttress roots**, and some **shrubs**, with **alternate**, **simple**, **entire** leaves that are often **coriaceous** or parchment-like. Flowers **actinomorphic**, **bisexual**, with 5 **sepals** and 5 free or **connate petals**, 5 to many **stamens**, **ovary superior** to semi-**inferior** of 2–5 fused or 3–4 free **carpels** and **locules**. **Inflorescence** a **raceme** or **panicle**. Fruit is a **nut** or **capsule**. There are 17 genera of 680 species found throughout the tropics but especially in western Malesia. Many are sources of commercially valuable **hardwood**.

diquat A **contact herbicide** that is used to control **broad-leaved** weeds and to desiccate foliage to facilitate the harvesting of potatoes. It breaks down rapidly but is moderately toxic to mammals.

Dirachmaceae (order **Rosales**) A **monogeneric** family (*Dirachma*) of small **trees** and **shrubs** with small, **alternate**, **serrate** to **dentate** leaves with persistent **stipules**. Flowers, terminal and borne singly, are **actinomorphic**, **bisexual**, with a 4- to 8-lobed **epicalyx**, 5–8 **sepals**, **connate** at the base, 5–8 free **petals**, 5–8 **stamens**, **ovary superior** of 5–8 fused **carpels** with 5–8 **locules**. Fruit a **capsule**. There are two species occurring in Somalia and the island of Socotra.

directed speciation An evolutionary trend in plants, in which successive **species** do not exhibit a continuum of **adaptation** to their environment, but appear distinctly different, as though in a series of evolutionary steps.

directional selection **Natural selection** that acts on a range of **phenotypes** for a particular **character** by shifting the mean phenotype toward a phenotypic extreme.

disafforestation *See* deforestation.

disassortative mating Mating between two individuals with different **phenotypes**.

disc A fleshy outgrowth from the **stamens** or **receptacle** of a **flower**; it often secretes **nectar**.

Discalis A genus, now extinct, of vascular plants (**Tracheophyta**) that lived during the Early Devonian epoch (416–397.5 million years ago). They had creeping stems up to 5 mm in diameter with many **branches** and also trailing or partly erect stems that were slightly smaller and also branched. All the stems bore spines. Sporangia (*see* sporangium) were disc-shaped (hence the name), about 3.7 mm across, with spines, and borne on stalks rising laterally from fertile stems; stems lacking sporangia had **circinnate** tips.

disc floret In a **capitulum** flower (e.g. in **Asteraceae**), the central part of the **inflorescence**. *Compare* ray floret.

discharge A measure of the amount of water flowing past a particular point, e.g. in a river (*see* gauging station) or from a well abstracting **groundwater**.

disclimax A plant community that replaces a **climax community** following a disturbance, in the **monoclimax** theory.

discontinuous distribution *See* disjunct distribution.

discordant drainage A **drainage pattern** that does not reflect the underlying geology, i.e. is discordant.

Discus rotundatus (round snail, rotund disc) A species of **snails** with flattened, red-brown, banded shells 2.5–6.0 mm high and 5.5–7 mm across, that inhabit moist, sheltered places and are found under stones, in dead wood, and in leaf litter, where they feed on **detritus** and **Fungi**. They sometimes form colonies. They occur throughout Europe and in parts of North America.

disintegration 1. The fragmentation of rocks as a result of **mechanical weathering**. **2.** The breaking up of a **clod** or soil **aggregate** by the action of wind-driven rain.

disjunct distribution (discontinuous distribution) The occurrence of a **taxon** in a limited number of locations that are separated by oceans, e.g. members of the family **Caricaceae** are most abundant in South America with a few occurring in Central America, but the genus *Cyclicomorpha* occurs in tropical Africa; ***Araucaria*** species occur in South America and Australasia.

disjunction During the **anaphase** stage of **mitosis** and **meiosis**, the separation of **homologous chromosomes** and their movement toward the **spindle**.

disomy The **diploid** condition.

dispersal mechanism The strategies or structures by which a **sessile** or slow-moving organism disseminates its **propagules**.

dispersion 1. The separation of soil particles from **aggregates** or **peds**, allowing them to engage individually in chemical reactions. **2.** The lateral spreading of flowing water as it crosses a surface.

displacement activity Behaviour that appears irrelevant to the situation in which the animal performing it finds itself. It may result from **conflict**, or from frustration in an animal that is prevented from attaining a goal.

display Stereotypical behaviour, usually inherited, that animals use in communication, e.g. in courtship or threat.

disruptive coloration A pattern of skin colours that disrupts the outline of an animal when seen against the typical background, thus making the animal more difficult to see. The stripes of a tiger are disruptive.

disruptive selection Selection that results in divergence of the frequency of **alleles**, leading to the emergence of diverging extreme **phenotypes**, e.g. by planting seeds collected from the tallest and shortest plants in a population.

disseminule A part of a plant from which a new plant may arise.

distal Farthest from the point of attachment.

distichous In two ranks.

distyly **Heterostyly** in which there are two types of **flower** (morphs) that differ in the lengths of their **pistils** and **stamens**.

disulfoton An **organophosphate insecticide** and **acaricide** that is **systemic** and used to control aphids (**Aphididae**), leafhoppers, thrips (**Thysanoptera**), **spider mites**, etc. It inhibits the **enzyme** anticholinesterase and is toxic to mammals, birds, and aquatic animals.

disulphide bridge A **covalent bond** between two sulphur atoms. Disulphide bonds in **peptides** and **proteins** help stabilize the molecular structure.

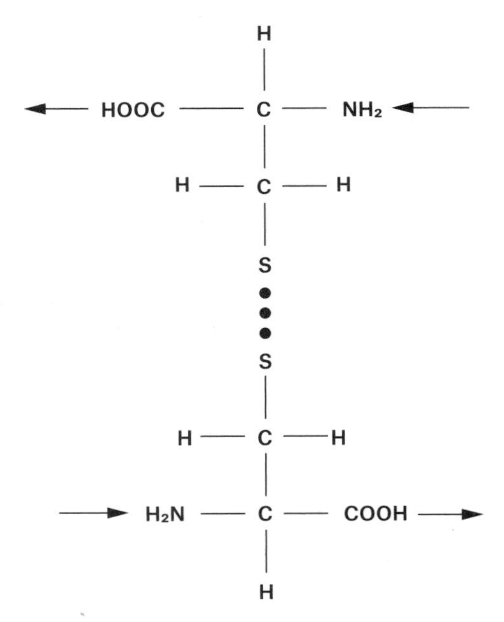

A disulphide bridge is a covalent bond that forms between two sulphur atoms on the side chains of a cysteine residue; it is important in the way the protein molecule folds to form its tertiary structure.

Ditylenchus destructor (potato rot nematode, potato tuber nematode, iris nematode) A species of nematodes (**Nematoda**), about 0.8–1.4 mm long, that feed on **starch** inside **tubers**, especially of potatoes. They can live in stored potatoes and sweet potatoes, and in **stolons**. They occur throughout temperate regions, wherever potatoes are grown.

Ditylenchus dipsaci (stem and bulb nematode, onion bloat) A species of nematodes (**Nematoda**), about 1.5 mm long, that are migratory **endoparasites**. They enter plants through wounds or **stomata** and release an **enzyme** that dissolves **cell walls**. They then live in the intercellular spaces in onion and garlic leaves and between the scales in onions, feeding on cell contents. This often causes the formation of **galls**. As the **bulb** swells, the nematodes migrate down the stem, which also swells and eventually collapses. As well as onions and garlic, *D. dipsaci* infests carrots, oats, some beans, strawberries, and certain ornamentals. It occurs in most temperate regions.

diurnal During the day, or at daily intervals.

diuron (DCMU, DMU) A **urea herbicide** that is applied to soil for total weed control on land not used to produce crops and to control **annual** weeds around tree fruit and nursery crops. It inhibits photosynthesis. It is harmful to fish and can cause eye irritation.

divergence A flow of air in which **streamlines** flow outward, away from a centre of high **atmospheric pressure**. This removes air, reducing the pressure at the centre.

divergent evolution The process in which groups of organisms descended from a common ancestor become increasingly different over successive generations, e.g. the divergence of **angiosperms** and **gymnosperms**, which both diverged from a common stem group. *Compare* convergent evolution.

Diversisporales An order of **Fungi** belonging to the **Glomeromycota** that live underground in mycorrhizal (*see* mycorrhiza) association with trees. They produce a variety of types of **spores**, hence the name.

diversity The number of species present in a community or area, usually with an assessment of the relative abundance of each species.

divide The boundary between adjacent **catchments**. In U.K. usage a divide is often called a watershed.

division In plant **taxonomy**, a rank equivalent to **phylum**; the term is falling from use, partly in favour of phylum, but also because the rank is not used in the classification of the **Angiosperm Phylogeny Group**.

DMS *See* dimethyl sulphide.

DMU *See* diuron.

DNA (deoxyribonucleic acid) A **nucleic acid** comprising a chain of the sugar deoxyribose to which are attached the pyrimidine **bases cytosine** (C) and thymine (T), and the purine bases adenine (A) and **guanine** (G). Pyramidines and purines form **base pairs**, A-T and C-G linking two strands of DNA, which form a double helix. DNA is the genetic material of almost all living organisms. *See* appendix: The Genetic Code.

DNBP *See* dinoseb.

DNOC *See* dinitro-*ortho*-cresol.

DNOPC *See* dinocap.

DNOSPB *See* dinoseb.

DNP *See* dinitrophenol.

DNSBP *See* dinoseb.

Doassansiales An order of **Fungi** belonging to the **Exobasidiomycetes** that are parasites of aquatic plants, causing smut diseases.

dodder *See* Convolulaceae, *Cuscuta*.

Dodoens, Rembert (1517–85) A Flemish physician and botanist who wrote a **herbal** with 715 illustrations, *Cruydeboek* (Plant Book), published in 1554; the book grew from 877 to more than 1500 pages in its final, 13th edition published in 1583, and remained the most widely used botanical reference for more than 200 years. Dodoens arranged plants in 6 groups according to their properties of 'species, form, name, virtue [i.e. usefulness], and temperament', expanding this in later editions eventually to 26 groups, by which time the work had 1309 illustrations.

dog days July and the first half of August in the Northern Hemisphere. This is the hottest part of the summer and the time when Sirius, the 'Dog Star', the brightest star in the sky, rises in conjunction with the Sun.

dog lichen A member of a genus (*Peltigera*) of about 90 species of **lichenized Fungi**, many with a cyanobacterial (*see* cyanobacteria) **symbiont**, that grow on soil, rock, trees, etc. They are able to fix atmospheric nitrogen (*see* nitrogen fixation).

dog stinkhorn *See Mutinus caninus*.

dogwood (*Cornus*) *See* Cornaceae.

Dokuchayev, Vasily Vasilyevich (1846–1903) A Russian soil scientist who devised a theory of soil formation, and a system of soil classification and of drawing soil maps. He is regarded as the father of soil science and many of the soil names he introduced remain in use, e.g. **chernozem**, **podzol**, **rendzina**, **solonetz**.

Dolichopodidae (long-legged flies) A family of flies (**Diptera**) most of which are small with long legs, covered with bristles and metallic blue-green or green in colour, and have large eyes. They are often found on flowers, feeding on nectar and sap, but adults are also predators of soft-bodied insects and the larvae are carnivorous. There are more than 7000 species distributed worldwide.

doliform Shaped like a barrel or jar.

dolomite 1. (pearl spar) A widely distributed mineral ($CaMg[CO_3]_2$), most of which forms by the process of dolomitization, involving a reaction between **limestone** and solutions containing magnesium (Mg). Dolomite is used in construction. **2.** (dolostone) A sedimentary rock formed by the dolomitization of limestone.

dolomitization *See* dolomite.

dolostone *See* dolomite.

domain The highest category in the taxonomical system based on comparisons of ribosomal RNA. There are three domains: **Bacteria**, **Archaea**, and **Eukarya**; members of one domain are not closely related to members of others. All plants belong to the domain Eukarya.

domatium A chamber on a leaf or **bract** that is inhabited by arthropods (**Arthropoda**), usually ants or mites, commonly consisting of a depression enclosed by leaf tissue or hairs.

domestic pigeon *See Columba livia*.

dominant The species that exerts the greatest influence on the characteristics of a community. Often this is the most abundant species.

dominant gene A **gene** in a **diploid** organism that always produces the same phenotypic **character** when one of its **alleles** are present, i.e. in the heterozygous (*see* heterozygosity) condition, as when two alleles are present, i.e. the homozygous (*see* homozygosity) condition. If gene *A* is dominant over gene *a* (the recessive), then the heterozygote (*Aa*) will produce the same **phenotype** as the homozygote (*AA*).

dormancy (hypobiosis) A resting condition, e.g. in seeds that are nor germinating or **buds** that are not growing.

dormin *See* abscisic acid.

dorsal **1.** In a plant, **abaxial**. **2.** In an animal, pertaining to the back, posterior, or upper side. **3.** The side of an organism farthest from the substrate; usually the upper side.

dorsifixed Describes **anthers** that are attached at the rear of the **filament**.

dorsiventral Having different upper (**dorsal**) and lower (**ventral**) sides.

Doryanthaceae (order **Asparagales**) A **monogeneric**, **monocotyledon** family (*Doryanthes*) of huge, **caespitose herbs** with **linear** to **lanceolate** leaves, arranged spirally and up to 2.5 m long, with a thin sheathing base. Flowers are large, **actinomorphic**, **bisexual**, with 3+3 **tepals** and 3+3 **stamens**, **ovary inferior** of 3 **carpels** and **locules**. **Inflorescence** a **thyrse** borne at the tip of an unbranched stem up to 6 m long. Fruit is a **loculicidal capsule**. There are two species found only in eastern Australia.

Dothideomycetes A class of **ascomycete fungi** most of which produce pseudothecia (*see* pseudothecium) and **bitunicate** asci (*see* ascus). Most species are **saprophytes** on decaying wood, leaves, or dung, or **endophytes**; some species are plant pathogens. There are 1300 genera with more than 19,000 species found worldwide.

double coconut (*Lodoicea maldivica*) *See* Arecaceae.

double fertilization In **angiosperms** and **Ephedraceae**, the method of reproduction in which the male **gametophyte**, comprising the **pollen grain** and **pollen tube**, contains three **haploid** nuclei: two sperm nuclei and one tube (vegetative) nucleus. The **tube nucleus** degenerates once the pollen tube has penetrated the **embryo sac** (the female gametophyte). Both sperm nuclei enter the embryo sac, one fertilizing the **ovum** to produce the **zygote**, the other uniting with the two **polar nuclei** present in the embryo sac to produce a triploid cell, which multiplies by **mitosis** to form the **endosperm**. In Ephedraceae, one sperm nucleus fertilizes the ovum nucleus, the other unites with an adjacent cell, but develops no further.

Douglas fir (*Pseudotsuga menziesii*) *See* Pacific coast forest.

doves *See* Columbidae.

down Lowland grassland that results from and is maintained by grazing. It occurs in Britain, usually on **chalk** or **limestone** soils, but occasionally on **acidic rock**.

downburst A downward rush of air from a **convection cell** in a **cumulonimbus** cloud that spreads to the sides when it reaches the surface, producing strong **gusts**.

downdraught A current of air that flows downward inside a **cumulonimbus** cloud, usually at less than approximately 18 km/h, but sometimes much faster in a **supercell** cloud.

down feather *See* feather.

downregulation *See* RNA interference.

downrush A very strong **downburst** often associated with a **cloudburst**. It consists of cold air dragged downward by falling snow and rain.

downwash Air that is carried to the ground by an **eddy** on the **lee** side of building or steep-sided hill.

downy mildew A plant disease caused by a water mould (**Oomycota**) of the family Peronosporaceae, or the pathogen itself. Infected plants usually have yellowish patches on the underside of leaves that expand and turn brown, and have a white or purple mould below the surface; these are sporangia (*see* sporangium) produced by the oomycete as it reproduces asexually. The Peronosporaceae contains 17 genera and more than 600 species, most of which cause downy mildew.

downy mildew of grapes *See Plasmopora viticola.*

downy woodpecker *See Picoides pubescens.*

drainage **1.** The movement of water across the land, at and beneath the surface, eventually to the sea. **2.** The removal of water from soil by gravity.

drainage density The average distance between the streams draining an area, calculated as the total length of all the streams divided by the area they drain.

drainage pattern The pattern formed by the relationship between the courses of the streams draining an area of land. This often reflects the type of underlying rock. There are nine basic patterns: dendritic; sub-dendritic; trellis; parallel; sub-parallel; radial; rectangular; annular; and pinnate.

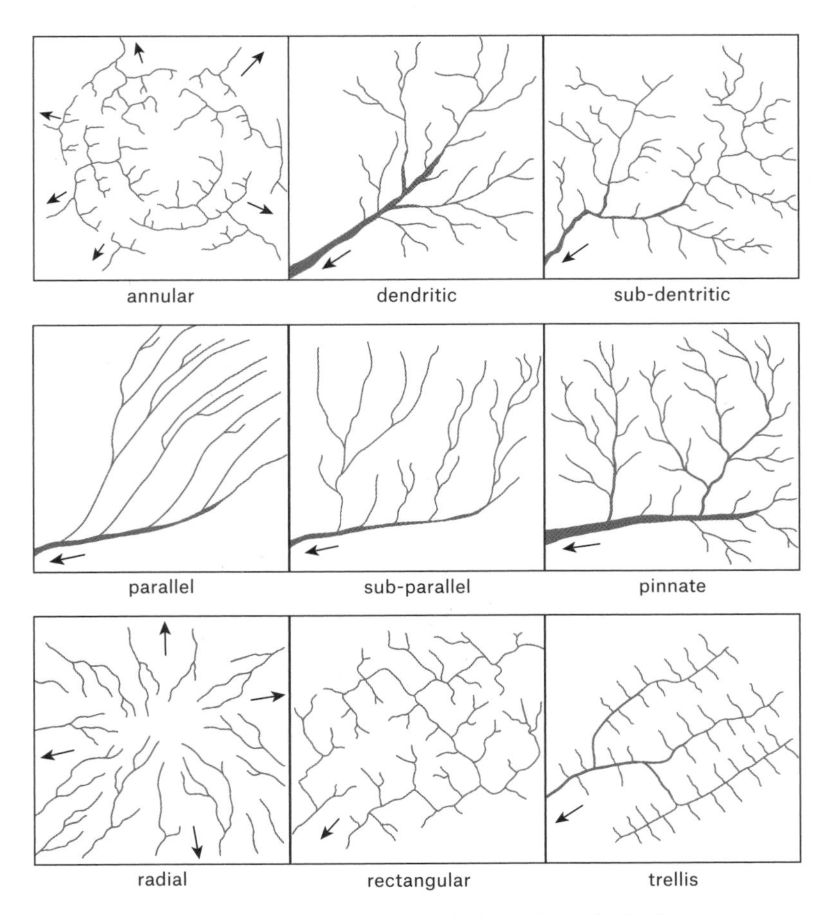

The nine basic patterns formed by water as it drains from the land.

drainage wind *See* katabatic wind.

drawdown The lowering of the **water table** in an area from which water is being abstracted through a well or borehole.

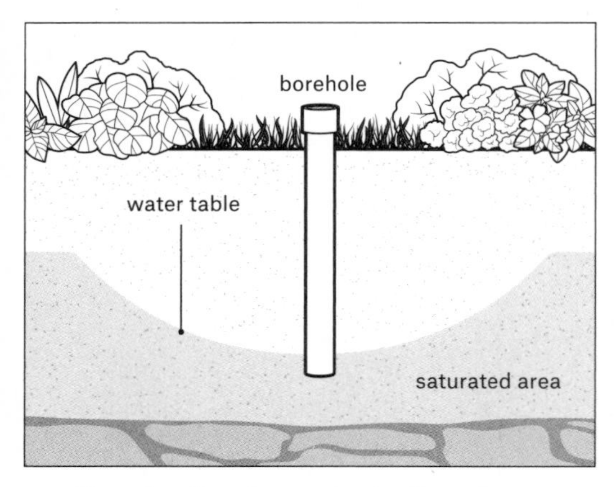

The extraction of groundwater through a borehole or well lowers the water table in the immediate vicinity.

drepanium A **cyme** in which there are flowers on only one side of the **axis**. The **inflorescence** is often flat and bent to the side.

Drepanophycus A genus, now extinct, of tree-like horsetails with an erect or arched stem up to 1 m long and true roots. It lived during the Devonian period (416–359.2 million years ago).

drey The nest of a squirrel.

drift A sediment deposited by a glacier.

Driloleirus americanus (giant Palouse earthworm, Washington giant earthworm) A pale pink earthworm (**Annelida**) that is known to grow to about 500 mm long and is rumoured to reach twice that length. The worm burrows to depths of more than 4 m, emerging at night to feed on plant material at the surface. It inhabits bunch-grass Palouse prairie in parts of Washington, Idaho, and Oregon. Once widespread and common it is now seldom seen and its survival is threatened by **habitat** loss and **competition** from invasive **exotic** species.

Driloleirus macelfreshi (Oregon giant earthworm) A pink earthworm (**Annelida**) that was first discovered in 1937 and is believed to be threatened by loss of **habitat**. It occurs only in Oregon, in the **acid soils** of coniferous forests. The worm grows up to 1.3 m long and 50–100 mm wide and when disturbed emits an odour reminiscent of lilies.

drizzle Precipitation that consists of liquid droplets all of approximately similar size, very close together, and smaller than 0.5 mm in diameter.

drone A male ant, bee, or wasp (**Hymenoptera**) that has the single function of mating with the queen, otherwise contributing nothing to the colony.

drone flies *See* Syrphidae.

drop *See Sclerotinia sclerotiorum.*

Drosera (family **Droseraceae**) A genus of **perennial**, occasionally **annual**, carnivorous **herbs** (sundews), often with **tubers** or **rhizomes**, with rosettes of leaves with long or circular **blades** bearing stalked glands that secrete a sweet substance that attracts insects, a sticky substance on which insects become trapped, and **enzymes** that digest the prey, and **sessile** glands that absorb the nutrient liquid. The stalks bend toward the centre of the leaf on contact. Trapped insects usually die within 15 minutes. Most of the nutrients acquired in this way are used in seed production; they are not essential to the plant's survival, but allow it to live on poor soils, usually in acid **bogs**. Flowers are **actinomorphic**, usually **pentamerous**, **ovary superior** of 1 **carpel**. Fruit is a **dehiscent capsule**. There are 110 species with **cosmopolitan distribution**, except for Antarctica, and especially well represented in Australia and New Zealand. Many are cultivated for ornament and some produce edible **corms** and substances with industrial uses.

Droseraceae (order **Caryophyllales**) A family of **perennial**, occasionally **annual**, carnivorous **herbs**, with leaves in **whorls** or spirally arranged, with or without **stipules**, and hinged to form a bilobed trap or bearing tentacles that trap insects (*see Drosera*). Flowers are **actinomorphic**, **hermaphrodite**, usually **pentamerous** occasionally 4-merous or up to 12-merous, free or fused **sepals**, free **petals** alternating with the sepals and often brightly coloured, usually 5 free **stamens**, **ovary superior** usually of 5 **carpels** with 1 **locule**. **Inflorescence** a terminal or lateral **cyme**. Fruit is a **dehiscent capsule**. There are 3 genera of 115 species with a worldwide distribution.

Drosophyllaceae (order **Caryophyllales**) A **monotypic** family (*Drosophyllum lusitanicum*) of carnivorous **perennial subshrubs** with **abaxially** coiled, **linear** leaves that have sticky tentacles on the underside. Flowers are **actinomorphic**, **hermaphrodite**, **pentamerous**, the **sepals** alternating with sulphur-yellow **petals**, 10 free **stamens** in 2 **whorls** of 5, **ovary superior** of 5 fused **carpels** and 1 **locule**. **Inflorescence** is terminal, a **cymose panicle** covered in glandular hairs. Fruit is a dry, **indehiscent capsule**. The plant occurs only in the southern Iberian Peninsula and Morocco.

drought A prolonged period during which the amount of precipitation falling over a particular area is markedly less than the amount that usually falls in that place over a similar period at that time of year. The length of time needed to define a drought varies from place to place.

Drude, Carl Georg Oscar (1852–1933) A German botanist, ecologist, and biogeographer, who described types of vegetation in terms of **formations**. He collaborated with H. G. A. **Engler** to produce *Die Vegetation der Erde* (Vegetation of the Earth) published between 1896 and 1923, and published *Die Ökologie der Pflanzen* (The Ecology of Plants) in 1913.

drumlin A smooth, oval, low hill, tapered at one end and rounded at the other, resembling a half-buried egg and made from glacial **drift**, usually **clay** but sometimes solid rock. Drumlins form beneath a glacier or advancing ice sheet and are aligned with the direction of flow. Drumlins usually occur in groups called drumlin fields or drumlin swarms.

drumlin field *See* drumlin.

drumlin swarm *See* drumlin.

drupe A fleshy fruit containing one or a few seeds, each surrounded by a stony layer. A plum is a drupe.

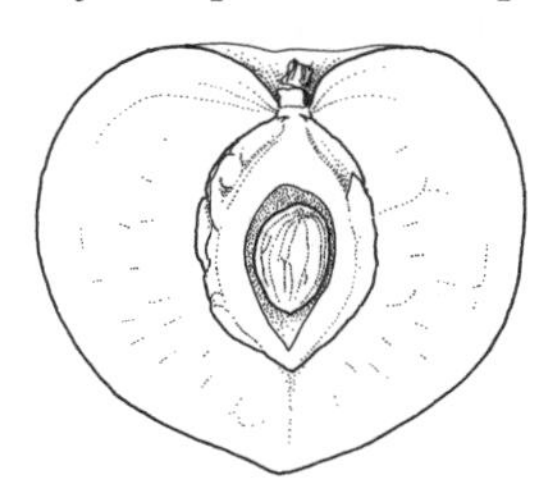

Drupe fruit (nectarine).

drupelet The individual small **drupes** of certain **aggregate fruits**, e.g. blackberry.

druse (sphaeroraphide) A mass of crystals, usually of calcium oxalate, that form in many species of algae (*see* alga), **angiosperms**, and **gymnosperms**, free inside cells or attached to **cell walls**. Druses are toxic and may deter **herbivores**.

dry adiabatic lapse rate (DALR) The rate at which a rising or subsiding body of dry air cools or warms adiabatically. It is 9.8°C/km.

dryad's saddle *See Polyporus squamosus*.

dry bubble disease A fungal disease that affects *Agaricus bisporus*, the most widely grown commercial mushroom. A mass of tissue appears among a cluster of mushrooms and overwhelms them, the **stipes** may split, and spots of necrotic tissue appear on the caps of infected mushrooms. The disease is caused by

Verticillium fungicola, which also parasitizes many wild fungal species.

dry-bulb temperature The air temperature registered by thermometer with a dry bulb that is exposed directly to the air. *Compare* wet-bulb temperature.

D

dry climate A **climate** in which the average annual precipitation is less than the **potential evapotranspiration**, so the ground is relatively dry most of the time, restricting plant growth.

dry deposition An equivalent of **acid precipitation** in which acid particles carried in dry air adhere to plant surfaces. This tends to be more harmful than **acid rain**.

dry haze **Haze** that contains no water droplets.

dry line (dewpoint front) A boundary that frequently forms in spring and summer over the Great Plains of North America between hot, dry air to the west and warm, moist air to the east. As the dry air advances it pushes beneath the moist air, raising it and often triggering the formation of huge **cumulonimbus** clouds. Dry line storms often produce **tornadoes**.

dry-matter production The dry weight of material produced by plants or animals in a unit area over a specified period.

Dryocopus pileatus (pileated woodpecker) A woodpecker that is about the size of a crow, 400–500 mm long with a wingspan of 660–750 mm. It is mainly black with a red crest, a white line on the throat, and white on the wings. It inhabits forest and feeds on insects, fruits, and nuts. It occurs in forests throughout North America.

dry snow Snow made from ice crystals that are linked directly, with no liquid water between them.

dry spell A period during which no precipitation falls, but that is shorter than a **drought**. In the United States, a dry spell is a period of two weeks or longer during which no measurable precipitation falls.

dry tongue A long, narrow protrusion of dry air into a region of moister air.

duckweed See *Wolffia*.

ductless gland See endocrine gland.

duff Plant material, e.g. leaves, needles, twigs, that has fallen to the ground.

Duke of Argyll's tea-plant See *Lycium*.

dulosis Slave-making behaviour found in certain parasitic species of ants that raid the nests of other ant species and remove pupae which they raise to forage for food that they give to their captors.

dulse The edible red seaweed *Palmaria palmata* (phylum Rhodophyta), in which the **thallus** is flattened and usually **dichotomously branching**. It grows in the intertidal zone.

Dumetella carolinensis (gray catbird) A species of grey birds with a black cap and black tail, 200–240 mm long with a wingspan of 220–300 mm, that inhabit dense shrub thickets, occasionally entering gardens. They feed on insects, spiders, and fruit, helping to control pests, especially **caterpillars**. They occur throughout much of North America.

dunce caps See *Conocybe*.

dunnock See *Prunella modularis*.

duplication The occurrence within a **chromosome** of more than one copy of a particular section.

duric horizon A subsurface **soil horizon** that is hard because it contains at least 10 percent of weakly **cemented** or indurated (*see* induration) nodules of **silica**, called durinodes.

duricrust A weathered soil deposit (*see* weathering) at or close to the surface, that may eventually form a hard mass. It forms primarily in subtropical environments and different minerals dominate particular types. *See* alcrete, caliche, ferricrete, silcrete.

Du Rietz, Gustaf Einar (1895–1967) A Swedish ecologist, botanist, and phytosociologist, who founded the Uppsala School of **Phytosociology** independently of the Zürich-Montpellier School established by Josias **Braun-Blanquet**; the two schools later merged.

durinode *See* duric horizon.

duripan (petroduric horizon) A **diagnostic horizon** made from material **cemented** by **silica**; if it occurs at the surface it is a **silcrete**.

durisols A group of soils that have a **duric horizon** within 100 cm of the surface. Durisols are a reference soil group in the **World Reference Base for Soil Resources**.

dusky salamander *See Desmognathus fuscus.*

dust Solid particles that are small enough to be raised and transported by wind.

dust mulch A **mulch** made by working the soil surface so intensively as to produce a very fine, **dust**-like texture. The technique is said to conserve moisture, but it is controversial.

dust whirl (dancing devil, desert devil, sand auger, sand devil) A small, rotating column of air that is strong enough to carry aloft **dust**, dry leaves, scraps of paper, and similar light items. It may be up to 100 m wide and rise to 1000 m.

Dutch elm disease A **vascular wilt** disease that can affect all elm (*Ulmus*) and *Zelkova* species, causing **branches**, then the tree, to wilt and die. It is caused by the **ascomycete** fungus *Ophiostoma ulmi* in Europe and North America early in the 20th century, *O. himal-ulmi* in the Himalaya, and *O. novo-ulmi* in Europe and North America during the later 20th century. The pathogen is spread by elm bark beetles (**Scolytidae**).

Dutchman's breeches (*Corydalis*) *See* Papaveraceae.

Dutchman's trousers *See Lamprocapnos spectabilis.*

dwarf bee (*Apis florae*) *See Apis.*

dwarf willow *See* arctic scrub.

dyad During the first division of **meiosis**, one of the products of the **disjunction** of a **tetrad**.

dyer's fucus *See* bladder wrack.

Dysaphis foeniculus (carrot root aphid, hawthorn aphid, parsley aphid) A species of aphid (**Aphididae**), native to Eurasia but now widely distributed in North America, that overwinters on hawthorn (*Crataegus* spp.), moving in spring to members of the **Apiaceae**, especially carrot, but also celery, fennel, parsley, and parsnip. It forms colonies on the root, just below ground level and feeds on the **taproots**; the aphids are often attended by ants. Infestations seldom cause severe damage, but may stunt growth.

Dysaphis plantaginea (rosy apple aphid) A species of aphid (**Aphididae**) that lays eggs in autumn in crevices in the **bark** or on leaf **axils** of apple trees. The eggs overwinter, emerging in spring as a generation of females, which produce the first of several generations of live young. Winged adults appear in midsummer. The aphids cause leaves to curl and often to turn bright red, often resulting in malformation of the fruit. The **honeydew** the aphids excrete attracts **sooty mould**.

dystrophic Describes standing water that is brown, owing to large amounts of **humus**, and usually shallow but if deeper the lower water will be depleted of oxygen.

E

E

-eae In plant **taxonomy**, the suffix used to indicate a **tribe**.

early blight of potato and tomato A disease that produces dark lesions on leaves; these grow, leading to **chlorosis** and significant **defoliation**, and may girdle tomato stems, killing the plant. The disease is caused by the **ascomycete** fungus ***Alternaria*** *solani*. The disease affects crops in parts of North America but is uncommon in Europe. *Compare* late blight of potato.

earthflow The movement of unconsolidated material down a hillslope.

earth tongues *See* Geoglossaceae, Neolectomycetes.

earwigs *See* Dermaptera, Forficulidae.

East African steppe floral region The area from southern Ethiopia to South Africa and westward to the Atlantic coast of northern Angola. There are about 150 **endemic** genera including *Saintpaulia*, several species of which are widely grown as African violets.

eastern box turtle *See Terrapene carolina.*

eastern coral snake *See Micrurus fulvius.*

eastern cottonwood (*Populus deltoides*) *See Populus.*

eastern diamondback rattlesnake *See Crotalus adamanteus.*

eastern garter snake *See Thamnophis sirtalis.*

eastern goldfinch *See Carduelis tristis.*

eastern gray squirrel *See Sciurus carolinensis.*

eastern hognose snake *See Heterodon platirhinos.*

eastern kingsnake *See Lampropeltis getula.*

eastern narrowmouth toad *See Gastrophryne carolinensis.*

eastern newt *See Notophthalmus viridescens.*

eastern phoebe *See Sayornis phoebe.*

eastern pigmy rattlesnake *See Sistrurus miliarius.*

eastern towhee *See Pipilo erythrophthalmus.*

Ebenaceae (order **Ericales**) A family of small, **evergreen trees** and **shrubs** with black **bark** and roots, **alternate** (occasionally **opposite**), **simple**, **entire**, **exstipulate** leaves, usually with **petioles.** Flowers **actinomorphic**, usually **unisexual** (plants **dioecious**) but occasionally **bisexual**, **trimerous** or **pentamerous** (occasionally 6- or 7-merous), fused **sepals**, **petals** fused into a tube with as many lobes as there are sepals, **staminate** flowers with **stamens** in 2 whorls with 2 or 4 times the number of stamens as sepals, **pistillate** flowers with a single whorl of **staminodes**, **ovary superior**

and **sessile** with as many **locules** as there are petals and sepals. **Inflorescence axillary**, sometimes a single flower. Fruit is a **berry**. There are 4 genera with 548 species in tropical and warm temperate regions. *Diospyros* species yield the timber ebony and several species have edible fruit including persimmons.

ebony (*Diospyros* spp.) *See* Ebenaceae.

ebracteate Lacking **bracts**.

eccrine gland (sweat gland) A gland in the skin of **Mammalia** that opens to the exterior and secretes a solution mainly of salt water (sweat). The **evaporation** of sweat cools the skin.

Ecdeiocoleaceae (order **Poales**) A **monocotyledon** family of **caespitose**, **monoecious**, rush-like **herbs** with **rhizomes**, the leaves reduced to sheaths. Flowers are flattened, **trimerous**, with 2 **whorls** of 3 **tepals** and either 2 sets each of 3 **stamens** or 2 or 3 superior **carpels. Inflorescence** 1 or a few **spikelets** each with several **glumes**. Fruit is a **nut** or **capsule**. There are two genera with three species occurring in southwestern Australia.

ecdysis The periodic shedding of the **exoskeleton**.

ecdysone A **hormone** that triggers moulting in insects.

ecesis The successful establishment of a migrating plant or animal in a new **habitat**.

Echinops (family **Asteraceae**) A genus of plants (globe thistles) with spiny leaves and globular blue or white **inflorescences**, each individual flower with an **involucre** of bristle-like **bracts**. There are about 120 species occurring from Europe to Central Asia and in the mountains of tropical Africa. Many are cultivated for ornament.

echinulate Covered with spines or small points.

ecocline (ecological gradient) A gradual change when moving from one **ecosystem** to another, when there is no sharp boundary between them.

ecological efficiency A measure of the flow of energy through an **ecosystem**, calculated as the proportion of the energy entering the system that is utilized within each **trophic level** and that passes between trophic levels.

ecological energetics The study of the flow of energy through **ecosystems**.

ecological factor *See* limiting factor.

ecological gradient *See* ecocline.

ecological indicator A type of organism or species that indicates by its presence or absence particular environmental conditions, e.g. many **lichens** are highly sensitive to air pollution, so their presence indicates clean air.

ecological isolation The separation of populations or groups of organisms that comes about because of changes in their environment. This may lead to reproductive isolation and eventually to the emergence of new species.

ecological pyramids (Eltonian pyramids) Graphical representations of the structure of an **ecosystem**, portraying the **trophic levels** as horizontal bars of equal thickness but different widths. The **producers**, usually green plants, form the lowest bar, with bars for the primary, secondary, and tertiary **consumers** arranged above them. The bar at each level is narrower than the bar below, so the structure forms a stepped pyramid. There are three pyramids. The pyramid of numbers represents the number of individuals at each level; the pyramid of **biomass** represents the biomass at each level; and the pyramid of energy represents the amount of energy available at each level. The pyramids were devised by the British zoologist Sir Charles Elton.

ecological system *See* ecosystem.

ecology The study of the interrelationships between individual and groups of organisms and between living organisms and the **abiotic** features of their environment. The German zoologist and naturalist Ernst Heinrich Haeckel coined the term *Ökologie* in 1866, from the Greek *oikos*, household.

economic injury level (EIL) The level of a pest infestation below which the cost of reducing the infestation would exceed that of the damage it causes.

ecosphere *See* biosphere.

ecosystem (ecological system) A discrete unit comprising living organisms and their **abiotic** surroundings that interact to form a stable system. Arthur Roy Clapham coined the term in 1930 and Arthur George **Tansley** popularized it.

ecosystem services The tangible benefits that people derive from natural processes that occur in functional **ecosystems**. These include provisioning, regulating, cultural, and supporting services. Provisioning services include the supply of food, raw materials, fuel, water, **genetic resources**, etc. Regulating services include plant **pollination** and the cleansing of air and water. Cultural services include recreational opportunities, and the appreciation of the aesthetic qualities of landscapes, seascapes, etc. Supporting services include the **hydrological cycle** and **biogeochemical cycles**.

ecotone A clearly defined and fairly narrow transition zone between two or more **ecosystems**.

ecotope The **habitat** in a **biogeocoenosis**.

ecotype A population of a widely occurring species that exhibits minor changes resulting from **adaptation** to local conditions.

ectal Outer; exterior.

ectexine *See* sexine.

ectocrine (exocrine, environmental hormone) A chemical substance that an organism releases into the **environment** and that affects another organism.

ectomycorrhiza A **mycorrhiza** in which the fungal **hyphae** do not penetrate the root of the host plant.

ectophloic siphonostele A **monostele** in which there is a cylinder of **phloem** around the **xylem**, both surrounding the **pith**.

ectopic recombination **Crossing over** between non-**homologous chromosomes**.

ectoplasm The outer, gel-like layer of **cytoplasm** containing many **microtubules** that lies immediately beneath the **cell membrane** in the cells of plants and some **Protozoa**.

ectotherm (heterotherm) An animal that maintains a fairly constant internal temperature by behavioural means, e.g. basking or seeking shade.

ectotrophic mycorrhiza A **mycorrhiza** in which the fungus forms a sheath with two layers around the roots of the host plant, the inner layer comprising a dense mesh of **hyphae**, called a hartig net.

edaphic Of or influenced by the soil.

edaphology The study of the influence of soils on living organisms, especially plants. It is one of the two major branches of soil science, the other being **pedology**.

eddy A turbulent flow of air, with frequent and erratic changes in direction.

edge effect A change in the number of species that are found in the area where two **habitats** overlap. The overlap area may contain species characteristic of both habitats, but since the overlap constitutes a habitat different from the others it will also support some species peculiar to itself. The area may, therefore, be rich in species, but since those species are ill-adapted to one or both of the adjacent habitats the rate of extinction is usually high, and the abundance of species attracts predators. Consequently, ecologists consider the edge effect to be a sign of ecological deterioration.

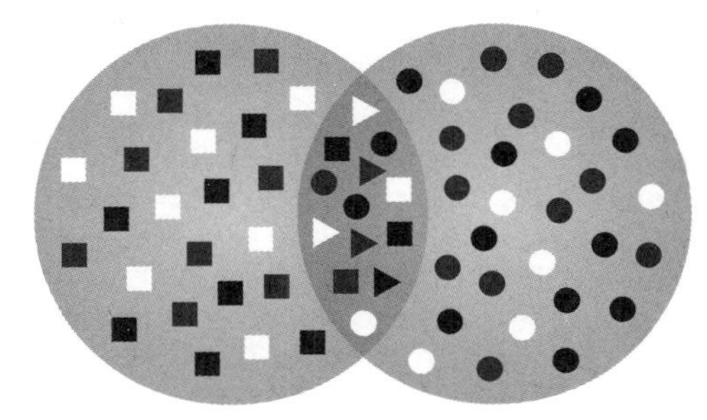

Edge effect. Where two habitats overlap and each supports different species (in this example one with black and white dots, the other with black and white squares), the area of overlap will support species from both habitats, plus species (black and white triangles) unique to the overlap area.

eel grass *See Zosteraceae.*

eelworms *See* Nematoda.

effective porosity **1.** The proportion of the **pore** space in a medium, e.g. rock or soil, that are interconnected and, therefore, capable of holding and transmitting fluids. If pore spaces are very small, as in **clay** soils, the **porosity** may be high, but **surface tension** binds the water tightly to the surface of clay particles. *Compare* absolute porosity, porosity. **2.** The proportion of the pore space through which **groundwater** is able to flow.

effective precipitation The amount of **precipitation** that is available to plants, i.e the amount of precipitation minus losses due to **evaporation**.

effective soil depth (ESD) The vertical distance from the soil surface to a layer of material that is impenetrable to plant roots. Soils are categorized by their effective soil depth as very shallow, ESD less than 25 cm; shallow, ESD 25–50 cm; moderately deep, ESD 50–90 cm; deep, ESD 90–150 cm; and very deep, ESD more than 150 cm.

effector molecules **Protein** molecules produced by **pathogens** that interfere with cell chemistry, thereby facilitating further infection.

effector-triggered immunity An immune response (*see* immunity) initiated by the detection of **effector molecules**.

effector-triggered susceptibility The infection of a plant by a **pathogen** that has been facilitated by the release of **effector molecules**.

effigurate Having a margin that lacks a definite form.

effused Spreading irregularly or loosely.

effused-reflexed Describes a fungal fruit body (*see* fruiting body) that is flat and spreading, with edges that curl upward or away from the substrate.

eft The juvenile, terrestrial stage in the life of a newt, especially in ***Notophthalmus viridescens***.

egestion The expulsion of undigested waste products.

egg wrack (knotted wrack, Norwegian kelp, rockweed) The brown seaweed (Fucaceae) *Ascophyllum nodosum* that grows, often abundantly, on sheltered rocky shores around the North Atlantic, and in sea lochs. The **thallus** is flattened and strap-like, with egg-shaped, air-filled **bladders**.

EIA *See* environmental impact assessment.

Eichhornia crassipes (water hyacinth) *See* Pontederiaceae.

EIL *See* economic injury level.

Eisenia fetida (brandling worm, red wiggler, red worm) A species of **epigean** earthworms (**Annelida**) that have retractable bristles on each segment with which they grip their surroundings as they move. When handled they can exude a foul-smelling liquid, hence the *fetida* in their name. They are rarely encountered in soil, preferring to live in decaying plant material, **compost**, and manure. They are native to Europe and have been introduced in every other continent except Antarctica and are bred commercially for sale.

E

Eisenia veneta (compost worm) A species of **epigean** earthworms (**Annelida**), very similar to ***Eisenia fetida***, about 100–200 mm long, that has a pink or yellowish body with dark red bands, usually with a paler **clitellum**. It inhabits decaying vegetation, soils rich in plant material, manure, and **compost** heaps. They are also known as *Dendrobaena veneta*. They are native to Europe but have been introduced widely elsewhere and are bred commercially for sale.

Eiseniella tetraedra (square-tail worm) A brown or yellow-brown earthworm (**Annelida**), 20–80 mm long with a distinct **clitellum**, that lives in mud, and beneath stones in ponds and rivers, and in wet places in woodlands, pastures, and gardens. It occurs throughout the world.

ektexine *See* sexine.

Elaeagnaceae (order **Rosales**) A family of **evergreen** or **deciduous shrubs**, **trees**, or climbers with leathery, **alternate**, **opposite**, or whorled, **simple**, **entire**, **exstipulate** leaves. Flowers **bisexual** or **unisexual** (plants **monoecious** or **dioecious**), **perigynous**, **perianth** often **petaloid** with tubular **hypanthium**, otherwise in a 2- to 4-lobed **whorl**, 4 **stamens** in 1 whorl or 8 in 2 whorls, **ovary superior** with 1 **locule**. Flowers **axillary**, solitary in clusters, **racemes**, or **spikes**. Fruit is an **achene**. There are 3 genera with 45 species occurring in northern temperate and warm tropical regions, Malesia, and Australia. Several species cultivated for ornament and some have edible fruits.

Elaeocarpaceae (order **Oxalidales**) A family of **trees** (some with **buttress roots**), **shrubs**, and **herbs** with **alternate**, **opposite**, spirally arranged, or occasionally whorled, usually **simple** leaves with **entire** or **serrate** margins with **stipules** or mucilaginous (*see* mucilage) hairs. Flowers **actinomorphic**, usually **hermaphrodite**, with 3–5 **valvate sepals**, 4 or 5 usually free, valvate **petals**, or no petals, 4 to many free **stamens**, **ovary superior** with 2 to many **locules**. **Inflorescence compound** or **simple**, **racemose** or **cymose** or sometimes of solitary flowers, **axillary** or terminal. Fruit is a **capsule** or **drupe**. There are 12 genera with 605 species occurring throughout the tropics, but not in mainland Africa. Several are cultivated for ornament and some produce edible fruit.

elaioplast A **leucoplast** that stores oils.

elaiosome A structure containing oil on the surface of a seed, usually as an attractant to ants.

Elaphe obsoleta (rat snake, Texas ratsnake, western rat snake) A species of slender colubrid snakes (**Colubridae**), 1.1–1.8 m long, with a triangular head, that are black except for a white chin. They have keeled (*see* keel) scales. The snakes inhabit a variety of **habitats** and are partly arboreal. They feed on rodents and other small vertebrates that they kill by constriction. They occur throughout eastern North America from southern Canada to Texas.

Elapidae (cobras, coral snakes, kraits, mambas, taipans) A family of venomous snakes that have short, rigid, grooved fangs at the front of the mouth through which they inject venom, in most species a neurotoxin (nerve poison). They are long, agile snakes, most **crepuscular**, and they feed mainly on small vertebrates. All are potentially deadly. There are 61 genera with 325 species distributed in tropical and subtropical regions of all continents except Europe.

elastic fibre Fibre composed of bundles of **elastin** that is found in the extracellular matrix of **connective tissue**.

elastic growth A reversible expansion of a **cell wall**.

elastin An elastic **protein** found in many parts of the bodies of vertebrates, which allows tissues to stretch and then resume their former shape.

elater One of the long, tubular cells with spiral thickenings found within the **spore**-bearing **capsule** in most liverworts (**Marchantiophyta**). In many species they assist spore dispersal.

Elateridae (click beetles, elaters, skipjacks, snapping beetles, spring beetles) A family of long (1–75 mm) beetles (**Coleoptera**) that right themselves if they fall on their backs by jack-knifing in a sudden movement that throws them into the air with a clicking sound. Most are black or brown with long antennae (*see* antenna) that are usually comb-like. *Phyrophorus* species (fireflies) are bioluminescent as larvae and adults. Adult elaterids feed on plant **sap**. Larvae (wireworms) are long, cylindrical, and tough. They live in rotten wood, soil, and among leaf litter. Some are serious pests, feeding on roots, others are predators or feed on dead wood. There are about 10,000 species with a worldwide distribution.

elaters *See* Elateridae.

Elatinaceae (order **Malpighiales**) A family of **annual** or short-lived **perennial herbs** and **subshrubs** with **opposite** or whorled, **entire** or **serrate** leaves with small **stipules**. Flowers usually small and inconspicuous, **actinomorphic**, **bisexual**, with 2–5(–6) free or partly **connate sepals**, 2–5 free **petals**, **ovary superior** of 2–3(–5) **carpels** with 2–5 **locules**. Flowers solitary in leaf **axils** or in **cymes**. Fruit is a **capsule**. There are 2 genera with 35 species occurring worldwide in tropical and temperate regions.

elder (*Sambucus*) *See* Adoxaceae, *Sambucus*.

electrode potential *See* oxidation-reduction potential.

electron carrier (redox carrier) A chemical compound that acts as a donor and acceptor of electrons or protons (hydrogen nuclei) in an **electron-transport chain**.

electron-transport chain A series of **electron carriers** that transport electrons and/or protons (hydrogen nuclei) by a sequence of **redox reactions** that occur in mitochondria (*see* mitochondrion) and in the processes of **photosynthesis** and **respiration**.

Eleocharis (family **Cyperaceae**) A genus of sedges (spikesedges, also called spikerushes), **caespitose annual** or **perennial herbs** with leaves reduced to sheaths and photosynthesizing stems growing from a **rhizome**. Some species are aquatic, most with submerged, branching stems; some able to switch between **C3** and **C4 pathways** of **photosynthesis**. **Inflorescence** is a many-flowered **spikelet**. There are about 150 species with worldwide distribution. **Tubers** of *E. dulcis* are Chinese water chestnuts.

elephant ear *See Xanthosoma*.

elevation head *See* hydraulic head.

elfin woodland A type of tropical **montane forest** that occurs on exposed sites or those with an extreme **climate**. It comprises dwarfed and distorted trees. *See also* kampfzone, krummholz.

elliptic *See* elliptical.

elliptical (elliptic) Of a leaf, oval, widest in the middle, and with a pointed tip.

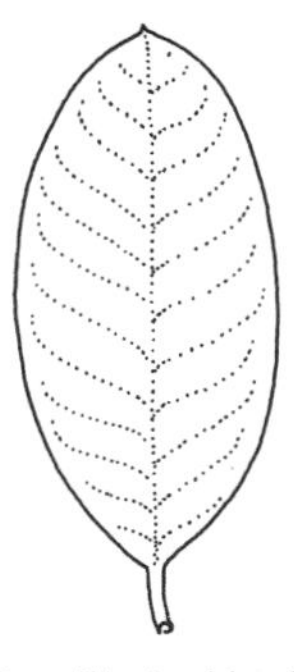

An elliptical leaf has pointed ends and is widest in the middle. This is a rhododendron leaf.

elm (*Ulmus*) *See* Ulmaceae.

El Niño A change in the distribution of surface **atmospheric pressure** over the South Pacific Ocean at intervals of two to seven years, bringing a weakening or even reversal of the easterly trade winds and the Equatorial Current that they drive. Warm water floods eastward and the water around Indonesia cools. It produces widespread effects on the weather. *See* teleconnections.

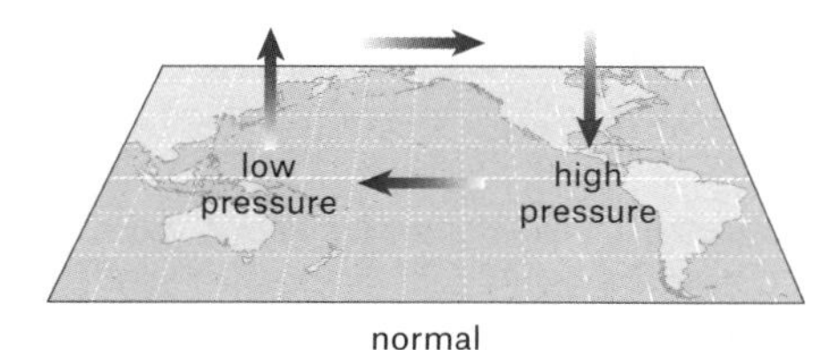

At average intervals of two to seven years, a change in the distribution of surface pressure over the equatorial South Pacific causes a change in wind strength and direction and a weakening of the westerly Equatorial Current.

El Niño-Southern Oscillation *See* ENSO.

ELR *See* environmental lapse rate.

Eltonian pyramids *See* ecological pyramids.

eluvial horizon (eluvial zone) A **soil horizon** rich in materials deposited by **eluviation**. It forms the lower part of the A horizon, designated Ae or E, and is typically pale in colour.

eluvial zone *See* eluvial horizon.

eluviation The removal by water of materials and substances from surface **soil horizons** and their deposition at a lower level in the **soil profile**. *Compare* leaching.

eluvium Soil material that is moved downward through the **soil profile** by the process of **eluviation**.

elytra *See* elytron.

elytron (pl. elytra) The hard fore wing of a beetle (**Coleoptera**) or **earwig**. It is often distinctively marked and in some species brightly coloured. The elytra are raised when the insect flies to allow free movement of the hind wings.

emarginated Having a notch at the tip, or at the edge of a fungal **gill** nearest the **stipe**.

Embden-Meyerhof pathway *See* glycolysis.

Emblingiaceae (order **Brassicales**) A **monotypic** family (*Emblingia calceoliflora*), consisting of a low-growing, probably short-lived, **perennial shrub**, its younger parts covered with stiff hairs. Leaves approximately **opposite** with **entire** or **undulate** margins. Flowers are solitary in leaf **axils** and are held upside down. They are **zygomorphic**, hypogynous (*see* hypogyny), the **calyx** of 5 fused **sepals** is split **adaxially**, 2 **connate petals**, the **ovary** is attached to the underside of the **androgynophore**. Fruit is **indehiscent**. The plant occurs only in western Australia.

Embothrium (family **Proteaceae**) A genus of **evergreen**, occasionally **deciduous trees** and **shrubs** with **simple**, **lanceolate** leaves that produce dense bunches of 4-lobed tubular, red occasionally white or yellow flowers that reflex, exposing the **stamens**. There are two to eight species occurring in southern South America. *Embothrium coccineum* (Chilean firebush) is widely cultivated for ornament.

embryo 1. A young plant that has developed from an **ovum**; in **seed plants** (**Spermatophyta**) the embryo is contained within the seed. **2.** A young animal that is developing from an egg (ovum) during the time it is contained within the egg membranes or within its mother's body.

embryogenesis *See* embryogeny.

embryogeny (embryogenesis, embryony) The process of forming an **embryo**.

embryony *See* embryogeny.

Embryophyta (Metaphyta) A subkingdom comprising those plants that produce **embryos** from multicellular reproductive organs, i.e. **Bryophyta**, **angiosperms**, and **gymnosperms**. Since these are predominantly terrestrial they are sometimes called land plants.

embryophyte A plant that produces **embryos** by means of multicellular reproductive organs, i.e. **Bryophyta** and **Spermatophyta**.

embryo sac (megagametophyte) In **angiosperms**, the female **gametophyte**, formed by the division of the **nucleus** of the **haploid megaspore** and consisting of six haploid cells lacking **cell walls** and two haploid nuclei. It is where the **ovum** is fertilized and the **embryo** develops.

emergence marsh The upper region of a **salt marsh**, between mean high-water level and the mean level of spring tides. Usually, it is submerged no more than 360 times a year and for less than 30 minutes during hours of daylight.

emergent Describes an aquatic plant with roots below water and stems that rise above the surface.

emigration *See* migration.

emissary sky A sky that is covered by patches of **cirrus**; it is an emissary of the wind and rain that will soon arrive.

enantiostyly A **flower morphology**, developed in at least ten **angiosperm** families, both **dicotyledons** and **monocotyledons**, in which flowers are mirror images of each other, with the **style** deflected to the left or right of the **axis**. There are two forms: monomorphic enantiostyly in which both forms occur on the same plant, and dimorphic enantiostyly in which the two forms occur on different plants. ↗

enation 1. A leaf-like structure that grows laterally from a stem. 2. An outgrowth from a leaf symptomatic of a viral disease.

Encarsia formosa A species of wasps (**Hymenoptera**), about 0.6 mm long, that is a **parasitoid** of ***Trialeurodes vaporariorum***. Females have a black and yellow abdomen and opalescent wings. Males are black and larger than females. Females lay eggs in third **instar nymphs** or **pupae** of the hosts. *Encarsia formosa* is used in **biological control**.

Enchytraeidae (potworms) A family of **Oligochaeta**, comprising white, transparent, segmented worms, about 25 mm long and 0.7–1.5 mm wide, resembling tiny earthworms, that inhabit moist, acid environments. They feed on decaying organic matter, do not feed on living plants, and their burrows improve soil structure.

Enchytraeus buchholzi (whiteworm, grindal worm) A species of earthworms (**Annelida**) that are white, about 25 mm long and 6 mm wide, and that feed on decaying organic matter. They often thrive in **compost** heaps and they are also bred for use as fishing bait.

Encyrtidae A family of brown or black wasps (**Hymenoptera**), 0.5–7.0 mm long, in which the middle legs are adapted for jumping. About half of all species are **parasitoids** or egg predators of scale insects (**Coccidae**), others of various insects and arachnids (**Arachnida**). Many species are polyembryonic (*see* polyembryony), producing from 10 to 1000 or more young from a single egg. ↗

endarch Describes primary **xylem** that develops outward from the **axis**. *Compare* exarch, mesarch.

endemic Describes a **taxon** that occurs only in a particular region, the size of the region depending on the taxonomic rank of the taxon, e.g. a family is likely to be endemic to a larger area than a genus.

endexine *See* nexine.

endive (*Cichorium endiva*) *See Cichorium.*

endobiotic Growing within a living organism.

endocarp *See* pericarp.

endocrine gland (ductless gland) A gland that secretes **hormones**.

endocytosis The process by cells absorb molecules by engulfing them in an **invagination** of the **cell membrane** that is then detached.

endocytosymbiosis A type of **endocytosis** in which the engulfed cell forms a symbiotic (*see* symbiosis) relationship with the cell that engulfed it.

endodermis A layer of tissue that forms a boundary between the **cortex** and **stele** in the roots and shoots of **Pteridophyta** and the roots but not usually stems of **Spermatophyta**. It is surrounded by the **Casparian strip**.

endogeic Describes an earthworm that makes horizontal burrows and lives and feeds below ground.

endolithic Living inside rock, coral, or an animal shell.

endomitosis A doubling of the number of **chromosomes** in a cell that fails to divide, leading to a type of **polyploidy** called endopolyploidy.

endomycorrhiza A **mycorrhiza** in which the fungal **hyphae** penetrate the cells of the host plant.

endoparasite *See* parasite.

endoperidium The inner of the two layers of the **peridium** found in certain **basidiomycete fungi**.

endophloeodal Growing inside **bark**.

endophyte A plant that lives inside another plant and is not a parasite.

endoplasm The inner **cytoplasm** of plant and some protozoan (*see* Protozoa) cells, containing the principal **organelles**.

endoplasmic reticulum (ER) A complex network of **tubules**, **cisternae**, and **vesicles** in the **cytoplasm** of the cells of all **eukaryotes**. There are three types of endoplasmic reticula. Rough ER synthesize **protein**, smooth ER synthesize **lipids** and **steroids**, and sarcoplasmic ER regulate calcium levels.

endopleura *See* tegmen.

endopolyploidy *See* endomitosis.

endopterygote Describes an insect (**Insecta**) in which the wings develop inside the body of the larva until the final moult, when the larva pupates and undergoes **metamorphosis**.

endorheic basin *See* closed drainage basin.

endorheic lake A lake that loses water only by **evaporation**.

endoskeleton A skeleton that is entirely inside the body.

endosperm In the seeds of many **angiosperms**, a **triploid** structure that surrounds and nourishes the **embryo**.

endospore A type of resting cell to which certain **Bacteria** (e.g. ***Bacillus*** and *Clostridium*) reduce themselves when environmental conditions threaten the survival of the **vegetative cell**. Endospores remain viable for thousands of years and when favourable conditions return the cell resumes its vegetative state. Despite the name, an endospore is not a **spore**.

endosulfan An **organichlorine insecticide** and **acaricide** that is persistent and bioaccumulates (*see* bioaccumulation). Its production and use are being phased out throughout the world.

endosymbiont A **symbiont** that lives inside the body of its host.

endosymbiosis **Symbiosis** in which one **symbiont** lives inside the body of the other.

endosymbiotic theory The theory, supported by compelling evidence, that **chloroplasts** and mitochondria (*see* mitochondrion) originated as independent **Bacteria** that became incorporated into the **cells** of **eukaryotes** as endo (inside) **symbionts**. Both chloroplasts and mitochondria possess their own **DNA**, which is different from the host's nuclear DNA, and both synthesize **proteins** and **enzymes** needed for their own functioning.

endothecium 1. In **angiosperms**, a layer of tissue lining the **lumen** of the **anther** that secretes substances necessary to the maturation of the **pollen grains**. 2. In mosses (**Bryophyta**) a cylinder of tissue surrounding a **columnella** of sterile cells in the **sporangium**.

endotherm An animal that maintains an approximately constant body temperature by means of internal mechanisms, e.g. shivering, sweating, panting. Birds (**Aves**) and mammals are endotherms.

Endothia parasitica *See Cryphonectria parasitica.*

endotoxin A substance present in the **cell walls** of Gram-negative (*see* Gram reaction) **Bacteria** that is toxic to animals.

endotrophic mycorrhiza A **mycorrhiza** in which the fungal **hyphae** penetrate the root of the host plant without altering the **morphology** of the root.

endozoochory Dispersal of seeds or **spores** by being ingested and excreted by animals.

endrin An **organochlorine insecticide** and **rodenticide** that is toxic to aquatic animals and harmful to humans if it enters domestic water supplies. Its use is banned in the European Union, United States, and many other countries.

energy flow The transfer of energy from organism to organism through a **food chain** or **food web**.

energy of activation *See* activation energy.

Engler, Heinrich Gustav Adolf (1844–1930) A German plant taxonomist and phytogeographer (*see* phytogeography) who devised a system of plant classification (Engler system) that is still used, and who collaborated with C. G. O. **Drude** in producing *Die Vegetation der Erde* (Earth's Vegetation), a major work on plant geography.

Ensete ventricosa (Abyssinian banana) *See* Musaceae.

ensiform Shaped like a sword.

ENSO (El Niño-Southern Oscillation) A complete cycle of **El Niño** and **La Niña**, associated with a **southern oscillation**.

Entandrophragma cylindricum (sapele) *See* Meliaceae.

Entandrophragma utile (utile) *See* Meliaceae.

Enterobacteriaceae A large family of Gram-negative (*see* Gram reaction), rod-shaped **Gammaproteobacteria** that are **facultative anaerobes**. Some occur in water and soil, and many form part of the gut flora of animals. The family also includes parasites and pathogens or animals and plants.

enterotoxin A bacterial toxin that affects the intestinal mucosa of animals, causing diarrhoea and gastroenteritis.

entire Of a leaf, having an undivided margin.

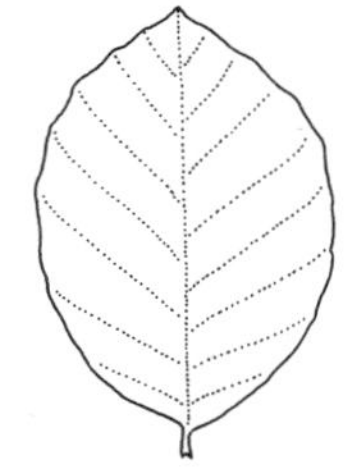

A leaf with an entire margin.

entisols Embryonic mineral soils with no visible **soil**

horizons. These are at the first stage of developing **soil profiles** and occur on **flood plains**, sand dunes, steeply eroding slopes, and recent **aeolian** or volcanic ash deposits. Entisols comprise an order in the U.S. Department of Agriculture **soil taxonomy**.

entomochory Dispersal of seeds or **spores** by insects.

E

entomopathogenic Able to cause disease in insects.

entomopathogenic fungus A fungus (*see* Fungi) that parasitizes insects and usually kills them. Such fungi are potential agents for **biological control**.

entomophilous Describes plants with **flowers** adapted to pollination by insects.

Entomophthorales An order of **Fungi** most of which are parasites of insects, some parasitizing **Nematoda**, **Tardigrada**, mites (**Arachnida**), and some living as **saprotrophs**. The **sporangium** acts as a single **conidium**, discharging its **spores** explosively.

Entomophthora muscae A species of **Fungi** that parasitizes many families of flies (**Diptera**) including houseflies (*Musca domestica*). **Spores** falling on a fly form a **conidium** on its body; within hours **hyphae** from the conidium penetrate the **cuticle** and enter the body cavity, digesting the contents and altering a section of the brain that causes the fly to climb upward and behave in a way that disperses fungal spores most effectively.

Entorrhizomycetes A class of **basidiomycete fungi** that survive in soil as **teliospores** and infect the roots of rushes (**Juncaceae**) and sedges (**Cyperaceae**), forming **galls**. There are four species.

entrainment Mixing between a body of air and the air surrounding it such that some air from each becomes incorporated in the other.

Entylomatales An order of **Exobasidiomycetes** that cause smut diseases in **dicotyledons**. There is one genus (*Entyloma*) with 160 species found worldwide.

environment The sum of the biological and physical conditions in which an organism lives.

environmental hormone *See* ectocrine.

environmental impact assessment (EIA, environmental impact statement) A document in which a person or organization proposing an industrial development, major project, or legislation affecting the **environment** sets out the anticipated consequences for the biological, physical, and aesthetic environment, and for human health and well-being. In most countries, the preparation and acceptance of an environmental impact assessment is required by law and is a condition of permission to proceed.

environmental impact statement *See* environmental impact assessment.

environmental lapse rate (ELR) The rate at which the air temperature decreases between the surface and the **troposphere** as this is measured in a particular place at a particular time. The cooling is not adiabatic (*see* adiabatic cooling and warming).

environmental variance The part of the difference between **phenotypes** that is due to **adaptation** to different environmental conditions.

enzyme A **catalyst** produced by a living **cell**.

eocyte A heat-loving, sulphur-metabolizing member of the Crenarchaeota, one of the divisions of the **Archaea**, that, on the basis of similarities in their **proteins**, may be more closely related to **Eukarya** than to other members of the Archaea. If so, the **three-domain system** may have to be abandoned in favour of a two-domain system in which Eukarya are included in Archaea. This proposal is known as the eocyte hypothesis. ↗

eocyte hypothesis *See* eocyte.

Ephedraceae (order **Pinales**) A **monogeneric** family (*Ephedra*) of **xeromorphic shrubs** and small **trees** with grooved, jointed stems that are the principal site of **photosynthesis**, the leaves reduced to scales that are soon shed. The main **branches** are whorled or **opposite**. Plants usually **dioecious**, occasionally **bisexual**. Small **cones** are borne along the shoots. There are 65 species occurring in drier temperate to tropical regions. Several species yield the drug **ephedrine**.

ephedrine An **alkaloid** derived from *Ephedra* (**Ephedraceae**) that has a variety of medicinal uses.

ephemeral Short-lived, completing its life cycle in a short time.

ephemeral stream A stream that flows only after heavy rain or snow-melt.

ephemerophyte An **ephemeral** plant.

epibiotic Growing on the surface of another living organism.

Epiblema uddmanniana (bramble shoot moth) A moth with a 15–20 mm wingspan and a prominent brown blotch on its grey forewings that flies in June and July. It is widely distributed in Europe and around the Mediterranean, and feeds on ***Rubus*** species.

epicalyx An **involucre** of **bracts** that resemble **sepals** but lie outside the **calyx**. ***Hibiscus*** species are divided into four major taxonomic groups on the basis of the **morphology** of the epicalyx.

Epichloë typhina A species of **ascomycete fungi** that is a pathogen of grasses (**Poaceae**). It enters through stems or leaves, possibly through wounds inflicted by sap-sucking insects, and causes the disease choke of grasses.

epicotyl The part of a **seedling** lying above the **cotyledons** that will develop into the shoot.

epicuticular wax Wax deposited on the surface of a **cuticle**.

epidermis The outermost layer of tissue; in plants and many invertebrates it is one cell thick, in vertebrates it is much thicker.

epigamic Describes a character with which an animal seeks to attract a partner for mating.

epigeal 1. Describes a germinating seed in which the **cotyledons** emerge above the ground surface carried on a **hypocotyl**. **2.** Describes a structure (e.g. **stolon**) that grows across the ground surface.

epigean (epigeic) Describes an organism that crawls across the surface or lives among leaf litter.

epigeic *See* epigean.

epigene Occurring or produced at the Earth's surface.

epigenetic Describes a change in **gene** expression caused by a mechanism other than a change in the **DNA**.

epigenetic drainage (superimposed drainage) A **drainage pattern** that formed over an earlier land surface that stood high above the present surface. Subsequent **erosion** and river incision lowered the surface, leaving a drainage pattern that bears no relation to the underlying geologic structure.

epigenetics The study of the way environmental and developmental cues determine **gene** expression.

epigeous Describes a plant that grows on the ground surface.

epigynous Describes a **flower** in which the **stamens**, **calyx**, and **corolla** are inserted near the tip of the **ovary**.

epilimnion The upper layer of warm water that forms in summer in lakes that become thermally stratified. Water in the epilimnion circulates. The epilimnion is usually shallower than the **hypolimnion**.

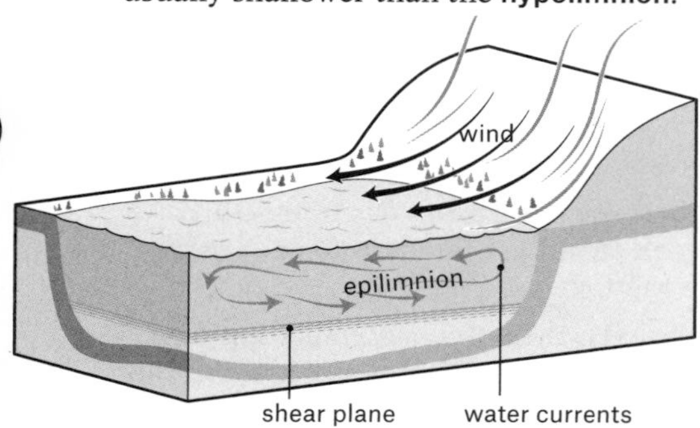

In summer, some lakes become thermally stratified. The wind then drives currents that cause the water to circulate above a shear plane separating the warmer water from the cooler water below.

epilithic Attached to or growing on the surface of rock.

epinastic growth *See* epinasty.

epinasty (epinastic growth) The more rapid growth on the upper or **adaxial** side of part of a plant than on the lower or **abaxial** side, causing that part to bend over. It is induced by **auxin**, **ethene**, and some **herbicides**, and may be a sign of disease, in which case there is a pronounced droop of the leaves.

epineuston The organisms that live on the surface of water.

epipedon A **soil horizon** that lies below the surface.

epipelic Growing on the surface of mud.

epipetalous Borne on or attached to the **petals**.

epiphloeodal Attached to or growing on the surface of **bark**.

epiphragm A temporary structure with which a **snail** seals the **aperture** of its shell to avoid **desiccation** during a time when the animal is inactive.

epiphyllous (folicolous) Growing on the surface of leaves.

epiphyte A plant that grows on the surface of another plant, using it for mechanical support but not obtaining nourishment from it.

epiphytotic An outbreak of a plant disease.

episepalous Borne on or attached to the **sepals**.

epistasis The prevention by an **allele** of one **gene** (the epistatic gene) of the expression of all alleles of another gene. *Compare* hypostasis.

epistatic gene *See* epistasis.

epitheca The older, outer half of the **frustule** of a **diatom**.

epithelium A layer of **cells** that lines the internal surface of certain organs, e.g. the developing **ovary**, the **resin** canal in **gymnosperms**.

epixylous Growing on the surface of wood.

epizoic Describes animals that attach themselves to the surface of other animals but are not parasites.

Epsilonproteobacteria A class of **Proteobacteria** most of which live in the digestive tracts of animals.

equatorial division During **metaphase** of **mitosis** and **meiosis**, the breaking of a **chromosome** into two equal halves, which become part of the two **daughter nuclei**.

equatorial plane *See* spindle.

equatorial plate During **metaphase** of **mitosis** and **meiosis**, an arrangement of the **chromosomes** so they lie in a single plane at the centre (equator) of the **spindle**.

Equigard *See* dichlorvos.

equilibrium level (level of zero buoyancy) The height at which rising air becomes neutrally buoyant (*see* buoyancy) and ceases to rise.

equilibrium species A species typical of a stable **environment** that maintains its access to resources by producing relatively few offspring and caring for them. *See K*-selection.

equinox One of the two dates each year when at noon the Sun appears directly overhead at the equator and, therefore, the Sun is above the horizon for 12 hours and below it for 12 hours everywhere in the world. These dates are 20–21 March and 22–23 September. *Compare* solstice.

Equisetites hemingwayi The earliest known member of the family Equisetaceae and the direct ancestor of the **extant** genus ***Equisetum***. It lived during the Carboniferous period (359.2–299 million years ago).

Equisetopsida *See* Sphenopsida.

Equisetum (family Equisetaceae) The only surviving genus of horsetails, **vascular cryptogams**, i.e. plants that reproduce by **spores**. They have hollow, photosynthetic stems, **whorls** of reduced, scale-like leaves that usually do not perform **photosynthesis**, **branches** in whorls, and apical **cones** producing spores that germinate into prothalli (*see* prothallus) bearing the sex organs. There are 25 species with worldwide distribution except Antarctica, usually in damp **habitats**. Horsetails are **living fossils.**

equitant Describes the arrangement of leaves that overlap at the base to form a flat, fan-like shape, but are not fused, e.g. the leaves of ***Iris***.

ER *See* endoplasmic reticulum.

ergastic Describes metabolic products of **cells** that are not part of the **protoplasm**.

ergatogyne In some ant species (**Hymenoptera**), an individual that is intermediate between a worker and a queen, having a **gaster** larger than that of a worker. It is able to function as an additional reproductive, or to replace the queen.

ergosome *See* polysome.

ergot *See Claviceps purpurea.*

Erica (family **Ericaceae**) A genus of mainly **calcifuge evergreen shrubs** (heath, heather) and some **trees** that have small, leathery, needle-like leaves and **campanulate** flowers with a **disc** usually borne in terminal **umbels** or **spikes**, sometimes **axillary**. There are about 860 species occurring from Europe to the Near East and mountains in Africa, at least 660 **endemic** to South Africa, where *Erica* is the largest genus of the **fynbos**.

Ericaceae (order **Ericales**) A family of **evergreen** or **deciduous shrubs**, **herbs**, and **trees**, with some scramblers and climbers, and some parasites lacking **chlorophyll**. Leaves are **simple**, **alternate** and arranged spirally or **opposite** or whorled, **entire** or **serrate**, sometimes needle-like, **exstipulate**. Flowers **actinomorphic**, usually **hermaphrodite**, with 4–5 or somtimes 2–9 fused **sepals**, 3–9 fused or 4–5 free **petals**, 2–8 fused or 5 free **stamens**, **ovary** usually **superior** of 4–5 or sometimes 1–14 fused **carpels**. **Inflorescence** is a terminal or **axillary raceme**. Fruit is a **capsule**, **berry**, or **drupe**. There are about 126 genera with 3995 species with worldwide distribution, but rare in tropical lowlands. Many species are cultivated for ornament, e.g. ***Erica***, ***Rhododendron***, ***Arbutus***, and *Pieris*.

Ericales An order of woody plants comprising 25 families with 346 genera and 11,545 species. *See* Actinidiaceae, Balsaminaceae, Cyrillaceae, Clethraceae, Diapensiaceae, Ebenaceae, Ericaceae, Fouquieriaceae, Lecythidaceae, Marcgraviaceae, Mitrastemonaceae, Pentaphylacaceae, Polemoniaceae, Primulaceae, Roridulaceae, Sapotaceae, Sarraceniaceae, Sladeniaceae, Styracaceae,

Symplocaceae, Tetrameristaceae, and Theaceae.

Erigeron (family **Asteraceae**) A genus of **annual**, **biennial**, but mainly **perennial herbs** (fleabane), with **simple** leaves in a **radical** rosette or **alternate**. Flower heads are on stalks, solitary or in **panicles**, with numerous **involucre bracts**. Ray **florets** are female, numerous, and in two or more rows; **disc florets bisexual**, tubular, 5-toothed. Fruit is an **achene** with a **pappus** of bristles. There are about 390 species occurring throughout temperate regions, but especially in America. Many are cultivated for ornament.

Erinaceus europaeus (hedgehog, European hedgehog) A mammal that is unmistakable because the hairs of its **pelage** are completely replaced by several thousand spines, about 22 mm long, covering the whole of its upper surface apart from its face. The spines are pale brown with a dark band near the tips. Hair on the underside of the body is brown. The hedgehog is about 160–260 mm long, males larger than females, and has a **plantigrade** gait. It inhabits woodland edges, hedgerows, sand dunes, and suburbs, entering gardens, and feeds almost entirely on invertebrates, with occasional bird eggs and chicks. It occurs throughout Europe and western Asia.

erineum A **hypoplasia** of **trichomes**, often accompanied by the accumulation of pigments, caused by mites.

Eriocaulaceae (order **Poales**) A family of tropical and subtropical, **annual** and **perennial**, mainly aquatic **herbs** (pipeworts) with **rhizomes**. Some non-aquatic species have leafy stems up to 4 m tall or large trunks with many **adventitious** roots. Many aquatics have leafy floating stems. Submerged leaves usually **linear** or **filiform** in rosettes arranged spirally. Flowers usually **unisexual** in **monoecious** species with males and females mixed in the flower head or with males in the centre surrounded by females. A few species **dioecious**. Flowers are small, **actinomorphic**, **trimerous** or 2-merous, with free **sepals**, **petals** usually fused into a tube, 1 or 2 whorls of 3 or 2 **stamens**, **ovary superior** of 2 or 3 fused **carpels** and 2 or 3 **locules**. **Inflorescence** a **spike** of 10 to more than 1000 flowers at the end of a leafless **peduncle**, or up to 100 spikes in an **umbel**. Fruit is a membranous, **loculicidal capsule**. There are 6 genera with 1160 species found throughout the tropics and extending into temperate regions. Many are gathered as everlasting flowers.

Eriophyes similis (plum pouch-gall mite) A species of **gall**-forming, worm-like mites (**Eriophyidae**), about 0.5 mm long, that are active in summer, feeding on ***Prunus*** species, especially blackthorn, forming galls on leaves, most densely around the leaf margins.

Eriophyidae (gall mites) A family of mites (**Arachnida**) that are parasites, many triggering the formation of **galls** on a wide variety of plants. The mites are very small, worm-like with two pairs of legs, and yellow or pink in colour. They disperse mainly by wind. They are serious pests, although some species are used in the **biological control** of invasive plants. There are more than 200 genera, with at least 3600 species and probably many more.

Eriosoma lanigerum (woolly aphid, woolly apple aphid) A species of red-purple aphids (**Aphididae**) with bodies covered in woolly masses of pale blue wax. They overwinter as eggs, or as young **nymphs** inside root **galls**, emerging in spring as females that lay eggs in crevices on the **bark** of host species. **Apterous** nymphs feed on new growth before turning into winged forms that migrate to other host plants, where they feed on wounds in **branches** and stems before moving to the roots. The aphids infest apple, pear, quince, and several other tree species. They occur throughout temperate regions in both hemispheres.

Eriostemon (family **Rutaceae**) A genus of **shrubs** with **alternate**, **simple** leaves bearing oil glands. Solitary white, blue, or pink flowers (wax flowers) have a large **calyx** of 5-pointed **sepals**, 5 spreading, waxy **petals** usually forming a star shape, 10 **stamens**. There are 2 species, **endemic** to Australia, *E. australasius* and *E. banksii*; 30 species of wax flowers formerly included in the genus have been transferred to the closely related genus *Philotheca*.

Erithacus rubecula (European robin, robin) A **passerine** bird, 125–140 mm long with a wingspan of 200–220 mm, that has a pale brown back and wings, buff underside, and an orange breast. Formerly classed as a member of the **Turdidae**, the robin is now placed in the Muscicapidae (fantails and flycatchers). It inhabits woodland and in Britain it is common in parks and gardens. It feeds on invertebrates and seeds. The robin is very aggressive to other members of its own species, and is renowned for being unafraid of humans, often approaching people disturbing the soil in the hope of taking earthworms. It occurs throughout Eurasia.

ermine moths *See* Yponomeutidae.

erodibility The extent to which a soil is susceptible to **erosion**. This depends on the ability of the soil to resist the impact of raindrops on the surface, and the shearing action of flowing water in **gullies** and **rills** on **clods**, which depends on the rate of water flow and the size of soil particles.

erosion 1. The removal of surface material through the action of water, wind, moving ice, and **creep**. **2.** That part of the process of **denudation** that involves the breaking down, dissolving, and transport of surface material.

erosion rate The rate at which **erosion** occurs, measured in **bubnoff units** (B). Glacial **abrasion** = 1000 B; soil **creep** in temperate maritime conditions = 1–5 B; **solifluction** = 25–250 B; erosion resulting from human activity, e.g. agriculture = 2000–8000 B.

erosion surface 1. (planation surface) A gently rolling land surface that is the product of a very long period of **erosion**. **2.** A surface that has been cut into rock or sediment by the action of water, ice, or wind.

eruciform **Caterpillar**-like; having a cylindrical body with true legs on the thorax and prolegs in the hind region.

erumpent Bursting out.

Erwinia A genus of **Enterobacteriaceae**, comprising Gram-negative (*see* Gram reaction), rod-shaped, **motile Bacteria**, which includes several plant pathogens.

Erwinia amylovora A species of **Enterobacteriaceae** that causes fire blight, a contagious, systemic disease affecting members of the **Rosaceae**, especially apples and pears. Infected areas appear blackened and shrivelled, as though scorched. The disease is transmitted by insects, wind, and rain, and infects young shoots, opening leaves, and flowers. It occurs in North America and much of Europe.

Erwinia rhapontici A species of **Enterobacteriaceae** that is an opportunist pathogen of plants. It is a **facultative anerobe** that causes crown rot in rhubarb (*Rheum rhaponticum*) that also penetrates the root, pink seed (pink pea) that reduces the yield and quality of seeds, and internal browning of hyacinth **bulbs**. It occurs in water and soil in Europe, the Middle East, Japan, Korea, and North America.

Erysiphales An order of **ascomycete fungi** that are **obligate parasites**, many of which cause **powdery mildew**. The fungal **mycelium** grows across the surface of the host plant, extracting nutrients through **hyphae** that penetrate the **epidermis** as haustoria (*see* haustorium). There are about 100 species occurring worldwide.

Erythropalaceae (order **Santalales**) A family of **trees**, slender **shrubs**, and **lianas** with **alternate**, **simple**, **exstipulate** leaves on long **petioles**. Flowers **pentamerous**, **ovary inferior** of 3 **carpels**. **Inflorescence** a **cyme**. Fruit a **drupe**. There are 4 genera with 40 species occurring throughout the tropics but not in Madagascar or eastern Malesia.

Erythroxylaceae (order **Malpighiales**) A family of **evergreen** or **deciduous shrubs** and small **trees** with **alternate** occasionally **opposite**, **simple**, **entire** leaves with often **caducous stipules**. Flowers small, **actinomorphic**, **bisexual** or occasionally **unisexual** (plants **dioecious**), hypogynous (*see* hypogyny), with 5 **connate**, **valvate**, or **imbricate sepals**, 5 free, imbricate, caducous **petals**, 10 **stamens** in 2 whorls of 5, **ovary superior** of usually 3 sometimes 2 or 4 fused **carpels** with 2–4 **locules**. Flowers solitary or in a terminal **fascicle**. Fruit a **drupe**. There are 4 genera with 240 species of pantropical distribution. Leaves of *Erythroxylum coca* and *E. novaganatense* yield cocaine.

Escalloniaceae (order **Escalloniales**) A family of **trees**, **shrubs**, **subshrubs**, and **herbs**, with **opposite** or **alternate**, **simple**, **entire** or **dentate** leaves, usually with **petioles**. Flowers 4-, 5-, to 9-merous, **sepals** attached to a **hypanthium**, as many **petals** as sepals, usually 5 **stamens**, **ovary superior** of 2–4 fused **carpels**. **Inflorescence cymose** or **racemose**. Fruit a **capsule** or **drupe**. There are 9 genera with 60 species found in Réunion, the eastern Himalayas, and southern China to eastern Australia and New Caledonia. Many are cultivated for ornament.

Escalloniales An order of plants comprising 1 family with 9 genera and 130 species. *See* Escalloniaceae.

escape reaction A behavioural response to the presence of a predator in which an animal seeks to escape or disappear.

Eschscholzia californica (Californian poppy) *See* Papaveraceae.

esculent Edible for humans.

ESD *See* effective soil depth.

ESP *See* exchangeable sodium percentage.

essential element A chemical element that is necessary for the healthy growth of any plant. These fall into two groups: **macronutrients** and **micronutrients**.

essential oil (volatile oil, ethereal oil, aetherolea) A concentrated liquid derived from plants and containing strongly scented compounds.

EST *See* expressed sequence tag.

ester A chemical compound formed when hydroxyl (OH) is removed from an acid and hydrogen (H) from an **alcohol**, forming water (H_2O). Most oils and fats are **fatty acid esters** of **glycerol**.

estivation *See* aestivation.

estrus *See* oestrus.

estrus cycle *See* oestrus cycle.

etaerio *See* aggregate fruit.

ete *See* Amazon floral region.

ethane (ethylene) A compound (C_2H_4), gaseous at room temperature, that is produced naturally by plants and functions as a **hormone** regulating several processes including **germination**, **cell growth**, fruit ripening, **abscission**, and **senescence**.

ethanol (ethyl alcohol) A colourless compound (CH_3CH_2OH), liquid at room temperature, that is produced by **fermentation** of a sugar solution resulting from anaerobic **respiration**. Ethanol mixes with water and absorbs water vapour.

ethereal oil *See* essential oil.

ethnobotany The study of the human uses of plants.

ethyl alcohol *See* ethanol.

ethylene *See* ethane.

etiolation A process that occurs in plants that are grown in darkness. Having little **chlorophyll** they are pale green or yellow, and have rudimentary leaves and very long **internodes** so the stem is abnormally tall, a response that carries the shoot rapidly upward toward the light.

etioplast A **chloroplast** that has been kept in darkness. It has little or no **chlorophyll** and gives the plant a pale green or yellow colour.

Eubacteria In the **five-kingdom system** of **taxonomy**, a subkingdom in the kingdom **Bacteria** that contains the 'true' bacteria, distinguishing them from the **Archaea**. In the **three-domain system** the Archaea comprise one of the domains, all bacteria are placed in a second domain, and the term Eubacteria is not needed.

Eubacteriales One of the two principal orders of **Eubacteria**, comprising spherical or rod-shaped **Bacteria** that have no photosynthetic pigments and in which **motile** cells possess **peritrichous** flagella (*see* flagellum).

Eucalyptus (family **Myrtaceae**) A genus of mainly **evergreen trees** with usually **lanceolate**, **glaucous** and resinous leaves that develop through seedling, juvenile, intermediate, and adult forms. Flower **buds** comprise an **operculum** composed of fused **sepals**, **petals**, or both enclosing the many fluffy, coloured **stamens**. On opening the operculum is thrown off, so the flower has no petals. Fruit is a woody **capsule**. There are more than 700 species, most native to Australia where they dominate forests; a few occur in Indonesia and New Guinea, and one in the Philippines. Many are cultivated for their timber and for ornament.

eucarpic Describes a fungus in which only part of the **thallus** forms a **fruiting body**.

euchromatin **Chromosome** material that does not accept microscope stains in the **interphase** stage. *Compare* heterochromatin.

Eucommiaceae (order **Garryales**) A **monotypic** family (*Eucommia ulmoides*) comprising a **deciduous tree** with spirals of **ovate** to **elliptic**, **crenate** to **serrate**, **exstipulate** leaves with **petioles**. Flowers **unisexual** (plants **dioecious**), without **sepals** or **petals**, 5–12 **stamens**, **ovary** of 2 **carpels**, one of which aborts, with 1 **locule**. Fruit is a **samara**. The family occurs only in central China.

Eucryphia (family **Cunoniaceae**) A genus of **evergreen** (one species **deciduous**) **trees** and large **shrubs** with leaves **opposite**, **simple** or **pinnate** with 3–13 **leaflets**, with **stipules**. Flowers have 4 **petals** and many **stamens** and **styles**. Fruit is a woody **capsule**. There are seven species, two occurring in southern South America and five in eastern Australia. Many are cultivated as ornamentals.

Eukarya (Eukaryota) In the **three-domain system** of taxonomic classification, one of the domains containing all the **eukaryotes**, comprising the kingdoms **Animalia**, **Fungi**, **Plantae**, and **Protista**. In the older **five-kingdom system**, a superkingdom containing the Animalia, Fungi, Plantae, and **Protoctista**.

eukaryote An organism formed from one or more cells that have a distinct **nucleus** contained in a **nuclear envelope**. All animals, fungi, plants, and protists are eukaryotes. *See* Eukarya.

Euleia heraclei (celery fly, celery-leaf fly, celery-leaf miner) A small fruit fly (Tephritidae) with green eyes and wings strongly marked with dark patches that is active from early spring to late autumn. Its minute larvae damage celery and, to a lesser extent, parsnip plants by mining the leaves. A large infestation early in the year can destroy the foliage, halting growth; late-season attacks are less damaging.

Eulophidae A large family of **Apocrita**, most of which are black or with a bright metallic sheen and 1–3 mm long. Most species are **parasitoids** of the larvae of **leaf miners** and gall-forming insects, mites, and nematodes. Some species spin cocoons on or close to host larvae. There are 4472 species with a worldwide distribution.

Eumeces fasciatus (*Plestiodon fasciatus*, five-lined skink) A species of skinks (**Scincidae**), 130–220 mm long in which juveniles and young females have a black body with five yellow longitudinal stripes and a bright blue tail. The body colour later fades to the grey, olive, or brown shared by males; males have a red head and throat. The body is slender and the legs short. The skinks inhabit moist wooded areas and feed mainly on insects. They occur throughout most of temperate eastern North America.

Eumeces inexpectatus (*Plestiodon inexpectatus*, southeastern five-lined skink) A species of dark brown or black skinks (**Scincidae**) with five yellow longitudinal stripes and a bright blue to grey tail. The skink is 140–215 mm long. It occurs in a variety of **habitats** including **deciduous** forest and grassland, and feeds on invertebrates. It occurs throughout the southeastern United States.

Eumeces laticeps (*Plestiodon laticeps*, broadhead skink) A skink (**Scincidae**) that is 150–330 mm long with short legs and a streamlined body, a grey, black, or brown body with five pale longitudinal stripes, and mature males with a large, orange head. The young have bright blue tails. They inhabit mainly woodland and are partly tree-dwelling, and feed on invertebrates. They occur along the coastal plain of the southeastern United States.

eumelanin *See* melanin.

Euphorbia (family **Euphorbiaceae**) A genus of **annual** or **perennial herbs** and **deciduous shrubs** or **trees**, many **succulent**, all of which produce a poisonous, caustic **latex**, once used as a purgative which gives the plants their common name of spurge. Leaves are **opposite**, **alternate**, or in **whorls**, with **stipules** that in some species are missing or in the form of glands or spines. Flowers are tiny, **unisexual** (most plants **monoecious**), without **perianths**; several male flowers, each with 1 **stamen**, grouped within an **involucre**, the single female flower with 3 **carpels**. **Inflorescence** is a **pseudanthia**. Fruit is an explosive **capsule**. There are 2008 species occurring throughout most tropical and temperate regions. Many are cultivated.

Euphorbia pulcherrima (poinsettia) *See* Euphorbiaceae.

Euphorbiaceae (order **Malpighiales**) A family of **trees**, **shrubs**, **lianas**, and **annual** and **perennial herbs**, most of which produce a caustic **latex**. Leaves **simple** or palmately **compound**, **entire** to **dentate** or with deep lobes, usually **stipulate**. Leafless, **succulent** species superficially resemble cacti. Flowers **unisexual** (plants **monoecious** or **dioecious**); **sepals**, **petals**, **discs**, **staminodes**, and **pistillodes** may be absent. There are up to 1000 free or fused **stamens**, **ovary superior** of usually 2–5, sometimes 1 or up to 20 fused **carpels**. **Inflorescence** diverse, but in the **tribe** Euphorbieae it is a specialized **pseudanthia** known as a cyathia. Fruit usually an explosive **capsule** or **schizocarp**. There are 218 genera with 5735 species, occurring throughout the tropics and most temperate regions. Many are cultivated. *Manihot esculenta* (cassava, also called manioc and tapioca) is a staple tropical food, *Hevea brasiliensis* is the rubber tree, *Ricinus communis* is the castor oil plant, and *Euphorbia pulcherrima* is poinsettia.

Euphroniaceae (order **Malpighiales**) A **monogeneric** family (*Euphronia*) of **shrubs**

and some **trees** with **alternate**, **simple** leaves with small **stipules**. Flowers **zygomorphic**, **hermaphrodite**, the **calyx** with 5 unequal lobes, 3 free, **spathe**-like **petals**, 4 **stamens**, usually 1 **staminode**, **ovary superior**, **syncarpous**, of 3 **carpels** and 3 **locules**. **Inflorescence** an **axillary** or terminal **raceme**. Fruit a **capsule**. There are one to three species occurring in northern South America.

euploid Possessing any number of sets of **chromosomes**.

Eupomatiaceae (order **Magnoliales**) A **monogeneric** family (*Eupomatia*) of **glabrous**, aromatic **shrubs**, **trees**, or **herbs** with **rhizomes**. Leaves **alternate**, **simple**, **entire**, **exstipulate**, with short **petioles**. Flowers are strongly scented, **perigynous**, **actinomorphic**, **hermaphrodite**, with no **perianth**, a **calyptra** formed from an **amplexicaul bract** that falls to expose the **receptacle**, many **stamens** arising from the edge of the receptacle, the inner rows **petaloid staminodes**, **ovary** of many **carpels**. Flowers solitary or occasionally in pairs. Fruit is **berry**-like. There are three species occurring in New Guinea and eastern Australia. Some are cultivated for ornament.

Eupteleaceae (order **Ranunculales**) A **monogeneric** family (*Euptelea*) of **trees**, often with multiple trunks, with spirally arranged **simple**, **elliptical**, **dentate**, **exstipulate** leaves with **petioles**. Flowers **hermaphrodite**, with **bracts** but lacking **sepals** and **petals**, 6–19 **stamens**, **ovary superior** of 8–31 **carpels**. Flowers **axillary**, in clusters. Fruit a **samara**. There are two species occurring in temperate southeastern Asia.

Eupteryx melissae (banded sage hopper, chrysanthemum leafhopper, sage leafhopper) A small bug (**Cicadellidae**), about 3 mm long, with a pale body marked with dark spots that is active from late spring to early autumn and may overwinter. It jumps if disturbed. It lays eggs in leaf veins and **petioles**. These hatch into pale yellow larvae that develop dark bands as they go through five moults. Both **nymphs** and adults feed on the foliage of herbs (**Lamiaceae**) and some **Asteraceae**, e.g. chrysanthemum. They are native to Europe but now occur in many other parts of the temperate world.

Eurasian badger *See Meles meles.*

Eurasian blackbird *See Turdus merula.*

Eurasian collared dove *See Streptopelia decauocto.*

Eurasian high An area of high **atmospheric pressure** that develops over central Eurasia in winter and disappears in April.

Eurasian jackdaw *See Corvus monedula.*

Eurasian jay *See Garrulus glandarius.*

Eurasian siskin *See Carduelis spinus.*

Eurasian wren *See Troglodytes troglodytes.*

European earwig (*Forficula auricularia*) *See* Forficulidae.

European goldfinch *See Carduelis carduelis.*

European greenfinch *See Carduelis chloris.*

European hedgehog *See Erinaceus europaeus.*

European magpie *See Pica pica.*

European mole *See Talpa europaea.*

European red mite *See Panonychus ulmi.*

European red slug *See Arion rufus.*

European red spider mite *See Panonychus ulmi.*

European robin *See Erithacus rubecula.*

European starling *See Sturnus vulgaris.*

European toad *See Bufo bufo.*

Eurosiberian floral region The area that covers Eurasia between the **arctic and subarctic floral region** and the

Mediterranean, West and Central Asiatic, and **Sino-Japanese floral regions**, within the **boreal region**. It is divided into two parts along the line of the Ural Mountains. There are about 150 **endemic** genera in Europe but only about 12 in Siberia.

Eurotiomycetes A class of **ascomycete fungi** that comprises two subclasses: the Chaetothyriomycetidae, which form perithecia (*see* perithecium), and the Eurotiomycetidae, which form cleistothecia (*see* cleistothecium) or other structures that are not perithecia. Most members of the Chaetothryiomycetidae form sooty moulds; most Eurotiomycetidae are **mycobionts** in **lichens**.

Eurotiomycetidae *See* Eurotiomycetes.

Eurycea cirrigera (southern two-lined salamander) A salamander (**Salamandridae**), 60–120 mm long, that has a yellow, orange, or rust-coloured body with two dark longitudinal stripes ending in speckles on the tail. They inhabit moist areas and feed on small invertebrates. They occur in the eastern and southeastern United States.

Eurycea wilderae (Blue Ridge two-lined salamander) A species of lungless salamanders (**Amphibia**), 70–107 mm long, that are bright orange-yellow with two black longitudinal stripes. They mate in late winter in water and their larvae are aquatic. At other times they inhabit forests and streams, usually about 1200 m above sea level, feeding nocturnally on aquatic and terrestrial invertebrates. They occur in the mountains of North and South Carolina, Georgia, Tennessee, and Virginia.

eusociality An extreme form of sociality in which only one female produces offspring, all other members of the colony attending to her needs, caring for the young, constructing, maintaining, defending, and provisioning the nest, and performing such other tasks as are necessary.

eusporangium A sporangium in which the **initials** form a layer, so the sporangium is larger than a **leptosporangium** and holds more **spores**, and has a wall with several layers. Eusporangia are found in all vascular plants (**Tracheophyta**) apart from more advanced ferns (**Pteridophyta**).

eustele A **siphonostele** in which **vascular bundles** form one or two rings around the **pith**. This arrangement occurs in the stems of most seed plants (**Spermatophyta**) and in the roots of **monocotyledons**.

eutelic Having a fixed number of body cells when mature, the number being constant for all members of a species.

eutrophic Rich in nutrients.

eutrophication The enrichment of an aquatic **ecosystem** with plant nutrients, usually nitrates or phosphates, so the productivity of the ecosystem increases, sometimes triggering an **algal bloom**. When the algae (*see* alga) die, their aerobic decomposition may deplete the water of its dissolved oxygen, causing the asphyxiation of sensitive species.

euxinic Describes an aquatic **environment** in which the circulation of water is restricted, e.g. a swamp or thermally stratified lake, leading to a depletion of dissolved oxygen.

evagination Turning inside out, e.g. to expel the contents of a **vesicle**.

evanescent Soon disappearing; transitory.

evaporation The change in phase from liquid to gas.

evaporimeter An instrument that measures the rate at which moisture is evaporating. There are several types. The simplest comprises a graduated reservoir filled with water and sealed by a cork with a ring allowing it to hang. A U-tube at the bottom of the reservoir expands into an open, conical area of known dimension, covered with filter paper. As water evaporates from the filter paper, the water level in the reservoir falls.

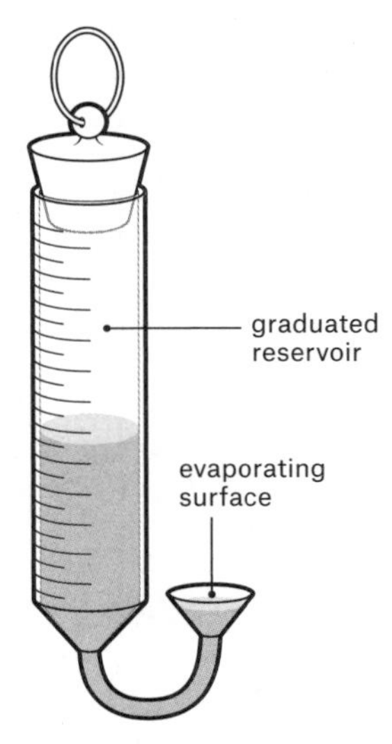

Evaporimeter. The funnel contains filter paper as an evaporating surface. As moisture evaporates from the paper, the level falls in the graduated reservoir.

evaporite A sedimentary rock formed by the precipitation of a salt as water evaporated from an enclosed area of sea, e.g. a lagoon, or a salt lake. Rock salt, **dolomite**, and **limestone** are evaporite rocks.

evapotranspiration **Evaporation** and **transpiration** combined. These are usually considered together because in open-air measurements it is very difficult to distinguish between the two sources of water vapour.

evening grosbeak *See Coccothraustes vespertinus.*

evening primose *See Oenothera.*

evergreen Describes a **shrub** or **tree** that bears leaves throughout the year. Although the **branches** are never bare, each leaf has a limited lifespan, so leaves are being shed constantly. Evergreen leaves are tougher and usually live longer than broad leaves.

evergreen forest A forest in which the trees do not lose all of their leaves at the same time. Such forest occur in all latitudes.

evergreen mixed forest A forest in which the **dominants** are **broad-leaved evergreen** and coniferous trees. Such forests are common in the Southern Hemisphere but less so in the Northern Hemisphere.

everlasting daisies *See Helichrysum, Helipterum.*

eversible Able to be turned inside out (everted).

evolute Having the margins unrolled and opened out. *Compare* involute.

evolutionary ecology A branch of **ecology** that takes account of the evolution of the constituent species, their life cycles, and the relationships among them.

evolutionary lineage A line directly linking a **taxon** with its ancestral taxon.

evolutionary rate The amount of evolutionary change that occurs over a specified time.

evolutionary stable strategy (ESS) A concept derived from **game theory** that identifies **characters**, combinations of characters, and behaviours that cannot be defeated, ensuring that the species possessing them cannot be replaced by a rival.

evolutionary tree *See* phylogenetic tree.

exalbuminous Describes a mature seed that contains no **endosperm**.

exannulate Lacking an **annulus**.

exarate Describes an insect (**Insecta**) **pupa** in which the appendages are free and able to move.

exarch Describes primary **xylem** in that develops inward from the outside, so the oldest strands are farthest from the **axis**. *Compare* endarch, mesarch.

Excavata (excavates) A group of single-celled **eukaryotes**, ranked as a kingdom by some authorities and as members of the **Protista** by others, many of which have a feeding groove along one side, hence their name. Most lack mitochondria (*see* mitochondrion) and they have two or four flagella (*see* flagellum). Some are free-living, others symbionts (*see* symbiosis), and some are **obligate parasites**, mainly of insects but some infecting vertebrate or invertebrate animals, or plants, and some parasitic groups contain **chloroplasts**. Excavates cause several serious diseases of humans, e.g. sleeping sickness, Chagas disease, and leishmaniasis.

excavates *See* Excavata.

exchangeable-cation percentage The percentage of a particular soil **cation** that is exchangeable (*see* cation exchange capacity).

exchangeable ions **Ions** that are adsorbed (*see* adsorption) onto sites on the surface of soil particles, and that are able to replace one another at those sites. Ions released by exchange into the soil solution are available as nutrients to plant roots.

exchangeable sodium percentage (ESP) The percentage of exchangeable **cations** accounted for by sodium. A **sodic soil** has an ESP of 6 percent or greater. A high ESP reverses **aggregation**, leading to **deflocculation.**

exchange capacity The total charge of all the **ions** in the soil **adsorption complex**.

exchange pairing Pairing of **homologous chromosomes** that allows genetic material to be exchanged between them.

excipulum The sterile tissue that forms the walls containing the **hymenium** of an **apothecium**, consisting of two layers, the **ectal** excipulum and medullary (*see* medulla) excipulum.

exclusion principle *See* competitive exclusion principle.

exclusive species In the phytosociological (*see* phytosociology) scheme devised by the school led by Josias **Braun-Blanquet**, one of the five classes of fidelity (*see* faithful species) that describe and classify plant communities. It is totally or almost totally confined to a particular community. *Compare* accidental species, indifferent species, preferential species, selective species.

excurrent **1.** Having a single, undivided, main stem with **branches** extending from it. **2.** Extending beyond the edge, e.g. a leaf midrib that extends beyond the leaf tip. **3.** Of a liquid, flowing outward.

exfoliation The weakening and flaking away of surface layers from a rock due to **weathering** or **erosion**.

exiguous Meagre, scanty.

exine The outer coat of a **pollen grain**. It resists decay and the overall shape of the grain and its surface markings (*see* colpus) are characteristic for a plant family, sometimes for a genus or even a species. Study of pollen grains preserved in sedimentary deposits, called palynology or pollen analysis, makes it possible to reconstruct past plant communities and, therefore, environments.

Exobasidiales An order of **Fungi** in the class **Exobasidiomycetes**, in which basidia (*see* basidium) are borne in a **hymenium**, but there is no true **basidiocarp**. They are **obligate parasites** of a wide range of plants.

Exobasidiomycetes A class of **basidiomycete fungi**, all of which are **obligate parasites** of vascular plants (**Tracheophyta**) and many of which cause the formation of **galls**. There are 7 orders and 1 group of unassigned species, with a total of 307 species.

exocarp *See* pericarp.

exocrine *See* ectocrine.

exocrine gland A gland that excretes its product through a duct leading to the external environment.

exocytosis A process in which a **vacuole** enclosed by a **membrane** fuses with the **plasma membrane**, allowing it to discharge its contents outside the cell.

exodermis In many **angiosperms**, the outermost layer of the **cortex**.

exoenzyme An **enzyme** that is released from a **cell**.

exogamy The tendency for distantly related **gametes** to fuse more readily than closely related ones.

exon That part of a **gene** that appears in the mature **RNA** after **transcription** and with all **introns** removed.

exoparasite *See* parasite.

exoperidium The outer of the two layers of the **peridium** found in certain **basidiomycete fungi**.

exopterygote Describes an insect (**Insecta**) in which the wings develop gradually and externally; the insect undergoes no pupal stage or **metamorphosis**, and its young are called **nymphs**.

exorheic lake A lake that loses water through outflow streams.

exoskeleton The hard, outermost layer of tissue found in some species of algae (*see* alga), e.g. stoneworts (**Charophyceae**), and the hard outermost layer of many invertebrate animals.

exosymbiont A **symbiont** that lives on the exterior of its host.

exothecium An external layer of tissue, often with **stomata**, surrounding the **sporangium** of mosses (**Bryophyta**).

exotherm *See* poikilotherm.

exotic Describes a species that has been introduced.

exotoxin A poison (toxin) secreted by a living organism.

exozoochory Dispersal of seeds or **spores** by being carried on the exterior of an animal.

exploitation competition (exploitative competition) **Competition** between species for a resource that is in limited supply. The more efficient competitor is more likely to succeed. *Compare* interference competition.

expressed sequence tag (EST) A single-stranded sequence of **DNA**, usually 200–500 **nucleotides** long, that is produced by sequencing an expressed **gene** from one or both ends.

expressivity The extent to which a **genotype** is expressed in the **phenotype**.

exserted Protruding, e.g. **stamens** that extent beyond the **corolla**.

exsiccation The drying out of a land area by processes that do not involve a reduction in **precipitation**. Draining a wetland will lead to exsiccation.

exstipulate Lacking **stipules**.

extant Living at present.

extinct Describes a **taxon** that has no currently living member.

extinction The disappearance of a **taxon** locally, regionally, or globally.

extirpation Rendering a species extinct throughout a part of its **range**.

extrafloral nectary (extranuptial nectary) A **nectary** that is not located in a **flower**. It may serve in defence, attracting insects that also eat herbivorous insects they encounter on the plant, and some carnivorous plants use nectar to attract insect prey.

extranuptial nectary *See* extrafloral nectary.

extratropical cyclone A **cyclone** that develops in middle latitudes, i.e. outside the tropics.

extremophile An organism that thrives under extreme conditions. Acidophiles inhabit acid environments (**pH** 1.0–5.0); alkaliphiles inhabit alkaline environments (pH >9.0); halophiles inhabit saline environments; piezophiles (barophiles) live under high pressure; psychrophiles grow best at temperatures lower than 15°C; thermophiles grow best at temperatures higher than 40°C; hyperthermophiles grow best at temperatures higher than 80°C; xerophiles grow in very arid conditions.

extrorse Describes **anthers** that are held away from the centre of the **flower** so they release their **pollen** away from the flower.

eye of storm The region of calm air and fairly clear sky that occurs at the centre of a deep **cyclone** and especially a **tropical cyclone**.

eye of wind The direction from which the wind is blowing, or the position on the horizon from which it appears to blow.

eyespot 1. An **organelle** containing **carotenoid** pigments found in certain unicellular algae (*see* alga). It is believed to be involved in **phototaxis**. **2.** (strawbreaker) A fungal disease of wheat caused by *Tapesia yallundae* (also called *Pseudocercosporella herpetrichoides*) that causes elliptical lesions on the lower part of the stem.

eyewall The mass of towering **cumulonimbus** cloud that surrounds the eye of a **tropical cyclone** or an **extratropical cyclone** generating **hurricane-force winds**.

F

F *See* coefficient of inbreeding.

F_1 The first filial generation produced by breeding from two parental (*P*) lines.

F_2 The second filial generation produced by breeding from **F_1** parents.

Fabaceae (order **Fabales**) A family of plants (legumes) formerly known as Leguminosae, that comprises 727 or 732 genera (depending on the classification) and between 19,000 and 19,700 species of **trees**, climbers, **shrubs**, and **herbs**. It is the third largest plant family after the **Orchidaceae** and **Asteraceae** and includes many plants of commercial value. The **inflorescence** is **racemose**, the **flowers** having 5 **petals** and **sepals**, the sepals often being united. There are typically 10 **stamens** (but more in the subfamily **Mimusoideae**) which are fused in some species. **Ovary superior** of 1 **carpel**. The fruit is a **pod**, usually containing several **seeds** and usually **indehiscent**. There are three subfamilies, distinguished by the structure of their flowers: Caesalpinoideae, Mimosoideae, and Papilionoideae.

Fabales An order of plants comprising 4 families of 754 genera and 20,080 species. *See* Fabaceae, Polygalaceae, Quillajaceae, and Surianaceae.

facies Appearance.

facilitated diffusion A mechanism driven by a **diffusion** gradient in which a carrier transports molecules across a **membrane**.

facultative Describes an organism that is able to adopt more than one mode of life, e.g. a facultative **parasite** can live either independently or parasitically.

FAD *See* flavin adenine dinucleotide.

Fagaceae (order **Fagales**) A family of **deciduous** or **evergreen** timber **trees** (beeches, oaks, sweet chestnuts) and a few **shrubs**, with **alternate**, occasionally whorled, **simple**, **entire** to pinnately lobed leaves with **scarious stipules**. Flowers **unisexual** (plants **monoecious**) with a 4- to 7-lobed **bract**-like **perianth**, up to 40 **stamens**, **pistillate** flowers in groups of 12–13 surrounded by a basal **involucre**, **ovary inferior** of 3–6 **styles** and **locules**. **Inflorescence** a **catkin** or small **spike** comprising flowers of only one sex (e.g. oaks) or with female flowers at the base of inflorescence of male flowers (e.g. sweet chestnuts). Fruit is a **nut**. There are 7 genera with 670 species with worldwide distribution. Many are cultivated.

Fagales An order of plants with 8 families of 33 genera and 1055 species. *See* Betulaceae, Casuarinaceae, Fagaceae, Juglandaceae, Myricaceae, Nothofagaceae, Rhoipteleaceae, and Ticodendraceae.

Fagus (family **Fagaceae**) A genus of **deciduous trees** (beeches) with broad **ovate**, **entire** or **dentate** leaves. Flowers **unisexual** (plants **monoecious**), female flowers in pairs, male flowers in wind-pollinated **catkins**. Fruit is a three-angled nut, single or in pairs held in a **cupule**. There are ten species occurring in the temperate Northern Hemisphere. Many are cultivated.

fair-weather cumulus *See* cumulus.

fairy ring A circle of dead grass, sometimes with toadstools, in a lawn or other grassy area. It is caused by a fungus, most often ***Marasmius*** *oreades*, which infects grass roots and expands outward from a central area.

faithful species (fidelity) A plant species that is completely or almost completely confined to a particular **plant association**. In the school of **phytosociology** led by Josias **Braun-Blanquet**, five fidelity classes are recognized: **accidental species**; **indifferent species**; **preferential species**; **selective species**; and **exclusive species**.

falcate Curved, like a sickle or scimitar.

falcato-secund Curved (**falcate**) to one side.

Falcon *See* methoxyfenoxide.

fall (rock fall) A type of **mass wasting** involving rocks of varying size that roll and bounce down a hillslope.

fallout The removal from the air of particles that fall by gravity.

fallow **1.** The practice of leaving an area of land uncultivated for all or part of a growing season. **2.** Land that has been left fallow.

fallstreak hole A hole that sometimes appears in clouds comprised of supercooled (*see* supercooling) water droplets. Droplets in part of the cloud freeze, other droplets freeze on to them, and they fall from the cloud, melting into raindrops as they descend, often forming **virga**, and leaving behind an area of cloud-free sky.

fallstreaks *See* virga.

fall wind *See* katabatic wind.

false bindweed *See Calystegia.*

false blusher *See Amanita pantherina.*

false branching In filamentous algae (*see* alga), an appearance of branching that results from one or both ends of a broken **filament** protruding from the sheath.

false chanterelle *See Hygrophoropsis aurantiaca.*

false cirrus Cloud resembling **cirrus** that is what remains of the top of a **cumulonimbus** cloud which has dissipated or from which it has become detached.

false cypress *See Chamaecyparis.*

false death cap *See Amanita citrina.*

false morel *See Gyromitra esculenta.*

family In plant **taxonomy**, a rank between **genus** and **order**; a family comprises one or more related genera, an order comprises one or more related families.

farinose Powdery or floury.

fasciation A thickening and flattening of shoots so it looks as though many shoots are pressed together to form a ribbon. Many undersized leaves or flowers may grow from the distorted stems. It is a rare condition that can also distort flower heads, roots, and fruit.

fascicle A bunch or cluster of leaves or **branches** all arising from a single point.

fascicular cambium **Cambium** that develops within **vascular bundles**.

fatiscent Cracked or gaping.

fatty acid A carboxylic acid with a long chain in which each **carboxyl** group links to a side-chain of carbon atoms, and hydrogen atoms are attached to some or all of the side-chain carbon atoms. If hydrogen atoms occupy all the carbon bonds the fatty acid is said to be saturated; if some carbon bonds are unoccupied it is unsaturated; if only one carbon bond is unoccupied it is monounsaturated; if more than two sites on the side-chain are unoccupied it is polyunsaturated.

F-box genes A group of **genes** that in plants code for approximately **amino acids** that form **proteins** involved in processes including leaf senescence, branching, self-incompatibility, and response to stress.

Fe *See* iron.

feather An outgrowth from the skin of a bird, made from **keratin**, that provides insulation and aids in flight. Some species also use feathers in **display**. A bird has several kinds of feathers. Contour (pennaceous) feathers have a central shaft (rhachis) with **barbs** on either side. Down feathers lie beneath the contour feathers and provide thermal insulation; some birds have powder down, which are modified down feathers that grow continuously, their tips disintegrating and releasing small keratin fragments (powder). Intermediate feathers (semiplumes) have a large shaft and down-like vanes. Filoplumes resemble hairs, with a fine shaft and a few vanes at the tip. Bristles have a strong shaft and vanes only around the base.

fecundity (fruitfulness) The extent of the ability to reproduce of an individual or population, measured as the number of eggs that develop in a mated female over a specified time.

fell Open mountainside with low-growing vegetation, from the Norse *fiall*, hill.

femur In **tetrapods**, the upper bone of the hind limb. In insects (**Insecta**), the third segment of the leg; the largest and most robust segment in most insects.

fen A **peat**-forming wetland area in which the vegetation receives water from both rainfall and the flow of **groundwater**, and where in summer the **water table** is below the surface. *Compare* bog.

fencerow A row of **trees**, **shrubs**, or **herbs** that provide resources for wildlife.

fenestrated Having small perforations or transparent areas.

fenitrothion An **organophosphate insecticide** and **acaricide** that is used to control aphids (**Aphididae**) and **caterpillars** in fruit crops, moths and weevils (**Curculionidae**) in peas, leatherjackets (**Tipulidae**) in cereals, and beetles (**Coleoptera**) in grain stores. It is of low toxicity.

fennel *See Foeniculum.*

fentins A group of three **fungicides** and **molluscicides** that contain tin and are moderately toxic.

fenugreek (*Trigonella foenum-graecum*) *See Trigonella.*

feral Describes a formerly cultivated plant or animal that is living in the wild.

feral pigeon *See Columba livia.*

fermentation (anaerobic respiration) A form of **respiration** that produces **adenosine triphosphate** (ATP) in the absence of oxygen. Commonly, it produces **ethanol** from **carbohydrates**.

fern *See* Pteridophyta.

fern frost **Frost** that forms patterns resembling fern fronds. It is frozen **dew** and most often seen in early morning on the windows of unheated rooms.

ferralic horizon A subsurface **soil horizon** that results from long and intense **weathering**, is at least 30 cm thick, and is enriched in iron, aluminium, manganese, titanium, and other oxides.

ferralitization *See* ferralization.

ferralization (ferralitization) A **leaching** process in tropical soils in which large amounts of iron and aluminium accumulate in the B **soil horizon**.

ferralsols Soils that have a highly weathered (*see* weathering) **ferric horizon** enriched in iron and aluminium. Ferralsols are a reference soil group in the **World Reference Base for Soil Resources**.

ferredoxin An iron and sulphur-containing **protein** with a low **redox potential** that functions as an **electron carrier** in **photosynthesis** and **nitrogen fixation**.

Ferrel cell In the **three-cell model** of the **general circulation**, the middle latitude cell that is bordered by the **Hadley** and **polar cells**. It is an indirect cell, driven by the direct cells on either side, and air moves through it in the opposite direction from that of the direct cells, rising at the **polar front** and subsiding where it meets the subtropical side of the Hadley cell. It was discovered by the American climatologist William Ferrel (1817–91).

F

ferric horizon A **soil horizon** more than 15 cm thick in which there is distinct red mottling or **concretions** caused by the segregation of iron.

ferricrete A **duricrust** dominated by sesquioxides of iron.

ferruginous Rust-coloured.

fertigation The application of **fertilizers** or other water-soluble substances through an irrigation system, i.e. *ferti*lization + irri*gation*.

fertility factor *See* conjugation.

fertilization The union of two **gametes** to produce a **zygote** during sexual reproduction.

fertilizer A substance rich in plant nutrients that is applied to soil to stimulate plant growth.

fetch The horizontal distance that air travels continuously across the surface.

F-factor *See* conjugation.

fibre An elongated plant cell with tapering ends and thick walls containing **lignin** that provides structural strength in stems, **branches**, and roots. Fibre cells are dead and the interior cavity is small. Dietary fibre is from the **cell walls** of edible plants and from the seeds and **sap** of certain plants.

fibril 1. A small fibre or thread-like structure. **2.** A trail of cloud sometimes seen extending from a **cumulonimbus**.

fibula In **tetrapods**, the posterior of the two bones of the hind limb.

Ficus (family **Moraceae**) A genus of mostly **evergreen trees**, **shrubs**, climbers, **stranglers**, **epiphytes**, and hemiepiphytes, many with aerial roots, and all of which have a white or yellowish **latex**. Leaves usually have paired **stipules** that leave scars on the twigs when they fall. Flowers are minute, **unisexual**, and inserted on a concave **receptacle** that forms a closed sphere. Pollination is by specialized fig wasps that enter through an **ostiole** to lay their eggs. Fruit is a **synconium**. There are about 850 species, known as figs or fig trees, occurring throughout the tropics with a few in warm temperate regions. Many are cultivated for their fruit, especially *F. carica*, the common fig.

Ficus benghalensis (banyan) *See* strangler fig.

fiddlehead *See* crozier.

fiducial point Any fixed position from which other positions can be measured. The fiducial point, also called standard temperature, for a barometer is the temperature at which that barometer gives a correct reading in latitude 45°. At any other latitude or temperature the barometer reading must be corrected.

field blewit *See Clitocybe.*

field capacity The amount of water a soil is able to retain after free **drainage** has removed any excess. It is usually measured as a percentage of the soil volume or of the weight of the soil after it has been oven-dried.

field layer In a plant community, the **herbs** and small **shrubs**.

field mouse *See Apodemus sylvaticus.*

field mushroom *See Agaricus campestris.*

field slug *See Deroceras reticulatum.*

fig (fig tree) *See Ficus*, Moraceae.

figwort weevil *See Cionus scrophulariae.*

filament 1. The stalk of a **stamen**, bearing the **anther**. **2.** Algal cells joined end to end. **3.** A strand of **protein** found in many types of **cell**.

filbert *See Corylus.*

Filicopsida (Polypodiidae) A group that comprises all the **leptosporangiate** ferns. The **sporophyte** is herbaceous or tree-like, with true roots, stems, and **simple** or **compound** leaves (fronds) arranged spirally. In most the stem is an underground or surface **rhizome**, usually covered in protective hairs or scales. Fronds are typically **circinate** when young. There are 300 genera with about 12,000 species found worldwide.

filiform Long, thin, thread-like.

filling An increase in **atmospheric pressure** at the centre of a **cyclone**.

filoplume *See* feather.

filter strip *See* buffer strip.

fimbriate Having a fringed margin.

fimiculous Growing on animal dung.

fine earth Soil consisting of particles smaller than 2 mm.

finger and toe *See Plasmodiophora brassicae.*

fipronil A broad-spectrum **insecticide** that is effective against a wide variety of pests, disrupting their nervous systems. Its use was banned in the European Union from 2014, except in greenhouses, because of concern over its contribution to declining bee populations.

fir (*Abies*) *See* Pinaceae.

fire blight *See Erwinia amylovora.*

firebush (*Bassia scoparia* f. *trichophylla*) *See Bassia.*

fire climax *See* pyroclimax.

fire scar A mark found among the **tree rings** of a plant that has survived fire. The scar indicates fire damage and its location makes it possible to determine when the fire occurred.

fire storm A wind storm that is generated by **convection** and **convergence** in a very hot fire.

fire weather Weather conditions that favour the ignition of dry material, causing a forest fire, in a specified area.

fishers *See* Mustelidae.

Fistulina hepatica (beefsteak fungus, beefsteak polypore, ox tongue) A species of **ascomycete fungi** which is a bracket fungus that resembles a slab of raw meat and bleeds a red juice when cut. It is edible and has been used as a meat substitute, but must be picked young and requires long cooking. It is common on living and dead oaks (***Quercus***) and sweet chestnut (***Castanea** sativa*) and found throughout Europe, North Africa, North America, and Australia.

fitness (adaptive value, selective value) The balance between the inherited advantages and disadvantages that determines the ability of an individual or **genotype** to survive and reproduce.

fivehorn smotherweed (*Bassia hyssipifolia*) *See Bassia.*

five-kingdom system A taxonomic system (*see* taxonomy) in which living organisms are classified in five kingdoms: **Animalia**; **Plantae**; **Fungi**; **Monera**; **Protista**. The system is widely used, but increasingly is being replaced by the **three-domain system.**

five-lined skink *See Eumeces fasciatus.*

fixation 1. A process by which inorganic molecules are incorporated into organic molecules, e.g. the fixation of carbon in **photosynthesis** and **nitrogen fixation** by **Bacteria**. **2.** Chemical reactions in soils that render soluble nutrient elements less soluble and, therefore, less available to plants. **3.** *See* gene fixation.

flabelliform Fan-shaped.

flaccid Limp; the condition of **cells** that lack water. *See* wilting.

Flagellariaceae (order Poales) A monogeneric, monocotyledon family (*Flagellaria*) of scrambling or climbing lianas that arise from sympodial rhizomes. Leaves distichous, grass-like, with a sheathing base and terminating in a sensitive tendril by which the plant attaches itself to its support. Flowers small, sessile, trimerous, bisexual, with 2 whorls of pale, membranous tendrils, 2 whorls of stamens, ovary syncarpous, superior with 3 locules. Inflorescence a large terminal panicle. Fruit drupe-like. There are four species occurring throughout the Old World tropics.

flagellum 1. A thread-like structure, one or more of which protrude from a cell and are used in locomotion. Bacterial and archaeal (*see* Archaea) flagella rotate (at 200–1000 rpm), but differ in structure and are not thought to be homologous. Eukaryotes have flagella that undulate and are more complex. There are two types: whiplash, which are smooth; and tinsel, which have many fine hair-like structures. **2.** A slender stem carrying much reduced leaves in some liverworts (Marchantiophyta).

Flagship *See* chloronicotinyls.

flannel flower (*Actinotus helianthi*) *See Actinotus.*

flask fungi Fungi that produce spores in perithecia (*see* perithecium) or ascocarps.

flat-backed millipede *See Oxidus gracilis.*

flat wrack (*Fucus spiralis*) *See* spiral wrack.

flatworms *See* Platyhelminthes.

flavescent Yellow or turning yellow.

flavin One of a group of yellow, water-soluble, light-sensitive pigments that influence the action of phytochrome and play a part in phototropism. They are coenzymes of flavoprotein. Riboflavin belongs to the group.

flavin adenine dinucleotide (FAD) A coenzyme which consists of riboflavin bound to the phosphate group of an ADP (adenosine diphosphate) molecule. It is an intermediate in oxidative photophosphorylation. Despite the name, it is not a nucleotide.

flavin mononucleotide (FMN) Riboflavin 5'-phosphate, a coenzyme that acts as a prosthetic group to several dehydrogenases, serving as an electron carrier.

flavonoids A range of pigment compounds responsible for most of the red, blue, and yellow colours in petals. They also filter ultraviolet (UV) radiation and are involved in bacterial nitrogen fixation. Some have an oestrogenic effect on animals and some, e.g. rotenone, are poisonous to some species and are used as insecticides.

flavoprotein A conjugated protein in which the prosthetic group is a flavin coenzyme. It acts as an electron carrier and is involved in many functions, including photosynthesis repair of DNA, and apoptosis.

flax (*Linum usitatissimum*) *See* Linaceae, *Linum.*

fleabane *See Erigeron.*

flea beetles A tribe (Alticini) of leaf beetles (Chrysomelidae), black or metallic green or blue, 2–3 mm long, that have enlarged hind legs, enabling them to jump when disturbed. They feed on all members of the Brassicaceae as well as potatoes, corn (maize), and other crops, and some ornamental plants, making round depressions in the upper surface of leaves that sometimes penetrate the leaf as a hole. There are about 10,000 species distributed worldwide.

flexistyly A type of dimorphism in which a flower changes its form, so at one time it is functionally male and at another time functionally female (*see* anaflexistyly, cataflexistyly), and the stigma moves vertically. This reduces the chance of self-pollination and may also prevent the **style** and **stigma** from interfering with the export of **pollen** from the flower. *See* herkogamy, heterostyly.

flexuous Wavy or bending in a zigzag shape.

flies *See* Diptera.

floccose Loose, cotton-like, or woolly.

flocculation 1. A process in which soil particles adhere to form **aggregates. 2.** A chemical process in which a **colloid** comes out of suspension and forms flakes.

floccus Tufts of hairs or **filaments** that give a woolly appearance.

flocking Among birds, the formation of a group with a social organization.

flood plain Part of a river valley that is periodically inundated and that is covered with unconsolidated **alluvium**.

flora All of the plant **species** that comprise the vegetation in a specified area or during a specified time period.

floral cup *See* hypanthium.

floral formula A convention that uses capital letters, numbers, and symbols to describe the structure of a **flower**. K = **calyx**; C = **corolla**; A = **androecium**; G = **gynoecium**; P = **perianth**. Numbers indicate the number of components (e.g. the number of **petals** in the corolla). If this exceeds 12 the symbol for infinity (∞) is used. If the number is enclosed in brackets it means the parts are fused. A single bracket placed horizontally above the number indicates that one **whorl** is fused to its neighbour. A line above the number following G indicates an inferior ovary, a line below indicates a superior ovary. The formula begins with ⊕ if the flower is **actinomorphic** and · · or ↑ if it is **zygomorphic**.

floral kingdom *See* floral region.

floral province *See* floral region.

floral region (floral kingdom, floral province) A geographic area that is defined by the distinctiveness of its **flora**, with particular regard for the number of **endemic** taxa. The most widely used system, and the one used here, was devised by Ronald **Good**.

floral tube *See* hypanthium.

floret One of the individual flowers of an **inflorescence**.

floridean starch *See* glycogen.

florigen (flowering hormone) A **messenger-RNA** that is produced in leaves in response to crossing a threshold in **photoperiod** or temperature and transported in **phloem** to the tip of the stem, where it initiates flowering.

floristics *See* phytogeography.

flow A type of **mass wasting** in which a large volume of soil or similar material moves down a hillslope. An avalanche is a flow of snow.

flower The sexual reproductive structure in an **angiosperm**, consisting of the male (**androecium**) and female (**gynoecium**) organs, often surrounded by **petals** forming the **corolla** and **sepals** forming the **calyx**. The male and female parts may be in the same flower or in separate flowers. *See also* imperfect flower, incomplete flower, inflorescence, perfect flower.

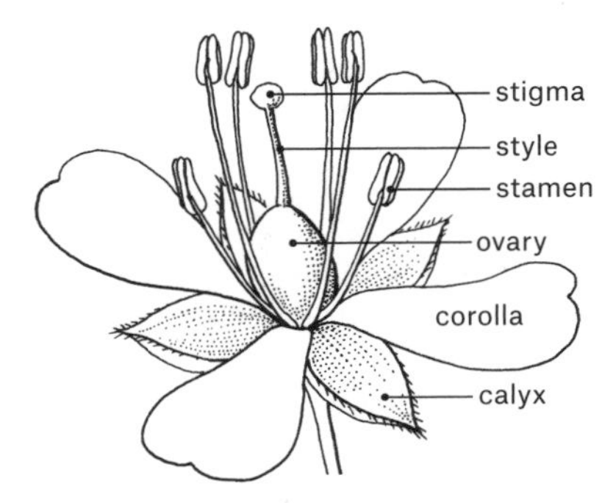

This drawing shows the female and male reproductive structures of a perfect flower.

flower bugs *See* anthocorid bugs.

flower flies *See* Syrphidae.

flower head *See* pseudanthium.

flowering hormone *See* florigen.

flowering nutmeg (*Leycesteria formosa*) *See Leycesteria.*

flowerpeckers (*Diglossa*) *See* nectar robber.

flukes *See* Platyhelminthes.

F

fluorescence Luminescence that occurs when an atom or molecule absorbs radiation at one wavelength, raising its electrons to a higher energy level from which they fall almost immediately, releasing radiation at a longer wavelength as they do so.

fluviatile Describes river-borne (fluvial) sediments.

fluvic horizon A thick, black **soil horizon**, at least 30 cm thick, at or close to the surface that has a low **bulk density** and is rich in organic matter.

fluvisols Soils formed on **alluvial** deposits and that have a **fluvic horizon** 25 cm below the surface and extending to more than 50 cm. Fluvisols are a reference soil group in the **World Reference Base for Soil Resources.**

fly agaric *See Amanita muscaria.*

flying buttress *See* buttress root.

FMN *See* flavin mononucleotide.

foehn wind (föhn wind) A warm, dry wind that blows, most commonly in spring, on the northern side of the European Alps, and by extension any wind of similar type, e.g. the **chinook.** It occurs when moist air is forced to rise to cross the mountains, cooling at the **saturated adiabatic lapse rate** and losing much of its moisture, then subsides down the opposite side and warms at the **dry adiabatic lapse rate.**

Foeniculum (family **Apiaceae**) A **monotypic** genus (*F. vulgare*, fennel), an erect, **glaucous, perennial herb** with hollow stems and finely dissected leaves that contain a pungent aromatic oil used in cooking and for flavouring. Flowers are small, yellow, and borne in a terminal **umbel.** It is native to the Mediterranean region but naturalized widely elsewhere and cultivated extensively.

fog **Stratus** cloud that extends to the surface and reduces horizontal visibility to less than 1 km. Typically, fog contains less than 1 g/m^3 of water.

fog drip Water that is deposited by **fog** on to tall vegetation and other structures and that drips to the ground.

fog droplet A particle of **fog**, which is a water droplet 1–20 μm in diameter.

föhn wind *See* foehn wind.

foliar diagnosis A technique for determining the **fertilizer** requirement of a particular crop that involves the chemical analysis of leaves taken throughout the growing cycle from plants receiving differing fertilizer treatments. This reveals the response of the plant to individual nutrients.

foliar nectary A **nectary** located on the leaf of a plant.

folic horizon A surface or shallow subsurface **soil horizon** that consists of well-aerated organic matter. It is more than 10 cm thick and contains more than 20 percent organic carbon (35 percent organic matter) by weight.

folicolous *See* epiphyllous.

foliose Leaf-like. *See also* lichen.

folivorous Describes an organism that eats foliage.

follicle **1.** A small gland, sac, or cavity. **2.** A dry fruit, derived from a single **carpel**, which dehisces (*see* dehiscent) along one side only.

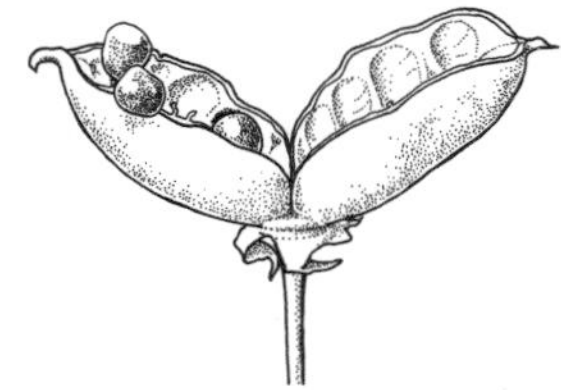

Follicle fruit (*Paeonia*).

Fonticulida A group of cellular slime moulds (**Acrasiomycetes**) in which the **fruiting body** is shaped like a volcano.

food chain A sequence of trophic relationships that reflect the transfer of energy from one **trophic level** to the next using single species, e.g. grass → rabbit → fox.

food-chain efficiency The energy value each animal in a **food chain** derives from the food it eats (i.e. the nutritional value discarding the inedible or indigestible parts). This reveals the proportion of the original energy supplied by sunlight for **photosynthesis** that passes to each **trophic level**.

food web A diagram that depicts the feeding relationships among the inhabitants of an **ecosystem**, by linking **consumers** with the species on which they depend. In effect, the diagram is a set of interconnected **food chains**. Despite its apparent complexity, the diagram

F

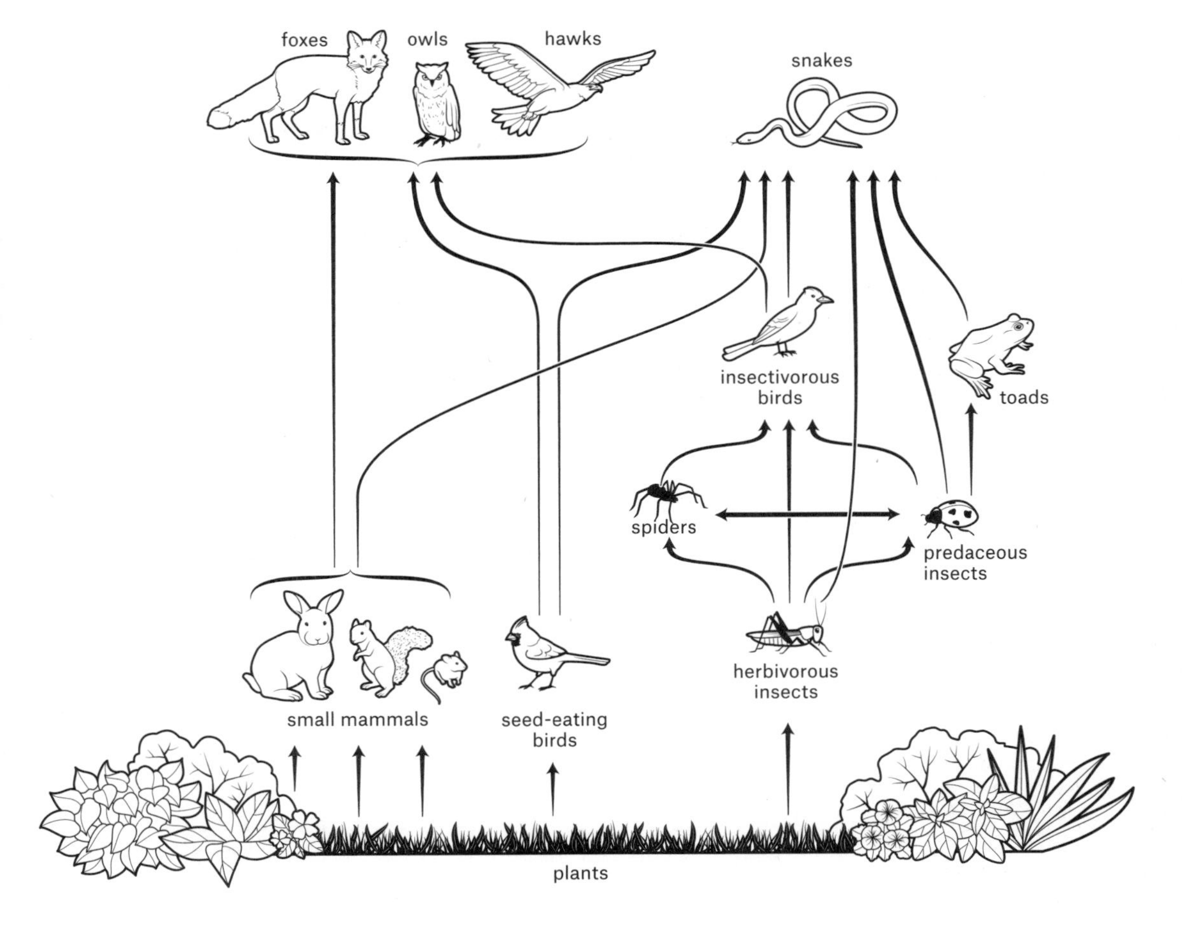

This food web diagram represents the relationships between organisms in an ecosystem, with the arrows indicating the food items on which each animal depends. Real ecosystems are much more complex than such diagrams suggest.

inevitably oversimplifies, not least because consumers eventually die and provide food for organisms at a different level.

foot The part of the **embryo** of a **sporophyte** that connects it to the **gametophyte**, from which it absorbs nutrients.

foraging **1.** Describes the method of **vegetative reproduction** of a plant that spreads by producing **runners**. In some plants that spread in this way the runners grow faster when they cross a surface with a low nutrient content and more slowly, producing roots, on more nutritious soil. **2.** Animal behaviour associated with obtaining and consuming food, including searching and hunting.

forb A non-woody plant other than a grass.

Forbes, Edward (1815–54) A Manx naturalist who toured Norway in 1833, compiling a botanical survey. In 1846 he published his major contribution to biogeography in the *Memoirs* of the Geological Survey 'On the Connexion between the Distribution of the Existing Fauna and Flora of the British Isles, and the Geological Changes which have Affected their Area'.

forceps Modified **cerci** borne at the tip of the abdomen in **earwigs** and in the earwig-like Japygidae. Insects use their forceps to grasp prey, and in attack and defence.

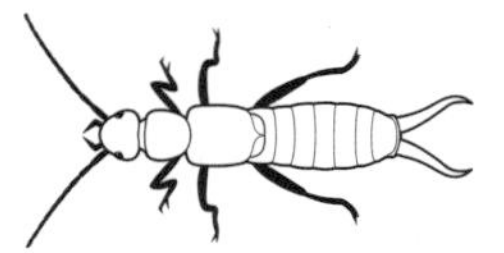

The forceps are the modified cerci at the end of the abdomen of this European earwig (*Forficula auricularia*). They are used in attack and defence, and to grasp prey.

forcipate Forked; pincer-like.

forecast period The period of time that is covered by a weather forecast.

forecast skill The accuracy of a weather forecast, measured on a scale from zero (completely incorrect) to one (completely correct).

forest A plant community comprising **trees** with crowns that touch to form a continuous **canopy**.

Forficula auricularia (European earwig) *See* Forficulidae.

Forficulidae (earwigs) A family of typical earwigs (**Dermaptera**), 25–32 mm long and light reddish brown to black. Most earwigs belong to this family, including *Forficula auricularia* (European earwig), which is both a pest of many cultivated plants but also a predator.

forget-me-not *See Myosotis*.

forked lightning A **lightning stroke** that is seen as a very bright, jagged line between two clouds or between a cloud and the surface.

formation A type of vegetation, e.g. sclerophyllous scrub, that is defined by its **growth form**, structure, and cover rather than the species of which it is composed.

formula of vegetation A scheme for recording vegetation quickly and precisely by means of a shorthand, using capital and lower case letters and numerals to denote **trees**, **shrubs**, and **herbs** that are tall, of medium height, or short, and that grow densely or sparsely, etc.

fossil The remains or traces of a once-living organism, regardless of their age.

fossil fuels Originally, a **fossil** was any object retrieved from below ground, so a fossil fuel was one obtained by mining. Today the term applies to fuels containing carbon and formed by the arrested decomposition of organic matter. Peat, lignite (brown coal), bituminous coal, anthracite, natural gas, and petroleum are fossil fuels.

founder effect The derivation of a population from a single individual or limited number of immigrants (the founder). Since the founder represents only a proportion of its **gene pool** the descendant gene pool will be smaller than the original; **natural selection** acting on the restricted gene pool will produce **gene** combinations different from those in the ancestral population, leading to rapid speciation.

Fouquieriaceae (order **Ericales**) A **monogeneric** family (*Fourquieria*) of spiny, or woody, **xeromorphic**, **succulent shrubs** and small **trees** with **entire**, **exstipulate** leaves borne singly or in groups. Red or yellow flowers are **actinomorphic**, **bisexual**, with 5 **imbricate sepals**, 5 hypogynous (*see* hypogyny) **petals** fused into a tube, 10 or more hypogynous **stamens**, **ovary superior** of 3 fused **carpels** and 1 **locule**. **Inflorescence** a terminal or **axillary spike**, **raceme**, or **panicle**. Fruit is a **capsule**. There are 11 species occurring in southwestern North America.

fovea A small pit or depression.

foveate Having foveae (*see* fovea).

Fowler's toad *See Bufo fowleri.*

foxglove *See Digitalis*, Scrophulariaceae.

fra *See* fractus.

fractus (fra) A species of cloud (*see* cloud classification) that consists of fragments detached from a parent cloud or that remain after the parent cloud has dissipated.

Fragaria (family **Rosaceae**) A genus of stoloniferous (*see* stolon) **herbs** with basal rosettes of leaves, each with three toothed **leaflets**. White or pink flowers with **pentamerous epicalyx**. The fruit (strawberry) is a swollen **receptacle** with tiny **achenes** on the surface. There are more than 20 species, with many **hybrids** and **cultivars**, distributed throughout northern temperate and tropical regions, and in Chile, and widely cultivated. *See* diageotropism.

fragic horizon A subsurface **soil horizon** at least 25 cm thick, with a higher **bulk density** than the overlying horizon and less than 0.5 percent organic carbon by weight. Plant roots and percolating water are able to penetrate only along faces between **peds**.

fragipan A subsurface **diagnostic horizon** with a high **bulk density** and little or no **cementation**. It is dense, compact, and brittle and occurs in **acid soils**.

fragmentation **Vegetative reproduction** in which the plant breaks into separate pieces (fragments), each of which develops into a new plant.

Francoaceae (order **Geraniales**) A family of **perennial**, rhizomatous (*see* rhizome) **herbs** with **alternate**, **simple**, **exstipulate**, **entire** or dissected, sheathing leaves, with **petioles**. Flowers sometimes **zygomorphic**, plants **hermaphrodite**, more or less **tetramerous**, **hypanthium** free or absent, 4(–5) **sepals**, **corolla** of 2 or 4–5 **imbricate** or **contorted petals**, 4 or 8 **stamens** alternating with 4 or 8 **staminodes**, **ovary superior** of 4 occasionally 2 **carpels** and **locules**. **Inflorescence** a terminal **raceme** or **panicle**. Fruit a **capsule**. There are two genera with two species occurring in Chile.

frangipani *See Plumeria.*

Frankeniaceae (order **Caryophyllales**) A **monogeneric** family (*Frankenia*) of clump-forming **shrubs** or cushion-like **subshrubs**, with stems containing salt glands. Leaves **opposite**, **decussate**, **simple**, **entire**, **exstipulate**, often inrolled at the margin and united by a sheathing membrane around the stem. Flowers **actinomorphic**, **hermaphrodite** or **unisexual** (plants **gynodioecious**), 4–5(–6) fused **sepals** and **petals**, 3–6(–24) **stamens** in 2 **whorls**, in **pistillate** flowers reduced to **staminodes**, **ovary superior** of 1–(2–4) fused **carpels** with 1 **locule**. **Inflorescence** terminal or **axillary**, solitary or in a dichasial (*see* dichassium) **cyme** with 2 free or 4 fused **bracts**. Fruit a longitudinal **dehiscent capsule**. There are 90 species with a scattered but worldwide distribution in warm, dry regions. Some species grown for ornament.

Frankia A genus of **Actinobacteria** that form **filaments** resembling fungal **mycorrhizae** and live in soil. They convert atmospheric nitrogen to ammonia (NH_3) and some inhabit **root nodules** on non-leguminous (actinorhizal) plants.

frankincense *See* Burseraceae.

Frankliniella occidentalis (western flower thrips) A species of brown, yellow, or red thrips (**Thysanoptera**) in which adult males are about 1.0 mm long and females 1.4 mm. Most are female and reproduce by **parthenogenesis**, but also lay from 40 to more than 100 eggs in flowers, fruit, or foliage, where the yellow, red-eyed **nymphs** feed for their first two **instars** then fall to the ground to complete two more instars. The thrips are attracted to brightly coloured flowers. They feed on more than 500 species of plants and cause considerable damage. The species originated in North America but now occurs in Europe, South America, and Australasia.

frass Fine, powdery material left by herbivorous insects, consisting of insect excrement mixed with plant material.

Fraxinus (family **Oleaceae**) A genus of medium or large, mostly **deciduous** (a few tropical species **evergreen**) **trees** (ash) with **opposite**, **pinnate**, occasionally **simple** or in **whorls** of 3, leaves. Flowers are reduced, mostly wind-pollinated, and borne in **racemes**. Fruit is a **samara**. There are 63 species occurring throughout the temperate Northern Hemisphere, a few tropical. They are widely grown for ornament and for their timber.

Fraxinus ornus (south European flowering ash, manna ash) *See* mannitol.

free atmosphere The whole of the atmosphere that lies above the top of the **planetary boundary layer**.

free-central placentation **Placentation** in which the **ovules** are borne on a central growth from the base of the **ovary**. *See* axile placentation, basal placentation, parietal placentation.

freezing 1. The change of phase from liquid to solid. **2.** The temperature (0°C) at which pure water freezes at 100 kPa pressure.

freezing drizzle **Drizzle** that consists of supercooled (*see* supercooling) droplets which freeze on contact with surfaces at below freezing temperature.

freezing fog **Fog** which forms when the air temperature, and therefore the ground and other surfaces, is below freezing. **Fog droplets** freeze on contact with surfaces, depositing a layer of ice.

freezing index The cumulative number of days when the air temperature is below freezing.

freezing level The lowest height above sea level at which the air temperature is 0°C.

freezing nuclei Small, airborne particles onto which supercooled (*see* supercooling) water droplets will freeze.

freezing rain Rain that consists of supercooled (*see* supercooling) water droplets. These freeze almost instantly on impact against surfaces that are at below freezing temperature.

fresh breeze Wind of 9–11 m/s. *See* appendix: Beaufort Wind Scale.

friable Describes a soil that crumbles easily.

friction layer (surface boundary layer) The lowest part, about 10 percent, of the **planetary boundary layer**, where friction with the surface ensures that air is thoroughly mixed.

Fringilla coelebs (chaffinch) A bird about 150 mm long with a wingspan of 245–285 mm, in which the male has a rust-red body, pale pink underside, blue-grey crown and neck, and an olive rump; females are dull brown. They inhabit woodland, but are common in urban and suburban gardens. They feed mainly on seeds, but also eat insects, and the young are fed exclusively on insects. They occur throughout Eurasia and are resident except in the colder parts of their range, and have been introduced to New Zealand and South Africa.

fringing forest (gallery forest) A forest that extends along a river bank from **tropical rain forest** into adjacent **savanna**.

frogs *See* Anura, Ranidae.

frog spawn The eggs of frogs (**Anura**).

frog storm (whippoorwill storm) In spring in North America, the first bad weather to follow a spell of fine, warm weather.

frond A large **compound** leaf, e.g of a palm or fern.

frons The front part of the head of an invertebrate animal.

front The boundary between two **air masses**.

frontal depression *See* frontal wave.

frontal fog (precipitation fog) **Fog** that forms at a **front**, where warm air is rising against cooler air, and its water vapour is condensing.

frontal precipitation Precipitation that falls from clouds which develop along **fronts**.

frontal system A structure comprising a **cold**, **warm**, and occluded (*see* occlusion) **front**, with their associated **isobars** and **cold** and **warm sectors**, as these are depicted on a weather map.

frontal thunderstorm A **thunderstorm** that develops when warm, moist air rises up a **front** and becomes unstable (*see* conditional instability).

frontal wave A wave that forms on the **polar front** and develops into a **frontal system** with a **cyclone** (frontal depression) around the wave crest.

frost Ice crystals that form on and coat a surface.

frost day A day on which **frost** occurs.

frost heaving An upward movement of the ground surface caused by the formation below the surface of a lenses of ice up to 30 mm thick. The vertical displacement is approximately equal to the combined thickness of all the lenses.

frost hollow (frost pocket) A small, sheltered depression surrounded by higher ground, where at night cold air subsides down the hillsides and accumulates, so in winter the nighttime temperature is more often below freezing in the hollow than it is elsewhere.

frostless zone Part of a hillside that remains free from **frost** when frost forms in the valley. It happens because cold, dense air sinks downhill to produce frost at lower levels, and is replaced by warmer air.

frost pocket *See* frost hollow.

frost point The temperature at which water vapour turns directly to ice.

frost shattering *See* frost wedging.

frost smoke **Steam fog** consisting of ice crystals.

frost wedging (congelifraction, frost shattering, gelifraction, gelivation) The shattering of rock that occurs when water freezes and expands in crevices along planes of weakness or in **pore** spaces.

frozen fog Low **stratus** in which the water droplets have frozen.

frozen precipitation Any type of precipitation that reaches the ground as ice or snow.

fructification A fungal **fruiting body** or the process of forming one.

fructose (D-fructose, fruit sugar, levulose) A **monosaccharide** produced by many plants. It is often linked with **glocuse** to form **sucrose**. Fructose is obtained commercially from sugar cane, sugar beet, and corn (maize).

frugivore An animal that feeds on fruit.

fruit In **angiosperms**, one or more ripened ovaries (*see* ovary) and their contents, including seeds, and in some cases other tissues with which they are combined. Pea and bean pods, ears of maize, cereal grains, and tomatoes are all fruits.

fruit body *See* fruiting body.

F

fruitfulness *See* fecundity.

fruiting body (fruit body, sporocarp) In **Fungi** and **slime moulds**, the multicellular structure that bears the structures which produce **spores**. Mushrooms and toadstools are fruiting bodies.

fruit moths *See* Tortricidae.

fruitlet-mining tortrix moth *See Pammene rhediella.*

fruit sugar *See* fructose.

fruit tree red spider mite *See Panonychus ulmi.*

fruit tree tortrix moth *See Archips podana.*

frustule The wall of a **diatom**, made from silica (SiO_2).

fruticose **Shrub**-like, with woody **branches**, or resembling a shrub, e.g. certain **lichens**.

Fuchsia (family **Onagraceae**) A genus of **deciduous** or **evergreen shrubs** and small **trees** with leaves **opposite** or in **whorls** of 3–5, **lanceolate**, **serrate** or **entire**. Flowers with 4 long, slender **sepals** and 4 shorter **petals**, **ovary inferior**, pollinated by hummingbirds. Fruit is an edible **berry**. There are about 110 species, most occurring in South America, with some in Central America and from New Zealand to Tahiti. They are widely cultivated for ornament.

Fucus serratus *See* serrated wrack.

Fucus spiralis *See* spiral wrack.

Fucus vesiculosus *See* bladder wrack.

fugacious Appearing only briefly; soon disappearing.

fugitive species *See* opportunistic species.

Fujita tornado intensity scale A six-point scale for classifying **tornadoes** according to the damage they inflict.

RATING	WIND SPEED (km/h)	DAMAGE
Weak		
F-0	64–116	slight
F-1	117–180	moderate
Strong		
F-2	182–253	considerable
F-3	254–331	severe
Violent		
F-4	333–418	devastating
F-5	420–512	incredible

fuliginous Dusky, matt black, soot-like.

fuller's earth A highly absorptive **clay** that consists principally of expanding clays, e.g. **montmorillonite**. It is used industrially as an absorbent.

fuller's teasel (*Dipsacus fullonum*) *See Dipsacus.*

fulvic acid A **humic acid** that can be extracted from **humus**. It is yellow (Latin *fulvus* means yellow) and soluble in strong acid. It reacts strongly with metals, increasing their solubility in water.

fulvous Reddish brown, reddish yellow, tawny.

fumaric acid A dicarboxylic acid ($HO_2CCH{=}CHCO_2H$), with a fruity taste, that is an intermediate in the **citric-acid cycle** and that occurs in **bolete fungi** and in certain **lichens** (e.g. **Iceland moss**).

fumulus A layer of cloud that is so tenuous as to be barely visible.

fundamental niche The **niche** that a population of a species would occupy in the absence of **competition**.

fundatrix (pl. fundatrices) A female, born by **parthenogenesis**, which founds a population, e.g. of aphids (**Aphididae**).

fungal sheath The fungal tissue that surrounds the root of the host plant in a **mycorrhiza**.

Fungi (Mycota, Eumycota) The taxonomic kingdom in the **domain** (or superkingdom in older classifications) **Eukarya** that comprises **heterotrophic eukaryotes**. Fungal **cell walls** are made from **chitin** and fungi consist of **hyphae** that form a **mycelium**. Reproductive hyphae form **fruiting bodies**, which produce and release **spores**. Fungi live as **saprotrophs**, **symbionts**, and **parasites**, and may be may be single-celled, filamentous, or plasmodial (*see* plasmodium). Fungi occur worldwide and there may be as many as 5 million species. Fungi can be large; an ***Armillaria** ostoyae* honey fungus in Oregon occupies 8.9 km^2, probably weighs more than 600 tonnes, and may be 2400 years old.

fungicide A chemical compound that kills **Fungi**.

fungicole A plant that grows on **Fungi**.

Fungi Imperfecti (Deuteromycota) **Fungi** that are known only in their asexual forms and therefore cannot be classified confidently. About 25,000 species fall into this group.

fungus gnats Small flies (**Diptera**), 1.5–3 mm long with clear or pale grey wings, long legs, and antennae (*see* antenna) longer than the head, and with **mandibles** adapted for tunnelling and gnawing. Females lay eggs in soil or plant debris and can produce several generations a year. Larvae are about 6 mm long and legless, and feed on plant roots and **Fungi**; adults are pollinators and also disperse fungal **spores**. Depending on temperature the flies develop from egg to adult in three to four weeks and adults live about eight days. They can damage seedlings and houseplants. There are many species in several families, found worldwide.

funicle (funiculus) The stalk of an **ovule**.

funiculus *See* funicle.

funnel cloud A cloud shaped like a funnel that develops in a **mesocyclone** and extends downward through the **cumulonimbus** cloud to emerge at the base. As it extends it also narrows, increasing its angular velocity, i.e. wind speed around the core, in order to conserve its angular momentum. If it touches the ground it becomes a **tornado**.

funnelling The acceleration that occurs when the wind is forced through a narrow passage, e.g. along a street with tall buildings on both sides.

furbelows The brown seaweed *Saccorhiza polyschides*, which is an **annual** but that grows to 3–5 m. The **stipe** is flattened, with frilly margins and curled at the base, and the **holdfast** is bulbous and has a knobbly surface.

furca *See* furcula.

furcipulate Resembling pincers.

furcula (furca) The springing organ of springtails (**Collembola**).

furfuraceous Covered with small scales.

Fusarium A genus of **ascomycete fungi**, most of which are **saprotrophs**, but with others that are plant pathogens or that produce toxins harmful to animals.

fuscous Dark or black, without lustre.

fusiform Long, with tapering ends, like a spindle.

fusion *See* coalescence.

future-natural Describes the community that would develop were all direct human influence removed.

fynbos Sclerophyllous vegetation, similar to **chaparral**, that is found in Cape Province, South Africa.

G

gaging station *See* gauging station.

gaining stream (influent stream) A stream that receives water from **groundwater**, through a **spring** or **seep**, increasing its flow.

galactolipid A **lipid** containing **galactose**.

galactose A **monosaccharide sugar** that is less sweet than **glucose**. It usually occurs as a component of a larger molecule.

gale Wind of 17–21 m/s. *See* appendix: Beaufort Wind Scale.

galericulate With a cap or cover.

gall (cecidium) An outgrowth or swelling on the roots, stems, or leaves of a plant, induced by bacterial or fungal infection, or by attack from certain species of **mites**, nematodes (**Nematoda**), or insects (**Insecta**). Some galls are beneficial to the plant (e.g. nitrogen-fixing **root nodules**), some do little harm, and others (e.g. **crown gall** and **clubroot**) are disease symptoms.

gallery forest *See* fringing forest.

gall gnats *See* Cecidomyiidae.

gall midges *See* Cecidomyiidae.

gall mites *See* Eriophyidae.

galochrous White, like milk.

galvanotaxis A change in the direction of locomotion of a **cell** or organism in response to an electrical stimulus.

game cropping The controlled killing of game animals for meat or other products.

gametangiophore *See* gametophore.

gametangium A **haploid** organ or **cell** on which **gametes** develop in **Protista**, algae (*see* alga), **Fungi**, and plant **gametophytes**. A male gametangium is usually known as an **antheridium**, a female one as an **archegonium**.

gamete A **haploid** cell that unites with another haploid cell of the opposite sex or **mating type** in the process of **fertilization**.

game theory The representation of relationships within a community and the physical or behavioural **characters** of species as contests in which participants seek to gain advantages. Attaching numerical values to gains and losses allows the contests to be modelled mathematically. The approach has led to important insights into ecological structures and animal behaviour.

gametic disequilibrium *See* linkage disequilibrium.

gametic equilibrium *See* linkage equilibrium.

gametocyte A **cell** that produces **gametes** by **meiosis**.

gametogenesis The formation of **gametes**.

gametophore (gametangiophore) The **thallose** or leaf-like structure bearing the gametangia (*see* gametangium) in mosses (**Bryophyta**) and ferns (**Pteridophyta**).

gametophyte In the life cycle of plants (*see* alternation of generations), a **haploid** stage arising from a haploid **spore** produced by **meiosis** from a **diploid sporophyte**, in which **gametes** are produced by **mitosis**. The gametophyte is the dominant, visible generation in mosses (**Bryophyta**) and liverworts (**Marchantiophyta**).

gametophytic self-incompatibility **Self-incompatibility** in which the development of the **pollen tube** ceases before **fertilization** can occur. It is the **genotype** of the **gametophyte** that determines the compatibility of **gametes**.

gamma diversity (regional diversity) The number of species present in a set of data referring to an inventory of a **habitat**. *See* alpha diversity, beta diversity.

gamma-hexachlorocyclohexane *See* lindane.

Gammallin *See* lindane.

Gammaproteobacteria A class of **Proteobacteria** that includes many serious pathogens.

gammaxene *See* lindane.

gamopetalous Having the **sepals** fused at the base or completely.

gamosepalous Having the **petals** fused at the base or completely.

gap analysis A technique used to identify **ecosystems** that are in need of **conservation**. The **ranges** of several endangered or rare species are plotted, each on a separate map, and the maps are laid one on top of another. The set of maps is then laid over a map showing protected areas and reserves. This reveals areas that remain unprotected.

Garcinia mangostana (mangosteen) *See* Clusiaceae.

garden millipede *See Oxidus gracilis.*

Garden Organic The British charity (non-profit) that is dedicated to promoting organic horticulture and providing advice and assistance to organic gardeners and growers. ↗

garden pea (*Pisum sativum*) *See Pisum.*

garden slug *See Arion hortensis.*

garden snail *See Cornu aspersum.*

garlic (*Allium sativum*) *See Allium.*

garlic snail *See Oxychilus alliarus.*

garrigue A form of secondary vegetation that is widespread around the Mediterranean and derived from **evergreen mixed forest**. It comprises aromatic **herbs** and dwarfed **shrubs**, many belonging to the **Fabaceae** or **Lamiaceae**.

Garrulus glandarius (Eurasian jay) A species of corvids (**Corvidae**), about 340 mm long with a wingspan of 550 mm, that are mainly reddish brown, with bright blue spots with black speckles on their otherwise black wings, black beaks, black moustache stripes, and black tails with a white patch near the base. They inhabit areas with dense foliage and trees, but sometimes feed on the ground. They eat nuts and acorns, and cache food for winter, mostly remembering where they left it. By leaving some uneaten the birds help disperse the seeds. They occur throughout Eurasia. ↗

Garrya (family **Garryaceae**) A genus of **evergreen shrubs** with **coriaceous**, **opposite**, **simple**, **ovate**, **entire** leaves with a short **petiole**. Flowers **unisexual**, **ovary inferior**, borne in pendulous **catkins**. Fruit is a **berry**. There are about 18 species occurring in western North and Central America and the Caribbean. Several are cultivated as winter-flowering ornamentals known as silk tassel or tassel bush.

Garryaceae (order **Garryales**) A family of **dioecious**, **evergreen shrubs** and **trees** with **opposite**, **simple**, **coriaceous**, more or less **entire** leaves, sub-**sessile** or with small **petioles**. Male flowers **actinomorphic** with 1 **perianth whorl** of 4 **tepals** alternating with 4 **stamens**. Female flowers with reduced perianth, **ovary** of 2 occasionally 3 **carpels** with 1 **locule**. **Inflorescences** terminal or **axillary** and **catkin**-like, males up to 30 cm long, females shorter and broader. Fruit is a **berry**. There are 2 genera with 17 species occurring in western North America, Central America, the Caribbean, and Eastern Asia. Some cultivated for ornament.

Garryales An order of woody plants comprising 2 families with 3 genera and 18 species. *See* Eucommiaceae and Garryaceae.

gaseous exchange The flow of gases into and from a living organism, e.g. in **photosynthesis** and **respiration**.

gaster In **Hymenoptera**, the abdomen apart from the first segment, which is included in the thorax and separated from the rest of the abdomen by the **petiole**.

gastrolith A stone that is ingested by an animal and that lodges in the **gizzard** where it helps break up food items.

Gastrophryne carolinensis (eastern narrowmouth toad) A species of toad-like **Amphibia** that have smooth, moist skin, unlike most toads. They grow to about 50 mm and have a fold of skin above their eyes that they can move forward to sweep away insects settling on their eyes. They are light to dark brown or red. They burrow in loose, moist soil and are found in sheltered, moist places. They feed mainly on ants, but also on other insects, and occur throughout the southeastern United States.

Gastropoda (slugs, snails, limpets, sea slugs) A class of asymmetrical molluscs (**Mollusca**) that have a true head, unsegmented body, and large, flat foot. Most have a well-developed **radula** with which they feed on soft plant material. During its development the **mantle** cavity and visceral mass of a gastropod rotates sideways through 180 degrees (the process is called torsion) so in the adult the anus is situated on the upper side of the body. The shell, in those gastropods that possess them, is in one piece and coiled spirally, at least in the young. More than two-thirds of all gastropods are marine; the remainder inhabit fresh water or live on dry land, but always in moist places. There is no agreement on the number of **fossil** and **extant** gastropod species, but there are more than 62,000 named species living at present.

gas vacuole A **vacuole** containing a gas. Gas vacuoles are found in certain aquatic **Bacteria** and **cyanobacteria** and apparently give the cell **buoyancy**.

gated pipe A surface irrigation pipe that has gates covering the holes, allowing greater control of the water used.

gauging station (gaging station) An installation close to a river containing instruments that monitor river flow or the level of **groundwater**.

Gaultheria *See* methyl salicylate.

Gause principle *See* competitive exclusion principle.

Geissolomataceae (order **Crossomatales**) A **monotypic** family (*Geissoloma marginatum*), which is a bushy **shrub** with **opposite**, **decussate**, **simple**, **entire**, **ovate** to suborbicular, **coriaceous** leaves with **connate** pairs of **petioles** and small **stipules**. Flowers single, **axillary**, **bisexual**, **actinomorphic** 4-merous, with a single **perianth whorl**, 4 **tepals**, 8 **stamens** in 2 whorls of 4, **ovary superior** of 4 fused **carpels** and 4 **locules**. Fruit a **loculicidal capsule**. The family occurs in Cape Province, South Africa.

geitonogamy **Pollination** by **pollen** from a different **flower** on the same plant.

gelatinous lichen A **lichen** in which a cyanobacterium (*see* cyanobacteria) is the **phycobiont**.

gelifluction (congelifluction) The slow, downslope movement of surface material, lubricated by water, over frozen ground. It is the equivalent of **solifluction** and occurs only in the **active layer**.

gelifraction *See* frost wedging.

gelisols A group of soils that occur in very cold climates. There is **permafrost** within 2 m of the surface and, because of the low temperature, soil development and the decomposition of organic matter proceed very slowly. Gelisols are an order in the U.S. Department of Agriculture **soil taxonomy**.

geliturbation (congeliturbation, cryoturbation) Any movement of soil or other surface material that is caused by **frost**, including **frost wedging** and **gelifluction**.

gelivation *See* frost wedging.

Gelsemiaceae (order **Gentianales**) A family of **shrubs** or **lianas** with **opposite**, **simple**, **entire**, **lanceolate** to broadly **ovate** leaves with **stipules** between the short **petioles**. Flowers more or less **actinomorphic**, **pentamerous**, usually heterostylous (*see* heterostyly), **sepals** sometimes fused, bright yellow, white, or pink **corolla** trumpet-shaped, **ovary superior** of 2 **carpels** and 2 **locules**. **Inflorescence** an **axillary cyme** of 1 or a few flowers. Fruit is a **capsule**. There are 2 genera with 11 species occurring throughout the tropics. Some have medicinal uses and some cultivated as ornamentals.

gemma 1. A **propagule** in the form of a modified organ of the parent plant, e.g. a **bulbil**. **2.** In **Fungi**, a thick-walled **spore** formed from a vegetative **hypha**. **3.** In **Bryophyta** and **Marchantiophyta**, a structure that functions in **vegetative reproduction**.

gemmation *See* budding.

GenBank A database that holds all publicly available **DNA** sequences.

gene The basic unit of heredity, comprising a segment of **DNA**, or **RNA** in some viruses, occupying a fixed **locus** on a **chromosome**, and that has a particular effect on the **phenotype** when it is transcribed. Genes may mutate (*see* mutation), producing different versions (**alleles**).

genealogical species A definition of a **species** as a group of organisms that are more closely related to each other than they are to organisms outside the group.

gene bank A place where hereditary material is conserved and stored in a viable state. It holds material from endangered plants and **cultivars** that are no longer in commercial use. When dried to a moisture content of about 4 percent and held at 0°C, many seeds remain viable for up to 20 years (but *see* recalcitrant seed). Pollen can also be stored, but it does not remain viable for so long.

gene centre *See* centre of diversity.

gene conversion A natural process whereby one member of a **gene family** acts as a template for correcting other members of the family.

gene duplication The copying of a **gene** twice, the two copies lying side by side on the same **chromosome**.

gene family A group of identical or very similar **genes** that result from **gene duplication** and usually occur on the same **chromosome**. They may function in concert or some may be suppressed (*see* gene silencing), becoming **pseudogenes**.

gene fixation A **gene frequency** in which all members of a population are homozygous (*see* homozygosity) for a certain **allele** at a particular **locus**.

gene flow The movement of **genes** within an interbreeding group of organisms that results from mating.

gene frequency For a given population, the number of loci (*see* locus) at which an **allele** occurs expressed as the proportion of the loci at which it could occur.

gene library A collection of **DNA** fragments that ideally includes all the genetic information for a particular **species**.

gene network *See* gene regulatory network.

gene pool The total number of **genes** possessed by all the members of a population of sexually reproducing organisms.

G

general adaptation An **adaptation** that fits an organism for a broad environmental condition, e.g. a plant leaf. The particular type and shape of leaf is a special adaptation to a more restricted set of conditions.

general circulation (atmospheric circulation) The overall movements of the atmosphere whereby heat is transported away from the equator, and winds, clouds, and precipitation are generated. *See* three-cell model.

generation time The time required for a **cell** to complete a full life cycle.

gene regulatory network (gene network, transcription network) A collection of segments of **DNA** that interact indirectly, through **RNA**, with each other and with other contents of a cell. Such networks regulate responses to the external environment, e.g. the **circadian rhythm** in plants.

gene sequencing The suite of laboratory procedures used to determine the order in which **nucleotides** occur in a segment of **DNA** of any length up to a whole **genome**. Sequences from different organisms can then be compared.

gene silencing An **epigenetic** process of gene regulation in which the activity of a **gene** is reduced or the gene is 'switched off' completely. It occurs at the **transcription** or post-transcription level, and modifies interactions in the **gene regulatory network**.

genetic Pertaining to ancestry.

genetic code The sequences of **codons** and the **amino acids** for which they encode. *See* appendix: The Genetic Code.

genetic drift Random fluctuations in **gene frequency** throughout a population such that offspring are not perfect genetic representatives of their parents. Drift occurs in all populations.

genetic engineering (genetic modification) Splitting and rejoining **DNA** to form **hybrids** that bypass the constraints ordinarily limiting the exchange of genetic material, in some cases to the extent of combining **genes** from only distantly related organisms. It is used in research and also to tailor an organism for a particular purpose or application.

genetic erosion The loss of **genes** that occurs when an introduced and highly adaptable **cultivar** replaces and eventually threatens the survival of local crop varieties that represent the genetic base of the crop **species**.

genetic load The average number of **lethal mutations** per individual in a population.

genetic map (linkage map) A diagram prepared for a **species** or population that shows the positions on **chromosomes** of known **genes** or **genetic markers** determined by the frequency of **recombination** between markers on **homologous chromosomes**; the greater the frequency of recombination the farther apart the markers are and vice versa.

genetic marker A **gene** or segment of **DNA** occupying a known position on a **chromosome** that can be used to identify a **species** or individual.

genetic modification *See* genetic engineering.

genetic polymorphism The presence in a population of two or more **genotypes** in a frequency that recurring **mutations** cannot explain. It may be balanced (**balanced polymorphism**), when **allele** frequencies are in equilibrium at a specified **locus**, or transient (**transient polymorphism**), when a mutation is spreading through the population.

genetic resources The **gene pool** of organisms that is available for human exploitation.

genetics The study of heredity and **genes**.

genetic system The arrangement of **genes** in a **species** and the method by which they are transferred from parents to offspring.

genetic variance The part of the variation in **phenotypes** that is attributable to variations in **genotype** among members of a population.

geniculate With a sharp bend, like a knee.

genome The total genetic information carried by a single (i.e. **haploid**) set of **chromosomes**.

genome duplication The doubling of a **genome** in **autopolyploidy**.

genome obesity The possession of an extremely large **genome**, as in certain plants, e.g. **Liliales**, many members of which have more than 100,000 million **base pairs**, when 1000 Mbp is more typical of plants. It is often due to the unchecked expansion of **retrotransposons**.

genotype The genetic constitution of an organism, as contrasted with its **phenotype**.

Gentianaceae (order **Gentianales**) A family of **annual** and **perennial herbs** with a few large **trees** or woody **lianas**, and several **saprophytes** that lack **chlorophyll**. Leaves **opposite**, occasionally **alternate** or whorled, **entire** or occasionally **saccate**. Flowers **bisexual**, hypogynous (*see* hypogyny), 4- or 5-merous, rarely 3- to 16-merous, usually **actinomorphic**, **sepals** fused, sometimes **calyx zygomorphic**, **corolla** tubular to **campanulate**, as many **stamens** as corolla lobes, **ovary superior** of 2 **carpels** and 1–2 **locules**. **Inflorescence** a terminal or **axillary cyme**, occasionally **raceme**, **spike**, or head. Fruit is a dry **capsule** or **berry**. There are 87 genera with 1655 species with worldwide but mainly temperate distribution. Many have medicinal uses, several cultivated as ornamentals.

Gentianales An order of plants comprising 5 families of 1118 genera and 16,627 species. *See* Apocynaceae, Gelsemiaceae, Gentianaceae, Loganiaceae, and Rubiaceae.

gentle breeze Wind of 4–5 m/s. *See* appendix: Beaufort Wind Scale.

genus In **taxonomy**, a group of related **species**, representing a taxonomic rank between species and **family**.

geobotanical anomaly A local concentration of plant species that may indicate the presence of an ore deposit or a source of **hydrocarbons**.

geobotanical exploration The use of plant **indicator species** or **assemblages** to identify areas that may contain deposits of metals.

geocarpic Describes a plant that produces fruit below ground, e.g. *Arachus hypogaea*, peanut.

geofrutices (geoxylic plants) Plants that produce a massive, woody, branched or unbranched, underground stem. Geofrutices occur in open areas where fires are frequent.

Geoglossaceae (earth tongues) A family of **ascomycete fungi** that produce small, dark, club-shaped **ascocarps** 20–80 mm tall containing brown **ascospores**. They occur widely in temperate regions, among decomposing vegetation in grassland in Europe, and in forests in North America.

geologic erosion (normal erosion) **Erosion** of soil in a natural environment, supporting natural vegetation, and undisturbed by human activity, that results from the action of wind, water, ice, freezing, thawing, etc.

Geomalacus maculosus (Kerry slug, Kerry spotted slug) A dark grey or brown round back **slug** (**Arionidae**) with yellow spots, 70–80 mm long, with about 25 rows of small projections along either side of the body. It has an internal shell. It inhabits moist environments with **acid soils**, is mainly nocturnal, and feeds on mosses (**Bryophyta**), **lichens**, liverworts (**Marchantiophyta**), and **Fungi**. It has a **disjunct distribution** in Ireland, northwest Spain, and Portugal, and is protected.

geophyte In the life-form classification devised by Christen **Raunkiær**, a terrestrial plant with underground storage organs, e.g. **bulbs**, **corms**, **rhizomes**, **tubers**, that allow it to survive unfavourable periods.

Georgefischeriales An order of **Expobasidiomycetes** that cause smut diseases. Most occur among grasses.

Georgia jumper *See Amynthas gracilis.*

geostrophic wind The wind above the **planetary boundary layer**, which blows parallel to the **isobars** because the **pressure-gradient force** (PGF) acting toward the centre of low pressure is precisely balanced by the deflection due to the **Coriolis effect** (CorF).

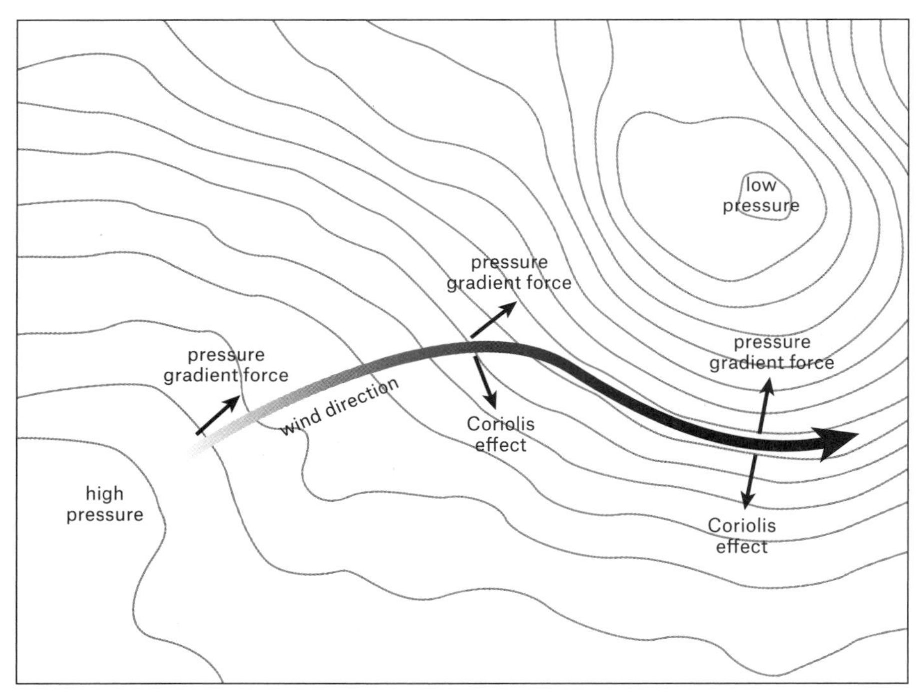

As air moves in response to the pressure gradient force (PGF), acting at right angles to the isobars, the Coriolis effect (CorF) deflects it to the right. When the two forces are in balance the air flows parallel to the isobars. That is the geostrophic wind.

geostrophic wind level (gradient wind level) The height at which the wind becomes geostrophic or gradient, at 500–1000 m.

geotaxis A change in the direction of locomotion of a **cell** or organism in response to gravity.

geotropism (gravitropism) A movement or change of direction by a plant or fungus in response to gravity. Positive geotropism is directed downward, e.g. a **taproot**, negative geotropism is directed upward, e.g. a shoot.

geoxylic plants *See* geofrutices.

Geraniaceae (order **Gerianales**) A family of **shrubs**, **herbs**, and **geophytes**, most **perennial** but some **annual** or **biennial**, with **alternate**, **simple** or lobed, deeply divided, or **pinnate** leaves, usually with **petioles** and **stipules**. Flowers **bisexual**, radially or bilaterally symmetrical, **pentamerous**, bracteolate (*see* bracteole), **perianth** of **sepals** and **petals** usually brightly coloured, 5, 10, or 15 **stamens**, sometimes as **staminodes**, **ovary superior** of 5 fused **carpels**. Fruit a **schizocarp**. There are 7 genera with 805 species occurring through warm temperate and temperate regions. *Pelargonium* and ***Geranium*** are widely cultivated as ornamentals.

Geraniales An order of plants comprising 5 families of 17 genera and 836 species. *See* Francoaceae, Geraniaceae, Greyiaceae, Ledocarpaceae, Melianthaceae, and Vivianiaceae.

Geranium (family **Geraniaceae**) A genus of **annual**, **biennial**, and **perennial herbs** with **palmate**, broadly circular leaves. Flowers **actinomorphic**, **pentamerous**, **petals** white, pink, purple, or blue. Fruit is a **capsule**, in some species with the five seeds attached to a central column resembling a bird's bill that is exposed when the capsule springs open, giving the plants their common name, cranesbills. There are 422 species occurring throughout temperate regions, especially in the eastern Mediterranean region, and on tropical mountains. Many are cultivated for ornament.

geranium aphid *See Acyrthosiphon malvae.*

geranium sawfly *See Protemphytus carpini.*

gerbils *See* Muridae.

germinal furrow *See* colpus.

germinal selection A theory proposed by August Weismann (1834–1914) that **natural selection** acts on the **germ plasm** to retard the spread of **mutations**.

germination The commencement of growth of plants, **Bacteria**, and **Fungi** following a period of **dormancy** spent as **seeds** or **spores**, usually triggered by an improvement in external conditions, e.g. a rise in temperature, increase in moisture and oxygen.

germ line The cells from which **gametes** are derived.

germ plasm Hereditary material that passes to offspring in **gametes** and gives rise to **somatic cells**.

germ plasm bank A repository for **germ plasm** where it may be conserved.

germ pore A thin-walled area on a **spore** or **pollen grain** from which the **germ tube** or **pollen tube** emerges.

germ tube The **filament** that emerges from a germinating **spore**. It grows and develops by **mitosis** to form **hyphae**.

Gerrardinaceae (order **Huerteales**) A **monogeneric** family (*Gerrardina*) of small **trees** and **shrubs** with **alternate**, **entire** or **serrate** leaves with **stipules**. Flowers **actinomorphic**, **pentamerous**, free **hypanthium**, 5 **imbricate sepals** in 1 **whorl**, 5 **sessile petals**, **ovary superior** or partly **inferior**, **syncarpous** with 1 **locule**. **Inflorescence** a **cyme**. Fruit a **berry**. There are two species occurring in the tropics and subtropics of southern and eastern Africa.

Gesneriaceae (order **Lamiales**) A family of **herbs**, **epiphytes**, **lianas**, and **trees** with weak stems, some species **monocarpic**. Many have only one enlarged **cotyledon**, 1 m or more long, and no foliage. Where leaves are present they are **opposite**, sometimes **alternate**, whorled, or forming a rosette, **entire** or toothed, usually covered in hairs. Flowers with 5 **sepals** free or fused into a tube, **corolla** in a tube, more or less **actinomorphic** or strongly **zygomorphic**, 2 or 4 **stamens**, **ovary superior**, semi-**inferior**, or inferior usually with 1 **locule**. **Inflorescence** an **axillary** or terminal **cyme**. Fruit is a **capsule**. There are 147 genera with 3311 species with a largely tropical distribution. Many cultivated for their flowers, known as African violets and gloxinias, and some for medicinal use.

Gessner, Conrad (1516–65) A Swiss polymath best known as a botanist, who in 1541 published *Historia plantarum*, a dictionary of plants, and in 1561 *De hortis Germaniae*, describing French, German, Italian, and Swiss gardens, and prominent gardeners. His uncompleted works were gathered together and published posthumously in Frankfurt as *Opera botanica* between 1751 and 1771 in 2 volumes.

giant bee (*Apis dorsata*) *See Apis.*

giant Palouse earthworm *See Driloleirus americanus.*

giant polypore *See Meripilus giganteus.*

giant puffball *See Calvatia gigantea.*

giant willow aphid *See Tuberolachnus salignus.*

gibberellins A group of **plant hormones** which stimulate leaf and shoot growth, usually affecting the whole plant so they do not induce bending movements. Gibberellins also occur in certain **Fungi** and fungal gibberellins cause **bakanae disease** in rice.

Gigantopteridales A group of plants, probably **polyphyletic** but sometimes ranked as an order, that lived during the Late Permian epoch (260.4–251 million years ago). They grew up to 50 cm tall, with a woody stem with spines, and leaf structures resembling fern fronds when young, but more like leaves when mature. They resembled **angiosperms** but are not known to have produced flowers.

gilgai The undulating relief found on a very small scale in soils containing high concentrations of **clay minerals**, e.g. **montmorillonite**, that swell and shrink in response to wetting and drying. The effect can be large enough to fracture pipes and move poles and fence posts out of alignment.

gill In the fruit body (*see* fruiting body) of an **agaric**, a membranous structure that holds the **hymenium**. It is composed of **blades** arranged radially, usually situated on the underside of the **pileus**.

The gill is the blade-like structure beneath the pileus of a mushroom or other agaric.

ginger (*Zingiber officinale*) *See* Zingiberaceae.

Ginkgo (family **Ginkgoaceae**) A genus of **deciduous**, **dioecious trees** (maidenhair tree) with fan-like, **alternate**, 2-lobed leaves with **dichotomous venation**. Males produce naked pairs of **catkin**-like clusters of **stamens**. Female plants produce a naked **ovule** that develops into two seeds that have a foul-smelling, fleshy, outer layer and a hard, edible, inner layer. There is one **extant** species, *G. biloba*, which is a **living fossil** discovered in eastern China that has changed little since 270 million years ago. It is widely cultivated and has many medical uses.

Ginkgoaceae (order **Ginkgoales**) A **monotypic** family (***Ginkgo** biloba*) of large, **deciduous trees** with fan-shaped leaves that have **dichotomous venation**.

Ginkgoales An order of plants that contains only the **monotypic** family **Ginkgoaceae**.

ginseng (*Panax* spp.) *See* Araliaceae.

girdle band *See* connecting band.

girdling Cutting across the **phloem** tissue of a plant, thus preventing the downward movement of substances needed to sustain the roots. Complete girdling of a tree, by severing the phloem around the entire circumference of the trunk, kills the tree.

Gisekiaceae (order **Caryophyllales**) A **monogeneric** family (*Gisekia*) of prostrate or erect **herbs** with **opposite** or whorled, **entire**, **exstipulate** leaves with **petioles**. Flowers are **actinomorphic**, **biexual**, **pentamerous**, with 5 **ovate** to **lanceolate tepals**, 5–20 free **stamens**, and 3–15 fused or 4–5 free **carpels**. **Inflorescence** is an **axillary umbel**-like **cyme** or diffuse terminal cyme. Fruit is a cluster of **mericarps**. There are five species occurring in Africa and Asia.

gizzard A strongly muscular section of the alimentary canal where food items are broken into small fragments.

glabrous Smooth, hairless.

glacial till (till) Sediment deposited by glacial ice with no contribution from liquid water.

gladiate Sword-like.

Gladiolus (family **Iridaceae**) A genus of **monocotyledon herbs** growing from **corms** usually with unbranched stems, each producing one to nine narrow, **ensiform** leaves with a longitudinal groove, enclosed in a sheath, the lowest leaf a **cataphyll**. Flowers **bisexual**, each subtended by 2 **bracts**, **tepals** fused into a tube, the **dorsal** tepal arching over the 3 **stamens**, **ovary** with 3 **locules**. **Inflorescence** is a long, one-sided **spike** with each flower enclosed in its own **spathe**. Fruit is a longitudinally **dehiscent capsule**. There are about 260 species, 10 occurring in Eurasia, the remainder in sub-Saharan Africa. Many are cultivated as ornamentals.

glaebule *See* caliche.

glandular hair *See* glandular trichome.

glandular trichome (glandular hair) A plant stalk (**trichome**) bearing a **gland** at its tip.

glasshouse millipede *See Oxidus gracilis*.

glasshouse-potato aphid *See Aulacorthum solani*.

glass snails *See* Vitrinidae.

glasswort *See Salicornia*.

glaucous Bluish green or sea-green.

glaze (clear ice) A layer of solid, clear ice that covers surfaces.

Gleason, Henry Allan (1882–1975) An American ecologist who in 1917 published his own **individualistic hypothesis** in opposition to the **climax theories** proposed by Frederic Edward **Clements** and later by Arthur George **Tansley**.

gleba Tissue that bears **spores** in the **fruiting bodies** of certain **Fungi**, e.g. puffballs.

glenoid Relating to a socket.

gley The end product of anaerobic, waterlogged soil, in which iron compounds are reduced (*see* reduction). The soil is often grey and mottled with rust-red colours.

gleying (gleyzation) The process of forming **gley**.

gley soil A soil subjected to prolonged waterlogging and that contains **gley soil horizons**. The name is used in the classification devised by the Soil Survey for England and Wales.

gleysols A group of soils that show evidence of **gleying** within 50 cm of the surface. Gleysols are a reference soil group in the **World Reference Base for Soil Resources**.

G

gleyzation *See* gleying.

gliadins *See* gluten.

Glinka, Konstantin Dimitrievich (1867–1927) A Russian soil scientist who was a student and colleague of Vasily Vasilyevich **Dokuchayev**. He conducted soil surveys of most of Russia.

Globodera pallida (white potato cyst nematode) A species of pale-coloured cyst nematodes (**Nematoda**) that are sedentary **endoparasites**. They feed on many **Solanaceae** species, but especially potatoes. The tan-coloured **cysts** persist in the soil for more than ten years. Juveniles emerging from the cysts penetrate plant roots and establish feeding sites in the **stele**. The nematode is dispersed in soil and contaminated plant material. Infestations cause severe crop losses. The nematode occurs throughout most of Europe, western Asia, South America, but is uncommon in North America.

Globodera rostochiensis (golden nematode, golden eelworm, yellow potato cyst nematode) A cyst nematode (**Nematoda**) that feeds on plants of the **Solanaceae**, especially potatoes and tomatoes, forming **cysts** containing eggs on their roots. Infestation causes reduced growth, **chlorosis**, and **wilting**. The nematode originated in South America but now occurs throughout Europe and North America.

globose Spherical.

globular protein A **protein** in which at least one **polypeptide** chain is folded into a spherical shape.

glochid A short, barbed hair (e.g. in **Cactaceae**).

Gloeocapsa (glow caps) A genus of single-celled **cyanobacteria** in which clusters of cells are embedded in mucilage. The colonies are usually spherical, often brightly coloured (hence the common name) and occur in masses worldwide on wet rocks and tree **bark**.

Gloger's rule The rule proposed by the German zoologist C. W. L. Gloger that many animals that live in wet climates are dark-coloured and those living in **dry climates** are pale-coloured. There are many exceptions to the rule, but so far as it is true it may be because wet areas tend to be darker in colour than dry areas, so dark and pale animals are better camouflaged.

Glomerales An order of **Fungi** belonging to the **Glomeromycota**. They are **biotrophs**, most exchanging nutrients with their hosts by means of arbuscular (*see* arbuscule) **mycorrhizae**.

Glomeromycota A **phylum** of **Fungi** that are **obligate symbionts**, forming arbuscular (*see* arbuscule) **mycorrhizae** with the roots of their hosts, which include most tropical trees and herbs. They are not known to reproduce sexually. There are 10 genera with about 150 species.

gloom The condition in which smoke or dense cloud reduces the intensity of daylight, but horizontal visibility remains good.

Glossopteris An extinct **genus** of woody, seed-bearing, **gymnosperms** that grew as **shrubs** or **trees** with **lanceolate** or tongue-shaped **deciduous** leaves, that lived in middle and high latitudes of **Gondwana** during the Permian (299–251 million years ago) and that contributed greatly to the Permian coal deposits of the Southern Hemisphere.

glossy pillar *See Cochlicopa lubrica.*

glottis The vocal cords and the spaces between them.

glow caps *See Gloeocapsa.*

gloxinia *See* Gesneriaceae.

glucan (glucosan) A **polysaccharide** in which **glucose** units are linked by **glycosides.**

gluconeogenesis The metabolic process, occurring in all living organisms, by which **glucose** is synthesized from non-**carbohydrate** precursors, including **amino acids, fatty acids, glycerol,** and intermediates in the **citric-acid cycle.** Gluconeogenesis is the opposite of **glycolysis.**

Gluconobacter oxydans A genus of **Alphaproteobacteria** that are oval or rod-shaped, non-**motile**, Gram-negative (*see* Gram reaction) **aerobes** that occur worldwide in soil and also in flowers, fruits, fruit juices, soft drinks, and alcoholic beverages. They can survive high-concentration sugar solutions and low **pH**, and partly oxidize **carbohydrates** and **alcohols**, causing bacterial rot of apples and pears.

glucosan *See* glucan.

glucose (dextrose, D-glucose, grape sugar) The simplest **monosaccharide sugar**, $C_6H_{12}O_6$. It is one of three sugars synthesized by plants (*see* fructose, galactose) as a product of **photosynthesis** and the primary source of energy for **cells.**

glucoside A **glycoside** formed when **glucose** is hydrolyzed (*see* hydrolysis) or broken down by **fermentation** or reactions catalyzed by **enzymes.**

glucosinolates A group of about 100 water-soluble **glucosides** formed as secondary metabolites by plants in the order **Brassicales**, especially the family **Brassicaceae** (cabbages, Brussels sprouts, etc.). The glucosinolate sinigrin gives these plants their bitter flavour. When hydrolyzed (*see* hydrolysis) by myrosinase **enzymes** glucosinolates yield oils that give mustard and horseradish their piquancy. ↗

glume In **Poaceae**, one of the pair of **bracts** that subtends each **spikelet.** In **Cyperaceae**, a bract that subtends the **inflorescence.**

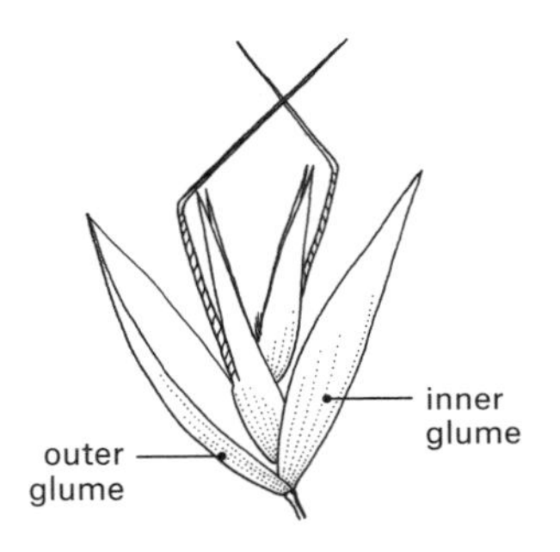

gluten A **protein** found in wheat, barley, and rye grains that gives dough its elasticity. Gluten is composed of two classes of proteins, gliadins, which are soluble, and glutenins, which are insoluble, joined to **starch** in the **endosperm.**

glutenins *See* gluten.

glycerides (acetylglycerols) A group of **esters** formed from **glycerol** and one to three **fatty acid** molecules; depending on the number of these glycerides are designated mono-, di-, or triglycerides. Vegetable oils and animal fats are mainly triglycerides, but **enzymes** break them down to monoglycerides, diglycerides, and free fatty acids.

glycerine *See* glycerol.

glycerol (glycerine) A trihydroxy **alcohol** ($CH_2OHCHOHCH_2OH$) produced by a reaction catalyzed by the **enzyme** lipase that occurs when **carbohydrates** are exhausted and fats supply energy for

respiration; a fat is converted to a **fatty acid** and glycerol. **Phosphorylation** of glycerol produces intermediates in **glycolysis**.

glycine The smallest **amino acid** found in **proteins**.

glycogen (floridean starch) A long chain of **glucose** molecules with many side branches that is used as a storage sugar in **red algae** (Rhodophyta) and in some **Bacteria**, **cyanobacteria**, and **Fungi**, but not in plants.

glycolate cycle (C2 cycle, oxidative photosynthetic cycle) A complex **metabolic pathway** in which glycolate produced in **chloroplasts** is converted to **glycine** in **peroxisomes**, and into **serine** in mitrochondria (*see* mitochondrion). The cycle recovers some of the carbon fixed by **photorespiration**.

glycolysis (Embden-Meyerhof pathway) A sequence of anaerobic reactions in which **glucose** is converted to **pyruvic acid** with the formation of two molecules of **adenosine triphosphate** (ATP) and two molecules of nicotinamide adenine diphosphate hydrogen (NADH).

glycoprotein A **conjugated protein** in which the **prosthetic group** is a **carbohydrate**.

glycoside A compound comprising a **sugar** molecule joined to another molecule by a **covalent bond**.

glyoxysome A specialized **peroxisome** found in plant and fungal **cells**. It contains **enzymes** that trigger the breakdown of **fatty acids** and other enzymes involved in the synthesis of **sugars** by **gluconeogenesis**.

glyphosate A broad-spectrum, **systemic herbicide** that is used to control grasses and **broad-leaved** herbs and woody plants, and to desiccate foliage on crops prior to harvest. It is only mildly toxic to mammals.

Gnetaceae (order **Gnetales**) A **monogeneric** family (*Gnetum*) of **gymnosperms**; they are **evergreen**, **dioecious lianas**, **trees**, and **shrubs** that, unlike other gymnosperms, possess **vessel elements**, placing the plants ambiguously between gymnosperms and **angiosperms**. Leaves **opposite**, **exstipulate**, **simple**, **elliptic**, **entire**, with **stipules**. Male **strobilus** with a **perianth** and **stamen**, female strobilus with perianth and 1 **ovule**. Fruit **drupe**-like. There are 30 species with a somewhat **disjunct** tropical distribution.

Gnetales An order of plants comprising 3 families of 3 genera and 96 species. *See* Ephedraceae, Gnetaceae, and Welwitchiaceae.

gnetophyte A group of **gymnosperms** comprising 3 families, each with 1 genus, and 68 species of **trees**, **shrubs**, and **lianas** (**Gnetaceae** and **Ephredaceae**) and turnip-like plants (**Welwitschiaceae**) that have many similarities with **angiosperms**. The grouping is probably artificial and their **fossil** record is poor.

gnotobiotic Describes a **culture** of microorganisms in which the precise composition is known.

goat moths *See* Cossidae.

goat nut *See* Simmondsiaceae.

goatsbeard *See Tragopogon*.

goat tang The **perennial** red seaweed *Polyides rotundus* found in rock pools and below the low-water mark. The **thallus** is cylindrical, **dichotomously branched**, and tough, the **holdfast** disc-shaped.

golden eelworm *See Globodera rostochiensis*.

golden nematode *See Globodera rostochiensis*.

goldenrod *See Solidago*.

golden trumpet *See Allamanda*.

goldfinch *See Carduelis carduelis*.

Golgi body (Golgi apparatus, Golgi complex) A system of **cisternae** surrounded by **vesicles** found in almost all **eukaryote** cells and involved in the packaging of many metabolic products. *See also* dictyosome.

Golgi, Camillo (1843–1926) An Italian physician and pathologist who devised a method of staining nerve and **cell** structures that revealed many details of cell **organelles** including the **Golgi body**. He shared the 1906 Nobel Prize in Physiology or Medicine with Santiago Ramón y Cajal for their work on the structure of the nervous system.

Gomortegaceae (order **Laurales**) A **monotypic** family (*Gomortega keule*), which is an aromatic **tree** with **opposite**, **simple**, **entire**, **exstipulate** leaves. The small flowers are **actinomorphic**, **perianth** of **sepals**, usually 7, sometimes 5–9 free **tepals**, usually 10, sometimes up to 13 free **stamens**, **ovary syncarpous inferior** of 2–6 **carpels** arranged spirally, with 2–3 **locules**. **Inflorescence** a terminal or **axillary raceme**. Fruits are **drupelets**. The family occurs only in Chile.

gonad A gland in which sperm or eggs form.

Gondwana A former supercontinent that comprised what are now South America, Africa, India, Australia, and Antarctica.

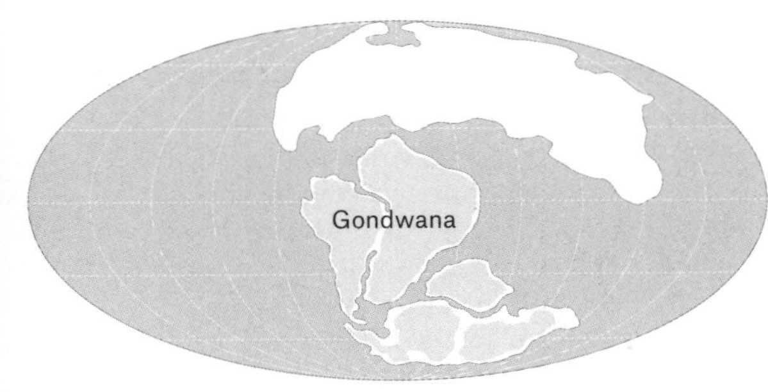

The supercontinent that comprised all of the present continents of the Southern Hemisphere as it appeared about 200 million years ago, at the end of the Triassic period.

It resulted from the breakup of **Pangaea**, and existed from about 510–180 million years ago.

gonidium 1. One of the single-celled algae (*see* alga) in a **lichen thallus**. **2.** In certain filamentous **Bacteria**, a **motile cell**. **3.** In some **cyanobacteria**, a motile **spore**. **4.** In some **green algae (Chlorophyta)**, a **vegetative cell** that undergoes division.

Good, Ronald d'Oyley (1896–1992) An English botanist and biogeographer who devised a scheme to divide the world into **floral regions**, describing it in *The Geography of the Flowering Plants*, first published in 1948 and still the standard reference on plant geography. The floral regions described here are taken from Good.

Goodeniaceae (order **Asterales**) A family of **annual** or **perennial**, **glabrous** or **pubescent** to **tomentose herbs** and **shrubs**, with a few climbers and small **trees**. Leaves leathery and much reduced in some species, **alternate**, **opposite**, or whorled, **simple**, **entire** or **serrate** or **dentate**, **exstipulate**, **sessile** or with **petioles**. Flowers **zygomorphic**, occasionally **actinomorphic**, **bisexual**, usually **pentamerous**, **corolla** tube often split longitudinally with the 5 lobes regular or with an upper lip of 2 and lower of 3 with lateral wings, 5 **stamens**, **ovary inferior** to semi-inferior, of 2 **carpels**. Fruit is a **drupe**, **nut**, or **capsule**. There are 12 genera with 430 species, Australian except for 1 pantropical genus (*Scaevola*). Some cultivated as ornamentals.

gooseberry (*Ribes uva-crispa*) *See Ribes.*

gooseberry aphid *See Aphis grossulariae.*

gooseberry sawfly *See Nematus ribesii.*

gorge wind *See* mountain-gap wind.

Gossypium (cotton) *See* Malvaceae.

Goupiaceae (order **Malpighiales**) A **monogeneric** family (*Goupia*) of **evergreen trees** with **alternate**, **coriaceous**, **simple**, **entire** or toothed leaves with **caducoud stipules**. Flowers **actinomorphic**,

bisexual, with 5 **connate sepals**, 5 **petals**, 5 **stamens**, **ovary superior** of 5 **locules**. **Inflorescence umbel**-like and **axillary**. Fruit is a **berry**-like **drupe**. There are two species occurring in central and northern South America. They yield valuable timber.

GPP *See* primary productivity.

grackles *See* Icteridae.

gradate sorus (graduate sorus) In some ferns (**Pteridophyta**), a **sorus** in which the sporangia (*see* sporangium) develop in sequence in a specific direction, most commonly from margin toward the centre. *Compare* mixed sorus, simple sorus.

gradient wind A wind that is not geostrophic but that blows parallel to the **isobars**. It is much more common than the **geostrophic wind** and occurs where the isobars are curved, so the moving air is subjected to a **centrifugal** force (inertia) that acts away from the centre of the low-pressure, so the wind is affected by three forces rather than two.

gradient wind level *See* geostrophic wind level.

graft 1. To take a part of one organism and implant it in a different position on the same organism or on another organism. **2.** A piece of tissue transferred in this way.

grain sizes *See* particle sizes.

graminoid Grass-like; the grass family (**Poaceae**) was formerly known as Gramineae.

Grammatophyllum (family **Orchidaceae**) A genus of **epiphytes** that includes *G. speciosum* (giant orchid), the world's biggest orchid, with stout stems up to 2 m long, pseudobulbs 2.5 m long, and big **panicle**-like **racemes** of flowers. Groups of giant orchids form clumps weighing up to 1 tonne. There are 11 species occurring from Malaysia to the Pacific Islands. They are widely cultivated.

Gram reaction A response to a laboratory procedure devised in 1884 by the Danish scientist Hans Christian Gram, in which **Bacteria** are killed, stained, e.g. with crystal violet, treated with a solvent, e.g. acetone or **ethanol**, then exposed to the Gram stain (or Gram's stain). Bacteria that readily lose their colour are said to be Gram-negative; those that retain their colour are Gram-positive. The difference reflects fundamental differences in the structure of the **cell wall**, indicating two distinct types of bacteria.

granadilla *See* Passifloraceae.

granite A pale-coloured, coarse-grained, **igneous** rock consisting of at least 20 per-cent quartz, with alkali feldspar, mica, and other minerals.

granite moss One of two genera of mosses (**Bryophyta**), *Andreaeobryum* and *Andreaea* that grow on **granite** rock faces in arctic and mountainous regions.

granulose Consisting of or covered by small grains.

granum Part of the internal structure of a **chloroplast**, consisting of 5–30 **thylakoids**. A typical chloroplast contains 40–80 grana.

Granville wilt *See Ralstonia solanacearum.*

grape phylloxera *See* Phylloxeridae.

grape sugar *See* glucose.

grapevine (*Vitis vinifera*) *See* Vitaceae.

grapevine leafroll A disease of grapevines (*Vitis vinifera*) in which leaves curl into cup shapes and the main veins remain green. The disease delays fruit ripening and causes a significant reduction in yield, and the vines lose vigour. Leafroll is caused by ten species of **Closteroviridae** and occurs worldwide.

grapevine yellows A group of diseases caused by **phytoplasmas** in which leaf veins and whole leaves become discoloured, necrotic, and curl downward, shoots are stunted, and grapes shrivel. The diseases occur worldwide.

Grapholita funebrana (plum fruit moth) A drab coloured tortrix moth (**Tortricidae**) with a wingspan of 10–15 mm. Females lay eggs on the underside of ***Prunus*** fruits and the **caterpillars**, pink with brown heads and up to 12 mm long, feed inside the fruit, finally leaving to pupate in soil, on tree **bark**, or in dead wood. The moth occurs throughout Eurasia and North Africa, but is not established in North America.

grass bugs *See* Miridae.

grass frog *See Rana temporaria.*

grasshopper nematode *See Mermis nigrescens.*

grassland A major global vegetation type, comprising areas where the natural vegetation is dominated by grasses (**Poaceae**). It occurs where the average rainfall is between those of forests and deserts. Grazing and fire destroy tree seedlings, thereby extending grassland to form a **plagioclimax** in some formerly forested areas. Tropical grassland is known as **savanna**. Temperate grassland is known as **pampas**, **prairie**, or **steppe**.

grass minimum temperature The lowest temperature registered by a thermometer set at the level of the tops of grass **blades** in short turf in the open over a specified period.

grass snake *See Natrix natrix, Opheodrys aestivus.*

grass temperature The temperature registered by a thermometer that is set level with the tops of grass **blades** in short turf, in the open.

graupel (soft hail) Precipitation in the form of opaque, usually spherical pellets of ice, 2–5 mm in diameter, that flatten or shatter when they strike a hard surface.

gravel Mineral particles, 2–60 mm in size.

graveolant Emitting a strong, often unpleasant, odour.

gravitational water Water that moves through soil by the force of gravity and that must be allowed to drain from the soil before the **field capacity** can be determined.

gravitropism *See* geotropism.

gravity wind *See* katabatic wind.

Gray, Asa (1810–88) An American botanist and taxonomist who popularized the study of botany and who supported, but not uncritically, Charles **Darwin**'s theory of evolution by **natural selection**.

gray catbird *See Dumetella carolinensis.*

grazing food chain *See* grazing pathway.

grazing pathway (grazing food chain) A **food chain** in which green plants are consumed by **herbivores**, which are consumed by **carnivores**.

greasewood *See* Sarcobataceae.

great crested newt *See Triturus cristatus.*

greater bulb fly *See Merodon equestris.*

great grey slug *See Limax maximus.*

great spotted woodpecker *See Dendrocopos major.*

great tit *See Parus major.*

great white butterfly *See Pieris brassicae.*

green algae A **paraphyletic** group of more than 7000 species of primarily aquatic **algae** that contain two types of **chlorophyll**. *Ulva*, sea lettuce, belongs to the group, but not all members are green.

green anole *See Anolis carolinensis.*

greenfinch *See Carduelis chloris.*

greenfly *See* Aphididae.

green frog *See Rana clamitans clamitans.*

green grass snake *See Opheodrys aestivus.*

greenhouse effect The warming of the **troposphere** that is due to the **absorption** and re-radiation of infrared radiation from the Earth's surface by molecules of water vapour and, to a much lesser extent, of other **greenhouse gases**, without which the surface temperature would be 30–40°C lower than it is.

G

greenhouse gas A gas that absorbs and immediately re-radiates infrared radiation from the Earth's surface, causing the **greenhouse effect**. The principal greenhouse gases are water vapour (H_2O), carbon dioxide (CO_2), nitrous oxide (N_2O), methane (CH_4), ozone (O_3), chlorofluorocarbons (CFCs), and hydrofluorocarbons.

greenhouse planarian *See Bipalium kewense.*

greenhouse slug *See Lehmannia valentiana.*

greenhouse whitefly *See Encarsia formosa, Trialeurodes vaporariorum*, beet pseudo-yellows virus, potato yellow vein, tomato infectious chlorosis virus.

green manure Fast-growing plants that are sown in bare soil following the harvesting of a crop in order to take up remaining nutrients that might otherwise be lost by **leaching**, smother weeds, and prevent soil **erosion**. They are dug into the soil the following spring to improve soil structure and release their stored nutrients.

green peach aphid *See Myzus persicae.*

green pug moth *See Pasiphila rectangulata.*

green tree frog *See Hyla cinerea.*

green worm *See Allolobophora chlorotica.*

greges *See* grex.

Grew, Nehemiah (1641–1712) An English plant anatomist and physiologist who published in 1682 a four-volume work *Anatomy of Plants*, with 82 plates and 7 appendixes, most dealing with botanical chemistry. He was the first person to extract **chlorophyll** from plant tissue, he coined the terms **parenchyma** and **radicle**, and he wrote the first detailed description of **pollen**. Carolus **Linnaeus** named *Grewia* (**Malvaceae**) in his honour.

grex (pl. greges) **1.** A group of **cultivars** derived from the same **hybrid** parents. **2.** A **pseudoplasmodium** formed by cellular slime moulds (**Acrasiomycetes**) and resembling a slug.

grey-brown podzolic An eluviated (*see* eluviation), free-draining soil with a B **soil horizon** enriched in **clay**. It develops beneath temperate woodlands. *See* alfisols.

grey field slug *See Deroceras reticulatum.*

grey garden slug *See Deroceras reticulatum.*

grey mould *See Botrytis cinerea.*

grey squirrel *See Sciurus carolinensis.*

grey worm *See Aporrectodea caliginosa.*

grindal worm *See Enchytraeus buchholzi.*

Grisebach, August Heinrich Rudolf (1814–79) A German botanist and phytogeographer who introduced the concept of the floral province, describing this in *Vegetation der Erde* (1872).

Griselinaceae (order **Apiales**) A **monogeneric** family (*Griselinia*) of **dioecious**, **evergreen shrubs**, small **trees**, climbers, and some **epiphytes**, with **simple**, **entire** or strongly **spinose** leaves. Flowers **actinomorphic**, **petals** absent in some **pistillate** ones. **Inflorescence** a **panicle** or **raceme**. Fruit is a **berry**. There are six species occurring in New Zealand and South America. Some cultivated for ornament.

grisette *See Amanita vaginata.*

groening's slime *See Lycogala epidendrum.*

gross primary productivity *See* primary productivity.

Grossulariaceae (order **Saxifragales**) A **monogeneric** family (*Ribes*) of mainly **evergreen shrubs**, some prostrate or scrambling, with **alternate**, **simple**, often toothed, often **palmately lobed** leaves with **stipules** attached to **petioles**. Flowers **actinomorphic**, **bisexual** or **unisexual** (plants **dioecious**), with tubular or **rotate hypanthium**, usually 5, sometimes 4 free **sepals**, as many **petals** as sepals, 4–5 **stamens**, **ovary inferior** of 2 joined **carpels** with 1 **locule**. **Inflorescence** a **raceme**. Fruit is a **berry**. There are 150 species occurring throughout the temperate Northern Hemisphere and also in the Andes. Many cultivated for their edible fruits, e.g. blackcurrant (*R. nigrum*), redcurrant (*R. rubrum*), gooseberry (*R. uva-crispa*), etc.

ground beetles *See* Carabidae.

ground frost **Frost** that forms when the air temperature is above freezing but the ground surface is below freezing.

ground inversion *See* surface inversion.

ground mesophyll (ground parenchyma) The collective term for the **bundle sheath**, **palisade mesophyll**, and **spongy mesophyll**.

ground parenchyma *See* ground mesophyll.

ground pine *See Ajuga.*

ground rattlesnake *See Sistrurus miliarius.*

ground streamer A column of air that becomes ionized from the ground up at the start of a **lightning stroke**, from the point on the ground toward which the **stepped leader** is descending.

groundwater Water that fills all the pores between mineral soil particles in the saturated zone of the soil, and flows slowly downslope.

group translocation (PEP group translocation, phosphotransferase system, PTS) A method by which **Bacteria** transport **sugar** across their **cell membranes** using phosphoenolpyruvate (PEP) as a source of energy.

grove snail *See Cepaea nemoralis.*

growth The increase in size of a **cell**, organ, or organism as a result of cell division or the enlargement of cells.

growth form **1. Morphology**, especially as this reflects **adaptation** to the environment. **2.** The way the size of a **population** changes, as this appears on a graph.

growth regulator A chemical compound that modifies the rate of growth of a plant or insect.

growth retardant A chemical compound that inhibits activity in the **meristem** below the **apex**, thereby reducing stem elongation. Growth retardants are used to prevent **lodging** in cereals.

growth ring *See* tree ring.

growth substance A chemical compound, other than a nutrient, that modifies the growth of a plant.

Grubbiaceae (order **Cornales**) A **monogeneric** family (*Grubbia*) of **evergreen**, heath-like **shrubs** with **opposite**, **linear-lanceolate** to linear leaves. Flowers usually less than 1 mm across, **actinomorphic**, **tetramerous**, **epigynous**, with 1 **perianth whorl**, 8 **stamens**, **ovary inferior** of 2 **carpels** with 2 **locules**, and later 1 following breakdown of the **septum**. **Inflorescence** a **sessile**, **axillary dichasium**. Fruit is a **syncarp**. There are three species occurring in Cape Province, South Africa.

Guamatelaceae (order **Crossosomatales**) A **monotypic** family (*Guametala tuerckheimii*), which is a sprawling, **evergreen shrub** with **opposite**, **simple**, **serrate** leaves, white to **tomentose** on the underside, with **stipules**. Flowers **hermaphrodite**, with **bracts**, **pentamerous**, free **hypanthium** short when present, when absent **petals** inserted in mouth of the

G

calyx tube, distinct calyx and **corolla**, 10 **stamens**, **ovary** of 3 **carpels**. **Inflorescence** is a **raceme**. Fruit a **follicle**. The plant occurs in Central America.

guanine A **purine** base ($C_5H_5N_5O$) that occurs in both **DNA** and **RNA**.

guano The accumulated droppings of birds, bats, or seals that is collected and marketed as a **fertilizer** rich in nutrients, especially phosphate.

guanosine phosphate A **nucleotide** of **guanine** that is used to construct **DNA** and **RNA**.

guard cell One of a pair of cells in the **epidermis** that surround each **stoma**; changes in the **turgor** of the guard cells opens and closes the stoma.

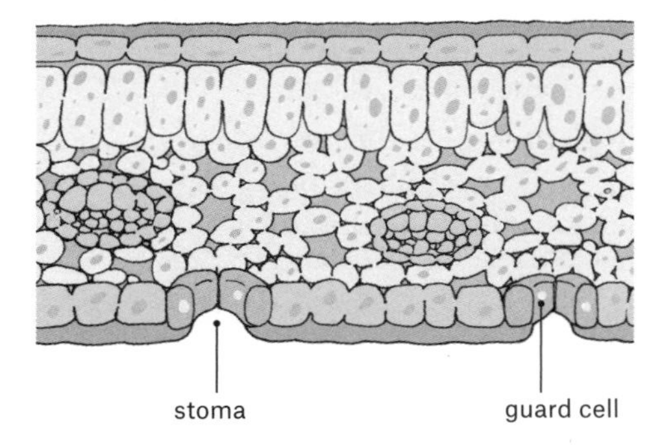

In this view of the cells in a cross-section of a leaf, the two indentations or pits on the lower surface are stomata, each with two guard cells that control their opening and closing. Here, the guard cells have released moisture, causing them to shrink and allowing the stomata to open. When the guard cells absorb moisture they swell, closing the stomata.

guava (*Psidium guajava*) *See* Myrtaceae.

guerrilla growth form The distribution of a plant that spreads by means of **rhizomes** or **stolons** that are long, but often short-lived, so they appear and disappear, growing between other plants, and the plants that grow from them emerge unpredictably, like a guerrilla force. *Compare* phalanx growth form.

Guiacum officinale (lignum-vitae) *See* Zygophyllaceae.

guild A group of species in which all the individuals employ similar techniques to exploit similar resources.

gully A type of soil **erosion** that develops where torrential rain forms a channel by scouring away soil and surface sediments. If it forms on a hillside it is called a valley-side gully, and if it forms on a valley floor it is an arroyo. Gullies often begin as **rills**.

gum A substance, often a **polysaccharide**, produced by many plants, especially woody ones, and often exuded when the **bark** is cut or the plant otherwise damaged. Some gums are water-soluble, others swell in water into a viscous **colloid** that is insoluble in organic solvents.

gum duct A channel between cells in **angiosperms** lined with **epithelium** that secretes **gum** into the **lumen** of the duct.

gummosis The production of **gum**.

Gumuia A genus, now extinct, of vascular plants (**Tracheophyta**) that lived during the Early Devonian epoch (416–397.5 million years ago). It had an apparently **sympodial** stem with sporangia (*see* sporangium) borne laterally and terminally on fertile stems. **Fossil** specimens were found in Yunnan, China, and first described in 1989.

Gunneraceae (order **Gunnerales**) A **monogeneric** family (*Gunnera*) of **perennial**, rarely **annual herbs** that arise from **stolons** or **rhizomes**, with symbiotic (*see* symbiosis) ***Nostoc*** **cyanobacteria** in the root and rhizome **parenchyma**. Stems are covered with triangular scales. Leaves **alternate**, **simple**, in rosettes, **dentate**, **crenate** or lobed, and small or up to 3 m wide (in *G. manicata*, which has **petioles** up to 2.5 m long). Flowers small or minute, somewhat **zygomorphic**, **bisexual** or unisexual (plants **monoecious**, **dioecious**, or **polygamomonoecious**), with 2 or sometimes 3 free, **valvate sepals**, 2 or sometimes 3 **petals**, 1–2 **stamens**, **ovary inferior** of 2 **carpels** with 1 **locule**. **Inflorescence** a **spike**-like, terminal or **axillary panicle**. Fruit is a **drupe** or **nut**. There are 40–50 species occurring around the South Pacific and in Africa and Madagascar. Some grown for ornament.

Gunnerales A family of plants comprising 2 families with 2 genera and about 45 species. *See* Gunneraceae and Muyrothamnaceae.

gust A sudden, sharp, but brief increase in the wind speed.

gust front (pressure jump line) An area immediately ahead of an advancing storm, where warm air is being drawn into the base of the cloud, producing strong wind **gusts**.

gustnado A small **tornado** that forms in the **gust front** ahead of a **supercell** storm.

guttation The extrusion of **xylem** sap, sometimes containing other substances, from **hydathodes** at the edges or tips of leaves of some vascular plants (**Tracheophyta**). It occurs when the **transpiration** rate is low, e.g. at night when **stomata** are closed, but the soil is moist so water enters roots and accumulates, generating **root pressure**.

Guttiferae *See* Clusiaceae.

gymnocarpy 1. In conifers, having the seed projecting from the flesh of the **cone**. **2.** In a fungal **fruiting body**, having the tissue bearing **spores** exposed during the whole of its development.

gymnosperm A seed plant (**Spermatophyta**) in which the **ovules** are borne naked on the **cone** scales (*compare* angiosperm). The group includes the conifers, **Cycadales**, **Ginkgoales**, and **Gnetales**.

Gymnosporangium juniperi-virginianae A species of **basidiomycete fungi** that is a **pathogen** of several trees, but especially of apple and crab-apple trees (***Malus***) and eastern red cedar (***Juniperus** virginiana*), causing cedar-apple rust. Symptoms begin with bright yellow spots on leaves that later turn orange or red, followed by the appearance of black spots, and in *J. virginiana* by the formation of red-brown **galls**. The fungus occurs in North America.

gynaecium *See* gynoecium.

gynobasic Describes a **style** that arises near the base of a deeply lobed **ovary**.

gynodioecious Describes a **dioecious** species in which **hermaphrodite** and female flowers occur on different plants.

gynoecium (gynaecium) The female reproductive organs of a **flower**, i.e. the **carpels**.

gynomoecious Describes a **monoecious** species in which **hermaphrodite** and female flowers occur on the same plant, but separately.

gynophore The stalk of a **gynoecium**.

gynostemium The **column** that forms when the **stamens** and **pistil** of a flower fuse into a single structure.

gypsic horizon A **soil horizon**, at least 15 cm thick, that contains 15 percent or more **gypsum**; if it contains more than 60 percent gypsum it is said to be hypergypsic.

gypsisols A group of soils that have a **gypsic horizon** within 100 cm of the surface, or more than 15 percent **gypsum** in the uppermost 100 cm. Gypsisols are a reference soil group in the **World Reference Base for Soil Resources**.

gypsobelum *See* love dart.

gypsum An **evaporite** mineral ($CaSO_4$ $2H_2O$) that is very insoluble and is the first mineral to precipitate from evaporating sea water. It forms less commonly in volcanic regions when sulphuric acid reacts with **limestone**. A clear, transparent variety of gypsum is called selenite, a fibrous variety is satin spar, and a fine-grained variety is alabaster.

Gyromitra esculenta (false morel, turban fungus) A species of **ascomycete fungi** with a **fruiting body** that has a red-brown **pileus**, 50–150 mm across, that is deeply lobed, resembling a brain, and a creamy white **stipe** 20–50 mm tall. It occurs in sandy soil in conifer woodland, especially in upland areas, throughout Europe and is widespread in North America. The fungi are very poisonous and have caused many deaths.

gyrose Curved, sinuous.

Gyrostemonaceae (order **Brassicales**) A family of herbs, shrubs, and small trees with **alternate**, **simple**, **entire** leaves, **sessile** or with **petioles** and small **stipules**, often **succulent**. Plants **monoecious** or **dioecious**, flowers more or less **actinomorphic**, with 4- to 8-lobed or entire, persistent cup-shaped **calyx** and no **corolla**. **Staminate** flowers with 7–100 **stamens**, **pistillate** flowers with **superior ovary** of 1 to many **carpels**. **Inflorescence** a terminal or **axillary raceme** or **panicle**. Fruit is a **schizocarp**, **syncarp**, or **achene**. There are 5 genera with 18 species occurring in Australia.

H

H *See* hydrogen.

ha *See* hectare.

haar A cold **fog** from the sea.

habit The form or shape of a plant.

habitat The area in which an organism or community live.

habitat action plan (habitat conservation plan) A management plan that defines objectives and targets for the **conservation** of a specified area of **habitat**, together with an action plan for realizing them.

habitat conservation plan *See* habitat action plan.

habitat enhancement A deliberate change made to an area of **habitat** with the aim of improving its quality for some or all of its species.

habitat fragmentation The division of an area of **habitat** into a number of smaller areas. *See* SLOSS debate.

habitat restoration The removal of **exotic** species from an area and the reintroduction of **native** species in order to allow the original community to become re-established.

Hadley cell Part of the **general circulation** of the atmosphere, first described by the English meteorologist George Hadley (1685–1768), in which warm, moist air rises close to the equator, losing much of its moisture as it does so to form towering clouds, moves away from the equator close to the **tropopause**, then subsides around latitude 30°, warming adiabatically as it does so (*see* adiabatic cooling and warming) and reaching the surface as warm, dry air. This results in the arid climates of the subtropical deserts. There are several Hadley cells in each hemisphere.

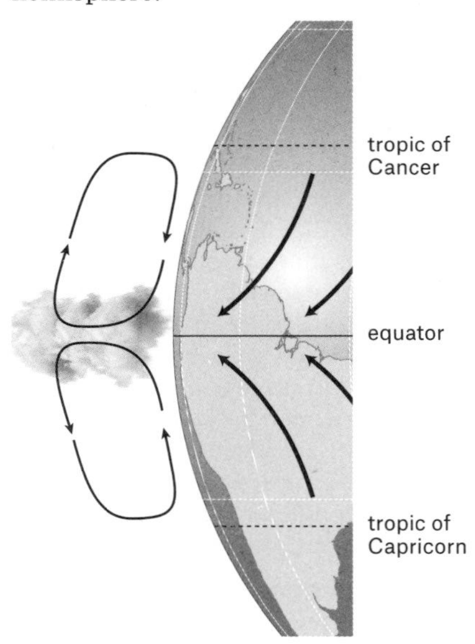

Hadley cell. A vertical cell in which air rises at the equator, moves away from the equator at high altitude, and subsides over the subtropics.

Haeckel, Ernst Heinrich (1834–1919) A German naturalist who strongly supported Charles **Darwin**'s theory of evolution by **natural selection** and did much to promote it in Germany. He also coined the term **ecology** (*Ökologie*), in 1866.

haematophagy Feeding on blood.

Haemodoraceae (order **Commelinales**) A family of **monocotyledon**, **perennial herbs** with **rhizomes**, **corms**, or **bulbs**, and red roots. Leaves **distichous**, **ensiform**, **linear**, or **acicular**, **glabrous** or hairy, with a sheathing base. Flowers **actinomorphic** or somewhat **zygomorphic**, **bisexual**, of either 3+3 free or basally fused **tepals**, or tepals as a tube with 6 free, **valvate** lobes, 3 or 6 **stamens adnate** to the tepals, **ovary superior** or **inferior** of 3 fused **carpels** with 3 **locules**. **Inflorescence** terminal and variable but often a **raceme** or **panicle**. Fruit is a **capsule**. There are 14 genera with 116 species occurring in tropical and warm temperate regions. Several species cultivated for ornament; *Anigozanthos manglesii* (kangaroo paw) is the state emblem of Western Australia.

Haemorhous mexicanus (house finch) A bird that is 125–150 mm long with a wingspan of 200–250 mm. It is brown or grey, with a square-tipped tail, the colour varying somewhat with the season. Adult males have a red face, upper breast, and rump. The birds are gregarious and inhabit urban and suburban gardens and parks, and more open **habitats** in part of the range, feeding on seeds, berries, and insects, and frequently visiting feeders. They occur throughout North America south of southern Canada and in Mexico.

Haemorhous purpureus (purple finch) *See Carpodacus purpureus.*

haem protein *See* cytochrome.

hail Precipitation in the form of hailstones, which are approximately spherical ice pellets that form inside large **cumulonimbus** clouds, the size of the stones increasing with the depth of the cloud. Most hailstones are 5–50 mm in diameter, but can be larger, and large hailstones consist of alternate layers of opaque and clear ice.

hail cannon A device, e.g. a large gun, situated on the ground that fires upward to create a shock wave inside a **cumulonimbus** cloud, typically firing at one- to ten-second intervals as the cloud passes, in the hope of disrupting the formation of hailstones. There is no convincing evidence that it works.

hail day A day on which **hail** falls.

hail region One of the 13 regions into which the United States is divided, ranked by the frequency and intensity of the hailstorms they experience.

hailshaft A visible column of falling hailstones beneath a **cumulonimbus** cloud.

hailstreak A narrow strip of ground that is completely covered by hailstones.

hailswath An area of ground that is partially covered by hailstones.

hairy marshmallow (*Althaea hirsuta*) *See Althaea.*

hairy snail *See Trochulus hispidus.*

hairy woodpecker *See Picoides villosus.*

Hales, Stephen (1677–1761) An English clergyman and physiologist, who was the first person to describe **transpiration**, in his book *Vegetable Staticks*, published in 1727. Hales also discovered that the pressure drawing water into plant roots varied through the day and according to the temperature, and noted that leaves absorb light, which he found was necessary to their growth.

half-inferior Describes the position of an **ovary** when the lower part is embedded in the **pedicel** and the upper part exposed.

Halloween lady beetle *See Harmonia axyridis.*

halophile *See* extremophile.

halophilic Preferring salt-rich conditions.

Halophytaceae (order **Caryophyllales**) A **monotypic** family (*Halophytum ameghinoi*), a **monoecious, annual herb** with **alternate, exstipulate** leaves. Flowers **unisexual, axillary racemes** of female flowers produced first, followed by **spikes** of male flowers above them. Female flowers of 1 small **bract** and 2 **bracteoles**, with no **perianth, ovary superior** of 3 fused **carpels**. Male flowers with **bracts**, with 4 **tepals** alternating with 4 **stamens**. Fruit a **nutlet**. The family occurs in Argentina.

halophyte A plant that is adapted to thrive in salty soils and salt air.

Haloragaceae (order **Saxifragales**) A family of small **annual** or **perennial herbs**, many aquatic or amphibious, and small **trees** with leaves that are **opposite, alternate**, or whorled, narrow and **entire** to **serrate** in terrestrial genera, and **dimorphic** in wetland genera, with **pinnate** submerged leaves and broad, entire to serrate **emergent** leaves. Flowers **bisexual** or **unisexual** (plants **monoecious** or **dioecious**), **actinomorphic**, 2-, 3-, or 4-merous, usually 4 **sepals** and **petals** but petals sometimes absent, 8 **stamens** in 2 **whorls, ovary inferior** with 4 **locules. Inflorescence** a **spike** or flat-topped **corymb**. Fruit an **indehiscent nut** or **schizocarp**. There are 8 genera with 145 species occurring worldwide, but especially in Australia. Some *Myriophyllum* species (water milfoil) cultivated as aquarium or pond plants, but can become invasive.

Haloragis (family **Haloragaceae**) A genus of perennial **herbs** (raspwort) with **opposite** or **alternate, entire, dentate**, or lobed leaves. Flowers **unisexual** or **bisexual**, 2–4 **sepals** united in a tube, 2–4 keeled (*see* keel) **petals**, often incurving and hairy on the outside, often absent in female flowers, twice as many **stamens** as petals, 2–4 **carpels. Inflorescence** a terminal **raceme** or **panicle**. Fruit is a **nut**. There are three species occurring in Australasia and the Pacific Islands.

halosere A plant **succession** that develops in a saline environment, e.g. a **salt marsh**.

haltere The modified hind wings of true (two-winged) flies (**Diptera**). Shaped like drumsticks, the halteres help the fly sense movement and direction while it is flying.

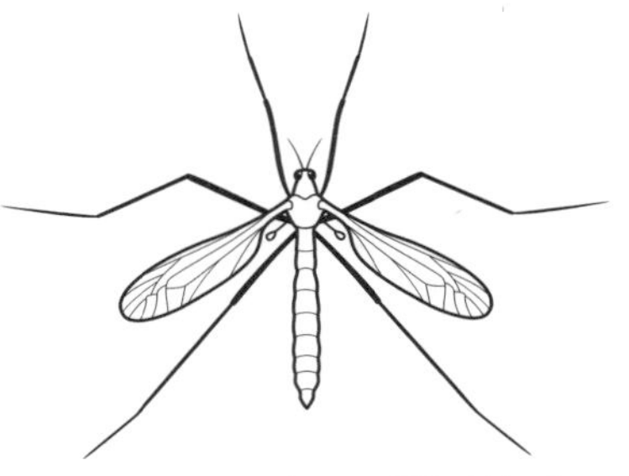

The halteres are the modified hind wings, shaped like drumsticks, of this cranefly.

hamada (hammada) A desert surface that consists of loose rocks and exposed **bedrock**.

Hamamelidaceae (order **Saxifragales**) A family of **trees** and **shrubs** with **alternate**, rarely **opposite, simple, ovate** to **elliptical** or **palmate, entire** or **serrate** leaves, usually with **stipules**. Flowers **actinomorphic** or **zygomorphic, bisexual** or **unisexual** (plants **monoecious** or **andromonoecious**), usually **epigynous** but some hypogynous (*see* hypogyny), 4–5 or up to 10 **sepals** or none, 4–5 **petals** or none, 4–5 or up to 10 **stamens**, sometimes with many **staminodes, ovary superior** to **inferior** of 2 fused **carpels. Inflorescence** usually a **spike**, sometimes a lax spike, or spherical aggregation of **sessile** flowers. Fruit is a **capsule**. There are 27 genera with 82 species occurring throughout tropical and temperate regions, but not South America. Many cultivated for ornament, timber, and medicinal or industrial products. *Hamamelis virginica* is witch hazel.

Hamamelis virginica (witch hazel) *See* Hamamelidaceae.

hamate With the tip hooked.

hammada *See* hamada.

hamsters *See* Cricetidae.

handkerchief tree (*Davidia involucrata*) *See* Cornaceae.

Hanguanaceae (order **Commelinales**) A **monogeneric** family (*Hanguana*) of **monocotyledon**, **dioecious herbs** with **linear** to **lanceolate** leaves, spirally arranged with a sheathing base. Flowers **actinomorphic**, **unisexual**, **trimerous**, male flowers with 6 **stamens** and rudimentary **ovary**, female flowers with 6 **staminodes** and **superior ovary** with 3 **locules**. **Inflorescence** terminal, branched **spicate**. Fruit is a **berry**. There are ten species occurring in Sri Lanka, southeastern Asia, and northern Australia.

haplobiont A plant that exists only as either a **gametophyte** or a **sporophyte**, i.e. it experiences only one of the two generations of most plants. *See* alternation of generations.

haplocheilic Describes a **stoma**, found in some **gymnosperms**, in which the **guard cells** and **subsidiary cells** are derived from different mother cells. *Compare* syndetocheilic.

haploid Having one set of **chromosomes**.

haplontic Describes a life cycle in which all stages are **haploid** apart from the **zygote**.

haplostele (ectophloic protostele) A **stele** in which a cylinder of **phloem** surround a central core of **xylem**. The stele is usually surrounded by an **endodermis**.

haplotype Closely linked (*see* linkage) **genetic markers** on the same **chromosome** that tend to be inherited together (i.e. they are not easily separated during **recombination**).

hapteron A swelling or outgrowth from the stem by which a plant is attached to its substrate. *See* holdfast.

hardening *See* acclimatization.

hard frost *See* black frost.

hardpan A **soil horizon**, typically in the middle or lower part of the **soil profile**, that is hardened due to **induration** or by a variety of **cemented** materials.

hardwood The wood from an **angiosperm tree**, or the tree itself.

hardy Describes a plant that is able to survive adverse conditions, usually of weather.

Hardy-Weinberg law The law stating that in an infinitely large, interbreeding population, in which mating is random and in which there is no selection, migration, or **mutation**, the frequencies of **genes** and **genotypes** will remain constant across the generations. If **alleles** A and a occur in a **diploid** population with frequencies of p and q respectively, the three possible genotypes AA, Aa, and aa will occur with frequencies of p^2, $2pq$, and q^2 respectively.

hares *See* Leporidae.

harlequin chromosomes Sister **chromatids** that stain differently so that when examined microscopically one appears darker than the other.

harlequin coral snake *See Micrurus fulvius*.

harlequin ladybird *See Harmonia axyridis*.

Harmonia axyridis (Asian lady beetle, harlequin ladybird, Halloween lady beetle, multicoloured Asian lady beetle) A large ladybird (**Coccinellidae**), 5.5–8.5 mm long, that is most commonly red or orange with black spots, black with 4 red spots, or black with 2 red spots, but there are many other forms and the beetles may have 0–19 spots. It is native to Asia but was introduced for **biological control** of aphids (**Aphididae**) and scale insects (**Coccidae**) in the United States and Europe and is now established widely. It is a voracious predator that attacks native coccinellids and it sometimes congregates in swarms that may enter homes.

Harpalus rufipes (strawberry seed beetle) A black ground beetle (**Carabidae**) with fine, yellowish hairs on the **elytra** and reddish brown legs and antennae (*see* antenna). Adults are 11–17 mm long. They lay eggs from late summer to autumn in soil among weeds. Larvae then feed, pupating in the soil. Adults remove the seeds from fruits, usually damaging the surrounding flesh. Larvae also eat seeds, but do not damage the surrounding fruit.

hartig net *See* ectotrophic mycorrhiza.

harvestmen *See* Arachnida.

hastate Of a leaf, shaped like a spear head, but with two lower lobes at right angles.

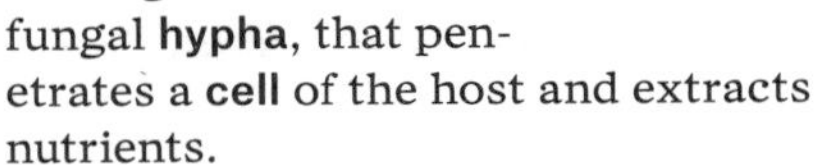
A hastate leaf.

haulm The stem of a cultivated plant.

haustorium The highly modified root or stem of a parasitic plant, or an outgrowth from a fungal **hypha**, that penetrates a **cell** of the host and extracts nutrients.

Hawaiian floral region The area that covers the Hawaiian islands within the **Palaeotropical region**. The most isolated of all floral regions, about 20 percent of its genera and 50 percent of its species are **endemic**.

hawthorn aphid *See Dysaphis foeniculus.*

haze A reduction in horizontal visibility due to **haze droplets** or other **aerosol** particles. Visibility in a haze is greater than 2 km.

haze droplet A water droplet less than 1 µm in diameter that forms by **condensation** on to a **hygroscopic nucleus** when the **relative humidity** (RH) is greater than about 80 percent. The droplet remains suspended in the air and as the RH rises above 90 percent the number of droplets increases, producing **haze**.

hazel *See* Betulaceae, *Corylus.*

hazel sawfly *See Croesus septentrionalis.*

heart and dart moth *See Agrotis exclamationis.*

heartwood The dead wood at the centre of a tree trunk or large **branch**. As **secondary growth** adds a new layer of **xylem** each year, the old xylem cells die. Their contents decay but the cells retain their **lignin** and fill with waste products, altering their colour and making heartwood easy to distinguish from **sapwood**.

heat capacity (thermal capacity) The amount of energy that must be supplied to a substance as heat in order to raise the temperature of that substance. If measured in relation to a unit mass of the substance it is known as the specific heat capacity; if measured in relation to a unit amount of the substance it is known as the molar heat capacity.

heath *See Erica.*

heather *See Calluna; Erica,* Ericaceae.

heath forest A type of **tropical rain forest** dominated by trees with small, sclerophyllous leaves (*see* sclerophyllous vegetation) that develops on **silica**-rich **podzolic soils**.

heathland A lowland plant community dominated by ericaceous **shrubs** (*see* Ericaceae) that develops on acid, **podzolic soil**.

heat island An area within which the air temperature is markedly higher than it is in the surrounding area.

heat lightning Flashes of **lightning** that are not followed by the sound of **thunder**. They are seen most often on warm summer nights and are silent because they occur more than about 10 km from the observer and all the sound waves have been either refracted upward by the change in air density with height, or absorbed by the air.

heat low *See* thermal low.

heat thunderstorm A **thunderstorm** that occurs on a warm, humid, summer afternoon when the ground has been heated strongly, rendering the air unstable (*see* instability).

H

heat wave A period, usually lasting several days or weeks, during which the average temperature is higher than is usual for that place at that time of year.

Hebe (family **Plantaginaceae**) A genus of **evergreen shrubs** and small **trees**, closely related to ***Veronica***, with leaves in 4 rows of **opposite**, **decussate** pairs. Flowers **perfect**, **corolla** with 4 lobes, 2 **stamens**. **Inflorescence** is a **spike**, **raceme**, or **panicle**. Fruit is a **capsule**. There are about 100 species occurring in New Zealand (where it is the largest plant genus), Australia, islands of the South Pacific, and South America. Many species cultivated for ornament. Their flowers attract butterflies.

hebetate With a blunt or soft tip.

hectare (ha) A unit of area; 1 ha = 2.471 **acres**, 1 acre = 0.4047 ha.

Hedera (family **Araliaceae**) A genus of **evergreen**, woody, climbers or creeping plants (ivy), most with dimorphic (*see* dimorphism) leaves; juvenile leaves **palmate**, adult leaves **entire**, **cordate**. Flowers with 5 small **petals** borne in **umbels**. Fruit is a **berry**. There are 12–15 species occurring in Europe, northwestern Africa, Macaronesia, and central, southern, and eastern Asia. Many insects and birds depend on their flowers and fruits, and deer eat the leaves. Some varieties of *H. helix* are cultivated for ornament, but the wild plant is invasive.

Hedera helix (ivy) *See Hedera*, Araliaceae.

hedge accentor *See Prunella modularis.*

hedgehog *See Erinaceus europaeus.*

hedgerow (living fence, live fence) A linear barrier of **shrubs** and small **trees** that defines a field boundary and confines farm livestock, and that also provides **habitat** for a variety of wild plants and animals.

hedge sparrow *See Prunella modularis.*

heirloom plant (heirloom variety, heirloom vegetable) A **cultivar** that was formerly grown widely but that is no longer considered commercially viable, and that is maintained by enthusiasts or by growers in isolated communities.

heirloom variety *See* heirloom plant.

heirloom vegetable *See* heirloom plant.

hekistotherm A plant of high latitudes, able to tolerate extreme cold.

Helianthemum (order **Cistaceae**) A genus of **evergreen shrubs** and **subshrubs** with **opposite**, **simple**, **oval** leaves. Flowers have 5 **petals**. There are about 110 species occurring throughout the Northern Hemisphere. Many are cultivated for ornament, and known as rock rose, sunrose, or rushrose. Many lepidopteran larvae feed on them.

Helianthus (order **Asteraceae**) A genus comprising 52 species of **annual** and **perennial herbs**, native to North America, in which the ray florets, if present, are sterile, and the **disc florets** usually have a **cauducous pappus** of two scales. *Helianthus annuus* (sunflower) is an annual, with lower leaves **opposite**, **ovate**, or heart-shaped, and upper leaves **alternate**. It is widely cultivated. *Helianthus tuberosus* (Jerusalem artichoke) is a perennial with opposite leaves, larger on the lower stem. It is cultivated for its edible **tubers**.

Helichrysum (family **Asteraceae**) A genus of **annual** or **perennial herbs** and **shrubs**

with usually **alternate**, **entire**, **oblong** to **lanceolate** leaves. The papery, yellow, white, brown, or pink **involucre bracts** extend beyond the flowers and resemble **petals. Disc florets** mostly **bisexual** and tubular. Fruit is a **nut** or **achene**. There are about 600 species, occurring in Africa, Madagascar, Australasia, and Eurasia. Many are cultivated for their flowers (everlasting daisies), which retain their colour for a long time.

heliciform (helicoid) Spirally coiled, i.e. forming a helix.

helicoid *See* heliciform.

Heliconiaceae (order **Zingiberales**) A **monogeneric** family (*Heliconia*) of **herbs** with **rhizomes** and aerial shoots or **pseudostems** formed from overlapping leaf sheaths. Leaves **distichous** and **simple**. Flowers **zygomorphic**, **bisexual**, fused **tepals** and **stamens** form a tube, 5 fertile **stamens**, 1 **staminode**, **ovary inferior** of 3 fused **carpels** and 3 **locules**. **Inflorescence** comprises **racemes** of flowers in the **axils** of the large, coloured **bracts**. Fruit is **drupe**-like. There are 100–200 species, most occurring in tropical America, some in Indonesia and the Pacific Islands. Some cultivated for ornament.

heliophyte A plant that is adapted to strong sunlight.

heliosis *See* solarization.

heliotropic (phototropic) Describes growth or movement of a **sessile** organism, but especially a plant, toward or away from sunlight.

heliotropic wind A slight but steady change in wind direction that occurs through the day as the Earth turns and the area most strongly heated by the Sun moves.

heliotropism *See* phototropism.

Helipterum (family **Asteraceae**) A genus of mainly **annual herbs** with **alternate** leaves. Flowers have soft, papery, white, yellow, brown, or black **petaloid involucre bracts**, **disc florets** often black. Fruit is an **achene** with a feathery **pappus**. There are 35 species occurring in South Aftrica and Australia; several species formerly assigned to this genus have been transferred to others. Several species are cultivated for their flowers (everlasting daisies).

Helix aspersa *See Cornu aspersum*.

Helleborus (family **Ranunculaceae**) A genus of **acaulescent**, **perennial**, **evergreen herbs** with **sympodial**, branched **rhizomes**, leaves growing from the base or **alternate**, divided or lobed. Flower **bisexual** with 5 persistent **sepals** that supply most of the colour. There are about 20 species occurring in much of Europe into the Caucasus. The plants contain poisonous **alkaloids**. Many cultivated and known, because of their time of flowering, as Christmas roses or Lenten roses.

helminth A parasitic worm that lives inside its host.

helophyte A plant in which the **perennating bud** lies immersed in mud or soil below the water level but the aerial parts of the plant stand above the surface (e.g. ***Phragmites australis***, common reed). *Compare* hydrophyte.

Helotiales An order of **ascomycete fungi** that have cup- or disc-shaped apothecia (*see* apothecium). Most are **saprotrophs** living in **humus** and decomposing plant material, but some are serious plant pathogens, e.g. ***Monilinia fructicola*** and ***Sclerotinia sclerotiorum***. Other species form mycorrhizal (*see* mycorrhiza) associations with members of the **Ericaceae**. There are 10 families, with 501 genera and 3881 species.

helotism A relationship, found among some species of ants, in which one organism or colony enslaves another.

Helwingiaceae (order **Aquifoliales**) A **monogeneric** family (*Helwingia*) of

evergreen shrubs and **trees** with **alternate** to sub-**opposite**, **simple**, **serrate**, **ovate** to **linear-lanceolate** leaves with **petioles** and **caducous stipules**. Flowers small, **actinomorphic**, **unisexual** (plants **dioecious**), 3–5 free, **valvate petals** in 1 **whorl**, 3–5 **stamens**, **ovary superior** of 2–4 fused **carpels**. **Inflorescence cymose**, borne on the **adaxial** surface of leaf **blades**. Fruit is a **drupe**. There are three species occurring from the Himalayas to Japan.

hemelytron A forewing that is leathery, but with a membranous tip, typical of bugs (**Heteroptera**).

hemicellulose A group of **polysaccharides** found in plant **cell walls**, where they for a matrix in which **cellulose** fibres, and **lignin** in woody plants, are embedded. The structure and abundance of hemicelluloses vary widely according to the plant species and cell type.

hemicryptophyte A plant that produces its **perennating buds** at ground level and has above-ground stems which die back when conditions are unfavourable. It is one of the categories in the classification of life forms devised by Christen **Raunkiær**. There are three subcategories. *See* partial rosette plant, protohemicryptophyte, rosette plant.

hemimetabolous Describes an insect with young, called **nymphs**, that develop through a series of moults, with incomplete **metamorphosis**.

hemiparasite A parasitic plant that contains **chlorophyll** and performs **photosynthesis**, but that lacks roots, or has greatly reduced roots, and uses a host for support or as a source of water and additional nutrients other than carbon, which it obtains by means of haustoria (*see* haustorium) from the host's **xylem**. **Facultative** hemiparasites, also known as meroparasites, are sometimes or always able to complete their life cycle independently of a host; **obligate** hemiparasites cannot survive without a host.

Hemiptera (true bugs) An order of **Insecta**, 1–150 mm long, that do not go through a pupal stage, never have an eleventh abdominal segment or **cerci**, and have mouthparts developed into a **rostrum** adapted for piercing and sucking, or at certain stages in the life cycles of some species, no mouthparts. Most feed on **phloem sap**. Most have two pairs of wings, the forewings are hemelytra (*see* hemelytron). There are about 35,000 species with a worldwide distribution.

hemlock (*Tsuga*) *See* Pinaceae.

hemp (*Cannabis sativa*) *See* Cannabaceae.

hemp-leaved marshmallow (*Althaea cannabina*) *See Althaea*.

henna (*Lawsonia inermis*) *See* Lythraceae.

Hennig, Emil Hans Willi (1913–76) A German biologist who originated **phylogenetic systematics**. He worked in the German Democratic Republic while living in West Berlin and his book *Grundzüge einer Theorie der phylogenetischen Systematik* (Basis of a theory of phylogenetic systematics), published in the DDR in 1950, made little international impact until an English translation, *Phylogenetic Systematics*, appeared in 1960. It then rapidly achieved an authoritative status and modern **taxonomy** is based on its precepts.

Hepatophyta The liverworts, a **phylum** of non-vascular plants now known as **Marchantiophyta**.

heptachlor An **organochlorine insecticide** formerly used to control mites and insect pests, especially in maize (corn). It is persistent, toxic, and bioaccumulates (*see* bioaccumulation) and it is no longer used.

heptamerous With seven parts.

herb A small, non-woody, seed-bearing plant in which all the aerial parts die back at the end of each growing season.

herbage 1. The **herbs** on which domestic animals graze. **2.** Herbs that are grown as a crop.

herbal A book containing descriptions, usually with illustrations, of medicinal plants.

herbarium A collection of dried plants, together with written notes describing them, their ecology, and the habitat and location where they were found; used as a reference in plant classification.

herbicide A chemical compound that kills plants.

herbivore A **heterotroph** that feeds on green plants or other **autotrophs**.

herding The behaviour of mammals that form groups with a social structure.

heritability A measure of the extent to which a **phenotype** is determined genetically and, therefore, can be modified by breeding.

herkogamy The separation of **anthers** and **stigma** in order to reduce the chance of self-pollination, and possibly to reduce interference from **styles** and stigmas with the export of **pollen** from the flower.

Hernandiaceae (order **Laurales**) A family of **trees**, **shrubs**, and some **lianas** with **alternate**, usually **simple**, **entire** or **compound**, **exstipulate** leaves. Flowers **actinomorphic** or **zygomorphic**, **bisexual**or **unisexual** (plants **monoecious**), 4–8 free or 3–10 fused **tepals** in 1 or 2 **whorls**, 3–7 **stamens** in 1 or 2 whorls, **ovary inferior** of 1 **carpel**. **Inflorescence axillary**, occasionally terminal, in **corymbose** or **panicle**-like **cymes**. Fruit is a **nut** or **drupe**. There are 5 genera with 55 species of pantropical distribution.

herpokinetic mobility A snake-like motion, e.g. in some algal **filaments**.

hesperidium The **berry** of a citrus **fruit**, in which the fleshy part is divided into segments and enclosed in a skin.

heteranthery The production by a **flower** of 2 or more types of **stamen**. It occurs in plants that secrete no **nectar**, so **pollen** is food for pollinators as well as the carrier of male **gametes**. Heteranthery allows the flower to specialize, with separate pollinating and feeding stamens.

hetero- Different from.

heteroallelic mutant An **allele** that has **mutations** at different sites within the **gene**.

Heterobasidion annosum A species of **basidiomycete fungi** that causes the disease annosum foot rot. It occurs throughout the Northern Hemisphere and attacks at least 200 species of **broad-leaved** and coniferous trees. It is the most important forest pathogen in the Northern Hemisphere. ↗

heteroblasty A series of changes in the form of an organ during the development of a plant, most often observed in leaves, with a plant producing juvenile, transitional, and adult leaves, e.g. in some ***Acacia*** species. *Compare* homoblasty.

heterocaryon *See* heterokaryon.

heterochromatin **Chromosome** material that accepts microscope stains during the **interphase** stage.

heterochrony During growth, the development of individual features at different rates or with different timing of onset and offset from those that occurred in the ancestor. It is a major factor in the evolution of new forms. This may lead to **paedomorphosis** or **peramorphosis**.

heterocyclic Describes a molecule that includes a ring containing at least two different elements.

heterocyst A **cell** that specialized in **nitrogen fixation** in certain **cyanobacteria**.

Heterodera carotae (carrot cyst nematode, carrot root nematode) A cyst nematode (**Nematoda**) that feeds only on carrots. It is white, lemon-shaped, with lemon-shaped **cysts**, found attached to carrot roots, that later darken in colour. It occurs throughout Europe and has been recorded in the United States. Infestation reduces growth and causes stunting, with bronzing of leaves.

Heterodera cruciferae (brassica cyst nematode) A cyst nematode (**Nematoda**) that feeds principally on brassica crops (**Brassicaceae**, formerly called Cruciferae). Females are white with plump, almost spherical bodies, and the **cyst** is dark brown and spherical or lemon-shaped. Infestations cause stunted growth and reduced growth rates. It has a worldwide distribution.

Heterodera goettingiana (pea cyst nematode) A cyst nematode (**Nematoda**) that feeds on legumes (**Fabaceae**), especially peas and broad beans. Females are white with a lemon-shaped body. **Cysts** are brown. It occurs widely in Europe, and in North Africa, the Middle East, and China.

Heterodera schachtii (beet cyst eelworm, sugar beet nematode) A cyst nematode (**Nematoda**) that feeds on more than 200 species of plants, including sugar beet and brassicas. **Cysts** in the soil contain 500–600 eggs that hatch when stimulated by substances secreted by the roots of potential host plants. The larvae penetrate the root and start feeding. Females mate, lay eggs, and die, their bodies hardening and becoming cysts. The nematode occurs throughout Europe and has been recorded in North America, Russia, Turkey, Israel, Australia, and South Africa. ☍

Heteroderidae (cyst nematodes) A family of **Nematoda** that are sedentary, **obligate** parasites of plant roots. There are 18 genera of which 6, comprising 34 species, are **gall**-forming. Females are **globose** in most species, males are wormlike. Cyst nematodes are serious pests of several crops including potato, soybean, maize (corn), cucurbits, lettuce, tomatoes, onions, peas, sunflowers, and melons. Infestations cause poor development of leaves and fruit and yellowing of leaves, and swellings (galls or **cysts**) on roots.

heterodimer A **protein** that consists of pairs of **polypeptides** with different **amino acid** sequences.

Heterodon platirhinos (eastern hognose snake, hissing adder, puff adder, spreading adder) A species of colubrid snakes (**Colubridae**) with thick bodies, 500 mm–1.2 m long, with highly variable colouring, but usually with large, dark blotches against a paler background, giving them some resemblance to rattlesnakes. They have a wide head with an upturned snout. The snakes inhabit areas with loose, dry soils in which they burrow, but are also found in woodland and grassland. They feed mainly on frogs and toads, but will also eat small mammals, reptiles, birds, and insects. When threatened they flatten their necks, raise their heads, hiss, and strike. This behaviour accounts for their common names, but they do not bite and are quite harmless. They occur throughout most of the United States and southern Canada. ☍

heterogametic *See* sex chromosome.

heterogamous Having two different types of **flower** on the same plant, e.g. in the capitula (*see* capitulum) of some **Asteraceae** the **disc florets** are **bisexual** and the **ray florets** female.

heterogamy Reproduction involving two types of **gamete**.

heterokaryon (heterocaryon) A **cell** that contains two or more genetically different nuclei.

heterokont Having flagella (*see* flagellum) of different lengths, or an organism with flagella of different lengths. *Compare* isokont.

Heterokontophyta (Stramenopila) A **phylum** of **eukaryotes**, most of which are algae (*see* alga), but that also includes the

Oomycota. All members have a life cycle that includes a **motile** stage with two differently shaped flagella (*see* flagellum). There are more than 100,000 species.

heteromeric *See* dimer.

heteromerous 1. Composed of different parts. **2.** Having different numbers of parts, e.g. 4 **sepals** and 5 **petals. 3.** Of a **lichen**, having a stratified **thallus** with layers containing different types of tissue.

heteromorphic Existing in different forms (morphs).

heteromorphic self-incompatibility **Self-incompatibility** that is based on **heterostyly.**

heterophylly Having leaves of different shapes on the same plant.

heteropolysaccharide A **polysaccharide** comprising more than one type of **monosaccharide.**

Heteroptera A suborder of true bugs (**Hemiptera**) in which the head is continuous ventrally with the mouthparts. Most species possess stink glands. **Nymphs** resemble adults but are wingless. They occur in terrestrial, semi-aquatic, and aquatic **habitats** (e.g. water boatmen). Many terrestrial species feed on plants or seeds and some are pests, others are scavengers, predators of smaller **Arthropoda**, or parasites (e.g. bedbugs). There are more than 50,000 species with a worldwide distribution.

Heterorhabditis A genus of nematodes (**Nematoda**) comprising 13 species, all of which are **obligate** parasites of insects. Some are used in **biological control.**

heterosis (hybrid vigour) Increased growth, robustness, and fertility that occurs in **hybrids** as compared with the parental **homozygotes**; it is always linked to increased **heterozygosity.**

heterospory Producing **spores** of two types (**megaspores** and **microspores**) on the same plant. *Compare* homospory.

heterostyly In **angiosperms**, a **polymorphism** in which a **flower** has **anthers** and **styles** of different lengths, e.g. in pin-eyed (long style) and thrum-eyed (short style) *Primula vulgaris*. It ensures **cross**-pollination by visiting insects. Usually the anthers of one type are at the same level as the **stigma** of the other.

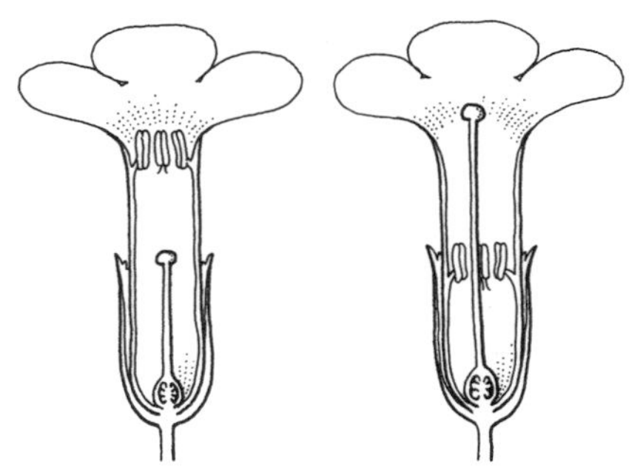

Heterostyly is a polymorphism that ensures cross-pollination, often by having anthers and styles of different lengths.

heterosymbiosis **Symbiosis** between different species.

heterothallic Describes taxa in which male and female reproductive organs occur in different individuals. Heterothallic yeasts (**Saccharomycotina**), have two **mating types**; after each cell division a mother cell changes type.

heterotherm *See* ectotherm.

heterotrichous In some algae (*see* alga), having prostrate **filaments** for attachment and erect filaments for **photosynthesis.**

heterotroph An organism that is unable to synthesize complex organic compounds from inorganic precursors and, therefore, obtains all its nutrients by consuming other organisms, which may be living or dead.

heterozygosity The presence of different **alleles** at a particular **gene locus.**

heterozygote A **diploid** or **polyploid** individual that has different **alleles** at a minimum of one **gene locus.**

Hevea brasiliensis (rubber tree) *See* Euphorbiaceae.

hexakisphosphate *See* phytic acid.

hexamerous Having parts in sixes.

hexaphosphoinositol *See* phytic acid.

Hexapoda A subphylum of **Arthropoda** that includes the **Insecta**, springtails (**Collembola**), Diplura (small, white, wingless, blind invertebrates), and Protura (minute, blind, wingless invertebrates).

hexose A **monosaccharide sugar** with six carbon atoms, $C_6H_{12}O_6$.

hexose monophosphate shunt (pentose phosphate shunt, phosphogluconate pathway) A series of metabolic reactions leading to the formation of deoxyribose and ribose sugars for the synthesis of **nucleic acids** and **nucleotides**, and nicotinamide adenine dinucleotide phosphate hydrogen (NADPH) from NADP for the synthesis of **fatty acids** and **steroids**.

hibernation A strategy for surviving the winter in a state of **dormancy**.

Hibiscus (family **Malvaceae**) A genus of **trees** and **shrubs** with **alternate**, **ovate** to **lanceolate**, **dentate** or lobed leaves. Flowers are bird-pollinated and showy, with 5 or more **petals** and a prominent **stamen** tube. Fruit is a **capsule**. There are several hundred species with a pantropical distribution. Several are cultivated as ornamentals.

Hicklingia A genus of plants, now extinct, that lived during the Middle Devonian epoch (397.5–385.3 million years ago) and is known from **fossils** found in Scotland. It grew in tufts with narrow, leafless stems, up to 17 cm tall, that branched **dichotomously** and bore sporangia (*see* sporangium) on short stalks on all sides and at the tip of the stem.

hickory *See Carya*, Juglandaceae.

high An area in which the **atmospheric pressure** is higher than it is in the surrounding area. *See* anticyclone.

high arctic tundra The northern section of the **tundra** where vegetation occurs mainly in marshy areas and does not cover all of the ground, except in sheltered places.

hill fog (upslope fog) **Fog** that forms when moist air is forced to rise up a hillside and cools adiabatically (*see* adiabatic cooling and warming).

hillock tundra Poorly drained **tundra** interspersed with small hillocks, about 25 cm high, where **drainage** is better and more plants occur.

hilum 1. The scar on a seed that marks its site of attachment to the plant. **2.** The site on a fungal **spore** where it was attached to the **sporophore**.

Himalayan honeysuckle (*Leycesteria formosa*) *See Leycesteria*.

Himantandraceae (order **Magnoliales**) A **monogeneric** family (*Galbulimima*) of aromatic **trees** with **alternate**, **simple**, **exstipulate** leaves with **petioles**. Flowers **actinomorphic**, **hermaphrodite**, **polypetalous**, with 4 or 6 leathery **sepals**, 7–9 **petals**, 15–40 **petaloid stamens**, 8–10 **staminodes**, **ovary superior** of 7–10 free **carpels** each with 1 **locule**. Flowers usually solitary. Fruit a **syncarpous**, **globose drupe**. There are two species occurring in the Celebes, New Guinea, and northeastern Australia. The **bark** has local medical uses and the plants contain **alkaloids** of pharmaceutical interest.

Himanthalia elongata *See* sea thong.

Hippophaë (sea buckthorn) *See* root nodule.

Hirudinea (leeches) A class of worms (**Annelida**) in which segmentation is not obvious, there is usually a sucker at each end of the body, a **clitellum**, and **chaetae** are absent. Most feed on decaying animal bodies and open wounds, and about 10 percent feed on blood. There are nearly 700 species, of which 90 are terrestrial, with a worldwide distribution.

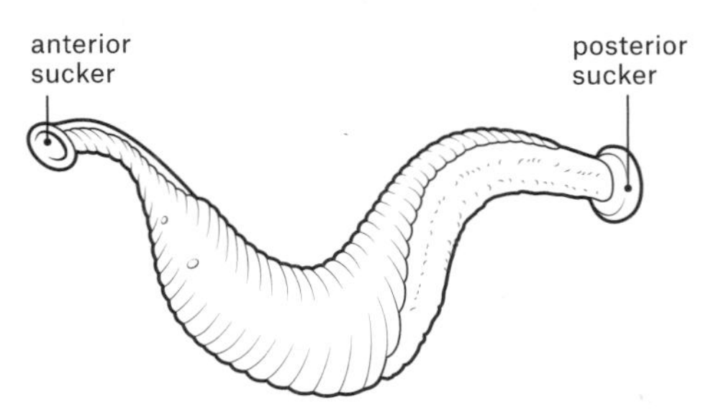

A leech (Hirudinea) is a worm, in most species with a sucker at each end.

Hirundo rustica (swallow, barn swallow) A species of migratory **passerine** birds, 170–190 mm long with a 320–350 mm wingspan, that have a blue head, back, and wings, **rufous** forehead, bib, and throat, white underside, curved, pointed wings, and a deeply forked tail. They thrive anywhere with access to water, open space for **foraging**, ledges for resting, and outbuildings for nesting. They feed on insects caught in flight, breed worldwide except for Antarctica and Australia, and winter in the tropics.

hispid With short, stiff bristles or hairs.

hissing adder *See Heterodon platirhinos.*

histic epipedon *See* humus.

histic horizon A surface or near-surface **soil horizon** that consists of poorly aerated organic material and that is saturated with water for at least one month in most years, unless it is artificially drained. It contains either 18 percent organic carbon by weight (30 percent organic matter) if there is 60 percent or more **clay**, or 12 percent organic carbon (20 percent organic matter) if there is no clay.

histisols A group of soils that consist mainly of organic matter. If they overlie unconsolidated mineral material they must be at least 40 cm thick, but they may be of any thickness if overlying solid rock. Histosols are a reference soil group in the **World Reference Base for Soil Resources.**

histochemistry The study of the chemistry of **cells** and tissues.

histology The study of **cells** and tissues.

histones A group of alkaline **proteins** that are the principal proteins in **chromatin** in **eukaryote** cells. They arrange **DNA** into **nucleosomes.**

hoar frost A thin layer of ice crystals that forms a white coating on exposed surfaces.

hobby farming *See* amenity horticulture.

hognut (*Carya glabra*) *See Carya.*

hogplum (*Ximenia*) *See* Ximeniaceae.

Holarctica A biogeographical region that includes all of the Northern Hemisphere to the north of latitude 15° N. The region is divided into the **Nearctic** and **Palaearctic.**

holdfast A structure that anchors a **sessile** organism to its substrate. It may be root-like or disc-shaped and may or may not possess suckers.

holistic (holological) Pertaining to the whole; in **ecology**, the study of complete **ecosystems.**

holly (*Ilex*) *See* Aquifoliaceae, *Ilex.*

hollyhock (*Althaea rosea*) *See Althaea.*

holocarpic Describes a fungus in which the entire **thallus** divides to form one or several **fruiting bodies.**

holocentric Describes a **chromosome** in which the properties of the **centromere** are distributed throughout the chromosome, so the entire chromosome acts as the centromere. *See also* acrocentric, metacentric, telocentric.

holocyclic Describes an insect that produces both viviparous (*see* vivipary) females and oviparous (*see* ovipary) females and males. A **fundatrix**, usually wingless, produces both winged and wingless viviparous females, which in turn produce wingless oviparous females and winged or wingless males. These mate and the female lays eggs that typically overwinter. *Compare* anholocyclic.

holoenzyme An **enzyme** consisting of an **apoenzyme** and its **cofactor**.

hologamete A **gamete** that comprises the entire cell of a protist (**Protista**).

holological *See* holistic.

holometabolous Describes an animal life cycle in which there are distinct juvenile and adult forms.

holomictic Describes a lake in which the water turns over completely at least once every year.

holomorph The whole fungus in **Ascomycota** and **Basidiomycota**, including both **anamorph** and **teleomorph**.

holoparasite A plant that is an **obligate** parasite, having little or no **chlorophyll**.

holophyletic Of a **taxon**, including all the descendants of their common ancestor.

holozoic Describes a method of feeding that involves ingesting organic material obtained from other organisms.

homeotherm (homoiotherm) An animal in which the body temperature varies only within narrow limits, being regulated metabolically (*see* endotherm) or behaviourally (*see* ectothrm).

home range The area within which an animal lives and obtains its food.

homoallelic mutant An **allele** that has more than one **mutation** at the same site within the **gene**.

Homobasidiomycetes *See* Agaricomycetes.

homoblasty Development of a plant that involves no changes in the form of organs, especially leaves. *Compare* heteroblasty.

homodimer A **protein** composed of a pair of identical **polypeptides**.

homogametic *See* sex chromosome.

homograft *See* allograft.

homoiohydry The ability of an organism to regulate the water content of its cells and tissues.

homoiomerous With all parts of the same type.

homoiotherm *See* homeotherm.

homokaryon A cell containing two or more nuclei that are genetically identical.

homologous Describes organs or other structures in unrelated **species** that have a similar evolutionary origin although they may serve different purposes, e.g. fertilization in **Bryophyta** by **motile gametes** is homologous to fertilization in **Spermatophyta**, where the gametes are non-motile but transferred by **pollen grains**.

homologous chromosomes **Chromosomes** that have identical linear sequences of **genes** and that pair during **meiosis**.

homologous recombination **Crossing over** between regions on **homologous chromosomes**.

homology Similarity of structures in different organisms that results from the descent of those organisms from a common ancestor. Homologies are evidence of relatedness between organisms. ↗

homomeric *See* dimer.

homomorphic self-incompatibility **Self-incompatibility** in plants where all the flowers have the same structure. It is achieved by genetic or biochemical mechanisms. There are two types: **gametophytic self-incompatibility** and **sporophytic self-incompatibility**.

homoplasy The occurrence of similar **characters** in distantly related taxa (*see* taxon) as a result of **convergent evolution** or **parallel evolution**.

Homoptera A suborder of **Hemiptera** with forewings that are not hemielytra (*see* hemelytron). They are the most destructive of all plant bugs, including aphids (**Aphididae**), leafhoppers (**Cicadellidae**), and scale insects (**Coccidae**). There are about 45,000 species.

homospory In **cryptogams**, the production of **spores** that are all of the same type and size. **Gametophytes** developing from them usually contain both female and male cells. *Compare* heterospory.

homosymbiosis **Symbiosis** between members of the same species.

homothallic Having male and female reproductive cells on the same **thallus**, in **Fungi** and algae (*see* alga).

homozygosity The condition in segments of **homologous chromosomes** in which identical **alleles** occur at one or more loci (*see* locus).

homozygote An individual in which identical **alleles** occur at one or more loci (*see* locus). The individual always breeds true at **homologous** loci that are homozygous (*see* homozygosity).

honeybees *See* Apidae, *Apis*.

honeydew A sticky, sugary liquid that is excreted by aphids (**Aphididae**), scale insects (**Coccidae**), bugs (**Hemiptera**), and some caterpillars (**Lepidoptera**) which feed on plant **sap**. In order to obtain sufficient **protein** from the sap, the insects are forced to excrete a large amount of surplus sugar. Honeydew often attracts **sooty mould**, some ants collect it directly from aphids, and some wasps and bees convert it to honeydew honey.

honey fungus *See Armillaria.*

honeysuckle *See* Caprifoliaceae, *Lonicera*.

hooded crow *See Corvus cornix.*

Hooke, Robert (1635–1703) An English experimental scientist with a very wide range of interests, who devised the compound microscope and made detailed and accurate drawings of the specimens he observed, publishing these in his book *Micrographia*, published in 1665. One of his drawings, of a section of **cork**, showed its porous structure; he called the compartments 'cells'.

Hooker, Sir Joseph Dalton (1817–1911) An English botanist and plant explorer who sailed on an expedition to Antarctica, subsequently publishing *Flora Antarctica* (1844–47), *Flora Novae Zelandiae* (1853–55), and *Flora Tasmanica* (1855–60). Between 1855 and 1857 he published *Flora of British India*. The many rhododendrons he collected and introduced into Britain became popular ornamentals. He edited the fifth and sixth editions of **Bentham**'s *Handbook of the British Flora*, which became known as 'Bentham and Hooker'. He was appointed assistant director of the Royal Botanic Gardens, Kew, in 1855, and succeeded his father as director in 1865.

Hooker, Sir William Jackson (1785–1865) An English botanist and authority on **cryptogams** who was the first director of the Royal Botanic Gardens, Kew.

hop (*Humulus lupulus*) *See* Cannabaceae.

hophornbeam (*Ostrya* spp.) *See* Betulaceae.

Hoplocampa flava (plum sawfly) A sawfly (**Symphyta**), 4–6 mm long, that lays eggs on the flowers of ***Prunus*** species, mainly in early summer. The larvae feed inside the developing fruit, which may fall. The larvae pupate in a **coccoon** in the soil. Adults feed on **nectar** and **pollen**. It occurs throughout Europe.

Hoplocampa testudinea (apple sawfly) A sawfly (**Symphyta**) that lays eggs in the **receptacles** of apple blossoms. These hatch soon after the **petals** fall and the larvae burrow into the developing fruits all the way to the core, migrating from apple to apple, leaving large holes contaminated with wet, brown **frass**. Badly damaged apples fall, those damaged only by scars remain on the tree until harvest. It occurs throughout Europe.

hop stunt *See* viroid.

horizontal gene transfer (lateral gene transfer) The movement of **genes** between organisms other than by reproduction. It is the usual means by which **Bacteria** acquire resistance to antibiotics.

hormogonium In certain **cyanobacteria**, a short section of a **filament (trichome)** that becomes detached and acts as a **propagule**.

hormone A substance produced in small amounts by specialized cells that affects other cells to which it is conveyed.

hornbeam *See* Betulaceae and *Carpinus*.

hornwort *See* Anthocerotophyta.

horotely An average rate of evolutionary change within a given **taxon**. *Compare* tachytely.

horse chestnut (*Aesculus hippocastanum*) *See Aesculus*, Sapindaceae.

horse latitudes Regions located in latitudes of about 30° in both hemispheres where warm, stable air is subsiding on the poleward sides of **Hadley cells** and winds are light and variable, and sometimes the air is calm. Sailing ships could be becalmed in such areas and if supplies of fresh water ran low horses, carried as cargo, might die.

horse mushroom *See Agaricus arvensis.*

horsetail (*Equisetum*) *See* Sphenopsida.

hortic horizon An **anthropedogenic horizon** that develops from long, deep **cultivation** (Latin *hortus*, garden), involving the application of **fertilizers** and organic matter. It is dark in colour and has an average organic content of at least 1 percent.

horticultural oil *See* insecticidal oil.

hot lightning **Lightning** that starts forest fires because the **lightning stroke** is sustained for long enough to ignite dry material. *Compare* cold lightning.

hot tower A narrow column of air that is rising rapidly by **convection** and is surrounded by air that is rising much more slowly, or subsiding. Hot towers generate large **cumulonimbus** and violent storms.

house finch *See Haemorhous mexicanus.*

houseleeks (*Sempervivum*) *See* Crassulaceae.

house martin *See Delichon urbicum.*

house sparrow *See Passer domesticus.*

house wren *See Troglodytes aedon.*

hoverflies *See* Syrphidae.

Howard, Luke (1772–1864) An English industrial chemist and meteorologist who devised the first practical scheme for classifying clouds (*see* cloud classification). Most of the names he gave to the cloud genera remain in use today.

Huaceae (order **Oxalidales**) A family of **shrubs**, **lianas**, and **herbs** with **alternate**, 2-ranked, **simple**, **entire** leaves wth **caducous stipules**. Flowers **actinomorphic**, **bisexual**, with usually 5, sometimes 4 free, **valvate sepals** or **sepals connate** forming a closed **calyx**, usually 5, sometimes 4 free or **sessile petals**, twice as many **stamens** as petals, **ovary superior** of 5 **carpels** with 1 **locule**. Flowers **axillary**, solitary or in clusters. Fruit is a **capsule**. There are two genera of three species occurring in tropical Africa.

hub gene A **node** with many branches leading from it in a **gene network**.

Huerteales An order of plants comprising 4 families of 6 genera, with 24 species. *See* Dipentodontaceae, Gerrardinaceae, Petenaeaceae, and Tapisciaceae.

Huia A genus, now extinct, of two known species of vascular plants (*see* Tracheophyta) that lived in the Early Devonian epoch, about 410 million years ago. The **sporophyte** had dichotomous (*see* dichotomous branching) and pseudomonopodial (*see* monopodial) leafless stems, and sporangia (*see* sporangium) as downward-curving spirals in terminal **spikes** on the fertile stems.

Humboldt, Friedrich Wilhelm Heinrich Alexander, Freiherr von (1769–1859) A German naturalist, geologist, vulcanologist, mining engineer, and biogeographer who spent the years 1799–1804 exploring South America accompanied by the French botanist Aimé **Bonpland** (1773–1858), returning to Europe with more than 30 cases of botanical specimens. In *Ideen zu einer Physiognomik der Gewächse* (Ideas on a physiognomy of plants, 1806) Humboldt proposed the concept of biogeography.

humic acid A mixture of organic compounds that can be extracted from soil using dilute alkali and precipitated with acid (in contrast to **fulvic acid**, which is soluble in acid).

humic gley soil (humic gleysol) A dark-coloured **gley soil** that has no free calcium carbonate near the surface. It is continually or intermittently moist, with or without a covering of **peat**, and has a gleyed horizon (*see* gleying).

humidity The amount of water vapour (not liquid water or ice crystals) that is present in the air. It can be expressed as **absolute humidity**, **mixing ratio**, **specific humidity**, or **relative humidity**.

humidity index The extent to which the amount of water that is available to plants exceeds the amount required for healthy growth, calculated as $100W_S/PE$, where W_S is the water surplus and PE is the **potential evapotranspiration**.

humification The formation of **humus** from decaying organic material through the action of **saprotrophs**. It is essentially an **oxidation** process involving the breakdown of large, complex molecules into simpler organic acids that may subsequently be mineralized into inorganic forms that can be absorbed by plant roots.

humin Those organic compounds in soil that do not dissolve in a diluted alkali solution.

H

Humiriacae (order **Malpighiales**) A family of **evergreen trees** with **buttress roots** and smooth, peeling **bark**. Leaves **alternate**, **simple**, **entire** to **serrate** with **petioles** often swollen. Flowers **actinomorphic**, **bisexual**, with 5 **connate sepals**, 5 **petals**, 10–30 **stamens** in 1 or 2 **whorls**, **ovary superior** of 4–5 or up to 8 fused **carpels** with 1 **locule**. **Inflorescence** usually **axillary**, **cymose**. Fruit **ovoid** to **globose**, **drupe**-like. There are 8 genera with 50 species occurring in tropical America and West Africa.

Humulus lupulus (hop) *See* Cannabaceae.

humus 1. A dark brown, amorphous substance consisting of decomposed organic matter found in soils that are aerated for at least part of the year. **2.** A surface **soil horizon** classed as either **mor** or **mull**. It is called a histic epipedon in the U.S. Department of Agriculture **soil taxonomy**. **3.** Informally, any organic matter present in soil.

hurricane Originally, a **tropical cyclone** occurring in the North Atlantic or Caribbean, but nowadays often applied to any tropical cyclone.

hurricane-force wind Wind of more than 33 m/s. *See* appendix: Beaufort Wind Scale.

hyaline Transparent or translucent.

hyaloplasm The ground substance of **cell cytoplasm**.

Hyalopterus pruni (mealy plum aphid) A pale green aphid (**Aphididae**) mottled with darker green, 1.5–2.6 mm long, and with antennae (*see* antenna) 0.5–0.75 the length of the body. Individuals are usually covered with white waxy meal. Eggs overwinter on ***Prunus*** and hatch in spring, forming colonies on the underside of leaves. Winged forms emerge in summer and migrate to grasses and reeds near water. In autumn the aphids move back to *Prunus*, though in smaller numbers. Large infestations can cause significant damage, although they do not cause leaves to curl, and the aphids excrete **honeydew**, which attracts **sooty mould** and inhibits **photosynthesis**. It occurs throughout North America and Europe.

hybrid An individual that results from breeding between parents with **genomes** sufficiently different for them to be considered separate **species** or **subspecies**. Hybrids may be fertile or sterile, but sterile plant hybrids may reproduce vegetatively.

hybrid speciation The appearance of a new **species** as a result of hybridization (*see* hybrid), usually due to **polyploidy**, although sometimes **diploid** species may hybridize to produce new diploid species. For example, hybridization between ***Helianthus*** *annuus* and *H. petiolaris* (both diploid) produced three diploid species: *H. anomalus*, *H. deserticola*, and *H. paradoxus*.

hybrid swarm A continuous series of morphologically different **hybrids** resulting from the hybridization of two species followed by the **cross-breeding** and backcrossing (*see* backcross) of subsequent generations.

hybrid vigour *See* heterosis.

hybrid zone A region in which **hybrids** of geographically **subspecies** occurs.

Hydatellaceae (order **Nymphaeales**) A **monogeneric** family (*Trithuria*) of very small, often **annual**, **monocotyledon** aquatic plants with **simple**, **linear**, **entire** leaves with only 1 **vein**. Flowers reduced, **unisexual** (plants **monoecious** or **dioecious**), of 1 **stamen**, **ovary** with 1 **locule**. **Inflorescence** a **capitulum** of 2–4 **bracts** upon a short **scape**. Fruit splitting into 3 **valves** or an **achene**. There are ten species occurring in India, Australia, and New Zealand.

hydathode A specialized tissue in the **epidermis** of leaves terminating in pores that exude water from the **xylem** during **guttation**. It probably evolved from modified stomata (*see* stoma).

Hydnoraceae (order **Piperales**) A family of leafless root parasites that lack **chlorophyll**. They have two types of roots: horizontal **rhizome**-like roots and haustorial (*see* haustorium) roots growing from them. Large, solitary flowers arise from the roots and are subterranean, partly above ground, or above ground, **actinomorphic**, **bisexual**, 3–4 fused **tepals**, 3–4 **stamens** or stamens fused with tepals into a tube (tepalostemon). Fruit **baccate**, with about 90,000 seeds. There are two genera with seven species occurring in Costa Rica, South America, Africa, Madagascar, and the Arabian Peninsula. They are possibly the strangest plants in the world.

hydragric horizon A subsurface **anthropedogenic horizon** that develops through wet **cultivation** (Greek *hydros*, water). It is more than 10 cm thick and often overlies a **buried soil**.

Hydrangeaceae (order **Cornales**) A family mainly of **shrubs**, but with a few **herbs**, with **rhizomes** and that climbs by means of **adventitious** roots. Leaves usually **opposite**, occasionally **alternate**, **simple**, **crenate** or **serrate**, but sometimes **entire** or palmately lobed. Flowers sometimes very dimorphic (*see* dimorphism), with fertile **perfect flowers** and sterile flowers with an enlarged **calyx**, 4–5 free or up to 10 fused **sepals** and **petals**, 2–5 times more **stamens** than petals, or rarely the same number or more than 200. **Ovary syncarpous, inferior** or partly so, of 3–5 free, 2–7 fused, or 12 **carpels. Inflorescence** usually a terminal **cyme**, sometimes **capitate** in **xeromorphic** species. Fruit is a **capsule** or **berry.** There are 17 genera with 190 species, most occurring in warm temperate regions, some tropical. Some with medicinal uses, many cultivated as ornamentals, e.g. *Hydrangea*, ***Philadelphus***.

hydrarch succession *See* hydrosere.

hydration The chemical combination of a substance with water.

hydraulic conductivity *See* permeability.

hydraulic gradient The change in the **hydraulic head** of **groundwater** over a horizontal distance.

hydraulic head 1. The elevation of a water body above a datum level. **2.** The potential energy possessed by a unit weight of water at a specified point. The hydraulic head comprises the elevation head, which is the height above a datum level, the pressure head, determined by the **atmospheric pressure**, and the velocity head, which is a measure of the rate of flow.

hydric Describes an area that is extremely wet.

hydrocarbon A chemical compound that contains hydrogen and carbon.

Hydrocharitaceae (order **Alismatales**) A family of **monocotyledon annual** or **perennial** aquatic **herbs**, with **rhizomes** or erect stems and basal roots. Leaves in rosettes in rhizomatous genera, otherwise arranged spirally or in **whorls**, usually submerged, sometimes floating but rarely **emergent**, **simple**, **linear** to **orbicular.** Flowers usually **actinomorphic** and **unisexual** (plants **dioecious**) or irregular and **bisexual** or unisexual (male and female flowers on separate plants), **perianth** in 1 or 2 series of 3 or occasionally 2 free segments, the inner series showy and **petal**-like, many **stamens**, **ovary inferior** of 2–20 **carpels** and in some genera a long **hypanthium. Inflorescence sessile** or **scape**-like. Fruit membraneous, **dehiscent** or **indehiscent**, with 1 seed. There are 18 genera with 116 species occurring worldwide. Several species cultivated as aquarium plants, some troublesome weeds.

hydrochory Dispersal of seeds or **spores** by water.

hydrocollapsibility The propensity of certain soils, e.g. **loess**, to experience sudden mechanical failure when the ground beneath them is close to saturation with water. Collapse is due to the combination of high **porosity** and weak **cementation.**

hydrogen (H) The simplest chemical element, its atom consisting of a nucleus of one proton with one electron. Hydrogen is essential for life, being the primary component of water (H_2O), and present in **carbohydrates**, **hydrocarbons**, **proteins**, etc.

hydrogenase An **enzyme** that catalyzes reactions in which hydrogen is added to a substrate.

hydrogen bond A chemical bond that occurs because the electron in the hydrogen atom is held only weakly, allowing it to

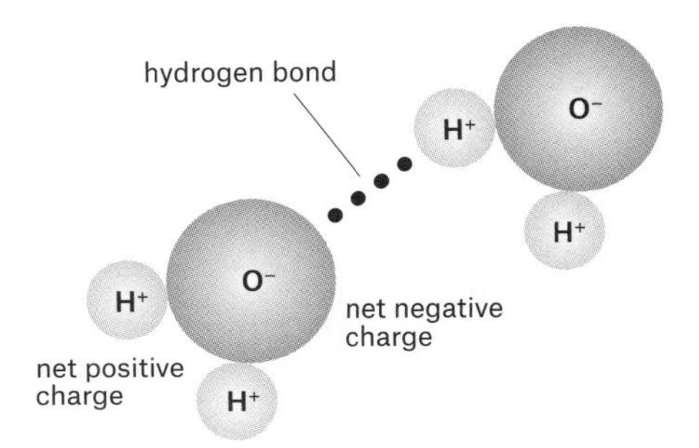

The net positive charge on one hydrogen atom is attracted to the net negative charge on the oxygen atom of a nearby water molecule, forming a weak bond between the two molecules.

form an **ionic bond** with another atom. Water molecules link by hydrogen bonds in the liquid and solid phases, but in water vapour the molecules have sufficient energy to break these bonds. Many organic molecules are held together by hydrogen bonds, including **DNA**.

hydroid In some mosses (**Bryophyta**), a thin-walled, elongate cell, usually with diagonal, porous, end walls, that conduct water, analogous to a **tracheary element**. *Compare* leptoid.

H

hydrolase An **enzyme** that catalyzes reactions in which a substrate is hydrolyzed (*see* hydrolysis).

Hydroleaceae (order **Solanales**) A **monogeneric** family (*Hydrolea*) of **herbs** and **shrubs** with erect or prostrate, **succulent** or woody stems, often bearing 1 or 2 spines up to 3 cm long at **nodes**. Leaves **alternate**, **ovate** to **linear**, **entire** to **serrulate**, **glabrous** to **pubescent**. Flowers **actinomorphic**, **bisexual**, **pentamerous**, hypogynous (*see* hypogyny), **sepals lanceolate** to **cordate**, bright blue to purple or white **corolla campanulate**, **ovary** of 2 occasionally 3–4 **carpels**. Fruit a **globose capsule**. There are 12 species occurring in tropical and warm temperate regions.

hydrological cycle The movement of water on a global scale from the surface, by **evaporation** and **transpiration**, through the air, and back to the surface as **precipitation**.

hydrological drought A **drought** in which the **water table** falls markedly.

hydrology The study of the movement of water between the atmosphere, land, and oceans, but especially its movement on land through surface waters, e.g. rivers and lakes, and **groundwater**.

hydrolysis 1. A reaction in which the addition of a molecule of water causes a substance to split into two parts. **2.** In

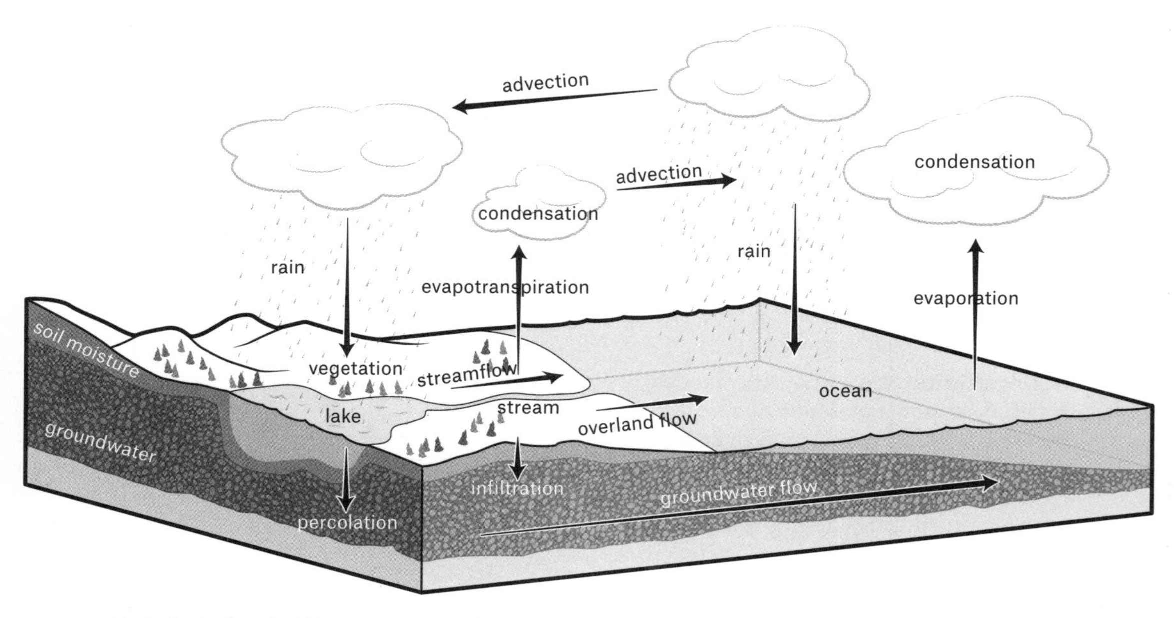

Hydrological cycle. Water evaporates from the ocean and land and is transpired by plants. It is advected as vapour, condenses to form clouds, and falls as precipitation. Precipitation falling on land returns to the atmosphere or ocean.

soils, the dissociation of water molecules ($H^+ + OH^-$) on the surface of minerals and the migration of H^+ into the crystal lattice, causing an imbalance that causes nutrient elements, e.g. calcium (Ca^-), magnesium (Mg^-), and potassium (K^-), to move out, into the soil solution.

hydrometeor All the liquid water and ice that is present in the atmosphere, including all forms of precipitation.

hydromuscovite *See* illite.

hydronasty A movement in a plant or plant organ triggered by a change in **humidity**.

hydrophilic Water-soluble or capable of being wetted.

hydrophily **Pollination** resulting from the transport of **pollen** by water.

hydrophobic Insoluble in water or incapable of being wetted; water-repellant.

hydrophyte A plant that grows in water or in very wet conditions, with its **perennating bud** at the bottom of open water. Leaves may be finely divided and submerged or large and floating and **aerenchyma** is usually present. *Compare* helophyte.

hydroponics The cultivation of plants in a liquid nutrient solution.

Hydropotes (Chinese water deer) *See* Cervidae.

hydrosere (hydrarch succession) A plant **succession** that begins on ground submerged under fresh water.

hydrosphere All of the water which exists at or close to the Earth's surface.

Hydrostachyaceae (order **Cornales**) A **monogeneric** family (*Hydrostachys*) of **annual** and **perennial herbs** that are submerged aquatics with almost no stem and a **holdfast** of **adventitious** roots that anchors them to rocks in fast-flowing water. A rosette of spiral, deeply **pinnate** leaves, covered with **enations**, arises from the holdfast, **inflorescence spikes** rising from the centre of the rosette, or from the pinnae (*see* pinna), suggesting that what appear to be leaves may be modified stems. Plants are **dioecious** or **monoecious**. Flowers are reduced, **sessile** and each subtended by a small **bract**, lacking **sepals** and **petals**, male flowers with 1 or 2 **stamens**, female flowers with **superior ovary** of 1 **carpel** and **locule**. Fruit is a **capsule**. There are 20 species occurring in central and southern Africa and Madagascar.

hydrostatic skeleton **Coelomic** fluid held under pressure that maintains the shape of a soft-bodied animal and provides muscles with a structure against which to contract.

hydrotaxis The movement of an organism in response to moisture, toward (positive) or away from (negative) the source of the stimulus.

hydrotropism The growth of a plant organ in the direction of moisture.

hydrous mica *See* illite.

hydroxide A substance, e.g. a mineral, that contains the **anion** comprising an oxygen atom bonded covalently with a hydrogen atom and carrying a negative charge: OH^-.

hydroxyl A compound consisting of an oxygen atom bound covalently to a hydrogen atom: –OH. A hydroxyl group is hydroxyl bonded to another substance (R–O–H); an **alcohol** is an organic compound containing a hydroxyl group. The neutral form of the **hydroxide ion** is a hydroxyl radical; hydroxyl radicals are highly reactive and, consequently, short-lived.

hyetogram A chart that records the amount and duration of rainfall at a particular place.

hyetograph An instrument that measures rainfall and records it as a line on a graph attached to a rotating drum.

hyetography The study of annual rainfall, its geographic distribution, and variations in it.

hygrometer An instrument that measures **humidity.**

hygrophanous Translucent or watery in appearance.

hygrophilic *See* hygrophilous.

hygrophilous (hygrophilic) Growing in or inhabiting moist environments.

Hygrophoropsis aurantiaca (false chanterelle) A species of **basidiomycete fungi** in which the **fruiting body** is initially convex then funnel-shaped, 20–80 mm across, with a **stipe** about 50 mm tall. The fruiting body is bright orange or peach-coloured, resembling a chanterelle (***Canthatellus***), but has **decurrent gills** unlike those of the true chanterelle. *Hygrophoropsis aurantiaca* is a **saprotroph**, common in forests throughout Europe and North America.

hygroscopic Capable of absorbing water from the surroundings, including from the air.

hygroscopic capacity *See* hygroscopic coefficient.

hygroscopic coefficient (hygroscopic capacity) The ratio, expressed as a percentage, of the weight of water to the weight of soil that a mass of dry soil will absorb and retain in equilibrium when the soil is in contact with saturated air.

hygroscopic moisture (hygroscopic water) Water that soil particles absorb from the air and that is bound very tightly to soil particles, so it is not available to plants.

hygroscopic nucleus A **cloud condensation nucleus** consisting of a substance that absorbs water, swelling as it does so.

hygroscopic water *See* hygroscopic moisture.

hygrotaxis The movement of an organism toward water.

Hyla chrysoscelis (Cope's gray tree frog) A species of tree frogs (**Bufonidae**), 30–50 mm long, usually grey in colour but may be brown or green. They inhabit woodland close to water. Their larvae feed on plant material and the adults on invertebrates. They occur throughout North America south of southern Canada. *Hyla chrysoscelis* is indistinguishable from *H. versicolor* (also called Cope's gray tree frog).

Hyla cinerea (green tree frog) A species of bright green, sometimes reddish brown tree frogs with a yellow or white underside and a yellow or white stripe on either side of the body. They are 340–620 mm long, with smooth skin and long legs. They inhabit forest areas close to water and feed on insects. They occur in the central and southeastern United States and are popular as pets.

Hyla squirella (squirrel tree frog) A species of tree frogs, about 25–40 mm long, that are highly variable in appearance. They are usually green, sometimes brown or yellow, plain or spotted, some with stripes, and some with bars between the eyes. They inhabit moist areas and can be found in gardens and feed on insects. They are mainly nocturnal but sometimes emerge by day when it is raining. They occur throughout the coastal plain of the southeastern United States.

Hyla versicolor *See Hyla chrysoscelis.*

Hylocichia mustelina (wood thrush) A thrush closely related to ***Turdus migratorius*** (American robin), 180–215 mm long with a 300–400 mm wingspan, that has a cinnamon-coloured crown and nape, a dull brown back, wings, and tail, and a white underside with dark spots on the breast, flanks, and sides. They also have a white ring round the eye. They inhabit **deciduous** and mixed forests, feeding on invertebrates and fruit. They breed on the eastern side of North America from southern Canada to Florida, and winter in Central America.

hymenium In a fungal **fruiting body**, the layer of tissue containing the cells that will develop into **spore**-bearing structures.

hymenomycetes A group of **basidiomycete fungi**, including **bracket fungi** and toadstools, in which the **hymenophore** is either exposed or covered only by a **velum**. The group was formerly ranked as a class, but is now known to be polyphyletic (*see* polyphyletism).

hymenophore In a fungal **fruiting body**, the structure that bears the **hymenium**.

Hymenoptera (ants, bees, sawflies, wasps) An order of insects that is divided into two suborders: **Apocrita**, comprising those with a constricted waist, and **Symphyta** (sawflies) comprising those without. Hymenopterans have two pairs of wings that can be linked together in flight, mouthparts adapted for chewing or modified into a **proboscis** for sucking, fairly long antennae (*see* antenna), and an **ovipositor** modified for piercing, sawing, or stinging in defence or to paralyze prey or hosts. They undergo full **metamorphosis**, usually pupate inside a **coccoon**, and pupae are **adecticous** and usually **exarate**. Larvae are **herbivores**, parasites of other insects including plant pests, or live in nests constructed and tended by adults. There are at least 103,000 species with a worldwide distribution.

Hymenoscyphus pseudoalbidus A fungus (**Ascomycota**) that, together with its **anamorph** *Chalara fraxinea*, is the cause of **ash dieback disease**. *Hymenoscyphus pseudoalbidus* forms during summer on the **petioles** of ash trees (***Fraxinus** excelsior* and *F. angustifolia*) and produces **ascospores** that are dispersed by wind.

hypanthium (floral cup, floral tube) An enlargement of the **receptacle** forming a cup-like structure in which the **sepals**, **petals**, and **stamens** are fused at the base and surround the **gynoecium** and fruit.

hypergenesis *See* hyperplasia.

hypergypsic horizon *See* gypsic horizon.

Hypericaceae (order **Malpighiales**) A family of small **trees**, **shrubs**, and **annual** and **perennial herbs** with **opposite**, **simple**, **oblong** leaves with no **petioles**. Flowers 4-merous or **pentamerous** with many **stamens**, **ovary** hypogynous (*see* hypogyny). Fruit is a **berry** or **drupe**. There are 9 genera with 560 species of worldwide distribution. *Hypericum*, with 370 species, is St John's-wort.

hyperparasite A parasite that parasitizes another parasite.

hyperplasia (hypergenesis) A proliferation of cells, in plants often the result of pest action.

hyperthermophile *See* extremophile.

hypertonic Having a higher solute concentration (i.e. **osmotic pressure**) on one side of a **membrane** than on the other. *Compare* hypotonic, isotonic.

hypha (pl. hyphae) A thread-like **filament** that is the basic structural unit in most **Fungi** and **Actinobacteria**.

Hyphomycetes A class of **Fungi Imperfecti** in which the vegetative stage is usually a well-developed **mycelium**, often called a mould, and the reproductive structures develop directly on the **hyphae** and are not enclosed. Consequently, their **spores** are carried by air currents, and most air contains them. Some species are predators of **Nematoda**, **Rotifera**, and other organisms. There are 1480 genera and more than 11,500 species, found worldwide.

hypobiosis **Dormancy**.

hypocotyl Part of the **plumule** or seedling that is located below the **cotyledon** and above the **radicle**.

hypodermis 1. The supportive layer of cells lying beneath the **epidermis** in the leaves and other organs of certain plants. It often contains **sclerenchyma** which adds to its strength. **2.** The layer of cells that lies beneath the **cuticle** in many invertebrates and secretes the **chitin** from which the cuticle is made. **3.** In vertebrates, the deepest layer of the skin.

H

hypogeal 1. Describes a germinating seed in which the **cotyledons** emerge below the ground surface. **2.** Describes a structure (e.g. a **rhizome**) that grows below the ground surface but parallel to it.

hypogean (hypogeic) Describes an organism that crawls across the surface.

hypogeic *See* hypogean.

hypogeous Describes a plant that grows on the ground surface.

hypogynous *See* hypogyny.

hypogyny (adj. hypogynous) The condition in which the **ovary** of a **flower** is superior.

hypolimnion The cooler, lower water in a thermally stratified lake.

hypolith A photosynthesizing organism that lives on the underside of rocks in a hot or cold desert, where it is protected from scouring by wind-blown sand and ultraviolet (UV) radiation, and where trapped moisture provides water for **photosynthesis**.

hypolithon A community of **hypoliths**.

hyponasty The bending upward of a leaf or other part of a plant due to the more rapid growth of the underside.

hyponeuston The organisms that live on the underside of the surface of water.

hypophloeodal Living or growing beneath **bark**.

hypostasis A **gene** (hypostatic gene) that cannot be expressed because of the expression of an **allele** of another gene.

hypostatic gene *See* hypostasis.

hypothallus 1. In some **lichens**, a layer of fungal **hyphae** lying beneath the **thallus** and extending beyond it. **2.** In sporangia formed by **Myxogastria**, the substrate from which the **sporangium** or its stalk arises.

hypotheca The younger, inner half of the **frustule** of a **diatom**.

hypotonic Having a lower solute concentration (i.e. **osmotic pressure**) on one side of a **membrane** than on the other. *Compare* hypertonic, isotonic.

Hypoxidaceae (order **Asparagales**) A family of **monocotyledon**, **perennial**, **geophyte herbs** with **rhizomes** or **corms**. Leaves **linear** to **lanceolate**, usually **sessile**. Flowers usually **actinomorphic**, **bisexual**, 2-, 3-, or 4-merous, free **tepals** in **whorls** or fused into a tube, often green and hairy on the outside and coloured on the inside, 3+3 **stamens**, **ovary inferior** of 3 **carpels**. **Inflorescence axillary** on a **scape**, **bracteate**, **spicate**, **corymbose**, occasionally **capitate**, **umbel**-like, or solitary. Fruit a **capsule**. There are 7–9 genera with 100–200 species, occurring in the seasonal tropics.

hysiginous Red in colour.

hyssop (*Hyssopus officinalis*) *See Hyssopus*.

Hyssopus (family **Lamiaceae**) A genus of **perennial herbs** and **evergreen shrubs** with **linear**, aromatic leaves. Flowers in **whorls** on a **spicate inflorescence**, the **calyx** with 5 nearly equal teeth, the **corolla** 2-lipped. There are 10–12 species occurring from the eastern Mediterranean to central Asia. *Hyssopus officinalis* (hyssop) is widely cultivated as a culinary and medicinal plant.

I

IAA *See* indole-acetic acid.

ianthinus Blue to purple in colour.

Icacinaceae A family of **shrubs**, **trees**, and **lianas** that are not placed in an order. The lianas have **tendrils** that are modified **branches** or **inflorescences**. Leaves **alternate**, sometimes **opposite**, **simple**, **entire**, occasionally **serrulate** or palmately 3- to 7-lobed. Flowers small, **actinomorphic**, **bisexual** or **unisexual** (plants **dioecious**), usually 5- but sometimes 4-merous, **calyx** cup-shaped, **petals** free or fused, as many **stamens** as petals or **corolla** lobes, **ovary superior**. Inflorescence **racemose**, occasionally **cymose**. Fruit is a **drupe**. There are 24 or 25 genera with 149 or 150 species occurring throughout the tropics.

ice-crystal haze **Haze** that contains only ice crystals.

ice day A day during which the air temperature fails to rise above freezing and when ice on water surfaces does not melt.

ice fog **Fog** that forms when warm water is suddenly exposed to air below freezing temperature and evaporating moisture changes directly to ice crystals.

Icelandic low A semipermanent are of low **atmospheric pressure** located between Iceland and Greenland, at about 60°–65° N.

Iceland moss The **lichen** *Cetraria islandica*, which has an erect, tufted **thallus**, with flattened, reddish brown **branches**, so it somewhat resembles a moss. It grows abundantly in northern regions, especially in Iceland. It is edible, but little used.

ice pellets Precipitation in the form of transparent or translucent ice particles less than 5 mm in diameter.

ice period The duration of surface snow, from the first fall to the melting of the last patches.

ice plant (*Carpobrotus edulis*) *See* Aizoaceae.

ice storm A storm of **freezing rain** that deposits a thick layer of clear ice on exposed structures, e.g. trees, telephone and power lines, and radio masts.

Ichneumonidae (ichneumon wasps, scorpion wasps) A family of wasps (**Apocrita**) all of which are parasites of other insects, each of the many subfamilies specializing in a particular group of hosts. Species vary greatly in size and colour, but most are slender and wasp-like, with antennae (*see* antenna) at least half as long as the body and many with **ovipositors** longer than the body. Unlike wasps, ichneumons do not sting in self-defence although they may make stinging motions if handled. There are at least 60,000 species, distributed worldwide.

ichneumon wasps *See* Ichneumonidae.

Icteridae (New World blackbirds, cowbirds, grackles, meadowlarks, orioles) A family of black, brown, orange, red, and yellow birds, many with white wing markings. They range from 150 mm to 520 mm in size, inhabit forests, scrub, grassland, and marshes, and feed on insects, small vertebrates, and fruit. Some breed in colonies and many are migratory. Cowbirds (five *Molothrus* species) are brood parasites (*see* brood parasitism) except for the bay-winged cowbird (*M. badius*), which is parasitized by the screaming cowbird (*M. rufoaxillaris*). Despite their common names, icterids are not closely related to Old World blackbirds, larks, or orioles. There are 31 genera with 111 species, distributed throughout the Americas.

I

Icterus bullockii (Bullock's oriole) A species of orioles (**Icteridae**) with marked **sexual dimorphism**. Males are 170–190 mm long and during the breeding season have black and orange plumage. Females are slightly smaller and a dull yellow. They inhabit open woodland and areas beside rivers and feed on insects, including some pests. It occurs throughout most of western North America, migrating to Central America in winter.

Icterus galbula (Baltimore oriole) A species of migratory birds that spend summers mainly in the eastern United States and winters in the tropics. They are 170–200 mm long, males slightly larger than females, adults with a black head, back, and beak, bright orange breast, rump, and underside, and black wings with orange and white bars. Females are olive or orange. They live in forests, grassland, and suburban areas, and feed on insects, especially **caterpillars**, including many pest species, and some small fruits and nectar.

ideal free distribution The distribution of **foraging** animals among patches of resources that results when the animals are free to move where they wish and are aware of differences in the amount of resources in the patches. The animals will tend to congregate in the richest patches.

idioblast An isolated plant **cell** that differs from those around it and that contains no living material.

idiomere *See* chromomere.

igapo *See* Amazon floral region.

igneous Describes rock that has crystallized from a **magma**. It is one of the three main groups of rock types. *See* metamorphic rock, sedimentary rock.

ileum The final section of the intestine in mammals (not to be confused with **ilium**).

Ilex (family **Aquifoliaceae**) A genus of **trees** and **shrubs** (holly) with mostly **evergreen**, **alternate**, leathery, glossy leaves, typically **serrate** or with spiny teeth. Plants **dioecious**, usually with male and female flowers on separate plants. Flowers inconspicuous, usually **tetramerous**. Fruit is a **drupe** (though commonly described as a **berry**). There are 400–600 species widely distributed in the tropics and temperate regions. *Ilex aquifolium* is cultivated as an ornamental. Yerba maté is made from the leaves of *I. paraguariensis*.

ilium The **dorsal** section of the pelvis in **tetrapods** (not to be confused with **ileum**).

illite (hydromuscovite, hydrous mica) A common **clay mineral** formed by the **chemical weathering** of muscovite or feldspar. It has an overall negative electrical charge.

illuvial horizon A **soil horizon** containing material deposited by **illuviation**.

illuviation The movement of soil materials by water, vertically or horizontally, and their deposition usually in a lower **soil horizon**.

imago The fully developed adult in **Pterygota**.

imbibition **Absorption** of liquid by porous tissue, causing it to swell.

imbricate Overlapping, like roof tiles.

imidacloprid *See* chloronicotinyls, neonicotinoid.

imine A compound containing the imino group NH.

imino acid A compound that contains both **imine** and **caroxyl** functional groups, both attached to the same carbon atom.

immature soil A soil that lacks a clearly defined **soil profile** because it has not existed long enough for one to develop.

immigration *See* migration.

immission The receipt of a substance, e.g. a pollutant, from a distant source.

immobilization The conversion, by living organisms, of a chemical element or compound from an inorganic form available to plant roots to an unavailable organic form, i.e. its removal fro the reservoir of soil nutrients.

immunity A resistance, inherited or acquired, to a **pathogen** or its products.

imparipinnate Of a leaf, **pinnate** with a single terminal leaflet.

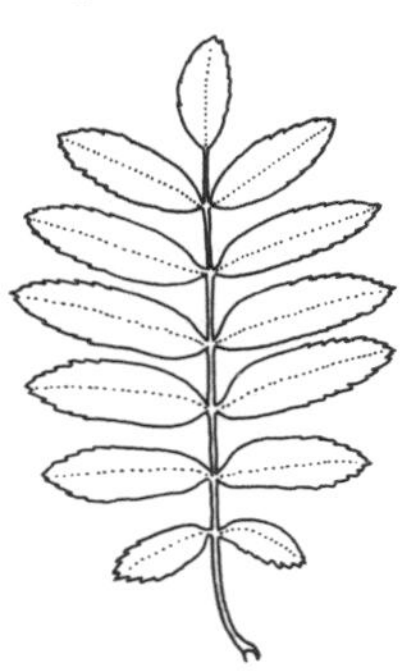

An imparipinnate leaf.

Impatiens (family **Balsaminaceae**) A genus of **annual** and **perennial herbs** with **entire** leaves with a water-repellant **cuticle** on the upper surface. Flowers with 3 **sepals**, the lowest **petaloid** and forming a conical tube ending in a spur, the upper 2 smaller but also petal-like, 5 **petals**, the uppermost concave and projected forward or upward. The generic name and the common name touch-me-not refer to the way the ripe **capsules** open explosively when touched, scattering their seeds. The plants are also known as jewelweed. Many are cultivated for ornament. *Impatiens walleriana* is busy lizzie.

imperfect cycle A **biogeochemical cycle** in which the element spends a prolonged period in soil or sedimentary rocks, during which it is unavailable to living organisms. *Compare* perfect cycle.

imperfect flower A **flower** that possesses a single set of reproductive organs, so is either male or female.

imperfect state The asexual state of a fungus.

imperilled species A plant or animal species, the survival of which is a matter for concern.

impervious soil A soil that will not permit the passage of water.

imported cabbage worm *See Pieris rapae.*

inactive front (passive front) A **front** or section of a front that has very little cloud or precipitation associated with it.

inbreeding Mating between close relatives, thereby tending to increase **homozygosity**.

inbreeding depression A decline in vigour that often occurs in the offspring of closely related parents, usually because of the expression of **recessive genes** due to **homozygosity**.

Inca lily (*Alstroemeria*) *See* Alstroemeriaceae.

inceptisols A group of soils that have one or more **soil horizons** in which mineral deposits have been either weathered (*see* weathering) or removed. They are at an early stage of forming horizons and lack a distinct **soil profile**. Inceptisols comprise an order in the U.S. Department of Agriculture **soil taxonomy**.

included flower A **flower** in which the **corolla** encloses the **stamens**.

included phloem **Phloem** tissue that is enclosed by **secondary xylem**.

incomplete dominance **Partial dominance** in which the **phenotype** of a **heterozygote** is intermediate between those of the corresponding **homozygotes**.

incomplete flower A **flower** that lacks one or more of the basic flower parts (*see* floral formula).

inconsequent drainage (insequent drainage) A **drainage pattern** that is unrelated to the underlying rocks.

incumbent Leaning on or resting against a support.

indehiscent Describes fruits that do not open to release their seeds.

independent assortment (random assortment) The distribution of **genes** in **gametes**, where **alleles** occur independently of each other. *See* Mendel's laws.

indeterminate (polytelic) Describes an **inflorescence** with **monopodial** growth; the terminal **bud** grows continually, flowers develop laterally along the **axis**, and a terminal flower never forms.

index of abundance An estimate of the size of an animal population that is calculated from the number attracted to a bait or the number caught for a specified amount of effort.

Indian floral region The area covering the whole of the Indian subcontinent, part of the **Palaeotropical region**. It contains about 150 **endemic** genera, most of which are **monotypic**.

Indian matting plant (*Cyperus papyrus*) *See* Cyperaceae.

indicator species A species with a narrow tolerance for certain environmental factors, so its presence is indicative of certain conditions.

indifferent species In the phytosociological (*see* phytosociology) scheme devised by the school led by Josias **Braun-Blanquet**, one of the five classes of fidelity (*see* faithful species) that describe and classify plant communities. Indifferent species are not rare, but have no particular affinity for any community. *Compare* accidental species, exclusive species, preferential species, selective species.

indigenous *See* native.

indigo bunting *See Passerina cyanea.*

individualistic hypothesis The hypothesis proposed by Henry Allan **Gleason** that the vegetation in an area changes in response to a continuously varying environment. Consequently, every plant community is to some degree unique.

indole-acetic acid (IAA) A carboxylic acid with the **carboxyl** group linked to an indole ring by a methylene group (CH_2). It is the most studied of the **auxins**, produced mainly in **buds** and young leaves, and inducing **cell growth** and cell division and the development of plant organs.

inducible enzyme (adaptable enzyme) An **enzyme** that is produced only when a suitable substrate is present and the enzyme is of immediate use. *Compare* constitutive enzyme.

indumentum A covering of hairs (**trichomes**).

induration The process, which may or may not involve **cementation**, of forming a **soil horizon** or **hardpan** that is hard, brittle, and has a high **bulk density**.

indusium A flap or scale that protects a fern (**Pteridophyta**) **sorus**. As the sporangia (*see* sporangium) mature the indusium shrivels allowing the **spores** to escape.

industrial melanism *See Biston betularia.*

infarctate Solid or turgid.

infection The invasion of tissues by a **pathogen**, or a disease caused by such an invasion.

inferior Describes an **ovary** that is inserted below the other flower organs.

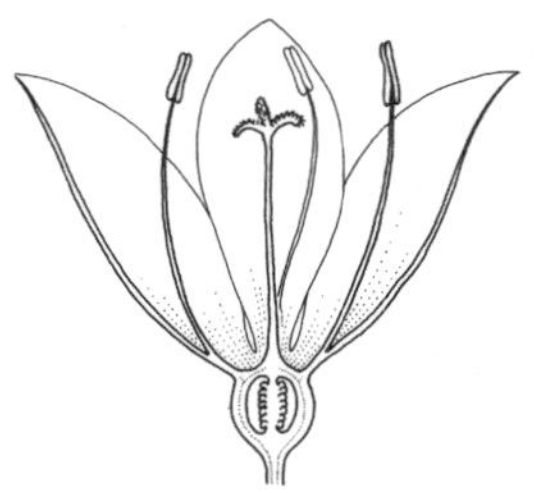

The inferior ovary is inserted below the other reproductive organs.

infiltration The downward movement of water through soil.

infiltration capacity The highest rate at which a soil or rock is able to absorb falling rain.

inflorescence A flowering structure that consists of more than one flower, and often of many. *See* catkin, corymb, cyme, panicle, raceme, spadix, spike, thyrse, umbel.

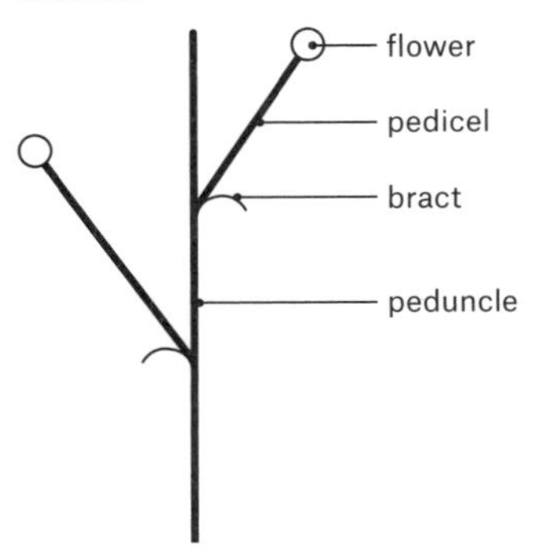

The standard schematic way to represent each type of inflorescence. Every inflorescence comprises flowers, each attached by a pedicel and subtended by a bract, and the flowers arise from a peduncle.

influent stream *See* gaining stream.

infructescence A group of fruits derived from the ovaries of an **inflorescence** and often retaining its structure.

infuscate Brown in colour.

ingroup The group of organisms being considered in the production of an evolutionary tree. *Compare* outgroup.

initial A **meristem** cell that is dividing repeatedly. Each division produces one cell which joins the specialized plant tissue and one cell that retains its meristemic ability to divide.

initiator The **transfer-RNA** (t-RNA) that transports methionine in **eukaryotes** and N-formyl-methionine in **prokaryotes** and that binds to the small unit of a **messenger-RNA** (m-RNA) to form an initiation complex, allowing the synthesis of a **peptide** chain to proceed.

ink cap The common name for the **fruiting body** of any **agaric** fungus in the genus *Coprinopsis* in which the **gills** undergo **autolysis** and dissolve into a black liquid resembling ink.

inoculation The addition to a soil of very small amounts of a substance, e.g. natural soil to sterilized soil, **rhizobia** to soil growing legumes (**Fabaceae**).

inoperculate Lacking an **operculum**.

inquilism A relationship between two organisms in which one lives inside the body of the other.

Insecta (insects) A class of **Arthropoda** that have three pairs of legs, a body comprising a head, thorax, and abdomen, two pairs of wings derived from outgrowths of the wall of the thorax (but reduced to one pair in **Diptera** and secondarily lost in some species), typically one pair of antennae (*see* antenna) and one pair of large compound eyes. Insects inhabit every **environment** except for the high arctic and Antarctica. About 1 million species are known with many more to be identified, meaning there are more species of insects than of all other animals combined. ☞

insecticidal oil (horticultural oil) An oil, usually derived from petroleum but sometimes from plants, that is made into an emulsion with water and sprayed on crops to control insects and mites, and in some formulations to control diseases, e.g. **powdery mildew**. The oils have little effect on beneficial insects, are safe to use, and cheap. ☞

insecticidal soap A soap or detergent that is mixed with water to make an approximately 2 percent concentration and sprayed on plants to kill arthropods (**Arthropoda**). Soaps used this way are believed to disrupt arthropod **cell membranes**. They are simple and safe, but must coat the pest in order to be effective. ☞

insecticide A chemical compound that kills insects.

insectivorous plants (carnivorous plants) Plants that trap and digest insects and other arthropods (**Arthropoda**). Such plants occur in **habitats** where nutrients are scarce, e.g. **bogs**. ☞

insects *See* Insecta.

insequent drainage *See* inconsequent drainage.

insolation The amount of solar radiation that reaches a unit area of the Earth's surface.

insolation weathering *See* thermal weathering.

instability The tendency of a rising body of air to continue rising because it is cooling adiabatically (*see* adiabatic cooling and warming) more slowly than the **environmental lapse rate**, so it is always warmer, and less dense, than the surrounding air.

instar An insect larva that is between moults of its **exoskeleton**, or between its final moult and pupation or emergence in its adult form.

insular climate The climate of a region, e.g. an oceanic island or peninsula, where the influence of the ocean exceeds that of any neighbouring continent.

integrated pest control *See* integrated pest management.

integrated pest management (integrated pest control, IPC, IPM) An approach to pest control that identifies those species that require control, determines a level of infestation at which action must be taken, and involves management practices that minimize the probability of infestations occurring. ☞

Integro *See* methoxyfenoxide.

integument 1. The one or more layers of tissue covering an **ovule** with a single opening, the **micropyle**. **2.** The **testa**. **3.** Any outer covering or skin.

intensification An increase in a **pressure gradient**, resulting in a strengthening of the wind, that occurs over hours or a few days.

intention movement The first step in a behavioural sequence that indicates the intention of the animal to perform the full ritual.

intercalary deletion *See* deletion.

interception 1. The abstraction of **groundwater** that would otherwise be lost, e.g. because it was flowing seaward close to the coast. **2.** The capture of falling rain by plant and other raised surfaces from which it evaporates without reaching the soil. **3.** The capture of direct sunlight by objects that cast shadows.

intercropping (interplanting) The growing of two or more crops in the same field or bed simultaneously, either in alternate rows or mixed, in order to optimize the use of space or reduce pest infestations. *See* companion planting.

interference competition Competition between species for a resource that is in limited supply, in which one competitor denies the other access to the resource by dominating the space in which the resource is located. *Compare* exploitation competition.

interflow (throughflow) The lateral flow of water through upper **soil horizons**, usually following heavy rain. If the flow occurs at shallow depth it may emerge above the surface near the bottom of slopes; it is then called return flow.

intergrade A soil or **soil horizon** with a composition or characteristics that are transitional between two soils or horizons that do not share a common origin.

intermediate feather *See* feather.

intermediate filament **Filaments** present in most **eukaryote** cells from multicellular organisms, associated with the **plasma membrane** and providing support for it.

intermediate rock An **igneous** rock that has a chemical composition intermediate between those of an **acidic rock** and a **basic rock**.

intermittent stream A stream that ceases to flow during periods of dry weather.

International Code of Nomenclature for Algae, Fungi, and Plants A set of rules applied in the formal naming of plants that aims to ensure that a single, unambiguous name is allotted to each **taxon**. The current edition was adopted in July 2011 by the 18th International Botanical Congress.

International Union for Conservation of Nature (IUCN, World Conservation Union) An organization based in Switzerland that promotes and initiates scientifically based **conservation** measures. Its members include more than 1200 organizations in about 160 countries and it has official observer status at the United Nations General Assembly.

internode The part of a plant stem that lies between two **nodes**.

interphase That part of the **cell cycle**, sometimes misleadingly called the resting phase, during which **chromosomes** are replicating and there is other intense activity, but there is no evidence of the cell **nucleus** dividing.

interplanting *See* intercropping.

interseeding Sowing plant seeds into standing vegetation.

interspecific competition *See* competition.

interstitial Describes the small spaces between particles.

interstitial fauna Animals that inhabit the spaces between sand grains.

intine The inside layer of a **spore** or **pollen grain**.

intraspecific competition *See* competition.

intraspecific parasitism A type of **brood parasitism** in which the parasite lays its eggs in the nest of another bird of the same species.

intrazonal soil A soil that has a distinct **soil profile** which reflects the dominant influence of a particular factor of age, **parent material**, or topography over the influence of vegetation or **climate**.

Intrepid *See* methoxyfenoxide.

introgression (introgressive hybridization) A flow of **genes** from one **species** into the **gene pool** of another.

introgressive hybridization *See* introgression.

intromit Insert.

intromittent organ An organ that is inserted, e.g. to transfer sperm for internal fertilization.

intron A segment of **DNA** that is removed from a **gene** during **transcription** and, therefore, is absent from the **messenger-RNA**.

introrse Describes **anthers** that release their **pollen** into the centre of the **flower**.

inulin A **polysaccharide** ($C_6H_{10}O_5$) that is a storage compound found in many plants, especially in the roots, **rhizomes** and **tubers** of many members of the **Asteraceae**. When hydrolyzed (*see* hydrolysis) it forms **fructose**.

invagination Infolding, to form a pocket.

invasive species 1. Any species that colonizes and comes to dominate a **habitat**. **2.** An **exotic** species that is capable of harming native species, the environment, or human health.

inversion 1. (genetic) A change in the arrangement of genetic material in which a segment of a **chromosome** is excised, rotated through 180 degrees, and reinserted in its original position, but with the **genes** in reverse order. **2.** *See* capping inversion, moisture inversion, precipitation inversion, radiation inversion, surface inversion, temperature inversion. **3.** *See* paracentric inversion, pericentric inversion.

inversostyly A **flower polymorphism** in which either the **styles** and **stamens** point in opposite vertical directions, i.e. if the styles point upward the stamens point downward and vice versa, or both styles and stamens point downward. It was first described in 2005 in *Hemimeris racemosa* (**Scrophulariaceae**).

invertase *See* sucrase.

inverted repeat Two **genes** that are identical, but with the **nucleotide** sequences in opposite order. Inverted repeats are common in **chloroplast DNA**; in lettuce and spinach it is 24,400 **base pairs** long and in maize it is 22,500 base pairs long.

invisible drought A **drought** that persists despite precipitation falling because when losses by **evaporation** and transpiration are deducted the amount is insufficient to recharge **aquifers** depleted during a preceding drought.

involucel A **whorl** of **bracteoles**.

involucre A structure that envelops and protects a plant organ, e.g. an **involucel** situated below an **inflorescence** or leaves that protect the reproductive organs in a leafy liverwort (**Marchantiophyta**).

involucrellum A dark structure surrounding the **ostiole** in many **Fungi**.

involute 1. Having the margins rolled up. *Compare* evolute. **2.** *See* vernation.

iodine cycle The flow of iodine (I) from the land, through living organisms, to the sea, to the air, and back to land. Decomposition of organic matter deposits iodine in marine sediments. Bacterial activity releases it into seawater as iodate (IO_3). Iodate is absorbed by certain marine algae (*see* alga), which release it as methyl iodate (CH_3IO_3). This enters the air, enters into several reactions, and finally returns to land as dry particles or dissolved in rainwater.

ion An atom that has acquired an electrical charge through the loss or gain of an electron. The loss of an electron produces a cation with a positive charge; the gain of an electron produces an anion with a negative charge.

ion exchange The reversible replacement of one **ion** by another in a solution.

ionic bond A chemical bond that forms when an electron moves from one atom to another, leaving the donor atom with a positive charge and the acceptor atom with a negative charge. Electrical attraction then holds the atoms together.

IPC *See* integrated pest management.

IPM *See* integrated pest management.

Ipomoea batatas (sweet potato) *See* Convolulaceae.

Iridaceae (order **Asparagales**) A family of **monocotyledon**, **deciduous**, with a few **evergreen**, **perennials** and a few **annual herbs** with **rhizomes**, **corms**, or **bulbs**. Leaves mainly **distichous**, sometimes **ensiform**, **cauline** or growing from the base, with an **imbricate** sheathing base. Flowers **actinomorphic** or **zygomorphic**, **bisexual**, usually with a large **perianth** of 2 **whorls** of 3 **petaloid tepals**, 3 occasionally 2 **stamens**, **ovary inferior**, occasionally **superior** of 3 fused **carpels**. **Inflorescence** usually **bracteate**, either terminal, 1- to many-flowered **umbel**-like, monochasial (*see* monochasium) **cyme**, or **spicate**, occasionally **panicle**-like, or a single, almost **sessile** flower. Fruit is a **capsule**. There are 66 genera with at least 2035 species with worldwide distribution. Many cultivated for ornament, e.g. ***Crocus***, ***Iris***, ***Crocosmia***, ***Gladiolus***.

iridescence Shimmering colours that are caused by the diffraction of light by molecules at the surface of a film of oil, ice crystals, or water molecules when these are all of about the same size.

iridoid A class of **secondary metabolites** with a very bitter taste that help defend plants against infections and **herbivores**. They are found in many medicinal plants and have a wide variety of pharmaceutical uses. They consist of monoterpenes (*see* terpene) derived from **isoprene**.

Iris (family **Iridaceae**) A genus of **perennial**, **monocotyledon herbs** with **rhizomes** or **bulbs**, with leafy shoots, usually flattened, and fan-shaped **inflorescences** of 1 or more flowers in which the 3 **sepals** (the 'falls') usually arch downward and are larger than the 3 **petals** (the 'standards'), which stand upright. The fruit is a **capsule**. There are 260–300 species occurring throughout the northern temperate zone. Many are cultivated for ornament.

iris nematode *See Ditylenchus destructor.*

iron (Fe) An element that is an essential trace nutrient for plants, taking part in reactions involving the transfer of electrons, e.g. in **phosphorylation** and **oxidative phosphorylation**.

iron pan An indurated (*see* induration) **soil horizon** in which the principal cementing material is iron oxide. It usually occurs at the top of the B horizon.

ironwood *See Carpinus.*

irradiance The rate at which solar energy passes through a unit surface area perpendicular to the radiation.

irragric horizon An **anthropedogenic horizon** that develops over a long period through irrigation (Latin *irrigare*) with water rich in sediment. The **clay** content is higher than that of the underlying soil and clay and **carbonates** are distributed evenly.

irregular *See* zygomorphic.

irrigation The provision of water to augment the amount available to crops naturally.

irrigation efficiency The ratio of the irrigation water utilized by a growing crop to the amount of water supplied.

irritability The magnitude of the response of an organism or organ to an external stimulus.

irruption A sudden and rapid increase in the size of a population.

Irvingaceae (order **Malpighiales**) A family of large **trees** with **buttress roots**. Leaves **alternate**, **simple**, **entire**, **ovate** to **elliptical**, with large **stipules** and short **petioles**. Flowers **actinomorphic**, **bisexual**, **pentamerous**, with free or slightly fused **sepals**, free **petals**, 10 free **stamens** in 1 **whorl**, **ovary superior**. **Inflorescence** a terminal or

axillary panicle. Fruit is a **samara**. There are three genera of ten species occurring in Africa and southeastern Asia.

isabelline Drab or grey.

ischium A rearward projection on the **ventral** side of the pelvis in **tetrapods**.

isidium An outgrowth of the **thallus** in some **lichens**, comprising both fungal hyphae (*see* hypha) and algal cells (*see* alga), that may break free, then acting as a **propagule**.

isobar A line drawn on a map that links places with the same **atmospheric pressure**.

isobaric map A map that shows the distribution of **atmospheric pressure** at a specified height above sea level.

isobaric slope *See* pressure gradient.

isobifacial *See* Acoraceae.

isodrosotherm A line drawn on a map that links places where the **dewpoint temperature** is the same.

isogamy **Fertilization** involving the fusion of similar **gametes** that occurs in some green algae (**Chlorophyta**), **Fungi**, and **Protozoa**. *Compare* oogamy.

isogeneic (syngeneic) Describes a **graft** in which the **scion** and **stock** are genetically identical.

isohel A line drawn on a map that links places that experience the same number of hours of sunshine.

isohume A line drawn on a map that links places of equal **humidity**.

isohyet A line drawn on a map that links places that receive the same amount of rainfall.

isokont Having flagella (*see* flagellum) of different lengths. *Compare* heterokont.

isolation species concept *See* biological species concept.

isoline A line on a map that links places with similar characteristics.

isomer One of several chemical compounds, all of which have the same molecular formula but different structural formulae, consequently differing in their physical and chemical properties.

isomerase An **enzyme** that catalyzes reactions involving the conversion of one **isomer** into another.

isomorphic Describes an **alternation of generations** in which the **sporophyte** and **gametophyte** are morphologically similar.

isomorphous substitution The substitution of one atom for another during the formation of a mineral, without causing any significant change in the crystal structure. In **clay minerals** this leads to the mineral having a negative charge, which is balanced by **cations** in the soil solution.

isophylly Having leaves that are all the same shape on the same plant.

Isopoda (pill bugs, slaters, woodlice) An order of crustaceans most of which have flattened bodies, lack a **carapace**, have a flexible, leathery **exoskeleton**, and are adapted for crawling. They have compound eyes without stalks and exchange gases by means of abdominal appendages (pleopods) that in terrestrial species resemble lungs. They inhabit many environments from the ocean floor to dry land. Most are scavengers or omnivores and some are parasites. There are about 10,000 species, of which about 4500 are marine, 500 freshwater, and 5000 live on land.

isoprene (2-methyl-1,3-butadiene) A colourless, volatile liquid produced by many trees and some legumes (**Fabaceae**). It is the structural base for other compounds (e.g. **terpenes**) and helps leaves combat heat stress. Released into the air, strong sunlight decomposes it, releasing ozone and contributing to the formation of photochemical smog.

isoprenoids *See* terpenoids.

isotach A line drawn on a map that links places that experience winds of the same speed.

isotherm A line drawn on a map that links places with the same air temperature.

isotonic Having an equal solute concentration (i.e. **osmotic pressure**) on both sides of a **membrane**. *Compare* hypertonic, hypotonic.

isotope A variety of a chemical element in which the atoms have the same number of protons and electrons as all other atoms of that element (so the atomic number is the same), but a different number of neutrons (so the atomic weight is different).

isotope hydrology The use of **isotopes** to identify and date bodies of water.

isotropic Possessing properties that have similar values in all directions.

Iteaceae (order **Saxifragales**) A family of **trees** and **shrubs** with small, **alternate**, **simple**, **elliptical** to **lanceolate** leaves with **petioles** and **deciduous** or persistent **stipules**. Flowers small, **actinomorphic**, **bisexual** or **unisexual** (plants **polygamonoecious**), with 5 persistent **connate sepals**, 5 persistent **valvate petals**, **5 stamens**, **ovary superior** to semi-**inferior** of 2 **carpels** and 2 **locules**. **Inflorescence axillary**, **racemose** or **paniculate**. Fruit is a **capsule**. There are 2 genera of 21 species scattered through temperate and tropical regions. A few cultivated as ornamentals.

iteroparity The condition of having more than one reproductive cycle during an organism's lifetime.

IUCN *See* International Union for Conservation of Nature.

ivy (*Hedera helix*) *See* Araliaceae, *Hedera*.

Ixioliriaceae (order **Asparagales**) A **monogeneric**, **monocotyledon** family (*Ixiolirion*) of **geophytes** with **bulb**-like **corms**. Leaves **alternate**, **linear**, with sheathing bases. Flowers **actinomorphic**, **bisexual**, with 3+3 equal **petaloid tepals**, 3+3 **stamens**, **ovary inferior** of 3 **carpels** and 3 **locules**. **Inflorescence** a **thyrse** with a few to many flowers, or **umbel**-like. Fruit is a **capsule**. There are three species occurring from Egypt to Central Asia.

Ixodidae A family of hard ticks (**Arachnida**) with bodies protected by a shield. Many are vectors of diseases. They have a single **nymph** stage. After mating, adult females feed, swelling as they do so, then fall from their hosts and lay eggs. There are 14 genera with 702 species found worldwide.

Ixonanthaceae (order **Malpighiales**) A family of **shrubs** or tall **trees** with **alternate**, **simple**, approximately **elliptical**, **entire** or **serrate** leaves with small, inconspicuous, **caducous stipules**. Flowers small, **actinomorphic**, **bisexual**, **pentamerous**, **sepals** more or less **imbricate**, 5, 10, 15, or 20 **stamens** in 1 **whorl**, **ovary superior** or slightly **inferior**, **syncarpous**, of 5 or 2 **carpels**. **Inflorescence** a terminal or **axillary cyme** or **raceme**. Fruit a **capsule**. There are 4–5 genera of 21 species with a pantropical distribution.

J

Jacaranda (family **Bignoniaceae**) A genus of **shrubs** to tall **trees** with **opposite**, **bipinnate**, sometimes **pinnate** or **simple** leaves. Flowers with a 5-lobed, blue to lilac **corolla**, and **staminodes** longer than the **stamens**. **Inflorescence** a **panicle** or cluster. Fruit is a **capsule**. There are 49 species occurring throughout tropical and subtropical America and the Caribbean. Many are cultivated as ornamentals.

jack dam *See* check dam.

jackdaws *See* Corvidae, *Corvus monedula*.

jackfruit (*Artocarpus*) *See* Moraceae.

jack pine (*Pinus banksiana*) *See* serotiny.

Japanese maple (*Acer palmatum*) *See Acer*.

Japanese umbrella pine (*Sciadopitys verticillata*) *See* Sciadopityaceae.

japweed (wireweed, strangleweed) The brown seaweed *Sargassum muticum* (Phaeophyta) that is believed to have originated in Japan and that was first detected in Britain in 1973. It is now established on European and North American coasts, growing below the low-tide mark.

Japygidae *See* forceps.

jarrah dieback *See Phytophthora cinnamomi*.

jasmine (*Jasminum*) *See* Oleaceae.

jasmonic acid A **hormone** that helps plants defend against insect attacks and **pathogen** infections, and also regulates fertility and fruit ripening.

jays *See* Corvidae.

jelly fungi *See Auricularia*.

Jerusalem artichoke (*Helianthus tuberosus*) *See Helianthus*.

jet-effect wind *See* mountain-gap wind.

jet stream A winding ribbon of fast-moving air in the upper **troposphere** or lower **stratosphere**. It is typically hundreds of kilometres wide, several kilometres deep, and extends for thousands of kilometres.

jewelweed *See Impatiens*.

jezebels *See* Pieridae.

JH *See* juvenile hormone.

jimsonweed (*Datura stramonium*) *See Datura*.

Joinvilleaceae (order **Poales**) A **monogeneric**, **monocotyledon** family (*Joinvillea*) of **perennial herbs** with **distichous**, **linear** to **lanceolate** leaves with open sheaths. Flowers **actinomorphic**, **bisexual**, **trimerous**, **ovary superior**. **Inflorescence** terminal, **paniculate**. Fruit is a **drupe**. There are two species occurring in the Malay Peninsula, Indonesia, and the Pacific Islands.

jojoba *See* Simmondsiaceae.

jonquil (*Narcissus*) *See* Amaryllidaceae.

J-shaped growth curve A curve on a graph that depicts the change in population density of a species that enters a new **environment** with abundant resources. The population size and density grow rapidly until an environmental factor checks it, and the population crashes. This pattern is typical of the population cycles of some insects and the sudden appearance and disappearance of **algal blooms**.

Juan Fernández floral region The area around the island of Juan Fernández and the Desventuradas Islands, off the coast of Chile, and part of the **Neotropical region**. More than 60 percent of the plants are **endemic**.

Juglandaceae (order **Fagales**) A family of **deciduous** sometimes **evergreen trees** (walnut, hickory, pecan, wingnut) with **alternate** occasionally **opposite**, **pinnate**, **exstipulate**, usually aromatic leaves. Flowers **bracteate**, **unisexual** (plants **monoecious**), **perianth** typically 4-lobed but often reduced or absent, 3–40 free **stamens**, **ovary inferior** of 2 fused **carpels** with 1 **locule**. **Staminate** flowers in pendulous, **catkin**-like **inflorescences** on previous year's growth, **pistillate** flowers in erect or pendulous **spikes** on new growth. Fruit a **drupe**-like **nut** or 3-winged and **samara**-like. There are 7–10 genera with 50 species occurring in northern temperate regions, Malesia, and central and South America. Many cultivated for timber (e.g. ***Juglans***), nuts, and oils.

Juglans (family **Juglandaceae**) A genus of tall, **deciduous trees** with **compound** leaves. **Inflorescences** are **catkin**-like, **staminate** and **pistillate** flowers occurring on the same tree, male flowers pendulous, female flowers erect. The fruit is a **drupe**-like **nut**. There are 21 species occurring throughout the northern temperate region. Many are cultivated, especially *J. regia* (Persian walnut), *J. nigra* (black walnut), and *J. cinerea* (butternut).

jugular Relating to the neck or throat.

jujube (*Zizyphus jujuba*) *See* Rhamnaceae.

jumbo red worm *See Dendrodrilus rubidus.*

jumper *See Dendrodrilus rubidus.*

jumping plant lice *See* Psyllidae.

jumping red worm *See Dendrodrilus rubidus.*

Juncaceae (order **Poales**) A **monocotyledon** family of **perennial** occasionally **annual**, grass-like **herbs** (rushes) with erect or horizontal **rhizomes** or erect stems. Leaves in 3 or sometimes 2 ranks, **linear** or **filiform**. Flowers **actinomorphic**, usually **bisexual**, with small **tepals** in 2 **whorls** of 3, usually 6 **stamens** in 2 whorls alternating with the tepals, **ovary superior** of 3 **carpels** and usually 3 **locules** but occasionally 1. **Inflorescence** terminal or lateral, with many flowers, compound, of open **panicles** or head-like or **spike**-like. Fruit is a **capsule**. There are 7 genera with 430 species with worldwide distribution.

Juncaginaceae (order **Alismatales**) A **monocotyledon** family of **annual** or **perennial herbs** with **rhizomes** found in marshy places; usually **emergent** but some floating. Flowers usually **actinomorphic**, **bisexual** or **unisexual** (plants **monoecious**, **dioecious**, or **polygamous**), 1–4 **tepals** or 6 in 2 **whorls**, 1, 4, or 6 **stamens**, **ovary superior** of 1, 3, or 6 free or partly joined **carpels**. **Inflorescence** a **spike** or **raceme**. Fruit a **schizocarp** splitting into **achenes**. There are 3 genera of 15 species with a cosmopolitan, mainly coastal distribution.

Junco hyemalis (dark-eyed junco) A sparrow-like bird, 130–175 mm long with a wingspan of 180–250 mm, that has a grey or brown back and wings, grey head, neck and breast, and a white underside. It inhabits woodland areas with abundant ground cover throughout most of temperate North America.

Juncus (family **Juncaceae**) A genus of usually **perennial**, tufted, **monocotyledon herbs** (rush) with **entire** leaves with open sheaths growing from the base, occasionally 1–3 leaves **cauline**, **blade terete** or flat, with channels, compressed, or reduced to **mucro** on the sheath, often with **auricles**. Flowers **bisexual** rarely **unisexual**, **tepals** in 2 **whorls**, 3–6 **stamens**, **ovary** with 1 or 3 **locules**. Flowers solitary or in clusters, with 1 papery **bract**, sometimes 2 papery **bracteoles**. Fruit is a **capsule**. There are about 300 species with worldwide distribution. Some cultivated as pond ornamentals.

June berry *See Amelanchier.*

juniper *See Juniperus.*

Juniperus (family **Cupressaceae**) A genus of **evergreen**, coniferous **trees** and **shrubs** (juniper) with needle-like (often juvenile) or scale-like, **imbricate** (mature) leaves. Plants are **monoecious** or **dioecious**. Female **cones** have 3–8 fleshy scales that fuse together to form a blue, occasionally orange or reddish brown, **berry**-like structure containing up to 12 seeds. These cones are often aromatic and used as a spice. There are 50–67 species distributed throughout the Northern Hemisphere. Some cultivated for timber or fruits.

Jussieu, Antoine-Laurent de (1748–1836) A French physician, botanist, and taxonomist who introduced a scheme of **natural classification**, describing 15 classes, 100 families, and many genera, all of which he saw as forming a continuous series with no clear boundaries.

jute The fibre obtained from the **phloem** tissue (**bast**) of *Corchorus* species (**Malvaceae**) and used to make hessian and strong cordage.

juvenile hormone (JH) A **hormone** secreted by insects (**Insecta**). It regulates development, advancement to the next stage being determined by the amount of JH present, the less JH there is the more adult the insect becomes. In many female insects it also regulates the production of eggs.

juvenile water Water formed in **magma** and released into the air; juvenile water has never before been in the atmosphere.

K

K *See* potassium.

kame A mound of sand and gravel, with steep sides and signs of it having collapsed near the margins. It is a glacial deposit that accumulated in a depression in the surface of a stagnant glacier that lay over a depression in the underlying surface and collapsed when the ice melted.

kampfzone A type of **subalpine forest** that grows between the **timber line** and **tree line** and consists of dwarfed or prostrate **trees**.

kangaroo paw (*Anigozanthos manglesii*) *See* Haemodoraceae.

karrikins A group of plant **growth regulators** that are found in smoke from burning plant material and that trigger seed **germination**. Karrikins from forest fires stimulate the germination of seeds on the ground.

karyogamy (caryogamy) The fusion of two cell nuclei to form a **zygote**.

karyotheca *See* nuclear envelope.

karyotype The number and appearance of the full complement of **chromosomes** in a **eukaryote** cell.

kastanozems Soils that have a **mollic horizon** more than 20 cm below the surface, and calcium compounds concentrated within 100 cm of the surface. Kastanozems are a reference soil group in the **World Reference Base for Soil Resources**.

katabaric (katallobaric) Describes meteorological phenomena that are associated with a decrease in **atmospheric pressure**.

katabatic wind (drainage wind, fall wind, gravity wind) A wind that blows down a slope. It comprises cold, dense air moving under the force of gravity.

katallobaric *See* katabaric.

keel (carina) Any sharply ridged structure resembling the keel of a boat, e.g. the two lower petals of a pea (*Pisum*) flower, which are fused, forming a boat-shaped structure around the reproductive organs.

keeled slug *See Tandonia budapestensis.*

kelp Any of about 30 genera of large, brown seaweeds (Laminariales) that grow below the low-tide level, often forming kelp 'forests' that support a wide variety of organisms. The name referred originally to the ash obtained by burning these seaweeds, which was used as **fertilizer** and in making glass and soap.

Kentucky bluegrass (*Poa pratensis*) *See Poa.*

keratin A group of **proteins** that form the base for hair, wool, nails, claws, and other structures arising from the **epidermis**.

Kerry slug *See Geomalacus maculosus.*

Kerry spotted slug *See Geomalacus maculosus.*

keystone species A species that exerts a strong influence on the character of an **ecosystem**, such that its removal destabilizes the system.

Khaya (mahogany) *See* Meliaceae.

kidney An organ in vertebrates and some invertebrates which controls the excretion or retention of water and the excretion of metabolic wastes.

killing frost A fall in temperature to below freezing that kills all but the hardiest plants.

kinase An **enzyme** that catalyzes reactions that transfer phosphate groups from a donor molecule, e.g. **adenosine triphosphate** (ATP) to a **substrate**.

kinesis A change in its speed of movement of a **cell** or organism in response to a stimulus, the direction of movement being random and unconnected to the direction of the stimulus.

kinetic energy The energy of motion, defined as the amount of work that a moving body could do if it were brought to rest. It is equal to $\frac{1}{2}mv^2$, where m is the mass of the moving body and v its speed. For a rotating body it is $\frac{1}{2}I\Omega^2$, where Ω is the angular velocity and I the moment of inertia. *See* potential energy.

kinetin A plant **hormone** that is a degradation product of animal **DNA**, first isolated from the sperm of herrings. In the presence of **auxin** it stimulates cell division and is sometimes used to induce the formation of **callus** in tissue **cultures** and to regenerate shoot tissues.

kinetochore The **protein** site at which a **chromosome** attaches to the **mitotic spindle** during **mitosis**. It is a transient structure that disappears once the chromosomes have segregated.

king bolete *See Boletus edulis.*

King Charles's apple *See* oak-apple gall.

kingdom In **taxonomy**, one of the major groups in which organisms are placed. In the **three-domain system**, kingdoms are ranked below the domains. In the **five-kingdom system**, the kingdoms are: **Bacteria**, **Protista**, **Animalia**, **Fungi**, and **Plantae**.

king sago palm (*Cycas revoluta*) *See* Cycadaceae.

Kirkiaceae (order **Sapindales**) A **monogeneric** family (*Kirkia*) of **deciduous trees** with **alternate** leaves typically clustered at the ends of shoots; leaves **compound**, **imparipinnate** with 20–30 pairs of **opposite**, **lanceolate** to suborbicular, **entire** to **serrate leaflets**. Flowers **actinomorphic**, **tetramerous**, usually dimorphic (*see* dimorphism), **unisexual** (plants **monoecious** or **polygamonoecious**), 4 **sepals**, 4 **petals**, 4 **stamens** alternating with the petals, **ovary superior** of 4–8 **connate carpels** each with 1 **locule**. **Inflorescence** an **axillary thyrse** of **dichotomous cymes**. Fruit a **schizocarp**. There are eight species occurring in tropical and southern Africa and Madagascar. *Kirkia acuminata* grown in South Africa as a hedge and for its wood.

Kiwi vine (Kiwi fruit, *Actinidia chinensis*) *See* Actinidiaceae.

kleptoparasitism **Parasitism** in which the parasite steals food from others.

knot (kt) A measure of speed, equal to 1 nautical mile (nm) per hour. Wind speed is often reported in knots. 1 kt = 1.852 km/h = 1.15 mph.

knotted wrack (*Ascophyllum nodosum*) *See* egg wrack.

Koeberliniaceae (order **Brassicales**) A **monogeneric** family (*Koeberlinia*) of **xerophyte shrubs** with spiny, photosynthesizing stems. Leaves, appearing only in the rainy season, are **simple**, **elliptic**, **entire**. Flowers **bracteate**, **tetramerous**, with 4 free **sepals** in 2 **whorls**, 4 free, overlapping **petals** in 1 whorl, 8 **stamens**, **ovary** with 2 fused **carpels**. **Inflorescence** is a short **axillary raceme**. Fruit a **berry**. There are two species occurring in central and southwestern North America and Bolivia.

Köppen climate classification A scheme for classifying climates according to their average summer and winter temperatures and aridity. The scheme was first proposed in 1918 by the German meteorologist Wladimir Peter Köppen (1846–1940) and the final version appeared in 1936. It is the scheme most widely used by geographers.

kraits *See* Elapidae.

Krameriaceae (order **Zygophyllales**) A **monogeneric** family (*Krameria*) of prostrate **perennial**, **hemiparasite herbs** and **shrubs** with small, **alternate**, **simple** or 3-foliate, **elliptical** to **linear**, **entire** leaves that are **sessile** or with **petioles**. Flowers inverted, pea-like, **zygomorphic**, **bisexual**, 4- or 5-merous, with 4 or 5 free unequal, elliptic, coloured **sepals** in spirals the lowest the largest, 5 (rarely 4) dimorphic (*see* dimorphism) **petals**, the 2 lower ones modified into fleshy oil-secreting **glands**, 4 **stamens**, **ovary superior** of 2 **carpels** one aborting early and 1 **locule**. **Inflorescence** a solitary, **axillary** flower or terminal **raceme** or **panicle** with each flower subtended by a pair of **bracteoles**. Fruit a **capsule**. There are 18 species occurring in southwestern North America, Central America, South America, and the Caribbean. *Krameria lappacea* (rhatany) cultivated for dyes, cosmetics, and an ingredient of toothpaste.

Krebs' cycle *See* citric acid cycle.

krummholz Literally, crooked wood (German), a term which describes the stunted, gnarled, often dwarfed trees, mainly conifers, that grow in the **kampfzone**.

***K*-selection** Evolutionary selection that maximizes the competitive strength of species living in stable environments. The resulting lifestyle involves producing few offspring that undergo a prolonged period of development, and a high survival rate for offspring. *K* stands for the **carrying capacity** of the **environment** for populations with an **S-shaped growth curve**.

kt *See* knot.

kurtosis A measure of the extent to which a distribution has a peak.

***k*-value** A measure of the number of individuals of a species that are lost at each stage in its life cycle, i.e. the 'killing power' of the environment. It is calculated as $k = \log_{10}a_x - \log_{10}a_{x+1}$, where a_x is the number of individuals at the beginning of each stage (time x), and a_{x+1} is the number at time x + 1, at the end of the stage.

Kylar *See* daminozide.

L

Labiatae *See* Lamiaceae.

Laboulbeniomycetes A group of **ascomycete fungi** all of which are **obligate parasites** of insects (**Insecta**) and other arthropods (**Arthropoda**) and live on the outside of their hosts. There is no true **mycelium**, but a small **haustorium** penetrates the host **integument**. **Fruiting bodies** are usually less than 1 mm across. There are more than 2000 species found worldwide.

Laburnum (family **Fabaceae**) A genus of small, **deciduous trees** with **trifoliate**, clover-like leaves and yellow, pea-like flowers borne in pendulous **racemes**. There are two species occurring in southern Europe. They are widely cultivated as ornamentals and for their wood. All parts of the plant are poisonous.

Laccaria laccata (deceiver, waxy laccaria) A species of **agaric fungi** with a mushroom-like **fruiting body**, the **pileus** up to 60 mm across and irregular **gills**. The colour can be pink, red, orange, or brown, the variability of its appearance accounting for its common name deceiver. It occurs throughout Europe and North America in **heathland** and woodland, and is edible but said to be tasteless.

Lacertilia ('true' lizards) A suborder of reptiles (**Reptilia**) that have overlapping scales but are members of neither the order Rhynchocephalia (tuatara) nor the suborder **Serpentes** (snakes). They have well-developed legs, although some species have lost their legs in the course of their evolution, and long tails capable of **autotomy**, and external ears. Otherwise lizards have no distinguishing features, since they are defined as not being snakes. There are more than 5000 species distributed on all continents except Antarctica.

lacewings *See* Neuroptera.

Lacistemataceae (order **Malpighiales**) A family of **shrubs** and small **trees** with **alternate**, **simple**, more or less **elliptical**, **coriaceous**, **entire** leaves with short **petioles**. Flowers small, more or less **zygomorphic**, **hermaphrodite** or **unisexual** (plants **monoecious** or **andromonoecious**), 1 fused or 2–6 free **sepals** or none, **petals** absent, 1 **stamen**, **ovary superior** of 2–3 **carpels** with 1 **locule**. **Inflorescence racemose** to densely **spicate**. Fruit is a **capsule**. There are 2 genera of 14 species occurring in Jamaica and Central and northwestern South America.

Lactarius (milk cap) A genus of mycorrhizal (*see* mycorrhiza) **agaric fungi** that exude a white liquid when the **fruiting body** or **gills** are damaged, hence the common name. The fruiting bodies are typically more than 100 mm across, with a **pileus** that appears woolly. There are about 400 species with a **cosmopolitan distribution**. Some are edible.

Lactuca (family **Asteraceae**) A genus of **annual**, **biennial**, and **perennial herbs** (lettuce) with underground **stolons**, **taproots**, or tuberous roots, leaves spirally arranged, basal leaves usually in a rosette,

sessile or with **petioles**, **entire** or **pinnatisect**. **Inflorescence** a **corymb**-like, **spike**-like, or pyramidal **panicle**. Fruit is an **achene**. There are about 100 species with worldwide distribution but especially in temperate Eurasia. Many are weeds. Cultivated lettuce is *L. sativa*.

lacuna 1. A gap between cells that is filled with air. **2.** *See* leaf gap.

lacunose With a pitted or indented surface.

lacustrine deposit An accumulation of sediment deposited on the bed of a lake.

LAD *See* last appearance datum.

lady beetles *See* Coccinellidae.

ladybirds *See* Coccinellidae.

ladybugs *See* Coccinellidae.

lady cows *See* Coccinellidae.

lady in a bath *See Lamprocapnos spectabilis.*

lady's mantle *See Alchemilla.*

Laelapidae (bee mites) A family of mites comprising about 90 genera, of which 43 are parasites of **Arthropoda**, 35 are parasites of mammals, and 10 are free-living predators dwelling in the soil. The genera *Euvarroa*, *Varroa*, and *Tropilaelaps* are **obligate** parasites of bees (***Apis***) and cause much harm. There are about 1300 species found worldwide.

laevigate Smooth, as though polished.

LAI *See* leaf-area index.

lake breeze A cool **breeze** that blows on warm days from a lake toward the shore, when **convection** over land reduces the surface **atmospheric pressure** and air moves shoreward from over the lake to replace it.

lake effect A modification in the characteristics of air as it crosses a large lake that is entirely surrounded by land. Air reaching the **lee** shore is moister than on the windward side, and the temperature more moderate.

lake-effect snow Snow that falls, often heavily, on the **lee** side of a large lake entirely enclosed by land. Air chilled by contact with the cold ground is warmed as it crosses the unfrozen lake and accumulates moisture. When it reaches the cold ground on the lee side its moisture condenses and falls as snow.

lake forest A coniferous forest, transitional between the **boreal forest** to the north and the **deciduous** forest to the south, that occurs in eastern North America.

Lamarck, Jean-Baptiste Pierre Antoine de Monet, chevalier de (1744–1829) A French naturalist and botanist who proposed a theory of evolution based on the inheritance by progeny of characteristics acquired by one or both parents during their lifetime, a theory later adopted by Trofim **Lysenko**.

lamb's lettuce *See Valerianella.*

lamella (pl. lamellae) **1.** A leaf or leaf **blade**. **2.** One of the **membranes** comprising a **thylakoid** disc. **3.** One of the membranes of a fungal **gill**.

Lamiaceae (order **Lamiales**) A family, formerly known as Labiatae, of **annual** and **perennial herbs**, **shrubs**, and **trees**, with some climbers, with **opposite**, occasionally **alternate** or in **whorls**, **simple**, toothed or deeply divided leaves. Flowers **zygomorphic** or secondarily **actinomorphic**, **gamopetalous**, **calyx** with usually 4 sometimes 5 teeth or lobes, or 2-lipped. **Corolla** usually strongly 2-lipped, usually 4 **stamens**, **ovary superior** of 2 **carpels** with 2 or 4 **locules**. **Inflorescence cymose**. Fruit is a **schizocarp**, **berry**, or **drupe**. There are 236 genera of 7173 species occurring worldwide. Many species cultivated for timber (e.g. *Tectona*, teak), as culinary and medicinal herbs, e.g. mint (***Mentha***), sage (***Salvia***), thyme (*Thymus*), marjoram

(***Origanum***), rosemary (*Rosmarinus*), basil (*Ocimum*), catmint (*Nepeta*), and for ornament.

Lamiales An order of plants that comprises 24 families or 1059 genera and 23,810 species. *See* Acanthaceae, Bignoniaceae, Byblidaceae, Calceolariaceae, Carlemanniaceae, Gesneriaceae, Lamiaceae, Lentibulariaceae, Linderniaceae, Martyniaceae, Mazaceae, Oleaceae, Orobranchaceae, Paulowniaceae, Pedaliaceae, Phrymaceae, Plantaginaceae, Piocospermataceae, Schlegeliaceae, Scrophulariaceae, Stilbaceae, Tetrachondraceae, Thomandersiaceae, and Verbenaceae.

lamina A flat structure, like a sheet.

Laminaria digitata *See* oarweed.

laminarin A **polysaccharide** storage product found in **brown algae** (Phaeophyta).

laminate 1. Broad and flat. **2.** Comprising thin layers.

Lamprocapnos spectabilis (family **Papaveraceae**) A species of **perennial herbs**, formerly known as ***Dicentra*** *spectabilis*, that have fleshy stems arising from **rhizomes**, bearing **compound** 3-lobed leaves, and **racemes** of heart-shaped flowers with pink outer and white inner **petals**. The plants, native to eastern Asia, are cultivated as old-fashioned bleeding heart, also known as Venus's car, lady in a bath, Dutchman's trousers, and lyre-flower.

Lampropeltis getula (eastern king snake, chain kingsnake, common kingsnake) A species of **Colubridae** with seven subspecies that vary in size and appearance. They are 610 mm–1.5 m long, glossy, black, with yellow or occasionally white bars across the body and yellow or white spots on the head. They inhabit forests, grasslands, and urban areas, and feed mainly on other snakes, including venomous species, but also on other small animals. They help control populations of venomous snakes and are therefore beneficial for humans, and are popular as pets, being docile when handled. They occur throughout most of North America.

Lanariaceae (order **Asparagales**) A **monotypic**, **monocotyledon** family (*Lanaria plumosa*) of **perennial herbs** that are densely covered with branching, woolly hairs. Leaves spirally arranged or **distichous** and **linear**. Flowers **actinomporphic**, **bisexual**, with 3+3 fused **tepals**, 3+3 **stamens**, **ovary inferior** of 3 **carpels** and **locules**. **Inflorescence** a dense **corymbose panicle**. Fruit is a **capsule**. The plant is **endemic** to the **fynbos** of Cape Province, South Africa.

Lanarkshire disease *See Phytophthora fragariae.*

lanate With matted hairs, giving a woolly appearance.

lanceolate Of a leaf, broad and tapering to a point, like the blade of a lance.

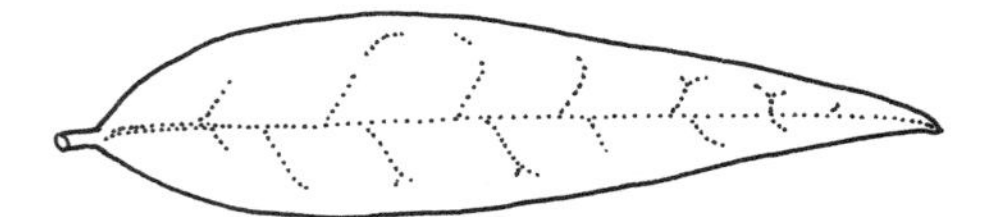

A lanceolate leaf.

land and sea breezes Breezes that blow in summer on the shores of the sea and large lakes. Dry land warms and cools more rapidly than water. During the day air over land becomes warmer than air over the water, causing it to rise by **convection** and drawing cooler air from over the water shoreward as a sea breeze. At night the land becomes cooler than the water and the reverse occurs, with a land breeze blowing toward the sea or lake.

land plants *See* Embryophyta.

landrace A local crop variety or **cultivar** that has developed largely naturally and that is maintained by **cultivation**.

landscape The visible features of an area of land.

landscape ecology The study of the **ecology** of landscapes for the purpose of informing planning decisions relating to landscape architecture.

landslide *See* slide.

landslide dam *See* debris dam.

land tortoises *See* Testudinidae.

La Niña The opposite of **El Niño**, when the southeasterly trade winds strengthen over the equatorial South Pacific, intensifying the South Equatorial Current and deepening the pool of warm water around Indonesia. *See* ENSO.

lanose Woolly.

lapse rate The rate at which the temperature of a body of air decreases as the air rises. Lapse rates vary according to local conditions and the **humidity**. *See* dry adiabatic lapse rate, environmental lapse rate, saturated adiabatic lapse rate.

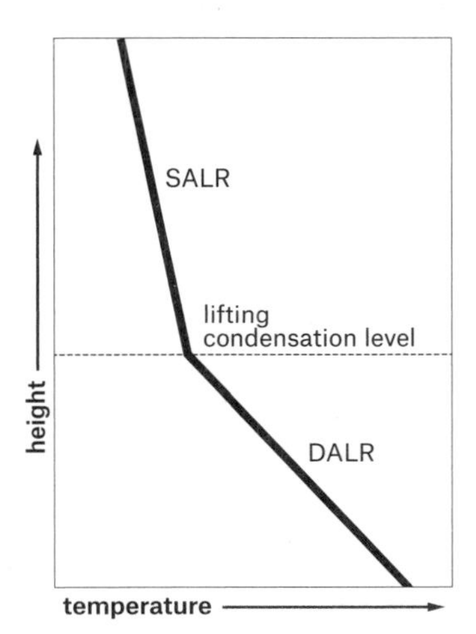

The lapse rate is the change of temperature with height. As the diagram shows, dry air cools at the dry adiabatic lapse rate (DALR), which is more rapid than the saturated adiabatic lapse rate (SALR), the change from DALR to SALR occurring at the lifting condensation level.

LAR *See* leaf-area ratio.

larch *See Larix*, Pinaceae.

Lardizabalaceae (order **Ranunculales**) A family of **deciduous** or **evergreen lianas** with leaves **alternate**, or in **fascicles**, **compound**, **pinnate**, with 3 **leaflets**, or **palmate** with up to 9 leaflets, **exstipulate**. Flowers **actinomorphic**, usually **unisexual** (plants **monoecious** or **dioecious**), **trimerous**, 3, 4, or 6 free, **petaloid sepals**, 6 free **petals** in 2 **whorls**, 6 free or 3–8 **connate stamens**, **ovary superior** of 3–6 (sometimes 3–12 free **carpels**. **Inflorescence axillary raceme**, **corymb**, or **umbel**. Fruit is a **follicle**. There are 7 genera of 40 species occurring in southeastern Asia and in Chile. Some species used medicinally, others cultivated for ornament.

large bulb fly *See Merodon equestris.*

large cabbage white *See Pieris brassicae.*

large fruit-tree tortrix moth *See Archips podana.*

large narcissus fly *See Merodon equestris.*

large red slug *See Arion rufus.*

large willow aphid *See Tuberolachnus salignus.*

Larix (family **Pinaceae**) A genus of large, **deciduous**, coniferous **trees** (larch) with soft, needle-like leaves borne singly and spirally arranged on long shoots or in dense clusters on short shoots. **Cones** are erect, oval, and persistent. There are 10–15 species occurring in the cooler parts of the temperate Northern Hemisphere. Some grown for timber.

larkspur *See Delphinium.*

Larrea tridentata (creosote bush) *See* Zygophyllaceae.

larva A juvenile animal that is **motile** and finds its own food between the time it hatches from an egg and undergoes **metamorphosis** to become adult, usually of a very different form.

larviparous Describes an animal that reproduces by producing larvae rather than laying eggs.

larynx In **tetrapods**, an expanded region of the **trachea** where it joins the **pharynx**, which contains folded membranes (vocal cords) that produce sound by vibrating as air is forced across them.

last appearance datum (LAD) Historically, the last recorded occurrence of a **taxon**.

late blight of potato *See Phytophthora infestans.*

latent heat Energy that is absorbed and released when a substance changes phase between solid, liquid, and gas. For water, the latent heat of freezing and melting is 334 joules per gram (J/g); of **condensation** and vaporization 2501 J/g; and of **deposition** and **sublimation** 2835 J/g (the sum of the latent heats of condensation/vaporization and freezing/melting). The **absorption** and release of latent heat does not alter the temperature of the substance, i.e. the heat is hidden, or latent.

L

lateral gene transfer *See* horizontal gene transfer.

lateral meristem **Meristem** tissue in the **cambium**.

laterite A **weathering** product that consists of **clay minerals**, some **silica**, and hydrated aluminium and iron oxides and hydroxides. It forms in humid tropical environments.

latex A liquid produced by many plants and some **agaric fungi** containing many substances in solution or suspension. Latex is often white but may be clear, red, or yellow. Certain **dicotyledons** produce latex containing caoutchouc (rubber).

Lathyrus (family **Fabaceae**) A genus of **annual** and **perennial** climbing **herbs** with **alternate**, **pinnate** leaves, each commonly with 1 or 2 pairs of **leaflets** and a terminal **tendril**. Flowers are pea-like, borne in clusters. There are 159 species occurring throughout temperate regions. Many are cultivated as sweet peas and vetchlings.

laticifer A **cell** or group of cells containing **latex**.

latifoliate **Broad-leaved**.

latitudinal vegetation zones Geographical belts, parallel to the equator, in which the vegetation is characteristic of a climatic regime, e.g. savanna grassland occurs in the seasonal tropics, approximately 10°–20° N and S, tropical rain forest occurs in the humid equatorial zone.

latosol A soil that is red in colour and composed of fine mineral grains.

lattice structure The regular, three-dimensional arrangement of atoms in a crystal.

Lauraceae (order **Laurales**) A family of large, **evergreen trees**, **shrubs**, and climbing parasites with **alternate** to **opposite**, **simple**, **entire**, **coriaceous**, **exstipulate** leaves. Flowers **bisexual** or **unisexual** (plants **monoecious** or **dioecious**), **perianth** of **whorls** of 3 free or 2–4 fused **tepals**, 9 free or 3–12 fused **stamens**, some reduced to **staminodes** or absent, **ovary superior** of 1 **carpel**. **Inflorescence umbel**-like, **racemose**, or **cymose**. Fruit a **berry** or **drupe**-like. There are about 50 genera of 2500 species of pantropical distribution. Some are commercially important, e.g. *Persea americana* (avocado pear), *Laurus nobilis* (bay laurel), *Cinnamomum* (source of cinnamon and camphor).

Laurales An order of plants comprising 7 families of 91 genera with 2858 species. *See* Atherospermataceae, Calycanthaceae, Gomortegaceae, Hernandiaceae, Lauraceae, Monimiaceae, and Siparunaceae.

Laurasia The northern supercontinent that resulted from the breaking apart of **Pangaea** during the Late Triassic epoch (228–199.6 million years ago) along the line of the North Atlantic Ocean and **Tethys Sea**. Laurasia comprised what became North America, Greenland, and Eurasia. *Compare* Gondwana.

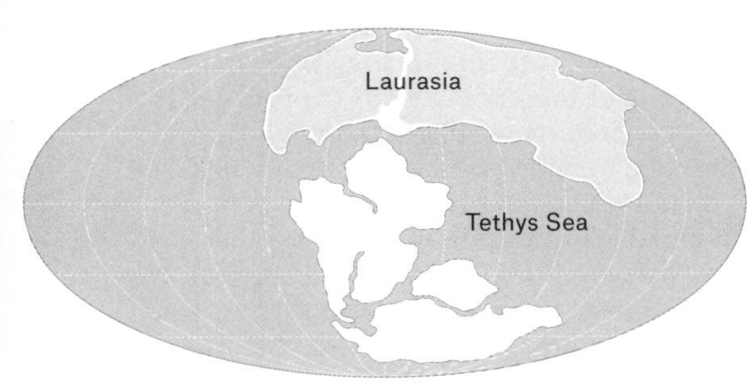

A map of the world as it was 200 million years ago, at the very end of the Triassic period. Laurasia comprised what are now all the continents of the Northern Hemisphere.

Laurus (family **Lauraceae**) A genus of small, **evergreen trees** with **alternate**, **simple**, **entire**, **elliptical** to **ovate** or **lanceolate**, leathery, **glabrous**, aromatic leaves. Flowers borne in **axillary umbels**. Fruit is a **berry**. There are three species, *L. nobilis* (bay laurel), *L. novocanariensis* (Canaries laurel), and *L. azorica* (Azores laurel). Bay laurel was used in ancient Greece to make laurel wreaths.

Laurus nobilis (bay laurel) *See Laurus*, Lauraceae.

Lavandula (family **Lamiaceae**) A genus of **xerophyte shrubs** and **perennial herbs** with **simple**, **pinnate**, or pinnately toothed leaves, usually bearing **indumentum** containing **essential oils**. Flowers with a tubular **calyx** and tubular **corolla**, usually of 5 lobes, borne in **whorls** held on **spikes**. There are 39 species occurring throughout northern warm temperate regions. Several cultivated as ornamentals, culinary herbs, or for their perfumed oils. *Lavandula angustifolia* is lavender.

lavender (*Lavandula angustifolia*) See *Lavandula*.

laver The edible seaweed *Porphyra umbilicalis* (purple laver), a **red alga** (Rhodophyta) that grows attached to rocks in the intertidal zone and that is widely cultivated. Its **thallus** is one cell thick. It has a high mineral content.

Lawsonia inermis (henna) *See* Lythraceae.

Lawson cypress (*Chamaecyparis lawsoniana*) *See Chamaecyparis*.

layer cloud A **stratiform** cloud resembling a sheet.

LC50 (lethal concentration fifty) The aerial concentration of a substance that will kill 50 percent of a group of test animals during a single exposure, typically of one to four hours.

LD50 (lethal dose fifty, median lethal dose) The size of the single dose of a substance that within a specified time will kill 50 percent of a group of test animals exposed to it by any route other than inhalation.

leachate The solution that forms as water percolates through soil or any other permeable medium.

leached soil Soil from which a substantial proportion of soluble materials, including plant nutrients, have been removed by **leaching**.

leaching The removal of soil materials in solution.

leader The actively growing tip of a main shoot, especially of a **tree**.

leaf An organ borne at a **node** on a plant stem, subtending an **axillary bud** in the **petiole**. It is usually green, thin, and expanded, and comprises a petiole and **lamina**, and is the principal site of **photosynthesis**.

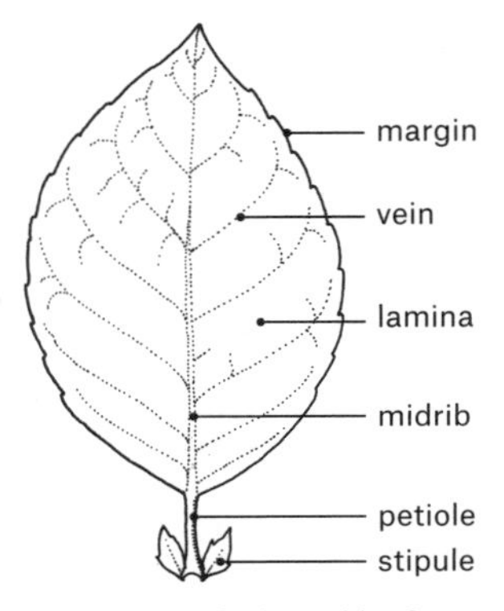

The parts of a broad leaf.

leaf-area index (LAI) The total area of all the leaves of a plant that are exposed to sunlight as a proportion of the ground area beneath the plant. An LAI of 3 means the exposed leaf area is 3 times greater than the area of ground below.

leaf-area ratio (LAR) The total area of all the surfaces of a plant that perform **photosynthesis** as a ratio of the unit dry weight of the plant, in m^2/g. It is the proportion of the plant **biomass** devoted to **photosynthesis**.

leaf beetles *See* Chrysomelidae.

leaf blister *See Taphrina.*

leafbugs *See* Miridae.

leaf curl *See Taphrina.*

leaf gap (lacuna) A break in the **stele** of a plant stem, filled with **parenchyma** tissue, through which a leaf grows.

leaflet Each individual small leaf in a **compound** leaf, i.e. a leaf that does not subtend a **bud** in its **petiole axis**.

leaf miner The larva of an insect that feeds on leaf tissue, inhabiting the inside of the leaf, where it is protected from predators and has access to cells with the least **lignin**. Its tunnels are visible as pale marks on the leaf.

leaf-rollers *See* Tortricidae.

leaf trace A strand of **vascular tissue** that grows through a **leaf gap** connecting a leaf to the plant stem.

leaf weevils *See Phyllobius.*

least willow *See* arctic-alpine species.

leatherjacket *See* Tipulidae.

leather-winged beetles *See* Cantharidae.

Lecanicillium longisporum A species of **ascomycete fungi** that is an **obligate parasite** of insects (**Insecta**); it produces sticky **spores** that adhere to the bodies of insects, then invade and eventually kill the host. The fungus is marketed commercially as an agent of **biological control** for aphids (**Aphididae**).

Lecanicillium muscarium A species of **ascomycete fungi** that is an **obligate parasite** of insects (**Insecta**); it produces sticky **spores** that adhere to the bodies of insects, then invade and eventually kill the host. The fungus is marketed commercially as an agent of **biological control** for whiteflies (**Homoptera**, especially ***Trialeurodes vaporariorum***, and thrips (**Thysanoptera**).

Lecanora *See* lecanorine.

lecanorine Describes apothecia (*see* apothecium) that are surrounded by a margin of tissue the same colour as the **thallus**. Such apothecia occur in *Lecanora* species of **lichens**.

Lecanoromycetes A class of **ascomycete fungi** in which the **fruiting body** is usually a typical **apothecium**, and which live as **mycobionts** in **lichens**; the great majority of **lichenized** fungi belong to this class.

lecideine (lecideoid) Describes apothecia (*see* apothecium) that lack a margin of thalline (*see* thallus) tissue, although parts of the thallus may be attached as a veil or collar, but that have a strongly darkened rim (the proper margin) formed from the upper edge of the **excipulum**. Such apothecia occur in *Lecidia* species of **lichens**.

Lecidia *See* lecideine.

lecidioid *See* lecideine.

lecotropal Horseshoe-shaped.

lectin A member of a group of **proteins** that bind **carbohydrates**. They occur in plants, especially **legumes**, certain **Fungi** and **lichens**, and also some fish and molluscs. Some lectins are poisonous, e.g. ricin from the castor oil plant (*Ricinus communis*).

Lecythidaceae (order **Ericales**) A family of **trees**, **shrubs**, and **lianas** with large, **simple**, sometimes **dentate** leaves,

distichous or spirally arranged in clusters at the tips of twigs. Flowers **actinomorphic** or **zygomorphic**, **bisexual**, 3–6 sometimes 2–6 fused **sepals**, 3–6 sometimes 3–18 **petals** or petals absent or replaced by **staminodes**, **ovary inferior** of 2–6 fused **carpels**. Pollination is by bats. **Inflorescence** a **spike**. Fruit a **capsule** or **drupe**-like. There are about 25 genera with 340 species occurring throughout the tropics. Some cultivated for timber or ornament. *Bertholletia excelsa* is the source of Brazil nuts.

Ledocarpaceae (order **Geraniales**) A family of small **shrubs** with usually **opposite**, **simple** and dissected or **compound** and **entire**, **exstipulate** leaves. Plants **hermaphrodite**, flowers **pentamerous**, **calyx** and **corolla** with 1 **whorl**, 10 **stamens**, **ovary** of 3 or 5 **carpels**. **Inflorescence racemose** in **corymbs** or flowers solitary. Fruit is a **capsule**. There are 2 genera with 11 species occurring in the American tropics.

lee Describes the side of an object that is sheltered from the wind.

leeches *See* Hirudinea.

lee depression A **cyclone** that forms in a westerly air flow on the **lee** side of a mountain range which is aligned north–south, e.g. the Rocky Mountains.

leek (*Allium ampeloprasum*) *See Allium.*

lee waves (standing waves) Undulations that develop in stable air on the **lee** side of a mountain.

legume A plant belonging to the family **Fabaceae**, or the fruit of such a plant. This is usually a **dehiscent pod** containing several seeds, e.g. pea, bean.

Lehmannia valentiana (three-banded garden slug, greenhouse slug) A species of translucent, yellow-grey or yellow-violet **slugs**, usually with 2 pairs of dark bands around the **mantle** and sometimes 3, that grow to 50–75 mm long. They are nocturnal, live in moist places, and feed on wood and plant material. They are pests in greenhouses. They occur in Europe, North and South America, Africa, Oceania, and Japan.

lek An area in which several male animals **display** in order to attract females for mating.

lemma The lower of the two **bracts** below each **floret** in grasses (**Poaceae**). *See also* palea.

lemmings *See* Cricetidae.

Lenten rose *See Helleborus.*

Lentibulariaceae (order **Lamiales**) A family of insectivorous **herbs** most of which are **rosette plants**. Some are aquatic. Flowers **zygomorphic**, **bisexual**, 2, 4, or 5 **sepals**, **corolla** 2-lipped, 2 **stamens**, sometimes 2 **staminodes**, **ovary** of 2 fused **carpels** with 1 **locule**. **Inflorescence** an erect **scape** with **bracts** and **bracteoles** bearing 1 flower or a **raceme**. Fruit is a **capsule**. In *Pinguicula* (80 spp.) leaves are flat with a sticky **adaxial** surface; in *Genlisea* (30 spp.) leaves tubular, forked, with insect traps on the fork **branches**; in ***Utricularia*** (bladderwort; about 220 spp.) leaves may or may not be present but the vegetative parts of the plant consists of **bladders** with trapdoors that open when 1 of 4 sensitive hairs is stimulated. There are 3 genera with 330 species with world-wide distribution.

lentic Pertaining to a freshwater **habitat**.

lenticel A blister-like pore in the stem of a woody plant that allows air to reach the underlying tissues. Lenticels also occur on fruits, e.g. as specks on apples and pears.

lenticular Biconvex, like a lens.

lenticular cloud (wave clouds) A series of lens-shaped (lenticular) **altocumulus** clouds seen on the **lee** side of a mountain.

leopard moth *See Zeuzera pyrina.*

leopard slug *See Limax maximus.*

Leopold, Aldo (1887–1948) An American ecologist and conservationist who, in his book *Sand County Almanac* published in 1949, argued that all living organisms are linked in a community to which humans also belong.

Leotiomycetes A class of non-**lichenized Fungi** belonging to the **Pezizomycotina** that are variable in form, but most of which have a small **apothecium** with an exposed **hymenium** and **unitunicate ascus** with no **operculum**, but an apical pore for the release of **spores**. Many species are plant pathogens, others are **endophytes**, **mycorrhizae**, parasites of other fungi, or **saprophytes**. They occur worldwide but with a patchy distribution.

Lepidobotryaceae (order **Celastrales**) A family of **evergreen trees** and large **shrubs** with **alternate**, **simple**, **entire** leaves with **caducous stipules**. Flowers **actinomorphic**, **unisexual** (plants **dioecious**), with 5 free, **imbricate sepals**, 5 free, imbricate **petals**, 10 **stamens** in 2 **whorls**, **ovary superior** of 2–3 **carpels**. **Inflorescence** a terminal **raceme** or **spicate panicle**. Fruit a **capsule**. There are two genera with two to three species with a scattered distribution in West Africa, and Central and South America.

Lepidodendrales An order, now extinct, of tree-like plants (the name is from the Greek for 'scale tree'), related to **Lycopsida**, that flourished in tropical swamps during the Pennsylvanian epoch (318.1–299 million years ago). They grew 30–40 m or more in height and formed extensive forests. Their stems were more than 1 m in diameter, rarely branched, but with a crown of **bifurcate branches** bearing clusters of spirally arranged, grass-like leaves. As the plant grew, leaf scars on the outer surface of the stem formed a distinctive diamond pattern that makes the **fossils** easy to identify.

lepidoid Scaly.

Lepidophloios kilpatrickense The earliest known member of the **Lycopsida**. It lived during the Pennsyvanian epoch (318.1–299 million years ago).

Lepidoptera (butterflies, moths) An order of **holometabolous Insecta** that have wings covered in minute, overlapping scales, which produce a wide variety of colours and patterns. Wingspan ranges from less than 5 mm to 250 mm. Most adults have mouthparts modified to form a **proboscis** they hold coiled beneath the head when not in use; a few species have chewing mouthparts. Butterflies have antennae (*see* antenna) with knobs or hooks at the tip; moth antennae are often **bipinnate**. Butterflies tend to be **diurnal** and rest with their wings folded vertically; moths tend to be nocturnal and rest with their wings folded horizontally. The distinction is largely artificial, however, and there are many exceptions. Larvae (**caterpillars**) have chewing mouthparts and almost all feed on plants; some are serious pests. There are more than 157,000 known species and at least as many yet to be described, distributed worldwide.

Lepidosaphes ulmi (apple bark louse, apple mussel scale, apple comma scale, mussel scale, oystershell scale) A scale insect (**Diaspididae**) up to 4 mm long, marked with brown bands on the waxy upper side and cream on the underside, shaped rather like a mussel shell, with short antennae (*see* antenna), and lacking eyes and legs. The female lays 20–100 white eggs beneath her body then dies, her scale darkening through the winter. The eggs hatch in spring and the larvae (crawlers) disperse to find a place to settle where the **bark** is thin enough for them to pierce the **vascular tissue** to feed. They are pests of more than 150 species of woody plants, including apple, pear, ***Prunus*** species, and currants, and a severe infestation can kill the plant. It occurs worldwide.

lepidote Covered in small scales.

Lepiota A genus of **basidiomycete fungi** in which the **fruiting body** is mushroom-like, most with a **pileus** less than 100 mm across and a slim **stipe**, with white or cream **gills**. They are **saprotrophs**. Most species prefer **calcareous soils**, rich in **humus**, and occur in woodland, grassland, or on sand dunes. There are about 400 species distributed worldwide. All are toxic and some contain **amatoxins**.

lepis A tiny, flat scale, attached at its centre and with an irregular margin. The fruits of *Myrialepis*, a genus of climbing palms (**Arecaceae**), have many (myriad) small scales (lepis).

Lepista *See Clitocybe.*

Leporidae (hares, rabbits, cottontails) A family of mammals in which the hind legs are long and adapted for jumping, the tail is reduced, and the ears are long. They are 250–700 mm long and females are usually larger than males. Rabbits are adapted for digging and the young are born in burrows, blind and naked; some species swim well and are semi-aquatic. Hares do not make burrows, and their young are born with their eyes open and with a full **pelage**. Cottontails do not dig burrows but may use burrows excavated by other animals. All members of the family are **obligate herbivores**. They occur in a wide variety of **habitats**. There are 11 genera with 54 species, found in most continents, but introduced in Australia, New Zealand, Java, and South America.

leprose Covered in powder, scurfy.

Leptinotarsa decemlineata (Colorado beetle, Colorado potato beetle, potato bug, ten-striped potato beetle, ten-striped spearman) A leaf beetle (**Chrysomelidae**), about 10 mm long, that has bright yellow or orange **elytra** with 5 brown, longitudinal stripes. Females lay orange eggs, about 1 mm long, in batches of about 30 on the underside of leaves. These hatch after 4–15 days into larvae that feed on the leaves, going through 4 **instars** before becoming prepupae that fall to the ground and pupate. Adults emerge after a few weeks and return to the plant to continue feeding and to mate. Depending on temperature there may be two or three generations a year. Adults of the final generation overwinter in the soil, emerging in the spring to lay eggs; if there is no host crop nearby the beetles are able to fly several kilometres. This is a major pest of potatoes and aubergines worldwide. It is not resident in Britain or Ireland, but adults occasionally arrive, carried by the wind from France while on their migrating flights, or in imported crops.

leptodermous With thin walls or skin.

leptoid In some mosses (**Bryophyta**), an elongate cell through which nutrients flow, analogous to a **sieve cell**. *Compare* hydroid.

leptokurtic A distribution, typical of wind-dispersed **propagules**, in which some points occur far from the origin but most are close to it.

leptosols Weakly developed soils in which there is solid rock or a layer with more than 40 percent calcium carbonate within 25 cm of the surface, or that have less than 10 percent fine-grained material in the upper 75 cm. Leptosols are a reference soil group in the **World Reference Base for Soil Resources**.

Leptospermum (family **Myrtaceae**) A genus of **evergreen shrubs** and small **trees** (tea tree) with **alternate**, **simple**, small leaves. Flowers solitary, with 5 **petals**. Fruit is a **capsule**. There are 80–86 species, most **endemic** to Australia, 1 in New Zealand, 2 in Southeast Asia, and 1 endemic to Malaysia. Several are cultivated for ornament.

Leptosphaeria coniothyrium A species of **ascomycete fungi** that causes cane blight in a wide range of plants including apples, strawberries, roses, ***Rubus*** spp., ***Ribes*** spp., and stone fruits. Infected stems weaken and may collapse. The fungus occurs worldwide.

Leptosphaeria maculans A species of **ascomycete fungi** that causes blackleg disease in brassicas (**Brassicaceae**). The fungus has two stages in its life cycle, as a **teliomorph** producing **ascospores** and as an **anamorph** (sometimes called *Phoma lingam*) producing pycnidia (*see* pycnidium). It begins as a **saprotroph** on stem residues left after harvest and penetrates the following crop, becoming a parasite. Blackleg disease produces **cankers** on stem bases, grey lesions on leaves, and rotting of the root, the cankers causing most crop losses.

leptosporangium A **sporangium** that develops from a single **initial**, has a wall with only 1 layer of tissue, and contains about 64 **spores**. Leptosporangia are found only in ferns (**Pteridophyta**). *Compare* eusporangium.

lessivage The **eluviation** of insoluble soil particles to a deeper layer.

lethal concentration fifty *See* LC_{50}.

lethal dose fifty *See* LD_{50}.

lethal mutation A **mutation** to a **gene** that causes the death of the organism carrying it.

lettuce (*Lactuca sativa*) *See Lactuca.*

lettuce infectious yellows virus (LIYV) A species of ***Crinivirus*** that causes yellowing or reddening, stunting, and rolling of leaves, with vein clearing. Leaves may become brittle. The disease affects lettuce (***Lactuca*** *sativa*), beet (*Beta vulgaris*), marrows (*Cucurbita pepo*), melons (*C. melo*), carrots (***Daucus*** *carota*), other cucurbits, and a number of wild plants. It is transmitted by a whitefly (***Bemisia tabaci***) and occurs in North America.

lettuce root aphid *See Pemphigus bursarius.*

leucoplast A colourless **plastid** that is involved in storage.

level of zero buoyancy *See* equilibrium level.

levulose *See* fructose.

Leycesteria (family **Caprifoliaceae**) A genus of **deciduous shrubs** with **opposite** leaves and flowers that are **actinomorphic**, **pentamerous**, and funnel-shaped, the **inflorescence** a pendulous **raceme**. Fruit is a **berry**. There are six species occurring in the Himalayas and the mountains of China. *Leycesteria formosa* is widely cultivated as Himalayan honeysuckle, also known as flowering nutmeg.

liana (liane) Any woody or wiry, free-hanging, climbing plant (vine) that roots in the ground and uses another plant, usually a **tree**, solely for support.

liane *See* liana.

lichen A composite organism consisting of a fungus, the mycobiont, living symbiotically (*see* symbiosis) with an **alga** or cyanobacterium (see **cyanobacteria**), the phycobiont. A lichen is classified by its fungal partner. Depending on the species, the lichen **thallus** may be **crustose**, **foliose**, or **fruticose**.

lichenicolous Growing on **lichens**.

lichenized Describes a fungal species (**Fungi**) that lives as a mycobiont in a **lichen**.

lichen woodland The northern edge of the **boreal forest**, bordered to the north by the **tundra**, where the woodland is open and park-like with a ground layer dominated by **lichens**. The lichen woodland is sometimes called the **taiga**.

lichen zone An area with a distinctive **lichen** population that relates to the quality of the air. The zones are categorized from 1 (algae only) to 10 (many species and pure air).

Lichinomycetes A class of **ascomycete fungi** most members of which are **mycobionts** of **lichens** in which the **phytobionts** are **cyanobacteria**.

Liebig, Justus von (1803–73) A German chemist who helped systematize organic and agricultural chemistry. In 1840, in *Die organische Chemie in ihrer Anwendung auf Agrikulturchemie und Physiologie* (Organic chemistry in its application to agricultural chemistry and physiology), he showed that plants take up nutrients in the form of simple chemicals and that nutrient deficiencies in soils may be remedied by the application of inorganic **fertilizers**. He maintained that plant growth is limited by the availability of the scarcest essential nutrient.

lifting condensation level The altitude at which the temperature of rising air falls to the **dewpoint temperature** and water vapour starts to condense. This marks the height of the base of clouds.

ligand 1. An atom, ion, or molecule that acts as an electron donor in bonding to a metal atom. **2.** A substance, usually a small molecule, that binds to a specific site on a **cell membrane**. **3.** *See* chelation.

ligase An **enzyme** that catalyzes a reaction joining two large molecules by forming a new chemical bond.

light air Wind of 1–2 m/s. *See* appendix: Beaufort Wind Scale.

light breeze Wind of 2–3 m/s. *See* appendix: Beaufort Wind Scale.

light-dependent stage (light reactions) During **photosynthesis**, those reactions that require light. Light photons impact **chlorophyll** molecules, which absorb their energy, releasing one electron for each photon absorbed, and the released electron attaches to an adjacent molecule, causing the release of another electron. Electrons move along an **electron-transport chain**, providing energy for **photophosphorylation** and **photolysis**. The H^+ from the breakdown of water attaches to nicotinamide adenine dinucleotide phosphate (NADP), converting it to NADPH, and the OH^- passes one electron to the chlorophyll, restoring its neutrality. Hydroxyls combine to form water, releasing oxygen ($4OH \rightarrow 2H_2O + O_2\uparrow$).

light-independent stage (dark reactions) During **photosynthesis**, those reactions that do not require light as a source of energy provided sufficient **adenosine triphosphate** (ATP) and **nicotinamide adenine dinucleotide phosphate** plus hydrogen (NADPH) are available. The reactions involve the reduction of carbon dioxide and the synthesis of sugar through the **Calvin cycle**.

lightning An electrical discharge that partly neutralizes a separation of charge which has accumulated between the top and bottom of a cloud, between two clouds, or between a cloud and the ground.

lightning channel The path that a **lightning stroke** follows; it is about 20 cm wide.

lightning stroke A flash of **lightning**, comprising the **stepped leader**, **dart leader**, and **return stroke**.

light reactions *See* light-dependent stage.

lignicolous Growing on **decorticate** wood.

lignified Describes cells that contain large amounts of **lignin**, which stiffens them.

lignin A constituent of **cell walls** in almost all terrestrial plants that anchors **cellulose** fibres and cements them together. After cellulose, lignin is the most abundant natural polymer in the world, and the only natural polymer that does not consist of

carbohydrate monomers. Its composition varies according to species, but comprises three different phenyl propane monomers that are cross-linked in complex ways.

lignum-vitae (*Guaiacum officinale*) *See* Zygophallaceae.

ligulate Possessing **ligules**, strap-like.

ligule 1. In most grasses (**Poaceae**) and sedges (**Cyperaceae**), a fringe of hairs or membrane that is part of the leaf, appearing where the leaf **blade** and sheath meet. It varies in length and may have a smooth or uneven margin. **2.** In some **Asteraceae**, the strap-shaped **corolla** of a **ray floret**.

Ligustrum ovalifolium (privet) *See* Oleaceae.

lilac *See Syringa.*

Liliaceae (order **Liliales**) A **monocotyledon** family of **perennial geophytes** with **bulbs** or **rhizomes**, and **alternate**, usually **sessile**, sometimes sheathing leaves. Flowers often large, **actinomorphic** or slightly **zygomorphic**, **bisexual**, with 6 free **tepals** in 2 **whorls**, 3+3 **stamens**, **ovary superior** of 3 **carpels** and **locules**. **Inflorescence** a **raceme** or **umbel**-like. Fruit is a **capsule** or **berry**. There are 19 genera with 610 species occurring throughout northern temperate regions. Many cultivated as ornamentals (lilies).

Liliales An order of **monocotyledon geophytes** that comprises 11 families of 67 genera and 1558 species. *See* Alstroemeriaceae, Campynemataceae, Colchicaceae, Corsiaceae, Liliaceae, Melanthiaceae, Petermanniaceae, Philesiaceae, Rhipogonaceae, and Smilacaceae.

Lilioceris lilii (lily beetle, red lily beetle, scarlet lily beetle) A bright red beetle (**Coleoptera**), about 8 mm long with a black head and legs. Its orange or red eggs, about 1 mm long, hatch into orange larvae with black heads that grow to 8–10 mm. Adults overwinter in sheltered places, emerging in spring to feed, mate, and lay eggs on the underside of leaves at intervals until early autumn. The beetles produce one generation a year. Both adults and larvae feed on ***Lilium*** and *Cardiocrinum* lilies and on fritillaries (*Fritillaria*) and can defoliate them. The beetle is native to Eurasia, but was introduced to southern England in the 19th century. It now occurs throughout most of the temperate Northern Hemisphere.

Lilium (family **Liliaceae**) A genus of **perennial**, **geophyte**, **monocotyledon herbs** that overwinter as **bulbs**. In some species the bulb develops into a **rhizome** with many small bulbs along it, and other species develop **stolons**. Flowers are large, often fragrant, with 6 free **tepals**, and **superior ovary**. **Inflorescence** is a **raceme** or **umbel**-like. Fruit is a **capsule**. There are about 100 species occurring throughout northern temperate regions, with some extending to The Philippines. Many are cultivated for their flowers (lilies).

lily *See* Liliaceae, *Lilium*.

lily beetle *See Lilioceris lilii.*

Limacidae (keel-back slugs) A family of **slugs** that have a distinct **keel** on the posterior part of the body, sometimes extending forward to the **mantle**. The family includes some of the largest terrestrial slugs, not all of which are pests. There are about 100 species occurring throughout the northern **Palaearctic**.

Limax maximus (great grey slug, leopard slug) A keel-back slug (**Limacidae**) that grows to 100–200 mm long. It is grey or brownish grey with dark blotches or spots and has a short **keel** on its tail, long and slender tentacles, and the reproductive pore close to the base of the right upper tentacle. The **clitellum** is about one-third the length of the slug and always has black spots. It inhabits damp, shady places, often close to human dwellings. It is nocturnal, feeding on **Fungi** and dead plant matter, and it is also a predator of other slugs, which it pursues, but also feeds on crop plants, making it a serious pest. It occurs widely in Europe and North America.

lime 1. The fruit of ***Citrus*** *aurantifolia*. 2. The alkaline compound calcium oxide (CaO), quicklime, or calcium hydroxide ($Ca[OH]_2$), slaked lime, or a material rich in calcium carbonate ($CaCO_3$) and other alkaline compounds, that is used to correct soil acidity and sometimes as a **fertilizer** to supply magnesium. 3. *See Tilia.*

Limeaceae (order **Caryophyllales**) A **monogeneric** family (*Limeum*) of **herbs** and **subshrubs** with spirally arranged, **exstipulate** leaves. Flowers **hermaphrodite**, with 5 **sepals**, 5 **petals** alternating with the sepals or absent, 8 **stamens** in 2 **whorls**, **ovary superior** of 3 united **carpels** with 1 or 3 **locules**. **Inflorescence** an **axillary cyme**. Fruit is a **capsule**. There are two species occurring in southern Africa, Ethiopia, and southern Asia.

limestone A sedimentary rock consisting principally of **calcite** and/or **dolomite**.

limestone forest A type of **tropical rain forest** that grows on **limestone** hills. There are few large **trees**, but many small trees and **shrubs**.

limiculous Growing in mud.

limiting factor (ecological factor) Any environmental condition that is close to the **limits of tolerance** for a specified species.

limits of tolerance The maximum and minimum boundaries to the range of environmental factors, e.g. light intensity and duration, temperature, water accessibility, availability of particular nutrients, within which a species can survive.

Limnanthaceae (order **Brassicales**) A family of soft, **annual herbs** with **alternate**, **pinnate** or 1–3 **pinnatisect**, **exstipulate** leaves. Flowers **actinomorphic**, **hermaphrodite**, **trimerous** or **pentamerous**, **sepals valvate** and fused, **petals** free, twice as many free **stamens** as petals, **ovary** of 3 or 5 fused **carpels**. **Inflorescence** a **raceme**. Fruit is a **schizocarp**. There are one or two genera (*Limnanthes, Floerkia*) with eight species occurring in temperate North America. *Limnanthes douglasii* widely cultivated as an ornamental.

limnic Pertaining to fresh water.

limnology The study of freshwater **ecosystems**.

limonacho *See* Achatocarpaceae.

Limonium (sea lavender) *See* Plumbaginaceae.

limpets *See* Gastropoda.

Linaceae (order **Malpighiales**) A family of **annual** and **perennial herbs**, **subshrubs**, and **shrubs** with **alternate** or **opposite**, usually **sessile**, **entire** leaves with small **stipules** or **exstipulate**. Flowers **actinomorphic**, **bisexual**, **pentamerous** or **tetramerous**, often heterostylous (*see* heterostyly), with 5 **imbricate sepals**, 5 free, **caducous petals**, 5 or 4 **stamens** alternating or opposite the petals, sometimes with additional **staminodes**, **ovary superior** of 2–5 **carpels** with 3–5 **locules**. **Inflorescence** a **cymose panicle**. Fruit a **capsule** or **schizocarp**. There are 10–12 genera with 300 species with worldwide distribution. Several species cultivated as ornamentals or for medicinal products. *Linum usitatissimum* is flax.

lindane (gamma-hexachlorocyclohexane, gammaxene, Gammallin) An **organochlorine insecticide** that was used to treat seeds, soil, and **parasites** of pets and farm livestock. It is persistent and toxic, and its agricultural use is banned.

linden *See Tilia.*

Linderniaceae (order **Lamiales**) A family of **annual** and **perennial herbs**, some aquatic, with **opposite**, **decussate**, sometimes in whorls or a rosette, occasionally **alternate**, **entire** or **serrate** leaves. Flowers with 2-lipped **corolla**, upper lip usually 2-lobed, lower lip 3-lobed and larger, 4 **stamens** or 2 fertile and 2 **staminodes**, **ovary** with 2 **locules**. Flowers **axillary** and solitary or in a terminal **raceme** or axillary **fascicle**. Fruit a **capsule**. There are about 14 genera with 195 species occurring throughout the tropic and warm temperate regions.

linear Of a leaf, several times longer than it is wide and not sharply pointed at the tip.

A linear leaf.

ling (*Calluna vulgaris*) *See Calluna*.

linkage An association between **genes** due to their presence on the same **chromosome**. The closer together they are the more closely linked they are, and the less likely to be separated during **crossing over**. All the genes on a chromosome comprise a linkage group.

linkage disequilibrium (gametic disequilibrium) The non-random association of **alleles** at two or more loci (*see* locus), not necessarily on the same **chromosome**, in a population.

linkage equilibrium (gametic equilibrium) The condition in which the frequency of **haplotypes** in a population is equal to the product of the frequencies of the **genetic markers** in each haplotype.

linkage group *See* linkage.

linkage map *See* genetic map.

Linnaeus, Carolus (1707–78) A Swedish botanist who devised a system of **taxonomy** and promoted **binomial classification**. He published his first classified list of plants, animals, and minerals in 1735 as *Systema Naturae*. He gave more details of his botanical classification, based on the number and arrangement of **stamens** and **pistils** in *Genera Plantarum* (1737). His most important botanical work was *Species Plantarum* (1735), which is still the starting point for botanical nomenclature.

linseed (*Linum usitatissimum*) *See Linum*.

Linum (family **Linaceae**) A genus of **herbs** with narrow, **linear**, **sessile** leaves. Flowers have 5-clawed, **caducous petals**, borne in **cymes**, some heterostylous (*see* heterostyly). There are 200 species occurring throughout temperate and subtropical regions. Some grown for ornament, *L. usitatissimum* is grown for fibre (flax) and oil (linseed oil).

Linum usitatissimum (flax) *See Linum*, Linaceae.

linuron A **systemic urea herbicide** that is applied to soils to kill grasses and **broad-leaved** weeds prior to their emergence and around cereal and vegetable crops, and ornamental **bulbs**. It is of low toxicity to mammals, but toxic to aquatic animals and its use is controlled.

Linyphiidae (money spiders) A family of very small, black or brown spiders (**Araneae**) which construct sheet webs in vegetation, the spider hanging on the underside of its web and running to catch prey that becomes trapped. The spiders have eight eyes in two rows of four and **chelicerae** with many teeth. The young disperse by **ballooning**. There are more than 4300 species found worldwide.

lipid A member of a group of **hydrophobic** compounds that includes fats, oils, waxes, **phospholipids**, and **steroids**. They store energy, act as **hormones** and vitamins, and contribute to the structure of cells.

lipoprotein A water-soluble **conjugated protein** which has a **lipid** as the **prosthetic group**.

liquid limit *See* Atterberg limits.

Liriodendron (family **Magnoliaceae**) A genus of **trees** with distinctive, 4-lobed leaves with a long **petiole**. Flowers, with 3 **sepals** and 6 **petals**, resemble tulips, hence the common name. Fruit is an aggregate of **samaras**. There are two species with **bicentric distribution**, *L. tulipifera* (tulip tree, tulip poplar) occurring in eastern North America and *L. chinense* in eastern China.

lithic contact The boundary between soil and the **parent material**.

lithification The formation of **rock** by the compression and cementing of mineral particles.

lithocyst A cell with a **cystolith**.

lithomorphic soils Soils that have a shallow **soil profile** and in which organic material lies directly above solid **bedrock**. This is a major group in the classification developed by the Soil Survey for England and Wales.

lithophyte A plant that grows on the surface of rocks.

Lithops (family **Aizoaceae**) A genus of **glabrous**, **succulent** plants (living stones) that have 2 massive, bulbous, fleshy, **opposite** leaves that are almost fused, white or yellow flowers arising from the split between them. All but the upper parts of the leaves are buried, the exposed surface being sufficiently translucent for light to penetrate, and the leaf surfaces are coloured to resemble stones. There are 37 species occurring in southern Africa. Many are cultivated.

lithosere A plant **succession** that begins on a surface of bare rock.

lithosols **Azonal soils** that are too stony or too shallow to be cultivated, with a layer of solid rock or **laterite** close to the surface, or that are deeper but consist mainly of stones and gravel. Such soils occur mainly in deserts and mountainous regions. Lithosols are a reference soil group in the **World Reference Base for Soil Resources**.

lithosphere The upper layers of the solid Earth, comprising the rocks of the oceanic and continental **crust** and the uppermost part of the mantle, where the rock is brittle. The lithosphere is broken into blocks, called tectonic plates, which move in relation to each other, producing the processes described as plate tectonics.

lithotroph A 'rock-eater', i.e. an organism that obtains energy by oxidizing inorganic (usually mineral) substances.

little tree worm *See Satchellius mammalis*.

littoral Pertaining to the shore of a sea or lake.

live fence *See* hedgerow.

live oak Any **evergreen** oak (*Quercus*), especially sand live oak (*Q. geminata*) and southern live oak (*Q. virginiana*), which (despite the name) is the state tree of Georgia.

liver A large organ in vertebrates and some invertebrates that arises from the intestine and is involved in the detoxification of blood, storage of food substances including **glycogen**, synthesis of **proteins**, production of **hormones**, and production of bile, which aids digestion.

liverwort A plant belonging to the **Marchantiophyta**.

living fence *See* hedgerow.

living fossil A **taxon** with no living relatives and that is otherwise known only from **fossils**. ***Ginkgo*** and ***Metasequoia*** are living fossils.

living stones *See Lithops*.

lixisols A group comprising all soils that have an **argic horizon** as the B **soil horizon** within 100–200 cm of the surface, other than **albeluvisols**, **alisols**, **acrisols**, and **luvisols**. Lixosols are a reference soil group in the **World Reference Base for Soil Resources**.

LIYV *See* lettuce infectious yellows virus.

lizards *See* Lacertilia, Squamata.

llanos The **savanna** grasslands in the Orinoco Basin of Venezuela.

LNR *See* local nature reserve.

loam A soil that comprises approximately equal parts of **clay**, **silt**, and **sand** particles. Loam retains moisture and plant nutrients well and is considered an ideal horticultural and agricultural soil. A slight imbalance favouring one of the components produces **clay loam**, silty loam, or sandy loam. *See* soil textural triangle.

loamy sand Soil that contains 70–90 percent **sand**, of which at least 18–22 percent is coarse sand, the percentage of **silt** plus 1.5 times the percentage of **clay** is equal to 15 or more, and the percentage of silt plus twice the percentage of clay is equal to less than 30. *See* soil textural triangle.

Loasaceae (order **Cornales**) A family of **annual** to **perennial herbs** and **shrubs**, climbers, and small **trees**, most covered with stinging hairs. Leaves **opposite** or **alternate**, **entire** or lobed or **pinnatisect** or **palmate**. Flowers **actinomorphic** with usually 5 but sometimes 4–8 free **sepals** and **petals**, many **stamens**, sometimes reduced to **staminodes**, **ovary inferior** or partly **superior** of 3–5 **carpels** usually with 1 **locule**. **Inflorescence** usually a **bracteate** terminal or **axillary raceme**. Fruit is a **capsule**. There are 14 genera with 265 species occurring in America, Africa, and the Marquesas Islands.

Lobaria pulmonaria *See* tree lungwort.

lobate With lobes, or resembling lobes.

lobed Of a leaf margin, undulating, forming lobes.

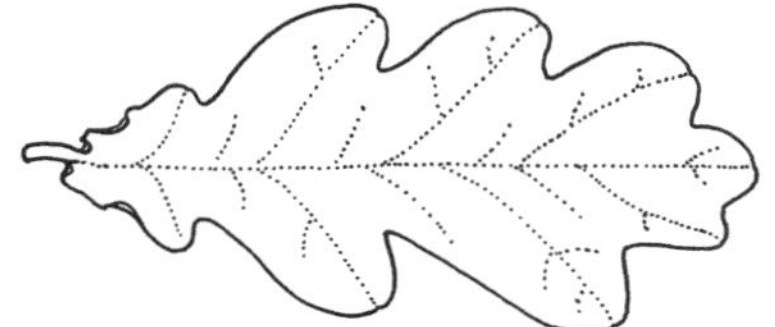

A leaf with a lobed margin.

Lobelia (family **Campanulaceae**) A genus of **annual** and **perennial herbs** and **shrubs** with **alternate**, **simple**, **entire** leaves. Flowers with 2-lipped **corolla** with 5 lobes. Fruit is a **capsule**. There are 360–400 species occurring throughout tropical to warm temperate regions. Many cultivated for ornament.

lob worm *See Lumbricus terrestris.*

local diversity *See* alpha diversity.

local nature reserve (LNR) In the United Kingdom, a **nature reserve** that is designated by a local government, usually on land it owns or leases, and that is of local but not national **conservation** value. Local authorities often delegate the management to voluntary conservation organizations.

loci *See* locus.

locule A small cavity or hollow chamber, e.g. in a plant **ovary**.

loculicidal Describes a **pod** that opens by splitting longitudinally along its midrib.

locus (pl. loci) The place on a **chromosome** where a particular **gene** is located. There is one gene at each locus; if there are several **alleles** of a gene, only one can appear at each locus.

locusts *See* Acrididae.

lodgepole pine (*Pinus contorta*) *See* serotiny.

lodging The flattening of a cereal crop due to the permanent bending of the plant stems due to severe weather, pest or fungal infestation, or a metabolic imbalance.

Lodoicea maldivica (double coconut) *See* Areaceae.

loess A unconsolidated sediment consisting mainly of quartz particles 0.015–0.05 mm in size that has been deposited by the wind. It shows little or no development of **soil profiles**. The soil is very fertile, making excellent agricultural land.

loganberry leafhopper *See Ribautiana tenerrima.*

Loganiaceae (order **Gentianales**) A family of **annual** and **perennial herbs**, **shrubs**, **trees** and **lianas** with **opposite**, **entire** leaves. Flowers **actinomorphic**, **bisexual** or **unisexual** (plants **dioecious** or **gynodioecious**), **tetramerous** or **pentamerous**, **sepals** fused or absent, **corolla** usually a narrow tube, **ovary superior** or semi-**inferior** of 2 fused **carpels** with 2 free or 1 or 3 fused **locules**. **Inflorescence** usually **cymose**, sometimes cincinnate (*see* cincinnus). Fruit a **capsule**, **drupe**, or berry-like. There are 13 genera of 420 species with a pantropical distribution. *Strychnos* produces several toxic **alkaloids**, e.g. strychnine and **curare**. It and several other species have medicinal uses.

loment A dry, **dehiscent schizocarp** in some **legumes** that breaks open at constrictions between its seeds, e.g. in tick trefoils (*Desmodium* spp).

London plane (*Platanus acerifolia*) *See* Platanaceae.

longevity The length of time for which an individual lives or a **taxon** remains **extant** if this is greater than the average for members of that **species** or other taxon.

Longidorus (needle nematodes) A genus of very large nematodes (**Nematoda**), 2–8 mm long, that occur throughout the world, mostly in sandy soil. They feed at or immediately behind the root tip of host plants, often causing the formation of **galls** and the reduction of the root system. They are pests of a wide range of crops including grapevines, corn (maize), mint, strawberries, lettuce, onion, and celery. There are many species.

long-legged flies *See* Dolichopodidae.

long-tailed bushtit *See Aegithalos caudatus.*

long-tailed field mouse *See Apodemus sylvaticus.*

long-tailed shrews *See Sorex.*

long-tailed tit *See Aegithalos caudatus.*

long-terminal repeat (LTR) A **DNA** sequence that is repeated hundreds or thousands of times. LTRs occur in **retrovirus** and **retrotransposon** DNA that flanks functional **genes**.

long waves *See* Rossby waves.

Lonicera (family **Caprifoliaceae**) A genus of **deciduous**, some **evergreen shrubs** and woody climbers with **opposite**, **simple**, **oval** leaves. Flowers **pentamerous** with a **corolla** that is tubular below and 2-lipped, or regular. Fruit is a **berry**. There are about 180 species occurring throughout northern temperate regions. Many cultivated for their sweet-scented flowers (honeysuckle) or for hedging.

loosestrife *See Lythrum.*

Lophiocarpaceae (order **Caryophyllales**) A family of **herbs**, **trees**, **shrubs**, and **lianas** with **alternate**, spiral, **entire**, usually **exstipulate** leaves. Flowers with 4 or 5 or up to 10 **imbricate sepals**, no **petals**, 4–5 or 5–50 **stamens**. Fruit an **achene** or **capsule**. There are two genera with six species occurring in Africa and to western India.

Lophopyxidaceae (order **Malpighiales**) A **monotypic** family (*Lophopyxis maingayi*), which is a **liana** with leaf **tendrils**. Leaves **alternate**, **simple**, **serrate**, with **stipules**. Flowers **actinomorphic**, **unisexual** (plants **monoecious**), 5 free or slightly **connate sepals**, 5 free **petals**, 5 **stamens** alternating with **petaloid staminodes**, **ovary superior** of 5 sometimes 4–5 **carpels** and **locules**. **Inflorescence** a cluster or **axillary panicle**. Fruit a **samara**. The plants occur from Malesia to the Solomon and Caroline Islands.

Loranthaceae (order **Santalales**) A family of parasitic, usually **evergreen shrubs** and some small **trees** that grow on the **branches** or less frequently roots of other **dicotyledons** attached by haustoria (*see* haustorium), sometimes with runners over the surface of the host. Leaves **opposite** or **alternate** or in **whorls**, **simple**, **entire**, **exstipulate**. Flowers **bisexual**, 4–5 free or 3–9 united, **valvate petals**, as many **stamens** as petals, **ovary inferior**. **Inflorescence** a 3-flowered **cyme**, **raceme**, **umbel**, or flowers solitary. Fruit usually a **berry**. There are 68 genera with 950 species occurring throughout the tropics.

lorica A hard, open sheath containing silica (SiO_2) that surrounds some unicellular algae (see alga).

losing stream A stream that has a permeable bed (*see* permeability) through which it loses water.

lotus fruit (*Zizyphus lotus*) *See* Rhamnaceae.

love dart (gypsobelum) A sharp dart, 1–30 mm long and made from calcium carbonate or **chitin**, that mature adults of certain terrestrial **slug** and **snail** species stab into each other during mating. The dart is coated with a **hormone** that triggers reactions that protect the sperm.

low An area of low **atmospheric pressure**.

low arctic tundra The southernmost part of the arctic **tundra**, where vegetation covers most of the ground, with a mosaic of communities reflecting local conditions.

lower atmosphere The atmosphere extending from the surface to the **tropopause**.

lowering The emergence of a mass of cloud from base of a **cumulonimbus** cloud, thereby lowering the cloud base.

Lowiaceae (order **Zingiberales**) A **monocotyledon**, **monogeneric** family (*Orchidantha*) of **perennial herbs** with **rhizomes**. Leaves **distichous**, **lanceolate**, **entire**, with a **petiole**. Flowers **zygomorphic**, **bisexual**, with 3 fused **sepals**, 3 **petals**, 5 free **stamens**, **ovary inferior** of 3 fused **carpels**. **Inflorescence spike**-like. Fruit a **capsule**. There are 15 species occurring from southern China to Borneo.

LTR *See* long-terminal repeat.

Lumbricus castaneus (chestnut worm) A common and widespread **epigeic** earthworm (**Annelida**), 30–70 mm long, that is brown between the head and the orange **clitellum**. It lives in leaf litter and beneath logs.

Lumbricus rubellus (redhead worm, red earthworm) A species of red-brown or red-violet, **epigeic** and **endogeic** earthworms (**Annelida**), 25–105 mm long, that live in soils with a loose structure and rich in organic matter. They feed on the surface on largely decomposed material. The worms are native to Europe but have been introduced widely elsewhere and are invasive.

Lumbricus terrestris (common earthworm, dew worm, lob worm, nightcrawler) An **anecic** earthworm (**Annelida**), 100–250 mm long and sometimes longer, reddish in colour, that feeds on the surface at night, pulling fragments of vegetation into the entrance to its burrow where it partly decays before being eaten. It is native to Europe but has been introduced in most other parts of the world and is invasive in parts of North America.

lumen **1.** An open space or cavity inside an organ. **2.** The derived SI unit of luminous flux.

lunate Half-moon-shaped.

lung The respiratory organ in air-breathing vertebrates and terrestrial **Mollusca**. It contains many alveoli (*see* alveolus) through which gases are exchanged.

lupin *See Lupinus*.

Lupinus (family **Fabaceae**) A genus mainly of **perennial**, some **annual herbs** (lupins) with a few **shrubs**, with **palmately lobed** leaves with 5–28 **leaflets**. Flowers pea-like, borne on an erect **spike**. Fruit is a **pod**. There are about 280 species, most occurring in North America, but some in South America, southern Europe, Africa, and Australasia. Some grown to feed cattle, others cultivated for their showy flowers.

luvisols Soils with an **argic horizon** that has a **cation exchange capacity** of 24 $cmol_c$/kg, and an illuvial (*see* illuviaton) accumulation of **clay**. Luvisols are a reference soil group in the **World Reference Base for Soil Resources**.

luxury consumption The **absorption** by a plant of an essential nutrient in a quantity exceeding that required for growth and reproduction, resulting in the accumulation of the element in plant tissues.

lyase An **enzyme** that catalyzes reactions that add groups to a **substrate** or remove them from it, forming or breaking double bonds, commonly between carbon atoms (C=C) or between carbon and oxygen (C=O).

lychee (*Litchi chinensis*) *See* Sapindaceae.

Lycium (family **Solanaceae**) A genus of **deciduous**, **perennial shrubs** with thorns. Leaves **alternate**, **simple**, **entire**. Flowers with a funnel-shaped, 5-lobed **corolla**, solitary or borne in small clusters. Fruit is a **berry**. There are about 90 species occurring in dry and saline environments throughout temperate and subtropical regions. Many are cultivated for ornament as boxthorn, also called desert thorn, Christmas berry, and Duke of Argyll's tea-plant.

Lycogala epidendrum (wolf's milk, groening's slime) A species of slime moulds (**Myxogastria**) that produces aethalia (*see* aethalium) resembling small puff ball **Fungi** 3–15 mm across. They may be pink, grey, yellowish brown, or greenish black, and grow on damp, rotting wood. It occurs worldwide.

Lycoperdon (puffballs) A genus of **agaric fungi** in which the **fruiting body** is approximately spherical (puffballs) and contains a vast number of **spores** that are ejected explosively when the mature puffball explodes. There are about 50 species with a widespread distribution in woodlands.

lycopod *See* Lycopodiophyta.

Lycopodiophyta (lycopods) The oldest **phylum** of the **Tracheophyta** that includes club mosses, quillworts, and spikemosses. Lycopods have green, branching stems, small, scale-like leaves, **rhizomes**, and reproduce by **spores**. They first appeared during the Silurian period (443.7–418 million years ago) and during the Carboniferous period (359.2–299 million years ago) they formed forests of trees more than 30 m tall. ***Baragwanathia longifolia*** is the earliest known species.

lycopsid *See* Lycopsida.

Lycopsida Depending on the classification a **class**, **clade**, or subphylum of vascular plants (**Tracheophyta**) with true stems and leaves, and most with roots. The lycopsids include the **extant** lycopods (**Lycopodiophyta**), Sellaginellales, and Isoetales, and the extinct **Lepidodendrales**. They first appeared during the Devonian period (416–359.2 million years ago).

Lygocoris pabulinus (common green capsid) A bug (**Miridae**) about 6 mm long with long legs and antennae (*see* antenna) that lays overwintering eggs on woody plants in autumn. These hatch in spring and feed where they emerge before moving to herbaceous plants. Winged adults emerge in summer and lay eggs from which a second generation hatches, feeding until it lays the overwintering eggs.

Lygus rugulipennis (tarnished bug) A brown or green bug (**Miridae**) about 6 mm long with long legs and antennae (*see* antenna) that overwinters as an adult in plant debris and lays eggs in late spring in plant stems and flower **buds**. A second generation emerges in late summer and attacks many garden plants.

Lyngbya A genus of filamentous **cyanobacteria** in which each long, unbranching **filament** is enclosed within a rigid sheath and the **filaments** are capable of gliding motility. They do not form **heterocysts** or **akinetes**. They occur worldwide in water and on wet rocks and in wet soil, and in water they may float free or form mats.

lyre-flower *See Lamprocapnos spectabilis.*

Lysenko, Trofim Denisovich (1898–1976) A Ukrainian agronomist and biologist who aimed to improve crop yields by controlling environmental conditions to produce plant traits he believed could be inherited. He became powerful in Soviet science, but although he remained director of the Institute of Genetics within the USSR Academy of Sciences until 1965, his influence declined during the 1950s.

lysis The rupture and breaking down of a cell, commonly due to an **enzyme** reaction, **osmotic pressure**, or bacterial or viral infection.

lysogenic cycle *See* lysogeny.

lysogeny (lysogenic cycle) A mechanism by which a **bacteriophage** reproduces, through integrating its **prophage** into that of its host or forming a circular copy of its genetic material in the host **cytoplasm**. Every subsequent cell division reproduces the viral **genome**. *See also* lytic cycle.

lysosome A specialized **vesicle** found in **eukaryote** cells that contains digestive **enzymes** that break down a variety of materials. The enzymes are **proteins** made primarily in the **endoplasmic reticulum** and packaged by the **Golgi bodies.**

Lythraceae (order **Myrtales**) A family of **trees**, including mangroves, some with **pneumatophores**. Leaves usually **opposite**, occasionally **alternate** or whorled, **simple**, usually dimorphic (*see* dimorphism) in amphibious genera, usually **entire** sometimes **dentate**, **stipules** small or absent. Flowers **actinomorphic** to **zygomorphic**, usually **hermaphrodite**, with **campanulate** to tubular **hypanthium**, 4–6 sometimes 4–16 **valvate sepals** and **petals**, twice as many **stamens** as **petals**, **ovary superior** sometimes **inferior** of 2–4 or many **carpels** each with 1 **locule. Inflorescence axillary** or terminal **raceme**, **panicle**, or **cyme**, or flowers solitary. Fruit is a **capsule**. There are 31 genera of 620 species occurring throughout the tropics with a few in temperate regions. Some are cultivated, e.g. *Punica granatum* (pomegranate), *Trapa natans* (water chestnut), and *Lawsonia inermis* (henna).

Lythrum (family **Lythraceae**) A genus of **annual** and **perennial herbs** with leaves in **whorls** or **opposite**, usually **simple**, **exstipulate**. Flowers **actinomorphic**, **bisexual**, with 12 **stamens** in 2 whorls. Flowers usually solitary, or **inflorescence** a **spike** or **raceme**. Fruit is a **capsule**. There are 38 species occurring throughout temperate regions, mainly in wet ground. Several are cultivated for ornament as loosestrife.

lytic cycle (lytic response) A mechanism by which a **bacteriophage** reproduces inside the host **cell**, leading to the **lysis** of the cell. *See also* lysogeny.

lytic response *See* lytic cycle.

M

macadamia nuts *See* Proteaceae.

macchia *See* maquis.

mace (*Myristica fragrans*) *See* Myristicaceae.

machair Low-lying, stable grassland, rich in herbs, that has developed on blown **sand** behind coastal sand dunes, typically on the west coast of Scotland and Hebridean islands.

mackerel sky Long, parallel rows of units of **cirrocumulus** cloud, resembling the pattern of scales on a mackerel.

macroaggregates Relatively large **aggregates**.

macroburst A large, powerful, downward rush of air from the base of a **cumulonimbus** cloud.

macroconsumer A **consumer** that feeds on other organisms or organic matter. All animals are macroconsumers. *Compare* microconsumer.

macrocyclic lactone insecticides A group of **insecticides**, **acaricides**, and **nematicides** that are used to control **spider mites**, **leaf miners**, aphids (**Aphididae**), and beetles (**Coleoptera**) in fruit and vegetable crops, and as antihelminthics to treat parasitic worms in mammals. They are moderately toxic to humans.

macroevolution Evolution of taxa (*see* taxon) above the **species** level.

macrofossil *See* megafossil.

macromolecule A large molecule, usually formed by the polymerization of subunits. **Nucleic acids**, **proteins**, and **carbohydrates** are polymer macromolecules, **lipids** are non-polymer macromolecules.

Macronesian floral region The area that includes the Canary and Azores Islands and Madeira, off the coast of northwest Africa, part of the **boreal region**. The region is noted for the large number of **succulents** of the family **Crassulaceae**.

macronutrient An **essential element** that plants need in relatively large amounts. These include carbon, hydrogen, oxygen, nitrogen, sulphur, phosphorus, potassium, and calcium. *Compare* micronutrients.

macrophyll *See* megaphyll.

macropores Soil **pores** larger than 75 µm.

Macrosiphum euphorbiae (potato aphid) A species of pink, green, or mottled pink and green aphid (**Aphididae**), 2.5–3.5 mm long, that feeds on a wide range of plants, especially **Solanaceae**, but also sunflowers, brassicas, peas and beans, apples, maize (corn), and asparagus. **Nymphs** resemble the adults. Females overwinter on sprouting potatoes, lettuce, or weeds, and emerge in spring, feeding on wild plants before migrating in early summer to their summer hosts, forming colonies on the undersides of leaves, flowers, and **buds**. The species originated in North America but now occurs worldwide wherever potatoes are cultivated.

Macrosiphum rosae (rose aphid) A species of pink, purple, or green aphid (**Aphididae**), 2.5 mm long with long, dark legs that feed on roses and pyracantha. They may complete their entire life cycle on a single plant. The species originated in Europe but now occurs throughout most of North America.

macrospore *See* megaspore.

Madagascar floral region The area that includes Madagascar and Comoros, the Seychelles, the Mascarenes, and the surrounding sea area to the African coast, part of the **Palaeotropical region**. A high proportion of the genera are **endemic**.

made ground (made land) An area of land that has been constructed by draining an area and using landfill to build a surface, e.g. reclaimed marsh or shoreline.

made land *See* made ground.

M

madroña laurel (*Arbutus menziesii*) *See Arbutus*.

maerl An agricultural dressing made from shell-rich **sand** and seaweed; the term is Breton and should not be confused with **marl**.

maggot A legless insect larva.

magic mushroom *See Psilocybe*.

magma Molten rock containing silicates, **carbonates**, or sulphides with dissolved volatile compounds and crystals in suspension that forms by the partial melting of crustal or mantle rocks. It is the parent material for all **igneous** rocks and processes.

magnesium (Mg) An element that contributes to the structure of **chlorophyll** and **membranes** and that takes part in many **enzyme** reactions, especially those involving the transfer of phosphate groups. Magnesium deficiency can cause **chlorosis** and discoloration of leaves.

Magnolia (family **Magnoliaceae**) A genus of **deciduous** and **evergreen trees** and **shrubs** with **simple** leaves that open after the flowers. Flowers are solitary, large, with **petaloid tepals**, and **stamens** and **carpels** in an elongate **receptacle**; carpels tough to withstand pollination by beetles. Flower **bud** enclosed in a **bract** rather than **sepals**. Fruit is a **follicle**. There are about 225 species with a **disjunct distribution** from the Himalayas to Japan and western Malaysia, and eastern North America, Central America, the Caribbean, and South America. Many grown for ornament.

Magnoliaceae (order **Magnoliales**) A family of **deciduous** or **evergreen trees** and **shrubs** with **alternate**, **simple**, **entire** leaves (lobed in ***Liriodendron***), with **petioles** and often large, **caducous stipules**. Flower **buds** enclosed by a **bract**, arising from the **peduncle**, that falls of as the flower opens. Flower usually solitary, often large and showy, **actinomorphic**, **hermaphrodite**, occasionally **unisexual**, with an elongate **receptacle**, **perianth** of 2 or 3 **whorls** of free **petaloid tepals**, many free **stamens**, **ovary superior**, **apocarpous**. Fruit coalescent woody **follicles** or aggregates of **samaras**. There are 2 genera of 227 species occurring throughout eastern North America, Central and South America, and Asia from the Himalayas to Malesia. Many are grown for ornament, some for timber, some for medicinal use.

magpie *See* Corvidae, *Pica pica*.

mahogany (*Khaya*, *Swietenia*) *See* Meliaceae.

Mahonia (family **Berberidaceae**) A genus of **evergreen shrubs** with **pinnate** leaves with 5–15 **leaflets**. Flowers **trimerous**, with 5 **whorls** of similar **perianth** segments, 2 whorls of **stamens**, **ovary** of 1 **carpel**. **Inflorescence** a **raceme** or **panicle**. Fruit is a **berry** (Oregon grape). There are about 70 species occurring in eastern Asia, the

Himalayas, and North America. Many are cultivated for ornament.

maiden A tree growing from a seed, **graft**, or less commonly a sucker, that is a year old and has not been **coppiced** or **pollarded**.

maidenhair fern *See Adiantum.*

maidenhair tree (*Ginkgo biloba*) *See Ginkgo*, Ginkgoaceae, Ginkgoales.

maintenance evolution (normalizing selection, stabilizing selection) The stabilizing effect of **natural selection** in environments that change little over time. Natural selection favours those **phenotypes** best adapted to the **environment** and eliminates extreme phenotypic variants.

maize (*Zea mays*) *See Zea.*

maize smut *See Ustilago maydis.*

major gene A **gene** that has a pronounced effect on the **phenotype**.

Malabar spinach (*Basella rubra* and *B. alba*) *See* Basellaceae.

malathion An **organophosphate insecticide** and **acaricide** that inhibits the **enzyme** anticholinesterase and is used to control aphids (**Aphididae**), leaf hoppers, codling moth (***Cydia pomonella***), and mites. It is harmful to fish, but of low toxicity to mammals, although its toxicity is enhanced by prior exposure to **parathion**.

Malaysian floral region The area that includes the Malay Peninsula, Java, Sumatra, Sunda Islands, Borneo, the Philippines, Celebes and the Molucca Islands, and New Guinea and Aru, part of the **Palaeotropcial region**. Floristically, the region is possibly the richest in the world.

maleic hydrazide A plant **growth regulator** that is used to control weeds on roadside verges and amenity areas, and to prevent sprouting in stored potatoes and onions. It is environmentally harmless and of low toxicity to mammals.

Malesherbiaceae (order **Malpighiales**) A **monogeneric** family (*Malesherbia*) of **perennial** and a few **annual herbs** growing to **subshrubs** or **shrubs** with **alternate, simple, linear** or **lanceolate** to **ovate** or **obovate, entire, pinnatiparite, dentate,** or lobed leaves. Flowers **actinomorphic, bisexual**, 5 **sepals** and **petals** forming a floral tube, central **androgynophore**, 5 **stamens, ovary superior, syncarpous** of 3 **carpels. Inflorescence** a terminal **raceme** or **panicle**. Fruit is a **capsule**. There are 24 species occurring in South America.

male sterility A condition resulting from the total or partial failure of a plant to produce functional **anthers, pollen**, or male **gametes**. Cytoplasmic male sterility is due to abnormalities in mitochondria (*see* mitochondrion) or **chloroplasts.** Cytoplasmic-genetic male sterility is due either to abnormalities in the **cytoplasm** or the cell **nucleus**. Plant breeders producing **hybrid** seeds often induce male sterility in order to establish female lines.

mallee Sclerophyllous vegetation, 2–3 m tall, dominated by ***Eucalyptus***, that occurs in southern Australia.

malleus *See* articular bone.

Malpighiaceae (order **Malpighiales**) A family of **trees, shrubs, subshrubs, herbs,** and climbers with usually **opposite** sometimes whorled or **alternate, simple** leaves with **stipules**. Flowers **actinomorphic** to **zygomorphic**, usually **bisexual** occasionally **unisexual** (plants **dioecious**), with 5 usually **imbricate** occasionally **valvate sepals**, 5 usually imbricate **petals**, 10 **stamens, ovary superior** of 3 free, or 2 or 4 fused **carpels. Inflorescence** a terminal or **axillary raceme** or **panicle**, or flowers solitary. Fruit usually a **schizocarp**, sometimes a **drupe** or **berry**. There are 68 genera of 1250 species occurring throughout the tropics and subtropics. Some cultivated for ornament or medicinal use.

M

Malpighiales An order of plants that comprises 39 families of 716 genera and 15,935 species. *See* Achariaceae, Balanopaceae, Bonnetiaceae, Calophyllaceae, Caryocaraceae, Centroplacaceae, Chrysobalanaceae, Clusiacae, Ctenolophonaceae, Dichepetalaceae, Elatinaceae, Erythroxylaceae, Euphorbiaceae, Euphroniaceae, Goupiaceae, Humiriaceae, Hypericaceae, Irvingiaceae, Ixonanthaceae, Lacistemataceae, Linaceae, Lophopyxidaceae, Malpighiaceae, Malesherbiaceae, Medusagynaceae, Ochnaceae, Pandaceae, Passifloraceae, Peraceae, Phyllanthaceae, Picrodendraceae, Podostemaceae, Putranjivaceae, Quilinaceae, Rafflesiaceae, Rhizophoraceae, Salicaceae, Trigoniaceae, and Violaceae.

maltose (malt sugar) A **disaccharide** sugar that consists of two **glucose** units linked by a **glycoside** bond.

malt sugar *See* maltose.

Malus (family **Rosaceae**) A genus of **deciduous trees** and **shrubs**, usually thorny in wild forms, with **alternate**, **simple**, **serrate** leaves. Flowers **actinomorphic**, with 5 **sepals** and **petals**, many **stamens**, **ovary half-inferior** of 3–5 fused **carpels**. **Inflorescence** a **corymb**. Fruit is a **pome**. There are 30–35 species occurring throughout the temperate Northern Hemisphere. Many are cultivated as apples and crab apples.

Malvaceae (order **Malvales**) A family of **herbs** and **subshrubs** with some **shrubs** and a few **trees** with **alternate**, **simple**, **entire** or **dentate** leaves. Flowers with 5 **valvate sepals**, 5 **petals**, **ovary superior**, **syncarpous**. **Inflorescence** a terminal **raceme** or **axillary cyme**. Fruit is a **capsule**. There are 243 genera of 4225 species with a widespread tropical and temperate distribution. Several cultivated, e.g. *Gossypium* (cotton), *Abelmoschus esculentus* (**okra**), *Hibiscus rosa-sinensis*, and ***Abutilon***.

Malvales An order of plants that comprises 10 families of 338 genera and 6005 species. *See* Bixaceae, Cistaceae, Cytinaceae, Dipterocarpaceae, Malvaceae, Muntingiaceae, Neuradaceae, Sarcolaenaceae, Sphaerosepalaceae, and Thymelaeaceae.

mambas *See* Elapidae.

Mamestra brassicae (cabbage moth) A noctuid moth (**Noctuidae**) with brown, mottled forewings and a wingspan of 34–50 mm that is common throughout Eurasia. Its larvae are green khaki, or brown, with darker spots that grow to about 25 mm and feed on a wide variety of plants, not only brassicas.

Mammalia (mammals) A class of homeothermic (*see* homeotherm) vertebrates in which the head is supported on a flexible neck, the lower jaw is formed from the dentary bone and articulates with the squamosal, teeth are usually present, the heart has four chambers, a **diaphragm** separates the thoracic and abdominal cavities, except in Monotremata (echidnas and platypus). The egg is small and develops in the uterus, and young are born alive and nourished by milk secreted by mammae (giving the class its name), and the skin bears at least some hairs.

mammals *See* Mammalia.

MAMPs *See* microbe-associated molecular patterns.

mandible **1.** The lower jaw of a vertebrate. In birds the term is sometimes applied to both parts of the beak, as the upper and lower mandibles. **2.** In **Arthropoda**, one of the pair of mouthparts that are used to seize and cut food items.

maneb A **carbamate fungicide** that is sprayed on to foliage to control fungal diseases in a wide range of fruit, nut, vegetable, and other crops. It can cause eye and skin irritation but is otherwise of low mammalian toxicity.

mangal *See* mangrove forest.

manganese (Mn) An element that is involved in the **light-dependent stage** of **photosynthesis** and that also takes part in **enzyme** reactions which break down **carbohydrates**. Deficiency causes **chlorosis** and other discoloration, especially in young leaves.

Mangifera indica (mango) *See* Anacardiaceae.

mango (*Mangifera indica*) *See* Anacardiaceae.

mangold fly *See Pegomyia betae.*

mangosteen (*Garcinia mangostana*) *See* Clusiaceae.

mangrove *See* Rhizophoraceae.

mangrove forest (mangal) A type of **swamp** forest, up to 30 m tall, that develops in salt or **brackish** water along tropical and subtropical coasts. Mangroves (*see* Rhizophoraceae) are often the only trees.

Manihot esculenta (cassava, manioc, tapioca) *See* Euphorbiaceae.

Manila hemp (*Musa textilis*) *See* Musaceae.

man-induced turnover The increase in the flow of an element through a **biogeochemical cycle** that is due to human activity.

manioc (*Manihot esculenta*) *See* Euphorbiaceae.

manipulated altruism (social parasitism) A form of **parasitism** in which the parasite tricks the host into nurturing or feeding it at a cost to itself. **Brood parasitism** is a version of manipulated altruism.

man-made soils Soil that results from such operations as the restoration of mines and quarries. It is a major soil group in the classification devised by the Soil Survey for England and Wales.

manna *See* mannitol.

mannan A polysaccharide made from branched or linear polymers derived from simpler sugars that is an important constituent of **hemicellulose** in the **cell walls** of vascular plants (**Tracheophyta**). Mannans also act as storage compounds in some seeds, e.g. palm (*Phoenix dactylifera*) and coffee (*Coffea arabica*).

manna sugar *See* mannitol.

mannite *See* mannitol.

mannitol (manna sugar, mannite) A polyhydroxy **alcohol** (sugar alcohol), $C_6H_8(OH)_6$, found in some algae (*see* alga) and many vascular plants (**Tracheophyta**), that reduces **osmotic pressure**. A white, crystalline solid derived from **mannose**, mannitol was first isolated from the south European flowering ash, also called manna ash (*Fraxinus ornus*) and called manna because of its similarity to the Biblical food.

M

mannose An aldohexose **monosaccharide** found in a wide variety of organisms. In some plants, especially members of the **Fabaceae**, it largely replaces **glucose** as the building block for polysaccharides, then called **mannans**.

manoxylic Describes wood with much **parenchyma** tissue in wide **rays** and **tracheids** with thin walls, making it soft and spongy. It is typical of cycads (**Cycadales**). *Compare* pycnoxylic.

mantid flies *See* Neuroptera.

mantle (pallium) In **Mollusca**, a fold of skin that encloses a space (mantle cavity) containing the gills or lungs. In **snails** the mantle secretes material for the construction of the shell.

maple (*Acer*) *See* Sapindaceae.

maquis The French name for **drought**-resistant scrub that occurs in the Mediterranean region. It consists of **sclerophyllous vegetation** comprising small, **evergreen trees** and **shrubs**. The maquis has developed from **evergreen forest** as a result of repeated burning and grazing. It is known as macchia in Italy and matorral in Spain.

Marantaceae (order **Zingiberales**) A **monocotyledon** family of **perennial herbs** with **rhizomes** or **tubers**. Leaves **distichous** and sheathing, with **petioles**. Flowers usually in pairs that are mirror images of each other, inconspicuous, **bisexual**, with 3 free, occasionally **connate**, **sepals**, 3 **petals**, outer **whorl** of **androecium** with 1–3 **petaloid staminodes**, sometimes absent, inner whorl of 1 fertile often petaloid half-**stamen**, 1 hooded staminode, and 1 fleshy staminode, **ovary inferior** of 3 fused **carpels** with 3 **locules**. **Inflorescence** is terminal or lateral, simple or **spike**-like or **capitate**. Fruit is a **capsule**, occasionally a **berry** or **caryopsis**. There are 31 genera with 550 species occurring throughout the tropics, but not Australia. Some cultivated for food or ornament.

Marasmius A genus of **agaric fungi**, several of which grow in **fairy rings**. They are **saprotrophs**. The **fruiting body** is mushroom-like but tiny and inconspicuous. Some are edible, e.g. *M. oreades* (fairy ring mushroom, Scotch bonnet), which has a buff or tan **pileus** 10–50 mm across, white or pale tan **gills**, and a **stipe** 20–80 mm tall. There are about 500 species distributed worldwide.

marathon *See* chloronicotinyls.

marbled salamander *See Ambystoma opacus.*

marcescence 1. The retention by plants of organs that are dead, e.g. of leaves through the winter by beech (***Fagus*** spp.). **2.** In some **Fungi**, the revival when moistened, with the release of **spores**, of **fruit bodies** that have dried out and withered.

Marcgraviaceae (order **Ericales**) A family of **lianas**, **shrubs**, and small **trees**, some of them **epiphytes**, with **alternate**, **distichous** or in spirals, **simple**, often **glabrous** or leathery, **entire** or **crenate**, **exstipulate** leaves. Flowers **actinomorphic**, **bisexual**, 4 or 5 **imbricate sepals**, 5 free or fused **petals**, 3 to many **stamens**, **ovary superior**. **Inflorescence** a terminal **raceme**, sometimes **umbel**-like or **spicate**. Fruit **globose**, fleshy, with many seeds. There are 7 genera with 130 species occurring in tropical America.

Marchantiophyta A **phylum** of nonvascular plants, formerly called Hepatophyta, which includes all the liverworts, plants in which the ovoid or spherical **capsule** is often surrounded by a tubular **perianth**, has no lid, and bursts when ripe into four sections to release its **spores**. Liverworts may be moss-like, or with leaves, often lobed or segmented, in two or three rows, or thallose, i.e. showing no differentiation into stem and leaves. There are about 9000 species, occurring in many types of **habitat**, especially in moist conditions.

mares' tails Fibrous **cirrus** cloud that form long strands, curled at the ends.

marginal placentation Placentation in which the **ovules** lie along the fused margins of a **carpel**, e.g. in **legumes**.

marine material Soil **parent material** consisting of **sedimentary rock** formed from sediments deposited on an ocean floor.

maritime air Moist air at a more moderate temperature than air over continents in the same latitude, which acquires its characteristics by prolonged contact with the ocean surface.

maritime climate (oceanic climate) The type of climate that is associated with **maritime air**. Compared with a **continental climate**, the **diurnal** and seasonal temperate range is smaller and precipitation is greater.

marjoram (*Origanum*) *See* Lamiaceae.

marl A marine sediment that occurs in beds with other **oozes**, and that consists of 30 percent **clay** and 70 percent

microfossils, with at least 15 percent comprising **fossils** made from **silica**. It occurs in soils as a **lime**-rich clay. The term should not be confused with **maerl**.

marsh An area with a **mineral soil** that is waterlogged most of the time. Marshes commonly occur on the margins of lakes and river **flood plains**. In North America, any herbaceous wetland is called a marsh. *Compare* swamp.

marsh gas *See* methanogen.

marshmallow (*Althaea officinalis*) *See Althaea*.

marsh pitcher *See* Sarracenciaceae.

marshy tundra Marsh that occurs in areas of **tundra** where **drainage** is poor, usually because a layer of **permafrost** lies close to the surface. The marsh supports grasses, sedges, and dwarf willows.

martens *See* Mustelidae.

Martyniaceae (order **Lamiales**) A family of **annual**, occasionally **perennial**, **herbs**, often with **tubers**, or **shrubs**, all covered with sticky hairs, with **opposite**, occasionally **alternate**, broad, **cordate**, **palmately lobed** or suborbicular leaves. Flowers with 5 **sepals** free or fused into a tube or **spathe**, **corolla** tubular below and **campanulate** above with 5 lobes, 4 **stamens** or 2 with 2 **staminodes**, **ovary superior** of 2 fused **carpels**. **Inflorescence** a lax terminal **raceme**. Fruit is a **capsule**. There are 5 genera with 126 species occurring throughout tropical and subtropical America. Some cultivated for ornament.

Masarinae (pollen wasps) A subfamily of wasps (**Vespidae**), 10–20 mm long, that feed their larvae exclusively with **nectar** and **pollen**, carrying the food in their crops and regurgitating into cells in their nests, made from mud or sand, before laying one egg in each cell and sealing the cell. They are important pollinators. Most pollen wasps are brown with yellow, red, or white markings. They occur on all continents but are abundant in only a few places.

mass flow The downslope movement of sediment, loose rock, or scree under the force of gravity.

mass mixing ratio *See* mixing ratio.

mass movement *See* mass wasting.

massula 1. A mass of **pollen grains** developed from a single **pollen mother cell**. **2.** In some aquatic ferns (e.g. *Azolla*), a layer of **cytoplasm** that extends from the **tapetum** to enclose the **megaspores** and **microspores**; four massulae enclose the megaspore and several massulae each contain a number of microspores.

mass wasting (mass movement) The movement down a hillslope of surface material that has lost internal **cohesion**, usually because of a large increase in its water content. The main types of mass wasting are **creep**, **fall**, **flow**, and **slide**.

mast The fruit of certain forest trees, especially beech (***Fagus***) and oak (*Quercus*).

mastigoneme A hair-like projection from a **flagellum**.

mast year A year in which there is abundant **mast**; mast years usually occur at intervals separated by years of poor mast production.

mating type The equivalent of sex in sexually reproducing organisms that lack reproductive organs or structures, e.g. single-celled organisms. Individuals possess on the cell surface **proteins** that bind to complementary proteins or **polysaccharides** on the surface of other individuals, providing differences by which mating types are defined, and ensuring that **conjugation** takes place only between cells of different mating types. There may be more than two mating types within a species and cells can change their mating type, e.g. the ciliate protozoon *Tetrahymena thermophila* has seven mating types.

matorral *See* maquis.

mature soil A soil that passed through all the stages in its formation and has stabilized. It exhibits a clearly defined **soil profile** and accumulates plant nutrients from **weathering** of **parent material** and the decomposition of organic matter at approximately the same rate as it loses them, e.g. through **leaching**.

Maundiaceae (order **Alismatales**) A **monotypic**, **monocotyledon** family (*Maundia triglochinoides*) of submerged aquatic plants (seagrasses) with **rhizomes** that are pollinated under water. Flowers **imperfect**, **anthers sessile**, **carpels** ascidiate (*see* ascidium). Fruit is **drupe**-like. The plants occur only in southeastern Australia. This family has been removed from the **Juncaginaceae** and many authorities do not recognize the move.

M

maxilla 1. In vertebrates, the bone at the rear of the upper jaw that holds all the teeth apart from the incisors. **2.** In some **Arthropoda**, one of the pair of mouthparts behind the **mandibles** that are used to ingest food items.

maximum sustained yield *See* optimum yield.

maximum thermometer A thermometer that records the highest temperature it registered since it was last reset.

Mayacaceae (order **Poales**) A **monogeneric**, **monocotyledon** family (*Mayaca*) of small **perennial herbs** with spirally arranged, **simple**, **sessile** leaves. Flowers solitary, **actinomorphic**, **trimerous**, with 3 **stamens**, **ovary superior** of 3 fused **carpels**. Fruit is a **capsule**. There are four to ten species, most occurring in tropical America, one in Africa.

Mazaceae (order **Lamiales**) A family of **annual** or **perennial herbs** with **opposite**, **dentate** leaves. Flowers have a pronounced lower lip. Fruit **indehiscent**. There are 33 species occurring from Central Asia to Australasia.

mazaedium An **ascocarp** in which the contents form a powdery mass.

MCPA A widely used **herbicide** (2-methyl-4-chlorophenoxyacetic acid) that is used to control **broad-leaved** weeds in pasture and cereal crops. It is of low mammalian toxicity.

MCPP *See* mecoprop.

meadow A field in which grasses (**Poaceae**) and a variety of **herbs** are grown for pasture.

meadowlarks *See* Icteridae.

meadow mushroom *See Agaricus campestris*.

meadow saffron (*Colchicum autumnale*) *See* Colchicaceae.

meadow steppe The northern part of the Eurasian **steppe**, bordering forest, that is dominated by sod-forming and tussock grasses up to more than 1 m tall, with a wide variety of flowering **herbs**.

mealybugs *See* Coccidae.

mealy cabbage aphid *See Brevicoryne brassicae*.

mealy hairs Hairs that form a surface coating with the consistency of meal.

mealy plum aphid *See Hyalopterus pruni*.

mean temperature The averaged air temperature measured at a specified place over a specified period.

mechanical weathering The breakdown of rocks and minerals by physical processes, e.g. by the freezing of water in crevices, the formation and subsequent expansion of salt crystals in crevices, expansion due to solar heating, and the release of pressure as overlying rocks fall away.

mecoprop (MCPP) A **herbicide** that is used to control **broad-leaved** weeds in cereal crops and among fruit trees. It breaks down rapidly and is not environmentally harmful. It can cause illness in humans at high doses, but is harmless at normal background levels.

median lethal dose *See* LD_{50}.

medicinal rhubarb (*Rheum officinale*) *See Rheum.*

medieval woodland **Woodland** that is known from records to have existed prior to the 17th century.

Mediterranean floral region The area that includes the coastal regions bordering the Mediterranean and the islands of the Mediterranean. The region contains a high proportion of **endemics**, reflecting its distinctive **climate**.

Mediterranean forest **Evergreen trees**, both conifers and **broad-leaved**, that formed the **climax vegetation** in lands bordering the Mediterranean. Almost all of it has disappeared, to be replaced by **garrigue** and **maquis**.

Mediterranean scrub A collective name that describes **garrigue** and **maquis**.

medlar (*Mespilus germanica*) *See Mespilus.*

medulla **1.** The central part of a structure or organ. **2.** A tangled mass of fungal hyphae (*see* hypha). **3.** A layer of hyphae in a **lichen thallus**.

medullary rays (pith rays, wood rays) Plates of **parenchyma** tissue that extend through wood from the centre (medulla) of the trunk or **branch** to the **cortex**, crossing the **tree rings** at right angles. In cross-section the plates appear as thin lines radiating from the centre to the edge, like the rays of the Sun.

megafossil (macrofossil) A **fossil** large enough to be visible without the aid of a microscope.

megagametophyte *See* embryo sac.

megaphyll (macrophyll) A leaf, typical of seed plants (**Spermatophyta**) and ferns (**Pteridophyta**), that is usually large and usually has **leaf gaps** associated with **leaf traces**. *Compare* microphyll.

megasporangium A **sporangium** that contains **megaspores**; in **angiosperms** it is the **ovule**.

megaspore (macrospore) The larger of the two types of **spore** produced by a heterosporous (*see* heterospory) plant. In most plants it develops into the female **gametophyte**. *Compare* microspore.

megaspore mother cell *See* megasporocyte.

megasporocyte (megaspore mother cell) A **diploid cell** that undergoes **meiosis** to produce four **haploid megaspores**.

megasporophyll (macrosporophyll) In a heterosporous (*see* heterospory) plant, a **sporophyll** that bears megasporangia (*see* megasporangium).

meiosis (reduction division) The nuclear division occurring at some stage in the life cycle of all sexually reproducing organisms in which the number of **chromosomes** is halved and genetic material is exchanged between **homologous chromosomes**. The nucleus, and cell, divides twice to produce **haploid gametes** or sexual **spores**. ↗

meiosporangium A **sporangium** in which **spores** (meisospores) are produced by **meiosis**. In **ascomycetes** the **ascus** is the meiosporangium.

meiospore A **haploid zoospore** formed in a **meiosporangium**.

meiotic drive (segregation distortion) Any mechanism or process that results in the over-representation of some **alleles** in the **gametes** formed by **meiosis**. This breaks Mendel's law of segregation (*see* Mendel's laws), which states that each gamete has an equal chance of receiving either of a dividing pair of alleles.

meiotic spindle The **spindle** that forms in **eukaryote cells** during **meiosis**.

Melanerpes carolinus (red-bellied woodpecker) A species of woodpeckers that are 230–270 mm long with a wingspan of 380–460 mm. They are mainly pale grey with black and white bars on the back, wings, and tail. Males have a bright red patch on the crown, females on the nape; both males and females also have a red patch on the belly. They inhabit all types of forest and feed on fruits, nuts, seeds, tree sap, and invertebrates. They occur throughout the eastern United States.

Melanesia and Micronesia floral region The large area that includes the islands of Micronesia (Marianas, Carolines, Marshall, Kiribati, and Tuvalu) and Melanesia (New Guinea, Solomons, Vanuatu, and Fiji), part of the **Palaeotropical region**. The **flora** is largely derived from that of the neighbouring continents and consequently there are comparatively few **endemics**.

M

melangeophilous Growing in **alluvial soil** or **loam**.

melanic horizon A dark or black (Greek *melanos*, black) **soil horizon**, at least 30 cm thick, that is rich in organic matter derived mainly from the decomposition of grass (**Poaceae**) roots, with an average 6 percent or more of organic carbon.

melanin One of a group of dark pigments found in most animals; in mammals they occur in the skin and hair. There are two principal forms: eumelanin is black or brown; phaeomelanin (pheomelanin) is red or pink. Neuromelanin is a dark pigment found in the brain of some animals.

melanism The occurrence in an animal population of black individuals.

Melanthiaceae (order **Liliales**) A family of **monocotyledon perennial herbs**, most with **rhizomes** or **bulbs**, a few with **corms**. Leaves **dentate**, **cauline** or in a rosette at the base, or in a spiral, or in **pseudowhorls** at the top of the shoot. Flowers usually **actinomorphic**, **unisexual** or **bisexual** (plants **monoecious**, **dioecious**, or **polygamous**), with 3+3 **petaloid sepals**, or 3 **sepaloid** and 3 petaloid, or 4–10 (or 4–9) **tepals**, 3+3 free **stamens** or up to 24 in 6 **whorls**, **ovary syncarpous**, usually **superior**, of 3 **carpels**. **Inflorescence** a **raceme**, occasionally a **panicle**, **spike**, or **umbel**-like, or flowers solitary. Fruit is a **capsule**. There are 16 genera with 170 species, occurring throughout the northern temperate region. Several with medicinal uses, some cultivated as ornamentals.

Melastomataceae (order Myrtales) A family of **trees**, **shrubs**, **herbs**, **lianas**, and **epiphytes** with leaves **opposite**, **decussate**, **simple**, **entire** or **dentate**, **exstipulate**, usually with **petioles**. **Anisophyly** common, ant domatia (*see* domatium), usually at leaf bases, in some genera. Flowers **actinomorphic** or with **zygomorphic** androecia (*see* androecium), often with **bracteoles**, **bisexual**, with 4–5 sometimes 3–8 **imbricate** to **valvate calyx** lobes and free **petals** in an urn-shaped or **campanulate hypanthium**, usually twice as many **stamens** as **petals**, **ovary superior** or **inferior** of 4–14 fused **carpels** and **locules**. **Inflorescence** usually a terminal, **axillary**, or **cauline cyme**, but occasionally flowers solitary. Fruit is a fleshy **berry** or **capsule**. There are 188 genera with 5005 species occurring in the tropics and subtropics. Some cultivated as ornamentals, for their timber, or for their edible fruits.

Melchior, Hans (1894–1984) A German botanist who made many contributions to **taxonomy**, especially in the Melchior system, a classification of the **angiosperms**.

Meles meles (Eurasian badger) Stocky, nocturnal mustelids (**Mustelidae**) with short, strong legs and a short tail, 560–900 mm long, and distinguished by the two dark stripes from the muzzle to each ear. They are gregarious, living in family groups of up to 20 individuals depending on the availability of resources in a large communal burrow (sett); young badgers often establish their own setts close to the parental one. Badgers are territorial, territories 0.0025–1.5 km^2, and feed mainly on earthworms (**Annelida**), augmented by other invertebrates and small mammals. They occur throughout the **Palaearctic** in woodlands near to open fields, farmland, and hedgerows, preferring sites among trees, shrubs, or rocks that will conceal the entrances to their setts.

Meliaceae (order **Sapindales**) A family of **trees** and **shrubs** with **alternate**, spiral, usually **pinnate**, occasionally **bipinnate**, **exstipulate** leaves. Flowers **actinomorphic**, **unisexual** but appearing **bisexual** (plants **monoecious**, **dioecious**, or **polygamous**), with 4–5 free or 3–8 partly united **sepals**, 3–5 free or 3–7 fused **petals**, 5–10 **stamens**, **ovary superior**, **syncarpous**, of 2–13 **carpels** and **locules**. **Inflorescence** a **cymose panicle**. Fruit is a **capsule**, **berry**, or **drupe**. There are 50 genera of 615 species occurring throughout the tropical lowlands. Many are important timber trees, e.g. *Swietenia* and *Khaya* (mahogany), *Entandrophragma cylindricum* (sapele), *E. utile* (utile), or with medicinal properties, e.g. *Azadirachta indica* (neem tree).

Melianthaceae (order **Geraniales**) A family of **shrubs** and small **trees**, some with **rhizomes**, with **alternate**, usually **pinnate** leaves, some with a winged **rhachis**, with **stipules**. Flowers **actinomorphic** to weakly **zygomorphic**, **resupinate** in some genera, **bisexual** or **unisexual** (plants **polygamodioecious**), 4- or 5-merous, 4 or 5 **sepals** and free **petals**, as many or twice as many **stamens** and petals, **ovary superior** of 4–5 **carpels**. **Inflorescence** a terminal **raceme** or **spike**. Fruit is a **capsule**. There are two genera with eight species occurring in sub-Saharan Africa. Some have medicinal uses.

Meloidogyne hapla (northern root-knot nematode) A nematode (**Nematoda**) that causes **galls** to form on the roots of many plants. Its eggs, laid about 1000 at a time, are able to survive harsh winters, hence the 'northern' in its name. The closely related southern root-knot nematode (*M. incognita*) produces galls less than half the size. Severe infections reduce yields and distort root crops.

Meloidogyne incognita (southern root-knot nematode) *See Meloidogyne hapla.*

Meloidogyne naasi (barley-root nematode, cereal-root nematode) A nematode (**Nematoda**) that lives on the roots of barley and wheat, and also on a wide variety of other crops including grasses (**Poaceae**) and legumes (**Fabaceae**). Juveniles penetrate plant roots and feed there, causing root swelling, finally emerging and moving into the soil. The pest occurs throughout Europe and has been introduced to North America.

Melospiza melodia (song sparrow) A species of sparrows, 120–170 mm long, in which the head is brown with a white or grey stripe on the crown and an eyestripe, and heavily streaked bodies. They inhabit open areas with **shrubs**, woodland edges, and thickets, and feed on seeds, berries, and some insects. They occur over most of North America and are partially migratory in parts of their range.

membrane A film-like structure that separates a **cell** from its surrounding **environment** and encloses various components within the cell. Membranes form the structural bases for **enzymes**, and may form part of the enzymes, and they also serve as selective barriers (*see* selectively permeable, semipermeable).

Mendel, Gregor Johann (1822–84) An Austrian Augustinian monk at the monastery of Brünn (now Brno, Czech Republic) who, from 1856, experimented with peas he grew in the monastery garden which led to his discovery of the principles of heredity. He reported his findings to two meetings of the Brünn natural history society (Naturforschenden Verein) on

8 February and 8 March 1865 and they were published in 1866 ('Versuche über Pflanzen-Hybriden' [Experiments with plant hybrids]) in the society's proceedings.

Mendelian character A **character** that is inherited according to **Mendel's laws**.

Mendelian population An interbreeding population of organisms that share a common **gene pool**.

Mendel's laws Two laws of inheritance that were formulated by Gregor **Mendel**. Expressed in modern terms these were: 1. (law of segregation) When two members of a pair of **genes** segregate during **meiosis** each **gamete** has an equal probability of obtaining either member of the pair; 2. (law of independent assortment) Different segregating gene pairs behave independently (this was later found to apply only to pairs that are unlinked or linked only distantly (*see* linkage).

Menispermaceae (order **Ranunculales**) A family of climbers, **shrubs**, and small **trees** with **alternate**, sometimes **peltate**, **exstipulate** leaves. Flowers **unisexual** (plants **dioecious**), 6 **sepals** (sometimes more) in 2 rows, **petals** in 2 or 3 rows, 3–6 or many **stamens**, 1–32 free **carpels**. **Inflorescence** a **raceme**, **panicle**, or **cyme**. Fruit is a **drupelet**. There are 70 genera with 442 species with a pantropical distribution, usually in the lowlands. **Curare** is obtained from the **bark** of *Chondrodendron tomentosum*. Other species also with medical uses.

Mentha (family **Lamiaceae**) A genus of aromatic, **perennial**, occasionally **annual herbs** with creeping **rhizomes** or **stolons**, **opposite**, **oblong** to **lanceolate**, **serrate** leaves, often downy. Flowers with 4-lobed **corolla**, 4 **stamens**, 10–13 **sepals**. Fruit is a **capsule**. There are 25 species with many **hybrids**, occurring in temperate Eurasia and South Africa. They are widely cultivated for the many varieties of mint.

Menyanthaceae (order **Asterales**) A family of **perennial**, a few **annual**, **herbs**, some with **rhizomes**. Leaves **alternate**, sometimes in a basal rosette, **linear** to **elliptical**, or **reniform**, or heart-shaped, or suborbicular, or **sagittate**, or **palmategely compound**. Flowers **actinomorphic**, **bisexual** or **unisexual** (plants **monoecious**), often heterostylous (*see* heterostyly), with 5 **sepals**, **corolla** with 5 lobes, 5 **stamens**, **ovary superior** to semi-**inferior** of 2 **carpels** with 1 **locule**. **Inflorescence** a simple or branched **cyme**, **raceme**, head, or flowers solitary. Fruit is a **capsule**. There are 5 genera with about 58 species of worldwide distribution.

Mephitidae (skunks, stink badgers) A family of nocturnal mammals recognizable by their distinctive black or brown fur with a contrasting pattern of white stripes or spots on their faces, tails, or backs, or a bold white stripe running from their nose to tail. They are broad and squat with short limbs and strong claws that they use for digging. They are also renowned for their defensive habit of squirting with great accuracy a foul-smelling liquid from their **anal glands**, but not before giving ample warning. They live in a wide variety of **habitats**, sometimes near human habitations. They feed on invertebrates and small vertebrates, including pests, but suffer persecution because occasionally they attack poultry and they are significant vectors for rabies. There are 4 genera with 13 species. Skunks (*Mephitis*, *Conepatus*, and *Spilogale* species) occur throughout North and Central America and northern South America. Stink badgers (*Mydaus* species) occur in the Philippines and Indonesia.

mericarp After a **schizocarp** has split, one of the parts, i.e. one **carpel**. It contains one or more seeds.

Meripilus giganteus (giant polypore, black-staining polypore) A species of **polypore Fungi** in which the **fruiting body** is a clump of yellowish brown brackets, 500 mm to 2 m across and 200–800 mm tall, near the base of a tree, usually a **broad-leaved** species, less commonly on conifers. It is edible when young. The fungus penetrates and grows inside the roots of mature trees and can cause considerable damage. It occurs throughout the Northern Hemisphere.

meristele *See* dictyostele.

meristem Tissue that consists of **cells** which continue to divide indefinitely, thereby generating new growth. They occur at the tips (apices) of stems and roots **(apical meristem)** and in the **cambium (lateral meristem).**

meristoderm The outer layer of the **stipe** in some **brown algae** (Phaeophyta); it resembles **meristem** in its ability to grow continually, thereby replacing tissue lost by **abrasion** against the rocks on which the seaweed grows.

mermaid's cup (mermaid's wineglass) The genus of unicellular **green algae (Chlorophyta)** *Acetabularia* that have a **thallus** of **branches** fused into the shape of a cup or wineglass borne on an erect stalk up to 10 cm tall. It is one of the largest of all single-celled organisms, its **nucleus** being in the root-like base that anchors it to the substrate. There are 13 species, most found in subtropical waters.

mermaid's wineglass *See* mermaid's cup.

Mermis nigrescens (grasshopper nematode) A large nematode **(Nematoda)**, up to 60 mm long, that lays eggs on vegetation, where they are ingested by herbivorous insects, especially grasshoppers **(Orthoptera)**, earwigs **(Dermaptera)**, beetles **(Coleoptera)**, caterpillars **(Lepidoptera)**, and **Hymenoptera**. The eggs hatch inside the host, break through the wall of the gut, and enter the body cavity where they feed. After emerging from the host they drop to the soil, maturing in two to four months but remaining in the soil for two to three years.

merocoenosis *See* merotope.

Merodon equestris (greater bulb fly, large bulb fly, large narcissus fly, narcissus bulb fly) A species of hoverfly **(Syrphidae)** that closely resembles a bumblebee. It is hairy, orange, yellow, and black, with black legs, about 12 mm long, and feeds on **pollen** and **nectar** from ***Lilium*** and *Narcissus* flowers. Its larvae feed on the **bulbs** of lilies and narcissi.

meromictic Describes a lake in which part of the water is permanently stratified, usually because of a chemical difference that distinguishes the **epilimnion** and **hypolimnion.**

meromixis Genetic **recombination** in which only part of the **genome** is transferred from the donor **cell** to the recipient, forming a partial **diploid**. This most often occurs in **Bacteria** but it is also known in other organisms.

meroparasite *See* hemiparasite.

merotope A **microhabitat** within a larger **habitat**, e.g. a pebble on a river bank. The organisms occupying a merotope constitute a merocoenosis.

mesarch Describes **xylem** in which the first strands form at the centre and subsequent strands develop on both the inside and outside of them. *Compare* endarch, exarch.

mesic Describes an area that is neither extremely wet nor extremely dry.

meso- Middle.

mesocarp *See* pericarp.

mesocotyl The **axial** part of the **embryo** in grasses **(Poaceae)**, situated below the

coleoptile and formed by the fusion of part of the **cotyledon** with the **hypocotyl**.

mesocyclone Air that spirals upward inside a **supercell** cloud, turning cyclonically (*see* cyclonic). If the mesocyclone extends downward through the base of the cloud it becomes a **funnel cloud**.

mesogenous Describes a **stoma** and surrounding cells, found in some **angiosperms**, in which the **guard cells** and **subsidiary cells** are derived from the same **initial**. *Compare* perigenous. *See also* syndetocheilic.

mesophile An organism that thrives in moderate temperatures, typically 20–45°C.

mesophyll The layer of tissue in a leaf that lies between the **epidermis** and the **vascular tissue**. It is composed of **palisade parenchyma** and **spongy parenchyma** and its principal functions are **photosynthesis** and the storage of **starch**.

M

mesophyte A plant that grows in areas that are neither extremely wet nor extremely dry.

mesopores Soil **pores** that are 2–50 nm in size.

mesosoma In **Apocrita**, the thorax.

mesosome In **prokaryotes**, an infolding of the **cell wall** containing respiratory **enzymes**. It is often where **cytokinesis** commences and the **chromosome** is usually attached to it.

mesothermal climate A climate of middle latitudes in which the mean temperature in the coldest month remains higher than −3°C.

mesothorax The second segment of an insect **thorax**.

Mespilus (family **Rosaceae**) A genus of **deciduous shrubs** an small **trees** with **elliptical** leaves and **hermaphrodite** flowers with 5 **petals**. The fruit is a **pome** (medlar). There are two species, *M. germanica*, native to southeastern Europe, which has been cultivated since ancient times, and *M. canescens*, which occurs in North America and was first described in 1990.

messenger-RNA (m-RNA) A single-stranded molecule of **RNA** that is synthesized during **transcription** and that transmits genetic information from nuclear **DNA** to the **ribosomes**.

metabolic pathway The sequence of chemical reactions, catalyzed by **enzymes**, in which **metabolites** are synthesized, degraded, or transformed.

metabolism The chemical reactions, catalyzed by **enzymes**, that take place inside all living **cells** and that allow them to maintain their structure, grow, reproduce, and respond to environmental stimuli. *See* anabolism, catabolism.

metabolite A product of metabolism or intermediate in a **metabolic pathway**.

metabolome All of the **metabolites** present in a **cell**.

metabolon A complex of **enzymes** that channel intermediate metabolic products from one reaction to the next in a **metabolic pathway**.

metaboly An ability to change shape.

metacarpal One of the bones in the forelimb of a **tetrapod** that articulates with the **carpus** and the phalanges (*see* phalanx).

metacarpus The part of a tetrapod forelimb between the **carpus** and phalanges (*see* phalanx), comprising the **metacarpal** bones.

metacentric Describes a **chromosome** in which the **centromere** is in the centre. *See also* acrocentric, holocentric, telocentric.

metaldehyde A **molluscicide** ($[CH_3CHO]_4$) that is used to kill **slugs** and **snails**. It is mildly toxic to humans but slug pellets containing it are toxic to dogs and cats.

metamere *See* metameric segmentation.

metameric segmentation (metamerism) The division of an animal's body into distinct segments (also known as metameres or somites) through the repetition of organs and tissues. It is seen most clearly in **Annelida**.

metamerism *See* metameric segmentation.

metamorphic rock Rock that consists of an aggregate of minerals formed by the recrystallization of pre-existing rocks due to changes in pressure, temperature, or the content of volatile compounds.

metamorphism The processes that alter the characteristics of a rock by the recrystallization of its minerals, but without major changes to its chemical composition.

metamorphosis An abrupt and radical change in the form (Greek *morphe*) of an animal as it transforms from a larva to an adult.

metam-sodium *See* metham-sodium.

metaphase In **mitosis** and **meiosis**, the stage during which the **chromosomes** arrange themselves in the equatorial region of the **spindle**.

metaphloem **Primary phloem** that develops after the **protophloem** and completes its elongation. It persists but may be obscured by the **secondary phloem** which follows.

Metaphyta *See* Embryophyta.

Metarhizium flavoviride A species of **ascomycete fungi** that parasitizes bugs (**Hemiptera**), some beetles (**Coleoptera**), and some **Orthoptera**. It is being developed as a **mycoinsecticide** against grasshoppers.

Metasequoia (family **Cupressaceae**) A **monotypic** genus (*M. glyptostroboides*, dawn redwood), which is a **deciduous tree** with **opposite** leaves. Male **cones** are small (6 mm long), borne on long **spikes**; female cones are larger (15–25 mm diameter), **globose** to ovoid. Known only from **fossils** until it was discovered in 1948 growing in China, it is now widely cultivated for ornament, growing rapidly in moist ground and striking readily from cuttings.

metasoma In **Apocrita**, the abdomen.

metatarsal One of the bones in the hind limb of a tetrapod that articulates with the **tarsus** and the phalanges (*see* phalanx).

metatarsus The part of a tetrapod hind limb between the **tarsus** and phalanges (*see* phalanx), comprising the **metatarsal** bones.

metathorax The third segment of an insect **thorax**.

metaxenia The effect of **pollen** on the female organs of a plant, influencing the time of ripening and to some extent the size of the fruit and seed. Selecting the pollen used for **fertilization** can produce an early fruit crop and a uniform and short ripening period. *Compare* xenia.

metaxylem **Primary xylem** that develops after the **protoxylem** and before the **secondary xylem**.

Metazoa *See* Animalia.

meteoric water Water that falls from the sky; precipitation.

meteorological drought A **drought** that is defined meteorologically as a decrease in precipitation.

meteorology The study of the atmospheric phenomena and conditions that produce day-to-day weather, and the forecasting of future weather.

metham-sodium (metam-sodium) A **thiocarbamate** compound that is used to fumigate soil. It is injected below the surface where it decomposes, moving upward as a gas, methyl isothiocyanate, which disinfects the soil. It is widely used, and

M

breaks down quickly leaving no harmful residues.

methanogen A member of the **Archaea** that derives energy from using hydrogen to reduce carbon dioxide to methane ($CO_2 + 4H_2 \rightarrow 2H_2O + CH_4\uparrow$), emitting the methane, which bubbles to the surface as marsh gas. Methanogens inhabit swamps, marshes, and mud where there is no free oxygen.

methanogenic Producing methane (CH_4).

methanotroph An organism that obtains nourishment from methane (CH_4).

methoxyfenoxide An **insecticide** sold under the trade names Falcon, Intrepid, Integro, Pacer, Prodigy, Rimi, and Runner, that accelerates moulting in **caterpillars**, but is harmless to beneficial insects. It is widely used on cotton, fruit, leafy vegetables, and other crops. Inhalation can cause irritation.

M

methyl The chemical group $-CH_3$.

methylation The addition of a **methyl** group to a chemical compound. Methylation of **DNA nucleotides** alters the expression of a **gene**.

methyl bromide (bromomethane) A gas (CH_3Br) that was formerly used as an **insecticide** but is no longer used because it destroys stratospheric ozone.

methylotroph An organism that uses as its sole source of carbon compounds with a single carbon atom in their molecule, e.g. methane (CH_4) and methanol (CH_3OH), or compounds with many carbon atoms but no carbon bonds.

methyl salicylate (oil of wintergreen) An organic **ester** produced by many plants but especially *Gaultheria* species (**Ericaceae**) and commercially from ***Betula*** *lenta* (black birch, cherry birch), both native to eastern North America. The oil is used in liniments, as an antiseptic, and as a flavouring and perfume.

methylthiomethane *See* dimethyl sulphide.

Metteniusaceae A **monogeneric** family (*Metteniusa*) of tall, **evergreen trees** that has not yet been definitely assigned to an order. Leaves **alternate**, **simple**, **entire**, **exstipulate**, with **petioles**. Flowers are fragrant, **actinomorphic**, **bisexual**, with 5 **imbricate sepals**, **corolla** partly fused with 5 **reflexed** lobes, 5 **stamens**, **ovary superior**. **Inflorescence** is an **axillary cyme** with short **pedicels** bearing up to 4 **bracts**. Fruit is a **drupe**. There are seven species occurring in southern Central America and northwestern South America.

Mexican orange *See Choisya.*

Meyen, Franz (1804–40) A German physician for whom Alexander von **Humboldt** secured the position of professor of botany at the University of Berlin. Following in Humboldt's footsteps, Meyen spent 1830–32 travelling in South America. He developed his own ideas about phytogeography, which he explained in *Grundriß der Pflanzengeographie* (Outline of plant geography), 1834, in which he showed the influence of environmental factors on the vegetation type, using **isolines** to delineate floristic areas.

Mg *See* magnesium.

mice *See* Muridae.

microaggregates Soil **aggregates** that are smaller than 250 µm in size. They form by reactions between **clay minerals**, polyvalent **cations**, and soil organic matter, and their presence helps protect soil organic matter.

microbe An organism that is visible only with the help of a microscope.

microbe-associated molecular patterns (MAMPs) Molecules or fragments of molecules that are characteristic of particular **microbes** and can be used to detect their presence.

microbial (microbic) Pertaining to **microbes**.

microbial genetics The study of the **genetics** of **microorganisms**, widely practised in the study of evolution because microorganisms have short generation times and **cultures** occupy little space.

microbial pesticide A microorganism that is used to kill a specific pest.

microbic *See* microbial.

microbiology The study of **microorganisms**.

microbivore An animal that feeds on microorganisms.

microbody A small **vesicle** found in the **cytoplasm** of **cells**; **glyoxisomes** and **peroxisomes** are microbodies.

Microbotryomycetes A class of **basidiomycete yeast fungi** that cause **rust** and **smut** diseases. There are about 224 species.

microburst A strong **downdraught** that occurs beneath a weak **convection cell** at some distance from the centre of a **cumulonimbus** cloud.

Microchiroptera (bats) A suborder of bats that use echolocation to pursue prey and avoid obstacles, aided by specialized ears and in many species by modification of the nose that allow them to control the frequency and direction of their sound emissions. Most are insectivores, but some feed on fruit, nectar, fish, or mammalian blood. There are 16 families with about 760 species found worldwide except for the Arctic and Antarctica.

microclimate The climate of a small area, when this differs from the climate around it.

Micrococcus denitrificans *See* denitrifying bacteria.

microconsumer (decomposer) A **consumer** that obtains energy and nutrients by breaking down complex organic molecules in dead **protoplasm**. Most microconsumers are **Bacteria** or **Fungi**.

microevolution Evolution that occurs within **species**.

microfibril A very fine fibre, comprising a more or less crystalline aggregation of **glycoproteins** and **cellulose** found in **cell walls**.

microfilament A **filament** 0.4–0.7 nm in diameter, made from the **protein** actin, that occurs beneath the **cell wall** of **eukaryotes**. It is involved in cell motility, **cytokinesis**, and **cytoplasmic streaming**.

microfossil A **fossil** that is visible only under a microscope.

microhabitat An area of **habitat** where particular organisms may be found within a larger habitat, e.g. beneath the **bark** of a tree.

Micromonospora A genus of **Actinobacteria** that form filamentous structures similar to a fungal **mycelium**. They are Gram-positive (*see* Gram reaction), **spore**-forming **aerobes** that occur in soil and **compost**, where they help decompose organic matter. They also yield antibiotics.

micronutrient (trace element) An **essential element** that plants require in relatively small amounts. These include iron, manganese, zinc, **copper**, **chlorine**, **boron**, molybdenum, and **cobalt**.

microorganism A **bacterium** (**Bacteria**), fungus (**Fungi**), **alga**, protozoon (**Protozoa**), or **virus** that can be seen only with the help of a microscope.

microphyll A leaf with a single **vein** and usually with no **leaf gap** associated with the **leaf trace**. *Compare* megaphyll.

micropores Soil **pores** smaller than 30 μm.

micropylar *See* micropyle.

micropyle (adj. micropylar) A small opening in the surface of an **ovule** through which the **pollen tube** passes during **fertilization**.

microRNA (miRNA) A small, single-stranded molecule of **RNA**, usually 21–24 **nucleotides** long (21-mers to 24-mers), with a function that depends on its length, e.g. 21-mers are involved in the degradation of **messenger-RNA** and 24-mers in **gene silencing**.

microsome A **vesicle**-like structure formed from pieces of the **endoplasmic reticulum** when a **eukaryote** cell is broken up.

microspecies The descendants of a plant that reproduces by **apomixis**. They are genetically uniform and although they exhibit phenotypic (*see* phenotype) variation, this is much smaller than in members of a **species**.

microsporangium In heterosporous (*see* heterospory) plants, a **sporangium** that produces **microspores**. In **gymnosperms** and **angiosperms** the microsporangium produces a microsporocyte (microspore mother cell), which forms four microspores by **meiosis**.

microspore In heterosporous (*see* heterospory) plants, the smaller of the two types of **spore**, which develops into a male **gametophyte**.

microspore mother cell *See* microsporangium.

microsporocyte In a **microsporangium**, a cell that divides by **meiosis** to produce four **microspores**.

microsporophyll A leaf or leaf-like structure that bears microsporangia (*see* microsporangium), e.g. the male **cones** of conifers. The **stamens** of **angiosperms** are modified microsporophylls. *See* megasporophyll.

Microstromatales An order of **Exobasidiomycetes smut fungi** that form **yeast** cells. There are 3 families with 11 species.

Microteaceae (order **Caryophyllales**) A **monogeneric** family (*Microtea*) of poorly known **annual herbs** with spiral leaves and flowers in groups of 3 on a **racemose inflorescence**, with 4 or 5 **petals**, 5–9 free or 2–9 fused **stamens**. Fruit is an **achene**. There are nine species occurring in Central and South America.

microtherm A plant that occurs in cool temperate regions where the average temperature in the warmest month is 10–22°C and does not fall below 8°C in the coldest month.

microthermal climate (moist subhumid climate) A climate of middle latitudes in which the mean temperature in the coldest month is lower than −3°C.

microtine cycle The **density dependent**, cyclical fluctuations in population, involving mass migrations, that affect certain species, e.g. locusts and the lemming (*Myodes lemmus*).

microtubule A tube, made from the **protein** tubulin, 15–25 nm in diameter, that are part of the **cytoskeleton** and occur in large numbers throughout the **cytoplasm** of all **eukaryote** cells and as components of cilia (*see* cilium) and flagella (*see* flagellum). They are involved in cell motility, help maintain the shape of the cell, and form part of the **mitotic spindle**.

Micrurus fulvius (eastern coral snake, common coral snake, American cobra, harlequin coral snake) A venomous snake (**Elapidae**), 600–900 mm long, that has a black snout and red, black, and yellow or white bands around its body, the yellow or white bands separating the red and black ones. It inhabits open woodland with decaying logs and surface rocks, and spends much of its time buried. It feeds on snakes, including its own species, other small vertebrates, and insects. The snake occurs throughout southeastern and southern central United States and

Mexico. Although venomous, the snake is not aggressive and bites are rare.

middle arctic tundra The **tundra** vegetation that grows on level ground along coastal plains with a thin **active layer**, and areas that are waterlogged. It includes dwarf heaths, *Sphagnum* moss, and sedges.

middle lamella A membrane that separates two adjacent **cell walls** and cements them together. The membrane consists mainly of **pectins**. It is the first layer to form during cell division and forms the outer wall of the cell, shared with the adjacent cell. it gives the cell the strength to withstand the **pressure potential** inside the cell. *See also* primary wall, secondary wall.

midges Small flies (**Diptera**) that are found throughout the world except for deserts, Antarctica, and the high arctic. They belong to several dipteran families and there are many species. Some feed on **nectar**, trigger **gall** formation on plants, or parasitize other insects. Biting midges (no-see-ums, punkies, family Ceratopogonidae) feed on humans and other mammals.

midlatitude mixed forest A type of forest found in middle latitudes that contains both coniferous and **broad-leaved** trees. Some such forests are a true climatic **climax**, others form an **ecotone** between coniferous forest and **broad-leaved deciduous** forest. Midlatitude mixed forests occur in southern Brazil, Chile, Tasmania, northern New Zealand, and South Africa's Cape Province.

midlatitude westerlies The prevailing winds, blowing from west to east, in the middle latitudes of both hemispheres.

midrib The thick structure that runs along the centre of a leaf, **thallus**, or leaf-like structure of seaweeds (algae) and mosses (**Bryophyta**). It provides support and in true leaves it is a **vein**.

mignonette (*Reseda odorata*) *See* Resedaceae.

migration 1. The movement of an organism or its **propagules** outward from an area (emigration), inward (immigration), or the periodic movement of an animal in both directions (migration). **2.** The movement of a plant **migrule** in an area from which plants were recently cleared.

migrule A migrating plant **propagule**. *See* migration.

mildew A white or pale film or patch consisting of fungal **hyphae** covering a surface. In horticulture the term usually refers to powdery mildew (*see* Erysiphales).

milk cap *See Lactarius.*

milkweed *See Asclepias.*

millet *See Panicum.*

Mimosa (family **Fabaceae**) A genus of **herbs** and **shrubs** that have multi-**pinnate** leaves and **stipules** sometimes resembling thorns. Flowers are small, **actinomorphic**, with a **valvate corolla** and fewer than 10 **stamens** (a feature that distinguishes *Mimosa* from the closely related ***Acacia***. There are about 400 species occurring throughout the tropics. Some species, e.g. *M. pudica*, the sensitive plant, show remarkable sleep movements (*see* nictonasty) in which the leaves droop in response to light and darkness, and also when stimulated mechanically.

Mimosoideae (family **Fabaceae**) One of the three subfamilies of the Fabaceae (sometimes ranked as a family, Mimosaceae), comprising mostly tropical and subtropical **shrubs** and **trees** often with **bipinnate** leaves and regular flowers with 10 or more **stamens**. There are 82 genera with 3275 species.

Mimulus (family **Phrymaceae**) A genus of **annual**, **perennial**, or **evergreen herbs** or **shrubs** with paired leaves. Flowers with a 2-lipped, tubular or trumpet-shaped

M

corolla. **Inflorescence** a terminal or **axillary raceme**. There are about 150 species occurring mainly in western North America and Australia, but also in southern Africa and Asia. Several species are cultivated for ornament and known as monkey flowers because the shape of their flowers, or markings on them, are reminiscent of a monkey's face.

Mimus polyglottos (northern mockingbird) A species of grey-brown birds with white patches on the wings, in which males are 220–255 mm long and females 210–235 mm. They inhabit woodland edges, open land, roadsides, parks, and residential areas wherever there is somewhere tall to perch, and feed on seeds, berries, insects, and other small animals. They are renowned for their ability to imitate sounds. The birds occur throughout North America.

mineral **1.** A natural inorganic substance that has a characteristic chemical composition and a crystalline structure by which it can be identified. Rocks are made from minerals. **2.** Any substance, including those of organic origin, which is obtained by mining.

mineralization The conversion of an organic compound to an inorganic compound by the action of living organisms. *See* ammonification.

mineral soil A soil that consists mainly of mineral particles and has characteristics that are determined more by the mineral than by the organic content.

minimum thermometer A thermometer that records the lowest temperature it registered since it was last reset.

minimum tillage A management technique in which crop residues are left on the ground surface rather than being incorporated into the soil, in order to reduce soil **erosion** by minimizing the number of tillage operations.

mink *See* Mustelidae.

mint *See* Lamiaceae, *Mentha*.

mint bush *See Prostanthera*.

mint moth *See Pyrausta aurata*.

minute pirate bugs *See* anthocorid bugs, *Orius*.

mire Wet, muddy ground, consisting mainly of **peat**.

Miridae (capsid bugs, grass bugs, leafbugs, plant bugs) A family of bugs (**Hemiptera**), 2–11 mm long with overlapping wings that sometimes form a Y or X shape on the back. They may be brown, red, yellow, or black, and have long antennae (*see* antenna). Some feed on **nectar** and some are predators, but most feed on plants and are very destructive pests. There are more than 10,000 species found worldwide.

mi-RNA *See* microRNA.

Misodendraceae (order **Santalales**) A **monogeneric** family (*Misodendron*) of **deciduous** stem parasites that grow as small, shrubby plants mainly on **branches** of ***Nothofagus***, attached by a **haustorium** that sometimes extends beneath the **bark** and extends secondary shoots some distance away. Leaves **alternate**, **entire**, **exstipulate**, and sometimes reduced to scales. Flowers small, usually **unisexual** (plants **monoecious** or **dioecious**), male flowers with no **perianth** segments and 2 or 3 **stamens** arising from a central cushion, female flowers with 3 perianth segments and 3 **staminodes**, **ovary superior**. **Inflorescence** a **raceme**, **spike**, or cluster. Fruit is an **achene**. There are eight species occurring in cool temperate South America.

mis-sense mutant A **mutant** in which the **mutation** has altered a **codon** so that it encodes a different **amino acid**. Almost invariably this results in the formation of an unstable or inactive **enzyme**.

mist Precipitation comprising droplets 0.005–0.05 mm in diameter that fall very slowly. Horizontal visibility is reduced, but remains more than 1 km.

mistletoe *See Viscum.*

mistletoe cactus *See Rhipsalis.*

mites *See* Arachnida.

mitochondrial-DNA (mt-DNA) Circular molecules of **DNA** that occur in mitochondria (*see* mitochondrion). It is entirely separate from nuclear DNA, with a few exceptions is transmitted through the female line, and codes for particular **RNA** components of **ribosomes**. Plant mt-DNA evolves very slowly (unlike animal mt-DNA).

mitochondrion A semi-autonomous **organelle** that occurs in large numbers in the **cytoplasm** of all **eukaryote** cells. Most mitochondria are oval (sometimes thread-like or spherical) and about 2 μm long, with an outer and inner membrane folded into cristae (*see* crista). A mitochondrion has its own **DNA** and **ribosomes** and reproduces by **binary fission**. Mitochondria are the principal site of **adenosine triphosphate** (ATP) production, and of the **enzymes** involved in the **citric-acid cycle** and **oxidative phosphorylation**. ↗

mitogenic Able to trigger **mitosis**.

mitosis The process of nuclear division that takes place at cell division and results in the formation of two **daughter nuclei**, each of which is identical to the parent nucleus. ↗

mitotic spindle The **spindle** that forms in **eukaryote** cells during **mitosis**.

Mitrastemonaceae (order **Ericales**) A **monogeneric** family (*Mitrastemon*) of root parasites (*see* parasitism), lacking **chlorophyll**, commonly found on **Fagaceae**. There are no stems. Leaves **opposite**, **decussate**, and scale-like. Flowers are white, terminal, solitary, **actinomorphic**, **bisexual**, **perianth** with 4 lobes, **stamens connate**, **ovary superior** of 9–15 free or 9–20 fused **carpels** with 1 **locule**. Fruit a **capsule**. There are two species with a scattered distribution in southeastern Asia, Malesia, Central America, and northwestern South America.

mixed cloud Cloud that contains both water droplets and ice crystals.

mixed sorus In certain ferns (**Pteridophyta**) a **sorus** in which the sporangia (*see* sporangium) develop over a prolonged period and in no definite order. This is considered the most evolutionarily advanced type of sorus. *Compare* gradate sorus, simple sorus.

mixed woodland A **woodland** that contains both coniferous and **broad-leaved** trees, with the less abundant type comprising at least about 20 percent.

mixing ratio (mass mixing ratio) The ratio of the mass of any gas present in the air to unit mass of air without that gas, expressed in grams of the gas per kilogram of air without the gas. It is most often used to report **humidity**, as gH_2O/kg air.

Mixiomycetes A class of **Fungi** in the **Pucciniomycotina** that contains one genus, *Mixia*, of which only one species is known, *M. osmundae*. It has multinucleate **hyphae** and is a parasite of ferns in the genus *Osmunda*, causing yellow to brown leaf spots. The fungus is known only from Japan, Taiwan, and the United States.

Mn *See* manganese.

mobbing Behaviour in which a group of prey animals collaborate in harassing a predator.

mock orange *See Choisya, Philadelphus.*

moder A type of **humus** that is intermediate between **mull** and **mor**, with an accumulation of humus near the surface.

moderate breeze Wind of 6–8 m/s. *See* appendix: Beaufort Wind Scale.

moderately deep soil *See* effective soil depth.

modifier gene A **gene** that modifies the expression of another gene.

Mohs's scale of hardness A scale devised in 1812 by the German mineralogist Friedrich Mohs, that ranks the hardness of materials by their ability to scratch one another. The scale runs from 1 to 10, 1 being the softest.

1. talc	6. orthoclase
2. gypsum	7. quartz
3. calcite	8. topaz
4. fluorite	9. corundum
5. apatite	10. diamond

moist climate A climate in which the annual precipitation is greater than the **potential evapotranspiration**.

moist subhumid climate *See* microthermal climate.

M

moisture inversion A layer of air in which the **humidity** increases with height.

Moko disease *See Ralstonia solanacearum.*

molar heat capacity *See* heat capacity.

molar mass *See* molecular weight.

mold *See* mould.

mole *See Talpa europaea.*

molecular clock The concept that at the molecular level evolution proceeds at a constant rate, so that the amount of difference in the **amino acid** content of their **proteins** can reveal how much time has elapsed since two species diverged from their common ancestor.

molecular drive The concept that an inherited **mutation** can spread through a population until the point at which many individuals with that mutation appear to arise at once; i.e. evolution can be affected by changes within the **genome** that are not influenced by **natural selection**.

molecular evolution The substitution of one **amino acid** for another during the synthesis of **proteins** due to a **mutation** in a **gene**.

molecular weight (molar mass, relative molecular mass) The weight of a molecule, calculated as the atomic weight of each constituent atom multiplied by the number of each atom in the molecule. It is measured in Daltons: 1 dalton = 1/12 of the weight of an atom of the **isotope** carbon-12 (^{12}C). The amount of a substance equal to the molecular weight of that substance is 1 mole (mol); the mole is the international standard unit of amount of substance.

mole drain A **drainage** channel made in a soil by dragging a bullet-shaped device through the soil at the desired depth, the pressure from the device compacting the sides of the hole so it remains open for several years.

mollic horizon A well structured, dark-coloured surface **soil horizon** that contains at least 1 percent organic matter (0.6 percent organic carbon) and a **base saturation** of 50 percent or more.

Mollicutes A class of **Bacteria** that lack **cell walls**. Most are parasites of animals, causing serious diseases in humans, and of more than 300 species of plants. The class contains 4 orders, 6 families, 12 genera, and 2069 species.

mollisols Soils that have a deep **mollic horizon** above a **mineral soil** with a high **base saturation**. Mollisols comprise an order in the U.S. Department of Agriculture **soil taxonomy**.

Molluginaceae (order **Caryophyllales**) A family of **annual** or **perennial** slightly **succulent herbs** or **subshrubs**. Leaves **alternate**, occasionally **opposite** or whorled, **entire**, **exstipulate** or with membranous **stipules**. Flowers **actinomorphic**, **hermaphrodite** occasionally **unisexual** (plants

dioecious), with 5 free or 4 fused **tepals** and some with 5 free or 5–8 fused **petaloid staminodes**, 4–5 free or 3–8 fused **stamens**, **ovary** of 2–5 (sometimes 1) **carpels**. **Inflorescence** usually a terminal **cyme**, occasionally terminal or solitary flower. Fruit is a **capsule**. There are 9 genera of 87 species occurring throughout tropical, subtropical, and warm temperate regions.

Mollusca A **phylum** of **coelomate** invertebrate animals that are highly diverse in form. They include bivalves (e.g. mussels), cephalopods (octopuses, squid, etc.), **Gastropoda**, and others. Molluscs possess **bilateral symmetry** and **metameric segmentation** is uncommon. Some have shells that form internal skeletons, many have a single muscular foot. There are at least 50,000 species occurring throughout the world.

molluscicide A chemical compound that kills **slugs** and **snails**.

Moluthrus (cowbirds) *See* Icteridae.

molybdenum (Mo) An element that is an essential **micronutrient** for plants, forming part of the **enzymes** nitrogenase and also nitrate reductase, which catalyzes the reduction of nitrate (NO_3^-) to nitrite (NO_2^-). Deficiency inhibits growth, produces irregularities in leaves, and causes **chlorosis**.

monadelphous Describes **stamens** with **filaments** that are fused together, often forming a tube.

Monera In the **five-kingdom system** of **taxonomy**, a kingdom that contains all the single-celled **prokaryotes**. In the more widely used **three-domain system**, prokaryotes are placed in the domains **Archaea** and **Bacteria**.

money spiders *See* Linyphiidae.

moniliform Resembling a necklace or string of beads.

Monilinia fructicola (*Sclerotinia fructicola*) A species of **ascomycete fungi** that cause brown rot of stone fruits. The fungus overwinters on fruit previously shrivelled (mummified) by the infection and in **cankers**. Infection appears first as small brown spots on fruit that grow until the entire fruit is covered and rots. The fungus occurs in North, Central, and most of South America, southern Africa, Australia, and New Zealand, but is absent from the European Union.

Monimiaceae (order **Laurales**) A family of **trees**, **shrubs**, and scrambling **lianas** with **opposite** occasionally whorled, **simple**, **entire** or **serrate** to **dentate**, **exstipulate** leaves. Flowers **actinomorphic**, usually **unisexual** rarely **hermaphrodite**, **sepaloid**, **petaloid** or calyptrate (*see* calyptra) **perianth** of 3 to many free or **connate tepals**, few to very many (1800) **stamens**, **ovary superior**, **apocarpous**, of 1 or up to 2000 **carpels**. Flowers cauliflorous (*see* cauliflory) or in **axillary racemes**. Fruit is an aggregate of **drupelets** enclosed by or embedded in the **receptacle**. There are 22 genera of 200 species occurring in the tropical Southern Hemisphere.

monkey flowers *See Mimulus*.

monkey puzzle (*Araucaria araucana*) *See Araucaria*.

monocalcium phosphate *See* superphosphate.

monocarpic Producing only 1 **carpel**.

monocentric Describes a **thallus** with a single reproductive centre.

monochasium A **cymose inflorescence** that consists of a single **axis** terminating in a flower.

monoclimax The theory proposed by Frederic Edward **Clements** that plant communities develop toward a definite **climax** determined by climate.

monoclinous Having functional **stamens** and **pistils** in the same flower.

monocotyedon An **angiosperm** in which the **embryo** typically has 1 **cotyledon**; this is usually **amplexicaul** and, in common with the later leaves, with parallel **nervation**.

monoculture A single crop species or variety that is grown over a large area to the exclusion of other species or varieties.

monoecious Having male and female reproductive organs on the same individual (e.g. plant).

monogeneric Describes a **taxon** containing only one **genus**.

monohybrid A **hybrid** of two individuals that are identically heterozygous (*see* heterozygosity) for the **alleles** of a particular **gene**, i.e. $Aa \times Aa$.

monoicious Of a moss (**Bryophyta**), having a **gametophyte** that bears both antheridia (*see* antheridium) and archegonia (*see* archegonium) on the same **gametophore**.

monokaryon A fungal **hypha** or **mycelium** in which each cell contains only one nucleus.

monolete Describes a **spore** marked by a single line, showing where the **sporocyte** split into four. *Compare* trilete.

monomictic Describes a lake in which the water circulates freely during only one season.

monomorphic enantiostyly *See* enantiostyly.

monophyly The condition of taxa (*see* taxon) that are all descended from a common ancestor.

monophyodont Describes an animal that possesses a single set of teeth which are not replaced.

monoplanetism In some water moulds (**Oomycota**) the occurrence of only one type of **zoospore** with only one swarming period.

monopodial 1. Having a single **axis**. **2.** Branching in which **branches** arise laterally from a central stem.

Monopodial. Lateral branches grow from a single stem.

monosaccharide The simplest **sugar** with the formula $C_x(H_2O)_y$ where $x \geq 3$ and the monosaccharide is classified by the number of its carbon atoms; **glucose** is a **hexose** sugar.

monosomic genome A **diploid genome** that has only one copy of a particular **chromosome**, so its chromosome number is $2n - 1$.

monosporangium In **red algae** (Rhodophyta), a **cell** that produces **monospores**.

monospore In **red algae** (Rhodophyta), a non-**motile**, asexual **spore**.

monostele A **stele** that comprises a single vessel.

monosulcate Describes a **pollen grain** possessing a single **sulcus**.

monotelic inflorescence *See* determinate inflorescence.

monotypic Describes a **taxon** containing only one **species**.

monounsaturated *See* fatty acid.

monozygotic polyembryony *See* polyembryony.

monsoon forest Seasonal tropical forest that occurs in Asia, where the monsoon climate alternates extremely dry weather with very heavy rains.

Monstera (family **Araceae**) A genus of **herbs** and **evergreen** climbers with aerial roots some of which hook over **branches** of trees they use for support and others that grow into the ground. Leaves **alternate**, large, **coriaceous**, and often with holes. **Inflorescence** is a **spadix**. Fruit is a **berry**. *Monstera deliciosa* (cheese plant) is widely cultivated as a houseplant, and for its edible berries.

montane forest Forest that grows on a mountain.

month degrees An expression of the conditions for plant growth that is used in some **climate classification** schemes. It is calculated by subtracting 6°C from the **mean temperature** for each month, the remainders being the number of degrees by which the temperature is above or below 6°C, 6°C being the minimum temperature for the growth of many plants.

Montiaceae (order **Caryophyllales**) A family of **annual** to **perennial herbs**, often fleshy, with **alternate** or **opposite**, **simple** leaves. Flowers **actinomorphic**, **bisexual**, with 2 sometimes 9 free **sepals**, 2–19 (sometimes 1) free or partly fused **petals**, 1 to many **stamens**, **ovary superior**. **Inflorescence** an **axillary** or terminal **cyme**, **raceme**, **panicle**, **umbel**, or flowers solitary. Fruit is a **capsule**. There are about 22 genera with about 230 species occurring in temperate regions of America, Asia, Eurasia, and Australia. Some cultivated for ornament.

Montiniaceae (order **Solanales**) A family of **shrubs** and **trees** with **alternate**, **opposite**, or subopposite, **simple**, **entire** leaves. Flowers **actinomorphic**, 4 free **petals**, 4 **stamens** alternating with the petals, **ovary inferior**. **Inflorescence** in male flowers a terminal or **axillary panicle**; female flowers solitary or in pairs, terminal. Fruit is a **capsule** or **drupe**. There are three genera with five species occurring in Africa and Madagascar.

montmorillonite A **clay mineral** that is able to absorb large amounts of water by expanding without altering its mineral structure, which comprises layers, each with one sheet of octahedral aluminium crystals bounded by two sheets of tetrahedral silicon crystals. It forms by the decomposition of volcanic ash.

moonflower *See Datura.*

moor An area of **acid soil** with abundant **peat**, usually at a high elevation, dominated by low **shrubs**, especially of the **Ericaceae**, with some rough grassland and sedges.

mor A surface **soil horizon** made from organic matter in varying stages of decomposition. It is acid and contains no microorganisms other than **Fungi**. There is a sharp boundary between **humus** and the underlying **mineral soil**.

Moraceae (order **Rosales**) A family of **shrubs**, **trees**, **lianas**, and a few **herbs**, often with prickles or thorns, some lianas stranglers (*see* strangling fig). They produce a milky **latex**. Leaves **alternate**, spiral, or **distichous**, rarely **opposite**, **simple** occasionally **palmately lobed**, rarely **pinnate**, **entire** to **dentate**, **stipules** often **amplexicaul**. Flowers **actinomorphic** or one-sided in reduced female flowers, **unisexual** (plants **monoecious** or **dioecious**), 2–6 or more **tepals** or tepals absent, 1–4 **stamens**, **ovary superior** of 2 **carpels** with 1 **locule**. **Inflorescence** a **raceme**, **spike**, **globose** head, flattened, or a **syconium** enclosing many flowers (fig). Fruit is an **achene** or **drupe**. There are 39 genera with 1125 species occurring in tropical and warm temperate regions. Many cultivated for timber, fibres, medicinal products, or edible fruits, e.g. ***Ficus*** (fig), ***Morus*** (mulberry), and ***Artocarpus*** (breadfruit, jackfruit).

Morchella (morel) A genus of **ascomycete fungi** with **fruiting bodies** that are variable in appearance, but have a conical or ovate **pileus** covered with ridges and pits, and no **stipe**. They either form **mycorrhizae** with trees or live as **saprotrophs**. There are about 50 species occurring throughout the Northern Hemisphere. They are edible and highly prized, provided they are cooked correctly, but they are often found close to the poisonous false morel (*see Gyromitra esculenta*).

morel *See Morchella.*

Moringaceae (order **Brassicales**) A **monogeneric** family (*Moringa*) of **deciduous trees** or **shrubs**, many with swollen trunks. Leaves **opposite**, **compound**, 1–3 **imparipinnate.** Flowers **actinomorphic** to strongly **zygomorphic**, **pentamerous**, **hermaphrodite**, with 5 free **sepals**, 5 free **petals**, 5 **stamens** alternating with **staminodes**, **ovary superior** of 3 fused **carpels** with 1 **locule. Inflorescence** an **axillary thyrse.** Fruit is a **capsule.** There are 12 species occurring in Africa and Madagascar to India. *Moringa oleifera* is being developed as a leaf-vegetable crop plant.

morning glory *See Calystegia.*

morph The reproductive stage in **Ascomycota** and **Basidiomycota.**

morphactins A group of plant **growth regulators**, based on fluorine-carboxylic acid, that are used to inhibit **germination** of weed seeds, root and shoot growth in seedlings, and **apical dominance** resulting in increased branching and tillering, to promote stem elongation resulting in dwarfing, and prolongation of **bud dormancy**, and to alter the timing and sequence of flowering.

morphology The structure and form of an organism.

morphospecies A group of individuals that are very similar to each other, but differ from all other groups by virtue of their **morphology.** For convenience such a group can be studied as though it were a **species.**

Morus (family **Moraceae**) A genus of **deciduous**, some **evergreen**, **trees** with **alternate**, **simple**, often lobed, **serrate** leaves. The edible fruits (mulberries) are aggregates of **drupelets**, resembling blackberries. There are 10–16 species with a **cosmopolitan distribution.** Many are cultivated for ornament, fruit, and for their foliage, which is food for silk worms.

mosaic evolution Different rates of **adaptation** within the same lineage, e.g. members of a particular **taxon** might differ in the rates of change of their leaves, roots, etc.

mosaic virus Any **virus** produces angular patches of discoloration on leaves, reminiscent of a mosaic.

moschatel (*Adoxa moschatellina*) *See* Adoxaceae.

moss *See* Bryophyta, Bryopsida.

moss-gall *See* robin's pincushion gall.

moss piglets *See* Tardigrada.

mossy forest A tropical **montane forest** dominated by trees, most 10–15 m tall, with dense crowns and **branches** and trunks festooned with mosses, **lichens**, and liverworts (**Marchantiophyta**).

mother-in-law's tongue (*Sansevieria trifasciata*) *See Sansevieria.*

moths *See* Lepidoptera.

motile Capable of moving independently.

mottle Any **virus** disease that produces diffuse or round patches of discoloration on leaves.

mottled worm *See Aporrectodea icterica.*

mould (mold) A fungus that produces a prominent, woolly **mycelium** with visible conidia (*see* conidium).

mountain ash (*Sorbus aucuparia*) *See Sorbus.*

mountain bluebird *See Sialia currucoides.*

mountain breeze A **katabatic wind** that occurs in some mountain regions.

mountain climate A climate that differs from that of the surrounding region by virtue of elevation. Compared with the regional climate, the mountain climate is cooler, windier, and wetter below the **snow line** but drier above it, because the air has lost most of its moisture.

mountain-gap wind (canyon wind, gorge wind, jet-effect wind) A wind that occurs locally when air is funnelled (*see* funnelling) between two mountains and accelerates.

mourning dove *See Zenaida macroura.*

mouse-ear snail *See Mysotella myosotis.*

m-RNA *See* messenger-RNA.

MSY *See* optimum yield.

mt-DNA *See* mitochondrial-DNA.

mucilage A thick, glue-like substance, produced by many plants and some **microorganisms**, consisting of a complex of **carbohydrates**. It is slimy and gelatinous when wet but dries hard.

mucin *See* mucoprotein.

muck soil A soil that consists mainly of **humus**.

Mucomycotina A subphylum of **Fungi**, with 325 species, most of which are **saprotrophs**, **mycoparasites**, or plant pathogens; some cause disease in humans. They produce **moulds** on fruit, e.g. strawberries, and are among the fastest-growing of all fungi.

mucoprotein A **conjugated protein** in which the **prosthetic group** is an **oligosaccharide** and the molecule is more than 3–4 percent **carbohydrate**. Mucoproteins are very viscous and were formerly called mucins.

Mucorales (pin moulds) An order of **Zygomycetes Fungi** that typically produce a coenocytic (*see* coenocyte), **eucarpic mycelium** with upright **sporangiophores**. **Asexual reproduction** occurs continuously. Most are **saprotrophs**, common in soil and on dung and decomposing plant matter, others are parasites or pathogens. They cause damage to food. There are about 300 species found worldwide.

mucro A sharp point or tip.

mucronate Describes an organ, e.g. a leaf, with a sharp tip (**mucro**).

mud cracks *See* desiccation cracks.

mud rain Rain that contains fine soil particles.

Muehlenbeckia (family **Polygonaceae**) A genus of **evergreen** or **deciduous** mat-forming **shrubs** and vigorous climbers (maidenhair vine, wirevine) with **alternate**, **cauline**, **linear** to **orbicular**, **simple** leaves with a sheathing membrane that unites the **stipules**. Stems prostrate, erect, or **scandent**. Flowers **bisexual** or **unisexual** (plants **dioecious** or **polygamous**), **apetalous**, **sepals** enlarged and fleshy, solitary or in a terminal or **axillary raceme**. Fleshy **calyx** resembles a **berry** and is edible in some species. Fruit is a triangular **nut**. There are 11 species occurring in Australasia and South America. Several are cultivated for ornament.

mulberry (*Morus*) *See* Moraceae.

mulch A surface **soil horizon** made from loose material. It protects the soil from the impact of rain (*see* cap) and rapid temperature changes, conserves moisture by reducing surface **evaporation**, and suppresses weeds. Mulches may occur naturally or be applied.

mull A surface **soil horizon** that is well aerated, alkaline, and provides ideal conditions for the decomposition of organic matter and the formation of **humus**. Organic and mineral material are well mixed.

Müllerian mimicry A form of mimicry, first described in 1879 by the German-born Brazilian zoologist Fritz Johann Friedrich Müller, in which two species of animals resemble each other and both are distasteful to predators. Predators that taste a member of one species learn to avoid both.

multicoloured Asian lady beetle *See Harmonia axyridis.*

M

multifactorial *See* polygenic.

multilocular Having many small cavities (**locules**).

multiple allelism The existence of several **alleles** of a particular **gene**.

multiple land-use strategy The management of an area to allow several compatible activities or uses of the land to be practised, or with the aim of rendering different uses mutually compatible.

multiprotein complex *See* protein complex.

multiseriate In several rows.

multivalent Describes the association of three or more **chromosomes** during the **prophase** stage of **meiosis**.

Muntingiaceae (order **Malvales**) A family of **shrubs** and small **trees**. Leaves **alternate**, **simple**, **serrate**, asymmetric and **cordate** at the base, and with **stipule**-like structures at each leaf. Flowers **actinomorphic**, **bisexual**, with 5 fused, **valvate sepals**, 5 **imbricate**, **caducous petals**, many **stamens**, **ovary inferior** of 5 **carpels** and **locules**. Flowers solitary or in clusters. Fruit is a **berry**. There are three genera with three species occurring in tropical America. *Muntingia calabura* is an important pioneer tree in forests, and is sometimes cultivated in the tropics for ornament or its **bark** or edible fruit.

murein (peptidoglycan) A substance comprising **polysaccharide** units linked by short **amino acid** chains to form a rigid structure that is the principal component of **cell walls** in **Bacteria**.

Muridae (mice, rats, gerbils) An Old World family of small, burrowing, terrestrial, arboreal, and semi-aquatic rodents (**Rodentia**) that have a slender body, long, scaly tail, and a pointed muzzle with prominent **vibrissae**. They inhabit a wide variety of **habitats**. ***Apodemus sylvaticus*** is the field or wood mouse, *Mus musculus* the house mouse. There are about 260 genera with about 1150 species, making this the largest of all mammal families. They occur throughout Eurasia and have been introduced worldwide.

muriform Patterned like a brick wall.

Murraya (family **Rutaceae**) A genus of **trees** and **shrubs** with **pinnate** leaves and fruit that is a **berry**. There are 12 species occurring from eastern Asia to the Pacific Islands. Several cultivated for ornament and hedging. *Murraya koenigii* (curry leaf) is used in Indian cooking.

Musa (family **Musaceae**) A genus of very large **herbs** with **rhizomes** that have an erect **pseudostem** formed from overlapping leaf bases. Flowers **unisexual**, the males terminal and subtended by coloured **bracts**. Flowers pollinated by bats. The fruit is an elongated **berry** with many seeds; these are absent from edible varieties. There are 66 species occurring throughout the tropics. Cultivated varieties (banana, plaintain) are **triploid** or tetrapoloid (*see* tetraploidy) **hybrids**.

Musaceae (order **Zingiberales**) A family of **monocotyledon**, large or gigantic **herbs** with **pseudostems** formed from the overlapping bases of the huge leaves. Leaves arranged spirally, **entire**, with a distinct **petiole**. Because of their large size, leaves are often split longitudinally by the wind. Flowers **zygomorphic**, usually **unisexual**, female at the base of the plant, males in terminal clusters on the same plant, 6 **petaloid perianth** segments in 2 **whorls**, 5 **stamens**, rarely 6, 1 small **staminode**, **ovary inferior** of 3 fused **carpels** and **locules**. Pollination is by bats, also by sunbirds and tree shrews. Fruit is a **berry** with many seeds, which are absent in **triploid** and tetraploid (*see* tetraploidy) **hybrids**. There are 2 genera with 74 species occurring in Africa, the Himalayas to southeastern Asia, the Philippines, and Australasia. ***Musa*** species cultivated for food (banana, plantain) and fibre (*M.*

textilis yields abaca, Manila hemp); *Ensete ventricosa* (Abyssinian banana) grown for food and fibre.

Musci Mosses, a former class of plants that were formerly placed in the **Bryophyta** together with the liverworts and hornworts. These have now been separated, leaving the mosses as the only members of the Bryophyta.

muscicolous Growing on or among mosses (**Musci**).

mushroom An edible fungal fruit body (*see* fruiting body). *Compare* toadstool.

muskeg An area of wet, boggy (*see* bog) ground with poor **drainage** that occurs inside the **boreal forest** and is dominated by *Sphagnum* moss with scattered, stunted trees.

muskmelon yellows virus *See* beet pseudo-yellows virus.

muskrats *See* Cricetidae.

mussel scale *See Lepidosaphes ulmi.*

mustard oil bomb *See* myrosin cell.

Mustelidae (weasels, stoats, badgers, fishers martens, mink, otters, polecats, wolverines) A family of carnivorous mammals that range in body length from 114 mm to more than 1 m. They have long bodies, short legs, short ears, long, sharp canine teeth, well-developed **carnassials**, and a **plantigrade** or **digitigrade** gait. They inhabit a wide range of **habitats** and some (otters, mink) are semi-aquatic. There are 22 genera of 56 species distributed worldwide except for Antarctica, Australia, Madagascar, and many ocean islands.

mutagen Any agent that is capable of causing a **mutation**.

mutagenic Causing **mutation**.

mutant A **gene**, **cell**, or organism that carries a **mutation**.

mutation A change in the structure of a **gene** or set of **chromosomes** or in the amount of genetic material within an organism.

mutation rate 1. The number of **mutations** per **gene** per cell generation. **2.** The frequency with which mutations occur in a particular **species** or population.

Mutinus caninus (dog stinkhorn) A species of **agaric fungi** in which the **fruiting body** is 80–150 mm tall with a honeycombed **pileus** beneath a sticky, smelly **gleba** containing the **spores**, on a fragile **stipe** 10–15 mm wide. The gleba emits a foul odour that attracts insects, which disseminate the spores. The fungus is a **saprotroph** usually found in conifer forests and close to rotting wood, sometimes forming **fairy rings**. It occurs throughout western Europe and North America.

mutual inhibition competition type Direct **competition** between two species in which each inhibits the other.

mutualism An association between two species that benefits both.

mycelial cord (mycelial strand) A linear aggregation of fungal **hyphae** that are bound together in a matrix with the hyphae fused at intervals. They develop in response to a shortage of nutrients. Each cord consists of wide, empty hyphae surrounded by narrower sheathing hyphae. They are able to transport nutrients and water over long distances.

mycelial strand *See* mycelial cord.

mycelium A mass of fungal **hyphae**, comprising the vegetative part of a fungus.

Mycetozoa The **phylum** that includes all **slime moulds**. These may be **amoeboid**, amoebo-flagellate (*see* flagellum), or plasmodial (*see* plasmodium), amoeboid and amoebo-flagellate members able to form **cysts**. Under appropriate conditions they form **fruit bodies**.

mycobiont The fungal symbiont (*see* symbiosis) in a **lichen**.

mycocecidium A **gall** (cecidium) formed by the action of a fungus.

mycoinsecticide A **insecticide** derived from a fungus.

mycology The study of **Fungi**.

mycoparasite A parasite of **Fungi**.

mycorrhiza A relationship between a fungus and a plant from which both partners benefit. The fungal **mycelium** absorbs sugars from the plant, at the same time greatly extending the volume of soil to which the plant roots have indirect access, thereby increasing the efficiency with which the plant obtains soil nutrients. Mycorrhizae also appear to protect roots from disease. Some orchids (**Orchidaceae**) and pines (*Pinus*) cannot develop normally in the absence of their mycorrhizal partners. The relationship is more important to wild plants than to cultivated plants, because growers feed their plants.

M

mycosis A disease of animals including humans that is caused by a fungus.

Mycosphaerella brassicicola *See* ring spot.

mycotoxin A toxin produced by a fungus.

mycotrophic Describes a plant that is a partner in a **mycorrhiza**.

mycovirus A virus that infects **Fungi**.

Myodocarpaceae (order **Apiales**) A family of usually **evergreen shrubs** and **trees** with **alternate**, **compound pinnate** or **simple**, **entire** or **serrate** leaves. Flowers **actinomorphic**, **bisexual** or **unisexual** (plants **dioecious**), **calyx** with 4 or 5 teeth fused to the **ovary**, 5–10 free or 5–12 partly fused **petals**, as many free **stamens** as petals and alternate with them, ovary **inferior** of 5–10 sometimes 2–12 fused **carpels** with 2–5 free or 1–10 fused **locules**. **Inflorescence** a **raceme**, **umbel**, or head. Fruits **terete** with specialized oil ducts in the endocarp (*see* pericarp). There are 2 genera with 19 species occurring in New Caledonia, Malesia, and Queensland.

Myosotis (family **Boraginaceae**) A genus of **annual** or **perennial herbs** (forget-me-not) with **alternate**, **simple** leaves. Flowers with a 5-lobed **calyx** and 5-lobed **corolla**. Fruit is a **pod**. There are about 50 species occurring throughout temperate regions. Many are cultivated for ornament.

Myricaceae (order **Fagales**) A family of aromatic, usually **evergreen**, **shrubs** and small **trees**, many with **trichomes**. Leaves **alternate**, **simple** or **pinnatifid**, **entire**, **serrate**, or **dentate**. Flowers inconspicuous, **unisexual** (plants **monoecious** or **dioecious**), males usually with 2 **bracteoles**, 4 free or 2–20 fused **stamens**, females with 2–4 bracteoles, **ovary superior** of 2 fused **carpels** with 1 **locule**. **Inflorescence** an **axillary spike** resembling a **catkin**. Fruit is a **drupe**. There are 3 genera with 57 species with a **cosmopolitan distribution**. Some species with medicinal properties. *Myrica gale* (bog myrtle) used for flavouring, dyeing, and as an insect repellant.

Myrica gale (bog myrtle) *See* Myricaceae, root nodule.

Myristicacea (order **Magnoliales**) A family of **trees** and **shrubs**, most **evergreen**, with **alternate**, **entire**, **exstipulate** leaves, in many species secreting aromatic oils. Flowers inconspicuous, **actinomorphic**, **unisexual** (plants **dioecious** sometimes **monoecious**), with 3 free or 2–5 partly **connate sepaloid tepals**, **stamens** fused into a **synandrium**, **ovary superior** with 1 **locule**. **Inflorescence** a terminal **raceme** or **corymb**, sometimes flowers cauliflorous (*see* cauliflory). Fruit is a **dehiscent berry**. There are 20 genera with 475 species with a pantropical distribution. *Myristica fragrans* (nutmeg tree) is the source of nutmeg and mace.

Myristica fragrans (nutmeg tree) *See* Myristicaceae.

myrmechory Dispersal of **spores** or seeds by ants.

myrmecodomatium A **domatium** inhabited by ants.

myrmecophily A type of **mutualism** in which an organism accommodates or supplies food for ants.

myrosinase *See* glucosinolates, myrosin cell.

myrosin cell An **idioblast** containing granules (myrosin grains) of a family of hydrolyzing (*see* hydrolysis) **enzymes** (myrosinases) that, when the cell is ruptured, remove the **glucose** group from a **glucosinolate**, the resulting product rapidly reacting, in a process sometimes called the mustard oil bomb, to release mustard oils that are toxic to many invertebrate animals, **Fungi**, and parasitic plants. It is a plant defence system. Myrosin cells occur mainly in members of the order **Brassicales**, and especially in the **Brassicaceae**.

Myrothamnaceae (order **Gunnerales**) A **monogeneric** family (*Myrothamnus*) of aromatic, resinous, **glabrous shrubs** that are **resurrection plants**. Leaves are small, **opposite**, **dentate**, **simple**, **plicate**, or flat, with **stipules**; they fold up during dry periods and open again after rain. Flowers **actinomorphic** to slightly **zygomorphic**, **sessile**, **unisexual** (plants **dioecious**), **apetalous**, 3–4 free or 4 sometimes 3–8 **connate stamens**, **ovary superior** of 3–4 connate **carpels** and 3–4 **locules**. **Inflorescence** a terminal **spike** sometimes with a terminal flower. Fruit is a **follicle**. There are two species occurring in Africa and Madagascar.

myrrh (*Commiphora myrrha*) *See* Burseraceae.

Myrtaceae (order **Myrtales**) A family of **trees** and **shrubs** with mostly **opposite** or spirally arranged, **simple** leaves with glands secreting ethereal oils, usually **terpenes**. Flowers **actinomorphic** sometimes **zygomorphic**, **bisexual** or **unisexual**, **epigynous** occasionally **perigynous**, 4–5 **sepals** and **petals** usually free sometimes fused into a **calyptra** or **operculum**, many **stamens**, **ovary inferior** to semi-inferior. **Inflorescence** a **panicle**, but with many variants. Fruit a **capsule** or **berry**. There are 131 genera with 4620 species with a worldwide but mostly warm temperate distribution. Many cultivated, e.g. ***Eucalyptus*** for timber, oils, and ornament, *Syzygium aromaticum* (**clove**) and *Pimenta dioica* (allspice) for spices, *Psidium guajava* (guava) for fruit, *Myrtus communis* (common myrtle) for ornament.

Myrtales An order of plants comprising 9 families of 380 genera and 11,027 species. *See* Alzateaceae, Combretaceae, Crypteroniaceae, Lythraceae, Melastomataceae, Myrtaceae, Onagraceae, Penaeaceae, and Vochysiaceae.

Myrtus communis (common myrtle) *See* Myrtaceae.

Mysotella myosotis (mouse-ear snail, salt marsh snail) A species of terrestrial **snail** with a brown or yellow-brown shell up to about 8 mm high, that occurs in salt marsh **habitats** in parts of western Europe and on the west coast of North America.

myxamoeba An **amoeboid** cell formed by a **slime mould**.

myxobacteria (slime bacteria) **Deltaproteobacteria** that occur in soil and are capable of gliding motility, often seeking food in swarms, sometimes called wolf packs, consisting of many joined cells. Resting cells also join to form a **fruiting body** that is often brightly coloured and sometimes visible to the naked eye. **Vegetative cells** are rod-shaped and often embedded in slime. The organisms are **aerobes** that produce **enzymes** to digest food items, these enzymes often disrupting other bacteria or **Fungi**.

M

Myxogastria (myxomycetes, acellular slime moulds, plasmodial slime moulds) A class of **Mycetozoa**, formerly known as Myxomycota, comprising single-celled **eukaryotes** that, at different stages in their life cycle, exist as individual, **amoeboid** cells, a **diploid**, multinucleate, slimy, amorphous **plasmodium**, and conspicuous, often brightly coloured, **fruiting bodies**. Plasmodia and fruiting bodies are often large. They feed on **Bacteria**, **Fungi**, and particles of decaying organic material. There are 900–1000 species occurring worldwide on decaying wood, **bark**, dung, and in soil.

myxomycetes *See* Myxogastria.

Myxomycota *See* Myxogastria.

myxospore A **spore** produced by a member of the **myxobacteria**.

Myzus ascalonicus (shallot aphid) A species of aphid (**Aphididae**) in which adults are pale green or yellow, wingless forms 1.1–2.2 mm long and winged forms 1.3–2.4 mm long. They overwinter in sheltered places such as crop stores and greenhouses and colonize plants in summer. They have been recorded on more than 200 species of plants and are serious pests of onions, shallots, strawberries, lettuce, brassicas, and potatoes, as well as many ornamentals including most flowers grown from **bulbs**.

Myzus cerasi (cherry blackfly) A species of shiny, black aphids (**Aphididae**) that feed on the **sap** of ***Prunus*** trees, forming colonies on the underside of leaves in late spring and early summer. The leaves curl and deform and the **honeydew** the aphids excrete coats parts of the plant, providing a substrate for **sooty mould**.

Myzus ornatus (ornate aphid, violet aphid) A species of green or pale yellow aphids (**Aphididae**), 1.0–1.7 mm long, that feed on a wide variety of plants, including violets but also brassicas, cucurbits, onions, peas, strawberry, and many ornamentals. It sometimes occurs in colonies of other species and is found throughout the world.

Myzus persicae (green peach aphid, peach aphid) A greenish or yellow aphid (**Aphididae**) that overwinters as eggs on ***Prunus*** trees, especially peach, and that feeds on plants in more than 40 families, including many horticultural and field vegetable crops as well as **broad-leaved** weeds. In mild climates they can produce more than 20 generations a year, each generation maturing in 10–12 days. They can form very dense colonies, reducing the yield of root and foliage crops, and they transmit viral diseases. They occur throughout the world.

N

N *See* nitrogen.

Na *See* sodium.

Nabidae (damsel bugs, nabids) A small family of soft-bodied, elongate, flying, brown or yellow bugs (**Hemiptera**) with large eyes and long legs that are predators of insects smaller than themselves, including many crop pests, which they seize and hold with their forelegs. There are 20 genera with about 500 species.

nabids *See* Nabidae.

narcissus bulb fly *See Merodon equestris.*

nacreous With a pearly lustre, like mother-of-pearl.

NAD *See* nicotinamide adenine dinucleotide.

NADP *See* nicotinamide adenine dinucleotide phosphate.

naked bud A **bud** that is not protected by scale leaves.

naked cell A cell that lacks a **cell wall**.

naked flower A **flower** lacking both **sepals** and **petals**.

naked ladies (*Colchicum*) *See* Colchicaceae.

NAO *See* North Atlantic oscillation.

Narcissus (family **Amaryllidaceae**) A genus of **monocotyledon herbs** with **bulbs** that bear regular flowers singly or in groups of up to 20 at the tip of a leafless stem, with a papery **spathe** around each flower or flower group. Flowers have a central **corona** surrounded by **perianth** of an inner **whorl** of 3 **sepals** and outer whorl of 3 **petals**, the perianth segments united into a tube at the base, **ovary** with 3 **locules**. There are 50 species occurring in meadows and woodland in Europe, North Africa, and western Asia. Many are cultivated for ornament.

Nartheciaceae (order **Dioscoreales**) A **monocotyledon** family of **perennial herbs** with erect or creeping **rhizomes**. **Linear**, **ensiform**, or **lanceolate** leaves growing from the base, **distichous** or spirally arranged. Flowers **bisexual**, **actinomorphic** with 5 **trimerous whorls**, 3+3 free or **connate petaloid tepals**, 3+3 **stamens**, **ovary superior** of 3 **carpels** and **locules**. **Inflorescence** a terminal, **bracteate**, **spike**, **raceme**, or terminal **corymb**. Fruit is a **loculicidal capsule**. There are 4–5 genera with 41 species with a scattered distribution in northern temperate regions, northern South America, and western Malesia.

nascent In the process of forming.

nastic movement *See* nasty.

nasturtium (*Tropaeolum majus*) *See* Tropaeolaceae.

nasty (nastic movement) A response by a plant organ to a stimulus that is diffuse rather than directional, e.g. the folding of a *Mimosa* leaf when it is disturbed.

natant Floating.

national nature reserve (NNR) In the United Kingdom, an area that has been designated as a **nature reserve** by English Nature, the Northern Ireland Environment Agency, Natural Resources Wales, or Scottish Natural Heritage. It may be managed by the designating agency or delegated to a voluntary (non-profit) organization. In the United States such an area is known as a national wildlife refuge.

national park An area that is set aside in perpetuity for **conservation** and to which the public is admitted for recreational purposes compatible with the overall objective. It is under the direct control of the state. As defined in 1975 by the **International Union for Conservation of Nature**, a national park is a large area of land containing **ecosystems** that have not been materially altered by human activities, and including plant and animal species, landscape features, and **habitat** of great scientific interest, or of beauty, or recreational or educational interest.

national wildlife refuge *See* national nature reserve.

native (indigenous) Describes a species that occurs naturally in a region or at a particular site, and has not been introduced by people.

natric horizon A dense, brown or black, subsurface **soil horizon** that contains as much more **clay** than the overlying horizon as is found in an **argic horizon**, and a high content of exchangeable sodium and/or magnesium.

Natrix natrix (grass snake, ringed snake, water snake) A colubrid (**Colubridae**) snake that is dark green or brown with a yellow collar, usually 900 mm–1.1 m long but occasionally up to 1.9 m. It inhabits woodland edge and is often found close to water; it swims well. The snake feeds mainly on frogs and toads, but also eats small mammals and fish. It is harmless to humans. It occurs throughout Europe, western Asia, and North Africa, but is absent from Scotland and Ireland.

natural Describes a community of **native** plants and animals. There are five categories: **future-natural**, **original-natural**, **past-natural**, **potential-natural**, and **present-natural**.

natural area In the United Kingdom, United States, and Canada, an area that possesses unique or significant features that should be protected.

natural classification An arrangement of living organisms into groups on the basis of their evolutionary (i.e. genetic) relationships.

naturalized Describes a species that was introduced from elsewhere but is now established, maintaining itself and reproducing without assistance.

natural selection A process in which environmental conditions determine which individuals of a **species** reproduce most successfully, thereby transmitting their **genes** to subsequent generations.

natural woodland **Woodland** that comprises only **native** tree species.

nature reserve An area that is set aside for **conservation** and associated scientific research and education and that has strong legal protection against other uses. The public may or may not be admitted, or access may be restricted, e.g. to certain areas and to footpaths only. *See* local nature reserve, national nature reserve.

Nearctic The part of **Holarctica** that includes North America and Mexico.

near gale Wind of 14–17 m/s. *See* appendix: Beaufort Wind Scale.

near-natural community A plant community that consists mainly or entirely of **native** species and is developing through a natural **succession**, but that has been modified by human intervention, e.g. coppiced woodland (*see* coppice), **wood-pasture**.

necrosis Death of a circumscribed piece of tissue.

nectar A liquid, up to 60 percent sugar, that is secreted by a **nectary** either to attract pollinating insects or to reward animals living mutualistically (*see* mutualism) that protect the plant from **herbivores**.

nectar guide A pattern of markings on the **corolla** of some flowers that guide pollinating insects toward the **nectary**. Some are most clearly visible in ultraviolet light.

nectarine *See Amygdalus.*

nectar robber An animal that takes **nectar** from a flower without transferring **pollen**, e.g. certain bumblebees and other insects have short tongues unable to reach the **nectary** from the top of the flower, so bore through the **corolla** to reach it. Flowerpeckers (14 *Diglossa* species) are birds that specialize in nectar robbing, with modified bills they use to pierce flowers to reach the nectar.

nectary A gland that secretes **nectar**, usually located at the base of the **perianth** of **flowers** pollinated by insects, but elsewhere in some other plants, e.g. in leaves, stems, fruits, and in the spines of some cacti. *See also* extrafloral nectary, foliar nectary.

nectria canker *See Nectria cinnarbarina, N. galligena.*

Nectria cinnarbarina A species of **ascomycete fungi** that causes nectria dieback, also called nectria canker and coral spot, affecting a number of trees and shrubs including apple, ash, birch, lime (linden), maple, pear, and rose. The fungus grows as a **saprpophyte** on dead wood, but becomes a weak opportunist parasite if the host is damaged, causing **cankers** and causing dieback of twigs and **branches**. As branches die, pink pustules appear on them.

nectria dieback *See Nectria cinnarbarina.*

Nectria galligena A species of **ascomycete fungi** that causes nectria canker, also known as apple canker, a disease affecting more than 60 species of trees and shrubs including apple, ash, birch, holly, maple, pear, and walnut. The disease can cause the formation of **cankers** that weaken the adjacent tissue, making it susceptible to breaking, although the disease is seldom fatal. The fungus occurs worldwide.

needle A narrow, pointed leaf, typical of many coniferous species.

needle nematodes *See Longidorus.*

neem tree (*Azadirachta indica*) *See* Meliaceae.

Nelsonioideae *See* Acanthaceae.

Nelumbonaceae (order **Proteales**) A **monogeneric** family (*Nelumbo*) of **perennial** aquatic **herbs** with creeping **rhizomes** that produce **tubers**, and with **adventitious** roots. Leaves **alternate**, in groups of 3, 2 of which are scale-leaves and 1 **emergent** or floating, **peltate**. Flowers **actinomorphic**, **bisexual**, with a **perianth** of 2–5 outer **sepals** merging with 20–30 spirally arranged **petals**, 200–400 spirally arranged **stamens**, **ovary apocarpous** of many **carpels** with 1 **locule**. Flowers borne above the water surface on long **peduncles**. Fruit is an **indehiscent nutlet**. There are two species occurring in temperate eastern and central North America, tropical and subtropical eastern Asia, and Australia. *Nelumbo nucifera* (sacred lotus) is widely cultivated for its edible tubers, rhizomes, young leaves, and seeds, and for its religious significance.

nematicide A chemical compound that kills eelworms (**Nematoda**).

N

Nematoda (eelworms, roundworms, threadworms) A **phylum** of thin, thread-like worms, most of which are minute, but some are up to 50 mm long and 1 species grows to 13 m (it parasitizes sperm whales). The body is unsegmented, covered in tough **cuticle**, lacks cilia (*see* cilium), and its muscles run longitudinally so it can move only from side to side and is unable to raise itself. The head is not distinct from the body. There may be as many as 1 million species, of which more than 16,000 are parasites, occurring worldwide in every type of environment.

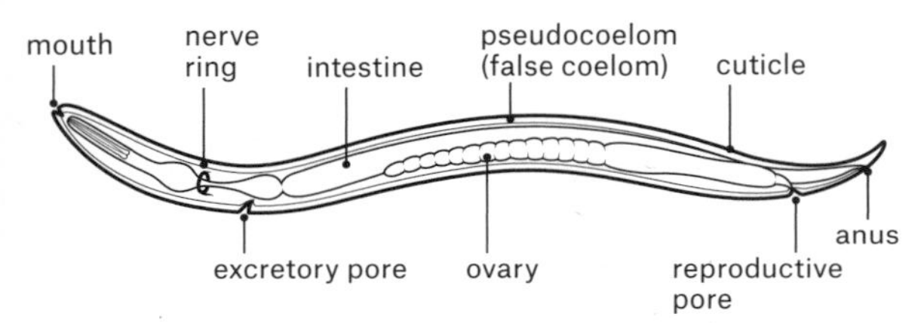

The basic anatomy of a nematode.

N

nematophagous fungi Carnivorous **Fungi** that trap and digest nematodes (**Nematoda**). Some live inside nematodes, others secrete adhesives to which nematodes stick, and others make rings that contract when a nematode tries to pass through. More than 160 species of nematophagous fungi are known; some are used in **biological control.**

Nematus ribesii (gooseberry sawfly) A sawfly (**Symphyta**) that is yellow with a black head and black markings on the thorax, males more black than females; females 5–7 mm long. **Caterpillars** are pale green with black spots and black heads, and up to 20 mm long. They feed on the leaves of gooseberry, redcurrant, and whitecurrant in spring and summer, and can cause severe **defoliation**.

Neocoenorrhinus germanicus (strawberry rhynchites) A species of shiny black weevils (**Curculionidae**) with straight antennae (*see* antenna). Adults are 2–3 mm long, larvae up to 2.5 mm. Adults emerge from hibernation in spring and feed on strawberry and ***Rubus*** plants, first on the leaves, then into blossoms, **petioles**, and **stolons**. They lay eggs in early summer on young tissues, usually petioles and tips of stolons, then puncture the tissues around the eggs, causing the plant tissue to die. Larvae burrow into the dead tissue and mature, pupating after a few weeks and reaching adulthood in two to three weeks. Loss of blossoms greatly reduces crop yields. The weevil occurs throughout Europe.

neofunctionalization The situation in which, following **gene duplication** or **genome duplication**, one of the resulting pair retains the original function while the other, having been liberated from constraints imposed by that function, acquires a new function.

neo-Lamarckism Any modern variant of the evolutionary theory, proposed by Jean-Baptiste **Lamarck**, that **characters** acquired during the lifetime of an individual can be inherited.

Neolectomycetes (earth tongues) A **monogeneric** (*Neolecta*) class of **ascomycete fungi** in which the **fruiting bodies** are upright, unbranched and sometimes lobed, yellow, orange, or yellow-green, and up to 70 mm tall. They have no close relatives, and are not related to members of the **Geoglossaceae**, also called earth tongues. They are said to be edible, but it is not known whether they are **saprotrophs**, parasites, or mutualists (*see* mutualism). The fungi occur in northern Europe, Asia, North America, and Argentina.

neonicotinoid A class of **systemic insecticides**, chemically allied to **nicotine**, that interfere with the nervous system of insects and are used to control sucking and piercing insect pests and fleas on dogs and cats. They are less toxic to mammals than **carbamate** and **organophosphate** products, but suspicions that they may harm bees led the European Union to rule that for two years from 1 December 2013, imidacloprid, clothianidin, and thiamethoxam may not be used for seed treatment, soil application in the form of granules, and as foliar sprays on crops attractive to bees.

neoplasm *See* tumour.

neospecies A new phylogenetic (*see* phylogeny) lineage that arises close to the edge of the range of its parent lineage due to random **genetic drift** or **adaptation** to environmental conditions that differ from those in the main part of the range.

Neotropical region The area that covers the tropical regions of the New World, from Central America and the Caribbean to southern Argentina and Chile, including the **Caribbean**, **Venezuela and Guiana**, **Amazon**, **south Brazilian**, **Andean**, **Pampas**, and **Juan Fernández floral regions**.

NEP *See* primary productivity.

Nepenthaceae (order **Caryophyllales**) A **monogeneric** family (*Nepenthes*) of **perennial**, **dioecious**, climbing, scrambling, or **epiphyte**, insectivorous **lianas** with **alternate**, **exstipulate** leaves. The **lamina** has wings on either side of a midrib, often attached at the base to the stem, and climbs by **tendrils** that are extensions of the midribs. The tip of the tendril then develops into a pitcher, 5–40 cm long, with a lid that opens as the pitcher matures. Insects attracted by colour or smell that enter the pitcher cannot escape because the sides are covered with very slippery wax scales; they fall to the bottom, drown and dissolve in the pool of liquid secretions, and the plant absorbs the released nutrients. Flowers are small, **actinomorphic**, **unisexual** (plants **dioecious**), **tetramerous**, the **perianth** of 3 or 4 **tepals**, 4–24 **stamens**, **ovary superior** of 4 fused **carpels** and **locules**. **Inflorescence** a **raceme** with **bracts** and **bracteoles**. Fruit is a **capsule**. There are 90 species occurring in the tropics from Madagascar to New Caledonia.

nephology The study of clouds.

nephridium An organ found in many invertebrates that is involved with regulating the water content of the body and excretion.

nereistoxin analogue insecticides A group of **insecticides** that are derived from a species of marine worm, *Lumbriconereis heteropoda*, and that are used to control **caterpillar** and beetle pests, including Colorado beetle (***Leptinotarsa decemlineata***). The best known is marketed as Cartap. They are of low toxicity to other organisms.

Nerium (family **Apocynaceae**) A **monotypic** genus (*N. oleander*, oleander), which is an **evergreen shrub** or small **tree** with leathery, **lanceolate**, **entire** leaves in pairs or **whorls** of 3. Flowers, often scented, with a 5-lobed **corolla**, borne in terminal clusters. Fruit is a **capsule**. Oleander has been cultivated for thousands of years as an oramental. All parts of the plant are very poisonous.

nervate Having nerves (**veins**).

nervation *See* venation.

nerve 1. In mosses (**Bryophyta**), a bundle of cells, resembling a leaf **vein**, near the centre of the leaf. **2.** *See* vein.

nest parasitism *See* brood parasitism.

net ecosystem productivity *See* primary productivity.

net primary productivity *See* primary productivity.

nettle *See Urtica*.

nettle leaf weevil *See Phyllobius pomaceus*.

network motif A gene circuit that occurs in a **gene regulatory network** more frequently than would be predicted by chance.

Neuradaceae (order **Malvales**) A family of **annual** and **perennial herbs** and **subshrubs** with **alternate, exstipulate, dentate** to **pinnatifid** leaves. Flowers with 5 free **sepals**, 5 free, **imbricate** or **contorted petals**, 10 **stamens**, 10 **carpels**. Flowers terminal, solitary. Fruit dry, **indehiscent**. There are three genera with ten species occurring in dry or desert areas from Africa to India. A few cultivated as ornamentals.

neural tube In chordate (**Chordata**) **embryos**, the precursor to the central nervous system, comprising the brain and spinal cord.

neuromelanin *See* melanin.

Neuroptera (ant lions, lacewings, mantid flies, owlflies) An order of **endopterygote Insecta**, up to 50 mm long with a wingspan of 5–150 mm, that have lace-like, membranous wings, simple, biting mouthparts, conspicuous antennae (*see* antenna), and large compound eyes. Larvae are grub-like with large, protruding **mandibles**. Both adults and larvae are predators of soft-bodied insects including aphids (**Aphididae**) and scale insects (**Coccidae**), although adults may also feed on **nectar** and **pollen**. There are more than 6000 species with a worldwide distribution.

Neuroterus quercusbaccarum *See* oak-spangle gall.

neurotoxin A poison (toxin) that alters the function of part of the nervous system.

neurotransmitter A chemical substance that contributes to the transmission of nerve impulses.

neutrality theory of evolution *See* neutral mutation theory.

neutral mutation theory (neutrality theory of evolution) The theory that many **mutations** have no significant effect on the **fitness** of their carrier (they are neutral) and, therefore, may become fixed in the **genome** at a random rate.

neutral soil Soil that has a **pH** between 6.6 and 7.3.

New Caledonia floral region The area that includes the Lord Howe and Norfolk Islands as well as New Caledonia, part of the **Palaeotropical region**. There are more than 100 **endemic** genera.

newts *See* Caudata, Salamandridae.

New World blackbirds *See* Icteridae.

New World rats and mice *See* Cricetidae.

New Zealand flatworm *See Arthurdendyus triangulatus.*

New Zealand flax *See Phormium.*

New Zealand floral region The area that includes both islands of New Zealand and the Kermadec, Chatham, Auckland, and Campbell Islands, part of the **Antarctic region**. There are about 30 **endemic** genera and many endemic species, but approximately 75 percent of the endemic species belong to genera that are not endemic.

New Zealand spinach (*Tetragonia tetragonioides*) *See* Aizoaceae.

nexine (endexine) The inner layer of the **exine** of a **pollen grain**.

niche The function an organism performs within an **ecosystem**.

Nicotiana (family **Solanaceae**) A genus of **annual** and **perennial herbs** with **exstipulate** leaves that vary in shape. Flowers **actinomorphic**, with 5 **sepals** and 5 **petals** fused into a tube, **ovary superior** of 2 fused **carpels**. Fruit is a **berry**. There are more than 70 species, occurring mainly in America but also in southwestern Africa, Australia, and islands in the South Pacific. All are known as tobacco, cultivated tobacco being *N. tabacum*, but several others grown as ornamentals.

nictitating membrane In many reptiles, birds, sharks, and some mammals, a membrane attached to a corner of the eye that can be drawn across the eye to reduce illumination of the retina.

nictotinamide adenine dinucleotide (NAD) A **coenzyme** comprising two **nucleotides** linked by their phosphate groups that transports electrons from one **redox** reaction to another. All cells use NAD in **respiration**.

nictotinamide adenine dinucleotide phosphate (NADP) A **coenzyme** that is involved in anabolic (*see* anabolism) reactions, e.g. the synthesis of **lipids** and **nucleic acids**. It is NAD with the addition of a phosphate group. It functions as a photon receptor in the **light-dependent stage** of **photosynthesis** and takes part in the reduction of carbon dioxide in the **light-independent stage**.

nicotine The active ingredient in tobacco, and derived from *Nicotiana* leaves, nicotine, as a 40 percent solution of nicotine sulphate, is used as an **insecticide** against aphids (**Aphididae**), thrips (**Thysanura**), and spider mites (**Tetranychidae**). It inhibits the **enzyme** anticholinesterase. Nicotine is readily absorbed through the skin and is very toxic, but it breaks down rapidly and is acceptable to organic growers.

nictonasty (sleep movement) A **diurnal** plant movement, e.g. the opening of flowers by day and their closing at night. *See* nasty.

nidicolous Describes a young bird that remains in the nest until it is capable of flying.

nidifugous Describes a young bird that leaves the nest very soon after hatching.

Nidulariaceae (bird's nest fungi) A family of **agaric fungi** in which the **fruiting body** is shaped like a bird's nest, 5–15 mm across, containing disc-shaped bodies containing **spores** and resembling eggs. The 'nest' is a splash cup; a raindrop striking it at a particular angle throws the 'eggs' out of the nest. The fungi are **saprotrophs** that grow on decaying wood. There are five genera and about six species found worldwide.

nightcrawler *See Lumbricus terrestris.*

nightingales *See* Turdidae.

night shine *See* tapetum.

Nile grass (*Cyperus papyrus*) *See* Cyperaceae.

nimbostratus (Ns) A genus (*see* cloud classification) of low, grey, fairly uniform clouds that often deliver steady rain or snow.

nimbus Latin for 'rain' and attached to the cloud names **cumulonimbus** and **nimbostratus** to associate them with rain.

nitic horizon A subsurface **soil horizon** that has a nutty structure and many shiny **ped** faces (Latin *nitidus*, shiny). It contains more than 30 percent **clay** and is at least 30 cm thick.

N

nitisols Soils that have a **nitric horizon** more than 30 cm below the surface, no evidence of **clay lessivage** within 100 cm of the surface, and a **cation exchange capacity** of less than 30 $cmol_c$/kg. Nitisols are a reference soil group in the **World Reference Base for Soil Resources**.

Nitrariaceae (order **Sapindales**) A family of **shrubs**, some with lateral **branches** modified as spines. Leaves **alternate** or in pairs at each **node**, **entire** or **pinnatifid**, with small or minute **stipules**. Flowers small, **actinomorphic**, **bisexual**, **pentamerous**, **sepals** fused, **petals** free, 10–15 **stamens**, **ovary superior** usually of 3 **carpels**. **Inflorescence** a terminal or **axillary**, dichasial (*see* dichasium) **cyme** or flowers solitary. Fruit is a **drupe**. There are 3 genera with 16 species occurring usually in arid regions from North Africa to eastern Asia, also southwestern Australia, and

Mexico. Fruits are edible and in Australia are important food for emus.

nitrate The **ion** NO_3^-. Most nitrate salts are soluble in water and are an important source of **nitrogen** for plants.

nitrification The oxidation of ammonia (NH_3) to nitrite (NO_2^-), or nitrite to nitrate (NO_3^-) by **Bacteria.**

nitrite The **ion** NO_2^-. It is an intermediate formed during both **nitrification** and **denitrification.**

Nitrobacter *See* Nitrobacteriaceae.

Nitrobacteraceae A family of Gram-negative (*see* Gram reaction), **motile** and non-motile **Bacteria** (order **Pseudomonadales**) in which the cells vary in shape, and typically use carbon dioxide as their only source of carbon. *Nitrosomonas* species obtain energy by oxidizing ammonia to nitrite; *Nitrobacter* species by oxidizing nitrites to nitrates. They occur worldwide in soil and aquatic environments.

nitrogen (N) An element that is an essential **macronutrient** for all living organisms. It forms part of all **proteins** and **nucleic acids**, and it is part of **chlorophyll.** Nitrogen deficiency causes **chlorosis**, small, thin, leaves, and short, thin shoots.

nitrogenase An **enzyme** used by nitrogen-fixing bacteria (*see* nitrogen fixation).

nitrogen cycle The movement of **nitrogen** from the atmosphere (*see* atmospheric compositon), to the surface where

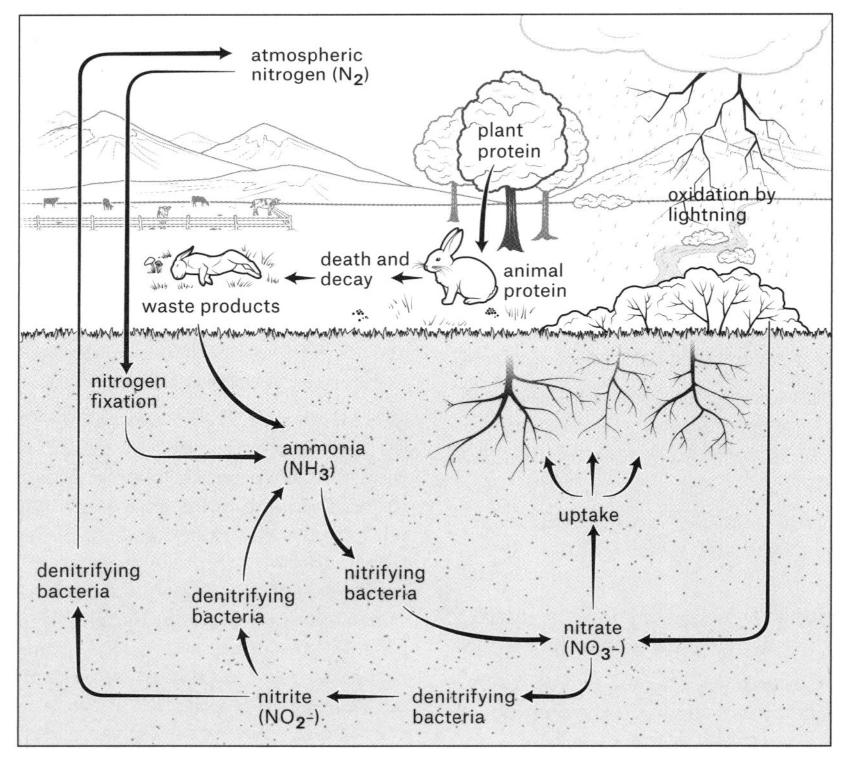

In the nitrogen cycle, nitrifying (nitrogen-fixing) bacteria and the energy of lightning convert N_2 to forms that can be absorbed by plants. Nitrogen is converted to plant protein, which becomes animal protein when heterotrophs consume the plants. The protein returns to the soil as wastes, and denitrifying bacteria convert it to NH_3 or to N_2.

Bacteria convert it to compounds plant roots can absorb (*see* nitrogen fixation), through living organisms where it forms an essential constituent of **proteins**, through conversion by **denitrifying bacteria** back into gaseous nitrogen (*see* denitrification), and so back to the atmosphere.

nitrogen fixation The conversion of gaseous nitrogen (N_2) into ammonia (NH_3) and then into other compounds that plant roots can absorb. This occurs when the energy of **lightning** causes nitrogen to react with water to form ammonia and nitrates (NO_3) and some nitrogen is fixed industrially to make **fertilizer**, but very much more is converted by the action of **nitrogen-fixing bacteria.**

nitrogen-fixing bacteria **Bacteria**, including **cyanobacteria**, that use the **enzyme** nitrogenase and energy from **adenosine triphosphate** (ATP) to combine gaseous nitrogen (N_2) with hydrogen (H) to form ammonia (NH_3). Some are free-living **aerobes**, e.g. ***Azotobacter***, ***Azomonas***, ***Bradyrhizobium***, ***Nostoc***, others free-living **anaerobes**, and ***Rhizobium*** species live symbiotically (*see* symbiosis) in the **root nodules** of legumes (**Fabaceae**) and certain other plants.

Nitrosomonas *See* Nitrobacteriaceae.

nivation **Erosion** that occurs beneath a covering of snow.

NNR *See* national nature reserve.

noble rot *See Botrytis cinerea.*

Noctuidae (army worms, cutworms, noctuids, owlets, underwings) The largest family of **Lepidoptera** of mainly nocturnal moths, some with **eyespots**, most with drab forewings, and some with coloured hind wings. Most are strongly attracted to light. **Caterpillars** are smooth and in many species they can feed on plants that are poisonous to other insects. The caterpillars pupate below ground and some (**cutworms**) live in the soil, feeding on the bases of plants such as brassicas and lettuce. There are more, and probably many more, than 35,000 species, distributed worldwide.

noctuids *See* Noctuidae.

node **1.** The point at which a leaf or leaves is attached to the stem of a plant. **2.** In a **phylogenetic tree**, the point at which a **taxon** divides; the sequential evolution of a taxon is indicated by adjacent nodes. **3.** A **gene** or product of a gene in a **gene network.**

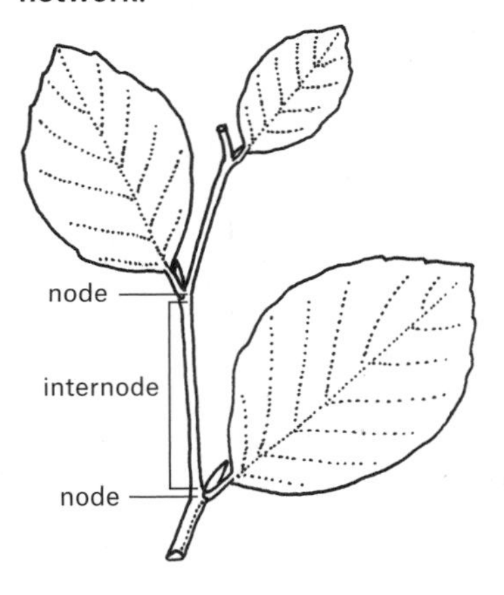

A node is a point at which a leaf is attached. The internode is the region of the stem between nodes.

node-based definition Under the **PhyloCode**, a name that associates a **taxon** with a **clade** that originates at a **node** on a phylogenetic (*see* phylogeny) tree.

nomogenesis The idea that evolution proceeds in a particular direction to some degree in response to sets of rules that are independent of **natural selection.**

non-cyclic photophosphorylation In the **light-dependent stage** of **photosynthesis**, two processes that result in the addition of a phosphate group to **adenosine diphosphate** (ADP) to form of **adenosine triphosphate** (ATP) and the formation of nicotinamide adenine dinucleotide phosphate hydrogen ($NADPH_2$). The hydrogen required for $NADPH_2$ is supplied by the breakdown of water (H_2O) and oxygen is released as a waste product. Electrons move from P_{680} and are taken up by P_{700}, but do not return to P_{680}, hence the process is non-cyclic.

non-essential amino acid An amino acid that an organism is able to synthesize for itself, so it is not an essential part of its diet.

non-exchangeable ions Ions that are held tightly, e.g. in mineral lattices, and cannot be displaced by other ions.

non-polar molecule A molecule in which charge is spread evenly because electrons are shared equally among the atomic nuclei. Non-polar molecules are soluble in fats but most are insoluble in water; some are able to dissolve polar molecules.

nonsense mutation A mutation resulting in the insertion of codon for which no transfer-RNA molecule exists, so the codon does not encode for any amino acid. Usually a nonsense mutation will terminate translation.

nor'easter (northeast storm) A storm with hurricane-force winds from the northeast that occurs in eastern North America.

normal erosion *See* geologic erosion.

normalizing selection *See* maintenance evolution.

North American pitcher plant *See* Saracenciaceae.

North and East Australian floral region The area that includes the forests of the eastern Northern Territories, Queensland, New South Wales, and Tasmania, part of the Australian region. There are many endemic species, including at least 260 on Cape York Peninsula.

North Atlantic Oscillation (NAO) A periodic change in the balance of atmospheric pressure between the Azores high and Icelandic low; when pressure is lower than usual over Iceland and higher over the Azores the NAO is said to be positive, and when pressure is higher over Iceland and lower over the Azores it is negative. The difference in pressure drives weather systems across the Atlantic, so the NAO has a major climatic influence.

northeast African highland and steppe floral region The area that includes Ethiopia, Eritrea, Somalia, Yemen, southern Saudi Arabia, and Socotra Island. There are about 50 endemic genera, about half of these on Socotra Island.

northeast storm *See* nor'easter.

northern cardinal *See Cardinalis cardinalis.*

northern crested newt *See Triturus cristatus.*

northern cricket frog *See Acris crepitans.*

northern dusky salamander *See Desmognathus fuscus.*

northern flicker *See Colaptes auratus.*

northern mockingbird *See Mimus polyglottos.*

northern root-knot nematode *See Meloidogyne hapla.*

Norwegian wrack (*Ascophyllum nodosum*) *See* egg wrack.

no-see-ums *See* midges.

nose leaf A fleshy modification of the nose in many Microchiroptera that is concerned with the emission of sounds for echolocation.

Nostoc A genus of filamentous cyanobacteria in which there are bead-like swellings along the filaments, and typically the filaments aggregate to form gelatinous colonies. They occur in wet soil, on wet rocks, and in fresh water, and perform

nitrogen fixation. Some species are **phycobionts** in **lichens**. *Nostoc punctiliforme* lives symbiotically in the leaf bases of *Gunnera*.

Nothia A genus, now extinct, of vascular plants (**Tracheophyta**) that lived during the Early Devonian epoch (416–397.5 million years ago). They had branching **rhizomes** and leafless aerial stems bearing sporangia (*see* sporangium). There is one known species (*N. aphylla*).

Nothofagaceae (order **Fagales**) A **monogeneric** family (*Nothofagus*, southern beeches) of **evergreen** or **deciduous trees** and **shrubs** with **alternate**, **simple**, **entire** or **serrate** leaves with **peltate stipules**. Flowers **unisexual** (plants **monoecious**), **staminate** flowers with **campanulate perianth**, 5–90 **stamens**, **sessile** or **pedunculate**, 1- to 3-flowered dichasia (*see* dichasium), **pistillate** flowers with small, **dentate** perianth, in a sessile or stalked **involucre**. Fruit is a **nut**. There are 35 species with a **disjunct distribution** from New Guinea to South America. Several cultivated for timber.

notochord (chorda dorsalis) In the **embryos** of **Chordata**, a flexible, rod-like structure that extends the length of the body, **ventral** to the **neural tube** and **dorsal** to the intestine. In some chordates it is retained throughout life as the main **axial** support to the body; in vertebrates it is wholly or partly replaced by the vertebral column.

Notophthalmus viridescens (eastern newt) A species of newts (**Salamandridae**) that are 7–9 mm long at hatching, changing after 2–5 months into a terrestrial **eft**, 34–45 mm long, that is orange or red with 2 rows of red spots outlined in black and has a dry skin. After 2–3 years the eft matures into an adult, 70–120 mm long, that is brown with a speckled underside and a slightly moist skin. The newts inhabit small ponds, lakes, and ditches in forests and can survive on land. They feed on small invertebrates. They occur throughout most of eastern North America.

nowcasting Issuing weather forecasts for up to two hours ahead.

NPP *See* primary productivity.

Ns *See* nimbostratus.

nucellus The tissue in a plant **ovule** that contains the **embryo sac**. The growing **embryo** may absorb it, or it may develop into the **perisperm** that feeds the embryo.

nuciferous Bearing **nuts**.

nuclear envelope (karyotheca, nuclear membrane, nucleolemma) In **eukaryote** cells, the membrane that separates the **nucleus** from the **cytoplasm**. It consists of 2 **lipid** layers, each 10 nm thick, separated by a perinuclear space 20–40 nm wide that is contiguous with the **lumen** of the **endoplasmic reticulum**. Nuclear pores in the nuclear envelope allow water and small molecules (e.g. **adenosine triphosphate**) to pass freely but regulate the passage of **proteins** and other large molecules.

nuclear membrane *See* nuclear envelope.

nuclear pore *See* nuclear envelope.

nuclear pore complex In **eukaryote** cells, a **protein complex** formed from a nuclear pore (*See* nuclear envelope) and its contents that regulates the movement of molecules between the cell **nucleus** and **cytoplasm**.

nucleic acid A **nucleotide** polymer produced in the **nucleus** and **cytoplasm** of living cells. There are two types, **DNA** and **RNA**, and they may be single- or double-stranded.

nucleoid The part of a **prokaryote** cell that contains the **DNA**.

nucleolemma *See* nuclear envelope.

nucleolus In the **nucleus** of an **eukaryote** cell, a structure, often spherical, that is composed of densely packed **fibrils** and granules of **proteins** and **nucleic acids**. It is not bound by a **membrane**. The nucleolus transcribes (*see* transcription) ribosomal RNA and assembles **ribosomes**.

nucleoprotein A **conjugated protein** in which the **prosthetic group** is a **nucleic acid**.

nucleoside A **glycoside** consisting of ribose or deoxyribose sugar bound to a **purine** or **pyramidine** base.

nucleosome A particle formed in isolated **chromatin** and consisting of **DNA** wrapped around a core of eight **histone** molecules.

nucleotide The unit structure of **nucleic acids**, comprising a **nucleoside** bound to a phosphate group.

nucleus The **organelle**, enclosed in a double membrane and containing the **chromosomes**, that is found in most non-dividing **eukaryote** cells. It disappears temporarily during cell division.

Nuctenea umbratica (orb-web spider, walnut orb-weave spider) A spider (formerly known as *Araneus umbraticus*) in which females grow up to 15 mm and males slimmer and up to 10 mm, with a flattened body that allows them to squeeze into small crevices, where they hide by day. At night the spider emerges and makes a web up to 700 mm across, with a line from the web to the hiding place, and after dark it waits in the centre for prey. Females are present all year, males mainly in summer. Both are dark-coloured with yellow or yellow-green flecks and small depressions on the upper surface of the **opisthosoma**. These spiders are common in greenhouses, where they help control pests. If handled, they can deliver a painful bite. They are common in gardens, occasionally entering houses, throughout Europe, North Africa, and southern Asia.

null allele *See* silent allele.

nuptial pad In male frogs (**Ranidae**), one of the thickened or horny pads on each thumb. They help the male to grasp the female during mating.

nut A dry, **indehiscent**, woody fruit.

nutcrackers *See* Corvidae.

nutlet A small **nut**.

nutmeg tree (*Myristica fragrans*) *See* Myristicaceae.

nutrient cycle A **biogeochemical cycle** in which the cycling element is an essential nutrient for living organisms.

nut sawfly *See Croesus septentrionalis.*

Nyctaginaceae (order **Caryophyllales**) A family of **trees**, **shrubs**, **lianas**, and **herbs** with **alternate**, **opposite**, or whorled, **simple**, **petiolate**, **exstipulate** leaves. Flowers **actinomorphic**, **bisexual** sometimes **unisexual** (plants **dioecious**), sometimes surrounded by coloured **bracts** resembling a **calyx**, **perianth** usually tubular, **petaloid**, no **petals**, usually 5 but 1 to many **stamens**, **ovary superior** of 1 **carpel**. **Inflorescence cymose** or **paniculate**. Fruit is an **achene**. There are 30 genera with 395 species occurring throughout tropical to warm temperate regions. Some cultivated for ornament, e.g. ***Bougainvillea*** *glabra* and *B. spectabilis* (bougainvillea), others have medicinal uses.

nyctigamous Describes a **flower** which opens at night.

nymph An insect larva that resembles the adult form. It moults as it grows, but it does not pupate.

Nymphaea (family **Nymphaeaceae**) A genus of **perennial** aquatic plants of shallow fresh water, with **rhizomes** and **peltate** or **cordate** leaves, usually floating. Flowers solitary, with coloured **sepals** and 3 to many **petals**, **ovary half-inferior**. Flowers often pollinated by beetles. There are 36 species occurring in tropical and temperate regions. Many are cultivated as ornamentals (water lilies).

Nymphaeaceae (order **Nymphaeales**) A family of freshwater aquatic, usually **perennial herbs** with long **rhizomes**, sometimes producing **tubers**. Leaves submerged, floating, or **emergent**, **simple**, **peltate**, **cordate** or **sagittate**, **ovate** to **orbicular**, **entire** or **dentate**, some very large (*see Victoria*). Flowers showy, solitary, **actinomorphic**, **hermaphrodite**, with 4–9 **petaloid sepals**, 3 to many **petals** or petals absent, **ovary inferior** or semi-inferior of 3–40 united or partly free **carpels**. Fruit is a **berry**-like **capsule**. There are 3 genera with 58 species with a worldwide distribution. Many cultivated as ornamentals. Some with edible seeds or rhizomes.

Nymphaeales An order of aquatic **herbs** with **rhizomes** that comprises 3 families of 6 genera and 74 species. *See* Cabombaceae, Hydatellaceae, and Nymphaeaceae.

Nyssaceae (order **Cornales**) A family of **trees** and **shrubs** closely related to the **Cornaceae** (dogwoods) and often included in that family. There are 5 genera with 22 species occurring mainly in eastern Asia, but also in Indo-Malesia and eastern North America.

O

O *See* oxygen.

oak *See* Fagaceae, *Quercus*.

oak-apple gall (King Charles's apple) A **multilocular gall** resembling a small, pink apple found on English (pedunculate) oak (*Quercus robur*). It is caused by the wasp *Biorhiza pallida*. The galls mature in summer, releasing wasps that reproduce sexually. Mated females lay their eggs on small roots. Larvae cause the formation of small galls from which an all-female, wingless generation of wasps emerges at the end of their second winter. They climb the tree and lay eggs in the bases of leaf **buds**, producing larvae that cause the oak-apple gall.

oak-marble gall A hard, brown **gall** rich in tannic acid, once used in dyeing and ink manufacture, found on English (pedunculate) oak (*Quercus robur*) and sessile (durmast) oak (*Q. petraea*) that is caused by the wasp *Andricus kollari*, introduced to Britain in 1830. The gall contains members of the all-female generation that emerge in spring to lay unfertilized eggs on **axillary buds** of Turkey oak (*Q. cerris*), where larvae form galls resembling ant pupae from which the sexual generation of wasps emerges the following spring, mates and the females lay eggs on *Q. robur* and *Q. petraea*.

oak moss The **fruticose lichen** *Evernia prunastri* found in mountain forests throughout the temperate Northern Hemisphere. It is used in the perfume industry as a scent component and fixative.

oak-nut Any **gall** that forms on an oak tree (*Quercus*).

oak-spangle gall A small, flat, red **gall**, about 1 mm across, found in late summer and early autumn on the underside of leaves of English (pedunculate) oak (*Quercus robur*) and sessile (durmast) oak (*Q. petraea*), often with high density (up to 100 per leaf). They are caused by the wasp *Neuroterus quercusbaccarum*, each gall containing one larva. When the leaves fall the galls remain attached, in spring releasing female wasps that lay unfertilized eggs on oak **catkins**, where the resulting larvae produce currant galls, resembling redcurrants, from which the sexual generation emerges in summer.

oak wilt *See Ceratocystis fagacearum*.

oarweed (tangle) The brown **alga** *Laminaria digitata*, which has a much branched **holdfast** resembling a root, a thick, cylindrical, smooth, flexible stem up to 2 m long, and an oar-shaped **blade** divided into strap-like ribbons. It grows at and just below the low-water mark and forms part of the offshore **kelp** forests that provide **habitat** for a wide range of species. Oarweed is also harvested for use as **fertilizer** and as a course of industrial chemicals.

oasis An isolated area within an arid region that contains sufficient water to support plants throughout the year. Oases most often occur in depressions where the **water table** is close to the surface. By analogy, the term is also applied to an area with one type of vegetation that is surrounded by land having a different type.

oasis effect The cooling effect, due to **evaporation** and the consequent **absorption** of **latent heat**, of an area of moist ground (the 'oasis') surrounded by dry ground (the 'desert').

obligate Describes an organism that is able to live only under particular environmental or other conditions. An obligate **aerobe** cannot survive in an **environment** lacking oxygen; an obligate parasite cannot survive without a host.

obligate parasite *See* parasite.

oblong leaf A leaf with parallel sides for most of its length.

obovate Describes an egg-shaped leaf with the **petiole** at the narrow end.

occipital Relating to the posterior part of the cranium.

occluded front *See* occlusion.

occlusion (occluded front) The final stage in **cyclogenesis**, in which the advancing cold air has lifted the warm air clear of the surface, thus separating it from the central low pressure. Occlusions may produce precipitation, but they are more often associated with a general drying out of the air and fine weather.

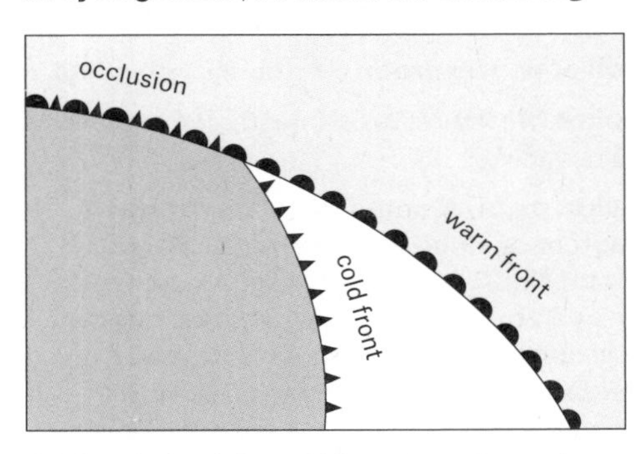

Cold air, behind the cold front, is undercutting the warm air, raising it aloft. The cold and warm fronts are then occluded, forming an occlusion, shown conventionally as alternate semicircles (warm front) and triangles (cold front).

oceanic climate *See* maritime climate.

Ochnaceae (order **Malpighiales**) A family of **trees**, **shrubs**, and **herbs** with **alternate**, **simple**, **crenate** to **serrate** sometimes **fimbriate** leaves with **stipules**. Flowers **actinomorphic** or slightly **zygomorphic**, **bisexual**, usually **pentamerous**, with 5 sometimes 3–4 or 6–10 **sepals**, 5 sometimes 3–4 or 6–12 free often **caducous petals**, 5 to many **stamens** in 1 to many **whorls**, inner whorls sometimes fused **staminodes**, **ovary superior** of 2–15 fused **carpels**. **Inflorescence** a terminal or **axillary raceme**, **panicle**, or **corymb**, or flowers solitary. Fruit is a **capsule** or **berry** or an aggregate of **drupelets**. There are 27 genera with 495 species occurring throughout the tropics, especially in Brazil. Some cultivated for ornament, timber, fibres, or medicinal use.

ochrea (ocrea) A protective sheath around the base of a stem formed from fused **stipules** or leaf bases.

ochric horizon A surface **soil horizon** that is either pale in colour, thin, or contains little organic carbon, or that is massive and very hard when dry.

ocrea *See* ochrea.

octa *See* okta.

Octoknemaceae (order **Santales**) A **monogeneric** family (*Octoknema*) of **trees** and **shrubs** with **alternate**, **petiolate**, **entire**, **exstipulate** leaves. Flowers **actinomorphic**, **unisexual** (plants **dioecious**), **trimerous**, **perianth** with distinct **calyx** and **corolla**, **staminate** flowers with **disc** and **pistillode**. **Pistillate** flower with **staminodes**, **ovary inferior**. **Inflorescence** an **axillary raceme**. Fruit is a **drupe**. There are 14 species occurring in tropical Africa.

Octolasion cyaneum (blue-grey worm, steel-blue worm, yellow tail worm) A species of blue-grey **endogeic** earthworms (**Annelida**), often with a lilac **dorsal** line and with a distinctive yellow tail, comprising the last four segments. It grows to 80–140 mm long and 5–8 mm wide. It lives in moist places, under stones and logs, and in gardens, pastures, and farmland. It is active to a depth of 400 mm. The worm occurs throughout western Europe and parts of North America, and has been introduced to Australasia.

Ocypus olens (devil's coach-horse beetle, cock-tail beetle) A black rove beetle (**Staphylinidae**), 25–28 mm long and covered with fine hairs, its abdomen strengthened with hardened plates, that raises its abdomen and opens its mouthparts when threatened. It has no sting but can deliver a painful bite and also defends itself by releasing a foul-smelling liquid from abdominal glands. It runs rapidly and is capable of flight, but seldom flies. It is a predator of invertebrates including earthworms and woodlice, and also feeds on carrion, seizing food items in its **mandibles**. The beetle is common throughout Europe and has been introduced to the Americas and Australasia.

Odocoileus virginianus (white-tailed deer, whitetail, Virginia deer) A deer (**Cervidae**), 1.5–2.0 m long and 800 mm–1.0 m tall at the shoulder, that is grey-brown in winter and red-brown in summer, with white behind the nose, around the eyes, and on the chin and throat, and a white tail. It inhabits forests, grasslands, swamps, farmland, and deserts, feeding on a variety of plants. It occurs in southern Canada and throughout the United States.

Odum, Eugene Pleasants (1913–2002) An American ecologist who pioneered the study of **ecosystems**, the dynamics of natural populations, and **ecological energetics**, with a special interest in the **ecology** of wetlands. His textbook *Fundamentals of Ecology*, first published in 1953, remained in use for many years.

Oenothera (family **Onagraceae**) A genus of **herbs** with spirally arranged, **dentate** or **pinnatifid** leaves. Flowers with 4 **petals**, usually yellow but sometimes white, purple, pink, or red. There are 79 species and many **hybrids** occurring in North America, but now naturalized in Europe and elsewhere. The unusual **chromosome morphology** has made *Oenothera* the subject of much genetic research. Many species cultivated as evening primrose, a name referring to the fact that in many species the flower opens very quickly in the evening.

oestrus (estrus) The stage in the **oestrus cycle** at which **ovulation** occurs.

oestrus cycle (estrus cycle) The sequence of stages by which a female mammal (other than most primates) is prepared for reproduction. The cycle is regulated by **hormones** and comprises **anoestrus**, **pro-oestrus**, **oestrus**, and **dioestrus**.

offshore wind A wind that blows from land toward the sea.

oidium A fungal **spore** that is produced asexually through the fragmentation of a **hypha**.

oil of wintergreen *See* methyl salicylate.

okra (*Abelmoschus esculentus*) *See* Malvaceae.

okta (octa) A unit used to report the extent of cloud cover in eighths of the total sky. The symbols used to depict cloud cover are usually in oktas, but can be converted to decimal fractions as: 1 okta = 0.1 or less; 2 oktas = 0.2–0.3; 3 oktas = 0.4–0.5; 4 oktas = 0.5; 5 oktas = 0.6; 6 oktas = 0.7–0.8; 7 oktas = 0.9 or overcast but with gaps in the cloud cover.

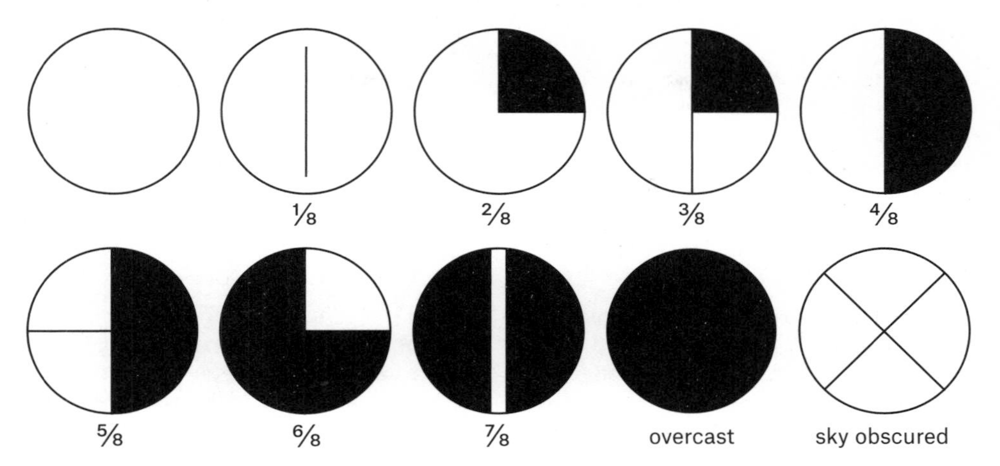

Okta. Cloud amount is conventionally reported as eighths of the sky, or oktas. These are the standard symbols used to show cloud cover on weather maps.

Olacaceae (order **Santalales**) A family of **evergreen trees, shrubs, lianas**, and root **hemiparasites** with **alternate** spiral or **distichous, simple, entire, exstipulate** leaves. Flowers **actinomorphic, bisexual** occasionally **unisexual** (plants **dioecious**), **calyx** with 3–6 lobes, 3–6 **petals**, as many or twice as many **stamens** as petals, **ovary superior**, semi-**inferior**, or inferior, of 2–5 **carpels. Inflorescence** an **axillary raceme** or **panicle**. Fruit is a **drupe** or **nut**. Depending on the classification there are 3 or 25 genera with 57 or about 250 species with a pantropical distribution.

old-fashioned bleeding heart *See Lamprocapnos spectabilis.*

old-field ecosystem An **ecosystem** that develops on abandoned farmland, through an old-field **succession**.

old-field succession *See* old-field ecosystem.

old-growth forest A North American term describing a forest in a late stage of a **succession**, that by implication has existed since prior to the arrival of Europeans.

old man's beard *See Usnea.*

Olea (family **Oleaceae**) A genus of **evergreen trees** and **shrubs** with small, **opposite, entire, elliptical** or **linear** leaves. Flowers with 4-lobed **calyx** and **corolla**, and 2 **stamens**, borne in **axillary racemes**. Fruit is a **drupe**. There are about 40 species occurring in tropical and warm temperate regions of Europe, Africa, southern Asia, and Australia. *Olea europaea* is the cultivated olive.

Oleaceae (order **Lamiales**) A family of **trees, shrubs**, and a few climbers and **herbs**, with usually **opposite, simple** to **imparipinnate, exstipulate, petiolate, entire** or **serrate** or **dentate** leaves. Flowers **actinomorphic, tetramerous**, 2 occasionally 4 **stamens, ovary superior** of 2 **carpels** and **locules. Inflorescence** is a compound **cyme**, usually terminal. Fruit is a **capsule, samara, berry**, or **drupe**. There are 24 genera with 615 species with a worldwide distribution. Many are cultivated, e.g. ***Olea*** (olive), ***Fraxinus*** (ash), *Jasminum* (jasmine), *Syringa* (lilac), *Forsythia*, and *Ligustrum ovalifolium* (privet).

oleander (*Nerium oleander*) *See Nerium*.

Olearia (family **Asteraceae**) A genus of **evergreen shrubs**, small **trees**, and **perennial herbs** with **alternate**, **simple**, **entire** or **dentate** leaves. Daisy-like flowers have a naked **receptacle**, **involucre bracts**. Fruit is an **achene**. There are 181 species occurring in Australasia and New Guinea. Many cultivated as daisy-bushes.

oleoresin A naturally occurring, semisolid mixture of an **essential oil** and a **resin**. Many oleoresins have a strong scent; spice oleoresins are widely used for flavouring and preserving food, but oleoresin capsicum, containing capsaicin (*see Capsicum*) is a powerful irritant used in pepper sprays.

oligo- Small or few.

Oligochaeta A class of **Annelida** comprising earthworms and freshwater worms. Oligochaetes have **metameric segmentation**, the segments possessing few **chaetae** (hence the name), and they lack **parapodia**, eyes, and tentacles. All are **hermaphrodites**. There are about 3500 species found worldwide.

oligomictic Describes a **meromictic** lake in which the water seldom mixes. Such lakes are common in the tropics where the surface temperature is 20–30°C.

oligopeptide A **peptide** consisting of three to ten **amino acids**.

oligosaccharide A **carbohydrate** consisting of two to ten **monosaccharides**.

oligotrophic Describes a soil or body of water that contains few plant nutrients.

oligotrophication The depletion of nutrients from a body of water or a reduction in the rate at which nutrients enter.

olive *See Olea*, Oleaceae.

olive knot A **gall** found on olive trees (*Olea*) caused by the **bacterium** *Pseudomonas savastonoi* that can kill twigs and **branches**.

olivillo *See* Aextoxicaceae.

Olpidium A genus of **Fungi** all of which are **obligate parasites**, most of plants but others of algae (*see* alga), other fungi, and invertebrates. The uniflagellate **zoospores** of plant parasites infect the roots of the host and form sporangia (*see* sporangium) and sometimes resting **spores** inside host cells. There are about 50 species occurring worldwide.

ombrogenous bog A **peat**-forming **bog** that forms above the **water table**. Its vegetation is isolated from the **mineral soil** below the peat, so it depends on rainfall for both water and mineral nutrients.

ombrotrophic Describes a **mire** that is fed by rain water.

Onagraceae (order **Myrtales**) A family of **perennial** or **annual herbs**, and a few **shrubs** and **trees**, with small, **caducous**, mostly **cauline** leaves that are **alternate** or **opposite** occasionally spiral or whorled, **simple**, **petiolate** to **sessile**, and **entire** or **dentate** to **pinnatifid**. Flowers **actinomorphic** or **zygomorphic**, **hermaphrodite** or occasionally **unisexual** (plants **gynodioecious**, **dioecious**, or subdioecious), usually **tetramerous**, opening morning or evening, **sepals valvate**, as many **petals** as sepals or petals absent, as many or twice as many **stamens** as sepals, **ovary inferior** of as many **carpels** and **locules** as sepals. **Inflorescence** an **axillary spike** or **raceme**, occasionally **panicle**, or flowers solitary. Fruit a **capsule**, **berry**, or **nut**. There are 22 genera with 656 species occurring worldwide. Some cultivated for ornament, e.g. ***Oenothera*, *Fuchsia***.

Oncothecaceae A **monogeneric** family (*Oncotheca*), that has not been assigned to an order, of **evergreen shrubs** and large **trees**, the entire plant **glabrous**, with **simple**, **entire**, **coriaceous** leaves borne in clusters near the ends of branches. Flowers about 2 mm across, **actinomorphic**, **bisexual**, with 5 **sepals**, **corolla** up to

O

2 mm long, 5 **stamens**, **ovary superior** of 5 **carpels**. **Inflorescence axillary**, branched. Fruit is a **drupe**. There are two species occurring in New Caledonia.

onion (*Allium cepa*) *See Allium.*

onion bloat *See Ditylenchus dipsaci.*

onion fly *See Delia antiqua.*

onion yellows phytoplasma A disease that produces yellowing and twisting of leaves, caused by ***Phytoplasma** asteris*, which is transmitted in the saliva of leafhoppers.

onshore wind A wind that blows from the sea toward the land.

oogamy **Fertilization** involving the fusion of a large, non-**motile**, female **gamete** with a small, usually motile, male gamete. *Compare* isogamy.

oogenesis The formation of eggs (ova) and, in **angiosperms**, the **embryo sac**.

oogonium The female sexual organ in certain algae (*see* alga) and **Fungi**, corresponding to the **archegonium** in mosses (**Bryophyta**) and ferns (**Pteridophyta**), that contains one or more **oospheres**.

oomycetes *See* Oomycota.

Oomycota (oomycetes, water moulds, downy mildews) A class of **Protista** resembling **Fungi** in that they feed on decaying matter and are filamentous, but differing in that their **filaments** are **diploid** (**hyphae** are **haploid**) and their **cell walls** are made from **celluloses** but not **chitin**. Their name means 'egg fungi' and refers to the oogonia (*see* oogonium) females produce. They reproduce by means of **zoospores** that have two flagella (*see* flagellum), one of the whiplash and one of the tinsel type. There are more than 500 species found worldwide. Most are aquatic but some are serious plant pathogens.

oosphere A female **gamete** contained in an **oogonium**.

oospore A thick-walled resting **spore** that develops within an **oogonium** from a fertilized **oosphere**.

ooze Mud that is rich in calcium carbonate and **silica** derived from the shells of marine organisms. Oozes accumulate in the deep ocean.

open canopy Describes a **tree** community in which the tree crowns do not meet and overlap, exposing areas of ground to direct sunlight and precipitation.

open population A population within which **gene flow** occurs freely.

operator A segment of **DNA** at one end of an **operon** that is the binding site for a particular **repressor protein**. It controls the functioning of adjacent **cistrons**.

operculum A small lid, e.g. the **dehiscent** cap on the **capsule** of some mosses (**Bryophyta**).

operon A set of adjacent **structural genes**, under the control of an **operator**, whose **messenger-RNA** is synthesized in a single piece together with the genes that affect the **transcription** of the structural genes. Operons occur principally in **prokaryotes** but also in some **eukaryotes**.

Operophtera brumata (winter moth) A species of dull grey moths (**Lepidoptera**) in which adults are active throughout the winter, but that cause most crop damage in spring when the **caterpillars** feed on the leaves, blossoms, and developing fruit of many **trees** and **shrubs**, especially apple, pear, plum, cherry, roses, oak, beech, and many more. Males fly strongly, with a wingspan of 22–28 mm, but females cannot fly. The moth occurs throughout Europe and the Near East, and is now established in North America.

Opheodrys aestivus (rough green snake, grass snake, green grass snake) A species of arboreal, slender, bright green colubrid snakes (**Colubridae**) with a yellow belly, about 1 m long, that inhabit forest edges

O

near water. They feed on insects and spiders. The snakes occur throughout the southeastern United States.

Opiliaceae (order **Santalales**) A family of **evergreen shrubs**, **lianas**, and small **trees**, some of which are root parasites. Leaves **alternate**, **simple**, **entire**, **petiolate**. Flowers **actinomorphic**, usually **bisexual** some **unisexual** (plants **dioecious** or **gynodioecious**), usually **tetramerous** or **pentamerous** occasionally **trimerous**, **perianth** inconspicuous sometimes absent, as many **stamens** as **tepals**, **ovary superior**. **Inflorescence** a **panicle**, **umbel**, or **spike**, usually **axillary** or cauliflorous (*see* cauliflory). Fruit is a **drupe**. There are 11 genera with 36 species with a pantropical distribution.

opisthosoma The posterior part of the body of **Arachnida**.

opistoglyphous Describes a snake that has enlarged teeth to the rear of the upper jaw with smaller teeth in front of them. The enlarged teeth may be solid or with a groove to allow saliva to enter the prey.

opium poppy (*Papaver somniferum*) *See* Papaveraceae.

opportunist species (fugitive species) A species that is able to colonize an area rapidly when conditions are favourable. Typically, opportunistic species are smaller than **equilibrium species** and have a shorter life cycle.

opposite Describes the arrangement of leaves that arise in pairs, one pair at each **node**.

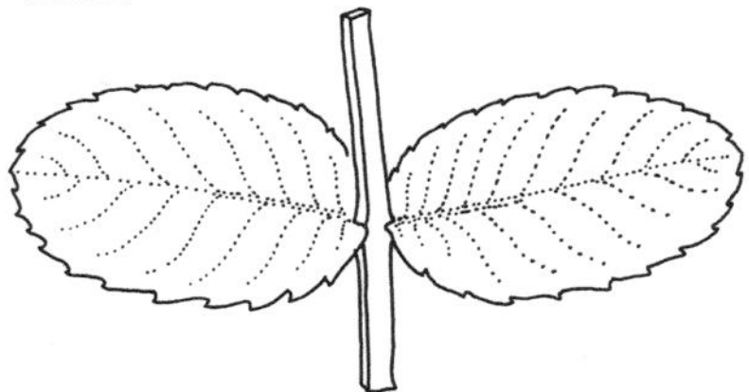

Opposite leaves.

optimum yield (maximum sustained yield, MSY) The largest number of individuals that may be removed from a population repeatedly without impairing the ability of that population to replace them. It is the theoretical point in the graph of a growing population where it reaches the greatest rate of increase. If the population follows an **S-shaped growth curve**, the optimum yield is equal to half the **carrying capacity**.

Opuntia (prickly pear) *See* Cactaceae.

orange-peel fungus *See Aleuria aurantia.*

orange tips *See* Pieridae.

orbicular Globular, circular, or disc-shaped.

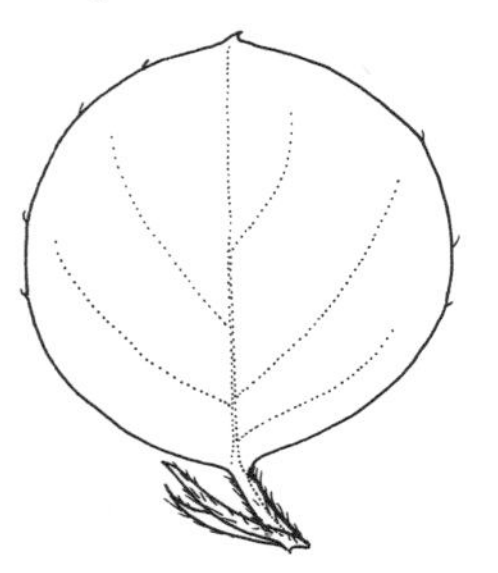

An orbicular leaf.

Orbiliomycetes A class of carnivorous **ascomycete fungi**, which trap and digest small invertebrates. They trap prey by means of a sticky **mycelium**, sticky knobs that project on stalks from **hyphae**, sessile adhesive knobs, adhesive columns, rings, and rings that constrict when a prey animal attempts to pass through. The fungi do not produce stromata (*see* stroma), but have brightly coloured or translucent, disc-shaped apothecia (*see* apothecium). There are 17 species found worldwide.

orbit The bones that form the eye socket in vertebrates.

orb-web spider *See Nuctenea umbratica.*

Orchidaceae (order **Asparagales**) The largest family of **angiosperms**, comprising **monocotyledon**, **perennial**, myccorhizal (*see* mycorrhiza) **herbs**, terrestrial or **epiphytes**, with some scramblers, climbers, or plants lacking **chlorophyll**, many with **rhizomes**, most tropical species with pseudobulbs. Most epiphytes with two types of root: flat and attaching to the substrate and long, dangling, and tangled aerial roots. Leaves usually **alternate** occasionally **opposite**, **distichous**, **simple**, **linear**, **lanceolate**, **ovate**, or **obovate**, sometimes reduced to scales or a single leaf. Flowers strongly **zygomorphic**, **bisexual** occasionally **unisexual** (plants **monoecious** or **dioecious**), with 2 **whorls** of 3 **tepals**, 3 **sepals**, and 3 **petals**, **ovary inferior**, usually of 3 **carpels** with 1 **locule**. **Inflorescence** lateral or terminal, often a **bracteate raceme**, sometimes a **spike** or **panicle**, or flowers solitary. Fruit is a **capsule** containing **dust**-like seeds that develop on associating with an appropriate fungus. Flowers are **ephemeral** to long-lasting and often fragrant (though some smelling of carrion). There are 880 genera with 22,075 species with a worldwide distribution.

orchid bees *See* Apidae.

order In **taxonomy** a rank above the level of **family**; a group of related families comprise an order. In animal taxonomy and some systems of plant taxonomy, a group of orders comprise a **phylum**. In the **Angiosperm Phylogeny Group** system (used here), orders are grouped into **clades**, clade being the highest rank below kingdom.

Oregon giant earthworm *See Driloleirus macelfreshi.*

Oregon grape *See Mahonia.*

oregano (*Origanum vulgare*) *See Origanum.*

organelle A persistent structure with a specialized function within a **eukaryote** cell. It is usually enclosed in a **selectively permeable membrane**. Mitochondria (*see* mitochondrion), **ribosomes**, **Golgi bodies**, **endoplasmic reticulum**, **vacuoles**, and the cell **nucleus** are organelles.

organic manure Decomposed material of plant and/or animal origin that is added to soil to supply plant nutrients and improve soil structure.

organic soil Soil that contains more than 60 percent organic matter, with an organic surface layer more than 50 cm thick, as defined in Britain; the U.S. Department of Agriculture defines an organic soil as one containing 20–30 percent organic matter, depending on the **clay** content. Usually the term applies to **peat**.

organochlorine A class of compounds that contain carbon, chlorine, and hydrogen, with very strong bonds between carbon and chlorine that prevent them from breaking down rapidly. They are insoluble in water but soluble in **lipids**, allowing them to accumulate along **food chains** (*see* bioaccumulation). A group of **insecticides** was based on them, but these are now banned or restricted in most countries.

organomercury A class of compounds that contain carbon and hydrogen, usually as a phenyl (C_6H_5) or methyl (CH_3) group, linked to mercury. They are used as seed dressings and sometimes to prevent fungal infestation of timber. Phenylmercury compounds are of low toxicity to vertebrates; methyl mercury compounds are highly toxic.

organophosphate A compound that is an **ester** of phosphoric acid. Organophosphates inhibit the **enzyme** anticholinesterase and are used as pesticides. They break down rapidly and do not accumulate along **food chains** (*see* bioaccumulation) and, consequently, are considered environmentally preferable to **organochlorine** compounds. Some are highly toxic to mammals.

Origanum (family **Lamiaceae**) A genus of **perennial herbs**, or small **deciduous** or **evergreen shrubs** with aromatic, **opposite**, **petiolate** leaves and small, tubular flowers with a 2-lobed **corolla** and 5-lobed **calyx**, borne in **spikes** or **corymbs** with conspicuous **bracts**. There are about 20 species occurring around the Mediterranean and in southeastern Asia. Several are cultivated as culinary herbs, e.g. *O. onites* (pot marjoram), *O. majorana* (sweet marjoram), and *O. vulgare* (oregano).

original-natural Describes a community that existed prior to any human intervention.

orioles *See* Icteridae.

Orius (minute pirate bugs) A genus of small bugs (**Hemiptera**) in which females are about 3 mm long and males rather smaller. Each day a female lays one to three eggs, 0.4 mm across, on the underside of leaves or in plant tissue. These hatch after five days into **nymphs** that go through five stages. Colour varies with species, but the nymphs always have red eyes. Nymphs feed on larvae of thrips (**Thysanoptera**). Adults feed on all thrips stages and also on aphids, spider mites, whiteflies, and the eggs of **Lepidoptera**. They also eat **pollen**, which allows them to maintain populations in the absence of prey.

ornate aphid *See Myzus ornatus.*

ornithophily Pollination by birds.

Orobranchaceae (order **Lamiales**) A family of more or less **succulent**, **annual**, **biennial**, or **perennial herbs** (broomrape), commonly with **rhizomes** or **tubers**, that lack **chlorophyll** and are root parasites. Leaves are absent or greatly reduced; where present they are small, **alternate**, spiral, **sessile**, **simple**, **entire**, **lanceolate** or **oblong** to **ovate**, **exstipulate**. Flowers **bracteate**, **zygomorphic**, **hermaphrodite**, 8 or 10 free or 6–10 fused, whorled **sepals** and **petals**, **ovary** of 2 **carpels** and 1 **locule**. **Inflorescence** a **raceme** or **spike**. Fruit is a **capsule**. There are 99 genera of 2060 species with a worldwide distribution.

orographic cloud Cloud that forms over high ground as a result of **orographic lifting**.

orographic lifting The forced rising of air as it crosses high ground.

orographic rain Rain that falls on the windward side of high ground from moist air that is forced to rise, cools adiabatically (*see* adiabatic cooling and warming), and its moisture condenses.

orthodox seed A seed that will survive being dried or frozen for prolonged periods, allowing it to be stored in a **seed bank**.

orthogenesis Evolutionary trends that appear to lead directly from ancestors to their descendants.

ortholog *See* orthologous.

orthologous Describes **homologous genes** (each called an ortholog) that become separated by **speciation**.

Orthoptera (crickets, grasshoppers, katydids, locusts) An order of **exopterygote Insecta** with hind legs adapted for jumping, toughened, leathery forewings, and membranous hind wings. They range in size from less than 5 mm long to more than 220 mm. Locusts and some katydids are serious pests. There are more than 20,000 species, distributed worldwide.

orthoselection A selective pressure that drives evolutionary change in a particular direction.

orthotropic Describes a movement of a plant or plant organ directly toward (positive) or away from (negative) a stimulus. *See* tropism.

orthotropous (atropous) Describes the position of an **ovule** that is upright, with the **micropyle** directly above the **funicle**.

ortstein A **hardpan** B **soil horizon** in **podzols** that is indurated (*see* induration) with mainly organic matter and iron (ferric) hydroxide as cementing materials. It tends to form immediately above the **water table**.

osier *See Salix.*

osmobiosis A type of **cryptobiosis** in which organisms tolerate an increase in the concentration of the solution that surrounds them.

osmometer An instrument used to measure **osmotic pressure**.

osmoregulation The mechanism by which an organism or cell controls its internal **osmotic pressure**.

osmosis The movement of a solvent, e.g. water, from a region of low solute concentration to a region of higher solute concentration across a **partially permeable membrane**.

osmotic potential (solute potential) The component of **water potential** that is due to the presence of solute molecules; it is equal to **osmotic pressure** but opposite in sign.

osmotic pressure The pressure that must be applied to a solution to prevent a solvent, e.g. water, from crossing a **partially permeable membrane** separating that solution from a more concentrated one; i.e. the pressure required to prevent **osmosis**. This pressure increases with increasing concentration of the solution.

osmotrophic Describes an organism that absorbs nutrients from a solution.

Osmundea pinnatifida *See* pepper dulse.

ossicle A small bone, especially one of the three bones of the inner ear.

ostiole 1. A small opening in an **alga** or fungal fruit body (*see* fruiting body) through which mature **spores** are released. **2.** In vascular plants (**Tracheophyta**) a small mouth, e.g. at the **apex** of a fig (***Ficus***), through which a female fig wasp enters to lay eggs, at the same time pollinating the flower contained within the fig.

ostracum The calcified part of the shell of an invertebrate animal. In the living animal it is covered by a layer of **protein**, forming a periostracum; this disappears after death.

Ostrya (hophornbeam) *See* Betulaceae.

Otiorhynchus rugostriatus (rough strawberry root weevil) A species of weevil (**Curculionidae**) that usually overwinters as larvae (in warm climates adults may hibernate), and emerges in early summer. Larvae are creamy white with brown heads and feed on roots. Adults are 6–8 mm long, red to brown with hairy **elytra**, and feed at night on **buds** and young shoots of a variety of plants. The species occurs throughout most of the world.

otic Relating to the ear.

Otiorhynchus sulcatus (black vine weevil) A species of weevil (**Curculionidae**) that is black and unable to fly, its **elytra** being fused. The adult is nocturnal, feeding on the edges of leaves of **broad-leaved** plants. Larvae are cream-coloured with a brown head, legless, and live in the soil feeding on roots and **cambium** at the base of plant stems. Herbaceous plants are most at risk, especially if growing in containers, and infestations can kill them.

otters *See* Mustelidae.

oubain A cardiac **glycoside** that inhibits the transport of sodium and potassium across **cell membranes**. It is found in ripe seeds and **bark** of certain African plants. Oubain has been used as an arrow poison; it also has medical applications, e.g. in treating heart failure.

outbreeding The **cross-breeding** of individuals that are not closely related.

outburst A sudden very heavy fall of precipitation produced by a strong downcurrent in a **cumulonimbus** cloud.

outgroup In **phylogenetic systematics**, a **species** that is chosen because it is the least related to those under consideration. Its inclusion makes it easier to distinguish apomorphic (*see* apomorph) and **plesiomorphic characters** that might otherwise remain undetected.

ovary The **gynoecium** of a plant.

ovate Widest at the base and narrower farther from the base.

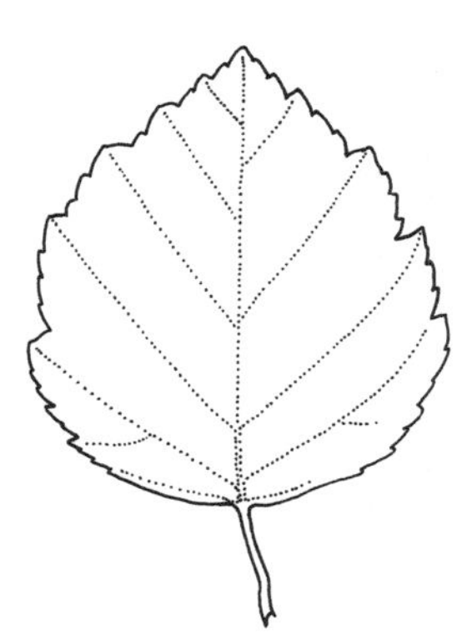

An ovate leaf.

overdispersion (contagious distribution) The situation in which a plant species is not distributed randomly but in a pattern with some densely populated and some empty areas.

overdominance *See* balanced polymorphism.

ovipary Animal reproduction in which the female lays eggs and the **embryos** develop outside her body, each egg developing into a young animal. *Compare* ovovivipary, vivipary.

ovipositor A specialized organ for laying eggs found in female insects (**Insecta**) and formed from outgrowths of the eighth and ninth abdominal segments. In sawflies (**Symphyta**) and ichneumons (**Ichneumonidae**) the ovipositor is very long and bears saw-like teeth used for cutting into plant tissue. In worker bees and sterile female wasps (**Apocrita**) the ovipositor is modified to form a stinging organ linked to a venom sac.

ovisac *See Pulvinaria vitis.*

ovovivipary Animal reproduction in which the female produces eggs which develop inside her body but separated from it by the egg membranes, and the **embryo** feeds on yolk. *Compare* ovipary, vivipary.

ovule In seed plants (**Spermatophyta**), the structure that following **fertilization** develops into the **seed**.

ovum An unfertilized female **gamete**, i.e. egg cell.

owlets *See* Noctuidae.

owlflies *See* Neuroptera.

owls *See* Strigiformes.

Oxalis (family **Oxalidaceae**) A genus of **annual** or **perennial herbs** with leaves that are **alternate**, **exstipulate**, and **palmate** with usually 3 (resembling clover) but up to 10 **obovate leaflets** with a notch at the top. In many species the leaflets fold down at night. Flowers usually solitary, **actinomorphic**, **bisexual**, with 5 free and persistent **sepals**, 5 usually fused **petals**, 10 **stamens**, **ovary superior** of 5 fused

carpels. Fruit is a **capsule**. There are 700 species with a worldwide distribution. Some cultivated and edible, e.g. *O. acetosella*, wood sorrel, but most poisonous; some are troublesome weeds.

oxaloacetic acid (oxalacetic acid) A dicarboxylic acid ($C_4H_2O_5^{2-}$) that is an intermediate in the **citric acid cycle**. It forms by the oxidation of **malic acid** and condenses with acetyl **coenzyme** A to form citric acid and coenzyme A. It is also a precursor in the synthesis of **amino acids**.

oxic horizon A fine-textured, mineral subsurface **soil horizon**, at least 30 cm thick, with a low content of unweathered material (*see* weathering) and a low **cation exchange capacity**. The lack of weatherable material means further weathering will release few plant nutrients.

Oxidalidaceae (order **Oxidales**) A family of small **trees**, **shrubs**, and climbers, but mainly **perennial** or sometimes **annual herbs**, often with **bulbs**, **tubers**, or fleshy roots, with **alternate**, **petiolate**, **simple**, **pinnate**, or **palmate** leaves, many with **leaflets** that fold down at night and in cold weather. Flowers often showy, rarely cleistogamous (*see* cleistogamy) and **apetalous**, **actinomorphic**, usually **hermaphrodite**, with 5 **sepals**, 5 **petals**, 10 **stamens** in 2 **whorls**, **ovary superior** of 5 free or fused **carpels** and 5 **locules**. **Inflorescence** is thyrsopaniculate (*see* thyrse, panicle), **cymose** or **racemose** as an **umbel**, **spike**, or head. Fruit is a capsule. There are 6 genera of 770 species (700 *Oxalis*) with a worldwide but mainly tropical and subtropical distribution.

Oxidales An order that comprises 7 families of 60 genera and 1815 species. *See* Brunelliaceae, Connaraceae, Cephalotaceae, Cunoniaceae, Elaeocarpaceae, Huaceae, and Oxalidaceae.

oxidase An **enzyme** that catalyzes reactions using molecular **oxygen** as an electron acceptor in the **oxidation** of a **substrate**.

oxidation A chemical reaction in which atoms or molecules gain oxygen, or lose hydrogen or electrons.

oxidation-reduction potential (electrode potential, redox potential) A measure of ease with which a substance will lose (oxidation) or accept (reduction) electrons, i.e. whether they are strong oxidizing or reducing agents. Oxidizing and reducing agents occur as couples.

oxidative phosphorylation A reaction that occurs during aerobic **respiration**, in which a phosphate group is added to **adenosine diphosphate** (ADP), converting it to **adenosine triphosphate** (ATP).

oxidative photosynthetic cycle *See* glycolate cycle.

oxidoreductase A group of **enzymes** that catalyze **redox reactions**.

Oxidus gracilis (flat-backed millipede, garden millipede, glasshouse millipede) A species of millipede (**Diplopoda**) that is light brown when young, becoming darker with age and sometimes with yellow borders. Most are 18–22 mm long, and with prominent antennae (*see* antenna). Males have 30 pairs of legs, females 31. They live for about two months and are intolerant of dry conditions. They are nocturnal. Originally tropical it is now widely distributed, especially in greenhouses, and can cause significant damage to plants if present in large numbers.

oxisols **Mineral soils** that have an **oxic horizon** within 2 m of the surface or **plinthite** close to the surface, and no **argic** or **spodic horizon** above the oxic horizon. Oxisols comprise an order in the U.S. Department of Agriculture **soil taxonomy**.

ox tongue *See Fistulina hepatica*.

Oxychilus alliarus (garlic snail) A pale yellow-brown, glossy glass **snail (Vitrinidae)**, 6–8 mm long, that gives off a strong smell of garlic when disturbed. Its shell is coiled, but without a tall spire. It occurs among plant litter, sometimes in gardens, and feeds on decaying plant material, algae (*see* alga), and moss. It is native to and widespread in Europe and also occurs in parts of North and South America.

oxygen (O) An element that is released into the air or water as a by-product of **photosynthesis** and that is essential for aerobic **respiration**, in which oxygen is the final hydrogen acceptor in a sequence of reactions that supply cells with energy.

oxygen cycle The cyclical flow of oxygen from the atmosphere through **respiration** by living **aerobes**, and its return to the atmosphere as a by-product of **photosynthesis**. Oxygen also enters the atmosphere as a result of the **photolysis** of water (H_2O) and nitrogen dioxide (NO_2) and by the **chemical weathering** of rocks. Chemical weathering also removes atmospheric oxygen through the oxidation of exposed minerals.

oyster fungus *See Pleurotus ostreatus.*

oyster plant (*Acanthus spinosus*) *See Acanthus.*

oystershell scale *See Lepidosaphes ulmi.*

oyster thief A seaweed, *Colpomenia peregrina*, that usually grows as an **epiphyte** on other seaweeds in sheltered rock pools on the middle and lower shore. Its **thallus** is a thin-walled, hollow sphere, yellowish green to olive-brown in colour with small brown spots, and usually 1–7 cm across. At low tide the spheres tend to fill with air, making them buoyant, so they rise on the incoming tide, lifting from the rocks any oysters (or other shells) attached to them.

P

P *See* phosphorus.

P *See* F_1, parental generation.

P680 A pigment (P) comprising two forms of **chlorophyll** *a* that absorbs light with a peak wavelength of 680 nm, in the red part of the visible light spectrum. It occurs in **chloroplasts** and is the energy trap for **photosystem II** in **photosynthesis**.

P700 A pigment (P) comprising **chlorophyll** *a* at the reaction centre of the molecule involved in **photosystem I** of **photosynthesis**. It absorbs light with a peak wavelength of 700 nm.

Pacer *See* methoxyfenoxide.

pachycaul With a thick stem.

Pacific coast forest The North American coniferous forest that extends from northern California to southern British Columbia, renowned for its giant trees, e.g. the big tree (*Sequoiadendron giganteum*), coastal redwood (*Sequoia sempervirens*), and Douglas fir (*Pseudotsuga menziesii*).

Pacific North American floral region The area that covers the western side of North America from southern Alaska and the Aleutian Islands to the Mexican Highlands, part of the **boreal region**. There are about 300 **endemic** species.

Paecilomyces fumosoroseus A species of **ascomycete fungi** that is a **mycoparasite** of insects. When a **blastospore** falls on an insect body it secretes an **enzyme** that dissolves a patch of the **cuticle**, allowing a **germ tube** to penetrate the body cavity, where the fungus grows. It can also penetrate through orifices in the cuticle. The fungus feeds on many species, especially of mites, and is used in **biological control**.

paedomorphosis A type of **heterochrony** in which the ancestral juvenile form is retained into adulthood, in some species (e.g. of **Cecidomyiidae**) allowing larval forms to reproduce.

Paeoniaceae (order **Saxifragales**) A **monogeneric** family (*Paeonia*) of **perennial herbs** or **shrubs** with **rhizomes**. Leaves **alternate**, **compound** with 3 to many **leaflets**, **exstipulate**, **linear** to broadly **elliptical**, **entire** or lobed. Flowers **actinomorphic**, **bisexual**, **bracteate**, with 5 sometimes 3–7 free, persistent **sepals**, 5–9 or more **petals**, many **centrifugal stamens**, **ovary** of 2–5 sometimes 1 or up to 8 **carpels**. Flowers large, showy, white, yellow, purple, red, or pink, solitary, usually terminal. Fruit a **follicle**. There are 33 species occurring in northern temperate regions, especially eastern Asia. Many widely cultivated for ornament (peonies) and for medicinal use.

painted turtle *See Chrysemys picta*.

Palaearctic The part of **Holarctica** that includes North Africa and Eurasia north of latitude 15° N.

palaeobotany The study of **fossil** plants.

palaeopolyploid A **diploid** organism descended from polyploid (*see* polyploidy) ancestors. Many plants are palaeopolyploid.

palaeosol *See* paleosol.

palaeospecies **Species** that are known only from **fossils**.

Palaeotaxus rediviva The earliest known species of yew (**Taxaceae**) that lived in the Late Triassic epoch (200 million years ago) and that during later periods of glaciation became confined to northern temperate zones. Some authorities consider it the ancestor of all **extant** yews.

Palaeotropical region The area that includes Africa, Asia south of the Himalayas, and the islands of the Pacific. It includes the **African-Indian desert, Sudanese park-steppe, northeast African highland and steppe, West African rain forest, East African steppe, South African, Madagascar, Ascencion and St Helena, Indian, continental Southeast Asia, Malaysian, Hawaiian, New Caledonia, Micronesian,** and **Polynesian floral regions.**

palate The roof of the mouth.

palea 1. The upper of the two **bracts** enclosing each **spikelet** in a grass (**Poaceae**) **inflorescence**. *See also* lemma. **2.** One of the bracts subtending the **receptacle** in a **capitulum** (*see* Asteraceae).

paleosol (palaeosol, relict soil) **1.** An ancient soil that has been buried beneath sediments or volcanic deposits. **2.** A soil formed in the distant past under climatic conditions and a type of vegetation markedly different from those of the present.

palisade mesophyll (palisade parenchyma) Tissue found in green leaves that consists of tightly packed, columnar, **parenchyma** cells resembling a palisade, each cell containing many **chloroplasts**, making this the primary site of **photosynthesis**. *See also* spongy mesophyll.

palisade parenchyma *See* palisade mesophyll.

pallium *See* mantle.

palmate (digitate) **1. Compound**, with 2 or more **leaflets** arising from the tip of a **petiole** or **rachis** and spreading like the fingers of a hand. **2.** Of leaf **veins** (*see* venation), diverging along the **lamina** from a point close to the top of the petiole, of veins all approximately of similar size.

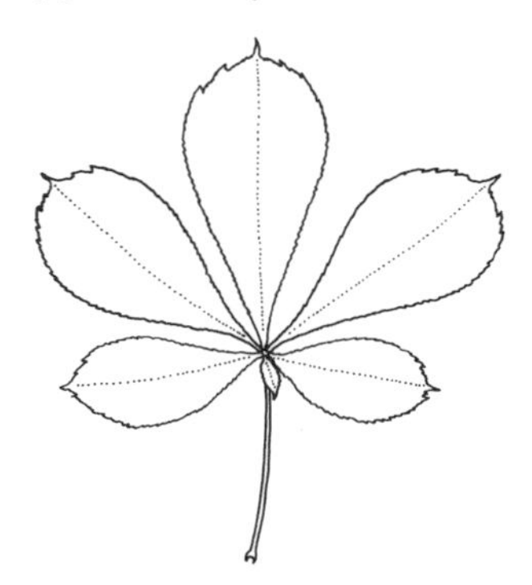

A palmate leaf has two or more leaflets that spread like the fingers of a hand.

palmately lobed Describes a leaf that is divided into lobes that spread like the fingers of a hand.

palmate newt *See Triturus helveticus.*

palmelloid Describes a colony of **microorganisms** that consists of non-**motile** cells embedded in a mucilaginous (*see* mucilage) matrix. Palmelloid colonies occur in some **cyanobacteria** and algae (*see* alga).

Palmer drought severity index (PDSI) A classification of **droughts** based on the extent to which the supply of water for plants departs from the average for that place and season.

Palouse prairie (bunch grass prairie) **Prairie** grassland that occurs in the Palouse region of Washington State, extending into Oregon, Idaho, and Utah in the United States. It comprises bunch grasses, e.g. blue-bunch wheatgrass (*Agropyron spicatum*), growing on deep, fertile soils developed over **aeolian** deposits. Much of the area now grows wheat.

palpigrades *See* Arachnida.

palsa A ridge or mound, 10–30 m wide, 15–150 m long, and 1–7 m high, consisting mainly of **peat** with a central lens of ice that forms in **mires** in **periglacial** environments.

palsa mire Peat-rich **tundra** with **palsas** covered in **lichens**. The high **albedo** of the palsas prevents them warming in summer, so the ice persists. In time, **erosion** removes the lichens, exposing the peat, which absorbs sunlight and warms, melting the ice, which forms a pool. Then the water freezes and a new palsa forms.

paludal Pertaining to **marsh**.

palynology *See* exine.

Pammene rhediella (fruitlet-mining tortrix moth) A species of dark brown tortrix moths (**Tortricidae**) with a wingspan of up to 10 mm that overwinter beneath loose **bark** as a cocoon, form **pupae** in spring and emerge as adults in early summer. Juvenile **caterpillars** have a black head and white body; the head turns brown as the larva matures. The caterpillars feed on the flesh of fruit, making the surface misshapen and rough-textured.

pampas Temperate grassland in South America, the largest area occurring in Buenos Aires Province, Argentina. Each individual area is known as a pampa (pl. pampas).

pampas floral region The area that includes Uruguay, southeastern Brazil, the Argentinian **pampas**, and western Argentina. The **flora** is dominated by the grasslands and there are only about 50 **endemics**.

PAMP-triggered immunity A range of responses that prepare plant tissues for attack, triggered by a pathogen-associated molecular pattern (PAMP) recognized by **pattern recognition receptors.**

pan A strongly compacted and indurated (*see* induration) **soil horizon**, usually below the surface.

panama hat plant (*Carludovica palmata*) *See* Cyclanthaceae.

Panax (ginseng) *See* Araliaceae.

Pandaceae (order **Malpighiales**) A family of **trees** and **shrubs** with **alternate**, **simple**, **entire** or **dentate** leaves. Flowers **actinomorphic**, **unisexual** (plants **dioecious**), 5 often **imbricate sepals**, 5 **petals**, 5–15 free **stamens** sometimes in 2 **whorls**, **ovary superior**. **Inflorescence** terminal or **cauline raceme**, or **axillary** or cauline to **ramiflorous** fasciculate (*see* fascicle) raceme. Fruit is a **drupe**. There are 3 genera of 125 species occurring in the tropics from Africa to New Guinea.

Pandanaceae (order **Pandanales**) A **monocotyledon** family of tall **trees**, **shrubs**, or climbers with stems marked by leaf scars and often with **prop roots**. Leaves long, narrow, **ensiform**, in 3 or 4 ranks sometimes twisted spirally. Flowers naked or with a vestigial **perianth**, **unisexual** (plants **dioecious**), **staminate** with a few up to several hundred **stamens**, **pistillate** with 1 to several **carpels** with 1 to many **locules**. **Inflorescence** unisexual, terminal sometimes lateral, **raceme** or **spike**. Fruit is a **drupe** or **berry**. There are 4 genera of 885 species occurring in tropical Africa, southern Asia, the Pacific Islands, and Australia. Some *Pandanus* spp. (screw pines) with edible fruit and **bracts**, others cultivated for fibres used in weaving and thatching, or for ornament.

Pandanales An order of **monocotyledon** plants comprising 5 families of 36 genera and 1345 species. *See* Cyclanthaceae, Pandanaceae, Stemonaceae, Truiridaceae, and Velloziaceae.

P

Pandora neoaphidis A species of **Fungi** that is a pathogen of aphids (**Aphididae**). It produces infective conidia (*see* conidium) that aphids encounter as they move about. Once infected, the insects emit volatile chemicals as distress signals; these alter the **foraging** behaviour of other aphids, thereby increasing the probability that they will contact fungal conidia. The fungus is used in **biological control**.

Pandorea (family **Bignoniaceae**) A genus of **lianas** and woody vines that have **decussate**, **compound**, **exstipulate** leaves. Flowers brightly coloured, with a 5-lobed **calyx** and **campanulate corolla**, borne in **cymes**. Fruit is a **capsule**. There are six species occurring in Malesia, Australia, and New Caledonia. Several cultivated for ornament, especially *P. jasminoides* (bower vine) and *P. pandorana* (wonga vine).

pandurate Shaped like a violin.

Pangaea A supercontinent that formed in the Late Permian epoch (260.4–251 million years ago) and began to break apart about 200 million years ago, into two sections, **Gondwana** and **Laurasia**.

panicle A compound **raceme**; more loosely, any complex, branched **inflorescence**.

A panicle is a compound raceme.

paniculate Having a **panicle** or an **inflorescence** resembling a panicle.

Panicum (family **Poaceae**) A **monocotyledon** genus of **annual** and **perennial** grasses, some of which grow up to 3 m tall. Flowers are borne in **panicles**, sometimes up to 60 cm long. There are about 450 species occurring throughout the tropics with a few in northern temperate regions. They include many important fodder, ornamental, and grain species (millet).

Panonychus ulmi (European red mite, European red spider mite, fruit tree red spider mite) A mite (**Arachnida**), less than 0.4 mm long, that feeds on **perennial trees** and **shrubs** by piercing leaves and consuming the cell contents. It can produce six to eight generations a year

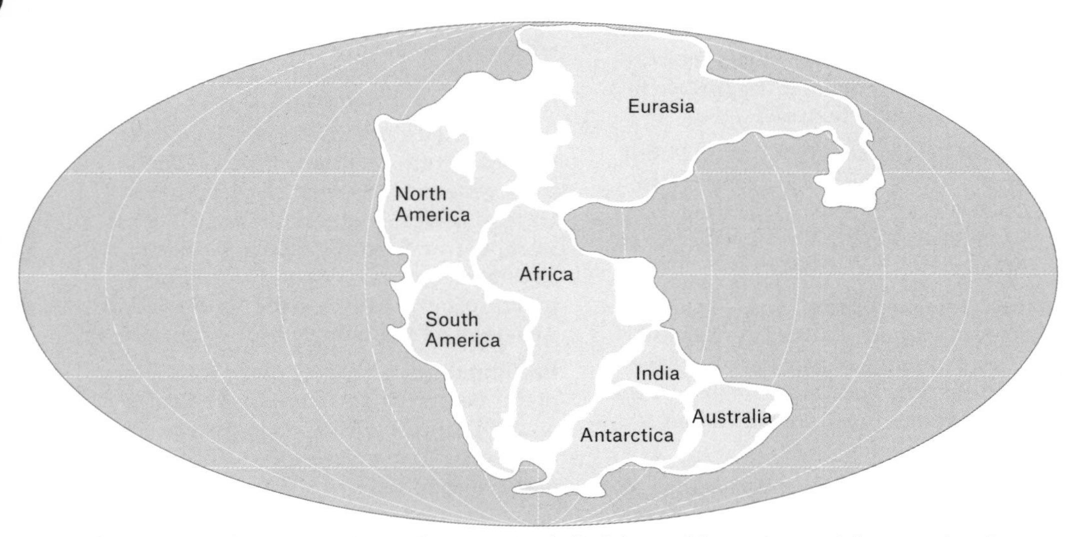

Pangaea was the supercontinent that comprised all of the world's continents. It began to break apart about 200 million years ago.

and is a serious pest of apple, pear, peach, and plum trees, brambles, cane fruit, and some ornamental plants. Adults are red with white spots at the base of the eight hairs on their back. It occurs throughout the world.

pansy *See Viola.*

Panthalassa The vast ocean that surrounded the supercontinent of **Pangaea**.

panther cap *See Amanita pantherina.*

Pantherophis guttatus (corn snake, red corn snake) A species of fairly large snakes (**Colubridae**) formerly known as *Elaphe guttata*, up to 1.2 m long, that are orange, reddish brown, or grey with up to 40 squarish brown or red blotches with black margins, and a spear-shaped marking on the head. They are not venomous, overpowering prey by constriction, and inhabit dry, exposed **habitats** with abundant mammal burrows. They often occur around buildings and sometimes enter human dwellings in search of prey. They feed on small mammals, birds, and lizards and are distributed throughout the eastern and southern-central United States and northern Mexico.

pantoporate Describes a **pollen grain** that has rounded pores covering its surface.

Papaveraceae (order **Ranunculales**) A family of **annual** and **perennial herbs** and a few **shrubs**, **geophytes**, and climbers, many producing **latex**. Leaves **alternate** or whorled, usually **pinnate** or **palmately compound**, **entire** but often lobed, dissected, or **bipinnatisect**, **exstipulate**. Flowers usually large, **bisexual**, hypogynous (*see* hypogyny) rarely perigynous (*see* perigyny), 2 free **sepals**, **petals** free, in 2 **whorls** of 2, many **stamens ovary superior** of 2 to many fused **carpels** and 1 **locule**. **Inflorescence racemose**, **cymose**, a **thyrse**, or flowers solitary. Fruit is a **capsule**. There are 44 genera of 760 species occurring mainly in temperate regions, also in South Africa and western South America. Many cultivated, e.g. *Papaver somniferum* (opium poppy), *Eschscholzia californica* (Californian poppy), *Corydalis* (Dutchman's breeches), and ***Dicentra*** (bleeding heart).

Papaver somniferum (opium poppy) *See* Papaveraceae.

papaya (*Carica papaya*) *See* Caricaceae.

paper reed (*Cyperus papyrus*) *See* Cyperaceae.

papilionoid Resembling or related to the pea-like flowers of the Papilionoideae subfamily of the **Fabaceae**.

papilla A small, rounded protrusion.

pappus A tuft of bristles or hairs derived from the **calyx** that is attached to a dry fruit and aids its dispersal by wind. It is found in many members of the **Asteraceae** and **Caprifoliaceae**.

papyrus sedge (*Cyperus papyrus*) *See* Cyperaceae.

paracentric inversion The end-to-end reversal of a section of a **chromosome** that does not include the **centromere**.

Paracoccus denitrificans *See* denitrifying bacteria.

Paracryphiaceae (order **Paracryphiales**) A family of **evergreen trees** and **shrubs** with sub-**verticillate**, **simple**, **exstipulate** leaves. Flowers small, **bisexual** or **staminate** (plants **andromonoecious**), with 4 **tepals**, usually 8 **stamens**, **ovary superior** of 1–8 **connate carpels**. **Inflorescence racemose**. Fruit is a **capsule**. There are 3 genera of 36 species occurring from the Philippines and New Guinea to New Zealand and New Caledonia.

Paracryphiales An order of plants with **tetramerous** flowers borne in **racemose inflorescences** comprising 1 family (**Paracryphiaceae**) with 3 genera of 36 species.

Paraglomerales An order of **arbuscular** mycorrhizal (*see* mycorrhiza) **Fungi**

belonging to the **Glomeromycota**, that occur only underground.

paralithic Describes soil material that is relatively unaltered and lacks the characteristics of any **diagnostic horizon**. Plant roots are able to penetrate only through cracks.

paralithic contact A boundary between soil and **paralithic** material that has no cracks or cracks penetrable by plants roots that are at least 10 cm apart.

parallel evolution Similar evolutionary trends that occur in descendants of a common ancestor, so the descendants are as alike as their ancestors.

parallel sequencing The simultaneous **gene sequencing** of many strands of **DNA**.

paralog *See* paralogous.

paralogous Describes **homologous genes** (each called a paralog) that become separated by **gene duplication** or **genome duplication**.

paramo Meadows between the **tree line** and **snow line** in the Andes that have a humid **arctic-alpine** vegetation with mosses and **lichens** and also scrub.

P

paramylum A **carbohydrate** resembling **starch** found as a storage product in certain **Protozoa** and algae (*see* alga).

paraná pine (*Araucaria angustifolia*) *See Araucaria*.

parapatric Describes **species** occupying adjacent but separate **habitats**.

parapatric speciation **Speciation** that occurs despite minor **gene flow** between subpopulations of the species.

paraphyletic Describes a **taxon** that includes some but not all of the descendants of an ancestral taxon.

parapodia Paired, muscular, lateral appendages bearing **chaetae** that extend from the body segments of bristleworms (Polychaeta) and some sea slugs. They are usually **biramous** but may be **uniramous**.

paraquat A fast-acting defoliant **herbicide** used to kill **broad-leaved** weeds and grasses. It is highly poisonous to humans.

parasexual cycle (parasexuality) In certain heterokaryotic (*see* heterokaryon) **Fungi** and other single-celled organisms, a form of **recombination** that is based on **mitosis** rather than **meiosis**, genetically distinct **haploid** nuclei fusing in the heterokaryon to produce **diploid** nuclei that multiply mitotically.

parasexuality *See* parasexual cycle.

parasite An organism that lives on (exoparasite) or inside (endoparasite) the body of a host, from which it obtains food, shelter, or some other necessity. Usually, but not always, this implies the host suffers some harm, ranging from very little to severe or fatal. An obligate parasite can live only parasitically, a **facultative** parasite can live either as a parasite or as a **saprotroph**, and a partial parasite is a facultative parasite that lives more efficiently as a parasite than as a saprotroph.

parasitic flies *See* Tachinidae.

parasitism A relationship between members of different species in which one individual (the parasite) lives on or inside the body of the other (the host), from which it obtains food, shelter, or some other benefit. The parasite is usually much smaller than its host. If the only resource the parasite obtains is food, it may be called a **biotroph**. Usually the parasite injures its host, but the effect may range from being undetectable to death.

parasitoid An organism that spends part of its life cycle as a **parasite** and part as a predator.

parasol pine (*Sciadopitys verticillata*) *See* Sciadopityaceae.

parathion An **organophosphate insecticide** and **acaricide** that is used against a variety of insect pests and mites. It is effective, but extremely toxic if inhaled or ingested. It is banned in many countries and may eventually be banned everywhere.

parchment bark *See Pittosporum.*

parenchyma 1. Plant tissue comprising unspecialized cells with air spaces between them. **2.** The cells from which an organ is constructed, rather than blood cells, **connective tissue**, nerve cells, etc.

parental generation (*P*) The generation comprising the parents of the **F_1** generation. The grandparental and great-grandparental generations are designated P_2 and P_3.

parent material (parent rock) The underlying rock from which a soil has developed.

parent rock *See* parent material.

Paridae (tits, titmice, chickadees) A family of small, very active birds, most 100–160 mm long, that are grey, brown, olive, or blue on the upper side and yellow, buff, or white on the underside. Many have crests. Their tails are short and square, their wings rounded. Those that feed on insects have narrow beaks, the beaks of those feeding on seeds are thicker. They are arboreal, inhabiting woodland and gardens. There are 8 genera and about 55 species distributed throughout most of the Northern Hemisphere and in Africa.

parietal placentation **Placentation** in which the **ovules** are in rows attached to the inner wall of the **ovary**. *See* axile placentation, basal placentation, free-central placentation.

paripinnate Describes a **pinnate** leaf with all the **leaflets** in pairs.

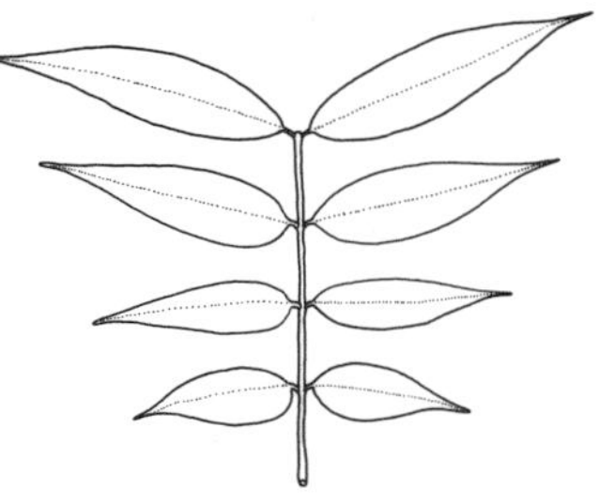

A paripinnate leaf.

park woodland **Woodland** with an **open canopy** growing above pasture.

parotoid gland *See* Bufonidae.

Parrotia (family **Hamamelidaceae**) A **monotypic** genus, *P. persica* (Persian ironwood), which is a **deciduous** small **tree** or **shrub** with an attractive **bark**, and **alternate**, **ovoid**, **simple**, **serrate** leaves that turn purple to red in autumn. Flowers are dark red, produced on bare stems in clusters subtended by white **bracts**, with 4 **sepals** and no **petals**. Fruit is a **capsule**. The species occurs in the forests of Iran, but is widely cultivated for ornament.

parsley aphid *See Dysaphis foeniculus.*

parsnip (*Pastinaca sativa*) *See* Apiaceae, *Pastinaca.*

parsnip aphid *See Cavariella pastinaceae.*

parthenocarpy The production of fruit without the **fertilization** of **ovules**. It may occur naturally or be induced and the fruit is seedless. *Compare* stenospermocarpy.

Parthenocissus (family **Vitaceae**) A genus of climbers with **palmately compound** leaves that climb by means of branched **tendrils** with sucker-like tips. Flowers tiny with 5 **petals**, borne in clusters. There are about 12 species occurring in the Himalayas, eastern Asia, and North America. Many cultivated as Virginia creeper.

parthenogenesis Reproduction in which a female **gamete** develops without having been fertilized by a male gamete, i.e. it is an incomplete form of sexual reproduction. Individuals that develop parthenogenetically are usually **diploid**, making them genetically identical to their mothers.

parthenospore (azygospore) A resting **spore** that develops by **parthenogenesis**. Parthenospores occur in certain algae (*see* alga) and **Fungi**. *Compare* zygospore.

partial dominance The occurrence in individuals heterozygous (*see* heterozygosity) for a particular **gene** of a **phenotype** that is intermediate between the two allelic (*see* allele) forms. It is usually a type of **incomplete dominance**, the individual resembling one parental **homozygote** more than the other. For example, when *Mirabilis jalapa* plants bred in a pure line with red petals are crossed with *M. jalapa* plants bred in a pure line with white petals, the first generation (**F_1**) offspring will have pink petals, the gene and its allele for red petals being incompletely dominant over the gene and allele for white petals.

partially permeable membrane (semipermeable membrane) A **membrane** that permits only solvent molecules to pass. Plant **cell walls** are partially permeable membranes. *See also* differentially permeable membrane, selectively permeable membrane.

partial parasite *See* parasite.

partial pressure In a mixture of gases, the pressure that can be attributed to one of the constituent gases. For example, the average surface **atmospheric pressure** is 100 kPa and air contains approximately 78 percent nitrogen and 21 percent oxygen, so the partial pressure of nitrogen is 78 kPa and that of oxygen 21 kPa.

partial rosette plant A **hemicryptophyte** that has a basal rosette of well-developed leaves with a few other leaves on stems above ground.

partial veil In the immature **fruiting bodies** of some **agaric fungi**, a membrane that connects with edge of the **pileus** with the **stipe**.

particle density The mass of soil particles per unit volume, usually expressed in grams per cubic centimetre.

particle sizes (grain sizes) The size of the particles that make up a sediment or **sedimentary rock**. The sizes of particles too small to measure directly are calculated as the diameter of a sphere with the same volume as the particle. There are two widely used scales: Udden-Wentworth (or Wentworth) and British.

Particle Sizes

	UDDEN-WENTWORTH (WENTWORTH)	BRITISH
boulder	>256 mm	>200 mm
cobble	64–256 mm	60–200 mm
pebble	2–64 mm	
gravel		2–60 mm
sand	62.5–2000 µm	
coarse sand		600–2000 µm
medium sand		200–600 µm
fine sand		60–200 µm
silt	4–62.5 µm	2–60 µm
clay	<4 µm	<2 µm

Parus bicolor (tufted titmouse) *See Baeolophus bicolor.*

Parus major (great tit) A bird belonging to the **Paridae**, 125–140 mm long with a blue-black crown, black head, neck, and throat, white cheeks, a broad black band along the underside, olive back, and yellow sides. It inhabits woodland and is a frequent visitor to garden feeders. It feeds on insects and other invertebrates, augmented by seeds and berries in winter. It occurs throughout much of Eurasia and part of North Africa.

Pasiphila rectangulata (green pug moth) A nocturnal moth (**Lepidoptera**) with a wingspan of 15–20 mm that is dark brown or green with dark bars and flies in midsummer. Its **caterpillars** are thick-bodied and green with a reddish stripe, and feed on the flowers of hawthorn and blackthorn, as well as several crop plants, including apple, pear, cherry, and quince. The moth is common and widespread in Europe and has been introduced to North America.

Passer domesticus (house sparrow) A stocky bird about 160 mm long with a brown back streaked with black and a pale underside. Males have a grey crown flanked by chestnut brown and black around their beak and on their throat. House sparrows inhabit farms and urban areas, feeding on the ground on seeds, nuts, fruit, as well as insects and other arthropods. They are native to Eurasia but occur worldwide.

Passeriformes (passerines, perching birds) An order of birds that have feet with three toes pointing forward and one backward, allowing them to perch. The birds are small- or medium-sized, varied in plumage, and with beaks adapted to a range of diets. There are about 5300 species, making this the largest group of birds.

Passerina cyanea (indigo bunting) A species of migratory birds, 115–130 mm long, that are brown with a little blue on the tail, but breeding males are brilliant blue with a dark blue crown. They inhabit woodland edges, field edges, and roadside and rail verges, often perching on power lines. They feed on invertebrates, seeds, and berries, and breed throughout eastern and western North America, wintering in Central America, northern South America, and the Caribbean.

passerines *See* Passeriformes.

Passifloraceae (order **Malpighiales**) A family of **trees**, **shrubs**, climbers, and **annual** or **perennial herbs** with **simple** or lobed, **petiolate** leaves. Climbers have modified **axillary inflorescences** as **tendrils**. Flowers **actinomorphic**, **hermaphrodite**, with 5 free or 3–8 fused **sepals** and **petals**, occasionally **apetalous**, 5 free or 4 fused, or 8 free or 8–10 fused **stamens**, **ovary superior** of 3 free or 2–5 fused **carpels**. Inflorescence a **cyme**, usually axillary, occasionally terminal or cauliflorous (*see* cauliflory). Fruit is a **berry**. There are 27 genera of 935 species occurring throughout the tropics with some in warm temperate regions. Many cultivated for their fruit (passion fruit, granadilla).

passion fruit *See* Passifloraceae.

passive absorption The **absorption** of water, nutrients, or other substances through the roots of a plant with no expenditure of energy by the plant.

passive chamaephyte A **chamaephyte** in which the shoots above ground die back and fall, to produce **buds** on horizontal axes at ground level.

passive front *See* inactive front.

Pastinaca (family **Apiaceae**) A genus of mostly **biennial herbs** with **pinnate** leaves and yellow flowers with tiny **sepals** or **asepalous**, 5 **petals**, and 5 **stamens**. borne in compound **umbels** with **bracts** or **bracteoles** that soon fall or are absent. Fruit is a **schizocarp**. There are 14 species occurring throughout temperate Eurasia and North Africa. *Pastinaca sativa* (parsnip) is cultivated for its edible **taproot**.

Pastinaca sativa (parsnip) *See* Apiaceae.

past-natural Describes a community with features that are derived directly from those of an **original-natural** community, with no human interference.

patagium A fold of skin between the fore and hind limbs in mammals, extended to aid gliding.

Patagonian floral region The area that includes Patagonia, Tierra del Fuego, the southern Andes, and the Falkland Islands. The **flora** are related to those of New Zealand.

patch dynamics The study of the proportion of feeding areas (patches) in a **habitat** that are occupied at a specified time by

P

members of the species of animals being studied.

patella The knee cap.

Patersonia (family **Iridaceae**) A **monocotyledon** genus of **perennial herbs** with leaves growing from a woody **rhizome** that sometimes forms a short trunk. Flowers open from a pair of **bracts**. They have 6 **tepals** in 2 **whorls** and 3 fused **stamens**. **Inflorescence** is a terminal cluster. Fruit is a **capsule**. There are about 20 species occurring mainly in Australia, also in Borneo and New Guinea. Several are cultivated for ornament.

pathogen A **microorganism** that is capable of causing disease.

pathogen-associated molecular patterns (PAMPs) Molecules or fragments of molecules produced by and characteristic of **pathogens** that can be detected by **pattern recognition receptors**. Most PAMPs are associated with **microbes** and are known as **microbe-associated molecular patterns**.

pattern recognition receptors (PRRs) **Proteins** located on the cell surface that identify pathogen-associated molecular patterns (PAMPs), e.g. bacterial and viral **DNA** and **RNA**, bacterial **carbohydrates** and **peptides**, triggering a response that prepares tissue for an impending attack.

Paulowniaceae (order **Lamiales**) A **monogeneric** family (*Paulownia*) of fast-growing, **deciduous trees** with large (up to 40 cm across), **opposite**, **simple**, **entire** leaves. Flowers **pentamerous**, resembling foxgloves borne in terminal, branched, **cymose inflorescences**. There are seven species occurring in temperate eastern Asia. Several are grown for ornament or timber.

pawpaw (*Carica papaya*) *See* Caricaceae.

Paxillus involutus (brown roll-rim, common roll-rim, poison pax) A species of **bolete fungi** in which the brown, funnel-shaped **fruiting body** is up to 60 mm tall, with a convex **pileus** up to 40–150 mm across with inrolled edges and **decurrent gills**, and a **stipe** 20–80 mm tall and up to 20 mm thick. It is common in woodlands and forms **mycorrhizae** with trees. It occurs throughout the Northern Hemisphere and has been introduced to Australia, New Zealand, and South America. Once thought to be edible, at least when cooked, it now known to be extremely poisonous.

PCR *See* polymerase chain reaction.

PDSI *See* Palmer drought severity index.

pea and bean weevil *See Sitona lineatus.*

pea aphid *See Acyrthosiphon pisum.*

peach *See Amygdalus.*

peach aphid *See Myzus persicae.*

pea cyst nematode *See Heterodera goettingiana.*

pea gall midge *See Contrarinia pisi.*

peak gust The highest speed of sustained wind or a **gust** that is recorded at a weather station during a period of observation, usually 24 hours.

pea midge *See Contrarinia pisi.*

pea moth *See Cydia nigricana.*

pear *See Pyrus.*

pear flower bud weevil *See Anthonomus pomorum.*

pear leaf curling midge *See Dasineura mali, Contrarinia pyrivora.*

pear leaf midge (*Dasineura pyri*) *See Dasineura mali.*

pearl spar *See* dolomite.

pear midge *See Contrarinia pyrivora.*

pear moss (common bladder moss) The moss (**Bryophyta**) *Physcomitrium pyriforme*, which forms pear-shaped **sporophyte capsules** in spring. It is distributed worldwide except for South America and Antarctica.

pear sawfly *See Caliroa cerasi.*

pear slug *See Caliroa cerasi.*

pear sucker *See Psylla pyricola.*

peat Organic matter that accumulates under waterlogged, mainly anaerobic conditions, where its decomposition proceeds very slowly. Bog mosses (*Sphagnum* spp.) are characteristic at such sites and, consequently, form a major component of peat.

peat moss Any of up to 350 species of *Sphagnum* moss (**Bryophyta**) found worldwide in wet, acid **habitats**, e.g. **bogs**, **moors**, etc. *Sphagnum* plants are branched. Their leaves lack **veins** and consist of green living cells and colourless, inflated, dead cells that readily fill with water, so the moss can hold up to at least 20 times its own weight of water.

peat podzol A **podzol** that has a surface **mor** horizon up to 30 cm thick, and usually an **iron pan** at the top of the B **soil horizon**.

peat soil A soil in which the O **soil horizon** is at least 40 cm thick, and often much thicker, and contains at least 65 percent organic matter, measured as dry weight. Peat soils are a major group in the classification developed by the Soil Survey for England and Wales. *See* histosols.

pebble In the Udden-Wentworth scale of **particle size**, a stone 2–64 mm in size.

pecan (*Carya illinoinensis*) *See Carya*, Juglandaceae.

pecking order The hierarchical social organization, based on dominance, found in many vertebrate and some insect species. An individual at a particular level in the hierarchy may threaten, or even peck, an individual at a lower level in order to gain prior access to a resource.

pectin A structural **heteropolysaccharide** found in the **middle lamella** and **cell walls**, especially in non-woody tissues and fruits of terrestrial plants. Reactions catalyzed by the **enzymes** pectinase and pectinesterase break pectin down in ripening fruit, softening the fruit, and in the **abscission** zone of the **petioles** of **deciduous** leaves. It is extracted commercially mainly from citrus fruits and used as a gelling agent, e.g. to help setting in jams and jellies.

pectinate Resembling a comb.

pectoral girdle The skeletal structure in vertebrates that provides support for the forelimbs.

ped The smallest structural unit of a soil, consisting of an **aggregate** of particles.

pedalfer A free-draining, **acid soil** that develops in a wet climate. Water moving downward to the **groundwater** leaches out aluminium (*al*) and iron (*fer*) from the soil **ped**.

Pedaliaceae (order **Lamiales**) A family of **perennial** or **annual herbs**, some **shrubs** or **trees**, with **opposite** sometimes **alternate**, **simple**, **entire**, lobed, occasionally **pinnatifid** leaves. Flowers **zygomorphic**, with 5 **connate sepals**, **corolla campanulate**, trumpet-shaped, or cylindrical, with 5 lobes, 4 **stamens** sometimes with 1 **staminode**, **ovary superior** usually of 2 fused **carpels** with 4 **locules**. Flowers solitary in leaf **axils** or in **axillary cymose** clusters. Fruit usually a **capsule**. There are 14 genera of 70 species occurring most in coastal or arid **habitats** in the Old World tropics. *Sesamum indicum* (sesame) is cultivated for its edible seeds.

pedate Describes a **palmate** leaf in which the lobes are divided.

pedicel A **flower** stalk.

pedipalps The second of the six pairs of appendages on the **prosoma** of arachnids (**Arachnida**). In arachnids with large

chelicerae the pedipalps have become walking legs; in others they are large and used to capture prey. All arachnids use their pedipalps to kill and manipulate prey, in self-defence, and for digging.

pedocal An **alkaline soil** that develops in **dry climates**, so although water drains freely it does not reach the **groundwater**.

pedogenesis (soil formation) The natural processes by which soil forms.

pedology The study of the formation, composition, and distribution of soils. It is one of the two main branches of soil science, the other being **edaphology**.

pedon A soil sampling unit that extends from the surface to the **parent material** and laterally by a sufficient amount to allow a complete exposure of the **soil profile**.

peduncle The stalk of an **inflorescence**.

pedunculate Pertaining to, or resembling, a **peduncle**.

Pegomyia betae (mangold fly) A species of grey-brown flies (**Diptera**) with orange legs that are up to 7 mm long. They fly in spring and lay small groups of white eggs, about 1 mm long, on the underside of leaves or **cotyledons**. These hatch into white, semi-transparent, legless maggots (beet leaf miners), 6–7 mm long, which burrow into the leaves and feed between the upper and lower surfaces before emerging to pupate in the soil, where they spend the winter. There can be two or three generations a year. They produce brown or silver blisters on leaves and inhibit growth; severe infestations can kill the plant. It occurs in North America, northern and central Europe, and Japan.

pelage In mammals, the hair covering the body; the coat.

pelagic Describes marine organisms that live in open water, and seabirds that spend most of their time at sea, coming ashore only to breed.

Pelargonium *See* Geraniaceae.

pelargonium aphid *See Acyrthosiphon malvae.*

pellicle *See* periplast.

pelosols Clay soils through which water percolates slowly and that have no gleyed (*see* gleying) **soil horizon** within 40 cm of the surface. They have a coarse, blocky structure and in dry weather they crack deeply. They are a major group in the soil classification devised by the Soil Survey for England and Wales.

peltate Shaped like a shield, with the **pedicel** near the centre.

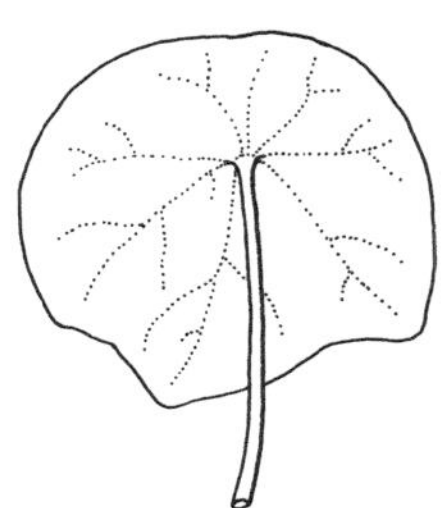

A peltate leaf.

Peltigera *See* dog lichen.

pelvis In vertebrates, the part of the **appendicular skeleton** that provides support for the hind limbs.

Pemphigus bursarius (lettuce root aphid) A species of aphids (**Aphididae**) that have antennae (*see* antenna) less than one-third the length of their bodies. They form gall-like colonies covered with powdery wax on the roots of **Asteraceae**, especially lettuce and chicory plants. Infested plants may fail to develop and the hearts fail to grow firm. Poplar trees (*Populus*) are the primary host.

Penaeaceae (order **Myrtales**) A family of **evergreen shrubs** and **subshrubs** with small, **opposite**, often **sessile**, **entire** leaves. Flowers **actinomorphic**, **hermaphrodite**, **perianth** of 4 **sepals** in 1 **whorl**, **petals** absent, 4 **stamens** alternating with the sepals, **ovary superior** of 4 **carpels** and 4 **locules**. Flowers borne singly but often crowded together in upper leaf **axils**, subtended by 2–4 coloured, leafy **bracts**. Fruit is a **loculicidal capsule**. There are 9 genera of 29 species occurring in eastern and southern Africa and St Helena. Some grown as ornamentals.

penicillin *See Penicillium.*

Penicillium A genus of **ascomycete fungi** in which the **thallus** comprises a branched network of multinucleate, septate (*see* septum) **hyphae** bearing **conidiophores.** The fungus occurs in soils worldwide in temperate regions. Most species are **saprotrophs**, forming **moulds**, and some cause disease. Different species are used in the production of cheeses and prepared meats, and *P. chrysogenum* yields the antibiotic penicillin.

pennaceous feather *See* feather.

Pennantiaceae (order **Apiales**) A **monogeneric** family (*Pennantia*) of **trees**, **shrubs**, and a few woody climbers, with **alternate**, **simple**, **entire** or **dentate** leaves. Flowers **actionomorphic**, **unisexual** (plants more or less **dioecious**), usually **pentamerous.** Fruit is a **drupe.** There are four species occurring in New Zealand, Norfolk Island, and northeastern Australia.

pennate diatom A diatom with **bilateral symmetry.**

penny bun *See Boletus edulis.*

Pentadiplandraceae (order **Brassicales**) A **monotypic** family (*Pentadiplandra brazzeana*) of **shrubs** and **lianas** with **alternate**, **simple**, **entire**, **exstipulate**, **petiolate** leaves. Plants **polygamous.** Flowers **pentamerous**, 10 free or 9–13 fused **stamens** (**staminodes** in female flowers), **ovary superior** of 3–5 **carpels** and **locules. Inflorescence axillary** or terminal, **racemose.** Fruit is a **berry.** The plant occurs in tropical West Africa. It is the source of two sweeteners, active ingredients being the proteins brazzein and pentadin.

pentamerous With parts in fives.

Pentaphragmataceae (order **Asterales**) A **monogeneric** family (*Pentaphragma*) of fleshy **herbs** with **distichous**, **exstipulate**, **serrate**, **dentate**, or **entire** leaves. Flowers **actinomorphic**, usually **hermaphrodite**, 5 free **sepals**, 4 or 5 usually united **petals**, 4 or 5 **stamens adnate** to the **corolla** tube, **ovary inferior**, **syncarpous**, of 2–3 **carpels** and **locules. Inflorescence** a **cyme.** Fruit is **baccate.** There are 30 species occurring from southeastern Asia to Malesia.

Pentaphylaceae (order **Ericales**) A family of usually **evergreen trees** and **shrubs** with **indumentum.** Leaves **distichous** or spirally arranged, **simple**, **entire** to **crenate** or **serrate**, **exstipulate.** Flowers usually **hermaphrodite** sometimes **unisexual** (plants **dioecious**), **pentamerous**, 5 persistent **sepals**, 5 **petals**, 5 to many **stamens**, **ovary** usually **superior** of 3–5 sometimes 2 or 6 fused **carpels**, each with 1 **locule. Inflorescence** an **axillary**, rarely terminal, **fascicle** or flowers solitary. Fruit a **berry**, **loculicidal capsule**, or **drupe.** There are 12 genera of 337 species occurring throughout the tropics and subtropics.

Pentathera *See Azalea.*

Pentatomidae (shield bugs, stink bugs) A family of bugs (**Hemiptera**), in which the brown or green **scutellum** is large, in some species almost covering the forewings and abdomen, and usually trapezoidal in shape. Some species are predators, but most feed on plants and are serious pests because they occur in large numbers and many are resistant to most **insecticides.** There are about 5000 species with a worldwide distribution.

Penthoraceae (order **Saxifragales**) A **monogeneric** family (*Penthorum*) of **herbs** with **stolons** or **rhizomes.** Leaves **alternate**, shortly **petiolate** or **sessile**, **lanceolate** to **elliptical**, **acuminate.** Flowers small, **actinomorphic**, **bisexual**, with 5 free or 5–8 fused **sepals** and **petals**, or **apetalous**, 10–16 **stamens**, **ovary** semi-**inferior** to **superior** of 5 free or 5–8 fused **carpels. Inflorescence** an **axillary** or terminal **cyme.** Fruit is a **capsule.** There are two species occurring in eastern and southeastern Asia and eastern North America.

pentose A **monosaccharide** with five carbon atoms. The **nucleotides** ribose and deoxyribose are pentose sugars.

pentose phosphate shunt *See* hexose monophosphate shunt.

Pentoxylon A genus, now extinct, of plants that grew in **Gondwana**, first appearing in the Late Permian epoch (260.4–251 million years ago) and becoming most abundant during the Jurassic and Early Cretaceous epochs (199.6–99.6 million years ago). The plants had branching stems with long, strap-like leaves at the ends of **branches**. Stems were eustelic (*see* eustele) with five or six wedges of **secondary growth** forming around a strand of primary **xylem** (giving the plant its name). **Cupules** bearing **ovules** were grouped in strobili (*see* strobilus) attached to short shoots.

peony *See* Paeoniaceae.

PEP *See* phosphoenolpyruvic acid.

Peperomia (family **Piperaceae**) A genus of **perennial herbs**, mostly **epiphytes**, with fleshy or **succulent**, **simple**, **entire** leaves. Flowers **hermaphrodite** with no **sepals** or **petals** and 2 **stamens**. **Inflorescence** a conical **spike**. The fruit is a **berry**. There are 845 species occurring throughout the tropics and subtropics, especially Central and northern South America. Some cultivated for their flowers or foliage.

PEP group translocation *See* group translocation.

pepper (*Capsicum*) *See* Solanaceae.

pepper dulse The small, red seaweed *Osmundea pinnatifida*, up to 8 cm long, with tufts of tough, flat fronds that are yellow-green on the upper shore but dark red-purple near the low-water mark. It branches alternately in one plane and grows in rock pools and on rocks. It has a strong, pungent smell and peppery taste, and is dried and used as a spice in some places.

peppered moth *See Biston betularia*.

pepper mild mottle virus (PMMoV) A **tobacco mosaic virus** that is the most serious viral pathogen of ***Capsicum*** plants. Symptoms vary, but include stunting, **chlorosis** of leaves, and deformation of fruits. It is transmitted by physical contact with contaminated objects or materials.

pepper tree (*Schinus molle*) *See* Anacardiaceae.

peptide A linear molecule made from two or more **amino acids** linked by **peptide bonds**. Depending on the number of amino acids, peptides are designated dipeptide (two), tripeptide (three), oligopeptide (three to ten), and **polypeptide** (more than ten).

peptide bond A chemical bond between a **carboxyl** group and **amino group** that links **amino acids** to form **peptides**.

peptidoglycan *See* murein.

Peraceae (order **Malpighiales**) A family of usually **evergreen trees** and **shrubs** with **alternate** rarely **opposite**, shortly **petiolate**, **simple**, **entire** leaves. Flowers **actinomorphic**, **unisexual** rarely **bisexual** (plants **dioecious** rarely **monoecious**), 4–6 free **sepals** or **calyx** fused with 2–4 sometimes 6 lobes, or **asepalous**, 5 free **petals** free or **apetalous**, 5–20 free or 2–5 fused **stamens**, **ovary syncarpous**, hypogynous (*see* hypogyny). **Inflorescence** an **axillary fascicle** or **racemose** or **paniculate**. Fruit is a **schizocarp**. There are 5 genera of 135 species with a pantropical distribution.

peramorphosis A type of **heterochrony** in which development takes much longer than in the ancestral form.

perched aquifer An **aquifer** that lies above a **confined aquifer**.

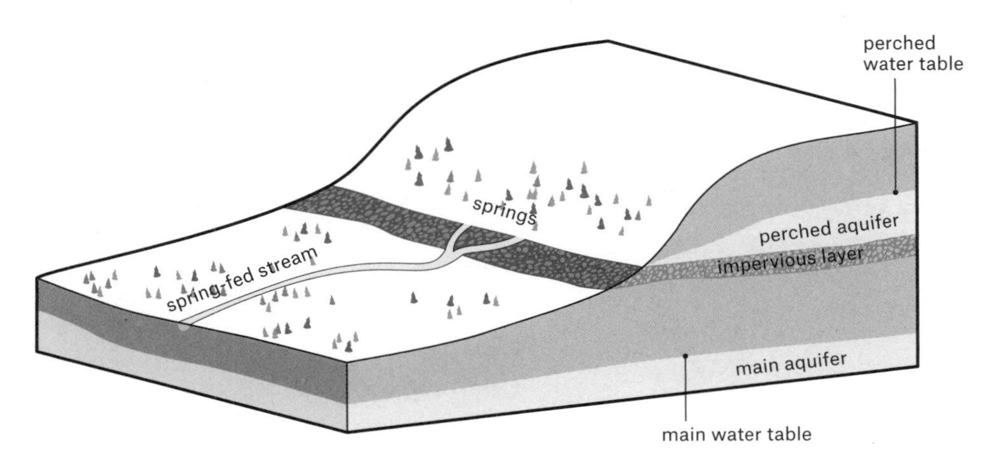

Perched aquifer. The main aquifer lies above a layer of impervious rock and below a layer of permeable material. A second impervious layer lies above the permeable material and a second aquifer is perched above that, with its own perched water table. Where the upper impervious layer intersects the ground surface, water from the perched aquifer emerges as springs that feed a stream.

perching birds *See* Passeriformes.

percolation The downward movement of water through a particulate medium, e.g. soil, especially if that medium is at or close to saturation.

percurrent *See* scalariform.

perennating bud A **bud** on the stem of a **biennial** or **perennial** plant that remains dormant through the season of no growth (winter or a dry season) and develops when favourable conditions return. *See* perennating organ.

perennating organ A part of a **biennial** or **perennial** plant that is modified to act as a store for energy, usually **carbohydrate**, or water during periods unfavourable for plant growth, e.g. winter or a dry season. **Bulbs**, **corms**, **pseudobulbs**, **rhizomes**, **tubers**, and **taproots** are perennating organs, as are **perennating buds**.

perennial Describes a plant that ordinarily lives for more than two seasons and when mature flowers annually.

perfect cycle A **biogeochemical cycle** in which an element enters the reservoir available to living organisms at approximately the same rate as it is removed; most perfect cycles involve a gaseous stage. *Compare* imperfect cycle.

perfect flower A **flower** that possesses a functioning **gynoecium** and **androecium**.

perfect state The state of a fungus when it is forming sexually produced **spores**.

perforation plate The end wall of a **vessel element**, which has openings to allow the passage of liquid.

perianth (perigonium) **1.** The outer part, i.e. **calyx** and **corolla**, of a **flower**. **2.** A sheath surrounding the **archegonium** of a moss (**Bryophyta**) or liverwort (**Marchantiophyta**).

pericarp The outer wall of a ripe **ovary** or fruit, consisting of a hardened or toughened outer layer, the exocarp, a middle layer, the mesocarp, which is often succulent or fleshy, and an inner endocarp that surrounds the seeds.

pericentric inversion The end-to-end reversal of a section of a **chromosome** containing the **centromere**.

perichaetium One of the enlarged leaves or **bracts** that surround the **archegonium** and **antheridium** of a moss (**Bryophyta**).

periclinal Parallel to a surface, e.g. describes the **cell wall** that is parallel to the plant surface. *Compare* anticlinal.

periclinal division Cell division in which the walls between **daughter cells** are **periclinal**, thus increasing the width or girth of the organ. *Compare* anticlinal division.

pericycle A thin layer of tissue, composed mainly of **parenchyma** or **sclerenchyma**, lying between the **endodermis** and **phloem** and forming the outermost layer of the **stele**. Lateral roots arise from the pericycle.

Peridiscaceae (order **Saxifrgales**) A family of **deciduous trees** and **shrubs** with **alternate**, **simple**, **entire** leaves. Flowers small, **actinomorphic**, **hermaphrodite**, with 4–5 or 7 free or 6 fused **sepals** the inner ones **petaloid**, **petals** absent, many **stamens**, **ovary superior** of 3–4 fused **carpels** with 1 **locule**. **Inflorescence** an **axillary raceme** or **fascicle**. Fruit is a **drupe** or **capsule**. There are 4 genera of 11 species occurring in tropical South America and West Africa.

peridium A membrane that encloses the **spores** in the **fruiting bodies** of certain **Fungi** and in **slime moulds**.

perigenous Describes a **stoma** and surrounding cells, found in some **angiosperms**, in which the **guard cells** and **subsidiary cells** are derived from different **initials**. *Compare* mesogenous. *See also* haplocheilic.

periglacial Describes an area adjacent to a present or former glacier or ice sheet, or to an **environment** in which freezing and thawing is or once was the predominant surface process.

perigonium *See* perianth.

perigynium 1. Any unusual structure surrounding the **pistil**. **2.** In sedges (***Carex***), a modified **bract** forming a sac enclosing the **achenes**. **3.** In mosses (**Bryophyta**) and liverworts (**Marchantiophyta**), a fleshy tube or cup that surrounds the **archegonium**.

perigynous Describes a **flower** that has the **calyx**, **corolla**, and **stamens** inserted around the edge of the cup-like **receptacle**.

perinuclear space *See* nuclear envelope.

periostracum *See* ostracum.

Periparus ater (coal tit) A bird belonging to the **Paridae**, which is 100–115 mm long with a black head, large white spot on the nape of its neck, white sides to the face, a black head, throat, and neck, a white bar on each wing, and pale underparts. It inhabits a variety of **habitats**, including gardens, and often forages in flocks. It feeds on seeds. It occurs throughout Eurasia and North Africa.

periphysis One of the short **filaments** that line the **ostiole** of the **perithecium** in certain **Fungi**.

periphyton Aquatic organisms that live clinging to submerged leaves, stems, or objects that protrude from the bottom.

periplasmic space In Gram-negative (*see* Gram reaction) **Bacteria**, the space between the inner and outer **cell membrane**.

periplasmodium A fluid resulting from the breakdown of the **tapetum** in some vascular plants (**Tracheophyta**) that surrounds the **sporocyte** and is absorbed by the **microspores**.

periplast (pellicle) A **protein**-rich layer inside a **cell membrane**.

perisperm Nutritive tissue derived from the **nucellus** in plants where the nucellus is not completely replaced by the **endosperm**.

perispore The outer covering of a **spore**.

peristome The mouth cavity of an invertebrate animal.

peristome teeth A set of structures, often delicate and thread-like, forming a ring around the rim of the **capsule** of a mature moss **(Bryophyta) sporophyte** that is revealed when the **operculum** falls away. The teeth respond to atmospheric **humidity**, closing the capsule when the air is moist and bending outward in dry air, allowing the **spores** to disperse.

perithecium An **ascocarp**, round or shaped like a flask, inside which **ascospores** form, to be discharged through a small pore.

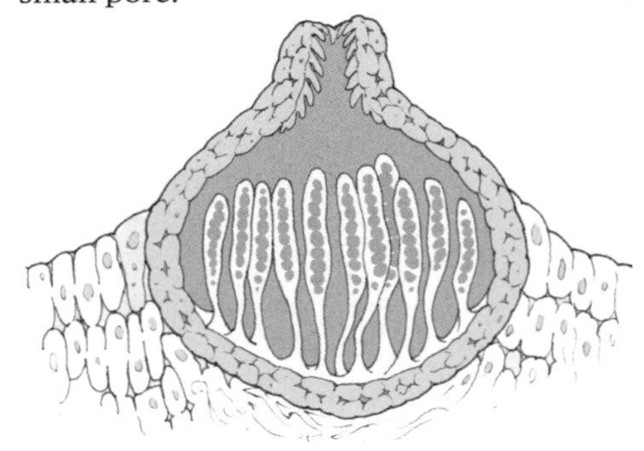

The ascospores form inside the flask-shaped ascocarp and are released through a small pore.

peritrichous Of flagella (*see* flagellum), distributed all over the cell.

periwinkle *See Vinca.*

permafrost (pergelisol) A layer of ground that is permanently frozen. Temperatures have remained below freezing through at least two consecutive winters and the intervening summer.

permanent drought **Drought** that typifies a desert, with very occasional precipitation and no permanent streams.

permanent wilting percentage *See* permanent wilting point.

permanent wilting point (PWP, permanent wilting percentage, wilting coefficient, wilting point) The amount of water present in a soil below which plants growing in that soil will lose **turgor**, i.e. wilt, and will fail to recover when wetted, i.e. wilting is permanent. It occurs when the force with which moisture in a drying soil adheres to soil particles exceeds the pressure exerted by plant roots. *Compare* temporary wilting.

permeability (hydraulic conductivity) The extent to which a medium permits the flow of fluids, i.e. the volume of fluid that passes through a unit cross-sectional area of a porous medium (*see* porosity) in a given period at a specified temperature, measured in units of metres per second or metres per day.

permease A member of a class of **proteins** that facilitate the passage of other substances across **cell membranes**.

Peromyscus leucopus (white-footed deer mouse) A rodent **(Rodentia)** with a body 90–100 mm long and a tail 65–95 mm long, with a pale or reddish brown head and back and white underside and feet. It inhabits warm, dry forests and scrub and feeds on seeds, nuts, berries, fruits, fungi, and insects. It occurs throughout most of the eastern United States.

Peronosporales One of the two principal orders of water moulds **(Oomycota)**, some of which are **saprotrophs** and others that are plant pathogens, e.g. ***Phytophthora*** and the organisms causing **downy mildew**.

peroxidase An **enzyme** that catalyzes the **oxidation** of compounds using hydrogen peroxide (H_2O_2) as an electron acceptor.

peroxisome A small **organelle**, 0.3–1.5 µm across and enclosed in a membrane, found in most **eukaryote** cells. They are made from **proteins** synthesized in **ribosomes** and reproduce by division, but they have no **genome**. Each peroxisome contains 50 or more **enzymes** involved in a number of metabolic reactions.

Persea (family **Lauraceae**) A genus of **evergreen trees** with spirally arranged or **alternate**, **simple**, **lanceolate** leaves. Flowers with 6 **perianth** segments and 9 **stamens**. **Inflorescence** is a **panicle**. Fruit is a **drupe**. There are about 150 species with a **disjunct distribution**, about 70 occurring in the New World tropics, 1 **endemic** to the Macaronesian Islands including Madeira and the Canary Islands, and about 80 in eastern and southeastern Asia. *Persea americana* is cultivated for its fruit, avocado pear.

Persea americana (avocado pear) *See Persea*, Lauraceae.

Persian ironwood (*Parrotia persica*) *See Parrotia.*

Persian walnut (*Juglans regia*) *See Juglans.*

persimmon (*Diospyros* spp.) *See* Ebenaceae.

perthophyte A parasitic fungus which obtains nutrients from dead tissue within a living plant.

pesticide A chemical compound that is used to kill pest organisms.

pesticide tolerance The concentration of a **pesticide** that is permitted by law to remain as a residue on a crop after it has been harvested.

petal A leaf-like structure, probably a modified leaf, found inside the **calyx** of a **flower**, a **whorl** of petals constituting the **corolla**. In insect-pollinated flowers petals are usually conspicuous and often brightly coloured; in wind-pollinated flowers they are small and inconspicuous.

petaloid Resembling a **petal**.

Petenaeaceae (order **Huerteales**) A **monotypic** family (*Petenaea cordata*) comprising small **trees** or **shrubs** with **cordate**, **petiolate**, **chartaceous**, **denticulate** leaves with minute, **caducous stipules**. Flowers with **valvate**, **lanceolate sepals**, **petals** absent, 8–12 **glabrous stamens**, **ovary superior**. **Inflorescence** an **axillary cyme** or **panicle**. Fruit is a **berry**. The species occurs only in northern Central America.

Petermanniaceae (order **Liliales**) A **monotypic**, **monocotyledon** family (*Petermannia cirrosa*), which is a **perennial** climber with a **rhizome**. The stem bears prickles and it climbs by **tendrils** opposite the leaves. Leaves **alternate**, **petiolate**, **lanceolate**, **entire**. Flowers small, **actinomorphic**, **bisexual**, with 3+3 **tepals**, 3+3 **stamens**, **ovary inferior** of 3 **carpels** with 1 **locule** rarely 3. **Inflorescence** is a terminal **cyme**. Fruit is a **berry**. The species occurs on the central eastern coast of Australia.

petiolate Possessing **petals**.

petiole 1. The stalk of a leaf. **2.** In **Apocrita**, the tight constriction ('waist') between the first and second segments of the abdomen.

petiolule The stalk of one of the **leaflets** in a **compound inflorescence**.

petrocalcic horizon An indurated (*see* induration) **calcic horizon** at least 10 cm thick that contains 50 percent or more calcium carbonate, which cements the layer, making it impenetrable to plant roots and impossible to dig.

petroduric horizon *See* duripan.

petrogypsic horizon A **soil horizon** at least 10 cm thick that is **cemented** by **gypsum**, which comprises at least 60 percent of the mineral content. It is impenetrable for plant roots. It is a **diagnostic horizon**.

petroplinthic horizon A rust-brown or yellowish **soil horizon** at least 10 cm thick that comprises a continuous layer of indurated (*see* induration) material, **cemented** mainly by iron. Organic matter is present only in traces, or not at all and the layer is impenetrable for plant roots.

P

Petrosaviaceae (order **Petrosaviales**) A **monocotyledon** family comprising one genus (*Petrosavia*) of leafless **saprophytes** that lack **chlorophyll**, and 1 genus (*Japanolirion*) with green, **linear** leaves at the base. Flowers **actinomorphic** with 5 **trimerous whorls**, **ovary superior** to semi-inferior. **Inflorescence** usually a **bracteate raceme**. Fruit is a **follicle**. There are two genera of four species with a scattered distribution in Japan, China, and western Malesia.

Petrosaviales An order of plants that includes only the family **Petrosaviaceae**, with two genera and four species.

Pezizaceae (cup fungi) A family of **ascomycete fungi** with **fruiting bodies** that are cup- or saucer-shaped, although that of ***Aleuria aurantia*** (orange peel fungus) resembles orange peel. There are 31 genera with 230 species, distributed worldwide.

Pezizomycetes A class of **ascomycete fungi** in which the **fruiting bodies** are cup-shaped and the **ascus** has an **operculum**. The Fungi are **saprotrophs**, mycorrhizal (*see* mycorrhiza), or plant parasites. The class includes the morel (***Morchella***) and truffles. There are more than 3700 species, most occurring in temperate regions.

Pezizomycotina The largest subphylum of **Ascomycota** comprising all the **Fungi** that produce a filamentous **mycelium** and **fruiting bodies**. There are more than 32,000 species, found worldwide.

PGA *See* phosphoglyceric acid.

PGF *See* pressure-gradient force.

pH A measure of acidity and alkalinity on a scale of 0–14, in which 7.0 represents neutrality; substances with a pH lower than 7.0 are acid, and those with a pH greater than 7.0 are alkaline. The value is calculated as the logarithm of the reciprocal of the concentration of hydrogen **ions** expressed in moles per litre ($pH = \log_{10} 1/H^+$). Household bleach has a pH of 13.0, sea water is about pH 8.0, lemon juice is about pH 2.0.

phaeomelanin *See* melanin.

Phaeophyta *See* algin, brown algae.

phaeozems A group of soils that includes all those with a **mollic horizon** other than **chernozems** and **kastanozems**. Phaeozems are a reference soil group in the **World Reference Base for Soil Resources**.

phage *See* bacteriophage.

phagocytosis The process by which a cell engulfs an external particle that it then digests by an **invagination** of the **cell membrane** that encloses the particle, then detaches inside the cell as the membrane closes behind it.

phalange *See* phalanx.

phalanx (phalange; pl. phalanges) One of the digits of a **tetrapod** limb.

phalanx growth form The distribution pattern that results when a plant spreads by means of **rhizomes** or **stolons** that are fairly short and often long-lived. They are closely spaced and the plants growing from them appear to advance along a front, like a Roman phalanx. *Compare* guerrilla growth form.

phallotoxins A group of at least seven toxic compounds present in the **fruiting bodies** of several species of ***Amanita*** **Fungi**, and especially ***A. phalloides*** (death cap). Ingestion by humans causes vomiting and diarrhoea, and may cause liver damage.

Phallus impudicus (common stinkhorn) A species of **agaric fungi** in which the **fruiting body** comprises a **peridium** enclosing a gelatinous **gleba** containing the **spores**. As the fruiting body matures the gleba emits a smell of carrion that attracts insects which disseminate the spores. The fruiting body is 100–300 mm tall, 40–50 mm wide, and topped with a

P

conical **pileus** 20–40 mm tall covered by the green-brown, slimy gleba. The inside of the stinkhorn is edible raw or pickled. The fungus occurs widely in Europe, North America, and parts of Asia, Africa, Australia, and South America. It is a **saprotroph**, usually found in woodland but sometimes in parks and gardens.

phanerophyte A plant that bears its **perennating buds** or shoot tips on shoots above ground. It is a category in the classification of life forms devised by Christen **Raunkiær**.

pharate The condition of an insect (**Insecta**) immediately prior to moulting. Its new **exoskeleton** has formed but the old one has not yet been shed.

pharmacopoeia A published list of medicines and health-care products and recipes for preparing them, nowadays those approved by governments or the World Health Organization. Originally the ingredients were obtained exclusively from plant, animal, and mineral sources. Most medicines were compounded from several ingredients; one derived from a single ingredient was called a simple.

P

pharynx The part of the vertebrate gut between the mouth and the oesophagus.

Phaseolus (family **Fabaceae**) A genus of **herbs** and climbers, many of which produce edible seeds (beans). There are 50–87 species, depending on the authority, occurring in tropical and warm temperate America, but now cultivated worldwide.

Phasmarhabditis hermaphrodita (rhabditid nematode) A nematode (**Rhabditidae**) that lives in soil, growing on decaying plant and animal matter and on slug faeces, where it consumes bacteria. The nematodes produce non-feeding juveniles with arrested development that enter slugs, especially ***Deroceras reticulatum***, where they resume their development and reproduce, killing the host. They are used extensively in **biological control.**

phasmid 1. One of a pair of sensory glands on either side of the tail of an eelworm (**Nematoda**). **2.** A stick insect or leaf insect of the order Phasmatodea.

Phaulothamnus *See* Achatocarpaceae.

pheasant's back *See Polyporus squamosus.*

Phellinaceae (order **Asterales**) A **monogeneric** family (*Phelline*) of **evergreen trees** and **shrubs** with **alternate**, **simple**, **entire**, **exstipulate** leaves crowded near the tips of **branches**. Flowers small, **unisexual** (plants **dioecious**), with 4–6 more or less **connate sepals**, 4–6 free, **valvate petals**, 4–6 **stamens**, **staminodes** in **pistillate** flowers, **ovary superior** of 2–5 fused **carpels** and **locules**. **Inflorescence** is an **axillary raceme** or **panicle**. Fruit is a **drupe**. There are 12 species occurring in New Caledonia.

Phellinus A genus of **agaric fungi** in which the brown, **cork**-like **fruiting bodies** are **resupinate**, **sessile**, and **perennial** brackets. The **hyphae** are yellowish brown and there are no **clamp connections**. The Fungi are **saprophytes** or parasites living on a wide variety of tree species. There are about 150 species distributed worldwide, many of which invade the vascular tissue and cause white rot.

phellogen (cork cambium) **Cambium** tissue that surrounds the **vascular cambium** in woody stems and **branches** experiencing **secondary growth**. Phellogen gives rise to **bark** and **cork**.

phenetic classification The classification of organisms on the basis of their physical similarities.

phenetic species concept A definition of the term **species** based on the observable similarities between organisms and clear differences between those and other organisms.

phenol (carbolic acid) A volatile compound (C_6H_5OH) produced from petroleum that is a precursor for many industrial products, including some **herbicides**.

phenology The study of seasonal events, such as changes in the date each year when plants produce leaves, flowers, and fruit, when leaves fall, the arrival and departure of migrants, nest-building, hibernation, etc.

phenols A group of compounds formed from a hydroxyl group (–OH) bonded to an aromatic **hydrocarbon**. The simplest is carbolic acid (C_6H_5OH). Phenols produced by plants include cannabinoids (*Cannabis sativa*), capsaicin (chilli peppers), salicylic acid (*Salix* spp.), and raspberry ketone (the aroma of raspberries).

phenophase A stage in the annual life cycle of a plant or animal that has an identifiable start and finish.

phenotype The observable features of an organism, i.e. the manifestation of its **genotype**.

phenotypic variance The total variation in an observable **character**.

pheomelanin *See* melanin.

pheromone A substance released into the **environment** by one animal that elicits a response in another animal of the same species. Many pheromones act as sexual attractants.

pheromone trap A trap that contains a natural or synthesized **pheromone** to attract insects, to detect their presence in the area or the density of their population.

Pheucticus ludovicianus (rose-breasted grosbeak) A species of birds, 180–220 mm long with a wingspan of 290–330 mm, in which the male has a black head, back, wings, and tail, white patches and red undersides on the wings, a red patch on the breast, and a white beak. They feed on fruit, seeds, and insects. Females are grey-brown with white markings and darker brown streaks. They breed in wooded areas and grassland throughout most of North America and winter in Central and northern South America and the Caribbean.

Pheucticus melanocephalus (black-headed grosbeak) A species of migratory birds, 150–200 mm long with a wingspan of 300–330 mm, a large, thick, conical, straw-coloured beak, males with a black head, black wings and tail with prominent white patches, an orange breast, and yellow underside, females brown with black streaks. They inhabit **deciduous** and **mixed woodland**, river banks, wetlands, and suburban areas, feeding on **snails**, other invertebrates, seeds, and berries. The birds spend most of the year in western North America and winter in Mexico.

Philadelphus (family **Hydrangeaceae**) A genus of mainly **deciduous**, a few **evergreen**, **shrubs** with **opposite**, **simple**, **serrate** leaves. Flowers, many sweet-scented, **tetramerous**, **pentamerous**, or **hexamerous** with many **stamens**. Fruit is a **capsule**. There are about 60 species occurring in North and Central America, Asia, and southeastern Europe. Many are cultivated for their flowers, known as mock orange because of their superficial resemblance and smell to those of ***Citrus*** species. They are also sometimes known incorrectly as syringa, ***Syringa*** being the genus to which the unrelated lilac belongs.

Philesiaceae (order **Liliales**) A **monocotyledon** family of **shrubs** and climbers with **distichous** or spirally arranged, **sessile** or shortly **petiolate** leaves. Flowers pendulous, **actinomorphic**, **bisexual**, with 6 **tepals** in 2 **whorls**, 3+3 **stamens**, **ovary superior** of 3 **carpels** with 1 **locule**. **Inflorescence axillary** or terminal. Fruit is a **berry**. There are two genera with two species occurring in southern Chile.

Philodendron (family **Araceae**) A **monocotyledon** genus mainly of climbers and **epiphytes** (some with a **symbiosis** with ants), with **alternate** leaves that are large, sometimes **pinnate**, often lobed, but that vary greatly in shape, and have juvenile and adult leaves with different shapes on the same plant. **Cataphylls** also occur. Some epiphytes begin life in the forest **canopy** and when they reach a certain size produce aerial roots that extend downward to the ground, after which they obtain nutrients from the soil and use the host only for support and access to light. Depending on the authority there are about 500 or 900 species occurring in tropical America. Many are cultivated as ornamentals.

Philomycidae A family of **slugs** in which the **mantle** is rounded and covers the entire body. The body is smooth and grows to 25–100 mm long. It inhabits moist locations in forests and cliff sides, sheltering beneath **bark** and in crevices, feeding on **lichen**, algae (*see* alga), and **Fungi**. They occur in parts of eastern Asia, eastern and central North America, Central America, and northern South America.

P

Philotheca *See Eriostemon.*

Philydraceae (order **Commelinales**) A **monocotyledon** family of **perennial herbs** with **rhizomes** or **corms**. Leaves crowded at the base, **distichous**, and **linear**, **ensiform**, or **terete**. Flowers solitary, **zygomorphic**, **sessile**, **bisexual**, with **petaloid perianth** of 4 **tepals** in 2 **whorls**, 1 **stamen**, **ovary superior** of 3 **carpels** and usually 1 **locule**. **Inflorescence** a simple or compound **spike**. Fruit is a **capsule**. There are four genera of five species occurring from Australia to southeastern Asia.

phloem In vascular plants (**Tracheophyta**), tissue that transports dissolved organic and inorganic nutrients to all parts of the plant.

phobotaxis A change in the direction of movement of a **motile** cell or organism that is made to avoid a stimulus.

Phoenix dactylifera (date palm) *See* African–Indian desert floral region.

Phoma lingam *See Leptosphaeria maculans.*

Phormidium A genus comprising about 200 species of filamentous **cyanobacteria** in which the sheathed **filaments** tend to form dense, leathery mats up to several centimetres across. They occur on wet rocks and wood, on aquatic plants, in wet soil and mud and are found worldwide, including the arctic.

Phormium (family **Xanthorrhoeaceae**) A **monocotyledon** genus of **evergreen**, **perennial herbs** with tough, **ensiform** leaves that grow up to 3 m long. Flowers small and tubular, borne in **panicles** at the top of stalks up to 5 m long. There are two species: *P. tenax* occurs in New Zealand and Norfolk Island; *P. colensoi* is **endemic** to New Zealand. Both are widely cultivated for their fibres and as ornamentals, and known as New Zealand flax.

Phorodon humuli (damson-hop aphid) A species of aphid (**Aphididae**) that overwinters on *Prunus*, emerging in late spring when some winged forms migrate to hop vines (*Humulus*); *P. humuli* is the main **limiting factor** in hop production. It is distributed worldwide.

phosphatase An **enzyme** that catalyzes reactions involving the **hydrolysis** of phosphoric acid **esters**.

phosphate rock *See* rock phosphate.

phosphoenolpyruvate *See* phosphoenolpyruvic acid.

phosphoenolpyruvic acid (PEP, phosphoenolpyruvate) An organic acid with four carbon atoms that has the phosphate bond with the highest energy of any compound found in living organisms.

Phosphoenolpyruvic acid is involved in **gluconeogenesis** and **glycolysis**, and is the **substrate** for carbon-dioxide **fixation** in plants with the **C4 pathway** of **photosynthesis**.

phosphogluconate pathway *See* hexose monophosphate shunt.

phosphoglyceric acid A 3-carbon molecule formed as an intermediate during **glycolysis** and that is also formed during the **Calvin cycle**, where it is the first stable product following the break of the unstable 6-carbon compound resulting from the combination of carbon dioxide with rubisco (**ribulose-1,5-biphosphate**); it is then often known as PGA.

phospholipids A class of **lipids** that are an important component of **cell membranes**. They are composed of **fatty acids**, **glycerol**, a phosphate group, and a **polar molecule**.

phosphorite *See* rock phosphate.

phosphorus (P) An element that is a plant **macronutrient**, absorbed from the soil solution as orthophosphate (PO_4^{2-}). Phosphorus is involved in the storage, transport, and release of energy through the **adenosine diphosphate-adenosine triphosphate** mechanism, central to **photosynthesis**, and in the synthesis of most **carbohydrates**. A deficiency causes intense green discoloration of leaves leading to **necrosis**.

phosphorus cycle The **imperfect cycle** by which **phosphorus** moves from rocks through living organisms and water, finally returning as sediments. Phosphorus enters the cycle by **weathering**, especially of the mineral apatite ($Ca_5[PO_4]_3[F,Cl,OH]$), which yields soluble phosphate (PO_4). Plant roots absorb phosphate as an essential nutrient, and **heterotrophs** obtain phosphorus in compounds contained in the food they consume. Phosphorus returns to the soil solution in urine and faeces, and through the decomposition of organic material. **Adsorption** onto particles, principally of **clay**, and reactions with minerals render phosphorus unavailable to plants and a proportion enter sediments that in time become sedimentary rocks.

phosphorylation A chemical reaction in which a phosphate group is added to a molecule.

phosphotransferase An **enzyme** that catalyzes reactions in which phosphate groups move between **substrates**.

phosphotransferase system *See* group translocation.

photoblastic Describes seeds that germinate in response to a stimulus from light.

photochemical reaction A chemical reaction triggered by the **absorption** of light energy, e.g. **photosynthesis**, **photophosphorylation**.

photochemical smog A form of air pollution that develops when ultraviolet radiation in very intense sunlight acts upon **hydrocarbons**, mainly in vehicle exhausts.

photodissociation A **photochemical reaction** that splits molecules into their constituent atoms. *See* photolysis.

photo-inhibition The prevention or retardation of a process in the presence of light.

photokinesis Movement or change in the speed or direction of movement of a **motile** organism or cell in response to a light stimulus.

photolithotroph A **phototroph** that oxidizes an inorganic substance, usually of mineral origin, in reactions that synthesize compounds.

photolysis (photodissociation) A sequence of chemical reactions driven by ultraviolet radiation in which molecules present in the atmosphere are broken down.

photomorphogenesis The influence of light on the growth and form of a plant. *Compare* skotomorphogenesis.

photonasty A response (*see* nasty) of a plant organ to the stimulus of light.

photoperiod The relative lengths of periods of daylight and nighttime darkness.

photoperiodism The response of an organism to changes in **photoperiod**, e.g. through the timing of flowering, setting of seed, leaf fall, etc.

photophosphorylation Part of the process of **photosynthesis** that involves forming **adenosine triphosphate** (ATP) using a proton gradient (hydrogen nuclei), similar to the **electron-transport chain** in **respiration**, created by energy from sunlight, i.e. **phosphorylation** powered by light photons (photo-).

photoreceptor A molecule that absorbs light, e.g. **chlorophyll**, **phytochrome**.

photorespiration A process that reduces the efficiency of the **light-independent stage** in the **C3 pathway** of **photosynthesis** when the concentration of carbon dioxide (CO_2) is below about 50 parts per million. The **enzyme** rubisco (**ribulose-1,5-biphosphate**) that catalyzes the fixation of CO_2 will accept either CO_2 or oxygen (O_2), so the gases compete; if O_2 is the more plentiful rubisco will add that rather than CO_2 to the next compound in the sequence, thereby altering the overall process with the resultant release of some of the CO_2 absorbed earlier. It is called respiration because it absorbs O_2 and releases CO_2, but it does so without yielding any energy.

photosynthesis A sequence of chemical reactions, powered by light energy absorbed by **chlorophyll**, in which green plants and some bacteria break down carbon dioxide (CO_2) and water (H_2O) and construct (synthesize) sugars. In green plants water serves both as hydrogen donor and a source of released oxygen, and the process can be summarized as:

$$6CO_2 + 6H_2O \xrightarrow[\text{light}]{\text{chlorophyll}} C_6H_{12}O_6 + 6O_2\uparrow$$

$C_6H_{12}O_6$ is **glucose**. *See* light-dependent stage, light-independent stage.

photosynthetic quotient The volume of oxygen released by **photosynthesis** expressed as a proportion of the carbon dioxide absorbed.

photosynthetic unit (PSU) A reaction centre where light for **photosynthesis** is absorbed, e.g. a **chlorophyll** molecule.

photosystem I (PSI) The sequence of reactions in **photosynthesis** in which **P700** and **accessory pigments** use light with a peak wavelength of 700 nm to reduce **nicotinamide dinucleotide phosphate** (NADP) to nicotinamide dinucleotide phosphate hydrogen (NADPH) and to produce **adenosine triphosphate** (ATP) through **photophosphorylation**. PSI was the first photosystem to be discovered, although its reactions occur after those of PSII.

photosystem II (PSII) The sequence of reactions in the **light-dependent stage** of **photosynthesis** in which **P680** and **accessory pigments** use light with a peak wavelength of 680 nm to dissociate (*see* photodissociation) water.

phototaxis A change in the direction of movement of a cell or organism in response to a change in light intensity.

phototroph An organism that obtains energy for its metabolism from light.

phototropism (heliotropism) A growth movement that occurs in response to light.

Phragmites (family **Poaceae**) A **monotypical**, **monocotyledon** genus of **perennial** grasses (*P. australis*, common reed) with **rhizomes**. Leaves are up to 50 cm long and 3 cm wide, the **ligule** a ring of hairs. The erect stems are up to 6 m tall. **Inflorescence** is a nodding **panicle** of slender **spikelets**, each **floret** having a tuft of long, silky hairs arising from the base. There are usually 3 **stamens**, the **ovary** is **glabrous**. Some authorities divide the genus into three or four species. *Phragmites* occurs in wetlands throughout temperate and tropical regions and can form extensive stands known as reed beds. It is widely cultivated for ornament and is used for thatching and purifying water (phytoremediation).

phragmoplast A structure that forms from the **mitotic spindle** during the **anaphase** and **telophase** stages of cell division in plants. At first it contains only **microtubules** but later acquires **Golgi bodies**, **ribosomes**, and **endoplasmic reticulum**. It acts as a scaffold for the **cell plate** during **cytokinesis**.

phreatic zone (zone of saturation) The region below the **water table** where all the soil **pores** are filled with water.

phreatophyte A plant with deep roots that obtains some of its water from the **phreatic zone**.

Phrymaceae (order **Lamiales**) A family of **annual** or **perennial herbs** and **shrubs** with **opposite**, **simple**, **petiolate**, **dentate** leaves. Flowers **zygomorphic** with 5 fused **sepals** and **petals**, and 4 **stamens**. Fruit is a **capsule** or **achene**. There are 13 genera with 188 species with a worldwide distribution, especially temperate western North America and Australia.

phycobilin A member of a group of **tetrapyrroles** that are **chromophores** in **cyanobacteria** and in the **chloroplasts** of **red algae** (Rhodophyta) and some other organisms.

phycobilisome An **organelle** that is the light-harvesting structure in **photosystem II** in **cyanobacteria** and **red algae** (Rhodophyta).

phycobiont The algal (*see* **alga**) or cyanobacterial (*see* cyanobacteria) **symbiont** in a **lichen**.

phycocyanin A blue **accessory pigment** found in the **chloroplasts** of many algae (*see* alga).

phycoerythrin A red **accessory pigment** found in the **chloroplasts** of many algae (*see* alga).

phycology (algology) The study of algae (*see* alga).

phycovirus A **virus** that infects and can replicate in algae (*see* alga).

phyletic evolution Change within an **evolutionary lineage** that is due to gradual adjustment to environmental conditions.

phyletic gradualism The theory that **macroevolution** is the result of **microevolution** continued over a very long period.

Phyllanthaceae (order **Malpighiales**) A family of **trees**, **shrubs**, and **herbs** with a few climbers, **succulents**, and aquatics. If present, **indumentum** simple, rarely **lepidote**, **stellate**, or **dendritic**. Leaves **alternate** or spiral, rarely fasciculate, whorled, or **opposite**, **simple**, usually **entire**; occasionally leaves absent. Flowers **actinomorphic**, **unisexual** (plants **monoecious** or **dioecious**), rarely **bisexual**, with 3–8 **sepals**, 4–6 free or 2–4 fused **petals** or **apetalous**, 3–10 free or 3–19 fused **stamens**, **ovary superior** of 2–5 free or 1 or 5–15 fused **carpels** and **locules**. **Inflorescence axillary** occasionally **cauline** or terminal; *Phyllocladus* spp. with **phylloclades**. Fruit is a **schizocarp**, **drupe**, or **berry**. There are 59 genera of 1745 species with a pantropical distribution, especially Malesia, and also in warm temperate regions.

phyllid A leaf-like structure, usually one cell thick, in a moss (**Bryophyta**) or liverwort (**Marchantiophyta**).

Phyllobius (leaf weevils) A genus of nine species of weevils (**Curculionidae**) that

P

have a wide **rostrum** and are covered in bright, metallic green scales. Adults feed on the foliage of fruit and nut trees and **hardy** ornamental trees.

Phyllobius pomaceus (nettle leaf weevil) A species of ***Phyllobius*** that is most often seen on the leaves of nettles (***Urtica***). Its larvae feed on roots and cause significant damage to strawberry plants.

phylloclade *See* cladode.

phyllocladium 1. A scale-like structure on the **pseudopodetium** of certain **fruticose lichens 2.** *See* cladode.

phyllode A flattened **petiole** that resembles a leaf and functions as one.

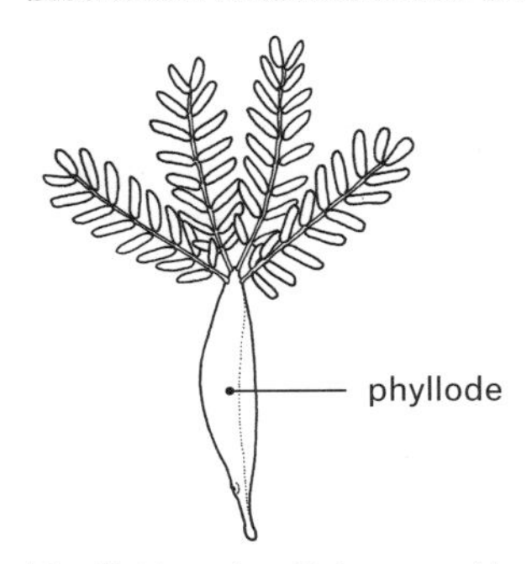

The flattened petiole resembles a leaf and functions as one.

phyllody The replacement of parts of a **flower** by leaf-like structures.

Phyllonomaceae (order **Aquifoliales**) A **monogeneric** family (*Phyllonoma*) of **glabrous**, **evergreen trees** and **shrubs** with **alternate**, **simple** leaves with small, somewhat **fimbriate stipules**. Flowers small, **actinomorphic**, **bisexual**, with 4–5 free **sepals** and **petals**, 5 **stamens**, **ovary superior** of 2 fused **carpels. Inflorescence** branched on the **adaxial** leaf surface near the tip. Fruit is a **berry**. There are four species occurring from Mexico to Peru.

phyllosphere The **microenvironment** for microorganisms provided by the surfaces of leaves or of all the above-ground surfaces of a plant.

phyllotaxis The arrangement of the leaves on a plant.

PhyloCode (International Code of Phylogenetic Nomenclature) A proposed set of rules governing phylogenetic (*see* phylogeny) nomenclature, initially to the level of **clade**. It does not mandate the use of other taxonomic ranks, although it does permit them.

phylogenetic species concept A definition of the term **species** as the smallest group of organisms that can be distinguished from other groups and that share a common ancestor.

phylogenetic systematics The study of genetic relationships among living organisms and their taxonomic (*see* taxonomy) classification based on evolutionary descent.

phylogenetic tree (evolutionary tree) A branching diagram, resembling a tree, that shows the inferred evolutionary relationships between biological taxa (*see* taxon). The order in which branches emerge reflects the order in which the taxa named at the ends of the branches appeared, and the branches indicate ancestral and descendant taxa.

phylogeny The evolutionary history of a **taxon.**

Phylloxeridae (phylloxera) A family of bugs (**Homoptera**) closely related to aphids (**Aphididae**) that form **galls** on a number of tree species. In the late 19th century grape phylloxera (*Daktulosphaira vitifoliae*), native to eastern North America, reached Britain and then continental Europe and almost destroyed the wine industry. The industry was saved by grafting (*see* graft) European vines on to stocks of an American species that was resistant to the stage in which the bug feeds on roots, causing deformities and fungal infections that can kill the plant.

phylum A taxonomic rank (*see* taxonomy) above class and below kingdom. In plant taxonomy, phyla are sometimes known as divisions. The phylum rank is not used in the **Angiosperm Phylogeny Group** classification.

Physcomitrium pyriforme *See* pear moss.

Physenaceae (order **Caryophyllales**) A **monogeneric** family (*Physena*) of **shrubs** and **trees** with **alternate**, **coriaceous**, **simple**, **entire**, **exstipulate** leaves. Flowers small, **actinomorphic**, **unisexual** (plants **dioecious**), **perianth** of 1 **whorl** of 5–9 **sepals**, 10–14 free or 8–10 or 14–25 partially fused **stamens**, **ovary superior** of 2 fused **carpels**. **Inflorescence** an **axillary raceme**. Fruit is a **capsule**. There are two species **endemic** to Madagascar.

physiognomy The structure and form of a natural community.

phytate *See* phytic acid.

phytic acid (hexakisphosphate, hexaphosphoinositol, phytate) In many plant tissues, especially bran and seeds, the principal form in which **phosphorus** is stored. Non-ruminant animals are unable to digest it, but ruminants (cattle, sheep, etc.) do so easily.

phytoalexin A substance produced by a plant that is toxic to invading organisms, especially **Fungi** and **Bacteria**.

phytochemical Any chemical compound produced by a plant.

phytochorion A geographic area across which the type of vegetation, and **taxa**, remain relatively constant.

phytochrome A pigment present in plants that responds to red and far-red light. There are two forms: P_R absorbs at a peak wavelength of 660 nm and is converted to the P_{FR} form, which absorbs at a peak of 730 nm. P_{FR} is the active form that initiates biological processes which include the induction of flowering, development of **chloroplasts** (but not the synthesis of **chlorophyll**), **germination**, **circadian rhythm**, leaf **senescence**, and leaf **abscission**.

phytoclimatology The study of the climate on plant surfaces and among growing plants.

phytogeography (floristics) The study of the geographic distribution of plants at different taxonomic (*see* taxonomy) levels.

phytohormone *See* plant hormone.

phytol An **alcohol** that is a major ingredient of **chlorophyll**.

Phytolaccaceae (order **Caryophyllales**) A family of **trees**, **shrubs**, climbers, and **herbs** with **opposite** or **alternate**, **petiolate**, **simple**, **entire** leaves, **exstipulate** or with very small **stipules** sometimes as spines. Flowers **actinomorphic**, **bisexual** rarely **unisexual** (plants **dioecious** or **monoecious**), **perianth** of 4 or 5 usually free and persistent segments, **stamens** hypogynous (*see* hypogyny), **ovary** usually **superior** of 1 to many united **carpels** each with 1 **locule**. **Inflorescence** a **raceme** or **spike**, rarely a **cyme**. Fruit is a **berry** or **nut** rarely a **loculicidal capsule**. There are 18 genera of 65 species occurring in tropical and warm temperate regions. Some cultivated for ornament or medicinal use.

phytoncide A substance produced by plants that has antimicrobial properties.

Phytonemus pallidus* ssp. *fragariae (strawberry mite) A subspecies of the cyclamen mite (*P. pallidus*, family **Tarsonemidae**), 0.25 mm long and usually barrel-shaped, that occurs on strawberry plants throughout the year, hibernating in winter in the crowns and emerging in spring, and starts feeding on the leaves as they are opening. The mites tend to conceal themselves among the leaf hairs. Mating occurs in summer and eggs are laid singly from March. Infested foliage becomes stunted and discoloured and productivity can be reduced substantially.

phytopathology The study of diseases that affect plants.

phytophagous Feeding on plants.

Phytophthora A genus of water moulds (**Oomycota**) most of which are host-specific parasites of **dicotyledons**, causing severe damage to a wide range of crops. They may reproduce sexually or asexually, although in many species sexual structures have not been observed. There are up to 500 species distributed worldwide.

Phytophthora cinnamomi A species of **Oomycota** that causes the disease **phytophthora die-back**, also known as dieback, jarrah dieback, cinnamon fungus, and root rot, in a wide range of plants including forest and ornamental trees, shrubs, herbs, grasses, and ferns (an unusually large number of hosts for a *Phytophthora* species). It lives in soil and in plant tissues and survives unfavourable conditions as **chlamydospores** which germinate when conditions improve, producing mycelia (*see* mycelium) and sporangia (*see* sporangium) that release **zoospores** which infect plant roots. Symptoms include **chlorosis** and wilting, and the disease is often fatal.

phytophthora die-back *See Phytophthora cinnamomi.*

Phytophthora fragariae A species of **Oomycota** that causes red stele, also known as red core and Lanarkshire disease, in strawberries and raspberries. The pathogen survives in the soil for up to four years and possibly longer as **oospores** which germinate to release **zoospores** that form **cysts** on the roots; **germination** tubes from the cysts then invade the vascular tissue of the root, which often turns red before dying back. After a time plant growth slows or ceases.

Phytophthora infestans A species of water mould (**Oomycota**) that causes late blight of potato and that also infects tomatoes and other members of the **Solanaceae**. It reproduces asexually by producing sporangia (*see* sporangium) continuously on the leaf surfaces of infected plants. The sporangia are dispersed by wind, but do not travel far. It also reproduces sexually. **Zoospores** emerge from the sporangia and enter host plants. The organism survives between crops either as **oospores** or as **mycelium** in infected tubers. The strain of *P. infestans* that caused the Irish Potato Famine and famine in Scotland is now thought to be extinct.

Phytophthora ramorum A species of water mould (**Oomycota**) that causes sudden oak death, also called ramorum dieback. Lesions appear on leaves and spread along the midrib, followed by lesions or **cankers** on twigs or stems, and wilting or dieback. The disease was first reported among oaks (***Quercus***) in the United States in the 1990s and in Britain in 2002. In Europe the disease also affects ***Rhododendron***, *Viburnum*, and ***Larix***.

phytoplankton Members of the **plankton** that perform **photosynthesis**, comprising mainly **diatoms** and in cool water and dinoflagellates (**Pyrrophyta**) in warm water. Phytoplankton form the base of aquatic **food chains**.

Phytoplasma A genus of **Mollicutes** comprising about 30 **clades** of Gram-positive (*see* Gram reaction) **Eubacteria** that occur worldwide. They are **obligate parasites**, most of which require an insect and a plant as hosts in their life cycle. A few harm their insect host, but most do not, and a few benefit the insect. In plants they invade and multiply in **phloem** tissue, interfering with plant development. They cause symptoms including yellowing of leaves and greening of other parts of the plant, reddening of leaves and stems, and the formation of witches' broom. They infect about 200 species of plants.

Phytoseiidae A family of predatory mites (**Arachnida**) most of which feed on other mites and small insects, and especially on spider mites; others feed on **honeydew** and **pollen**. The mites are 0.5–0.8 mm long and live in soil and leaf litter. There are more than 2200 species distributed worldwide.

Phytoseiulus persimilis A predatory mite (**Phytoseiidae**), about 0.5 mm long, that feeds almost exclusively on spider mites. It is native to the Mediterranean region but is used widely for **biological control** in greenhouses.

phytosociology The classification of plant communities on the basis of the species they contain. Josias **Braun-Blanquet** and his colleagues at Zürich and Montpellier developed the most widely used scheme, and G. E. **Du Rietz** and his colleagues developed a similar scheme at Uppsala.

phytotoxin A substance produced by a **pathogen** that is poisonous to plants.

Pica pica (magpie, black-billed magpie, European magpie) A black and white corvid (**Corvidae**), about 450 mm long with a wingspan of 520–620 mm, which has a glossy metallic sheen. It is believed to be among the most intelligent of all animals. It occurs in a wide range of **habitats** and is common in suburban areas, and feeds on insects, carrion, eggs, young birds, and plant material. It is distributed through Europe and most of Asia.

pickerel frog *See Rana palustris.*

pickleweed *See Salicornia.*

picocyanobacteria Very small **cyanobacteria**, less than 2 μm in size.

Picoides pubescens (downy woodpecker) The smallest North American woodpecker, 145–170 mm long with a wingspan of 250–310 mm. It is black and white with a white stripe above and below each eye; it closely resembles the hairy woodpecker (*P. villosus*), but is markedly smaller. It inhabits open, **deciduous** forest and woodland throughout North America.

Picoides villosus (hairy woodpecker) A woodpecker, 165–270 mm long with a wingspan of 445 mm, that closely resembles the downy woodpecker (***P. pubescens***); males have a red patch on the head. They inhabit forests and feed mainly on insects. They occur throughout North and Central America.

Picramniaceae (order **Picramniales**) A family of **trees** and **shrubs** with **alternate**, **compound**, **imparipinnate** leaves with **leafelts alternate** or **opposite**. Flowers small, **actinomorphic**, **unisexual** (plants **dioecious**), **trimerous** or **pentamerous**, **sepals** free or slightly fused, **petals caducous** or absent, as many **stamens** as petals, **ovary superior**, **syncarpous** with 2–3 **carpels** and 1–3 **locules**. **Inflorescence** a **linear** usually pendent **spike** with flowers **sessile**, rarely cauliflorous (*see* cauliflory). Fruit is a **berry** or **samara**. There are 3 genera and 49 species with a pantropical distribution. Some with useful timber or medicinal properties.

Picramniales An order comprising only the family **Picramniaceae**, of 3 genera with 49 species.

Picrodendraceae (order **Malpighiales**) A family of **trees**, **shrubs**, and **subshrubs** with **alternate**, **opposite**, or whorled, **simple** or **palmately compound**, **entire** or **dentate** leaves. Flowers **unisexual** (plants **monoecious** or **dioecious**), **apetalous**, 2–10 or more **sepals**, 2 or very many **stamens**, **ovary superior** of 2–5 **carpels** and **locules**. Male **inflorescence catkin**-like and **racemose** or a **thyrse**, or flowers solitary and **axillary**; female flowers solitary and axillary. Fruit is a **schizocarp** or **drupe**. There are 24 genera with 80 species occurring throughout the tropics, especially in Australia, New Guinea, and New Caledonia.

Pieridae (white butterflies, jezebels, orange tips, brimstones, sulphur butterflies) A family of butterflies (**Lepidoptera**) most of which have white or yellow wings. Some are migratory. There are more than 1000 species found throughout the world. Some are pests, especially of brassicas (**Brassicaceae**).

Pieris brassicae (cabbage butterfly, cabbage white, large cabbage white, great white butterfly, white cabbage butterfly) A butterfly (**Pieridae**) that has white forewings with black tips and with two black spots on the forewings of females; wingspan 50–65 mm. **Caterpillars** are hairy, and black with two longitudinal yellow stripes, and feed on wild and cultivated brassicas. It is common throughout Eurasia as far eastward as the Himalaya, and in North Africa, and has been introduced in South Africa and New Zealand.

Pieris rapae (small white butterfly, cabbage moth, cabbage white, imported cabbage worm, white butterfly) A butterfly (**Pieridae**) with creamy white forewings with black tips; females also have two black spots on the forewings; wingspan is 32–47 mm. Females lay eggs singly on the leaves of food plants. **Caterpillars** are green and live on the underside of leaves and are serious pests of a wide range of plants, especially brassicas. The butterfly occurs throughout Eurasia and North Africa and has been introduced to North America, Australia, and New Zealand.

piezophile *See* extremophile.

pigeons *See* Columbidae.

pigmy rattlesnake *See Sistrurus miliarius.*

pileated woodpecker *See Dryocopus pileatus.*

pileum The top of a bird's head from the beak to the nape.

pileus (cap) The leathery or fleshy structure in a fruit body (*see* fruiting body) that bears the **hymenium**, e.g. the cap of a mushroom.

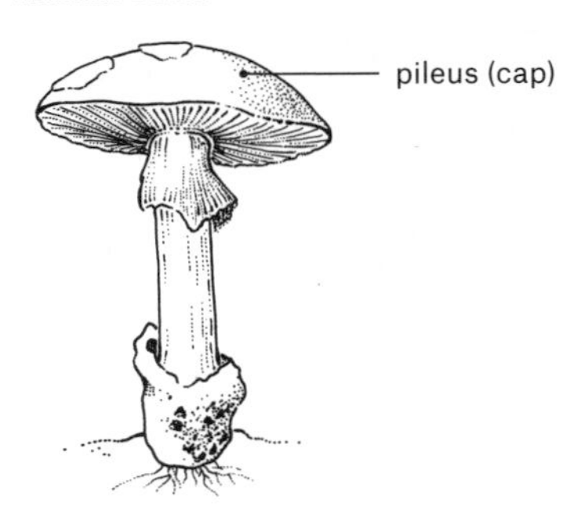

The pileus is the umbrella-shaped top of the typical mushroom or toadstool.

piliferous Having a hair (**trichome**).

piliferous layer (root-hair zone) The part of a root **epidermis** that produces abundant **root hairs**.

pill bugs *See* Isopoda, Porcellionidae.

pilose Covered with fine hairs.

pilus *See* conjugation.

Pimenta dioica (allspice) *See* Myrtaceae.

pimento (*Capsicum annuum*) *See Capsicum.*

Pinaceae (order **Pinales**) A family of **gymnosperm**, mostly **evergreen trees** and **shrubs** with spirally arranged needle-like leaves. Plants **monoecious**, female **cones** large, usually woody, with many scales in spirals. Male cones small and fall soon after pollination. There are 11 genera and 210 species occurring throughout northern temperate regions. Many are important timber trees, e.g. *Pinus* (pine), *Abies* (fir), ***Larix*** (larch), *Picea* (spruce), *Tsuga* (hemlock), ***Cedrus*** (cedar).

Pinales An order of conifers comprising 7 families with 68 genera and 545 species. *See* Araucariaceae, Cupressaceae, Pinaceae, Podocarpaceae, Sciadopityaceae, and Taxaceae.

pine (*Pinus*) *See* Pinaceae.

P

pineapple (*Ananas comosus*) *See* Bromeliaceae.

pine barren A pine forest developed on poor, sandy or marshy soil in which few or none of the trees attain their full size.

pine siskin *See Carduelis pinus.*

pin-eyed *See* heterostyly.

pink pea *See Erwinia rhapontici.*

pinks (*Dianthus plumarius*) *See Dianthus.*

pink seed *See Erwinia rhapontici.*

pink worm *See Dendrodrilus rubidus.*

pin moulds *See* Mucorales.

pinna 1. One of the **leaflets** of a **pinnate** leaf. **2.** The external ear in mammals.

pinnate 1. Describes a **compound** leaf in which **leaflets** lie on either side of a central **rachis**. **2.** Describes a **venation** in which the principal **veins** arise on either side of a **midrib**.

pinnatifid Of a leaf, **pinnate**, but not divided all the way to the **rachis**.

A pinnatifid leaf.

pinnatiparite Describes a **pinnate** leaf in which the divisions between **leaflets** extend from the margin to more than halfway to the **midrib**.

pinnatisect Describes a **pinnate** leaf in which the divisions between **leaflets** extend all the way to the **midrib**.

pinnule One of the smallest divisions of a fern frond.

pinocytosis A form of **endocytosis** in which particles brought into a cell by an **invagination** of the **cell membrane** are then contained within **vesicles**.

Pinus banksiana (jack pine) *See* serotiny.

Pinus contorta (lodgepole pine) *See* serotiny.

pioneer plant A plant species that becomes established early in a **succession** because it germinates readily on bare ground, grows rapidly, and produces abundant small seeds that disperse widely.

Piperaceae (order **Piperales**) A family of **trees**, **shrubs**, **lianas**, and **herbs** with some **epiphytes**; lianas climb by means of **adventitious** roots. Leaves **alternate**, **opposite**, spirally arranged, or at the base, **entire**, membranous or **succulent**. Flowers **actinomorphic**, **unisexual** (plants **dioecious** or **monoecious**), **asepalous**, **apetalous**, 2–6 free **stamens**, **ovary superior**. **Inflorescence axillary**, terminal, or opposite the leaves, **racemose**, **spicate**, each flower subtended by a small **bract**. Fruit **drupe**-like or berry-like. There are 5 genera with 3615 species with a pantropical distribution. Several cultivated; *Piper nigrum* (black pepper) is the source of peppercorns.

Piperales An order of mainly herbaceous plants comprising 4 families of 17 genera and 4090 species. *See* Aristolochiaceae, Hydnoraceae, Piperaceae, and Saururaceae.

Pipilo erythrophthalmus (eastern towhee) A sparrow, 170–230 mm long with a wingspan of 200–300 mm, in which adults have **rufous** sides, white underside, a long, dark tail with white edges, and red eyes (but white in the southeast of the range). The head, tail, and upper body are black in males and brown in females. The birds inhabit scrubland and feed on the ground on seeds, fruit, invertebrates, and small vertebrates, and they sometimes visit garden feeders. They occur throughout eastern North America.

Pipilo maculates (spotted towhee) A species of large sparrows with **rufous** flanks, white underside, red eyes, males with a glossy black head and back, females grey. They have white spots on the back and white wing bars. They breed in thickets in **chaparral**, nesting on the ground or in low **bushes**, and feed on insects, seeds, and berries. They occur in western North America and are partly migratory.

Piptoporus betulinus (birch polypore, birch bracket, razor strop fungus) A species of **agaric fungi** that grows as a bracket almost exclusively on dead or dying birch trees (***Betula*** spp.). The **fruiting body** is almost spherical and grey-brown, becoming flatter and browner on top and white beneath as it matures, when it is 100–250 mm across and 20–60 mm thick. It is a **polypore**. After its tree has died the fungus can live as a **saprotroph** on the decaying trunk. The fungus occurs throughout the Northern Hemisphere. Strips cut from the leathery surface were once used as razor strops, but it is also edible. Ötzi, the Iceman, the 5000-year-old mummy found in the Tyrol in 1991, carried two pieces of this fungus on a thong around his neck.

Piranga ludoviciana (western tanager) A species of birds, 165 mm long, in which adult males have a bright red face, yellow nape, shoulder, and rump, and black back, wings, and tail. Females are olive with a yellow head and dark wings and tail. They breed in forests, feeding on fruits and insects. They breed in western North America and winter in Central America.

Piranga olivacea (scarlet tanager) A species of birds, 160–170 mm long with a 250–290 mm wingspan, that are olive-green on the back and head, and yellow on the underside, with dark wings and tail. In the breeding season adult males are bright scarlet with black wings and tail. They inhabit **deciduous** or mixed forests, sometimes occurring in suburban areas with abundant trees. They feed on insects, which they catch in the treetops or on the ground. They breed in eastern North American and spend the winter in Central and western South America.

piscicide A chemical compound that kills fish.

pistachio (*Pistacia vera*) *See* Anacardiaceae.

Pistacia vera (pistachio) *See* Anacardiaceae.

pistil The female reproductive organ in a **flower**, comprising the **stigma**, **style**, and **ovary**.

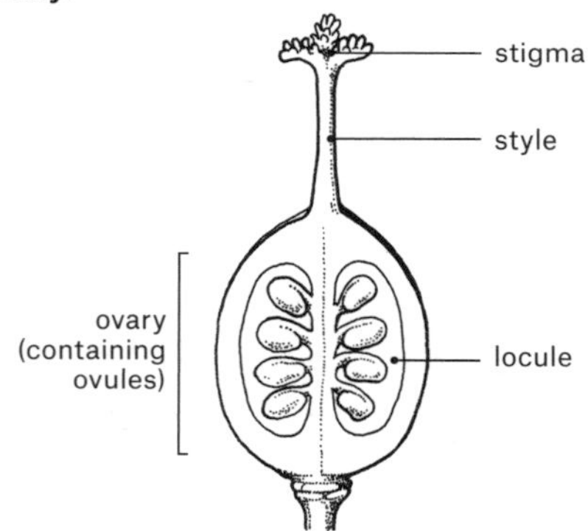

The pistil is the female reproductive structure of a flower, comprising the stigma, style, and ovary; the locule inside the ovary contains ovules.

pistillate Describes a **flower** that possesses a **pistil** but no **stamens**, i.e. the flower is female.

pistillode A sterile **pistil**, often reduced in size.

Pisum (family **Fabaceae**) A genus of **annual herbs** and climbers with leaf **tendrils**. Leaves **compound** with pairs of oval **leaflets**. Flowers **zygomorphic** with a large upright back **petal** and two smaller wing petals enclosing a **keel** petal. Fruit is a

pod. There are one to five species occurring in southwestern Asia and northeastern Africa. *Pisum sativum* (garden pea) is widely cultivated.

pit A small area of a **cell wall** where the **secondary wall** is thin enough for substances to pass, a pit in one cell usually aligned with one in an adjacent cell. *See* bordered pit, primary pit, secondary pit, simple pit.

pith Tissue that stores food at the centre of the stem of a non-woody plant, composed of **parenchyma** tissue.

pith rays *See* medullary rays.

Pittosporaceae (order **Apiales**) A family of **evergreen shrubs**, small **trees**, **lianas**, and scramblers with **alternate**, sometimes **opposite** or whorled, leathery, usually **entire**, **exstipulate** leaves. Flowers **actinomorphic** or weakly **zygomorphic**, **bisexual** rarely **uniexual** (plants **polygamous**), 5 free or slightly **connate sepals**, 5 mostly connate **petals**, 5 **stamens**, **ovary superior** of 2 sometimes 3–5 fused **carpels** with 1 or many **locules**. **Inflorescence umbel**-like, **corymbose**, **paniculate**, or flowers solitary. Fruit is a **loculocidal capsule** or **berry**. There are 6–9 genera with 200 species occurring in Old World tropical and temperate regions. Some cultivated for ornament, e.g. ***Pittosporum***.

Pittosporum (family **Pittosporaceae**) A genus of **evergreen shrubs** and small **trees** that have spirally arranged or whorled, leathery, **simple**, **entire** or **sinuate** rarely lobed, **exstipulate** leaves. Flowers often scented, **actinomorphic**, **bisexual**, **pentamerous**, **sepals** and **petals** united at the base, **ovary superior** of 2 or more fused **carpels**. **Inflorescence** an **umbel** or **corymb**, of flowers solitary. Fruit is a **capsule**. There are about 200 species occurring in tropical and subtropical Africa, Asia, and Australasia. Several species with valuable timber, others cultivated for ornament, as parchment bark or Australian laurel.

pit vipers *See* Crotalinae.

placenta The fused margins of the **carpel** to which the **ovules** are attached in a **flower**.

placentation The arrangement of the **ovules** within the **carpel** of a **flower**. *See* axile placentation, basal placentation, free-central placentation, parietal placentation.

placodioid Describes a **lichen thallus** that is attached to the **substrate** at its centre but lacks **rhizines** and that is free or **lobate** at the margins.

plaggen horizon A surface **soil horizon** more than 50 cm deep that results from manuring continued over many years. It may also contain fragments of pottery, builder's rubble, etc.

plaggic horizon An **anthropedogenic horizon** that has a uniform texture and usually consists of **sand** or **loamy sand** (Dutch *plag* means sod). The **base saturation** is less than 50 percent.

plagioclimax A stable **climax** arising from a **succession** that has been arrested or deflected directly or indirectly by human activity, so the climax is not the one that would have developed in the absence of human interference, even though it may consist entirely of **native** species.

plagiotropic Tending to grow horizontally.

planation surface *See* erosion surface.

planetary boundary layer (atmospheric boundary layer, surface boundary layer) The lowest part of the atmosphere, extending from the surface to about 500 m, in which the physical conditions are strongly influenced by the proximity of the land or sea surface.

planetary waves *See* Rossby waves.

planetary wind A wind that results entirely from the reaction of sunlight and the Earth's rotation.

plane tree *See* Platanaceae.

plankton Aquatic organisms that drift with the movement of tides and water currents and are only weakly capable of independent locomotion. **Phytoplankton** form the basis of the aquatic **food chain**. Zooplankton, including protozoons (**Protozoa**), small crustaceans, and larvae of larger animals, feed on phytoplankton.

planogamete A **motile gamete**.

planosols Soils that have a **soil horizon** that has been submerged beneath stagnant water for a prolonged period. Planosols are a reference soil group in the **World Reference Base for Soil Resources**.

planospore A **motile spore**.

plant association The basic unit in the Zürich-Montpellier School of Phytosociology classification developed by Josias **Braun-Blanquet** and his colleagues, consisting of a **faithful species** and others commonly present that give it coherence. In American and British **phytosociology**, a community characterized by its **physiognomy** as well as the species it contains.

P

plantation A stand of trees, comprising one or a few species, that is planted and grown as a commercial crop.

plant breeding The controlled reproduction of plants in order to emphasize particular desired **characters**.

plantigrade Describes a gait in which the entire sole of the foot makes contact with the ground, as in humans.

plant sociability A scheme for classifying the way members of a plant species are distributed within a community. In the **Braun-Blanquet** scheme the plants may grow singly; in groups or tufts; in troops, small patches, or cushions; in broad patches or carpets; or in large crowds or pure stands.

planozygote A **motile zygote**.

Plantae (Metaphyta) The taxonomic kingdom that contains all plants.

Plantaginaceae (order **Lamiales**) A family of **annual** and **perennial herbs** (plantains) and **subshrubs**, some **succulent** aquatics, some with a rosette of leaves around the base of 1 or more long **scapes** each bearing a terminal **inflorescence**, others with **alternate** or **opposite** leaves and many inflorescences all the way up the stem. Leaves **entire** or divided or lobed, **exstipulate**. Flowers **bisexual** or **unisexual** (plants **dioecious** or **gynodioecious**), **calyx** with 4 free or 3–4 fused **sepals**, **corolla scarious**, 4 free or 3–4 fused **petals**, 4 free or 1–4 fused **stamens**, **ovary superior** of 2 fused **carpels** with 1 or 2 **locules**. Inflorescence a **spike**, spherical head, or loose cluster with a central male flower. Fruit is a **capsule** or **nut**. There about 90 genera and 1900 species with a worldwide, mostly temperate distribution.

plantain *See Musa*, Musaceae, Plantaginaceae.

plant bugs *See* Miridae.

plant growth factor *See* plant hormone.

plant growth substance *See* plant hormone.

plant hormone (phytohormone, plant growth factor, plant growth substance) A substance other than a nutrient, **coenzyme**, **enzyme**, or product of detoxification, that is synthesized by a plant, transported actively or passively, and affects growth or some other physiological process.

Plantlife A British voluntary (non-profit) organization that is dedicated to the protection and growing of wild plants and fungi. Membership is open to all and it provides information and practical advice.

plaque A clear area that appears in an opaque **culture** of **Bacteria** due to **lysis** by a **bacteriophage**.

plasmalemma *See* cell membrane.

plasma membrane *See* cell membrane.

plasmid A circular molecule of double-stranded **DNA** present in a cell that is not associated with and can replicate independently of a **chromosome**. Plasmids occur in some **eukaryote** and **Archaea** cells and in all **Bacteria**.

plasmodesma (pl. plasmodesmata) A cytoplasmic (*see* cytoplasm) bridge, lined with a **cell membrane**, that connects adjacent cells.

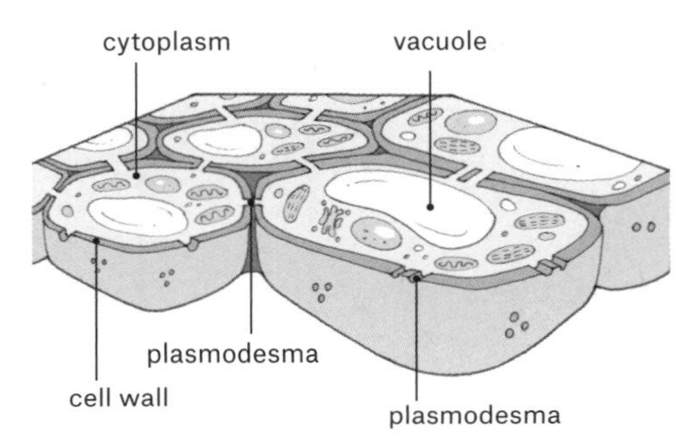

A plasmodesma is a bridge of cytoplasm that links adjacent cells.

plasmodial slime moulds *See* Myxogastria.

Plasmodiophora brassicae A species of **Protista** that is an **obligate parasite** of members of the **Brassicaceae**, causing the disease clubroot, also called finger and toe, in more than 300 species in 64 genera. The organism survives in soil for six to eight years as a dormant **cyst** and is activated by root exudates from brassica plants. Single cells of the pathogen then enter root hairs where they multiply. Infected plants have deformation (clubbing, although other pathogens also cause this) of roots, often as long, finger-shaped clubs, and the plant wilts. The pathogen occurs worldwide.

plasmodium 1. In acellular slime moulds (**Myxogastria**), a feeding structure that consists of a mobile mass of multinucleate, non-cellular **protoplasm**. **2.** A genus of parasites that cause malaria.

plasmogamy The fusion of the **cytoplasm** of two cells.

plasmolysis A process in which the **cytoplasm** in a plant cell shrinks, pulling the **cell membrane** away from the **cell wall**. It happens when the cell is in a **hypertonic** solution and loses water by **osmosis**.

Plasmopora viticola A species of **heterothallic Oomycota** in the order **Peronosporales** that causes downy mildew of grapes. It produces **oospores** that overwinter in the soil, germinating in spring and producing **zoospores** that are splashed by rain on to leaves and enter through stomata (*see* stoma). Yellow lesions appear on leaves and secondary infections follow. It is the most devastating disease of grapevines. It occurs worldwide where there is summer rainfall and summer temperature above 10°C.

plastic limit *See* Atterberg limits.

plastid An **organelle** found in plant cells, and one of the structures that distinguishes plant cells from those of other **eukaryotes**. Plastids are thought to have originated as endosymbionts (*see* endosymbiosis) and now serve many functions. They begin as protoplastids in **meristems** and differentiate depending on the type of cell in which they occur. **Chloroplasts**, **chromoplasts**, and **leucoplasts** are plastids.

plastocyanin A **protein** containing **copper** that acts as an **electron carrier** during **photosynthesis**, linking **photosystem I** and **photosystem II**.

plastoquinone A **quinone** molecule that acts as an electron acceptor in **photosystem II** of **photosynthesis**.

plastron The lower shell of a turtle or tortoise (**Chelonia**). *See also* carapace.

Platanaceae (order **Proteales**) A **monogeneric** family (*Platanus*) of **trees** (plane trees) with pale **bark** that tends to flake off in large pieces. Leaves **alternate**, **simple**,

P

usually **palmately lobed** or **entire** and **elliptical** to **oblong**, **dentate**, with large, leaf-like **stipules.** Flowers **actinomorphic**, **unisexual** (plants **monoecious**), **staminate** flowers with 3–4 free or 4–7 fused **sepals**, vestigial **petals**, as many **stamens** as sepals, **pistillate** flowers with 3–4 free or 4–7 fused sepals, **apetalous**, **ovary superior**, **apocarpous**, of 5–8 free or 3–5 or 9 fused **carpels. Inflorescence** a terminal head. Fruit is an **achene**. There are 10 species occurring in northern temperate regions. *Platanus acerifolia* (London plane) is widely grown as a street tree.

plate tectonics *See* lithosphere.

Platyhelminthes (flatworms) A **phylum** of **acoelomate**, dorso-ventrally flattened worms with **bilateral symmetry** but lacking **metameric segmentation**. They have no anus or blood-vascular system. Most are **hermaphrodites**. Many are parasites, e.g. flukes and tapeworms.

platyspermic Having bilaterally symmetrical, flattened seeds.

pleated sheet (beta sheet) A structure, usually found in fibrous **proteins**, where **polypeptide** chains are partly extended and the chains linked by bonds between NH and CO groups at each **peptide** bond.

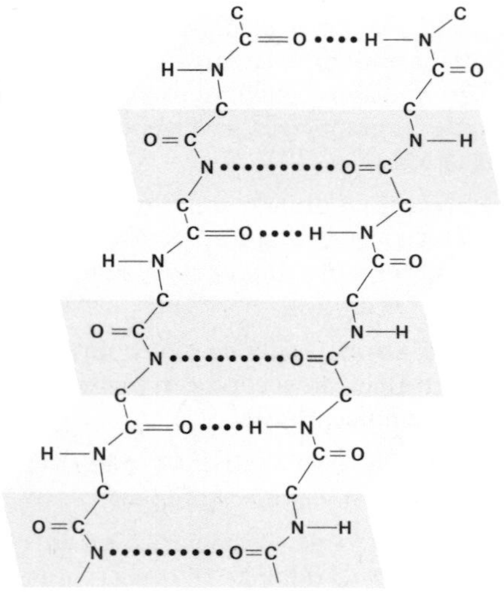

A pleated sheet.

plectenchyma Fungal tissue that consists of a mass of anastomosing hyphae (*see* hypha).

Plectonema A genus of filamentous **cyanobacteria** that occur worldwide in soil, where they help bind soil particles together, in water, and on wet surfaces. They perform **nitrogen fixation**.

plectostele A **protostele** in which the **xylem** appears in cross-section as plates surrounded by **phloem** tissue; in fact, the xylem and phloem form parallel, interwoven bands.

pleomorphic Able to exist in different forms.

pleopod *See* Isopoda.

Plesiocorus rugicollis (apple capsid) A bug (**Miridae**) about 6 mm long that lays eggs in late summer in the **bark** of apple trees. These hatch in spring and the insects feed on the leaves around flower **buds**, then on new leaves and the developing fruit. *Salix* (willow) is the natural host plant.

plesiomorphic Describes **characters** inherited from a common ancestor that are shared by different groups of organisms.

Plethodon chlorobryonis (Atlantic coast slimy salamander) A species of stocky, black or blue-black salamanders (**Amphibia**) with white or pale spots, 115–205 mm long, that occur in the eastern United States, mainly in forests. It secretes a mucus making it slimy to handle, and the mucus clings to the hands. The salamander is active except in winter, but shelters below ground in dry weather. It feeds on insects and arachnids (**Arachnida**), and sometimes on smaller salamanders.

Plethodon cylindraceus (white-spotted slimy salamander) A salamander

(**Salamandridae**), 114–210 mm long, that is slender, with a short nose and long tail, and shiny black with white spots on the back and a grey underside. It lacks lungs, breathing by **cutaneous respiration** and through mouth and throat membranes, and lives its entire life on dry land. If threatened it secretes a sticky mucus that clings to the hands. It inhabits **deciduous** forests with abundant leaf litter and feeds on invertebrates. It occurs in the Appalachian Mountains and elsewhere in the eastern United States.

pleurocarpous Describes a moss (**Bryophyta**) in which the arechegonia (*see* archegonium) are borne in **capsules** at the tips of lateral **branches**, rather than at the tips of main branches or stems.

Pleurocarpous mosses bear their female sex organs (archegonia) in capsules at the tips of lateral branches, rather than at the tips of main branches or stems.

Pleurotus ostreatus (oyster fungus) A species of **agaric fungi** in which the **fruiting body** has a broad **pileus** shaped like a fan or an oyster that is white, grey, tan, or brown and 50–250 mm across, with white or cream **gills**. It is a **saprotroph** feeding on decaying wood and is often seen growing on tree trunks, especially beech (***Fagus*** spp.), and it can cause a form of wood decay known as white rot. It is also carnivorous, feeding on nematodes (**Nematoda**). It occurs worldwide in temperate and subtropical forests, and is sometimes cultivated. It is edible and highly prized.

plicate 1. Wrinkled or folded. **2.** *See* vernation.

plinthic horizon A subsurface **soil horizon** that is rich in iron, contains little **humus**, and consists of a mixture of **clay**, **quartz**, and other minerals, and that changes irreversibly to irregular **aggregates** or a **hardpan** with repeated wetting and drying in the presence of oxygen.

plinthite A mixture of **clay** and **quartz**, with iron and aluminium oxides, and containing little **humus**, that develops through repeated **leaching** and **gleying**. On exposure to air it turns irreversibly into an ironstone **hardpan**.

plinthosols A group of soils that have a **plinthic horizon** within 50 cm of the surface. Plinthosols are a reference soil group in the **World Reference Base for Soil Resources**.

Plocospermataceae (order **Lamiales**) A **monotypic** family (*Plocosperma buxifolium*), which is a **shrub** or small **tree** with **opposite**, **entire**, **subsessile** leaves. Flowers **unisexual**, 5- to 6-merous, **corolla campanulate** or funnel-shaped, 5 **stamens**. **Inflorescence** a terminal **dichasium**. Fruit is a **capsule**. The tree occurs in Central America.

ploidy The number of sets of **chromosomes** in a cell **nucleus**.

Plumbaginaceae (order **Caryophyllales**) A family of **shrubs**, **lianas**, and **annual** and **perennial herbs**, many **halophytes** or **psammophytes**. Leaves spirally arranged, often with a basal rosette, **simple**, **entire**, **exstipulate**. Flowers **actinomorphic**, **hermaphrodite**, with **scarious bracts** sometimes forming an **involucre**, 5 fused, persistent **sepals**, 5 **petals** free, **connate** at the base, or fused into a tube, 5 **stamens**, **ovary superior** of 5 fused **carpels** and 1 **locule**. **Inflorescence** a **capitulum**. Fruit usually **indehiscent**, enclosed by the

calyx. There are 27 genera of 836 species occurring mainly from the Mediterranean region to Central Asia, scattered elsewhere. Some with medicinal properties, many cultivated for ornament, e.g. *Armeria* spp. (sea pink, thrift) and *Limonium* spp. (sea lavender).

Plumeria (family **Apocynaceae**) A genus mainly of **deciduous shrubs** and small, **pachycaul trees** (frangipani) that branch to produce a candelabra shape, with **alternate** leaves that vary in shape according to species. Flowers solitary, white yellow, pink, or red, **pentamerous** with partly **imbricate petals**. Flowers have a heavy fragrance, making them attractive to pollinating insects although they produce no **nectar**. There are 11 species occurring in tropical America. Many cultivated for ornament.

plum fruit moth *See Grapholita funebrana*.

plum pouch-gall mite *See Eriophyes similis*.

plum pox (sharka disease) A disease of ***Prunus*** species caused by the plum pox **virus** (**Potyviridae**), which is transmitted by aphids (**Aphididae**). It causes discoloration of blossoms and deformation and blemish in fruits. It occurs throughout Europe, and in North America, Argentina, Chile, South Africa, and China. It is a notifiable disease in Britain.

plum sawfly *See Hoplocampa flava*.

plumule The terminal **bud** of a plant **embryo**.

PMMoV *See* pepper mild mottle virus.

PMTV *See* potato mop top virus.

pneumathode A modified secondary root, made from **aerenchyma**, that is connected to an air chamber in the main root. Air entering the pneumathode is able to reach the root tissue. Many plants have pneumathodes. Tree ferns (order Cyatheales) have pneumathodes on the **stipe** or **rachis** of their leaves.

pneumatocyst A gas-filled, **bladder**-like swelling on the **stipe** of **brown algae** (Phaeophyta) that acts as a float.

pneumatophore A specialized air-breathing root, containing many pores, that projects above the surface of water-logged or strongly compacted soil, allowing the root to exchange gases with the air.

Poa (family **Poaceae**) A genus of **monocotyledon**, **annual** and **perennial** grasses with **glabrous** leaves, the **lemma** narrow, keeled (*see* keel), often with long, soft hairs arising from the base, usually with a **hyaline** tip and no **awn**. Flowers **monoecious**, a few **dioecious**, with 3 **stamens**. There are about 500 species occurring throughout temperate regions. *Poa annua* (annual meadow grass) may be the most widely distributed and commonest grass in the world. *Poa pratensis* is Kentucky bluegrass.

Poaceae (order **Poales**) A family, formerly known as Gramineae, of **monocotyledon**, **annual** and **perennial herbs** with **rhizomes** or **stolons**, or that are **caespitose**, a few woody. Fibrous roots often supplemented by **adventitious** roots growing from the stem, so each stem has its own root system, allowing the plant to grow to large size and a great age. It is thought that a tussock of *Festuca ovina*, 8 m across, could be 1000 years old. Leaves usually **distichous**, with a sheath that encloses the stem and a **blade**, absent in some species, that is long and narrow or **lanceolate** to **ovate**, and up to 5 m long. Flower a **spikelet** comprising an **axis** and 2-ranked scales, the lowest 2 (**glumes**) empty, the others (**lemmas**) forming the **floret**, glumes and lemmas often produced into a bristle (**awn**). Spikelets **bisexual** occasionally **unisexual**, mostly male or barren, plant **monoecious** sometimes **dioecious**, 3 or 1–6 or more **stamens**, **ovary superior**. **Inflorescence** a **panicle**, **spike**, or **raceme**. Fruit is a **caryopsis**, occasionally **berry**-

like or a **nut**. There are 707 genera of 11,337 species with a worldwide distribution. The family includes the cereal grasses, pasture and fodder grasses, bamboos, sugar cane, etc.

poached soil *See* puddled soil.

Poales A **monocotyledon** order of plants that comprises 17 families of 997 genera and 18,325 species. *See* Anarthriaceae, Bromeliaceae, Centrolepidaceae, Cyperaceae, Ecdeiocoleaceae, Eriocaulaceae, Flagellariaceae, Joinvilleaceae, Juncaceae, Mayacaceae, Poaceae, Rapateaceae, Restionaceae, Thurniaceae, Typhaceae, and Xyridaceae.

pocket rot A type of timber decay that occurs in discrete areas (pockets) surrounded by healthy tissue.

poculiform Shaped like a goblet.

pod A fruit characteristic of **legumes**, that splits down both sides, e.g. a pea pod.

podetium An upright structure, sometimes branched and sometimes cup-shaped, arising from the **thallus** of certain **lichens**, or part of the **apothecium**, usually bearing pycnidia (*see* pycnidium) and/or **ascocarps**.

Podocarpaceae (order **Pinales**) A family of coniferous, **evergreen shrubs** and **trees** with leaves spirally arranged sometimes **opposite**, scale- or needle-like or **linear** to **lanceolate**. Plants **monoecious** or **dioecious**. Male **cones catkin**-like with many **imbricate stamens**, female cones pendant, **pedunculate**, reduced to a few **bracts** or scales. Seeds enclosed in a fleshy structure (epimatium). There are 16 genera of 125 species occurring mainly in the Southern Hemisphere but scattered in Japan, Central America, and the Caribbean. Some important timber trees (*see Podocarpus*). The family also includes *Lepidothamnus fonkii* (Chilean pygmy cedar), the world's smallest conifer (grows to 60 cm).

Podocarpus (family **Podocarpaceae**) A genus of coniferous, **evergreen shrubs** and **trees** (some up to 40 m tall) with **alternate**, **linear** to **ovate** leaves. Plants **dioecious**. Seeds solitary, mature seeds covered with a fleshy to **coriaceous** layer formed from the enlarged **receptacle**. There are 100 species occurring in Southern Hemisphere temperate zones and extending into the tropics of both hemispheres. *Podocarpus totara* (totara) of New Zealand provides valuable timber and is of great importance in Maori culture.

Podostemaceae (order **Malpighiales**) A family of **annual** and **perennial herbs** with photosynthetic, **adventitious** roots, that grow submerged in fast-flowing water, including waterfalls. Stems often tiny but longer when flowering and sometimes up to 80 cm long and floating. Leaves **simple**, **linear**, sometimes lobed. Flowers **actinomorphic**, **bisexual**, **perianth** of 2–3 or 5 to many **tepals** or tepals absent, 1 to many **stamens**, **ovary superior** of 1–3 **carpels** each with 1 **locule**. **Inflorescence** a **spike** or **cyme** or flowers solitary. Flowers usually appear only when plants emerge in the dry season, or develop below the water surface and are self-pollinating. Fruit is a **capsule**. There are 48 genera of 270 species occurring throughout the tropics, especially in America.

podzolic soils Soils that have an acid organic surface layer and a black, dark brown, or yellowish subsurface **soil horizon** rich in iron, aluminium, organic matter, or some combination of these. They comprise a major group in the classification devised by the Soil Survey for England and Wales.

podzolization An advanced stage of **leaching** in which a soil has lost **humus**, iron compounds, and **clay minerals** from its surface **soil horizons**, some of these accumulating in the B horizon.

podzols Soils that form from a **parent material** rich in **quartz**. They usually have a coarse-textured, pale-coloured surface **soil horizon** containing little organic matter and from which iron oxides and **clay** have been removed by **leaching**, above a subsurface **spodic horizon**. Podzols are a reference soil group in the **World Reference Base for Soil Resources**.

Poecile atricapillus (black-capped chickadee) A short, plump bird with a black cap and bib, white cheeks, and dark grey back and wings with black and white streaks. They feed on invertebrate animals and berries in **deciduous** woodland, parks, and near the edge of wooded areas. They occur throughout most of North America.

poikilohydry The inability of an organism to control the water content of its cells and tissues, so this varies with the **humidity** of its environment.

poikilotherm (exotherm) An organism that does not regulate its temperature, which varies according to the temperature of its surroundings.

poinsettia (*Euphorbia pulcherrima*) *See* Euphorbiaceae.

point mutation A **mutation** at a single **locus**.

point snail *See Acicula fusca.*

poison ivy (*Rhus toxicodendron; Toxicodendron radicans*) *See Rhus.*

poison oak (*Rhus diversiloba; Toxicodendron diversilobum*) *See Rhus.*

poison pax *See Paxillus involutus.*

poison sumac (*Rhus vernix; Toxicodendron vernix*) *See Rhus.*

polar air An **air mass** that acquires its characteristics in the **anticyclones** over Siberia, northern Canada, and the Southern Ocean.

polar-air depression A **cyclone** that forms in the Northern Hemisphere when unstable (*see* instability) arctic or polar **maritime air** moves southward, along the eastern edge of a large north–south **ridge**.

polar auxin transport The **active transport** of **auxin** from cell to cell through a plant, always in a direction away from the shoots and toward the roots.

polar cell That component of the **general circulation** in which air subsides over the polar regions, moves away from the poles at low level, rises at the **polar front**, and flows poleward at high level. *See* three-cell model.

polar climate A climate of high latitudes where the mean air temperature remains below freezing throughout the year.

polar desert A region inside the Arctic Circle or Antarctic Circle where the annual precipitation is very low. The mean annual precipitation at the South Pole (snowfall converted to **water equivalent**) is about 25 mm and most of Antarctica receives less than 200 mm a year. Central Kalaallit Nunaat (Greenland) receives about 8 mm of water-equivalent a year.

polar easterlies The prevailing low-level winds in the **polar cells**, produced by the action of the **Coriolis effect** on air moving away from the poles.

polar flagella Flagella (*see* flagellum) arranged at one end of a rod-shaped cell.

polar front The boundary (**front**) between polar and tropical air that lies in middle latitudes at the meeting of the **polar cells** and **Ferrel cells**. The polar-front **jet stream** flows along the top of the front, close to the **tropopause**.

polarilocular Describes **spores** that comprise two cells connected by a channel or pore.

polarity The separation of a property into two contrasting types. In physics this may relate to the alignment of atoms or their arrangement in molecules, so one end carries a positive electromagnetic charge and the other end a negative charge. In evolutionary studies the term refers to the direction of evolution, contrasting whether character states are **primitive** or derived.

polar molecule A molecule that carries no charge overall, but in which one end of the molecule bears a small positive charge and the opposite end a small negative charge, so the molecule is a **dipole**. This arises because electrons are shared unequally among the constituent atoms. Water and ammonia (NH_3) have polar molecules.

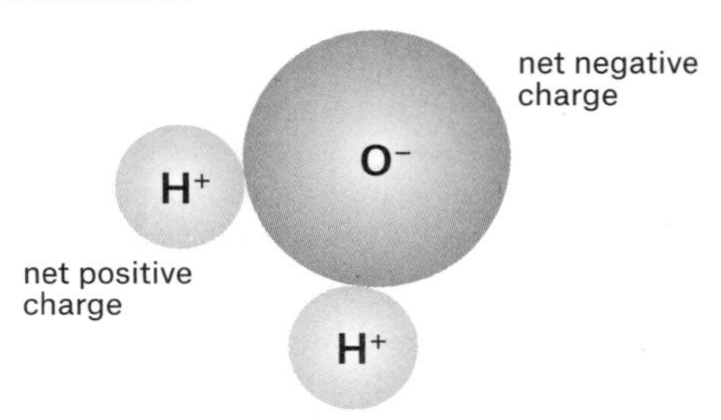

Polar molecule. The oxygen atom exerts a strong pull on the two (negative) electrons held by the two hydrogen atoms. This results in a small net negative charge on the oxygen side of the molecule and a small positive charge on the hydrogen side, while the molecule remains neutral overall. This is a water molecule and its polarity accounts for most of its properties.

polar nuclei In **angiosperms**, two **haploid** nuclei in the **embryo sac** that fuse with one of the sperm nuclei delivered by the **pollen grain** to form the triploid **endosperm**.

polar outbreak A protrusion of **polar air** into lower latitudes.

polar trough A **trough** in the upper **troposphere** over the poles that extends as far as the tropics.

polder A low-lying, level area of land that has been reclaimed from the sea.

polecats *See* Mustelidae.

Polemoniaceae (order **Ericales**) A family of **annual** and **perennial herbs**, and **shrubs**, **trees**, and climbers with leaves **alternate**, **opposite**, or in a **whorl**, **simple**, **linear** sometimes **coriaceous**, or linear-**pinnatisect** or **pinnate**, **exstipulate**. In climbers the **leaflet rachis** ends in branched **tendrils** each with a terminal hook. Flowers **actinomorphic** sometimes bilateral, **bisexual**, **pentamerous** occasionally 4- to 6-merous, with 5 **sepals**, narrowly tubular or **campanulate corolla**, as many **stamens** as corolla lobes, **ovary superior** of 3 sometimes 2 or 4 fused **carpels** each with 1 **locule**. **Inflorescence** usually a terminal **cyme**. Fruit is a **capsule**. There are about 18 genera of 385 species occurring in northern temperate regions, especially western North America, Central America, and the Andes. Many are grown as ornamentals, e.g. *Phlox*.

pollard To remove the upper part of a **tree**, usually at a height of about 2 m, in order to produce a crown of small **branches** that can be used for fuel, fencing, etc. (*see* coppice), but at a height where the foliage and young growth are beyond the reach of browsing animals.

pollen The mass of **pollen grains** produced by the **anthers** of **angiosperms** and the male **cones** of **gymnosperms**.

pollen analysis *See* exine.

pollen basket *See* corbiculum.

pollen grain (microgametophyte) In seed plants (**Spermatophyta**), a structure produced in a **microsporangium** that contains one **tube nucleus** (*see* pollen tube) and two sperm nuclei, all of them **haploid**, enclosed by an inner wall (the intine) rich in **cellulose** and a very tough outer wall (**exine**) made mainly from **sporopollenin**. A pollen grain is a **gametophyte** (*see* alternation of generations).

pollen mother cell The **microsporocyte** that gives rise to **pollen grains**, after undergoing two divisions by **meiosis** to produce four **microspores**, each of which becomes a pollen grain.

pollen rain The fall of **pollen grains** on to a particular place.

pollen sac The structure in seed plants (**Spermatophyta**) inside which pollen grains form.

pollen tube The tube that grows from a fertilized **pollen grain**, penetrates the **stigma** and **style**, and enters the **ovary**. Two sperm nuclei (male **gametes**) pass along the tube; when they reach the ovary the tube ruptures, releasing them. The pollen-**tube nucleus** then degenerates. *See* double fertilization.

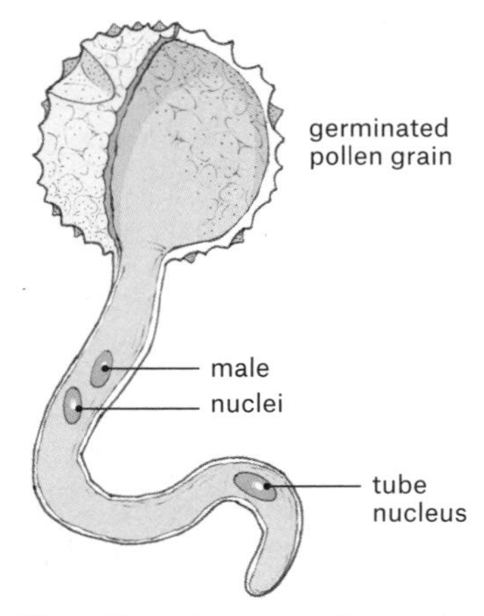

The pollen tube extends into the ovum from a germinated pollen grain. The tube possesses a nucleus, and two male nuclei from the pollen grain move along the tube and enter the ovum.

pollen wasps *See* Masarinae.

pollination The transfer of **pollen grains** from the **anthers** to the **stigma** of **angiosperms** or directly to the **micropyle** of **gymnosperms**, where one grain germinates releasing two **gametes** that fertilize the female gamete. (*See* double fertilization).

pollinium A mass of **pollen grains** that are transported together, e.g. in **Orchidaceae**.

pollywog *See* Amphibia.

polyacetylene A member of a group of more than 2000 organic polymers containing carbon-carbon triple bonds, many of which are produced by plants, **Fungi**, **lichens**, and some animals. More than 1100 are produced by plants in the family **Asteraceae**, and others by the **Apiaceae** (e.g. carrot) and **Araliaceae**. Polyacetylenes are poisonous to many animals, but some have therapeutic properties.

polyarch Describes **primary xylem** composed of many strands.

polycentric Having several centres, e.g. a polycentric **chromosome** has several **centromeres**; a polycentric **thallus** has several reproductive centres.

polyclimax The theory proposed by Arthur George **Tansley**, that a plant community develops through a **succession** to a **climax** that may be determined by climate, fire, soil, or some other factor.

polycyclic Describes a **stele** in which the **vascular bundles** appear in cross-section as concentric rings.

polyembryony The development of several **embryos** from a single **ovule**, an occurrence common in **gymnosperms**. Cleavage polyembryony (monozygotic polyembryony) results from the division of the **zygote**, adventive polyembryony from **somatic cells** in the **nucellus** or **chalaza**.

polyethism Division of labour in **eusocial** insects through the functional specialization of groups of members of the colony. This may involve **castes** with specialized forms (caste polyethism) or individuals changing their role as they age (age polyethism).

polyextremophile An **extremophile** that thrives under two or more extreme environmental conditions.

Polygalaceae (order **Fabales**) A family of **annual** and **perennial herbs** and **shrubs**, with some **trees** and **lianas**, and a few **saprophytes**. Leaves **simple**, **exstipulate**. Flowers usually **zygomorphic**,

hermaphrodite, subtended by a **bract** and 2 **bracteoles**, 5 occasionally 4–7 **sepals** often **petaloid**, **corolla** usually of 3 **petals**, usually 8 **stamens**, **ovary superior** of 2 sometimes 1 or 5 united **carpels** with 2 **locules. Inflorescence** an **axillary** or terminal **spike**, **raceme**, or **paniculate** or flowers solitary. Fruit is a **loculicidal capsule** or **nut**. There are about 21 genera of 965 species with a worldwide distribution.

polygamodioecious Describes a plant species with **perfect flowers**, but some individuals also bearing male (**staminate**) flowers and others bearing female (**pistillate**) flowers.

polygamous Describes a plant species in which male (**staminate**), female (**pistillate**), and **hermaphrodite** flowers occur on the same or different plants.

polygene A member of a group of **genes** that together control a particular **character**.

polygenic (multifactorial) Describes a **character** controlled by a group of **polygenes** acting in concert.

polygenic character A variable **character** whose particular form is determined by a group of **polygenes**.

Polygonaceae (order **Caryophyllales**) A family of **annual** and **perennial herbs**, **shrubs**, **trees**, climbers, and scramblers with leaves usually at the base, **alternate**, **opposite**, occasionally whorled, **simple**, sometimes reduced. Flowers small, white, pale gree, or pink, usually **hermaphrodite** occasionally **unisexual** (plants **dioecious** sometimes **monoecious**), subtended by a sheathing tube of fused **bracteoles**, 3–6 **tepals**, 6–9 free or 2–6 fused **stamens**, **ovary superior** of 3 sometimes 2 or 4 fused **carpels** with 1 **locule. Inflorescence** an **axillary** or terminal **paniculate**, **racemose**, spike-like, or **umbel**-like **thyrse**. Fruit is a triangular **achene** or **nut**. There are 43 genera of 1110 species with a worldwide distribution. Many cultivated for ornament or food. ***Rheum*** ×*hybridum* is rhubarb, *Fagopyrum esculentum* is buckwheat, *Rumex acetosa* is sorrel, *Rumex crispus* is the weed yellow dock.

polymerase An **enzyme** that catalyzes the repair and maintenance of **nucleic acids**.

polymerase chain reaction (PCR) A technique for producing many copies of a particular segment of **DNA** by means of a chain of **polymerase** reactions.

polymictic Describes a lake in which the water circulates continuously.

polymorphism The existence of two or more genetically distinct forms within an interbreeding population.

Polynesian floral region The area that includes the islands of the North and South Pacific to the west of New Caledonia and south of Hawaii eastward to about longitude 90° W. There are few **endemics**.

polypedon (soil individual) A soil unit that consists of two or more contiguous **pedons** within a single **soil series**.

polypeptide A long, unbranched chain comprising ten or more **amino acids** linked by **peptide bonds**.

polypetalous Having the **petals** free.

Polyphagotarsonemus latus (broad mite) A species of very small mites (**Tenthredinidae**) that feed on a wide variety of plants including important fruit crops and ornamentals, causing distortion of leaves and flowers and killing new growth. They are oval. Males are about 0.11 m long, females about 0.2 mm long, yellow to green with a paler stripe. Males have large hind legs, used in mating. They have a worldwide distribution.

polyphagous Feeding on many different kinds of plant or animal.

polyphyletism The presence in a **taxon** of members that are not all descended from the same common ancestor.

polyploidy The possession of more than the two sets of **chromosomes** found in

diploid organisms due to the replication of complete sets of chromosomes without nuclear division. The possession of three sets is called triploidy and written 3*n*, four sets is tetraploidy (4*n*), etc. *See also* allopolyploidy.

Polypodiidae *See* Filicopsida.

polypore (bracket fungus) A **basidiomycete** fungus that produces a **fruiting body** with tubes or pores on the underside. Some live in the soil and form **mycorrhizae** with trees, but most grow on the sides of trees and their fruiting bodies are known as conks.

Polyporus squamosus (dryad's saddle, saddle fungus, pheasant's back mushroom) A species of **basidiomycete bracket fungi** with a yellow or brown, **squamulose fruiting body**, 80–300 mm across and up to 100 mm thick, usually attached by a thick **stipe**. The fungus grows on dead logs or tree stumps and occurs in Europe, Asia, North America, and Australia. It is edible.

polyribosome *See* polysome.

polysaccharide A long, unbranched, **carbohydrate** polymer that consists of ten or more **monosccharides** linked by glycosidic (*see* glycoside) bonds.

polysome (ergosome, polyribosome) A cluster, line, or circle of **ribosomes** attached to a single **messenger-RNA** molecule. Many ribosomes read the m-RNA simultaneously, each contributing to the synthesis of the same **protein**.

polystele A **dictyostele**, typical of ferns (**Pteridophyta**) and **monocotyledons**, in which the **vascular bundles** appear in cross-section as single strands each surrounded by **phloem**.

polytelic *See* indeterminate.

polytomy 1. (phylogenetic pitchfork) A **node** in a **phylogenetic tree** from which more than two lineages emerge. This may result from insufficient data, making it impossible to determine how the lineages are related (called soft polytomy), or from several speciation events having occurred over a very short period so all the daughter lineages are equally closely related (hard polytomy). **2.** Several **branches** emerging from the **apical meristem** of a plant.

polytopic evolution *See* polytopism.

polytopism (polytopic evolution) The emergence of a new **taxon** in more than one place from parents of the same **species**.

polyunsaturated *See* fatty acid.

pome A **fruit** that develops inside the fleshy **receptacle**, with the **seeds** protected by a tough **carpel** wall. The carpel walls containing the seeds form the core, which is the true fruit. Apples and pears are pomes, as are the fruits of cotoneaster, hawthorn, medlars, etc.

pomegranate (*Punica granatum*) *See* Lythraceae.

ponding The accumulation of cold air in a **frost hollow**.

pond slider *See Trachemys scripta.*

pondweed *See* Potamogetonaceae.

Pontederiaceae (order **Commelinales**) A **monocotyledon** family of **annual** or **perennial**, submerged, floating, and **emergent** aquatic **herbs** with **determinate**, leafless, flowering stems and **indeterminate**, vegetative, leafy stems. Leaves **distichous**, rarely whorled, with a sheathing base and **simple blade**. Flowers more or less **zygomorphic**, **unisexual**, with usually 6 occasionally 3 or 4 basally **connate petaloid tepals**, 3 or 6 **stamens**, often dimorphic (*see* dimorphism), **ovary superior** of 3 fused **carpels** with usually 3 **locules**. **Inflorescence** a terminal **spike** or **umbel**-like **panicle**, or flowers solitary. Fruit is a **capsule** or **nutlet**. There are 9 genera of 33 species occurring in tropical and temperate regions, especially America. Some are cultivated ornamentals, e.g. *Eichhornia crassipes* (water hyacinth), which is also a noxious aquatic weed.

poor man's licorice *See Bulgaria inquinans.*

poor man's weatherglass (*Saccharina latissima*) *See* sea belt.

population genetics The study of inherited variation in time and space in a population of organisms.

Populus (family **Salicaceae**) A genus of **deciduous trees** with spirally arranged, **petiolate** leaves that vary greatly in shape, even on the same tree. In some the **petioles** are flattened causing the leaves to twist back and forth in the wind, so the tree seems to be trembling. Flowers wind-pollinated, borne in sessile or **pedunculate catkins**, are usually **dioecious** rarely **monoecious**, without **calyx** or **corolla**, **staminate** flowers with 4–60 **stamens** on a cup-shaped disc on the base of a scale attached to the **rachis** of the catkin, **pistillate** flowers comprise a single-celled **ovary**. Fruit is a **dehiscent capsule**. There are 25–35 species occurring throughout the Northern Hemisphere. Many grown for timber or ornament. *Populus tremula* is aspen, *P. tremuloides* is quaking or trembling aspen, *P. alba* is white poplar, *P. deltoides* is eastern cottonwood.

porate Having pores.

Porcellionidae (pill bugs, sowbugs, woodlice) A family of terrestrial isopods (crustaceans) that have a rigid **exoskeleton** in segments, prominent two-segmented antennae (*see* antenna), and seven pairs of legs. They live in moist soil and plant debris, and feed on decaying organic material. There are more than 3000 species, distributed worldwide.

porcini *See Boletus edulis.*

pore An air- or water-filled space completely surrounded by mineral or organic soil or rock particles that results from the packing of the particles.

pore space The total volume of all the interconnected **pores** in a soil or rock.

pore-water pressure The pressure that water contained in **pore spaces** exerts against the surrounding material. It is positive if the soil is saturated because of the **buoyancy** the water gives to the soil particles, zero when the pores are filled with air, and negative when the pores are partly filled with water because **surface tension** then has a suction effect that increases the shear strength of the soil.

porosity The proportion of the total volume of a medium, e.g. rock or soil, that is occupied by **pore** spaces. *See* absolute porosity, effective porosity.

porphin The chemical compound from which **porphyrins** are derived. It consists of four **pyrrole**-like rings linked by four CH groups; its formula is $C_{20}H_{14}N_4$.

porphyreus Purple.

porphyrin A group of chemical compounds derived from **porphin** that have a **tetrapyrrole** structure allowing them to bind metals into complexes. Porphyrins occur in many biological compounds including **chlorophyll** and **cytochromes.**

P

Portulacaceae (order **Caryophyllales**) A **monogeneric** family (*Portulaca*) of weedy, **succulent**, **annual herbs**, with more or less **terete** leaves, that use the **CAM pathway** of **photosynthesis**. Flowers with 2 **sepals**, usually 5 **petals**. **Inflorescence** terminal, **capitate**, with **involucre**. Fruit is a **circumscissile capsule**. There are 40–100 species with a worldwide distribution but especially tropical and subtropical America.

Posidoniaceae (order **Alismatales**) A **monocotyledon**, **monogeneric** family (*Posidonia*) of marine grasses with a creeping, **monopodial rhizome** covered with fibrous strands of old leaf sheaths. Leaves **linear** to **filiform** with persistent sheaths. Flowers **actinomorphic**, usually **bisexual**, lacking **tepals**, 3 **sessile stamens**, **ovary superior** of 1 naked **carpel**. **Inflorescence racemose** and **spike**-like. Fruit a buoyant **follicle**. There are 90 species occurring in the Mediterranean region and temperate Australia.

potamodromous Describes fish that undertake long migrations in fresh water.

Potamogetonaceae (order **Alismatales**) A **monocotyledon** family of usually **perennial** some **annual**, submerged or floating, aquatic **herbs** (pondweeds). The lower part of the stem creeping and **rhizome**-like, the upper part long, flexible, erect or floating. Leaves **alternate**, **opposite**, or in bunches, submerged leaves **simple**, **linear** to **orbicular**, floating leaves **lanceolate** to **ovate**, **entire** or **serrulate**. Flowers **bisexual** rarely **unisexual** (plants **monoecious** or **dioecious**), either regular with 2 or 4 free, **bract**-like scales, or a cup-like, 3-lobed sheath, or absent, with usually 1, 2, or 4 **stamens**, **ovary superior** of usually 4 free or partly united **carpels** each with 1 **anatropous** or **campylotropous ovule**, or of 1–8 free carpels each with an anatropous ovule. Fruit is a **drupe** or **berry**. There are 4 genera of 102 species with worldwide distribution, especially in temperate regions.

potassium (K) An element that is an essential **macronutrient** for plants. It helps control the **water potential** of cells and is involved in the movement of leaves and the action of **guard cells**. It is essential

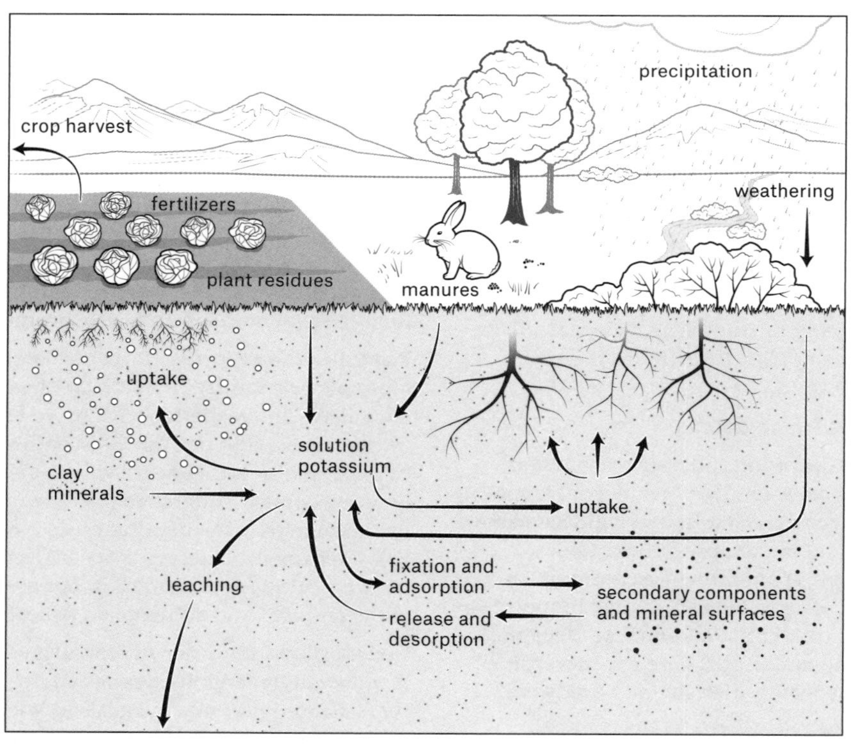

Potassium cycle. Potassium is an essential plant nutrient. Plants obtain it from the weathering of clay minerals in the soil, augmented by fertilizer and manure. It then cycles between living organisms and the soil solution.

primary One of the outermost flight **feathers** of a bird.

primary forest 1. A forest occupying a site that has been continuously forested for many centuries, in Britain since the last ice age, even though it may have been clear-felled provided the trees were replanted or regenerated naturally. **2.** A forest that is the natural **climax** and either has never been disturbed by human activity, or that has fully recovered from such disturbance.

primary growth Growth that is due to cell division and consequent expansion at the **apical meristem**.

primary mineral A mineral that has crystallized from a **magma** and undergone no further change.

primary phloem **Phloem** tissue that develops from the **procambium**. *See* metaphloem, protophloem.

primary pigments In **photosynthesis**, the pigment that releases the electrons that drive the sequence of reactions. In green plants the primary pigments are the P_{680} and P_{700} varieties of **chlorophyll** *a*.

primary pit A **pit** in a **cell wall** that results from the incomplete separation of **daughter cells** during **mitosis**. *See* bordered pit, secondary pit, simple pit.

primary production *See* primary productivity.

primary productivity (primary production) The rate at which **autotrophs**, principally green plants, produce **biomass** by **photosynthesis** or chemosynthesis. Gross primary productivity (GPP) is the total rate of biomass production per unit of ground or water-surface area, including the proportion consumed by **respiration** during the measurement period. Net primary productivity (NPP) is GPP minus the amount lost through respiration. Net **ecosystem** productivity (NEP) is NPP minus the amount consumed by **heterotrophs**.

primary sexual character *See* primary sexual organ.

primary sexual organ (primary sexual character) An organ that produces **gametes**; in **angiosperms** the **stamen** and **pistil** in the **flower**, in **gymnosperms** the **cone**.

primary structure The type, number, and linear sequence of **amino acids** that comprise a **protein**. *See* quaternary structure, secondary structure, tertiary structure.

primary succession (prisere) A **succession** that begins on recently cleared bare ground, following major earth movements, retreating glaciers, etc.

primary wall The second layer of a plant **cell wall** to form during cell division. It is made of **pectins**, **hemicellulose**, and **glycoproteins** and comprises a rigid structure of **microfibrils**. *See also* middle lamella, secondary wall.

primary woodland **Woodland** on a site that has been wooded continuously for many centuries, in Britain since the last ice age, even though it may have been clear-felled providing the trees were replanted or regenerated naturally.

primary xylem **Xylem** tissue that develops from the **procambium** during primary growth, beginning with the **protoxylem** followed by **metaxylem**.

primitive Describes a **character** or organism that is preserved unchanged from an ancestor.

primordium The earliest recognizable stage in the development of an organ.

Primulaceae (order **Ericales**) A family of **annual** and **perennial herbs** with leaves **alternate**, **opposite**, or whorled often as a basal rosette, **simple**, **linear** to **orbicular**, **entire** to **serrate**, rarely **pinnatisect**. Flowers **actinomorphic**, **bisexual**, often heterostylous (*see* heterostyly), usually **pentamerous** occasionally 7-merous or **trimerous**, **sepals** free or fused into a tubular to **campanulate calyx**, **corolla tubular** to campanulate, 5 **stamens**, **ovary superior** of 5 fused **carpels** with 1 **locule. Inflorescence scape**-like with a single terminal flower, or a terminal **umbel** or **raceme**, or an **axillary** cluster, or a solitary axillary flower. Fruit is a **capsule.** There are 58 genera of 2590 species with a worldwide distribution. Many cultivated for ornament, e.g. ***Cyclamen***, *Primula*.

prisere *See* primary succession.

privet (*Ligustrum ovalifolium*) *See* Oleaceae.

probability forecast A weather forecast that states the likelihood of a particular condition occurring.

proboscis A tubular projection from the head of an animal, in invertebrates used for feeding.

procambium Primary **meristem** tissue composed of groups of elongated cells at the growing tips of stems and roots that produces new **vascular bundles** as plant growth continues.

procumbent Lying along the ground; protruding forward approximately horizontally.

Procyonidae (raccoons, coatis, cacomistle) A family of small mammals with short legs and medium to long tails, most with broad faces, partly retractile claws, and a **plantigrade** or semi-plantigrade gait. All are more or less arboreal. They are omnivorous, feeding on plant material, small mammals, and birds. There are 6 genera with 18 species found throughout the Americas.

Prodigy *See* methoxyfenoxide.

producer An organism that is able to synthesize complex organic compounds from simple, inorganic precursors, i.e. an **autotroph**.

proestrus *See* pro-oestrus.

prokaryote An organism in which the cells lack a true **nucleus**, the **DNA** being present as a loop in the **cytoplasm**. There are no **chloroplasts** or mitochondria (*see* mitochondrion) and the **ribosomes** are small. Most prokaryotes are single-celled. They comprise two taxonomic **domains**, **Bacteria** and **Archaea**.

prolepsis The condition in which an **axillary shoot** grows from a **lateral meristem** that has branched from a terminal meristem following the formation of a **bud** or a period of **dormancy**. *Compare* syllepsis.

prolegs (false legs) Fleshy protuberances from the abdomen of an insect larva that are used for walking. Unlike true insect legs, they are not jointed and not developed from the thorax.

proline A water-soluble **imino acid**, present in all **proteins**, that accumulates in plants under stress and protects proteins and membranes from the adverse effects of high concentrations of inorganic **ions** and extremes of temperature.

promeristem The cells that initiate growth in **apical meristem**, and their immediate descendants.

promoter A **nucleotide** sequence in an **operon** that lies between the **operator** and the **structural gene**(s) and is a recognition site and point of attachment for **RNA polymerase**; it is the starting point for **transcription** but is not transcribed itself.

promycelium In smut fungi (**Ustilaginomycetes**), a structure of two or three cells that forms when a **teliospore** germinates, each cell of the promycelium then **budding** to produce several sporidia (*see* sporidium).

pronotum The **dorsal** hardened **cuticle** of the first segment of the thorax of an insect.

pro-oestrus (proestrus) The stage in the **oestrus cycle** when the reproductive organs become active.

propagule Any plant structure that is capable of giving rise to a new plant, especially one involved in **vegetative reproduction**, e.g. **bulb**, **corm**, **tuber**, etc.

prophage The **genome** of a **bacteriophage**.

propham *See* carbamate herbicides.

prophase The first stage in **mitosis** and both divisions of **meiosis**. In mitosis and meiosis I the **chromosomes** become visible; in prophase I of meiosis they also pair. In prophase II of meiosis and mitosis the nucleoli (*see* nucleolus) and **nuclear envelope** break down and the **chromatids** become shorter and thicker.

proplastid A small **organelle**, less than 1 μm across, that is enclosed in a double membrane, the outer membrane often extended into finger-like protrusions, but has little internal structure. Proplastids are found in **meristem** cells and are thought to develop into mature **plastids**.

prop root A root that arises from the stem of a **tree** some distance from the ground and extends to the side, providing mechanical support.

prosenchyma An obsolete term formerly applied to any tissue composed of elongated cells with tapering ends.

prosoma The anterior part of the body of **Arachnida**.

Prostanthera (family **Lamiaceae**) A genus of bushy, **evergreen shrubs** (mint bushes), most with glands that yield aromatic oils. Leaves **opposite**. Flowers with 2-lipped persistent **calyx**, **corolla** a short, wide tube, 4 **stamens**. **Inflorescence** a terminal **panicle** or **raceme**, or flowers solitary in **axils** of **bracts**. Fruit is a **nutlet**. There are about 90 species, all **endemic** to Australia. Several cultivated for ornament or their oils.

prostheca A stalk-like outgrowth of a bacterial cell that contains material from the **cell wall** with a core of **cytoplasm**. Prosthecae may serve to increase the surface area of the cell, thus aiding the ingestion of food particles.

prosthecate Possessing prosthecae (*see* prostheca).

prosthetic group The tightly bound, non-**protein** component of a **conjugated protein** or **enzyme** that is required for its function.

protandrous hermaphrodite A **hermaphrodite** species in which young, small individuals are male and older, larger individuals are female.

protandry The maturation of **anthers** before **carpels**. *Compare* protogyny.

Proteaceae (order **Proteales**) A family of **perennial trees** and **acaulescent shrubs** with clusters of lateral (proteoid) roots. Leaves often **coriaceous**, **compound**, dissected or lobed, **dentate**, or **simple** and **entire**, **exstipulate**. Flowers usually radially or bilaterally symmetrical, **bisexual** sometimes **unisexual** (plants **dioecious** or **monoecious**), 4 rarely 3 or 5 **tepals**, 4 rarely 3 or 5 **stamens**, **ovary superior** of 1 **carpel** rarely 2. **Inflorescence** very variable, a simple or branching **raceme** of pairs of flowers, or simple or compound and **racemose**. Fruit **dehiscent** or **indehiscent**, dry or fleshy. There are 80 genera of 1600 species occurring mainly in the Southern Hemisphere. *Macadamia* spp. and **hybrids** are the source of macadamia nuts; many cultivated for ornament, e.g. *Protea*, *Grevillea*, ***Banksia***.

Proteales An order of plants comprising 4 families of 85 genera and 1710 species. *See* Nelumbonaceae, Platanaceae, Proteaceae, and Sabiaceae.

protected area Land that is set aside, usually for wildlife **conservation**, with legal protection against other uses.

protective plant A plant that is grown to protect the soil from **erosion** or to shelter crop plants.

protein An organic polymer composed of **amino acids**. They serve as structural elements, **enzymes**, **hormones**, **genes**, etc.

protein complex (multiprotein complex) A group of two or more **polypeptide** chains that, depending on its component **proteins**, may have several catalytic functions.

Protemphytus carpini (geranium sawfly) A sawfly (**Symphyta**) that produces olive-green or greyish black larvae, up to 11 mm long, that feed in large numbers on the underside of leaves, especially of geranium, often causing serious **defoliation**. Adults are shiny, black, and 6–8 mm long. They are active in May and June and again in July and August, and produce two generations a year.

Proteobacteria A **phylum** of Gram-negative (*see* Gram reaction) **Bacteria**, many of which are free-living and move by means of flagella (*see* flagellum) or by gliding; others are non-**motile**, including the **Myxobacteria**. Many species perform **nitrogen fixation**; others are pathogens. There are six classes: **Alphaproteobacteria**; **Betaproteobacteria**; **Gammaproteobacteria**; **Deltaproteobacteria**; **Epsilonproteobacteria**; and **Zetaproteobacteria**.

proteome All of the **proteins** present in a cell.

proteroglyphous Describes snakes that have short, hollow fangs at the front of the mouth, often with small, solid teeth behind them.

prothallium *See* prothallus.

prothallus (prothallium) The short-lived **gametophyte** stage in the life cycle of a fern (**Pteridophyta**), where it is inconspicuous, heart-shaped, and about 2–5 mm across, bearing **rhizoids** and the reproductive organs. It develops from a germinating **spore**. The term is also applied to the gametophytes of some mosses (**Bryophyta**) and liverworts (**Marchantiophyta**). *See* alternation of generations.

Protista A diverse group of **eukaryotes** that were formerly classed as a kingdom including animal-like organisms (protozoa), plant-like organisms (protophyta), and fungus-like organisms (slime moulds and water moulds). All are unicellular, those that are colonial not being differentiated into tissues. The group is polyphyletic (*see* polyphyletism) and its members have been allocated to other groups in modern classifications.

protoconch The first shell of a molluscan (**Mollusca**) larva. As the animal grows the protoconch often survives at the tip of the larger shell spire.

protocorm 1. A mass of **parenchyma** tissue resembling a **tuber**, with **rhizoids** on its lower surface and containing symbiotic (*see* symbiosis) **Fungi** that protrudes through the **prothallus** of club mosses. **2.** In orchids (**Orchidaceae**), an **ephemeral** structure that develops from the germinating seed and from which the first true root and shoot emerge.

Protoctista In the **three-domain system** of **taxonomy**, a kingdom within the domain **Eukarya**. In the **five-kingdom system**, a kingdom within the superkingdom Eukarya. Protoctists are aquatic **eukaryotes** that are neither animals, plants, nor fungi. The kingdom includes naked and shelled amoebae, foraminiferans, zooflagellates, dinoflagellates, ciliates, **diatoms**, algae (*see* alga), **slime moulds**, slime nets, and protozoa. Single-celled protoctists were formerly placed in the **Protista**, but this ranking has been abandoned.

protoderm The outer layer of **apical meristem** tissue, from which the **epidermis** arises.

protogyny The maturation of **carpels** before **anthers**. *Compare* protandry.

protohemicryptophyte A **hemicryptophyte** in which the lowest leaves are scale-like or smaller than the other leaves, protecting the **bud**.

protonation The addition of a proton, i.e. a hydrogen nucleus H^+, to an atom, **ion**, or molecule, thereby forming the **conjugate acid**.

protonema 1. The juvenile form of a moss (**Bryophyta**) and some liverworts (**Marchantiophyta**), usually consisting of a branching green **filament**, but in *Sphagnum* (*see* peat moss) and rock moss (*Andreaea* spp.) a **thallus**; the protonema is not differentiated in thallose liveworts. **2.** An erect filament that appears following **germination** in stoneworts (**Charophyceae**).

protophloem The first **phloem** tissue to form from the **procambium**. It consists of thin-walled, narrow cells, and is followed by the **metaphloem**.

protoplasm The colourless, translucent, colloidal contents of a cell including the **cell membrane** but not the large **vacuoles**, masses of ingested material, or secretions. Protoplasm in the cell **nucleus** is often called nucleoplasm and protoplasm outside the nucleus is **cytoplasm**.

protoplast A bacterial, fungal, or plant cell with the **cell wall** removed.

protostele The more primitive of the two types of **stele** (*see* siphonostele), consisting of a central, cylindrical strand of **xylem** surrounded by **phloem**. There are three types of protostele: **actinostele**, **haplostele**, and **plectostele**.

Protosteliomycetes A class of acellular slime moulds (**Myxogastria**) in which the **myxamoebae** do not aggregate prior to fruiting and the feeding stage consists of groups of myxamoebae or small plasmodia (*see* plasmodium). These organisms are widely distributed and found on rotting wood and other plant material.

Prototaxites An organism that lived on land during the Silurian and Devonian periods (443.7–359.2 million years ago). It had a trunk-like structure up to 1 m in diameter and 8 m tall that apparently grew by the addition of outer layers. Despite its superficial resemblance to a tree and its name, 'first yew', *Prototaxites* was probably a fungus.

protoxylem The first **xylem** tissue to form from the **procambium** during the growth of **primary xylem**. It is followed by the **metaxylem**.

protovascular bundle A strand of the **procambium**.

Protozoa (sing. protozoon) A group of single-celled, heterotrophic (*see* heterotroph) **eukaryotes**, many of which are **motile**. Some classifications rank them as a subkingdom or **phylum** of **Animalia** or **Protista**, comprising protists capable of movement, but the group is now known to be polyphyletic (*see* polyphyletism). Protozoa occur worldwide in soil and aquatic environments. Some are predators of algae (*see* alga), **Fungi**, and **Bacteria**, some are herbivorous, others are prey of small invertebrates.

protozoon *See* Protozoa.

provirus The **genome** of a **virus** after it has been incorporated into the genome of its host.

provisioning services *See* ecosystem services.

proximal Closest to the point of attachment.

PRRs *See* pattern recognition receptors.

pruinose Powdery, covered in powder.

Prunella modularis (dunnock, hedge accentor, hedge sparrow) A brown, streaked bird, 35–140 mm long and resembling a house sparrow. It feeds on insects in woodlands and gardens, and occurs throughout most of temperate Eurasia. It is migratory in the northern part of its range.

Prunus (family **Rosaceae**) A genus of **deciduous** or **evergreen trees** and **shrubs** with **alternate**, **simple**, usually **lanceolate** leaves. Flowers **pentamerous**, with about 20 **stamens** and 1 **carpel**. Flowers solitary, or in **umbels** forming **racemes**. Fruit is a fleshy **drupe**. There are about 430 species occurring in northern temperate regions. Many are cultivated for ornament and for their fruit, which include plums, cherries, apricots, nectarines, and almonds.

psammon The microorganisms that inhabit the interstices between **sand** grains on a sea or lake shore.

psammophyte A plant that thrives growing in unstable sand, typically in deserts.

Pseudacris crucifer (spring peeper) A brown, olive, or grey frog (**Ranidae**), sometimes red or yellow, with a white or cream underside, 20–25 mm long, that has webbed feet, prominent disks on the tips of its digits, a dark cross on its back, and dark bands on its legs. It inhabits wooded marshes and **habitats** close to ponds and swamps. It climbs well but spends most of its time on the ground or concealed in leaf litter. It feeds on insects. The frog occurs in eastern North America.

Pseudacris feriarum (upland chorus frog) A species of brown, grey-brown, or red-brown frogs (**Ranidae**) with dark blotches, which grow to 20–40 mm long. They are nocturnal and secretive. They live in moist areas close to water and feed on insects. They occur throughout the southern and eastern United States.

pseudanthium (flower head) An **inflorescence** composed of **florets** that are reduced, in **Asteraceae** to a single **stamen** or **carpel**, but that are present in large numbers (up to 1000), so the pseudanthium itself may be large and appear to be a single flower. e.g. dandelion (*Taraxacum officinale*).

pseudobulb A swelling that serves as a storage organ in the stem of a **perennial** plant between leaf **nodes**. Pseudobulbs occur in many epiphytic (*see* epiphyte) orchids (**Orchidaceae**).

pseudocarp (false fruit, accessory fruit) A fruit in which the ripened **ovary** is combined with another structure, commonly the **receptacle**. A strawberry is a pseudocarp.

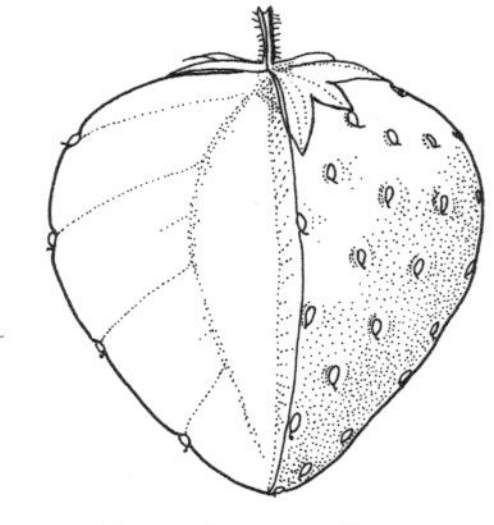

Pseudocarp fruit (strawberry).

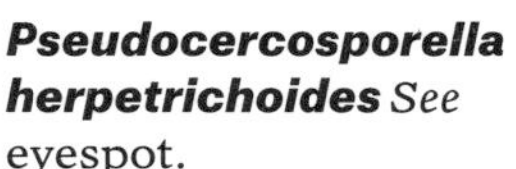

Pseudocercosporella herpetrichoides *See* eyespot.

pseudocoelom A second body cavity occupying the space between the inside of the body wall and the gut (the first body cavity).

pseudocyphella A very small pore or depression, visible as a white dot, on the surface of the **thallus** in some species of **lichen**.

pseudoendosperm Nutritive tissue in the seed of a **gymnosperm**. It is derived from the female **gametophyte** and is **haploid**, unlike the true **endosperm** of **angiosperms**, which is **triploid**.

pseudogamy A form of **asexual reproduction** that requires stimulation of the female **gamete** by the male gamete, but in which **fertilization** does not occur.

pseudogenes Genes that have been silenced in the course of evolution so they no longer have any function. They continue to mutate, however, at a constant rate.

Pseudolarix (family **Pinaceae**) A **monotypic**, **gymnosperm** genus (*P. amabilis*, golden larch), which is a **deciduous**, coniferous **tree** with dimorphic (*see* dimorphism) shoots, spirally arranged bright green leaves that turn golden yellow in autumn, spaced widely on long shoots and in dense **whorls** on short shoots. Male **cones** borne in **umbels** on spur shoots, female cones globular, resembling globe artichokes, with triangular scales. The plant is native to southeastern China but is cultivated widely for ornament.

Pseudomonadales One of the two principal orders of **Eubacteria**, comprising Gram-negative (*see* Gram reaction), spherical, spiral, or rod-shaped **Bacteria** in which the **motile** forms possess a single **flagellum**. *See* Eubacteriales.

Pseudomonas A genus of **Gammaproteobacteria** comprising 191 known species of Gram-negative (*see* Gram reaction), rod-shaped **aerobes** that do not form **endospores**. Most are **motile** with one or more **polar flagella**. They are common in soils and in aquatic environments, and are able to break down many organic compounds. Some are pathogens, e.g. ***P. syringae***, which infects a wide range of plants. Other species inhibit plant pathogens and are used in **biological control**, or detoxify pollutants and are used in **bioremediation**. *See* denitrifying bacteria.

Pseudomonas savastonoi *See* olive knot.

Pseudomonas solanacearum *See Ralstonia solanacearum.*

Pseudomonas syringae* pv. *phaseolicola The pathogenic variety (pv.) of *P. syringae* (*see Pseudomonas*) that causes halo blight of the bean *Phaseolus vulgaris* and several other **Fabaceae** crops. Lesions, soaked in water, appear on leaves, stems, **petioles**, or pods and greenish yellow haloes appear on leaves; in seedlings the disease causes **chlorosis**. Infection is spread by wind and rain, and the bacteria can enter through wounds. The disease occurs worldwide in temperate regions and in mountainous regions of the tropics.

Pseudomyrex ferruginea (acacia ant) *See* co-adaptation.

pseudoparenchyma In **red algae** (Rhodophyta) and certain **Fungi**, a mass of interwoven **filaments** or hyphae (*see* hypha) that superficially resembles **parenchyma**.

pseudoplasmodium A structure resembling a **plasmodium** formed by the aggregation of many **amoeboid** cells in cellular slime moulds (**Acrasiomycetes**).

pseudopodetium An erect, **fruticose** structure on the **thallus** of certain **lichens** that bears **ascocarps** if these are present. It may resemble a **podetium**, but develops from the thallus rather than the ascocarp.

pseudopodium A protrusion from the body of an **amoeba** by means of which the organism moves or ingests food items. It is usually withdrawn when no longer needed.

pseudorumination (refection) A type of feeding in some mammals in which food passes rapidly through the gut and is ingested again as it leaves the anus.

pseudoscorpions *See* Arachnida.

pseudostem A structure resembling a stem that develops from swollen leaf bases.

pseudothecium (pseudoperithecium) A fungal **ascocarp** that resembles a **perithecium**, but has **bitunicate** asci (*see* ascus) that are not regularly organized into a **hymenium**.

Pseudotsuga menziesii (Douglas fir) *See* Pacific coast forest.

pseudowhorl An arrangement of leaves that are attached almost, but not quite, at the same level so they resemble a **whorl**.

PSI *See* photosystem I.

PSII *See* photosystem II.

Psidium guajava (guava) *See* Myrtaceae.

Psila rosae (carrot fly) A small, black, shiny fly with a red head and orange legs, about 8 mm long that produces two or three generations a year. The larvae are creamy-white, 8–10 mm long, and feed on plant **taproots**, especially carrot but also other crops including parsnip, celery, and parsley. Brown scars appear on the outside of the root, which may then rot.

Psilocybe (magic mushroom) A genus of **agaric fungi** that produce psilocin and psilocybin, substances with psychedelic properties. The brown or yellowish **fruiting bodies** are small. The fungi are **saprotrophs** and there are about 40 species found worldwide.

Psilopsida (order Psilotales) A class of **Pteridophyta**, comprising the one family Psilotaceae with two living genera, *Psilotum* and *Tmesipteris*, which are ancient and primitive plants with forked stems bearing small, scale-like appendages (*Psilotum*) or flattened leaves (*Tmesipteris*). They have creeping **rhizomes**, lack true roots, and their **spore** capsules are fused in pairs (*Tmesipteris*) or threes (*Psilotum*). They live as **saprophytes** in association with a mycorrhizal (*see* mycorrhiza) fungus.

PSU *See* photosynthetic unit.

psychrometer A **hygrometer** that comprises two thermometers, one with its bulb wrapped in muslin partly immersed in a reservoir of water, so it acts as a wick. **Evaporation** from the wick cools the wet bulb and comparison between its temperature and the **dry-bulb temperature** gives the wet-bulb depression, from which the **dewpoint temperature** and **relative humidity** can be calculated, or read from published tables.

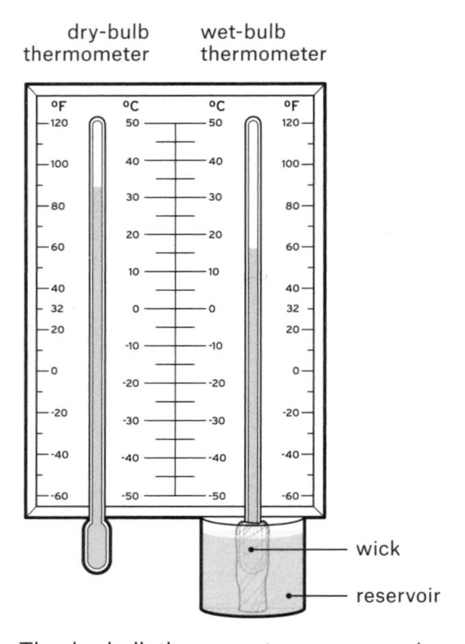

The dry-bulb thermometer measures air temperature, the wet-bulb thermometer, its bulb wrapped in a wick immersed in a reservoir of distilled water, measures the air temperature minus the energy lost by the latent heat of vaporization. The difference between the two (the wet-bulb depression) is used to calculate the dewpoint temperature and the relative humidity.

psychrophile *See* extremophile.

Psylla mali (apple sucker) A bug (**Psyllidae**) that feeds only on apple trees and that produces only one generation a year. The bugs overwinter as minute, straw-coloured eggs in leaf scars on fruit spurs, hatching in spring as **nymphs** with red eyes and a flattened, orange-brown body, becoming bright green after the second moult and maturing as adults, up to 3 mm long, four to six weeks after hatching. Large populations feeding on **sap** can cause browning of **petals** and in extreme cases the death of flower **buds**. They occur throughout the temperate Northern Hemisphere.

Psylla pyricola (pear sucker) A bug (**Psyllidae**), about 3 mm long with a black spot on the forewing, that resembles ***Psylla mali***. It overwinters as an adult and lays eggs in early spring on pear leaves and blossoms. The **nymphs** are orange-yellow with red eyes and feed on the blossoms, excreting **honeydew** that attracts **sooty mould**. There are usually three generations a year. They occur throughout the temperate Northern Hemisphere.

Psyllidae (psyllids, jumping plant lice) A family of bugs (**Homoptera**) in which the hind legs of adults are modified for jumping and the **rostrum** arises between the front pair of legs. Psyllids feed on sap and some cause **galls** to form, but most species feed on only one or a few plant species. Immature stages cannot jump and move little. There are more than 70 genera with about 800 species, found worldwide.

psyllids *See* Psyllidae.

Pteridophyta A **phylum** (division) of vascular **cryptogams** comprising the true ferns, flowerless plants in which a **spore**-bearing, non-sexual, **sporophyte** generation alternates with a sexual **gametophyte** generation. The sporophyte generation is large and bears leaves and roots. There are more than 230 genera with about 12,000 species with worldwide distribution.

Pteridospermaphyta *See* seed fern.

Pterocarya (family **Juglandaceae**) A genus of **deciduous trees** with **pinnate** leaves having 11–25 **leaflets**. Flowers are **monoecious** and borne in **catkins**. Fruit is a **nut** with two short, leafy wings (known as wing nuts). There are six species occurring from the Caucasus to Japan. Some cultivated for ornament.

pterochory Wind-dispersal of winged seeds.

Pterostylis (family **Orchidaceae**) A genus of terrestrial, **deciduous** orchids with fleshy leaves in a **radical** rosette. They produce one or more **tubers**. The green, white, or brown flowers have a **dorsal sepal** and upper two **petals** forming a hood and the lower two sepals forming a lip, often extending into a long tail. Fruit is a **capsule**. There are about 200 species occurring mainly in New Zealand, Australia, New Guinea, and New Caledonia. Many cultivated for ornament.

Pterygota The larger of the two subclasses of **Insecta** (*see* Apterygota), comprising insects that have wings, or that are secondarily wingless, i.e. they lack wings but are descended from winged ancestors. There are 29 orders and about 577,000 species with a worldwide distribution.

Ptilotus (family **Amaranthaceae**) A genus of mostly **perennial shrubs** and **herbs** with **alternate** or **opposite**, **entire**, **exstipulate** leaves. Flowers usually **actinomorphic**, **bisexual**, **perianth** segments partly fused, 1–5 **stamens**, **ovary superior**. **Inflorescence** an **axillary cyme** or flowers solitary. Fruit is a **nut** or **berry**. There are about 100 species, **endemic** to Australia. Several cultivated for ornament.

PTS *See* group translocation.

ptyxis *See* vernation.

pubescent Covered with down or soft hairs.

Puccinia graminis A species of **Basidiomycota** that causes black stem rust, a disease affecting many grasses (**Poaceae**) including wheat, barley, rye, and triticale. Reddish brown pustules appear on the stems, leaves, and other parts of infected plants, which then rupture to release a reddish mass of **urediospores**. Infected plants produce fewer **tillers** and seeds. Severely infected plants may die. *Puccinia graminis* requires two hosts, the other being barberry (***Berberis** vulgaris*), where the fungus produces black **teliospores**

that overwinter on stubble and other grasses.

Puccinia horiana A species of **Basidiomycota** that causes chrysanthemum white rust, a disease that affects *Chrysanthemum* species and some other members of the **Asteraceae**. Symptoms begin with green or yellow spots, up to 5 mm across, on the upper surface of leaves, especially young leaves. The spots grow larger and turn brown, and the leaves wither. The fungus occurs in Asia, North and southern Africa, Mexico, South America, and is widespread in Europe.

Pucciniomycetes (Pucciniales) A class of **basidiomycete fungi**, almost all of which are **obligate parasites** of animals, plants, or other fungi; they cause **rust** diseases in plants. The **mycelium** is **dikaryon** and sexual reproduction is through **basidiospores**. There are about 8000 species found worldwide.

Pucciniomycotina A subphylum of **basidiomycete fungi**, all of which are minute, but they are otherwise very diverse. There are about 215 genera with about 8400 species, 95 percent of which are in the class **Pucciniomycetes**.

puddingstone *See* conglomerate.

puddled soil (poached soil) Soil that has lost all structure due to trampling, tillage when wet, or the impact of raindrops.

puff adder *See Heterodon platirhinos.*

puffballs *See Lycoperdon.*

pulmonary Relating to the lungs.

Pulmonata (slugs, snails) An informal group of **Gastropoda** in which the **mantle** cavity is modified to form a lung. Most pulmonates have a coiled shell and **detorsion** has occurred in some. They are **hermaphrodites** and young develop directly, with no larval form. There are more than 20,000 species. Formerly regarded as a subclass, Pulmonata is now believed to be polyphyletic (*see* polyphyletism) and the name is no longer used taxonomically.

pulse 1. Any member of the family **Fabaceae**. 2. The edible seeds of a legume (**Fabaceae**).

pulverulent Covered with **dust** or powder.

Pulvinaria vitis (cottony grape scale, cottony maple scale, cottony vine scale, woolly vine scale) A scale insect (**Coccidae**) in which adult males are 1.5 mm long and adult females 5–7 m long, the dark brown scale circular, oval, or heart-shaped, and wrinkled. Adults appear in autumn and mate, the males then die and the females overwinter, beginning to grow and feed the following spring. Each female spins a white container (ovisac) in which she lays 1000 or more eggs over two to three weeks, then dies. The eggs hatch in early summer and the **nymphs** disperse to young wood. They reach adulthood in autumn. The insects infest gooseberry, currant, grapevine, apricot, and peach, and heavy infestations can cover large areas of the plant. They excrete **honeydew** that attracts **sooty mould**. The insect occurs widely throughout the **Palaearctic** and North America, and has been found in New Zealand.

pulvinate Swollen, convex, shaped like a cushion.

pulvinus 1. (geniculum) A thickening at the base of a **pinna** or leaf **petiole** that has a core of **vascular tissue**, allowing it to move water rapidly to and from **cell vacuoles**, thereby moving the leaf, e.g. in **nictonasty**. 2. In grasses (**Poaceae**), a thickened region at a stem **node** containing tissue that can bend and elongate, allowing geotropic (*see* geotropism) movement that raises the stem following **lodging**.

puna Vegetation that grows on high, arid plateaux along the western side of the Andes.

punctate Marked by pores or small depressions.

punctiform Marked with dots or points.

punctuated equilibrium The hypothesis that evolution involves long periods of stability (equilibrium) punctuated by short periods of rapid change.

Punica granatum (pomegranate) *See* Lythraceae.

punkies *See* midges.

pupa The stage in the life cycle of an insect during which it changes from its juvenile to adult form, involving a major reorganisation of its body structure. The pupa is usually immobile and covered with a hard shell (chrysalis) or silk (cocoon).

puparium A **pupa** that is formed from the **exoskeleton** of the final larval **instar**.

purine A **heterocyclic** organic compound ($C_5H_4N_4$) consisting of a **pyrimidine** ring fused to an **imidazole** ring. The two most important purines, adenine and **guanine**, are **nucleotide** bases. Caffeine and theobromine (the bitter taste of cocoa) are also purines. *See* DNA.

purple finch *See Carpodacus purpureus.*

puszta Grassland in Hungary, similar in character to the **prairie**.

Putranjivaceae (order **Malpighiales**) A family of **trees** and **shrubs** with leaves **distichous** rarely **opposite** or subopposite, **entire** to **dentate**. Flowers **unisexual** (plants **dioecious**), 3–7 usually **imbricate sepals**, **petals** absent, 2–20 or up to 50 **stamens**, **ovary superior** of 1–6 **carpels** and **locules**. **Inflorescence** an **axillary fascicle** sometimes **cauline**, subtended by inconspicuous **bracts**. Fruit **drupe**-like. There are 3 genera of 210 species occurring throughout the tropics especially in Africa and Malesia.

PVY *See* potato virus Y.

PWP *See* permanent wilting point.

pycnidium (pycnium) A spherical or flask-shaped, asexual **fruiting body** formed in certain **Pucciniomycetes Fungi**, in which conidia (*see* conidium) develop and are released through a pore.

pycniospore (pycnospore) A **haploid spore** formed in a **pycnidium** that functions as a male **gamete**.

pycnium *See* pycnidium.

pycnospore *See* pycniospore.

pygmy shrew (*Sorex minutus*) *See Sorex.*

pygostyle Several of the posterior vertebrae of a bird that are fused together to form the structure that supports the tail **feathers** and muscles.

Pyralidae (snout moths) A family of moths (**Lepidoptera**) with a wingspan of 13–40 mm. Most have **caterpillars** that bore into the stems, **buds**, flowers, or seeds of plants, some feed on dried plant material, and some feed on combs in beehives. The family includes **diurnal**, nocturnal, and **crepuscular** members. There are at least 6150 species found worldwide.

Pyrausta aurata (mint moth) A moth (**Lepidoptera**) with purple or brown wings, each with a single yellow spot, and a wingspan of 18–20 mm, that flies by day and by night. It has two generations a year, in early and late summer. Its **caterpillars** are 11–12 mm long, red or green with longitudinal rows of black spots, and feed on mints, marjoram, lemon balm, and other herbs. It occurs throughout northern and western Europe.

pyrene The hard stone of certain fruits, e.g. cherry.

pyrenoid A protein body, not enclosed by a membrane, in or beside the **chloroplasts** of certain algae (*see* alga) and hornworts (**Bryophyta**). It is where most of the chloroplast's **ribulose-1,5-biphosphate carboxylase** (rubisco) is concentrated, so it is also where carbon dioxide is concentrated during **photosynthesis**, and the pyrenoid is surrounded by plates of **starch**.

P

pyrethrins A group of **insecticides** derived from *Chrysanthemum* species, especially *C. cinerariifolium*. They act on contact and in doses insufficient to kill insects they deter them. Pyrethrins break down quickly and are of low toxicity to vertebrates.

pyrethroid insecticides A group of **insecticides** in which the synthesized active ingredient closely resembles **pyrethrum**. Pyrethroids are highly toxic to insects but of low toxicity to birds and mammals, but they are harmful to fish if they enter water directly.

pyrethrum An insecticide made from the dried flower heads of *Chrysanthemum* species, especially *C. cinerariifolium* and *C. coccineum*, which are then crushed and mixed with water.

pyriform Pear-shaped.

pyrimidine A **heterocyclic** compound ($C_4H_4N_2$) that is involved in the production of **proteins** and **starches** and the regulation of **enzyme** activity. The pyrimidines **cytosine**, thimine, and uracil are **nucleotide** bases. *See* DNA.

P

pyroclimax (fire climax) A **climax** that is controlled by repeated fires that burn off the surface vegetation and trigger a new **succession**. Interactions between fire and the trampling and grazing of large **herbivores** are also thought to have contributed to the formation of the world's major grasslands.

pyrophyte A plant that is adapted to fire or obtains a competitive advantage from it.

pyrrole A **heterocyclic** compound (C_4H_4NH) that is a colourless, volatile liquid. It is an ingredient of a range of substances including **porphyrins** and **chlorophyll**.

Pyrrophyta (Dinophyta, dinoflagellates) A **phylum** of protists (**Protista**), 0.01–2.0 mm in size, with two flagella (*see* flagellum), many with **cell walls**, and most with a **theca**. About half are photosynthetic (*see* photosynthesis) with **chlorophylls** *a* and *c* as well as golden brown pigments. Others are **heterotrophs** or mixotrophs (engulfing other organisms). Most reproduce asexually by cell division. Many are free-swimming with a spiral motion, others are **sessile**. There are about 2000 species found worldwide in freshwater and marine environments.

Pyrus (family **Rosaceae**) A genus of mainly **deciduous trees** and **shrubs**, some with thorns, leaves **alternate**, **simple**, **oval** to **lanceolate**. Flowers with 5 **sepals**, 5 **petals**, 20–30 **stamens**, 2–5 fused **carpels**. **Inflorescence** an **umbel**-like cluster. Fruit is a **pome**. There are about 20 species occurring in temperate Eurasia. Many cultivated for their fruit (pear).

pyruvic acid A **ketone** ($C_3H_4O_3$) that is the final product of **glycolysis** in aerobic **respiration** and is then oxidized to carbon dioxide and acetyl **coenzyme** A. In anaerobic respiration in plant cells pyruvic acid is converted irreversibly to **ethanol** and carbon dioxide (in animal cells it is converted to lactic acid). Acetyl coenzyme A can also catalyze the reaction converting pyruvic acid to **carbohydrates** by **gluconeogenesis** or to **fatty acids**.

Pythiales One of the two principal orders of water moulds (**Oomycota**) that produce coenocytic (*see* coenocyte) **hyphae** without septa (*see* septum). Some members are plant pathogens.

Pythium A genus of water moulds (**Oomycota**) of the order **Pythiales**, comprising more than 150 species, most of which are plant pathogens causing root rot in about 100 species of hosts. Infected plants are stunted and eventually may die.

pyxidium A **capsule** that dehisces around its circumference so the top falls off, e.g. in some **Amaranthaceae**.

Q

Quadraspidiotus perniciosus (San José scale) A golden brown scale insect (**Coccidae**), about 1 mm long, that feeds on about 200 species of plants, mostly **shrubs** and **trees** including apple, pear, and peach. Females lay eggs beneath the round, grey, wax scale, about 1.4 mm across, that they secrete. The eggs hatch into bright yellow crawlers, a female giving rise to about 10 crawlers a day for 40–55 days. The crawlers move or are carried by wind to new feeding sites. The insects cause a loss of vigour in the tree, slowing growth and reducing fruit production. It occurs throughout the world except for Antarctica.

quadrat An area marked out on the ground that is used in vegetation surveys. The sizes and shapes of quadrats vary according to the type of vegetation being sampled. A number of quadrats are set out in a pattern calculated to deliver a random sample, and all plants of the relevant species in each quadrat are counted.

quaking aspen (*Populus tremuloides*) *See Populus.*

quantasome A particle made from **lipids** and **proteins**, including several photosynthetic pigments and **electron carriers**, that is embedded in the surface of **thylakoid** discs in **chloroplasts**. There are two types. The smaller quantasome is thought to be the site of **photosystem I**, the larger of **photosystem II**.

quantitative inheritance The inheritance of a **character** whose manifestation occurs through the cumulative action of many **genes** each of which exerts only a small effect.

quantum evolution A sudden acceleration in the rate of change in an **evolutionary lineage**.

quantum speciation Rapid **speciation** that can occur in a small population isolated from its much larger ancestral population. It is also central to the hypothesis of **punctuated equilibrium**.

quartz (rock crystal) A silicate mineral (SiO_2) that is found in many **igneous** and **metamorphic rocks**. Its crystals are usually six-sided prisms with a six-faced pyramid at each end. Quartz is usually colourless and transparent, but can occur in a variety of colours, many of which are semi-precious stones.

quasi-stationary front A **front** that is moving at less than about 5 **knots** (9.25 km/h).

quassia An **insecticide** extracted from the wood and **bark** of *Quassia amara* (amargo, bitter ash, bitter wood), a **shrub** or small **tree** native to Central and South America, by steeping chippings in water. The active ingredient, quassin, is effective against aphids (**Aphididae**), Colorado beetle (***Leptinotarsa decemlineata***), **caterpillars**, and house flies. It also has medicinal uses.

Q

quaternary structure The final configuration of the **polypeptide** subunits that comprise a functioning **protein** molecule. *See* primary structure, secondary structure, tertiary structure.

queen The primary reproductive female in a colony of **eusocial** insects (**Insecta**).

Queensland arrowroot (*Canna edulis*) *See* Cannaceae.

Quercus (family **Fagaceae**) A genus of **deciduous** (oak) and **evergreen** (live oak) **trees** and **shrubs** with spirally arranged, **simple**, lobed, **serrate**, rarely **crenate**, or **entire** leaves, marcescent (*see* marcescence) in many deciduous species. Male flowers are **catkins**, female flowers solitary. Fruit is a **nut** held in a woody **cupule** (acorn). There are 600 species occurring throughout temperate and tropical regions. Many grown for timber and as ornamentals.

quick clay **Clay** that softens, sometimes behaving as a liquid, when disturbed, especially if it is saturated with water.

quick flow *See* surface runoff.

quicklime *See* lime.

quiescent centre A region in the **apical meristem** where cells are dividing slowly or not at all, but the cells will resume dividing should the **initials** be damaged.

Quillajaceae (order **Fabales**) A **monogeneric** family (*Quillaja*) of **evergreen trees** with **alternate**, spirally arranged, **simple**, **serrate**, **stipulate** leaves. Flowers **actinomorphic**, **bisexual**, 5 **valvate sepals**, 5 **petals**, 2 **whorls** of 5 **stamens**, **ovary superior** of 5 **carpels**. **Inflorescence** a terminal **cyme**. Fruit an asymmetrically lobed **follicle**. There are three species occurring in temperate South America.

quinine (*Cinchona*) *See* Rubiaceae.

quinoa *See* Amaranthaceae.

quinone 1. A group of compounds some of which act as **electron carriers** in mitochondria (*see* mitochondrion) and **chloroplasts**. **2.** A group of **fungicides** that includes chloranil used as a seed dressing, dichlone used as a foliar spray, and dithianon.

R

rabbits *See* Leporidae.

raccoons *See* Procyonidae.

raceme A **inflorescence** in which flowers are produced laterally from a main **axis** that continues to grow monopodially (*see* monopodial). The youngest flowers are at the **apex** or at the centre.

racemose Resembling or developing in the manner of a **raceme**.

Raceme.

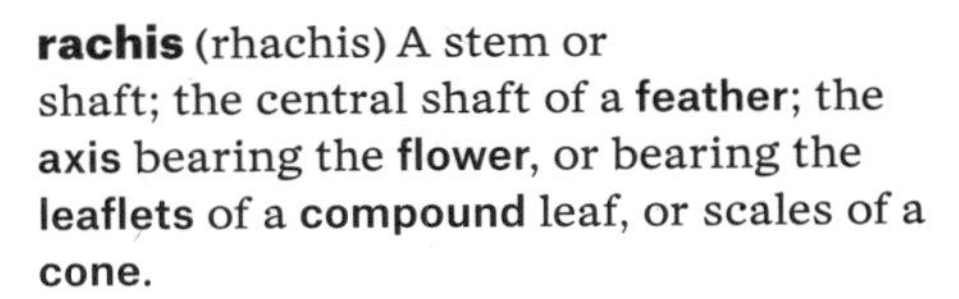

rachis (rhachis) A stem or shaft; the central shaft of a **feather**; the **axis** bearing the **flower**, or bearing the **leaflets** of a **compound** leaf, or scales of a **cone**.

radial symmetry An arrangement of body parts that locates them symmetrically around a central **axis**, allowing the animal to respond to stimuli from all directions without turning.

radiation cooling The fall in surface temperature that occurs at night, when the surface radiates its accumulated warmth while receiving no solar radiation to compensate, and in winter, when the hours of darkness exceed those of daylight.

radiation fog **Fog** that forms on clear nights when the air is moist. As energy radiates from the surface the surface-level air temperature falls; if it falls below the **dewpoint temperature** fog will form.

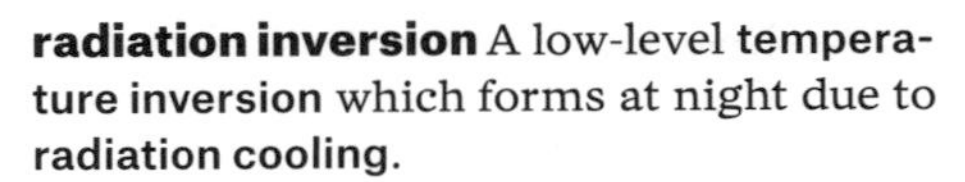

radiation inversion A low-level **temperature inversion** which forms at night due to **radiation cooling**.

radical Describes leaves that arise from the base of a stem or from a **rhizome**.

radicle The rudimentary root that is part of an **embryo**.

radius In **tetrapods**, one of the long bones of the forelimb; in primates it is the bone closest to the thumb.

radula A rasping tongue, comprising a strip of **chitin** that is constantly renewed, with rows of small teeth across its surface, found in most molluscs (**Mollusca**).

Rafflesiaceae (order **Malpighiales**) A family of stem or root parasites of the vine *Tetrastigma*, lacking **chlorophyll** and vegetative body resembling a fungal **mycelium** entirely inside the host. There are no stems; **bracts** subtending the flowers serve as leaves. Flowers **actinomorphic**, usually **unisexual** (plants **monoecious** or **dioecious**) occasionally also with **bisexual** flowers, **perianth** with 5, 10, or 16 **connate tepals**, 12–40 **extrorse anthers**, **ovary inferior** of 4–8 fused **carpels** with 1 **locule**. Flowers range in size from 8 cm to more than 1 m across, weigh up to 7 kg, and often smell of rotting flesh. Flowers solitary or in a variety of **inflorescences**. Fruit **baccate**. There are 3 genera of 20 species occurring in southern China, Assam, Bhutan, Thailand, and western Malesia.

rain **Precipitation** that consists of liquid drops 0.5–5.0 mm in diameter.

rain beetle *See Carabus violaceus.*

rain day A period of 24 hours beginning at 0900 Universal Time during which at least 0.2 mm of **rain** falls.

rain dove *See Zenaida macroura.*

raindrop A drop of liquid water that falls from a cloud. Raindrops are at least 0.5 mm in diameter and most are 2–5 mm, and they fall at an average 23–33 km/h.

rainfall inversion *See* precipitation inversion.

rain gauge An instrument that is used to measure the amount of rainfall during a specified period, usually one day. There are several types. The international standard gauge is a cylinder 20 cm in diameter mounted vertically with its top 1 m above ground level. Rain enters the cylinder through a funnel leading into a measuring tube with a cross-sectional area one-tenth that of the cylinder, so the amount of rainfall is one-tenth of the reading on the tube. In the tipping-buckets design, rain enters through a heated collecting funnel, passes down a tube to a second funnel, and from there falls into one of two 'buckets' mounted on a rocker. As a bucket fills it moves downward, finally making an electrical contact and emptying its contents, and the other bucket is positioned to collect the incoming water. The electrical contact is transmitted to a pen on a rotating drum.

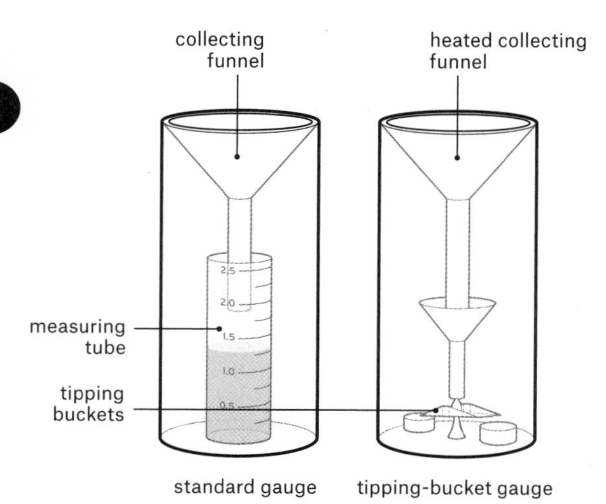

There are several versions of rain gauges. These two are the simplest.

rain shadow The area on the **lee** side of mountains that has a relatively **dry climate** because air approaching from the windward side of the mountains loses its moisture as it is forced to rise.

rainsplash The **detachment** of small soil particles by the impact of falling raindrops and their subsequent movement downslope.

rainwash The transfer of soil material by the action of rain. It includes **rainsplash** and **soilwash**.

rainy climate A climate in which the average annual precipitation is sufficient to sustain plants that are not adapted to arid conditions.

raised bog An **ombrogenous bog**, typically dome-shaped, that grows upward. The surface has hummocks, often with sedges and heathers, and pools, and different species of bog moss (*Sphagnum*) cover the waterlogged areas. The top of the dome (also called a cupola) may be 12 m above the **water table**.

Ralstonia solanacearum A species of **Bacteria**, formerly called *Pseudomonas solanacearum*, that causes **bacterial wilt**, also known as Granville wilt, Moko disease (in bananas), southern bacterial wilt, and southern wilt, in nearly 200 species of plants, but especially **Solanaceae**. The **bacterium** is an **aerobe**, rod-shaped, Gram-negative (*see* Gram reaction), and does not form **endospores.** It occurs worldwide, causing disease outdoors in tropical, subtropical, and temperate regions with moderate rainfall.

rambutan (*Nephelium lappaceum*) *See* Sapindaceae.

ramentum A thin scale on the young frond of a fern; most of the ramenta fall off as the frond opens, but some persist along the **rachis**.

ramet An individual member of a **clone**.

ramiflorous Flowering on the **branches**.

ramorum dieback *See Phytophthora ramorum.*

ramus A branch or a projection from a bone.

Rana catesbiana (American bullfrog) A frog (**Ranidae**) with a green back, pale underside, and brown blotches on the sides, 90–150 mm long with large individuals weighing 500 g, that occurs in the eastern United States and California, parts of South America, and Japan, and that has been introduced in Canada, Mexico, Jamaica, and Europe. It feeds on insects and small vertebrates including rodents and snakes.

Rana clamitans clamitans (*Lithobates clamitans clamitans*, green frog) A species and subspecies of green, greenish brown, or yellowish frogs (**Ranidae**) with yellow or white undersides, 75–125 mm long, with webbed toes, that live close to inland waters, and feed on invertebrates. They occur throughout eastern North America. Another subspecies, *R. clamitans melanota* (bronze frog) occurs in the southeastern United States.

Rana palustris (*Lithobates palustris*, pickerel frog) A species of frogs (**Ranidae**) that have rectangular spots on the back that may merge to form a single, long rectangle, and orange or yellow skin on the inside of the thighs. Their skin secretes toxins that can be irritating and that are lethal to small animals. The frogs are 450–750 mm long, females larger than males. They mainly inhabit streams in woodland, but in summer move some distance from water. They feed on plants as **tadpoles** and on invertebrates as adults. They occur throughout much of eastern North America.

Rana sphenocephala (southern leopard frog) A green or pale brown frog (**Ranidae**) with dark blotches and a pointed snout that is up to 90 mm long. It inhabits shallow fresh water with abundant vegetation and feeds on insects and other invertebrates. It occurs in the southeastern United States.

Rana temporaria (common frog, brown frog, grass frog) A frog (**Ranidae**), 75–80 mm long, with a wide head and short body, that has a brown or black triangular area around the ear. They are generally brown with black markings, but the colour is highly variable. It inhabits damp places not far from water and feeds on insects, insect larvae, and other invertebrate animals. It is valuable in helping control plant pests. The frog occurs throughout the **Palaearctic**.

range The area within which a species occurs.

Rangifer tarandus *See* antlers.

Ranidae (frogs) A family of amphibians (**Amphibia**) in which the hind legs are long and adapted for jumping, the hind feet are webbed, the tip of the tongue is notched, males often develop vocal sacs, and typically the skin is smooth and moist. Some species have vivid warning coloration and the skin secretes **batrachotoxin**. Frogs are air-breathing but also rely on **cutaneous respiration**. Fertilization is external and most breed in water; larvae (tadpoles, pollywogs) are aquatic, with gills. Most frogs live near water but some are arboreal. They occur on all continents except Antarctica. There are 14 genera and 363 species, but the family may be polyphyletic (*see* polyphyletism) and is currently being revised.

R

Ranunculaceae (order **Ranunculales**) A family of **perennial** and some **annual herbs** with a few **shrubs** and **lianas**, with **opposite** or spirally arranged, **simple**, **palmately lobed**, or **compound**, **exstipulate**, usually **petiolate** leaves. In some the petiole or the whole leaf forms a **tendril**. Flowers usually **actinomorphic**, **hermaphrodite**, typically with the parts arranged spirally along an elongated **receptacle**, or **perianth** segments in **whorls**, usually 5 or 3 to many **sepals**, often **petaloid**, true **calyx** and **corolla** often absent, many **stamens** with **extrorse anthers**. **Inflorescence** a terminal **raceme** or **cyme**, or flowers solitary. Fruit is a **berry**. There are 62 genera of 2525 species with a worldwide distribution. Many cultivated for ornament, e.g. ***Clematis***, *Trolius*, *Aconitum*, ***Delphinium***, ***Helleborus***; *Ranunculus* spp. are buttercups.

Ranunculales An order of plants comprising 7 families of 199 genera and 4445 species. *See* Berberidaceae, Eupteleaceae, Circaeasteraceae, Lardizabalaceae, Menispermaceae, Papaveraceae, and Ranunculaceae.

Rapateaceae (order **Poales**) A **monocotyledon** family of **perennial herbs** with thick **rhizomes**, and **distichous**, **simple**, often **ensiform**, rarely **petiolate** leaves. Flowers **actinomorphic**, **trimerous**, **bisexual**, with 3 **sepals**, 3 **petals**, 2 **whorls** of fused **stamens**, **ovary superior**. **Inflorescence** an **axillary capitulum** of 1–3 or up to 70 **spikelets**, subtended by **bracts** and borne on a leafless **scape**. Fruit is a **capsule**. There are 16 genera of 94 species occurring in tropical South America and Africa.

raphe 1. A ridge or seam on a seed, due to the **funicle** fusing to the **integument** of the **ovule**. **2.** A slit or groove along the axis of the cell in certain **diatoms**.

raphide A needle-like crystal, commonly sharp at one end and blunt at the other, usually of calcium oxalate, that is found inside a plant cell in more than 200 plant families. Plants accumulate them in response to a surplus of calcium and they may deter **herbivores**.

rarity The vulnerability of a species to extinction based on its relative abundance.

raspberry *See Rubus.*

raspberry aphid An aphid (**Aphididae**), *Amphorophora idaei* that is widely distributed and that feeds exclusively on raspberry (***Rubus*** *idaeus*). The wingless form is 2.6–4.1 mm long, pale green or yellowish green, with long legs and antennae (*see* antenna). Eggs hatch in spring and the aphids feed at the tips of leaf **buds**, moving later to the undersides of leaves. After two generations of wingless forms, winged forms appear in summer and migrate to other plants or other parts of the same plant. Eggs are laid in late autumn, usually not higher than 30 cm from ground level. The aphids act as vectors of a range of viral diseases.

raspberry beetle *See Byturus tomentosus.*

raspberry cane midge *See Reseliella theobaldi.*

raspwort *See Haloragis.*

rats *See* Muridae.

rat snake *See Elaphe obsoleta.*

Raunkiær, Christen Christensen (1876–1938) A Danish ecologist who devised a scheme for classifying plants according to the position of their **perennating buds** in relation to the soil surface. This, he believed, marked the **adaptations** plants had made as they spread away from the tropics, where he suggested the earliest **angiosperms** arose.

ravens *See* Corvidae.

ravine wind A wind that blows along a narrow valley or ravine.

raw gley soils Soils which form in material that has been waterlogged since its deposition. Such soils usually support no plants and occur in intertidal flats or saltings that are developing into **salt marshes**. They are a major group in the classification devised by the Soil Survey for England and Wales.

ray *See* medullary ray.

ray floret In a **capitulum** flower (e.g. in **Asteraceae**), a flower at the edge of the **inflorescence**, usually with only 1 petal.

Ray, John (1627–1705) An English naturalist who attempted to classify plants on the basis of their morphological differences and similarities. His three-volume *Historia generalis plantarum* (1686, 1688, 1704), each volume containing about 1000 pages, described more than 18,600 species, with information on their distribution, **ecology**, **germination**, growing habits, diseases, and, where appropriate, medicinal uses.

razor strop fungus *See Piptoporus betulinus.*

reaction centre The site, containing a type of **chlorophyll** *a*, at which the **absorption** of light energy triggers the transport of electrons for **photophosphorylation** in **photosynthesis**.

reading-frame shift The consequence of a **mutation** involving the insertion or excision of a **nucleotide**. In **transcription**, nucleotides are read in triplets, the reading frame being determined by a starting point. Adding or removing a nucleotide will cause the reading frame to shift one nucleotide forward or backward at that point, altering the composition of the triplets, and hence the **amino acids** produced, throughout the remainder of the **gene**.

reading mistake The insertion of an incorrect **amino acid** into a **polypeptide** chain during **protein** synthesis.

realized niche The **niche** that a species occupies when in the presence of **competition** for resources.

recalcitrant seed A seed that remains viable for only a short time, often one year or less, so it cannot be stored and the plant must be conserved by growing it. Many tropical plants produce recalcitrant seeds.

receptacle 1. The thickened part of the **peduncle** to which the parts of a **flower** are attached. In some plants the receptacle gives rise to the edible part of a **pseudocarp**. **2.** In algae (*see* alga), a swelling at the tip of a **branch** of the **thallus** that contains the **conceptacle**.

recessive gene A **gene** that is expressed in the homozygous (*see* homozygosity) condition, but masked by a **dominant gene** in the heterozygous (*see* heterozygosity) condition.

reciprocal cross **Cross-breeding** between organisms with different **genotypes** (call them A and B) in which male A × female B is followed by female A × male B (the reciprocal cross).

reciprocal genes Genes with no **alleles** that complement (i.e. reciprocate) each other.

reciprocal predation A relationship in which two species prey on each other.

recombinant A **genotype** carrying **gene** combinations that are not present in either parent due to **recombination**.

recombination The arrangement of **genes** in offspring in combinations different from those in either parent, and the assortment of **chromosomes** into different sets. It occurs through independent assortment (*see* Mendel's laws) during **crossing over**.

recombination frequency The number of **recombinants** divided by the number of progeny and expressed as a percentage.

recon The smallest segment of **DNA** that is capable of **recombination**.

recoverability *See* recreatability.

R

recreatability (recoverability, salvageability) The extent to which an **ecosystem** or natural community might recover or be re-established to a form indistinguishable from the original following a major disturbance.

rectiflorous Having the axes of the **florets** parallel to the axis of the **inflorescence.**

recurved Bent backward.

red algae (Rhodophyta) A group of algae (*see* alga), most of which are seaweeds consisting of membranous sheets of cells, although some are single-celled. Some deposit crystals of calcium carbonate in and around their **cell walls**, so they closely resemble corals. There are more than 5200 species, most common in tropical seas, but some found in fresh water. Some are edible; nori is obtained from *Pophyra* spp.

red and black froghopper *See Cercopsis vulnerata.*

red and black leafhopper *See Cercopsis vulnerata.*

red-bellied woodpecker *See Melanerpes carolinus.*

redbird *See Cardinalis cardinalis.*

red chestnut *See Aesculus.*

red clover (*Trifolium repens*) *See Trifolium.*

red core *See Phytophthora fragariae.*

red corn snake *See Pantherophis guttatus.*

redcurrant (*Ribes rubrum*) *See Ribes.*

red currant blister aphid *See Cryptomyzus ribis.*

red-eared slider *See Trachemys scripta.*

red earthworm *See Lumbricus rubellus.*

red fox *See Vulpes vulpes.*

red fucus *See* bladder wrack.

redhead earthworm *See Lumbricus rubellus.*

red-hot cattail (*Acalypha hispida*) *See Acalypha.*

red lily beetle *See Lilioceris lilii.*

Red List A list published by the **International Union for Conservation of Nature** that describes the status of species that are or may be at risk of extinction. It covers all mammals, birds, amphibians, sharks, reef-building corals, cycads, and conifers, and is being expanded to include reptiles, fishes, and some groups of invertebrates and plants.

redox carrier *See* electron carrier.

redox potential *See* oxidation-reduction potential.

redox reaction A chemical reaction that involves simultaneous **oxidation** and **reduction**, i.e. the oxidation of one substance involves the reduction of the other and vice versa.

red podzolic soils Soils formed by **podzolization** following a long period of **chemical weathering** and **leaching.** They resemble **podzols** but are more heavily weathered and contain a higher concentration of iron oxides, giving them their red colour. They occur in humid tropical environments.

red-seeded dandelion (*Taraxacum erythrospermum*) *See Taraxacum.*

red slug *See Arion rufus.*

redstarts *See* Turdidae.

red trout worm *See Dendrodrilus rubidus.*

reducing sugar A **sugar** that contains an aldehyde group (R-CHO, where R is a side chain) or is capable of forming one; they are responsible for the browning of certain foods. **Glucose** and **galactose** are reducing sugars.

reduction A chemical reaction in which atoms or molecules either gain hydrogen or electrons or lose oxygen.

reduction division *See* meiosis.

redundant cistron Among many copies of a **cistron** present on the same **chromosome** only one of which is expressed, one of the copies that is not expressed.

Reduviidae (ambush bugs, assassin bugs, three-legged bugs) A family of bugs (**Heteroptera**), 4–40 mm long, most with long legs and a prominent **rostrum** that fits into a ridged groove where it moves, making a rasping sound used to deter predators. Most reduviids are predators of other **Arthropoda**, seizing prey with the front legs. Three-legged bugs walk with their front legs raised. There are about 7000 species found throughout the world, but especially in the tropics.

red velvet mite *See Allothrombium fuliginosum.*

red wiggler *See Dendrodrilus rubidus, Eisenia fetida.*

red wiggler worm *See Dendrodrilus rubidus.*

red worm *See Eisenia fetida.*

reedmace *See Typha.*

refection *See* pseudorumination.

reflexed Bent back sharply.

refugium An area that has remained isolated from major changes, especially of climate, that have occurred elsewhere. Some of the plants and animals characteristic of a region survive adverse periods in refugia.

regeneration The regrowth of vegetation or re-establishment of communities following a disturbance that damaged or destroyed them.

regional diversity *See* gamma diversity.

regolith A general term describing unconsolidated surface material resulting from **weathering** and lying above unaltered, solid rock, e.g. rock fragments, sand, and other mineral grains.

regosols Soils that do not belong to any of the other reference soil groups in the **World Reference Base for Soil Resources.**

regular *See* actinomorphic.

regulating services *See* ecosystem services.

regulator gene In regulation by **operons**, a **gene** that, when transcribed (*see* transcription), produces a **protein** that switches off an **operator** gene and hence the operon. The regulator gene is not part of the operon and may not even be on the same **chromosome.**

reindeer *See* antlers.

reindeer moss (caribou moss) The **lichen** *Cladonia rangiferina*, a **fruticose** lichen with much-branched podetia (*see* podetium) up to 8 cm tall, that grows on well-drained soil, primarily in arctic **tundra.** It is an important food for reindeer (caribou, *Rangifer tarandus*).

relative humidity (RH) The ratio of the mass of water vapour in a unit mass of air to the mass required to saturate the air at that temperature, expressed as a percentage. This is the measure of **humidity** used in weather reports and forecasts.

relative molecular mass *See* molecular weight.

release factors **Proteins** that react to release a completed **polypeptide** from the **ribosome** when a termination **codon** is encountered during **transcription.**

relict soil *See* paleosol.

relief The surface features of a landscape expressed in terms of its vertical and horizontal dimensions, i.e. elevation, slope, and orientation.

remiges *See* remex.

remix (pl. remiges) One of the flight **feathers** of a bird, i.e. a **primary** or **secondary.**

removal time (residence time) The length of time that a molecule of a substance remains in a particular part of a **biogeochemical cycle**.

rendzina A **calcareous soil** that develops above **chalk**, **dolomite**, **limestone**, or unconsolidated calcareous material, and occasionally over **gypsum**. It contains gravel and stones, and so much calcium carbonate that it fizzes when dilute acid is dripped on to a sample.

reniform Kidney-shaped.

replaceability The ease and extent to which a damaged or lost **ecosystem** could be re-established. Ecosystems with low replaceability, i.e. it would be difficult or impossible to replace them, have a priority claim for **conservation**.

replacement ecology The planned establishment of a community that is entirely different from the original one, where it would be impossible to restore a site to its former condition. *Compare* restoration ecology.

replicase An **enzyme** that catalyzes the **replication** of **DNA** and **RNA**.

replication The formation of daughter molecules of **nucleic acid** from a parent molecule which serves as a template.

representativeness A measure of the extent to which a particular area is typical of the **ecosystem** in which it occurs.

repressor In regulation by **operons**, a **protein** produced by a **regulator gene** that switches off an **operator** gene and hence the operon.

reptant Creeping along the ground surface and taking root at intervals.

Reptilia (reptiles) A class of vertebrates that includes all **amniotes** other than **Aves** and **Mammalia**. Reptiles are **poikilotherms**, **extant** species with a body covered in scales, sometimes supported by **scutes** (some **fossil** reptiles were feathered). Young are air-breathing from the time of hatching and no larvae have gills; **ovovipary** is common. There are about 10,000 species found in all continents except Antarctica.

reptiles *See* Reptilia.

Resedaceae (order **Brassicales**) A family of **annual**, **biennial**, and **perennial herbs**, small **shrubs**, and climbers with leaves **alternate**, **stipulate**, **entire** or divided, sometimes absent. Flowers **zygomorphic**, usually **bisexual**, with 2–8 usually free **sepals** and **petals**, petals occasionally absent, 3–45 **stamens**, **ovary superior** of 2–7 more or less fused **carpels** with 1 **locule**. **Inflorescence** a **bracteate raceme** or **spike**. Fruit is a **capsule**, in one genus a **berry**. There are 3 genera of 75 species occurring in warm temperate and dry subtropical regions. *Reseda odorata* (mignonette) is cultivated for ornament and its oil used in perfumes.

Reseliella theobaldi (raspberry cane midge) A small, red-brown fly (**Diptera**) that overwinters in the soil and emerges in early summer. It lays eggs under the **bark** of new canes of ***Rubus*** crops. These hatch into pink larvae up to 4 mm long that feed on the canes for two to three weeks, then fall to the ground and pupate. A second generation emerges in later summer. The damage they cause allows infection by fungal diseases, especially raspberry cane blight. Summer-fruiting varieties are most at risk. The midge occurs throughout Europe.

reserve acidity The amount of lime that must be applied to raise the **pH** of an **acid soil** to 7.0 (neutrality).

reservoir pool A store of a nutrient element at a particular part of a **biogeochemical cycle**. Exchanges between a reservoir pool and an **active pool** are slow compared with the rate of flow through the active pool.

residence time *See* removal time.

resin A viscous liquid, consisting mainly of **terpenes**, that solidifies on exposure to air. Many woody plants secrete it but especially coniferous trees. It has many commercial uses in adhesives and varnishes.

resistance genes *See* R genes.

resorb Re-absorb; i.e. to metabolize substances or structures produced by the body, e.g. the resorbtion of non-viable foetuses by some mammals.

resource partitioning (differential resource utilization) The utilization of resources by ecologically similar species sharing a **habitat** in ways that avoid **competition**. They may rely on different resources or exploit the same resources but in different ways.

respiration 1. A sequence of chemical reactions in which food substances are degraded, with oxygen as the final acceptor in an **electron-transport chain** and **adenosine triphosphate** (ATP), carbon dioxide, and water as the products. The reactions are the opposite of those in **photosynthesis** and can be summarized as: $C_6H_{12}O_6 + 6O_2 \rightarrow 6CO_2 + 6H_2O$, with the release of about 3 MJ/mol of energy. *See also* fermentation. **2.** The transport of oxygen to tissues for respiration and of carbon dioxide from them.

respiration quotient (RQ) A dimensionless number (i.e. there are no units of measurement) calculated as the amount of carbon dioxide released divided by the amount of oxygen consumed in aerobic **respiration**.

resting phase *See* interphase.

Restio (family **Restionaceae**) A genus of **monocotyledon**, **evergreen**, **dioecious**, rush-like, **perennial herbs** with adult leaves reduced to sheaths. Flowers small, **actinomorphic**, **perianth** of two series of dry, often **hyaline** segments, 3 **stamens**, male flowers often with a rudimentary **ovary**, female flowers often with **staminodes**, **ovary superior** of 1–3 **carpels** and **locules**. **Inflorescence** a **spike**. Fruit **nut**-like or a **capsule**. There are 91 species occurring in Australia and South Africa. Some widely cultivated as structural garden plants.

Restionaceae (order **Poales**) A family of **monocotyledon**, **evergreen**, **caespitose**, usually **dioecious** plants with **rhizomes** or **stolons**. Stems photosynthetic, **dichotomous** or with **branches** in **whorls**. Leaves dimorphic (*see* dimorphism), juvenile plants with **blade**-like leaf, adult plants with leaves reduced to sheaths. Flowers **actinomorphic**, usually **unisexual** often with **sexual dimorphism**, or **bisexual**, **perianth** of 2 whorls of 3 **tepals**, 3 **stamens**, female flowers sometimes with **staminodes**, **ovary superior**. Flowers solitary subtended by 1 or 2 **bracts**, or in **spikelets** in a **spicate**, **racemose**, or **paniculate inflorescence**. Fruit is a **capsule** or **nut**. There are 58 genera of 500 species occurring in Africa, Madagascar, southeastern Asia, Australia, New Zealand, and Chile. Some cultivated for ornament.

restoration ecology The establishment of the original community on a disturbed site. *Compare* replacement ecology.

resupinate Upside-down.

resurrection plant A plant that is able to survive extreme and prolonged **desiccation**, becoming withered and apparently dead, but reviving when moistened. Many species behave in this way but the best known is *Selaginella lepidophylla* (rose of Jericho).

reticulate Having a network pattern.

reticulate evolution The emergence of many closely related **species**, usually through **polyploidy**; it is especially common in plants.

retrices *See* retrix.

retrix (pl. retrices) A tail **feather** of a bird.

retrotransposon A **transposable element** that contains **genes** allowing it to be copied into **RNA** then reinserted into the **genome** as **DNA**.

retrovirus An **RNA virus** that is able to make a **DNA** copy of itself and insert it into the **genome** of a host.

return flow *See* interflow.

reverse mutation (reversion) A **mutation** that cancels the effect of an earlier mutation, thereby restoring the ability of the affected **gene** to produce a functional **protein**.

reversion *See* reverse mutation.

revolute *See* vernation.

revolving storm A storm in which air circulates cyclonically (*see* cyclonic) around a centre of low pressure.

R genes (resistance genes) Plant **genes** that confer resistance to infection by **pathogens**.

RH *See* relative humidity.

rhabdites Rod-shaped structures in the **epidermis** of some **Turbellaria** species and certain other worms. Their function is unknown, but may be defensive; if the epidermis is irritated, rhabdites may be discharged and form a protective layer. The layer may also help in capturing prey.

R

Rhabditidae An order of nematodes (**Nematoda**), all of which are free-living parasites of animals or plants. There are 17 genera.

rhabditid nematode *See Phasmarhabditis hermaphrodita.*

Rhabdodendraceae (order **Caryophyllales**) A **monogeneric** family (*Rhabdodendron*) of **evergreen trees** and **shrubs** with **alternate**, **coriaceous**, **entire** leaves with **peltate** hairs on the underside. Flowers **actinomorphic**, **hermaphrodite**, **pentamerous** occasionally **tetramerous**, **sepals** small, **imbricate**, **petals** free, imbricate, 27–53 **stamens** in 3 **whorls**, **ovary superior** of 1 **carpel**. **Inflorescence** an **axillary raceme** or **racemose cyme**. Fruit is a **drupelet**. There are three species occurring in tropical South America.

rhachis *See* rachis.

rhagadiose Having deep chinks or cracks.

Rhamnaceae (order **Rosales**) A family of **trees**, **shrubs**, **lianas**, and **herbs**, often with spines. Leaves **alternate** occasionally **opposite**, **simple**, **entire** or **serrate**, **stipules caducous**. Flowers inconspicuous, **actinomorphic**, **bisexual** sometimes **unisexual**, sometimes **apetalous**, 4–5 **sepals**, or sometimes no **petals**, 4–5 **stamens**, **ovary superior** to **inferior** of 1–3 **carpels** with 2–3 free or 3–5 fused **locules**. **Inflorescence axillary** or terminal, **cymose** occasionally **racemose**. Fruit is a **drupe**, **nut**, occasionally **capsule** or **schizocarp**. There are 52 genera of 925 species with worldwide distribution. Some are dyeplants, others with medicinal properties. *Rhamnus catharticus* is buckthorn, *Zizyphus jujuba* is jujube, *Z. lotus* may be the lotus fruit.

rhatany (*Krameria lappacea*) *See* Krameriaceae.

rheotaxis A movement or change in direction in response to a water current.

rheotropism Movement or growth of a plant organ in response to a water current. *See* tropism.

Rheum (family **Polygonaceae**) A genus of **perennial herbs** usually with large, **deciduous** leaves growing from the base with a long **petiole**. Flowers small, **entomophilous**, **hermaphrodite**, **perianth campanulate** of 6 **sepaloid tepals**, 9 **stamens**. Fruit is an **achene**. There are about 60 species occurring in temperate and subtropical Asia. Several are cultivated. *Rheum ×hybridum* is rhubarb, *R. officinale* is medicinal rhubarb.

rhinarium The naked skin around the nostrils of a mammal.

Rhipogonaceae (order **Liliales**) A **monocotyledon**, **monogeneric** family (*Rhipogonum*) of **shrubs** and **lianas** that climb by means of their prickly stems. Leaves **alternate**, **distichous**, **opposite**, or **verticillate**. Flowers **actinomorphic**, **bisexual**, 6 free **tepals** in 2 **whorls**, 6 free **stamens**, **ovary superior** of 3 fused **carpels** with 3 **locules**. **Inflorescence** an **axillary** or terminal **spike**, **raceme**, or **panicle**. Fruit is a **berry**. There are six species occurring from New Guinea to New Zealand.

Rhipsalis (family **Cactaceae**) A genus of epiphytic (*see* epiphyte) cacti, known as mistletoe cacti, that may be erect, sprawling, or pendent, with **terete**, angular, or compressed, **succulent** stems. Flowers small, **actinomorphic**, with varying numbers of **perianth** segments, **stamens**, and **carpels**. Fruit is a **berry**. There are about 50 species occurring in tropical America, Madagascar, and Sri Lanka. Some are cultivated for ornament.

Rhizaria A large group of mainly single-celled, **amoeboid eukaryotes**. Most are marine, but they include many with shells that formed **limestone** and **chalk** rocks.

rhizina *See* rhizine.

rhizine (rhizina) In **lichens**, a structure formed from hyphae (*see* hypha) and resembling a root that usually serves to anchor the lichen to the substrate.

rhizobia Soil **Bacteria** that establish themselves in **root nodules** where they perform **nitrogen fixation**.

Rhizobiaceae A family of **Alphaproteobacteria** that are Gram-negative (*see* Gram reaction), usually rod-shaped, and **aerobes**. Many species are **rhizobia**. Others, e.g. ***Agrobacterium***, are plant pathogens. There are 8 genera with 40 species, found worldwide.

Rhizobium The largest genus of **Rhizobiaceae**, comprising **Bacteria** that occur worldwide in soil and in **root nodules** on plant roots, where they perform **nitrogen fixation**.

Rhizoctonia solani A species of **agaric fungi** that are **anamorphs** of *Thanatephorus* spp. They are **saprotrophs** but also opportunistic pathogens of a wide variety of plants, causing diseases such as damping-off, **canker**, and black scurf. The fungus occurs worldwide in soil and produces **basidiocarps** on decaying plant material.

rhizoid In mosses (**Bryophyta**) and liverworts (**Marchantiophyta**), a structure able to absorb water and minerals and that often serves to anchor the plant to the substrate.

rhizome An underground stem that creeps horizontally, bears roots and leaves, and usually persists from one growing season to the next.

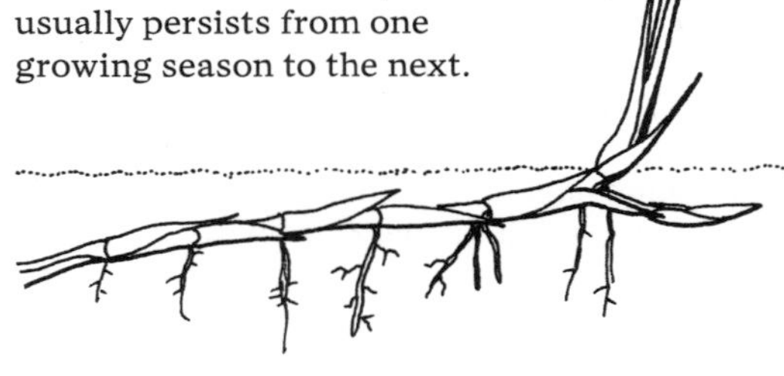

A rhizome is an underground stem.

rhizomorph A structure resembling a thick cord or thread that is composed of parallel fungal **hyphae**.

rhizomycelium In certain **Fungi**, a system of branched **hyphae** with few or no nuclei that form part of the **thallus**.

Rhizophoraceae (order **Malpighiales**) A family of **shrubs** and **trees**, some with prominent **stilt roots** and some with **pneumatophores**. Leaves usually **opposite** sometimes in **whorls**, **simple**, **entire**, **crenate**, or **dentate**, **stipules** conspicuous, **caducous**. Flowers **actinomorphic**, **bisexual** rarely **unisexual** (plants **monoecious**),

R

hypogynous (*see* hypogyny) to **epigynous**, 4–5 sometimes 5–16 persistent, **valvate**, **connate sepals**, 4–5 free or 5–16 fused **petals**, 8–10 or many free **stamens**, **staminodes** in **pistillate** flowers, **ovary** semi-**inferior** to inferior of 2–12 fused **carpels** and **locules**. **Inflorescence** a **cyme** or **raceme**, rarely flowers solitary. Fruit is a **berry** or **drupe**, or a dry, **indehiscent capsule**. There are 16 genera of 149 species with a pantropical distribution, mainly Old World. *Rhizophora*, *Bruguiera*, *Cassipourea*, and *Kandelia* spp. are mangroves.

rhizophore In spike mosses (Selaginellaceae), a leafless **branch** that grows downward from a fork in the stem and produces roots when it enters the soil.

rhizosphere The region of soil that surrounds and is affected by plant roots.

Rhododendron (family **Ericaceae**) A genus of **evergreen** or **deciduous shrubs** and **trees** with leaves **alternate**, **simple**, **entire** or **dentate**. Flowers tubular to **campanulate** with 5 lobes, 5–10 **stamens**. Fruit is an elongated **capsule**. There are more than 1000 species divided into 5 major and 3 minor subgenera, occurring mainly in the Himalayas and mountains of southeastern Asia, but some in Europe and North America and one in Australia. Many are cultivated for ornament. All are highly toxic as is honey from their flowers. ☠

rhoophilous Growing in creeks.

Rhopalosiphoninus ribesinus (currant stem aphid) A species of aphids (**Aphididae**) that feed on ***Ribes*** species, especially blackcurrant (*Ribes nigrum*) and redcurrant (*R. rubrum*). Males are wingless. It is **hygrophilous** and inhabits the lower shoots, near the ground and is most common in regions with a wet climate and in shady locations throughout Europe. It causes leaves to curl and brownish or red lumps to form on their upper surfaces.

rhubarb (*Rheum* ×*hybridum*) *See* Polygonaceae, *Rheum*.

Rhus (family **Anacardiaceae**) A genus of **shrubs**, small **trees**, and climbers (sumacs) with spirally arranged, usually **pinnate** sometimes **trifoliate** or **simple** leaves. Flowers **pentamerous**, borne in dense **panicles** or **spikes**. Fruit is a **drupe**, forming in dense clusters. There are about 250 species occurring in subtropical and temperate regions, especially Africa and North America. Some cultivated for ornament, but all produce a highly irritant exudate. *Rhus toxicodendron* (*Toxicodendron radicans*) is poison ivy, *R. diversiloba* (*T. diversilobum*) is poison oak, *R. vernix* (*T. vernix*) is poison sumac.

Rhytisma A genus of **ascomycete fungi** that are parasites of **deciduous** trees, causing tar spot on their leaves. There are about 18 species occurring throughout Europe and North America.

rib In vertebrates, a bone of the **axial skeleton** that articulates with a vertebra.

Ribautiana tenerrima (bramble leafhopper, loganberry leafhopper) A leafhopper (**Cicadellidae**), 3.0–3.5 mm long, that is mainly bright yellow. It is active in summer and found on ***Rubus*** species, including raspberries, loganberries, and cultivated brambles. It also feeds on apple and plum trees and hop vines. It causes mottling of leaves and reduces **photosynthesis**. The insect is very common throughout the temperate Northern Hemisphere and in Australasia.

Ribes (family **Grossulariaceae**) The only genus in the family, comprising **shrubs**, many species of which are grown for their fruit, e.g. redcurrant (*R. rubrum*), blackcurrant (*R. nigrum*), gooseberry (*R. uva-crispa*). There are about 150 species occurring throughout the temperate Northern Hemisphere and in the mountains of Central and South America.

ribitol A 5-carbon **alcohol** ($C_5H_{12}O_5$) formed by the **reduction** of **ribose**, that is an important part of the structure of **flavins**, **flavin adenine dinucleotide** (FAD), and **flavin mononucleotide** (FMN).

riboflavin Vitamin B_2, a compound formed from **ribitol** and **flavin** that occurs widely and is part of the structure of **flavin adenine dinucleotide** (FAD), and **flavin mononucleotide** (FMN).

ribonuclease An **enzyme** that catalyzes the **hydrolysis** of **RNA**.

ribose (ribulose) A **monosaccharide** ($C_5H_{10}O_5$) that is the **carbohydrate** component of **RNA**.

ribosome A molecule composed of two subunits, one of **RNA** (ribosomal RNA) and the other of several dozen **proteins**. Ribosomes are found in large numbers in all cells and are the sites of protein synthesis. The subunits join when a ribosome attaches to **messenger-RNA**, which receives **amino acids** from a molecule of **transfer-RNA**, from which proteins are assembled. Ribosomes are formed in the **nucleolus** but occur mainly in the cell **cytoplasm** either singly or in chains (**polysomes**), or attached to the **endoplasmic reticulum**, which is then called rough ER. Ribosomes are of two types: 70S ribosomes occur in **prokaryotes**, **chloroplasts**, and mitochondria (*see* mitochondrion); the larger 80S ribosomes in **eukaryotes**.

ribovirus Any **RNA virus** other than a **retrovirus**.

ribulose *See* ribose.

ribulose-1,5-biphosphate (RuBP) A compound formed joining a phosphate group from **adenosine triphosphate** (ATP) to ribulose phosphate (RuP). The double phosphate is then able to act as a carbon dioxide acceptor in the first stage of the **Calvin cycle**.

ribulose-1,5-biphosphate carboxylase (RuBisCo, rubisco) The **enzyme** that catalyzes the first step of carbon **fixation** in the **Calvin cycle**, which is the carboxylation of **ribulose-1,5-biphosphate** (RuBP). RuBisCo will also oxygenate RuBP, causing **photorespiration** to take the place of **photosynthesis**. Because it is present in every photosynthesizing plant and also assists in decomposition because it is not affected by temperature or pH, RuBisCo is believed to be the most abundant **protein** on Earth.

Ricinus (family **Euphorbiaceae**) A **monotypic** genus (*R. communis*, castor oil plant), a small **tree** with glossy, **alternate**, **palmate** with 5–12 segments, **dentate** leaves with long **petioles**. Flower **monoecious**, **apetalous**, **inflorescence paniculate**. Fruit is a bean-like **capsule** (castor bean), from which castor oil (a laxative but also industrial lubricant, e.g. Castrol) is obtained; it is also the source of the poison ricin. The plant occurs in Africa but is widely cultivated.

rictal bristle A stiff, modified **feather** with little or no **rhachis** found around the eyes and beaks of many insectivorous birds.

ridge A narrow protrusion from a centre of high **atmospheric pressure** which extends into a region of lower pressure.

rill A narrow, shallow channel that forms on slopes where **surface runoff** has cut into soil or soft rock. It is an early sign of **erosion** and may develop into a **gully**.

rill-wash Eroded material (*see* erosion) that flows intermittently along narrow channels or **rills**, carried by **surface runoff**.

rime ice A layer of white ice with an irregular surface.

Rimi *See* methoxyfenoxide.

ringed snake *See Natrix natrix*.

ringneck snake *See Diadophis punctatus*.

ring spot A disease of brassicas (**Brassicaceae**) in which circular brown or purple

R

patches, up to 25 mm across, speckled with concentric rings of black spots, appear on leaves, followed by the death of affected leaves. The disease is caused by the fungus *Mycosphaerella brassicicola*. It overwinters in crop debris and on wild plants.

riparian Pertaining to a river bank.

ripening A process in which **fruit** becomes softer and sweeter, often with a change in colour to advertise the increased palatability.

rithron The section of a river where the water is shallow, cold, and flows rapidly with a broken surface.

RNA (ribonucleic acid) A **nucleic acid** comprising a D-ribose sugar and the purine bases adenine and **guanine** and the pyrimidine bases **cytosine** and uracil. RNA occurs as **messenger-RNA**, ribosomal RNA, and **transfer-RNA**. *See* appendix: The Genetic Code.

RNA interference The regulation of the activity of **genes** by **small RNA** molecules. The binding of such a molecule to **messenger-RNA** that is subsequently degraded, or to **DNA** causing **gene silencing**, results in downregulation. Upregulation occurs when small RNAs interact with gene promoters.

RNA virus A **virus** in which the genetic material consists of **RNA**.

robin *See Erithacus rubecula.*

robin's pincushion gall (rose bedeguar, moss-gall) A crimson or red and green **gall** that forms on dog rose (*Rosa canina*) and field rose (*R. arvensis*), induced by the gall wasp *Diplolepis rosae*. The female lays her eggs in leaf **axils** or **buds**, although the galls appear to grow from the stem or twig. Each egg contains 1 larva, and up to 30 galls appear in June, each about the size of a pea.

rock An aggregate of **minerals** or organic matter that may be consolidated or unconsolidated. It forms by the crystallization of **magma**, forming **igneous** rock; by recrystallization, forming **metamorphic rocks**; by the accumulation of particles, forming **sedimentary rock**; and by the aggregation of **fossil** fragments **cemented** by **calcite** or fine mud, or as an aggregation of partially decayed organic matter, forming coal. To a geologist, soil is a type of rock.

rock bee (*Apis dorsata*) *See Apis.*

rock cress *See Arabis.*

rock crystal *See* quartz.

rock dove *See Columba livia.*

rock fall *See* fall.

rock moss (*Andreaea*) *See* protonema.

rock phosphate (phosphate rock, phosphorite) A **sedimentary rock** that is rich in phosphate, usually in the form of carbonate hydroxyl fluorapatite, $Ca_{10}(PO_4[CO_3])_6F_{2-3}$. It is processed to make phosphate **fertilizer**.

rock rose *See Helianthemum.*

rock tripe The **lichen** *Lasallia pustulata* that has conspicuous pustules or blisters across its surface that are **pruinose** near the centre. The lichen occurs on rocks mainly in upland regions and has been used for dyeing and as a survival food.

rockweed (*Ascophyllum nodosum*) *See* egg wrack; (*Fucus vesiculosus*) *See* bladder wrack.

Rodentia (rodents) An order of mammals that have only one pair of incisor teeth in each jaw. These grow continually throughout life and must be worn down by gnawing, but retain their chisel-like shape and sharp edges because they have a thick layer of enamel on the front but not on the back. There is a **diastema** to either side of the incisors and the canines and first premolars are absent. A rodent can draw a fold of skin into its diastema to divide the mouth into two sections; in some species these folds form cheek pouches. Rodents have claws and a **plantigrade** or semi-plantigrade gait. They vary in size from the South American capybara average weight 50 kg to the dormouse, weighing 23–43 g. There are about 1500 species of rodents found in all continents except Antarctica.

rodenticide A chemical compound that kills rodents (**Rodentia**), especially rats and mice.

rodents *See* Rodentia.

roe deer (*Capreolus capreolus*) *See* Cervidae.

roguing Removing inferior or infected plants from a stand by hand.

rooks *See* Corvidae.

rooster combs *See* sulphur fungus.

root The part of a vascular plant (**Tracheophyta**) that is usually below ground level and through which the plant absorbs water and nutrient minerals, and that anchors the plant.

root cap Tissue covering the tip of a root **apex**.

rooted tree A **phylogenetic tree** that usually includes an **outgroup**, thus allowing the common ancestor to be clearly identified.

root fly *See Delia radicum.*

root graft Roots of adjacent plants that become interconnected.

root hair A thin-walled **trichome** arising from a single **trichoblast** on the **epidermis** of a **root**. It increases the surface area of root available for the **absorption** of water and dissolved nutrients.

root-knot nematodes Sedentary **endoparasite** nematodes (**Nematoda**) in which upon hatching second-stage juveniles enter a plant root to feed. It grows and moults to a third-stage juvenile and if it is female it is no longer able to leave the root. It continues to the fourth juvenile stage and as it grows the cells around its head enlarge and form nurse cells, typically causing **gall** formation. The mature female lays up to several hundred eggs in a gelatinous egg mass from the posterior of her body. Her body remains inside the root and the eggs may be partly exposed. *Compare* cyst-forming nematodes.

root-lesion nematodes *See Pratylenchus.*

root nodule (actinorrhiza) A small growth, resembling a **gall**, on the root of certain plants caused by infection by symbiotic (*see* symbiosis) **Bacteria** that form colonies. The bacteria fix atmospheric **nitrogen** (*see* nitrogen fixation), much of which becomes available to the plant, and the colonies receive **carbohydrate** products of **photosynthesis** in return. Legumes (**Fabaceae**) have root nodules containing *Rhizobium* or ***Bradyrhizobium*** species and nodules containing **Actinobacteria** occur on a range of other plants including alder (***Alnus*** spp.), bog myrtle (*Myrica gale*), sea buckthorn (*Hippophaë* spp.), sumach (*Coriara* spp.), California lilac (*Ceanothus* spp.), etc.

root pressure **Osmotic pressure** that develops in the **xylem** of certain vascular plants (**Tracheophyta**) when soil moisture is high, causing fluid to rise and to exude from cut shoots.

root rot *See Phytophthora cinnamomi, Pythium.*

root-shoot ratio The ratio of the dry weight of the shoots of a plant to the dry weight of the roots. A plant with a high proportion of roots is better able to obtain water and nutrient minerals, while one with a high proportion of shoots is better able to obtain light for **photosynthesis**. The ratio is one indicator of the health of a plant.

root tuber *See* tuber.

Roridulaceae (order **Ericales**) A **monogeneric** family (*Roridula*) of small, insectivorous, **evergreen shrubs** with **alternate, linear** to **lanceolate, exstipulate, entire** or **laciniate** leaves with **capitate** hairs that secrete a resin. Flowers **actinomorphic, pentamerous, hermaphrodite, ovary superior** of 3 fused **carpels. Inflorescence** a terminal **raceme.** Fruit is a **loculicidal capsule.** There are two species **endemic** to southern Africa.

Rosa (family **Rosaceae**) A genus of mainly **deciduous**, prickly **shrubs** and climbers with **alternate, pinnate, stipulate, serrate** leaves. Flowers solitary, **pentamerous** (*R. sericea* **tetramerous**), with 5 or 4 **sepals** beneath the **petals**, many free **stamens**, many **superior ovaries** with free **carpels.** Fruit is an **achene** aggregated into a **berry**-like structure (rose hip) in a cup formed by the **calyx** tube. There are more than 100 species most occurring in Asia, but also in Europe, northwestern Africa, and North America. It is possibly the most widely cultivated of all plants, with many complex **hybrids**.

Rosaceae (order **Rosales**) A family of **deciduous** or **evergreen trees, shrubs, perennial herbs**, and climbers, many with prickles, with **alternate** rarely **opposite, simple** or **compound pinnate** or **palmate**, usually **stipulate** leaves. Flowers **actinomorphic**, usually **bisexual** occasionally **unisexual** (plants **dioecious**), 4- or 5-merous, **perigynous**, 5 free or 3–5 or 5–10 fused **sepals** and **petals** rarely absent, petal numbers increased in some **cultivars** by replacing **stamens** or **styles** with **petaloid** organs, 2, 3, or more times as many stamens as petals, **ovary superior**, semi-**inferior**, or inferior with many **carpels. Inflorescence** terminal or **axillary, racemose, cymose**, or **paniculate**, or flowers solitary. Fruit is a **drupe, drupelet**, aggregate of **achenes, capsule**, or **follicle**. There are 90 genera of 2520 species with a worldwide distribution. Many cultivated for ornament or for their fruits.

Rosales An order of plants comprising 9 families or 261 genera and 7725 species. *See* Barbeyaceae, Cannabaceae, Cynomoriaceae, Dirachmaceae, Elaeagnaceae, Moraceae, Rhamnaceae, Rosaceae, Ulmaceae, and Urticaceae.

rose aphid *See Macrosiphum rosae.*

rose bedeguar *See* robin's pincushion gall.

rose-breasted grosbeak *See Pheucticus ludovicianus.*

rosemary (*Rosmarinus*) *See* Lamiaceae.

rose of Jericho (*Selaginella lepidophylla*) *See* resurrection plant.

rose scale *See Aulacapsis rosae.*

rose shoot sawfly *See Ardis brunniventris.*

rose tip infesting sawfly *See Ardis brunniventris.*

Rossby waves (long waves, planetary waves) Waves with wavelengths of 2000–4000 km) which develop in ocean currents and in moving air in the middle and upper **troposphere**.

rosette plant 1. A plant with leaves that spread horizontally from a short stem, so they lie close to ground level, e.g. daisy (*Bellis perennis*). **2.** A **hemicryptophyte** in which leaves occur only at the base of the stem.

rostrum In true bugs (**Hemiptera**), an extension to the head, resembling a snout, that carries the specialized mouthparts used for piercing and sucking.

rosy apple aphid *See Dysaphis plantaginea.*

rosy-tipped worm *See Aporrectodea rosea.*

rotate Wheel- or disc-shaped.

rotenone An **insecticide** and **piscicide** extracted from the stems and seeds of several plants and used to kill a wide variety of arthropods (**Arthropoda**).

Rotifera (wheel animalcules) A **phylum** of spherical, pseudocoelomate (*see* pseudocoelcom) animals, most 0.1–0.5 mm across, that move by means of cilia (*see* cilium). They inhabit freshwater environments and films of water coating soil particles and feed on decomposing organic material, single-celled algae (*see* alga), and other small organisms; they are also cannibalistic. There are about 1800 species.

rotund disc *See Discus rotundatus.*

rough earth snake *See Virginia striatula.*

rough ER *See* endoplasmic reticulum.

rough green snake *See Opheodrys aestivus.*

rough strawberry root weevil *See Otiorhynchus rugostriatus.*

round back slugs *See* Arionidae.

round dance *See* dance language.

round snail *See Discus rotundatus.*

roundworms *See* Nematoda.

Rousseaceae (order **Asterales**) A family of **evergreen trees** and **lianas** with **alternate** or **opposite**, **serrate** leaves. Flowers **actinomorphic**, **bisexual**, with **valvate sepals**, 5–6 **stamens**, **ovary superior** of 3–7 **carpels** and **locules**. **Inflorescence** a **panicle**. Fruit is a **berry** or **capsule**. There are 4 genera of 13 species occurring in Mauritius and from New Guinea to New Zealand.

rove beetles *See* Staphylinidae.

rowan (*Sorbus aucuparia*) *See Sorbus.*

RQ *See* respiration quotient.

***r*-selection** A reproductive strategy favoured by **natural selection** for species living in an **environment** prone to rapid change and opportunists at the early stage of a **succession**. Such species maximize their intrinsic rate of increase (*r*) by producing large numbers of small seeds or offspring whenever conditions are favourable, thereby rapidly colonizing the **habitat**.

rubber buttons *See Bulgaria inquinans.*

rubber tree (*Hevea brasiliensis*) *See* Euphorbiaceae.

Rübel, Eduard August (1876–1960) A Swiss phytogeographer (*see* phytogeography) who assisted Josias **Braun-Blanquet** in developing the classification system devised by A. F. W. **Schimper** into the scheme used by the Zürich-Montpellier School of **Phytosociology**.

Rubiaceae (order **Gentianales**) A family mainly of small **trees** and **shrubs**, but also **annual** or **perennial herbs**, **lianas**, **epiphytes**, and **geofrutices**, with some **succulents** and aquatics, and some myrmecophiles (*see* myrmecophily). Leaves **opposite** sometimes whorled, **simple**, usually **entire**, **stipulate**. Flowers **bisexual** or **unisexual** (plants usually **dioecious**), **tetramerous** or **pentamerous**, **calyx adnate** to the **ovary**, **corolla** tubular, **actinomorphic** sometimes **zygomorphic**, **stamens epipetalous** as many as corolla lobes, ovary **inferior** of usually 2 but sometimes 5 or more **carpels**. **Inflorescence** variable, often a **thyrse**. Fruit a **berry**, **drupe**, or **capsule**. There are 611 genera of 13,150 species with a worldwide, mainly tropical, distribution. *Coffea* (coffee) is the most economically important genus, *Cinchona* is the source of quinine, *Gardenia* is cultivated for perfumery.

RuBisCo *See* ribulose-1,5-biphosphate carboxylase.

RuBP *See* ribulose-1,5-biphosphate.

Rubus (family **Rosaceae**) A genus of **shrubs**, **herbs**, and rambling plants, most with prickly stems and **compound** leaves of 3–7 **leaflets**. Flowers usually **hermaphrodite**, **pentamerous** with many **stamens** and **carpels**. Carpels sit on a conical **receptacle** and form clusters of 1-seeded **drupelets** aggregated into the fruit.

Classification of the genus is complicated because of the extent of **polyploidy**, **apomixis**, and hybridization (*see* hybrid), making it difficult to determine what is or is not a species, but there are hundreds, possibly thousands of species, with a **cosmopolitan distribution**, but mainly in northern temperate regions. They include blackberry, raspberry, dewberry, and cloudberry.

ruby-throated hummingbird *See Archilochus colubris.*

rubythroats *See* Turdidae.

ruderal Associated with human dwellings, or a plant that occurs around human dwellings or farms or that grows on waste ground.

rufous Reddish brown.

rufous hummingbird *See Selasphorus rufus.*

rugose Ridged or wrinkled.

Rumina decollata (decollate snail) A terrestrial **snail** with a light brown conical shell up to 45 mm long and 14 mm wide, **decollate** in adults because they deliberately chip off the tip of the shell. The snail is a predator of other snails and **slugs** and their eggs. It is native to the Mediterranean region but has been introduced into other parts of Europe, North, Central, and South America, and the Caribbean for **biological control**.

R

ruminate Describes **endosperm** that is pale and marked with an irregular pattern of dark lines.

runcinate Saw-toothed.

runner A **stolon** that produces roots at its **apex** from which a new plant grows.

Runner *See* methoxyfenoxide.

runoff *See* surface runoff.

Ruppiaceae (order **Alismatales**) A **monogeneric**, **monocotyledon** family (*Ruppia*) of usually **annual**, submerged **herbs** with lower stems **rhizomes** and upper stems floating. Leaves **alternate** or **opposite**, **simple**, **linear**, slightly **serrulate**. Flowers **actinomorphic**, **bisexual**, lacking **tepals**, 2 **stamens**, **ovary superior** of usually 4 free **carpels**. **Inflorescence capitate** with a terminal **spike**. Fruit is a **drupe**. There are up to ten species with a more or less worldwide distribution.

rush *See* Juncaceae, *Juncus.*

rushrose *See Helianthemum.*

rusts A group of fungal plant diseases that can affect most cultivated plants. They produce pale spots that develop into pustules bearing **spores** most often on leaves, but also on stems, flowers, and fruit. The pustules are often rust-coloured, but may also be black, white, brown, orange, or yellow, and there may be many pustules on a single leaf. Infection often, but not always, reduces the vigour of the plant. Rust diseases are caused by **obligate parasites** belonging to the **Pucciniomycetes**.

Rutaceae (order **Sapindales**) A family of mostly small, **evergreen trees**, also **shrubs** and woody climbers, most aromatic, with **alternate** rarely **opposite** or whorled, usually **compound** 1-foliate or 3-foliate, or **simple** and **entire** to **bipinnatisect** leaves. Flowers **actinomorphic** occasionally somewhat **zygomorphic**, **bisexual** rarely **unisexual** (plants **dioecious**), **pentamerous** or usually 4-, sometimes 2- or 3-merous, usually hypogynous (*see* hypogyny), usually 5 **sepals**, 5 **petals**, as many **stamens** as petals or sometimes 2 stamens and 3 **staminodes**, **ovary superior** of 4–5 free or 1–2 or 5–10 fused **carpels** each with 1 **locule**. **Inflorescence** variable. Fruit is a **capsule**, **schizocarp**, **drupe**, or **berry**, or in ***Citrus*** and related genera a **hesperidium**. There are 161 genera of 2070 species with a largely tropical distribution. *Citrus* species widely cultivated for their fruits, others for their oils.

S

S *See* sulphur.

Sabiaceae (order **Proteales**) A family of **evergreen** or **deciduous trees** and **lianas** with **alternate**, **simple**, **ovate** to **elliptical**, usually **acuminate**, **entire** to minutely **dentate**, **exstipulate**, **petiolate** leaves. Flowers **actinomorphic**, **bisexual**, 5- to 7-merous, 5 **epipetalous stamens**, **ovary superior** of 2 **carpels**. **Inflorescence** an **axillary cyme**, cymes aggregated into a **thyrse**, or flowers solitary. Fruit is a **drupe**. There are 3 genera with 100 species occurring from southeastern Asia to Malesia and in tropical America.

saccate Bag- or sac-like.

Saccharina latissima *See* sea belt.

Saccharomycetales *See* Saccharomycotina.

Saccharomycotina (Saccharomycetales) A subphylum of **ascomycete fungi** that consists of **yeasts** and comprises the single class Saccharomycetes. Most reproduce asexually by **mitosis**, producing a **bud** into which a second nucleus migrates. The bud grows and when it reaches adult size separates from the parent cell. These yeasts produce carbon dioxide and are used in **fermentation**, e.g. of bread and alcoholic beverages. Others cause disease in animals and some are plant pathogens.

saccharose *See* sucrose.

Saccharum (family **Poaceae**) A genus of large, robust, **perennial**, reed-like grasses with elongate, flat leaf **blades**. Flowers in pairs of **spikelets**, one **bisexual**, **sessile**, the other with a **pedicel**, **deciduous**, often with no **awn**, forming a **paniculate raceme**. There are two species (*S. spontaneum* and *S. robustum*) that occur in the wild in temperate and tropical southeastern Asia, and four (*S. barberi*, *S. edule*, *S. officinarum*, and *S. sinense*) that are **cultivars** which do not survive in the wild but have been accorded the status of species. Some cultivated for ornament. *Saccharum officinarum* is sugar cane.

sac fungi *See* Ascomycota.

sacred lotus (*Nelumbo nucifera*) *See* Nelumbonaceae.

saddle *See* clitellum.

saddle fungus *See Polyporus squamosus*.

Saffir-Simpson hurricane scale A five-point scale introduced in 1955 to extend the **Beaufort wind scale** to cover

Saffir-Simpson Hurricane Scale

CATEGORY	WIND MPH	WIND KNOTS	WIND KM/H	DAMAGE
1	74–95	64–82	119–153	some damage
2	96–110	83–95	154–177	extensive
3	111–129	96–112	178–208	devastating
4	130–156	113–136	209–251	catastrophic
5	157 or more	137 or more	252 or more	catastrophic

hurricanes. The wind speeds refer to sustained winds.

saffron A flavouring and intense yellow colouring obtained from the dried **stigmas** of ***Crocus*** *sativus*.

sage *See* Lamiaceae, *Salvia*.

sage leafhopper *See Eupteryx melissae*.

sagittate Shaped like an arrow head.

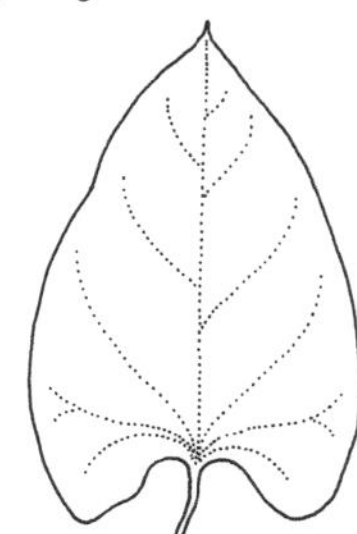
A sagittate leaf.

sago palm (*Cycas revoluta*) *See* Cycadaceae.

sailor beetles *See* Cantharidae.

St John's-wort (*Hypericum*) *See* Hypericaceae.

Saintpaulia (African violet) *See* East African steppe floral region.

salamanders *See* Caudata, Salamandridae.

Salamandridae (newts, 'true' salamanders) A family of amphibians **(Amphibia)** comprising animals with cylindrical bodies, well-developed legs and tails, lungs, and palatal teeth (teeth born on the **palate**). All species secrete skin toxins; newts are especially poisonous. Most salamandrids are less than 200 mm long and many are brightly coloured. All newts and most salamanders are aquatic as larvae. Adult salamanders are terrestrial, adult newts are either partly or wholly aquatic. Salamanders have smooth skin, newts have dry skin with a rough texture. Salamanders occur mainly in Europe, with a few species in the Middle East and North Africa. Newts occur in Europe, China, southeastern Asia, northwestern Africa, and North America. There are 15 genera with about 55 species.

Salicaceae (order **Malpighiales**) A family of **subshrubs** to **shrubs** and **evergreen** or **deciduous trees** with **opposite** or spiral, **simple**, **linear** to **orbicular**, **entire** to **serrate** or lobed leaves, with **stipules** often **caducous**. Flowers **unisexual** (plants **dioecious**), each subtended by a **bract**, **trimerous** to **hexamerous**, **sepals** and **petals** sometimes absent, up to 60 **stamens**, **ovary superior** of 2 free or 2–4 fused **carpels** with 1 **locule. Inflorescence** variable, often an erect or pendent **catkin**. Fruit is a **capsule**. There are 55 genera with 1010 species with a pantropical and temperate distribution. Many species of ***Salix*** are cultivated.

salic horizon A surface or shallow subsurface **soil horizon** at least 15 cm thick that is enriched in readily soluble salts, i.e. salts that are more soluble than **gypsum**. It is a **diagnostic horizon**.

Salicornia (family **Amaranthaceae**) A genus of **annual**, **halophyte**, **succulent herbs** (glasswort) and some woody **perennials** with small, scale-like leaves fused to the stem and to each other, with only the tips visible. Flowers solitary, **hermaphrodite**, 1–2 **stamens**, 1 **carpel**. There are up to 60 species with a **cosmopolitan distribution** on beaches, mud flats, **salt marshes**, and among mangroves. They use the **C4 pathway** of **photosynthesis** and their high internal suction pressure allows them to extract fresh water from strongly saline solutions. Some species edible pickled (pickleweed) or fresh (samphire).

Salix (family **Salicaceae**) A genus of **deciduous trees** and **shrubs** with leaves **alternate**, **simple**, **linear** to **lanceolate**, usually **serrate**, often **acuminate**, **stipules** sometimes prominent. Plants **dioecious**, flowers as **catkins**, without **calyx** or **corolla**, 2–10 **stamens**, **ovary** of 1 **carpel**. There are about 400 species occurring mainly on moist soils in northern temperate and cold regions. Many cultivated as willow, sallow, or osier. *See* arctic scrub.

salination *See* salinization.

saline-sodic soil Soil containing more than 15 percent exchangeable sodium, with a **pH** below 8.5, a **sodium-absorption**

ratio greater than 13, electrical conductivity greater than 0.4 siemens per metre, and a poor physical condition. The high pH and salinity inhibit the growth of most plants.

saline soil Soil containing sufficient salt to inhibit plant growth. It has a **pH** below 8.5, a **sodium-absorption ratio** below 13, electrical conductivity from 2 to more than 4 siemens per metre, and a normal physical condition.

salinization (salination) The accumulation in soil of soluble salts, usually by upward movement by **capillarity** from saline **groundwater** followed by **evaporation**.

Salix herbacea (least willow) *See* arctic-alpine species.

S alleles *See* S genes.

sallow *See Salix.*

SALR *See* saturated adiabatic lapse rate.

salsify *See Tragopogon.*

salt The product of a chemical reaction between an acid and a base, composed of cations (positively charged **ions**) and anions (negatively charged ions) bonded ionically to produce a molecule with no net charge. Common salt is sodium chloride (Na^+Cl^-).

saltation A motion in which a particle rises steeply, travels horizontally, and descends gently. It is one of the most important processes of particle transport in air and water.

Salticus scenicus (zebra spider) A species of spiders marked with distinctive black and white stripes, and with two very large eyes and six smaller ones. It does not build a web, instead stalking its prey and pouncing on it when it is within range. Prior to jumping, the spider attaches a silk thread to the substrate; if its leap misses, it climbs back up its thread. Zebra spiders are acutely aware of humans watching them, often raising their heads and altering their behaviour in response. The spiders are 5–7 mm long and prey on other spiders and insects, some much larger than themselves. They occur throughout **Holarctica**, often close to human habitations.

salt marsh An area of mud banks found in an estuary where it is immersed by tides, with a pattern of vegetation that forms zones related to the duration of each immersion. The plants are **halophytes** with varying tolerances for salt water.

salt marsh snail *See Mysotella myosotis.*

salt stress The inhibition of growth plants experience when exposed to excess salt. The stress arises through direct toxicity and through dehydration caused by the low **osmotic potential** of the soil solution.

saltwort *See* Bataceae.

Salvadoraceae (order **Brassicales**) A family of **shrubs** and small **trees**, some with thorns, with **opposite**, **coriaceous**, **simple**, **entire** leaves. Flowers **actinomorphic**, **hermaphrodite** or **unisexual** (plants **dioecious** or **polygamodioecious**), 2–4 free or 5 fused **sepals**, 4 free or 5 fused **petals** and **stamens**, **ovary superior** of 2 fused **carpels** with 1 or 2 **locules. Inflorescence** a terminal or **axillary raceme.** Fruit is a **berry** or **drupe**. There are 3 genera of 11 species occurring from Africa to southeastern Asia and western Malesia. Some are edible.

salvageability *See* recreatability.

Salvia (family **Lamiaceae**) A genus of **annual**, **biennial**, and **perennial herbs** and **subshrubs** with usually **entire** sometimes **dentate** or **pinnate** leaves. Foliage usually aromatic. Flowers with a tubular or **campanulate calyx, corolla** strongly 2-lipped, the upper lip often arched over the 2 **stamens. Inflorescence** a **raceme**

or **panicle**. Fruit is a **nutlet**. There are 700–900 species distributed widely in temperate and tropical regions. Many are cultivated as herbs or for ornament; all are known as sage.

samara A winged **nut** or **achene** with one seed.

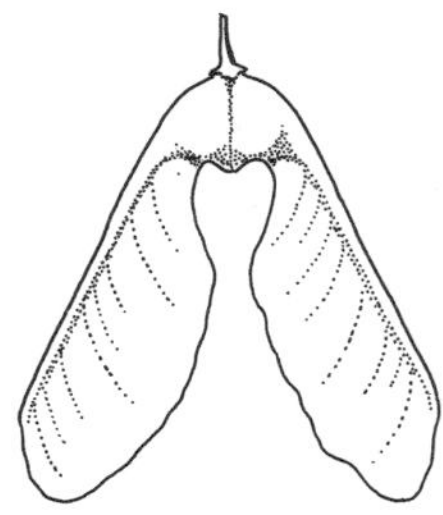

Samara fruit (maple, *Acer* spp.).

Sambucus (family **Adoxaceae**) A genus of **deciduous shrubs**, small **trees** (elder), and **perennial herbs** with **opposite**, **serrate**, **pinnate** leaves with 5–9 or rarely 3 or 11 **leaflets**, occasionally **bipinnate**, leaflets **ovate** to **lanceolate**. Flowers **actinomorphic**, **unisexual**, **calyx** small and inconspicuous, **corolla** with a short tube and 5 free or 3–5 fused lobes, as many **stamens** as corolla lobes, **ovary** with 3–5 **locules**. **Inflorescence** an **umbel**-like or **paniculate** cluster. Fruit is a **baccate drupe**. There are 5–30 species occurring widely in temperate and tropical regions. The fruit of *S. nigra* are edible; its fruit and flowers are made into cordial or wine.

samphire *See Salicornia.*

sand Mineral particles 62.5–2000 µm in size in the Udden-Wentworth scale or 60–2000 µm in the British scale. *See* particle sizes.

sandalwood tree (*Santalum album*) *See* Santalaceae.

sand auger *See* dust whirl.

sand devil *See* dust whirl.

sandstone (arenite) A **sedimentary rock** formed by the **lithification** of **sand** grains in a mud matrix and mineral cement.

sandwort *See Arenaria.*

Sansevieria (family **Asparagaceae**) A **monocotyledon** genus of **perennial herbs** with leaves as a rosette or **distichous**, species originating in arid regions with hard, sometimes **succulent** leaves, those from more humid tropical and subtropical climates with soft leaves. **Inflorescence** is a **raceme**. Fruit is a **berry**. There are about 70 species occurring in Africa, Madagascar, and Arabia. Fibres from *S. zeylanica* are used as bowstring hemp, *S. trifasciata*, with erect, strap-like, dark green, blotched leaves is grown as a houseplant, mother-in-law's tongue or snake plant.

Santa Ana A **foehn wind** that blows in southern California, especially in autumn and winter.

Santalaceae (order **Santalales**) A family of **trees**, **shrubs**, and **herbs**, most of which are root parasites or **epiphytes** that grow on tree **branches**. Leaves usually **alternate** sometimes **opposite**, **exstipulate**, sometimes scale-like or with **clades** resembling leaves. Flowers **actinomorphic**, **bisexual** or **unisexual** (plants **monoecious** or **dioecious**), **perianth** of 3–6 **tepals** in 1 **whorl**, as many **stamens** as tepals, **adnate** to them at the base, **ovary inferior** or semi-inferior of 2–5 **carpels**. **Inflorescence** a **spike**, **raceme**, or **capitulum**. Fruit is a **nut** or **baccate drupe**. There are 44 genera of 990 species with a worldwide but mainly tropical distribution. *Santalum album*, sandalwood tree, yields fragrant wood and oil.

Santalales An order of plants comprising 13 families of 151 genera and 1992 species. *See* Aptandraceae, Balanophoraceae, Coulaceae, Erythropalaceae, Loranthaceae, Misodendraceae, Octoknemaceae, Olacaceae, Opiliaceae, Santalaceae, Schoepfiaceae, Strombosiaceae, and Ximeniaceae.

sap The liquid that exudes from ruptured **vascular tissues** or **parenchyma**.

sapele (*Entandrophragma cylindricum*) *See* Meliaceae.

Sapindaceae (order **Sapindales**) A family of **trees**, **lianas**, **shrubs**, climbers, and a few **herbs** with **rhizomes**, with **alternate** sometimes **opposite**, **paripinnate**, **imparipinnate**, sometimes **bipinnate**, **ternate** or biternate, **palmately compound**, or **simple** leaves, **leaflets** alternate, opposite, or palmate, **entire** or **serrate**. Flowers **actinomorphic** sometimes **zygomorphic**, usually **tetramerous** or **pentamerous**, **unisexual** (plants **monoecious** or **dioecious**) or **bisexual** (plants **andromonoecious**), 5 sometimes 3–7 free **sepals** or in an **urceolate** occasionally **petaloid** tube, 4–5 sometimes 2–4 or 5–6 free **petals**, usually 8, sometimes 5–10 or up to 74 **stamens**, **ovary superior** with 1–3 free or up to 8 fused **locules**. **Inflorescence** a **panicle** or **thyrse**, rarely a **fascicle** or flowers solitary. Fruit is a **capsule**, **drupe**, **berry**, or **samara**. There are 140 genera of 1630 species with a worldwide distribution. Many cultivated for their edible **arils**, e.g. *Litchi chinensis* (lychee), *Nephelium lappaceum* (rambutan), for timber or ornament, e.g. ***Acer*** (maple), ***Aesculus hippocastanum*** (horse chestnut), or other products.

Sapindales An order of plants comprising 9 families of 471 genera and 6070 species. *See* Anacardiaceae, Biebersteiniaceae, Burseraceae, Kirkiaceae, Meliaceae, Nitrariaceae, Rutaceae, Sapindaceae, and Simaroubaceae.

sapling A young **tree**.

saponin A member of a group of **glycosides** that form colloidal solutions in water and foam when shaken. They occur in many plants but take their name from soapwort (*Saponaria*). They were once used as soap (Latin *sapo*) and *S. officinalis* is still used to clean delicate textiles. Saponins have a bitter taste and are toxic in large amounts.

Sapotaceae (order **Ericales**) A family of **trees**, **shrubs**, **lianas**, and **geophytes** with spirally arranged or **distichous**, **simple**, **entire** occasionally **dentate**, **stipulate** leaves. Flowers often nocturnal and bat-pollinated, **actinomorphic**, **bisexual** or **unisexual** (plants **monoecious** or **dioecious**), 2–11 **sepals** in 1 or 2 **whorls** or a spiral, as many **petals** as sepals in 1 whorl, 4–35 or up to 43 **epipetalous stamens** sometimes alternating with 0–8 or up to 12 sometimes **petaloid staminodes**, **ovary superior** usually of many fused **carpels** with 1–15 or up to 30 **locules**. **Inflorescence** an **axillary fascicle** or flowers solitary. Fruit is a **berry**. There are 53 genera of 1100 species with a pantropical distribution. Some important timber trees, others cultivated for **latex** or edible fruits.

saprobe *See* saprotroph.

saprophage An organism that feeds on dead organisms, thereby contributing to the recycling of nutrients. Most saprophages are **bacteria** or **Fungi**, but some are invertebrate animals, e.g. certain insect (**Insecta**) larvae.

saprophyte A **saprotroph** that is a plant or resembles one.

saprotroph (saprobe, saprovore) A **hetrotroph** that feeds on soluble organic compounds it obtains from dead plant or animal matter.

saprovore *See* saprotroph.

saprozoite A **saprotroh** that is an animal or resembles one.

sapwood Wood composed of active **xylem** tissue that surrounds the dead **heartwood**.

Sarcobataceae (order **Caryophyllales**) A **monogeneric** family (*Sarcobatus*) of thorny, **succulent shrubs** with **alternate**, **exstipulate**, **simple**, **entire** leaves without **petioles**. Flowers **unisexual** (plants **monoecious** or **dioecious**), male flowers without a **perianth**, 1–4 free **stamens**, in a **catkin**-like **inflorescence**, female flowers with **sepals**, **ovary** partly **inferior** of 2 **carpels**. Fruit enclosed in the fleshy perianth. There are two species occurring in saline **habitats** in southwestern North America. The wood, greasewood, is traditionally used to make tools and for firewood.

Sarcococca (family **Buxaceae**) A genus of **monoecious**, **evergreen subshrubs** with **alternate** leaves and inconspicuous but fragrant flowers borne in **axils**. Fruit is a **drupe**. There are 16–20 species occurring in the Himalayas and eastern and southeastern Asia. Several cultivated for ground cover or hedging, as sweet box or Christmas box.

Sarcolaenaceae (order **Malvales**) A family of **evergreen trees** and **shrubs** with **alternate**, **simple**, **entire**, **ovate**, **oblong**, **elliptical**, or circular leaves with **caducous stipules**. Flowers **actinomorphic**, **hermaphrodite**, often with **involucre** of **bracts**, 3 or 5 **sepals**, 5 or 6 **petals**, 6–12 or more than 20 **stamens**, **ovary superior** of 3–4 sometimes 1–3 or 5 united **carpels**. Flowers single or in pairs. Fruit is a **loculicidal capsule** or **indehiscent nut**. There are 8 genera of 60 species **endemic** to Madagascar.

sarcoplasmic ER *See* endoplasmic reticulum.

sarcotesta A fleshy seed coat (**testa**) in which seeds are embedded, e.g. in pomegranate and papaya, and in certain **gymnosperms**, e.g. ***Ginkgo***, **Cycadaceae**.

Sargassum muticum *See* japweed.

sarmentose Describes a straggling **shrub** with a **stolon**.

Sarracenciaceae (order **Ericales**) A family of **perennial**, insectivorous **herbs** with **rhizomes** (North American pitcher plant, cobra lily, marsh pitcher), in which each leaf is rolled into a tube with a ventral wing and a lid, forming a pitcher, sometimes containing water and with **nectar** glands that attract insects, which fall into the pitcher and drown. The pitchers are borne in rosettes, with short **petioles**. Flowers **actinomorphic**, **hermaphrodite**, **pentamerous**, 5 **sepals** in 1 **whorl** or sepals absent, 5 **petals**, 10–20 **stamens**, **ovary superior** of 3 fused **carpels** with 1 **locule**. **Inflorescence** terminal, of a single flower or **raceme** on an erect **scape**. Fruit is a **capsule**. There are 3 genera of 32 species. *Sarracenia* occurs in the eastern United States, *Darlingtonia* in the west, and *Heliamphora* in the highlands of Guyana. Some cultivated as a curiosity.

sarsaparilla *See* Smilacaceace.

saskatoon *See Amelanchier.*

Satchellius mammalis (little tree worm) An earthworm (**Annelida**) that grows to 24–41 mm long, with a dark upper surface from the first segment to the **clitellum**. It is found among leaf litter in many types of **habitat** but is seldom abundant. It occurs throughout western Europe.

satellite DNA A section of **DNA** with a base composition sufficiently different from that of most of the DNA in the **genome** that it can be separated by centrifuging. The difference is often due to the presence of long repetitive DNA sequences.

satellite virus A **virus** that can infect a host only when accompanied by a helper or master virus, in the absence of which it is unable to replicate.

satin spar *See* gypsum.

saturated *See* fatty acid.

saturated adiabatic lapse rate (SALR) The rate at which the temperature of saturated air changes as it rises or subsides. This varies because the **condensation** and vaporization of moisture release or absorb **latent heat** and consequently the SALR depends on the air temperature, ranging from about 5°C/km to 9°C/km, with an average value of 6°C/km. *See* adiabatic cooling and warming.

saturation The condition of air that can hold no more water vapour; the **relative humidity** is 100 percent. *See also* supersaturation.

saturation vapour pressure The vapour **pressure** at which the **boundary layer** above a water surface is saturated at a specified temperature.

saurian Pertaining to, or resembling, a lizard.

saurochory Dispersal of **spores** or seeds by snakes or lizards (Sauria).

Saururaceae (order **Piperales**) A family of **procumbent** or erect **perennial herbs** with **rhizomes** or **stolons**, and **alternate**, **simple**, **entire**, **stipulate** leaves, **cordate** at the base. Flowers **actinomorphic**, **bisexual**, without a **perianth**, 3, 6, or 8 **stamens**, **ovary superior** of 4 free or 3–4 fused **carpels** or semi-**inferior** of 3 or 4 carpels. **Inflorescence** is an **axillary spike** or **raceme** of up to 350 flowers, sometimes with an **involucre** of **petaloid bracts**. Fruit is a **schizocarp** or **capsule**. There are five genera of six species occurring in northern temperate regions. Several edible, with medicinal uses, or grown as ornamentals.

savanna (savannah) Tropical grassland dominated by grasses (**Poaceae**) with varying numbers of scattered trees and shrubs.

savannah *See* savanna.

savanna woodland **Savanna** in which trees and shrubs form an **open canopy**.

sawflies *See* Hymenoptera, Symphyta, Tenthredinidae.

saxicolous Growing on stones or walls.

Saxifragaceae (order **Saxifragales**) A family of **evergreen** or **deciduous**, **perennial herbs** with **alternate** rarely **opposite**, **simple**, sometimes **petiolate**, sometimes **pinnate** or **palmately lobed**, **peltate** rarely **compound**, sometimes **stipulate** leaves. Flowers usually **actinomorphic** rarely **zygomorphic**, **bisexual** rarely **unisexual** (plants **monoecious**), usually 5 free or 3–5 fused **sepals**, as many **petals** as sepals, 3–10 **stamens**, **ovary superior** or **inferior** of 2 free or 3 fused **carpels**. **Inflorescence** a terminal or **axillary raceme**, **cyme**, **spike**, **panicle**, head, or flowers solitary. Fruit usually a **capsule**. There are about 33 genera of 540 species occurring in northern temperate and arctic regions. Several grown as ornamentals, e.g. *Saxifraga* (saxifrage), *Astilbe*, *Bergenia*.

Saxifragales An order of plants comprising 15 families of 112 genera and 2500 species. *See* Altingiaceae, Aphanopetalaceae, Cercidiphyllaceae, Crassulaceae, Cynomoriaceae, Daphniphyllaceae, Grossulariaceae, Haloragaceae, Hammelidaceae, Iteaceae, Paeoniaceae, Penthoraceae, Peridiscaceae, Saxifragaceae, and Tetracarpaeaceae.

saxifrage (*Saxifraga*) *See* Saxifragaceae.

Sayornis phoebe (eastern phoebe) A dull grey-brown flycatcher, 140–170 mm long, that inhabits woodlands and feeds on insects, but also takes small fish and fruit. It occurs in North America from northern Canada to the southeastern United States.

Sc *See* stratocumulus.

scabrous Covered with small scales or bristles, rough to the touch.

scalariform (percurrent) Ladder-like.

scale (squama) **1.** A thin, small, plate-like, **adpressed** structure, e.g. a **cataphyll**. **2.** A flattened and much reduced leaf, typical of many conifers. **3.** An external covering to some single-celled algae (*see* alga).

scale insects *See* Coccidae.

scandent Climbing.

scape A leafless **pedicel**.

scaphoid Shaped like a boat.

scapula 1. The shoulder blade in mammals. **2.** The **dorsal** part of the **pectoral girdle** in other **tetrapods**.

Scarabaeidae (chafers, scarabs, tumblebugs) A family of beetles (**Coleoptera**) that are diverse in form but 2–150 mm

S

long and with distinctive antennae (*see* antenna) made from plates that the insect can make into a ball or fan out to detect different chemical signals. Many species have broad front legs, adapted for digging. Adults feed on plants and many are pests. Larvae are fleshy, C-shaped, and feed on roots. Most adults are nocturnal, but flower chafers and leaf chafers are active by day. There are about 28,000 species with a worldwide distribution.

scarabs *See* Scarabeidae.

scarious Dry and membranous.

scarlet lily beetle *See Lilioceris lilii.*

scarlet tanager *See Piranga olivacea.*

scavenging The removal of particles from the air by rain or snow.

Sceloporus undulatus (eastern fence lizard) A lizard, 90–190 mm long, that is grey, brown, or reddish in colour, males having a blue patch on the belly. They inhabit grassland and forest edges. They bask on fences, rocks, and logs. They feed mainly on insects, spiders, and centipedes, sometimes snails, as well as some plant material. If handled they may bite. They occur throughout much of the eastern and southern United States.

Scheuchzeriaceae (order **Alismatales**) A **monotypic**, **monocotyledon** family (*Scheuchzeria palustris*, Rannoch rush), which is a slender, grass-like, **perennial herb** with a **rhizome**. Leaves **distichous**, **linear**, with a sheathing base. Flowers **actinomorphic**, **bisexual**, with inconspicuous **tepals**, 6 **stamens** in 2 **whorls**, **ovary superior** of 3 occasionally 6 **carpels**. **Inflorescence** a terminal **raceme** with **bracts**. Fruit is a **follicle**. The plant occurs in marshy places and **bogs** in northern temperate and arctic regions.

Schimper, Andreas Franz Wilhelm (1856–1901) A German botanist and ecologist who made important contributions to the understanding of plant cells, but whose major work, *Pflanzengeographie auf physiologischer Grundlage*, first published in 1898 (first English language edition, *Plant Geography upon a Physiological Basis*, 1903), set out a classification of the world's vegetation based on the way plants are adapted physiologically to their environment.

Schinus molle (pepper tree) *See* Anacardiaceae.

Schisandraceae (order **Austrobaileyales**) A family of **evergreen** or **deciduous shrubs**, **trees**, and woody climbers with **alternate**, spiral, sometimes leathery, **petiolate**, **exstipulate**, **entire**, **denticulate**, sometimes **dentate** leaves. Flowers **actinomorphic**, **unisexual** (plants **monoecious** or **dioecious**), 5–24 **tepals**, 4–60 or up to 80 **stamens**, **ovary superior** of 12 to many **carpels**. Flowers solitary or aggregated in **axillary** clusters, occasionally cauliflorous (*see* cauliflory). Fruit an aggregate of **berries**. There are 3 genera of 92 species occurring in Sri Lanka, southeastern Asia to western Malesia, southeastern United States, Mexico, and Greater Antilles. Some cultivated for ornament.

Schisozosaccharomycetes A class of **Fungi** that contains the fission **yeasts**. These single-celled organisms grow at the cell tips and when they have doubled in size the cells divide at the centre.

schizocarp A dry, **dehiscent** or **indehiscent fruit** derived from two or more **carpels**, each of which matures as a single-seeded unit.

schizogony A form of **asexual reproduction** in certain parasitic **Protozoa** in which the **nucleus** divides prior to the cell dividing, producing **daughter cells** called merozoites that can either develop into **gametocytes** or enter new hosts and undergo further schizogony.

Schlegeliaceae (order **Lamiales**) A family of tall **trees**, **shrubs**, **epiphytes**, and **lianas** with leaves **opposite**, **simple**,

coriaceous, shortly **petiolate**, **elliptical**, **entire** or lobed, sometimes spiny resembling holly. Flowers more or less **actinomorphic**, **bisexual**, **calyx** 3- to 5-lobed **campanulate**, **corolla** 5-lobed tubular, 4 **stamens**, **ovary superior**. **Inflorescence** a terminal or **axillary raceme** or **cyme**. Fruit is a **berry**. There are 4 genera of 28 species occurring in Mexico and tropical South America.

Schoepfiaceae (order **Santalales**) A family of **trees**, **shrubs**, and **perennial herbs** that are root **hemiparasites** with haustorial (*see* haustorium) roots. Leaves **alternate**, **petiolate** or **sessile**, **exstipulate**, somewhat **coriaceous**, **entire**, **elliptical** to **ovate**, or **linear** or **lanceolate**, sometimes **acuminate** with a spine or prickle at the tip. Flowers **bisexual**, **heterostylous** (*see* heterostyly), **calyx** absent or a **calyculus**, 4–5 **connate** to **urceolate petals**, as many **stamens** as petals, **ovary** of 2 or 3 **carpels**. **Inflorescence** an **axillary**, **cymose** cluster, **umbel**, or **spike**. Fruit is a **nut**-like **achene**. There are 3 genera of 55 species occurring in Central and South America, and tropical southeastern Asia to western Malesia.

Sciadopityaceae (order **Pinales**) A **monotypic gymnosperm** family (*Sciadopitys verticillata*, Japanese umbrella pine, parasol pine), which is a coniferous tree with needle-like leaves fused in pairs throughout their length and borne in loose, umbrella-like **whorls** with scale-like leaves at the base of each whorl. The plant is **endemic** to central and southern Japan.

sciaphilic (skiaphilic) Shade-loving.

Scincidae (skinks) A family of terrestrial lizards that resemble **Lacertilia** but have wedge-shaped heads, long, streamlined bodies, small, weak legs and in some species lacking forelegs or all legs, and smooth scales. Most have long tails they can detach if a predator seizes them, and the tails may partially grow back. Skinks occur in a wide variety of **habitats** and many are burrowers. Most feed on insects and other small arthropods but some are herbivorous. There are more than 1500 species, occurring in temperate, subtropical, and tropical regions worldwide.

scion A shoot or other cutting that is grafted (*see* graft) on to another plant (the stock).

Scirpus (family **Cyperaceae**) A genus of **monocotyledon**, **perennial**, grass-like **herbs** (club rush) with **rhizomes** and solid stems, often triangular in section. Leaves usually arise from the stem base with a closed sheath and **blade** with no **ligule**. Flowers inconspicuous, **bisexual**, **perianth** a series of bristles or scales, **ovary superior**. **Inflorescence** a cluster of 2 to many **spikelets**. Fruit is an **achene**. There are about 120 species with a **cosmopolitan distribution** in wet **habitats**. *Scirpus lacustris* is the true bulrush.

Sciuridae (squirrels) A family of rodents (**Rodentia**) with species that are terrestrial, burrowing, arboreal, and some that glide by means of a **patagium**. They have four digits on the forelimbs and five on the hind limbs, all with sharp claws. Arboreal squirrels have long, bushy tails used for balance, and large ears, some with prominent tufts. Many ground-dwelling and burrowing squirrels have short, strong forelimbs they use for digging and most have tails less bushy than those of tree squirrels. Squirrels occur in almost every type of **habitat**. Most feed on seeds, nuts, and fruit, but some eat **Fungi**, **lichens**, insects, eggs, and small vertebrates. They are found worldwide except for some deserts, Greenland, Antarctica, Australia, southern South America, and some ocean islands. There are 51 genera and about 280 species.

Sciurus carolinensis (grey squirrel, eastern gray squirrel) A tree squirrel (**Sciuridae**) with a body 380–525 mm long and a bushy tail 150–250 mm long, that is grey often with shades of red or cinnamon. It inhabits forests and is common in urban and suburban areas. It feeds on nuts, seeds, **buds**, flowers, insects, frogs, and bird eggs. It occurs throughout eastern North America and was introduced to Britain in the 19th and early 20th centuries and is now naturalized and widespread. It was introduced to South Africa in about 1905.

Sciurus vulgaris (red squirrel, Eurasian red squirrel) A tree squirrel (**Sciuridae**) with a body 205–220 mm long and a bushy tail 170–180 mm long, that is chestnut in colour with white under parts and very prominent ear tufts, although the body colour is very variable. It inhabits forests, nesting in large trees, and feeds on nuts, and visits bird feeders. It occurs throughout Eurasia. Its population has declined in Britain due to disease and **competition** from the introduced grey squirrel (***Sciurus carolinensis***).

sclerenchyma Woody or fibrous tissue that provides support to a plant, formed from dead cells with walls strengthened with **lignin**, **hemicellulose**, and **cellulose**.

sclerite One of the hard sections of an insect **exoskeleton**.

sclerophyllous vegetation **Scrub** or **forest** dominated by **evergreen** plants with leaves that are thick, hard or leathery, and usually small, that occurs in climates with a hot, dry season.

S

sclerotinia disease *See Sclerotinia sclerotiorum.*

Sclerotinia fructicola *See Monilinia fructicola.*

Sclerotinia sclerotiorum A species of **ascomycete fungi** that is a pathogen known to infect more than 400 species of plants with the disease white mould, also called blossom blight, cottony rot, crown rot, drop, stem rot, sclerotinia disease, and watery soft rot. The fungus survives out of season as a **sclerotium** on or within plant tissue and in soil, germinating the following spring and producing apothecia (*see* apothecium) that release **spores**, which germinate when they encounter the leaves, roots, flowers, or fruit of a host, producing a **mycelium** that invades the plant and eventually appears on the surface as a white mould.

Sclerotium cepivorum A species of **ascomycete fungi** that causes the disease Allium root rot (also called white rot of onions), which can cause severe crop losses. The fungus dwells in the soil as a **sclerotium** until ***Allium*** root exudates stimulate it to produce **hyphae** that seek host roots; it infects only *Allium* species. The fungus occurs worldwide.

sclerotium A fungal resting body consisting of a compact mass of **mycelium** that resists unfavourable conditions, detaching from the fungus and remaining dormant in the soil.

Scolytidae (bark beetles, ambrosia beetles) A family of beetles (**Coleoptera**), most less than 5 mm long, with short, clubbed antennae (*see* antenna) and the head hidden beneath an extension of the covering of the first thoracic segment. The **elytra** are often grooved or incised toward the rear and used to shovel wood debris. The beetles excavate a chamber in or beneath tree **bark** in which they lay eggs; the large, fleshy larvae feed on wood, tunnelling away from the egg chamber. The elm-bark beetle (*Scolytus scolytus*) carries the fungus responsible for **Dutch elm disease**. Ambrosia beetles bore into the **sapwood** and their larvae feed on **ambrosia fungi** that develop on the walls of their tunnels. There are about 6000 species

scopa The specialized hairs between which female bees pack **pollen grains** together before carrying them back to the nest or hive.

scorpioid cyme A **cyme** in which the **axis** is coiled or curved like the tail of a scorpion.

scorpions *See* Arachnida.

scorpion wasps *See* Ichneumonidae.

Scotch bonnet (*Marasmius oreades*) *See Marasmius.*

scotch mist **Stratus** cloud that forms suddenly on high ground.

scramble competition **Competition** for a resource that is not present in a sufficient amount to satisfy the needs of all the competitors, but that is distributed evenly, so no competitor receives all it needs. In extreme cases, all the competitors die. *Compare* contest competition.

screaming cowbird (*Molothrus rufoaxillaris*) *See* Icteridae.

scree slope *See* talus.

screw pine (*Pandanus*) *See* Pandanaceae.

Scrophulariaceae (order **Lamiales**) A family of **annual** and **perennial herbs**, **shrubs**, and a few climbers, **trees**, and **lianas**, some parasitic, a few **epiphytes**, some aquatic, some **resurrection plants**. Leaves **alternate** or **opposite** occasionally whorled, **simple** to **pinnatisect**, reduced in aquatics. Flowers strongly **zygomorphic** to almost **actinomorphic**, sometimes **resupinate**, **bisexual**, usually **tetramerous** or **pentamerous**, **corolla** 2-lipped, strongly zygomorphic, usually 2-lobed upper and 3-lobed lower lips, 4 sometimes 5 **stamens**, **ovary superior** usually of 2 free **carpels**. **Inflorescence** a **raceme** or **thyrse** or flowers solitary. Fruit is a **capsule**, rarely a **berry**. There are 65 genera of 1800 species with a worldwide distribution. Many are cultivated for ornament. Heart drugs digoxin and digitalin are obtained from ***Digitalis*** (foxglove).

scrub Vegetation dominated by **shrubs**.

scud Tattered fragments of cloud below the main cloud base.

scute An enlarged bony plate embedded in the skin.

scutellum **1.** A structure, believed to be a modified **cotyledon**, that lies between the **embryo** and **endosperm** of a grass (**Poaceae**) seed. During **germination** the scutellum secretes **enzymes** involved in breaking down the endoperm. **2.** The most posterior of three **dorsal sclerites** on the **mesothorax** and **metathorax**, found in many winged insects.

Scytonema A genus of filamentous **cyanobacteria** with **filaments** showing **false branching**, due to loops of **trichomes** breaking through the **filament** sheath then breaking. The filaments form dark mats that are free-floating in water or attached to a substrate, forming erect tufts on wet surfaces and in soil. They perform **nitrogen fixation**.

sea belt (sugar-kelp, poor man's weatherglass) The yellowish brown seaweed *Saccharina latissima*, which has a root-like **holdfast** and a single, ribbon-like **blade** up to 5 m long and 20 cm wide with an undulating central region and wavy edges. When dry a sweet-tasting deposit sometimes accumulates on its surface, hence one of its common names. In high **humidity** the blade is soft and rubbery and in low humidity it is dry and brittle, hence its other common name.

sea buckthorn (*Hippophaë*) *See* root nodule.

sea lavender (*Limonium*) *See* Plumbaginaceae.

sea oak *See* bladder wrack.

sea pink (*Armeria*) *See* Plumbaginaceae.

sea slugs *See* Gastropoda.

seasonal drought A **drought** that occurs every year in climates with pronounced wet and dry seasons.

sea thong The brown seaweed *Himanthalia elongata*, found on the lower shore, which has a flattened **thallus** up to 3 cm across with a short stalk, from which

S

dichotomously branching thongs grow in autumn and winter, reaching up to 2 m by summer.

secondary An inner flight **feather** of a bird, on the trailing edge of the wing connected to the **ulna**.

secondary consumer A **carnivore** that preys upon **herbivores**.

secondary forest A forest that grows on land from which a previous forest has been cleared.

secondary growth (secondary thickening) The formation of new tissue in woody plants by the addition of successive layers through the repeated division of **cambium** cells, thereby increasing the girth of the **branch**, stem, or root. The layers remain visible as **tree rings**.

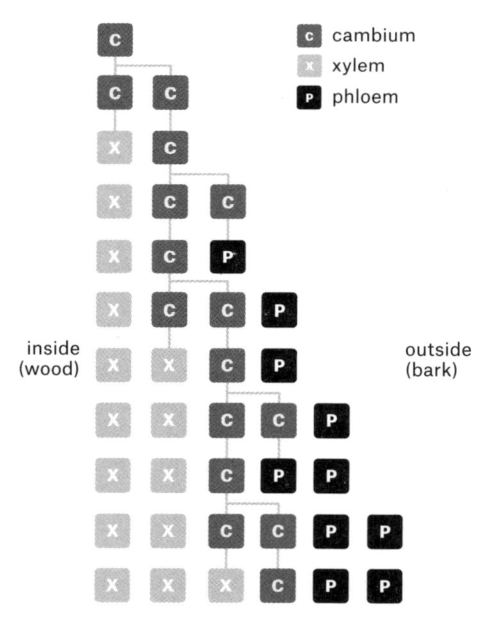

Secondary growth. Repeated division of the cambium cells produces xylem cells on the inside of the tree trunk or branch, phloem cells outside them, and more cambium cells.

secondary metabolite (secondary plant compound) A chemical compound produced by an organism's metabolic processes that serves no primary function, i.e. in growth or reproduction. Secondary plant metabolites often help protect the plant against **herbivores**. Some are used medicinally.

secondary mineral A **mineral** formed by the alteration of a **primary mineral** in an **igneous** rock.

secondary phloem **Phloem** tissue derived from the **vascular cambium** in plants that undergo **secondary growth**.

secondary pit A **pit** that develops in the adjacent **cell walls** of cells that were not originally connected or that are not sisters resulting from mitotic division. *See* bordered pit, primary pit, simple pit.

secondary plant compound *See* secondary metabolite.

secondary structure The configuration of a **protein** molecule that develops when the **primary structure** twists into an **alpha helix** or a folded **pleated sheet**. *See* primary structure, quaternary structure, tertiary structure.

secondary succession A **succession** that develops when a major environmental perturbation has disrupted a previous succession or a **climax community**, e.g. following fire, or the abandonment of **cultivation** or grazing.

secondary thickening *See* secondary growth.

secondary vein A leaf **vein** that branches from the primary vein running along the centre of the leaf.

secondary wall The third and final layer of a plant **cell wall** to form during cell division. It is made from **hemicellulose**, **cellulose**, and **lignin** and is very rigid. It helps the cell resist compression. *See also* middle lamella, primary wall.

secondary woodland Woodland that grows on land that has not supported woodland throughout history, or in Britain since the last ice age.

secondary xylem Xylem tissue derived the **vascular cambium** in plants that undergo **secondary growth**.

secretion 1. The process of discharging from a cell or **gland** a substance that is not a metabolic waste and that serves a useful purpose, e.g. an **enzyme** or **hormone**. **2.** Any substance that is secreted.

secund Arranged all on one side or curved to one side.

sedges *See Carex*, Cyperaceae.

sedimentary rock A rock formed by the compression of rock and mineral particles, often mixed with organic fragments, that have been transported and deposited on land or the seabed, or by the compression and cementing of compounds that were carried to the sea in solution and later precipitated.

Sedum (stonecrop) *See* Crassulaceae.

seed In seed plants (**Spermatophyta**), the body formed from a fertilized **ovule** from which a new plant emerges. It comprises an **embryo** plant with a store of food enclosed in a seed coat (**testa**). The food may be stored in the **cotyledons** or in **endosperm**.

seed bank A store in which seeds are held as a technique for conserving rare or endangered plants. Dried to a moisture content of about 4 percent and held at a constant 0°C, seeds of many species remain viable for up to 20 years.

seed fern (Pteridospermatophyta) A group of **seed plants** that first appeared during the Devonian period (416–359.2 million years ago) and became extinct early in the Eocene epoch (about 55 million years ago). The plants had fern-like foliage but reproduced by seeds.

seedling A young plant that has developed from a **seed**.

seed plants (spermatophytes) Plants that reproduce by means of seeds rather than **spores**, comprising the **angiosperms** and **gymnosperms**.

seed weevils *See* Bruchidae.

seep A place where the **water table** intercepts the ground surface and water soaks the ground. *Compare* spring.

segetal A plant that grows spontaneously on cultivated land.

segregation distortion *See* meiotic drive.

seismonasty A nastic response (*see* nasty) to vibration or touch, e.g. the sudden folding of the leaves of the sensitive plant (*Mimosa pudica*).

Selaginella lepidophylla (rose of Jericho) *See* resurrection plant.

Selasphorus rufus (rufous hummingbird) A hummingbird, 75–90 mm long, in which the adult male has a white breast, greenish back and crown, bronze-green **pileum**, bright orange collar, and iridescent red throat. The adult female has a metallic bronze-green back, dull white throat, and a duller pileum than that of the male. The birds feed on nectar from red, tubular flowers, e.g. honeysuckle, and also small insects. During the breeding season they inhabit forests. They breed in western and coastal North America and winter in Mexico.

selection coefficient (*s*) A means of comparing the **fitness** of two **genotypes** within a population, yielding a number between 0 and 1. If $s = 0$ there is no selection against the genotype; if $s = 1$ selection is total and the genotype will contribute nothing to the next generation.

selection pressure The influence of environmental conditions on evolution, exerted through **natural selection**. If the selection pressure is weak little

S

evolutionary change will occur, if it is strong evolutionary change will be rapid.

selectively permeable membrane A **membrane** that allows certain molecules to cross but prevents others. *See also* differentially permeable membrane, partially permeable membrane.

selective species In the phytosociological (*see* phytosociology) scheme devised by the school led by Josias **Braun-Blanquet**, one of the five classes of fidelity (*see* faithful species) that describe and classify plant communities. Selective species are common in a particular community and also occur occasionally in others. *Compare* accidental species, exclusive species, indifferent species, preferential species.

selective value *See* fitness.

selenite *See* gypsum.

selenotropism A tropic response (*see* tropism) to moonlight.

self-fertilization (selfing) The fusion of female and male **gametes** produced by the same plant.

self-incompatibility The mechanisms by which plants avoid **self-fertilization** and achieve outbreeding, thereby increasing genetic variability. There are two types: **homomorphic self-incompatibility** and **heteromorphic self-incompatibility.**

selfing *See* self-fertilization.

self-inverting soil *See* self-mulching soil.

selfish DNA A hypothesis that aims to explain the presence in **genomes** of apparently redundant **DNA** that is not translated into **protein**, proposing that selection favours any method by which DNA may replicate and rapid replication is best achieved by bypassing expression in the **phenotype**. The DNA replicates by spreading laterally, causing it to be duplicated at other loci (*see* locus); this is 'selfish' because the DNA confers no advantage on the organism carrying it and does not trigger the production of the materials from which it is made.

selfish genes The hypothesis that organisms act as agents for replication of **genes**, rather than genes existing to serve organisms, i.e. that **natural selection** operates at the level of the gene. It was expounded in *The Selfish Gene* by Richard Dawkins, published in 1976.

self-mulching soil (self-inverting soil) A **clay** soil in which the surface layers swell and shrink, producing deep crevices into which loose soil falls. Over time this repeated churning mixes the soil.

self-pollination The transfer of **pollen** from **anthers** to **stigmas** of the same **flower** or of different flowers but on the same plant.

self-sterility genes *See* S genes.

self-thinning The natural process in which the number of plants of a particular species in a specified area (the population density) decreases as the plants grow larger.

selva **Tropical rain forest**, especially that in the Amazon basin.

semelparity (big-bang reproduction) The condition of having only one reproductive cycle in the course of a lifetime.

semi-desert scrub A transitional type of vegetation found between true desert and an area supporting more abundant vegetation, e.g. **savanna.** Plants are more widely scattered than in the more vegetated region and **succulent** plants are more common.

semi-natural community A pattern of vegetation that has been altered significantly by human management or interference but that appears natural because of the length of time human influence has persisted, e.g. **heathland**, **down**.

semi-natural woodland Woodland that has been managed in the past, but which comprises mainly **native** species that have not obviously been planted.

semipalmate Describes the feet of a bird that are partially webbed, the webs not extending to the tips of the toes.

semipermeable Describes a membrane that allows solvent molecules to pass but prevents the passage of solute molecules. *See* partially permeable membrane.

semiplume *See* feather.

semi-slug A **slug** that lives on land and has a shell that is too small for the animal to retract into.

Sempervivum (houseleek) *See* Crassulaceae.

senescence The processes of deterioration, controlled by **hormones**, that terminate the life of an organ or organism. In plants it is often associated with flowering and fruiting and is accompanied by a reduction in the amount of **chlorophyll** and consequent reduction in **photosynthesis**.

sensible heat Heat that the skin can detect. *Compare* latent heat.

sensible temperature The temperature the body experiences. This may differ from the temperature registered by a thermometer, e.g. due to **wind chill**.

sensitive plant (*Mimosa pudica*) *See Mimosa*.

sepal A leaf-like organ in **angiosperms**, usually green, that covers a flower **bud** and that forms an outer **whorl** in the open flower. The set of sepals comprise the **calyx**.

sepaloid Resembling a **sepal**.

septate Having a **septum**.

Septoria A genus of **ascomycete fungi** that produce pycnidia (*see* pycnidium) and that cause leaf spot diseases and stem cankers on a wide variety of crops; septoria leaf blotch is a major disease of wheat in Britain. There are about 1072 species, distributed worldwide.

septum A partition.

Sequoiadendron giganteum (big tree) *See* Pacific coast forest.

Sequoia sempervirens (coastal redwood) *See* Pacific coast forest.

sere The sequence of changes that typically occur in a **succession**.

seriate Arranged in a row.

sericeous Silky; covered in silk-like hairs.

serine A polar (*see* polar molecule) **amino acid** (HO_2CCHCH_2OH) found in **proteins**.

serotiny The retention of seeds on a tree, often for several years, until a shock, most commonly the heat of a fire, triggers their release, when they fall on to ground enriched by ash from which competitors have been removed. Jack pine (*Pinus banksiana*), lodgepole pine (*P. contorta*), and many ***Eucalyptus*** species are serotinous.

serotonin A compound, 5-hydroxytryptamine ($C_{10}H_{12}N_2O$), derived from **tryptophan** that occurs in many plants including tomatoes, kiwi fruit, pineapples, bananas, and nettles; the richest sources are walnuts and hickory. It is the end product of reactions that prevent the accumulation of ammonia in drying seeds and serotonin may speed the passage of seeds through the digestive tract of animals that eat fruit.

Serpentes (snakes) A suborder of reptiles (**Reptilia**) that have long bodies, no legs, jaws that are attached by ligaments, allowing them to be opened very wide, and no external ears or eyelids, the eye surface being protected by a **spectacle**; it is the lack of visible ears and eyelids that distinguishes snakes from legless

S

lizards. The tongue is usually forked and protruded through a notch in the snout. Paired internal organs are positioned one in front of the other rather than side by side, and most species have only one lung. Snakes occur on all continents except for Antarctica and aquatic species occur in the Pacific and Indian Oceans. All are carnivorous, although a few species subsist mainly on eggs. There are about 500 genera and more than 2700 species.

serpentine barren An area of **scrub** or **heathland** that develops on soils with an excess of magnesium released by the **weathering** of serpentine rock.

serrate Having toothed margins, resembling a saw blade.

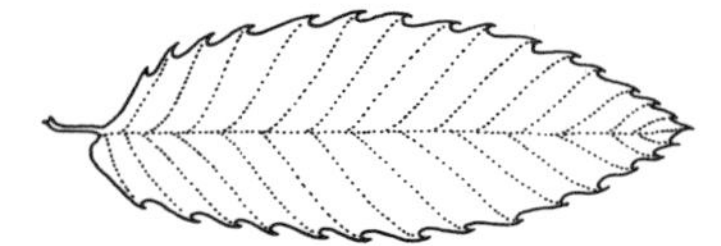

A leaf with a serrate margin.

serrated wrack (toothed wrack) An olive-brown, robust seaweed (*Fucus serratus*), about 60 cm long, that is very common on rocks near the low-water mark. The **thallus** is flattened and branched and the fronds are about 2 cm wide, with serrated edges, and repeatedly split in two.

serrulate Having very fine, tooth-like projections.

sesame (*Sesamum indicum*) *See* Pedaliaceae.

sesquioxides An oxide that contains three atoms of oxygen and two of another element. In soils, this describes the hydrated oxides of iron.

sessile **1.** Permanently attached to a substrate. **2.** Lacking a stalk.

seta (pl. setae) **1.** The stalk of the **capsule** of a moss (**Bryophyta**) or liverwort (**Marchantiophyta**). **2.** A bristle or stiff, hair-like structure.

Setchellanthaceae (order **Brassicales**) A **monotypic** family (*Setchellanthus caeruleus*), which is a **shrub** with tiny, **alternate**, **petiolate**, **exstipulate** leaves. Flowers **actinomorphic**, **bisexual**, 6 sometimes 5 or 7 fused **sepals**, 6 sometimes 5 or 7 **petals**, 60–76 free or 40–60 fused **stamens**, **ovary superior** of 3 fused **carpels** each with 1 **locule**. Flowers solitary in the leaf **axils**. Fruit is a **capsule**. The plant is **endemic** to Mexico.

Setophaga petechia (American yellow warbler, yellow warbler) A species of **passerine** birds, 100–180 mm long with a 160–220 mm wingspan, that are yellow with olive-green wings, males with red streaks on the underside and black streaks on the wings. They inhabit woodland and thickets and feed mainly on arthropods (**Arthropoda**). They breed throughout most of North America and winter in Central and South America.

setose Having setae (*see* seta).

settled Describes fine weather conditions that continue unchanged for at least several days and usually for longer than a week.

severe gale Wind of 21–24 m/s. *See* appendix: Beaufort Wind Scale.

severe storm A storm that endangers human life or damages property. The U.S. National Weather Service defines it as a storm with hailstones at least 19 mm across, winds gusting to 93 km/h or more, a **tornado**, or more than one of these.

sex chromosome A nuclear **chromosome** linked to the sex of the individual carrying it and that contributes to **sex determination** in offspring. The sex with a **homologous** pair of sex chromosomes is said to be homogametic (XX), the sex with a dissimilar pair of sex chromosomes is heterogametic (X or XY). **Gametes** from XX individuals are identical, all carrying the X chromosome; those from X or XY individuals produce two types

S

of gametes, one with an X chromosome and one with either a Y chromosome or no chromosome at all. The union of sexes therefore results in equal numbers of male and female offspring. In many plants the male is the heterogametic sex.

sex determination The process by which the sex of offspring is determined. In **monoecious** plants this is influenced and may be controlled by **hormones**. In some **diploid dioecious** plants (e.g. *Salix* spp.) the union of XX with X or XY **gametes** (*see* sex chromosome) determines sex, in others it is a single **gene** with two **alleles**.

sexine (ectexine, ektexine) The outer layer of the **exine** of a **pollen grain**.

sexual dimorphism The condition in which males and females of the same species differ in form (other than sexual characters), e.g. male birds are often more brightly plumed than females, and male mammals are often larger than females.

sexual reproduction Reproduction in which **haploid** nuclei (*see* nucleus) fuse.

S genes (S alleles, self-sterility genes) **Genes** that are involved in self-recognition in plant **self-incompatibility**. Female and male S genes occurring in a **supergene** are expressed in the **pistil** and **anther** respectively. The interaction of their **protein** products inhibits the development of a viable **embryo**.

shadbush *See Amelanchier.*

shade temperature The air temperature registered by a thermometer that is not exposed to direct sunlight. In direct sunlight a thermometer will absorb heat, directly raising its temperature to above that of the air, so it will give a false reading.

shallot (*Allium cepa*) *See Allium.*

shallot aphid *See Myzus ascalonicus.*

shallow fog Fog that extends no higher than 2 m above the surface.

shallow soil *See* effective soil depth.

sharka disease *See* plum pox.

sharp sand Sand that is composed principally of angular grains with only a small amount of other material.

sheath **1.** The base of a leaf **lamina** that encloses the stem. **2.** In **prokaryotes**, the **capsule**. **3.** *See* fungal sheath.

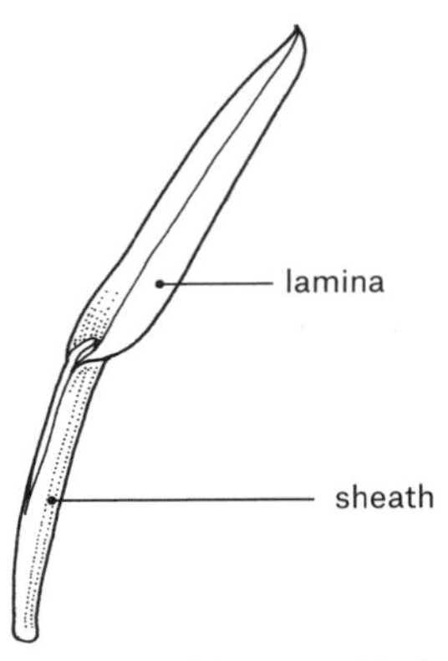

The base of the grass blade (lamina) encloses the stem, forming a sheath.

sheet erosion (sheet flow, sheet wash) The **detachment** of soil particles by the impact of raindrops and their transport across the surface as a sheet of material, rather than in defined channels. It results from rainfall so intense that small surface rivulets merge.

sheet flow *See* sheet erosion.

sheet lighting **Lightning** that is seen as a bright flash, but not as a **lightning stroke**, which is concealed either by intervening cloud or because it occurs between two regions inside a cloud.

sheet wash *See* sheet erosion.

shelter temperature (air temperature, surface temperature) The temperature registered by a thermometer located 1.25 m above ground level inside a suitable shelter, e.g. a **Stevenson screen**.

shield bugs *See* Pentatomidae.

shimmer An effect that is seen above a hot surface, e.g. of a road, caused by the refraction of light as it passes from cool to warm air. It is a mirage producing an inverted image of the sky.

shoot A **stem**, most of which is above ground.

short-day plant A plant that flowers when the period of nighttime darkness is longer than a critical threshold, and the period of daylight is shorter. The plant requires continual nighttime darkness, but moonlight and starlight are too dim to affect it.

short-horned grasshoppers *See* Acrididae.

shortwings *See* Turdidae.

shrews *See Sorex.*

shrinkage cracks *See* desiccation cracks.

shrinkage limit *See* Atterberg limits.

shrub A **perennial**, woody plant that is less than 10 m tall and that branches into several main stems close to or below ground level. It may be **deciduous** or **evergreen**, but the aerial parts do not die back at the end of each growing season. *Compare* herb, subshrub, tree.

Si *See* silicon.

Sialia currucoides (mountain bluebird) A small bird (Muscicapidae), 160–200 mm long, in which the male has a deep blue back and wings and paler blue underside and white belly; the female is blue-grey with a grey back, throat, and crown. They breed in high mountain meadows and winter in grassland at lower elevations, feeding on insects and berries. They occur throughout most of North America.

Sialia sialis (eastern bluebird) A small bird (Muscicapidae), 160–210 mm long with a wingspan of 250–320 mm, in which males have a bright blue head and wings, and a tan-coloured throat and breast; females have paler blue wings and tail and a grey crown and back. They inhabit open areas with scattered trees and are often seen in parks and gardens. They feed mainly on insects, augmented with berries in winter. They occur throughout most of North America.

Siberian ginseng (*Acanthopanax senticosus*) *See* Araliaceae.

Siberian high A large **anticyclone** that forms in winter over Siberia, centred south of Lake Baikal.

sibling species (aphanic species, cryptic species) A **species** that is almost identical in appearance to another, closely related species with which it is unable to breed.

sieve cells Long, narrow, tapering cells with pores (the sieve) through which nutrients move in solution from cell to cell. The sieve is distributed evenly throughout the **cell wall**, but there is no **sieve plate**. Sieve cells are found in the **phloem** tissue of **gymnosperms** and non-seed plants, but not in **angiosperms**.

sieve elements Long, narrow cells, ending in **sieve plates**, that are linked end to end to form **sieve tubes** in the **phloem** of **angiosperms**.

sieve plate The wall at either end of a **sieve element** in the **phloem** of **angiosperms**. The wall is perforated, forming a sieve, sometimes with strands of **cytoplasm** extending through the pores and facilitating nutrient transport between cells.

sieve tube A series of **sieve elements** in **angiosperms** or **sieve cells** in non-seed plants and **gymnosperms**, joined end to end, through which nutrients are transported in the **phloem**.

Sigillaria A genus, now extinct, of tall, tree-like but non-woody plants related to the **Lycopsida** (club mosses) with a stem, sometimes forked, strengthened by closely packed leaf bases just below the surface, and a plume of long, grass-like leaves at the top of the stem. The plants lived from the Carboniferous to the early Permian periods (359.2–270.6 million years ago).

sigmoid growth curve *See* S-shaped growth curve.

silcrete A **duricrust** dominated by **silica**.

silent allele (null allele) An **allele** that is not expressed in the **phenotype**.

silica **Silicon** dioxide (SiO_2). It occurs naturally in three forms: crystalline, cryptocrystalline, and amorphous hydrated. **Quartz** is crystalline silica, and the mineral content of **sand** consists of small quartz crystals; crystalline silica also forms other minerals. Cryptocrystalline silica forms a number of very finely crystalline minerals including chert, jasper, and flint. Amorphous hydrated silica forms minerals including opal. Some plants accumulate silica in the **cell walls** of their epidermal and vascular tissues, and it is found in the cell wall of **diatoms**.

silica-sesquioxide ratio The ratio of **silica** to **sesquioxides** in a soil. As silica is lost by **leaching**, the proportion of sesquioxides increases. This change is associated with a reduction in **cation exchange capacity** and water retention.

siliceous Containing **silica**.

silicle (silicula, silicule) A dry fruit that is at least as broad as it is long.

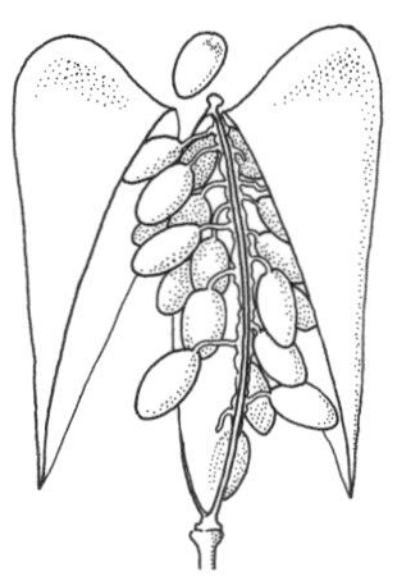

Silicle (shepherd's purse, *Capsella bursa-pastoris*).

silicon (Si) The second most abundant element in the Earth's **crust** (after **oxygen**), accounting for 27.7 percent of the crust by weight. It occurs most commonly as **silica**. Si is not counted as an essential nutrient, but it strengthens **cell walls**, thereby reducing water loss and inhibiting fungal infection, making leaves more erect, and reducing susceptibility to **lodging**.

silicula *See* silicle.

silicule *See* silicle.

siliqua *See* silique.

silique (siliqua) A dry, **dehiscent** fruit, formed from two fused **carpels** and containing many seeds, that is longer than it is wide, and that opens along either suture. It is found in many members of the **Brassicaceae**.

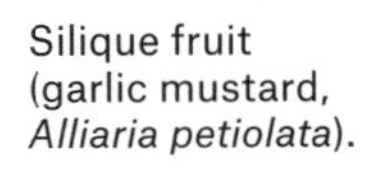

Silique fruit (garlic mustard, *Alliaria petiolata*).

silk tassel *See Garrya.*

silt A type of **soil texture** based on mineral particles measuring 4.0–62.5 µm on the Udden-Wentworth scale, or 2–60 µm on the British scale.

silty clay loam A soil in which the mineral composition is 27–40 percent **clay**, up to 20 percent **sand**, and 40–53 percent **silt**.

silver leaf *See Chondrostereum purpureum.*

silverleaf whitefly *See Bemisia tabaci.*

silviculture The management of a **forest** or **woodland** for the benefit of the entire **ecosystem**, with or without the commercial production of wood or timber.

Simaroubaceae (order **Sapindales**) A family of **trees** and **shrubs** with leaves **alternate** rarely **opposite**, **pinnate** sometimes with 1 **leaflet** rarely 3, **exstipulate** rarely **stipulate**. Flowers **actinomorphic**, **unisexual** (plants **dioecious** or

andromonoecious) sometimes **bisexual**, 3- to 8-merous, usually 5 **sepals** and **petals**, occasionally **apetalous**, usually twice as many **stamens** as petals, sometimes 5 reduced to **staminodes**, **ovary superior** of 2–5 free or 5–8 fused **carpels**. **Inflorescence** a terminal or **axillary thyrse**, **cyme**, or **raceme**. Fruit is **capsule**, **samara**, **drupe**, or **berry**. There are 19–22 genera of 110 species with a mainly tropical distribution, a few temperate. Some have medicinal or insecticidal properties or are used to make incense. *Ailanthus altissima* (tree of heaven) is widely grown for wood pulp.

simazine A **herbicide** applied before weeds emerge and used to control **broad-leaved** and grass weeds around crops with deep roots and fruit trees, and on paths and open ground. It degrades fairly slowly but is of very low toxicity to animals.

Simmondsiaceae (order **Caryophyllales**) A **monotypic** family (*Simmondsia chinensis*, jojoba or goat nut), which is a small, **evergreen**, **xerophyte tree** or **shrub** with **opposite**, **glaucous**, **coriaceous**, **elliptical** to **oblong**, almost **sessile**, **exstipulate** leaves. Flowers **actinomorphic**, **unisexual** (plants **dioecious**), **apetalous**, 10–12 **stamens**; female flowers with 5 **sepals**, **ovary superior** of 3 sometimes 4 fused **carpels**. Female flowers solitary, male **inflorescence** a **capitate** cluster. Fruit is a **capsule**. Jojoba oil is widely used in pesticides and cosmetics.

simple **1.** Describes a leaf that is not divided or lobed. **2.** *See* pharmacopoeia.

simple pit A **pit** in a **cell wall** between **parenchyma** cells that is not partly covered by an extension of the cell wall. *Compare* bordered pit.

simple sorus In certain ferns (**Pteridophyta**), a **sorus** in which all the sporangia (*see* sporangium) develop together. *Compare* gradate sorus, mixed sorus.

Sinadoxa *See* Adoxaceae.

single superphosphate *See* superphosphate.

sinistral coil A **snail** shell that coils counter-clockwise when viewed from above.

sink A natural reservoir in which substances or energy accumulate.

sinuate Wavy, curved, or indented.

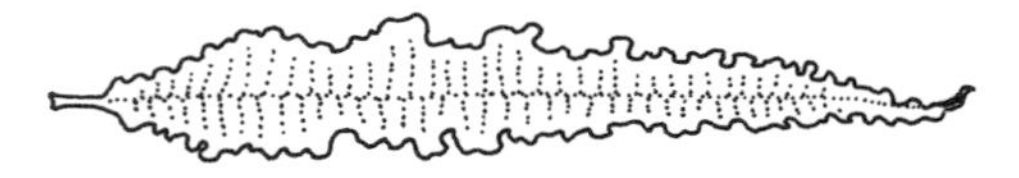

A leaf with a sinuate margin.

sinus A small space.

Siparunaceae (order **Laurales**) A family of **evergreen shrubs** and **trees** with leaves **opposite** or in **whorls**, **simple**, **entire** or **serrate**, **exstipulate**. Flowers usually **actinomorphic**, **unisexual** (plants **monoecious** or **dioecious**), **perianth** with 4–6 free or 4–7 fused **tepals**, 1–72 **stamens**, **ovary superior** of 3–30 free **carpels** with 1 **locule**. **Inflorescence** an **axillary cyme** or **fascicle**, or cauliflorous (*see* cauliflory). Fruit is a **drupelet** with a fleshy appendage. There are 2 genera of 75 species occurring in tropical America and West Africa.

siphonaceous (siphoneous) Describes an **alga** with a **thallus** that is not divided into compartments by septa (*see* septum). Typically there is a central **vacuole** surrounded by **protoplasm** in which there are nuclei and **chloroplasts** lining the **cell wall**.

siphonostele The more highly developed of the two types of **stele** (*see* protostele), in which the **xylem** tissue forms a cylinder surrounding **pith**, and there are often **leaf gaps** in the xylem. There are three types of siphonostele: **solenostele**, **dictyostele**, and **eustele**. A siphonostele may be **amphiphloic** or **ectophloic**.

siphuncle *See* cornicle.

siRNA *See* small interfering RNA.

S

siskin *See Carduelis spinus.*

sistentes *See* adelgids.

sister groups In **phylogenetic systematics**, two taxa (*see* taxon) connected at a single node.

Sistrurus miliarius (pigmy rattlesnake, ground rattlesnake, eastern pigmy rattlesnake) A pit viper (**Crotalinae**), 300–600 mm long, that is variable in colour but with a dark, vertical line through the eye and the side of the face, and dark, circular spots along the back. Its rattle is so small it sounds like the buzzing of an insect and is audible over only a short distance. It inhabits a variety of **habitats**, but all are close to water; the snake swims well. Its bite is serious but seldom fatal. The snake feeds on small vertebrates and invertebrates. It occurs throughout the southeastern United States.

site of special scientific interest (SSSI) In the United Kingdom, an area designated by a government authority as being of biological, ecological, geological, or other scientific importance. The owner of a designated site is required to notify the designating authority and obtain permission before undertaking operations that would alter its characteristics.

Sitona lineatus (pea and bean weevil) A species of weevil (**Curculionidae**), 3.5–5.5 mm long, usually metallic bronze or golden but sometimes green or blue, that feeds on **legumes**, especially peas, beans, and vetches, but also clovers, and that is a significant pest. Females lay eggs in the soil in spring. These hatch after about three weeks and the larvae find their way to **root nodules**, where they feed for six to seven weeks, by when they are up to 6.5 mm long and pupate in the soil, emerging as adults after about 17 days to feed on leaves of the hosts. They occur throughout the temperate Northern Hemisphere and have been introduced in South Africa and Australasia.

Sitta carolinensis (white-breasted nuthatch) A bird, 130–140 mm long with a 200–270 mm wingspan, that has a large head with a beak almost as long as the head and a glossy black cap, pale blue-grey back with a dark band at the shoulders, and black and grey wings with a white bar. They inhabit **deciduous** and mixed forests and visit garden feeders. They feed on nuts, seeds, and insects. The birds occur throughout most of North America and northern Mexico.

skiaphilic *See* sciaphilic.

skinks *See* Scincidae.

skip-jacks *See* Elateridae.

skotomorphogenesis The influence of darkness on the growth and form of a plant, e.g. stem elongation, lack of leaf expansion, undifferentiated **chloroplasts**. *Compare* photomorphogenesis.

skototropism A tropic response (*see* tropism) to darkness.

Skottsberg, Carl Johan Fredrik (1880–1963) A Swedish botanist who studied the vegetation of the far south, including the **floras** of Patagonia, Tierra del Fuego, the Falkland Islands, South Georgia, and Juan Fernández. He designed and directed the Gothenburg Botanical Garden and the Jardin Botanico 'Carl Skottsberg', founded in 1970 at Punta Arenas, Chile, is named in his honour.

skunks *See* Mephitidae.

SLA *See* specific leaf area.

Sladeniaceae (order **Ericales**) A family of **evergreen trees** with **alternate**, **petiolate**, **exstipulate**, **simple**, **ovate**, **lanceolate**, **oblong**, or **elliptical**, **entire** or **serrate** leaves. Flowers **hermaphrodite**, with 5 free **sepals**, 5 partly **connate petals**, 10 occasionally up to 13 **stamens**, **ovary superior** of 3 fused **carpels** with 3 **locules**. **Inflorescence** an **axillary** dichasial (*see* dichasium) **cyme**. Fruit is a **schizocarp**. There

S

are two genera of three species occurring in southeastern Asia and tropical East Africa.

slaked lime *See* lime.

slaking The breaking up of earth materials when they are exposed to water or air.

slaters *See* Isopoda.

sleep movement *See* nictonasty.

sleet 1. In Britain, a mixture of rain and snow falling together. **2.** In North America, small **raindrops** that freeze as they are falling.

slide (landslide) A more or less rapid displacement of surface material on a hillslope across one or more plane or curved surfaces.

slime mould A **eukaryote** organism that lives for much of the time as a single, **amoeboid** cell, but that aggregates with others to form a **plasmodium** or **pseudoplasmodium** and **fruiting bodies.** There are two groups, **Acrasiomycetes** (cellular slime moulds) and **Myxogastria** (acellular slime moulds). *See also* Oomycota.

slippery-moss snail *See Cochlicopa lubrica.*

SLOSS debate The discussion over whether a single, large **nature reserve** will support more species than a similar area divided into several smaller reserves, i.e. Single Large Over Several Small. This depends on the extent to which the species in the small reserves overlap. If there is little overlap several small reserves may be preferable, and small areas of **habitat** may protect species from the accidental introduction of parasites or predators.

slow worm *See Anguis fragilis.*

slug A land-dwelling, air-breathing gastropod (**Gastropoda**) that has no external shell, although it may have a greatly reduced internal shell. Like all gastropods, their bodies undergo **torsion**, but externally they are symmetrical. Most have two pairs of retractable tentacles, the upper pair sensitive to light and the lower pair to scent, behind the head a **mantle** with a respiratory opening, the anterior part of the mantle being modified to form a lung. The underside of the body forms a muscular foot that secretes a mucus containing fibres that help the animal move across inclined surfaces. Mucus covering the entire body helps prevent **desiccation**, but slugs must remain in damp environments and take shelter in dry conditions. Slugs are hermaphrodites. They obtain food by means of a **radula** and feed on a wide variety of plant and fungal material; some species are predators. There are about 5000 species distributed worldwide. *See also* Pulmonata.

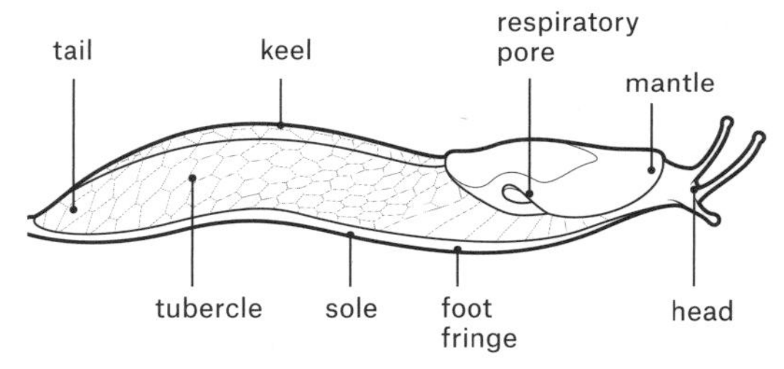

A slug is a terrestrial gastropod in which the shell is either absent or greatly reduced.

slug pellets *See* metaldehyde.

small ermine moth *See Yponomeuta padella.*

small hail Hailstones smaller than 5 mm in diameter, but that remain intact until they strike the ground.

small interfering RNA (siRNA) A type of **small RNA**, 20–25 **nucleotides** long, involved in **RNA interference.**

small RNA Molecules of **RNA**, usually 20–25 **nucleotides** long, that cause **RNA interference.**

small striped slug *See Arion hortensis.*

small white butterfly *See Pieris rapae.*

Smilacaceae (order **Liliales**) A **monogeneric monocotyledon** family (*Smilax*) of **lianas** and vines with prickly stems with spirally arranged, **petiolate**, **entire**, **coriaceous** leaves. Flowers erect, **actinomorphic**, **unisexual**, with 6 **tepals** in 2 **whorls**, 3–6 **stamens**, **ovary superior** of 3 fused **carpels** with 3 **locules**. **Inflorescence** an **umbel**. Fruit is a **berry**. There are 315 species occurring throughout the tropics and warm temperate regions. Some species edible, some the source of sarsaparilla.

smog 1. A mixture of smoke and **fog**, responsible for the 'pea soupers' that once occurred in winter in most industrial cities. **2.** *See* photochemical smog.

smokebush (smoke tree) *See Cotinus*.

smooth ER *See* endoplasmic reticulum.

smooth newt *See Triturus vulgaris*.

smudge *See Colletotrichum*.

smudge pot *See* smudging.

smudging Using oil-burning heaters, called smudge pots, to produce smoke that forms a layer close to ground level. The smoke traps warmth radiating from the surface at night, protecting valuable crops from frost.

smut fungi *See* Ustilaginomycetes.

snail A gastropod (**Gastropoda**) that has an external shell into which it can retract its body. Land-dwelling snails have a lung and breathe air, aquatic species breathe with gills. They have two pairs of retractable tentacles, the upper pair sensitive to light and the lower pair to scent. The underside of the body comprises a muscular foot. Snails are hermaphrodites. The obtain food by means of a **radula** and feed on a wide variety of material. There are many species, found throughout the world. *See also* Pulmonata.

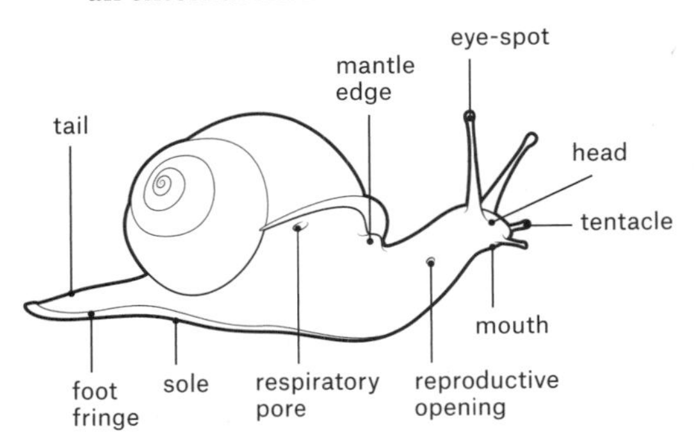

A snail is a terrestrial gastropod that has a shell in which it is protected from dessication.

snake eyes *See* Achatocarpaceae.

snake plant (*Sansevieria trifasciata*) *See Sansevieria*.

snakes *See* Serpentes, Squamata.

snapping beetles *See* Elateridae.

snout beetles *See* Curculionidae.

snout moths *See* Pyralidae.

snow **Precipitation** in the form of aggregations of ice crystals.

snow belt A strip of land on the **lee** side of a large lake, parallel to the shore and about 80 km wide, where more snow falls in winter than falls on the windward side. *See* lake-effect snow.

snowberry (*Symphoricarpus*) *See* Caprifoliaceae.

snow eater A warm, dry wind that removes snow by **sublimation**. The **chinook** is sometimes called the snow eater.

snow gauge A modified **rain gauge** that is used to measure the amount of snow fall.

snow line The boundary between ground that is covered by snow and ground that is snow-free, e.g. the edge of the permanent snow cover on a high mountain. The elevation of the snow line varies with latitude and with **aspect**.

snow worms *See* Cantharidae.

soboliferous Forming clumps.

social parasitism *See* manipulated altruism.

S

sodication An increase in the proportion of exchangeable sodium present in the soil. Sodium adsorbs on to **cation exchange** sites on soil particles. This causes **aggregates** to break up, leading to the closure of **pores**, making the soil impermeable.

sodic soil Soil containing more than 15 percent exchangeable sodium or sufficient sodium to inhibit plant growth. It has a **pH** greater than 8.5, a **sodium-absorption ratio** greater than 13, electrical conductivity of leas than 0.4 siemens per metre, and a poor physical condition.

sodium (Na) An element, found in all plants, that is not regarded as an essential nutrient but that is important in controlling **osmotic pressure** and can substitute for **potassium** in some metabolic functions. It is important for carbon-dioxide **fixation** in some C4 (*see* C4 pathway) and **CAM** plants.

sodium adsorption ratio (SAR) A measure of the extent to which sodium **cations** are absorbed at **cation exchange** sites at the expense of other cations. It is calculated as the amount of sodium (Na) present in relation to the amounts of calcium (Ca) and magnesium (Mg): SAR = $Na^{+}/\sqrt{½}\,(Ca^{2+} + Mg^{2+})$. A low sodium content implies a low SAR.

sodium-coupled transport The movement of a **metabolite**, e.g. **glucose**, into a cell against a concentration gradient that is accompanied by the movement of **sodium** across the **cell wall**. Sodium crosses cell walls easily because of the difference in concentration of solutions on either side (*see* osmosis), and the metabolite attaches to a carrier molecule that is also attached to the sodium.

soft hail *See* graupel.

soft rot Any plant disease in which tissues soften or liquefy, then becoming slimy and foul-smelling. Many **Bacteria** can cause soft rot, most being Gram-negative (*see* Gram reaction) species of ***Erwinia*** and ***Pseudomonas***.

soft scales *See* Coccidae.

softwood The wood of a coniferous tree or the tree itself.

soil association A group of soils that form a pattern of soil types characteristic of a geographic region. It is used as a mapping unit.

soil classification A scheme for arranging soil types in groups according to their distinguishing features, analogous to taxonomical systems (*see* taxonomy) used in biology. Most nations have devised their own systems. The two most widely used internationally are the **soil taxonomy** developed by the U.S. Department of Agriculture and the **World Reference Base for Soil Resources** developed in association with the Food and Agriculture Organization (FAO) of the United Nations.

soil complex A mapping unit that is used to display the geographic distribution of soil types. It is used where soils of different types are so mixed that it would be impossible or impractical to show each type separately. A soil complex is more precise than a **soil association**.

soil conservation Any system of land management which aims to prevent physical loss by soil **erosion** and chemical deterioration by loss of nutrients.

soil drainage The removal of excess water from the soil, e.g. by means of ditches to carry water into a stream and thereby preventing it from entering the land below, or by the installation of **mole drains** or **tile drains**.

soil fertility The ability of a soil to sustain crops, which is determined by the depth, structure, and texture of the soil, its ability to retain moisture without becoming waterlogged, and the content and availability of its plant nutrients.

S

soil fixation The processes occurring in the soil that convert certain chemical elements that are essential for plant growth from a soluble or exchangeable form to a form that is less soluble or non-exchangeable.

soil formation *See* pedogenesis.

soil grading curve A line on a graph that plots grain size on a horizontal, logarithmic axis against percentage on a vertical, arithmetic axis.

soil horizon A layer of soil that is fairly uniform in composition and appearance, lies more or less parallel to the soil surface, and is clearly differentiated from the layers above and below it. Soil horizons are labelled. The O horizon comprises organic material lying on and close to the surface. The A (or E for eluviated, *see* eluviation) horizon comprises a mixture of organic and mineral material from which compounds drain downward (eluviation) into the B horizon. The C horizon consists of material derived from the **weathering** of the **parent material** but without any further alteration. The R (or D) horizon comprises the underlying **bedrock**.

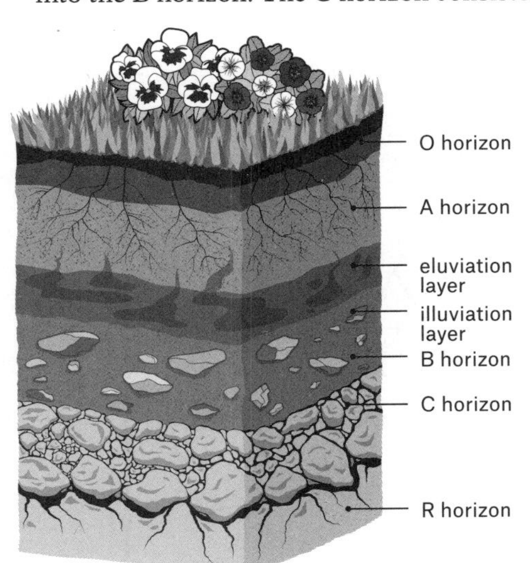

Soil horizon. The O horizon comprises surface organic material, the A horizon is rich in humus from which compounds drain downward (eluviation) into the B horizon. The C horizon consists of mineral material weathered from the bedrock (R horizon).

soil individual *See* polypedon.

soil macrofauna Larger animals that live in the soil. These include mice, moles, earthworms, spiders, and a wide variety of insects and insect larvae.

soil map A map that depicts the geographic distribution of soil types (*see* soil classification).

soil microfauna Soil-dwelling animals that can be seen only with the aid of a microscope. They are less than 0.1 mm in size and comprise mainly small **Arthropoda**, **Nematoda**, and **Protozoa**.

soil monolith A vertical slice cut through a soil from the surface to the subsoil, exposing its colours and layered structure, that is removed intact and used for teaching purposes.

soil morphology The features of a soil that can be observed in the field, e.g. the colour, texture, and changes through the **soil horizons** as revealed by a **soil profile**.

soil phase A description of a soil that is not part of a wider taxonomic scheme (*see* soil classification), but applies to a local variant, e.g. where water has deposited a surface layer of sediment, or part of the surface layer has been eroded by wind.

soil profile A vertical section that is cut through a soil from the surface to the **parent material**.

soil separates The mineral particles of the **fine earth** categorized according to size as **sand**, **silt**, and **clay**.

soil series A group of soils that developed from similar **parent material** under similar climatic conditions and supporting similar types of vegetation, and that have similar **soil profiles**. Soil series are a basic unit in soil mapping.

soil solarization *See* solarization.

soil structure A soil quality that derives from the way individual particles join to form **aggregates** and **peds**.

soil survey A systematic study of the soils of an area, involving their detailed examination, classification, and mapping.

soil taxonomy A system of **soil classification** that was developed by scientists of the U.S. Soil Survey, within the Department of Agriculture. It divides soils into 11 orders, which are subdivided into suborders, great groups, families, and soil series. These are defined by **diagnostic horizons**.

soil testing A series of procedures that evaluate the suitability of a soil for crop growing. The tests examine soil structure, texture, nutrient availability, water retention, **pH**, etc.

soil textural triangle A diagram that is used to classify soils by their textures, based on the percentages of **clay**, **silt**, and **sand** particles each type contains. In the illustration below, the triangle uses **particle sizes** defined by the USDA (United States Department of Agriculture); the triangle based on sizes defined by the Soil Survey of England and Wales is slightly different.

soil texture The mineral composition of a soil, comprising the relative proportions of sand, silt, and **clay** particles.

soil venting 1. The injection of air into soil or **groundwater** in order to stimulate **aerobic** organisms. **2.** A technique for removing contaminant gases from soil by pumping them out from a well dug to the affected depth.

soilwash The downslope movement of soil material carried by surface water flow.

soil-water zone (unsaturated zone, vadose zone) The region of the soil that lies between the surface and the **water table**.

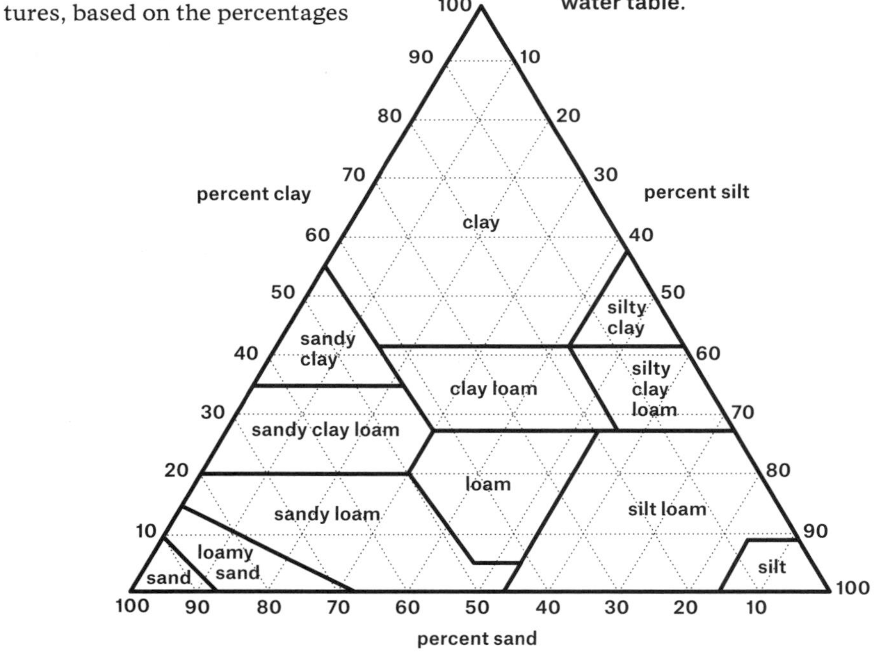

Soil textural triangle. The mineral component of soil comprises particles of sand, silt, and clay, with the texture of the soil depending on their relative proportions.

sol A continuous, homogeneous **dispersion** of solid particles in a liquid (a **colloid**), e.g. fluid mud.

Solanaceae (order **Solanales**) A family of **trees, shrubs, vines, lianas, epiphytes**, and **annual** and **perennial herbs** with **alternate**, usually **simple, exstipulate, entire** or dissected leaves. Flowers usually **actinomorphic** sometimes **zygomorphic, bisexual** or **unisexual** (plants **andromonoecious**), with 5 free or 4–7 fused **sepals** in 1 **whorl**, 5 sometimes 4–10 **plicate, imbricate** or **valvate petals, corolla** round and flat, **campanulate**, or tubular, 4 or 5 or up to 10 **stamens, ovary superior** of 2 fused **carpels** with usually 2 **locules. Inflorescence** terminal or **axillary, cymose**, or flowers solitary. Fruit a **capsule** or **berry**, sometimes a **drupe** or **schizocarp**. There are 102 genera of 2460 species with a worldwide but mainly tropical American distribution. Many species commercially important food crops, e.g. ***Solanum*** or ***Capsicum*** spp. (peppers), or grown for ornament.

Solanales An order of plants comprising 5 families of 165 genera and 4080 species. *See* Convolvulaceae, Hydroleaceae, Montiniaceae, Solanaceae, and Sphenocleaceae.

Solanum (family **Solanaceae**) A genus of **annual** and **perennial herbs**, vines, **subshrubs, shrubs**, and small **trees** in which the **corolla** lobes are spreading or **reflexed** and the fruit is usually a **berry**. There are 1500–2000 species with a worldwide but mainly tropical American distribution. Many are poisonous, e.g. *S. dulcamara* (nightshade), others are important food crops, e.g. *S. tuberosum* (potato), *S. melongena* (aubergine), *S. lycopersicum* (tomato).

solarization 1. (heliosis) The inhibition of **photosynthesis** at extremely high light intensity due to **photo-inhibition** and **oxidation** of the compounds involved. **2.** Scorching of leaves by intense sunlight. **3.** (soil solarization) Covering the ground with opaque sheeting to suppress weeds.

soldier beetles *See* Cantharidae.

solenoglyphous Describes snakes (**Serpentes**) that have long, hollow fangs which fold against the roof of the mouth when the mouth is closed. They are the only teeth in the upper jaw and are capable of injecting venom.

solenostele A type of **siphonostele** in which a central core of **pith** is surrounded by three cylinders of **phloem, xylem**, and phloem, and **leaf gaps** are scattered and not overlapping in cross section. A solenostele may be **amphiphloic** or **ectophloic**.

Solidago (family **Asteraceae**) A genus of **perennial herbs** growing from **rhizomes** or **caudices** with erect stems and **simple** often **serrate** leaves. Ray **florets** are **pistillate**, the 2–35 or up to 60 **disc florets bisexual. Involucre campanulate** to tubular, with **imbricate bracts. Inflorescence** a head of up to 1500 or more florets in **panicles**, sometimes **corymbose**. Fruit is an **achene**. There are about 100 species, most occurring in North, Central, and South America, but some Eurasian. Many are cultivated for their showy flowers (goldenrod).

solifluction (solifluxion) The downhill movement of surface material that is saturated with water.

solifluxion *See* solifluction.

soligenous mire A **mire** that receives water from both rain and **surface runoff**.

solodic soils A soil that was formerly saline but which has become leached (*see* leaching). The A **soil horizon** is slightly acid and the B horizon is enriched with **clay** saturated with sodium.

solodization The removal of alkalis from a soil by the **leaching** of sodium from the upper **soil horizons**.

solonchaks Soils that have a **soil horizon** more than 15 cm thick at or close to the surface that is enriched with soluble salts. Solonchaks often develop from recent

alluvial deposits. They are a reference soil group in the **World Reference Base for Soil Resources.**

solonetz A mineral soil that is undergoing the process of **solodization**. The A **soil horizon** is sandy and acid, and the B horizon is partly enriched with **clay** saturated with sodium.

soloth soils Soils that are similar to **solodic soils** but acidic throughout the **soil profile.**

solstice One of the two dates each year when the lengths of time the Sun is above and below the horizon are at their most extreme, because at noon the Sun is directly overhead at one or other of the **tropics**. These dates are 21–22 June and 22–23 December. *Compare* equinox.

solum (pl. sola) The part of a **soil profile** in which soil is developing and where most plant roots and soil animals occur.

solute potential *See* osmotic potential.

solution A homogeneous mixture of two or more substances, formed by dissolving a quantity of some components (solute) in a larger quantity of another (solvent), that can be separated only by boiling, condensing, or freezing. In soil, a solution forms by a **weathering** process in which weakly bonded minerals are detached by attaching to water molecules, which are polar (*see* polar molecule).

somatic cell In a multicelled organism, any cell that is not destined to become a **gamete**, i.e. a body cell.

somatic cell hybrid A **hybrid** cell that results from the fusion of two **somatic cells.**

somatic mutation A **mutation** that occurs in a **somatic cell.**

sombric horizon A subsurface **diagnostic horizon** in which **humus** has moved downward and the **base saturation** is less than 50 percent.

somite *See* metameric segmentation.

song sparrow *See Melospiza melodia.*

song thrush *See Turdus philomelos.*

sooty mould Black spots on leaves that are the mycelia (*see* mycelium) of **ascomycete fungi** growing on sugars in **honeydew**. Several genera of fungi cause sooty mould.

soralium A structure containing a mass of soredia (*see* soredium).

Sorbus (family **Rosaceae**) A genus of small **deciduous trees** and **shrubs** with **simple** or **pinnate** leaves. Flowers **pentamerous** with many **stamens**, **inflorescence** a compound **corymb**. Fruit is a **berry**. There are 100–200 species occurring throughout the temperate Northern Hemisphere. *Sorbus aucuparia* is rowan or mountain ash, *S. aria* is whitebeam, and *S. torminalis* is wild service tree.

Sordariomycetes One of the largest classes of **ascomycete fungi**, comprising more than 600 genera and at least 3000 species. It includes most of the ascomycetes that are not **mycobionts** of **lichens** and usually have flask-shaped asci (*see* ascus). These fungi are ubiquitous worldwide as plant pathogens and **endophytes**, parasites of arthropods (**Arthropoda**), mammals (**Mammalia**), and other Fungi, and as **saprotrophs.**

soredium A structure formed by certain **lichens** comprising one or more algal cells enmeshed in fungal hyphae (*see* hypha). Under certain conditions it may develop into a **thallus** and function in **vegetative reproduction**. A mass of soredia appear as fine grains or powder.

Sorex (shrews, long-tailed shrews) A genus of small mammals with long, mobile snouts, long tails, teeth with red tips, small ears, and small eyes. Their eyesight is poor and they rely mainly on hearing and smell. They feed on invertebrates, mainly earthworms and insects.

There are about 35 species occurring in northern Eurasia and North America. There are two species native to Britain. The common shrew (*S. araneus*) has a body length of 60–80 mm, a dark back and paler sides and belly. The pygmy shrew (*S. minutus*) has a body length of 40–60 mm and less contrasting colours.

sorocarp A **fruiting body**, produced by some **slime moulds**, that consists of an unenclosed mass of **spores** borne on a stalk.

sorrel (*Rumex acetosa*) *See* Pologonaceae.

sorus In ferns (**Pteridophyta**) and **Fungi**, a cluster of sporangia (*see* sporangium).

source 1. A rock from which later sediments are derived. **2.** (source region) The area, typically a continent or ocean, where an **air mass** forms.

source region *See* source.

South African region The area that includes only the **Cape floral region**.

south Brazilian floral region The area that covers the eastern coastal region, central uplands, eastern highlands, and Gran Chaco of Brazil, part of the **Neotropical region**. There are about 400 **endemic** genera as well as many endemic species.

southeastern five-lined skink *See Eumeces inexpectatus.*

southern bacterial wilt *See Ralstonia solanacearum.*

southern beech *See* Nothofagaceae.

southern cricket frog *See Acris gryllus.*

southern leopard frog *See Rana sphenocephala.*

southern oscillation A periodic change in the distribution of surface **atmospheric pressure** over the equatorial South Pacific Ocean, measured at Darwin, Australia, and Tahiti. It is associated with **El Niño** and **La Niña** events. *See* ENSO.

southern pill woodlouse *See Armadillidum nasatum.*

southern root-knot nematode (*Meloidogyne incognita*) *See Meloidogyne hapla.*

southern toad *See Anaxyrus terrestris.*

southern two-lined salamander *See Eurycea cirrigera.*

southern wilt *See Ralstonia solanacearum.*

southernwood (*Artemisia abrotanumis*) *See Artemisia.*

south European flowering ash (*Fraxinus ornus*) *See* mannitol.

south temperate oceanic-island floral region The area that includes the islands of the Southern Ocean, part of the **Antarctic region**. There is a high degree of floristic constancy between the islands, despite the great distances separating them.

Southwest Australian floral region The area that covers southwestern Australia, part of the **Australian region**. The floristic composition is very rich, with many **endemics**.

sowbugs *See* Porcellionidae.

spadix An **inflorescence** comprising a **spike** of flowers on a swollen **axis**.

spalling *See* thermal weathering.

Sparassis crispa (cauliflower fungus, brain fungus, wood cauliflower) A species of **agaric fungi** that grows at the bases of coniferous trees, especially Scots pine (*Pinus sylvestris*). It is weakly parasitic, but seldom causes serious harm. The **fruiting body** comprises creamy white, curled, leaf-like lobes, reminiscent of a cauliflower or brain. It is up to 400 mm across, 100–250 mm tall, and can weigh several kilograms. It is edible when young. It occurs throughout northern and central Europe, and North America, and is cultivated commercially in some countries.

Spadix.

S

spathe A large **bract** subtending a **monocotyledon inflorescence.**

spatulate With a broad, flattened end, resembling a spatula.

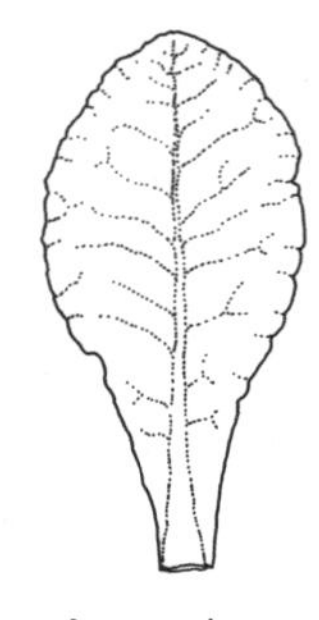
A spatulate leaf.

spawn A fungal **mycelium**, especially one used by mushroom growers to start a new **culture.**

special adaptation *See* general adaptation.

speciation The separation of populations of interbreeding organisms into independent groups that accumulate genetic differences until they lose the ability to interbreed and become distinct **species.**

species (sing. and pl.) A group of organisms that resemble each other more closely than they resemble non-members of the group. In **taxonomy**, an interbreeding group of organisms that are reproductively isolated from members of other groups; i.e. members of a species breed with other members to produce fertile offspring but either cannot breed with organisms outside the group or can do so, but the offspring are usually (but not always) infertile.

species turnover *See* beta diversity.

specific gravity The ratio of the density, i.e. the weight of a unit volume, of a substance to that of water at 4°C, expressed as a number (there is no unit).

specific heat capacity *See* heat capacity.

specific humidity The ratio of the mass of water vapour in the air to a unit mass of air including the water vapour.

specific leaf area (SLA) The area of one side of a fresh leaf divided by its oven-dried mass.

spectacle In snakes (**Serpentes**), a transparent scale that covers and protects the eye.

speculum A patch of brightly coloured, often iridescent patches on the **secondary feathers** of some birds.

speedwell *See Veronica.*

spell of weather A period of unchanging weather, usually of five to ten days.

spermatangium In **red algae** (Rhodophyta), the organ that produces spermatia (*see* spermatium).

spermatiophore A **hypha** bearing a **spermatium.**

spermatium A non-**motile** cell found in **red algae** (Rhodophyta) and some **ascomycete fungi** that functions as a male **gamete.**

spermatocyte *See* antherocyte.

Spermatophyta The **seed plants**, in some older classifications ranked as a **phylum** (division) in the kingdom **Plantae**, comprising the classes Cycadopsida (cycads), Ginkgoopsida (ginkgo), Pinopsida (conifers), Gnetopsida (**gnetophytes**), and Magnoliopsida (angiosperms). Alternatively, these are ranked as phyla, with Spermatophyta as a superphylum (superdivision).

spermatophytes *See* seed plants.

spermatozoid *See* antherozoid.

sphaeroraphide *See* druse.

Sphaerosepalaceae (order **Malvales**) A family of **deciduous trees** with **alternate, simple, coriaceous, entire** leaves. Flowers **actinomorphic, bisexual**, usually **tetramerous, 4 sepals** and **caducous petals**, 25–160 **stamens, ovary superior** of 2–4 sometimes 5 fused **carpels. Inflorescence** an **axillary raceme** with many **branches.** Fruit is a **berry.** There are 2 genera of 18 species **endemic** to Madagascar.

Sphagnum *See* peat moss.

Sphenocleaceae (order **Solanales**) A **monogeneric** family (*Sphenoclea*) of **annual herbs** with hollow stems and

alternate, **simple**, **entire**, **linear** to **elliptical** leaves. Flowers **actinomorphic**, **bisexual**, **pentamerous**, **epigynous**, **ovary** semi-**inferior** of 2 fused **carpels**. **Inflorescence** a terminal **spike**. Fruit is a **capsule**. There are two species with a pantropical distribution.

sphenoid A bone that forms part of the orbit in the vertebrate skull.

Sphenophyllum A genus, now extinct, of creeping or climbing plants, or understory **shrubs**, that lived from the Devonian to Permian periods (416–251 million years ago). It had jointed stems, **cuneate** leaves borne in **whorls** of 3, and terminal **sporophylls**.

Sphenopsida (Equisetopsida) A subphylum of the **Pteridophyta** that flourished in the coal swamps of the Carboniferous (359.2–318.1 million years ago) when one form, ***Calamites***, was tree-like and up to 30 m tall. Today the only surviving sphenopsid is ***Equisetum*** (horsetail).

spicate 1. Resembling a spike. **2.** Arranged in spikes, e.g. an **inflorescence**.

spicule A small spine or spike.

spider mites *See* Tetranychidae.

spiders *See* Arachnida.

spike A **racemose inflorescence** in which all the **florets** are **sessile**.

spikelet A small **spike**, consisting of an **axis**, 2 **bracts** or **glumes**, and 1 or more **florets**.

spikerushes *See Eleocharis.*

spikesedges *See Eleocharis.*

spillover Precipitation caused by **orographic lifting** that blows across the top of a mountain and falls on the **lee** side into what is usually a **rain shadow**.

Spikelet.

spinach *See* Amaranthaceae.

spinach stem fly *See Delia echinata.*

spindle In **eukaryote** cells an arrangement of **microtubules** that forms during the **metaphase** stage of cell division. **Chromosomes** attach themselves to the spindle, which moves them until they are aligned in a plane at the equator (the equatorial plane). During **anaphase**, the spindle separates the chromosomes and moves them to opposite poles of the cell.

spindle attachment *See* centromere.

spine A modified leaf that forms a sharp point.

Spinifex (family **Poaceae**) A **monocotyledon** genus of **dioecious**, **perennial** grasses with thick **rhizomes** and **stolons**, and silvery leaves. The **spikelets** of the **inflorescence** are **sessile**, male spikelets with 2 flowers borne in **spikes** with clusters of 4–6 with long **bracts**, female spikelets in dense heads with a pine-like **rachis**. There are three species occurring in eastern and southeastern Asia, the Pacific Islands, and Australia. The grasses are used to stabilize sand dunes.

spinney A small area of **woodland**, originally dominated by hawthorn (*Crataegus*).

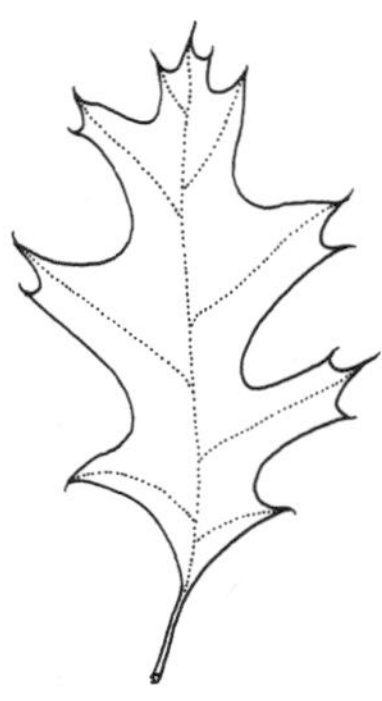
A leaf with a spinose margin.

spinose Bearing spines.

spiral wrack (flat wrack, twisted wrack) The brown seaweed *Fucus spiralis*, up to 40 cm long, found attached to rocks near the high-tide mark on sheltered or moderately exposed shores. It spends 90 percent of its time out of water and tolerates **desiccation**. The **thallus** lacks **bladders**. It is **hermaphrodite**.

Spiroplasma A genus of **Mollicutes** in which the cells are able to form spiral **filaments** and move with a corkscrew motion.

They are parasites of insects and plants, where they colonize **phloem** tissue. There are 37 species. *See* citrus stubborn disease, corn stunt disease.

splash erosion The **erosion** that occurs when falling raindrops impact bare soil, breaking up **aggregates** and propelling (splashing) soil particles up to 60 cm into the air and on to the surface up to 1.5 m away.

spodic horizon A subsurface **diagnostic horizon** in which aluminium and iron compounds have accumulated amorphously by **illuviation**, together with organic matter.

spodosols Soils that have a **spodic horizon**. They are coarse in texture (*see* soil texture), acid, and tend to form in humid, cool to temperate climates.

Spongospora subterranea A protozoon (**Protozoa**) that causes the disease powdery scab in potatoes. The organism begins as a **zoospore** with two flagella (*see* flagellum) that either unites with a different type of zoospore to produce a **zygote** that enters the plant through the root, or invades the root directly. Resting **spores** can remain viable in the soil for up to ten years. Infected tubers are marked by scabs on the skin; these do not affect the quality of the potato, but only its appearance. *Spongospora subterranea* occurs worldwide.

S

spongy mesophyll (spongy parenchyma) Tissue found in green leaves that consists of roughly spherical cells with large spaces between them filled with humid air and linked to air chambers beneath the stomata (*see* stoma). Spongy mesophyll cells contain **chloroplasts**, but fewer than are found in **palisade mesophyll**. Spongy mesophyll is the site for **gaseous exchange** for **photosynthesis** and **respiration**.

spongy parenchyma *See* spongy mesophyll.

spontaneous mutation A **mutation** that occurs naturally, rather than one induced chemically or by irradiation.

sporangiolum A **sporangium** that contains only one or a few **spores**.

sporangiophore A specialized **hypha** that bears a **sporangium**.

sporangiospore A **spore** that is formed in a **sporangium**.

sporangium A sac or enclosed space, made from one or more cells, in which **spores** form either by **mitosis** or, in most land plants and many **Fungi**, by **meiosis**.

spore 1. An asexual reproductive unit, usually consisting of a single **haploid** cell capable of developing into a new organism by **mitosis**. **2.** In **Bacteria**, a cell that functions as a **propagule** or as a dormant structure allowing the **bacterium** to survive adverse conditions.

sporeling A young plant that has grown from a **spore**.

spore mother cell *See* sporocyte.

spore print The pattern and colours of **spores** that appears when the **pileus** of a fungal **fruiting body** is placed on a sheet of paper, **gill**-side down, and left for several hours or overnight.

sporidium 1. A protozoan (*see* Protozoa) **spore**. **2.** A spore produced by a **promycelium**.

sporocarp *See* fruiting body.

sporocyte (spore mother cell) A **diploid** cell that divides by **meiosis** to produce four **haploid spores**.

sporodochium A compact **stroma**, shaped like a cushion and covered with **conidiophores**.

sporogenesis The formation of **spores**, or reproduction by spores.

sporogonium The **sporophyte** generation in mosses (**Bryophyta**) and liverworts (**Marchantiophyta**).

sporophore A structure that bears **spores**.

sporophyll A leaf that bears sporangia (*see* sporangium).

sporophyte The **diploid** generation in the life cycle of plants (*see* alternation of generations) that produces **haploid spores** by **meiosis**. These develop into the haploid **gametophyte**. In ferns (**Pteridophyta**), **angiosperms**, and **gymnosperms** the sporocyte is the dominant generation, i.e. the visible plant, and the gametophyte is small and comprises only the **pollen** and **embryo sac**. In **Bryophyta** and **Marchantiophyta** the gametophyte is the dominant generation.

sporophytic self-incompatibility A type of **homomorphic self-incompatibility** that is achieved genetically, by the **diploid genotype** of the **sporophyte** generation (*see* alternation of generations). **Pollen** will not germinate on the **stigma** of a flower that contains either of the two **alleles** in the sporophyte parent that produced the pollen. This occurs in **Brassicaceae**.

sporopollenin A polymer made from **fatty acids** and **phenol** compounds that is the principal component of the **exine** of **pollen grains**. It is said to be the most resistant organic substance known.

sport A **mutation** that causes an individual plant to differ from the species type.

sporulation The production and release of **spores**.

spotted asparagus beetle (*Crioceris duodecimpunctata*) *See* asparagus beetle.

spotted cucumber beetle (*Diabrotica undecimpunctata*) *See* bacterial wilt.

spotted laurel *See Aucuba.*

spotted towhee *See Pipilo maculates.*

spraing Brown rings and arcs in the flesh of potatoes resulting from disease.

sprayer A device that distributes a liquid under pressure through a nozzle or array of nozzles. It may be carried and powered by a hand pump or battery-powered electric motor, or operated from a tractor or an aircraft specialized for the purpose.

spreading adder *See Heterodon platirhinos.*

spring A place where the **water table** intercepts the ground surface and water emerges as a distinct trickle or stream. *Compare* seep.

spring amanita *See* destroying angel.

spring beetles *See* Elateridae.

spring peeper *See Pseudacris crucifer.*

springtails *See* Collembola.

spruce (*Picea*) *See* Pinaceae.

spur 1. A spike that projects outside a **flower**, usually from the base of a **perianth** segment. **2.** A short side **branch** that bears fruit. **3.** In conifers, a side branch that bears leaves.

spurge *See Euphorbia.*

squall A sudden, brief storm in which the wind speed increases up to 50 percent to at least 30 km/h, remains at that speed for at least two minutes, then slowly diminishes.

squall line A number of highly active **cumulonimbus** clouds that merge along a continuous line sometimes more than 900 km long, and advance at right angles to the line.

squama *See* scale.

Squamata (lizards, snakes) An order of reptiles (**Reptilia**) that have small, usually overlapping scales covering the body, a notched or forked tongue, and great flexibility of the bones at the rear of the skull. With more than 7000 species, the order includes 95 percent of all living reptiles. They are divided into two suborders, **Lacertilia** (lizards) and **Serpentes** (snakes), and occur on all continents except Antarctica.

S

squamosal The bone that forms the posterior part of the side of the vertebrate skull. In mammals it articulates with the lower jaw.

squamule *See* squamulose.

squamulose 1. Consisting of or bearing small scales (squamules). **2.** Describes a **lichen thallus** intermediate between **crustose** and **foliose**, that consists of small lobes with or without **rhizines**, and **dorsiventral** differentiation.

square-tail worm *See Eiseniella tetraedra.*

squarrose 1. With a rough or ragged surface. **2. Recurved** at the tip.

squirrels *See* Sciuridae.

squirrel tree frog *See Hyla squirella.*

S-shaped growth curve (sigmoid growth curve) The change that is often seen in the population density of a species entering a new **habitat**. At first the population increases slowly, but then much more rapidly as it exploits newly accessible resources, approaching an exponential rate of growth similar to the **J-shaped growth curve**. Then growth slows as **competition** for resources increases, until the population size and density stabilize. *See* carrying capacity, density dependence.

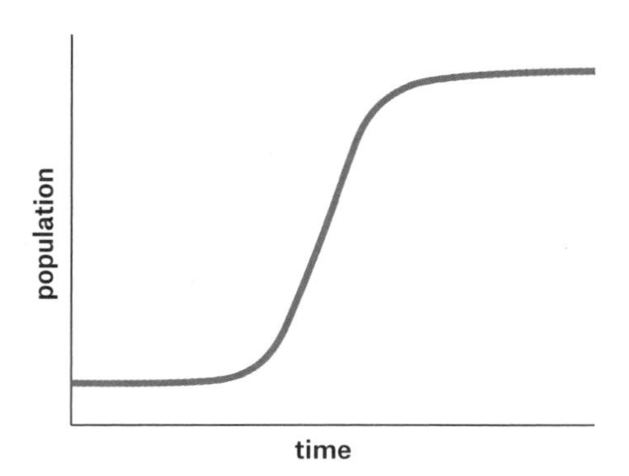

In the S-shaped growth curve the population initially grows very slowly. Growth then accelerates rapidly until the size of the population reaches a balance with the availability of resources to sustain it, when numbers stabilize.

SSP *See* superphosphate.

SSSI *See* site of special scientific interest.

St *See* stratus.

stabilizing selection *See* maintenance evolution.

stable air Air that is neutrally buoyant (*see* buoyancy). It remains at a constant height unless forced to rise and then sinks back to its former level.

Stachyuraceae (order **Crossosomatales**) A **monogeneric** family (*Stachyurus*) of **deciduous** or **evergreen shrubs** and small **trees**, some **scandent**, with leaves **alternate**, **simple**, nearly **orbicular** and **acuminate** to **linear**, **petiolate**, **serrate**, with small **stipules**. Flowers **actinomorphic**, **bisexual** or functionally **unisexual** (plants **dioecious**), hypogynous (*see* hypogyny), **tetramerous**, 4 free **sepals** and **petals**, 8 **stamens** in 2 **whorls**, **ovary superior** of 4 fused **carpels**. **Inflorescence** an erect or pendent **spike** or **raceme**, each flower subtended by 1 **bract** and 2 **bracteoles**. Fruit is a **berry**. There are five species occurring in southeastern Asia. Several cultivated for ornament.

stamen The male organ of a **flower**, comprising a **filament** with an **anther** at its tip; anthers are usually bilobed and united by a **connective**.

staminate Having **stamens**, i.e. male.

staminode A rudimentary or reduced **stamen** that produces no **pollen** but may have a **filament** expanded to form a **petal**-like structure, or modified to form a **nectary**, e.g. in witch hazel (*Hamamelis*).

standard 1. A plant that has been grafted (*see* graft) on to a strong stem so it stands without support. **2.** A **tree** that is allowed to grow to its full size. **3.** A tree that is large enough to be felled for timber.

standard atmosphere (standard pressure) The average sea-level **atmospheric pressure**, assuming that the air is a perfect gas at a temperature of 15°C and the acceleration due to gravity is 9.80655 m/s^2. The standard atmosphere is equal to a pressure of 1.013250×10^5 newtons per square metre (= 101.3250 kPa, 1.01325 millibars, 760 mm of mercury, 29.9213 inches of mercury).

standard pressure *See* standard atmosphere.

standard temperature *See* fiducial point.

standing cloud A stationary cloud.

standing crop *See* biomass.

standing waves *See* lee waves.

Staphyleaceae (order **Crossosomatales**) A family of **deciduous** or **evergreen shrubs** and **trees** with leaves **opposite**, **pinnate compound** with an odd number of **leaflets** that are usually **ovate** to **elliptical**, **acuminate**, **stipulate**, **crenate** or **serrate**. Flowers **actinomorphic**, **bisexual** sometimes **unisexual** (plants **dioecious**), hypogynous (*see* hypogyny), **pentamerous**, 5 free often **petaloid sepals**, 5 free **petals**, 5 **stamens**, **ovary superior** of 2–3 free or 4 fused **carpels**. **Inflorescence** a terminal or **axillary panicle**. Fruit is a **berry**. There are 2 genera of 45 species occurring in northern temperate regions, American tropics, and Malesia. Some cultivated for ornament.

Staphylinidae (rove beetles) A family of beetles (**Coleoptera**) with short **elytra** that leave more than half of the abdomen exposed. Many fly strongly. They are 1–35 mm long and most are shiny black, black and red, or metallic. They occur in damp places and most are predators of other invertebrates. There are about 58,000 species found worldwide.

starch (amylum) A **carbohydrate** that consists of **glucose** units linked by glycosidic (*see* glycoside) bonds. It is the principal storage carbohydrate in plants. There are two forms: typically, plants contain 20–25 percent **amylase**, which has linear molecules, and 75–80 percent **amylopectin**, with branched molecules.

starch sheath **Endodermis** containing grains of **starch**.

starling *See Sturnus vulgaris.*

startle response A rapid reaction to sudden danger that may cause an animal to make a threat **display** or flee; the air from an approaching swatter can set a cockroach running for cover in 20 milliseconds.

stasigenesis The situation in which an **evolutionary lineage** persists over a long period with little or no change and without splitting.

stationary front A **front** where the air on either side is moving approximately parallel to the line of the front. Consequently, the position of the front at ground level does not move, or moves only slowly and erratically.

statismospore A fungal **spore** that is not discharged explosively.

steam devil An almost vertical column of cloud that forms over the surface of a frozen lake.

steam fog Thin, wispy **fog** that forms when cold air moves across the surface of warmer water.

steel-blue worm *See Octolasion cyaneum.*

Stegnospermataceae (order **Caryophyllales**) A **monogeneric** family (*Stegnosperma*) of **shrubs**, **lianas**, and small **trees** with fleshy, **alternate**, **petiolate**, **entire**, **exstipulate** leaves. Flowers **actinomorphic**, **hermaphrodite**, hypogynous (*see* hypogyny), with 5 free **sepals**, 5 **petals**, 5 free or 5–10 fused **stamens**, **ovary** of 3–5 fused **carpels**. **Inflorescence** is a **thyrse**. Fruit is a **capsule**. There are three species occurring in Central America and the Antilles.

Steinernema A genus of **entomopathogenic rhabdite** nematodes (**Rhabditida**) that actively hunt insects by jumping onto passing prey, raising their bodies to attach themselves, or moving through the soil searching for prey. Each nematode species has its own hunting strategy and this dictates the insects it parasitizes. These nematodes are used in **biological control.**

stele In vascular plants (**Tracheophyta**), the central part of roots and stems that contains **vascular tissues**, sometimes with **pith** and **medullary rays**. There are two principal types: **protostele** and **siphonostele**, with subdivisions of each, and either may be a **monostele** or **dictyostele**.

stellate Star-shaped.

stem In vascular plants (**Tracheophyta**), the central **axis** that has **nodes** from which leaves and flowers arise, separated by **internodes**. Stems may be single or branching, and they often provide mechanical support. Stems are usually above ground and are positively phototropic (*see* phototropism) and negatively geotropic (*see* geotropism). *See also* rhizome.

stem and bulb nematode *See Ditylenchus dipsaci.*

stem flow Precipitation that is intercepted by plants and reaches the ground by flowing down their stems.

Stemonaceae (order **Pandanales**) A **monocotyledon** family of **perennial herbs** with a twining, creeping, or upright **habit**, and leaves **alternate**, **opposite**, or in **whorls**. Flowers **actinomorphic** rarely **zygomorphic**, **bisexual** or functionally **unisexual** (plants **monoecious**), **tetramerous** rarely **pentamerous**, 4 **tepals**, 4 **stamens**, **ovary inferior**, semi-inferior, or **superior** of 2 **carpels** with 1 **locule**. **Inflorescence** an **axillary** simple or compound **raceme**, **cymose** cluster, or **umbel**-like of several united clusters. Fruit **baccate** or a **capsule**. There are 4 genera of 27 species occurring from China and Japan to Australia and in southeastern United States.

Stemonuraceae (order **Aquifoliales**) A family of **evergreen shrubs** and **trees** with **alternate**, **simple**, **exstipulate**, **entire** leaves. Flowers **actinomorphic**, **bisexual** or **unisexual** (plants **dioecious**), usually **pentamerous**, **calyx** of 5 free or 4 or 6 fused lobes, 5 free or 4 or 5–7 fused **petals**, as many **stamens** as petals or **corolla** lobes, **ovary superior. Inflorescence cymose.** Fruit is a **drupe**. There are 12 genera of 95 species occurring throughout the tropics.

stem rot *See Sclerotinia sclerotiorum.*

stem tuber *See* tuber.

stenospermocarpy The production of fruit by a technique in which the plant is pollinated and fertilized normally, but the **embryo** is then aborted, resulting in fruits containing small, undeveloped seeds. Most seedless grapes are produced in this way. *Compare* parthenocarpy.

stephanokont A **spore** or **gamete** that has a circle of cilia (*see* cilium) at the anterior end.

steppe The temperate grassland of Eurasia, extending from the Danube Basin to China, and dominated mainly by **drought**-resistant, **perennial** grasses (**Poaceae**).

stepped leader The first stage of a **lightning stroke**, carrying charge away from the cloud.

stereotaxis *See* thigmotaxis.

sterile 1. Describes an organism that is incapable of reproducing. **2.** Describes land or soil that is incapable of sustaining plants, especially cultivated plants. **3.** Describes an **environment** that is totally devoid of living organisms.

sternum The breast bone in **tetrapods**; the bone at the centre of the ventral side of the thorax to which most of the ribs are attached.

steroids A group comprising hundreds of compounds made from 20 carbon atoms arranged in 4 rings fused together. They are synthesized by plants, animals, and fungi and include **phytosterols**, **brassinosteroids**, and certain **alkaloids**.

Stevenson screen A standard container for the **hygrometer**, **maximum thermometer**, and **mimimum thermometer** used at a weather station. It is a white-painted box with louvered walls of double thickness on four sides, standing on legs that raise it so the thermometers are 1.25 m above ground level.

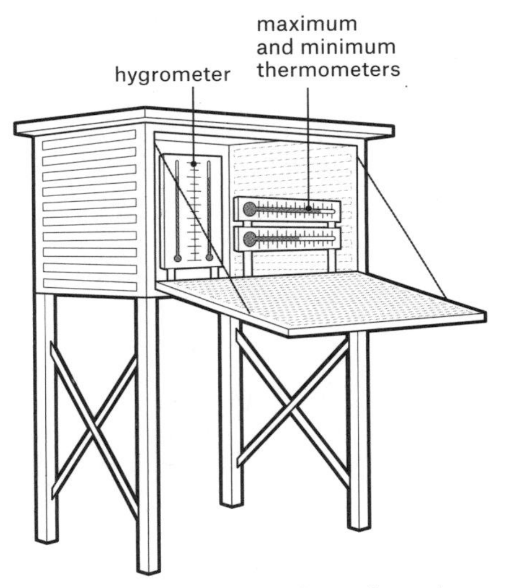

The Stevenson screen is the traditional, white-painted, louvred, standard container for meteorological instruments.

stigma 1. The part of the female reproductive organ in a **flower** on which **pollen grains** germinate. **2.** (pl. stigmata) A dark area on an insect wing.

stigma-height dimorphism A type of reciprocal **herkogamy** in which a plant species produces two types of **flower** that differ only in the length of the **style**. In the L-morph the style is long and in the S-morph it is short.

Stilbaceae (order **Lamiales**) A family of **shrubs**, some resembling ***Erica***, and **herbs**, with **linear** leaves with **recurved** margins in overlapping **whorls**. Flowers with 5 **sepals**, **corolla tubular** with 4 or 5 lobes, 4 **stamens**, **ovary superior** of 2 fused **carpels**, each flower subtended by a pair of **bracteoles**. **Inflorescence** a cluster or head. Fruit a **loculicidal capsule**. There are 11 genera of 39 species, most occurring in Cape Province, South Africa, also in tropical Africa, Madagascar, Arabia, and the Mascarene Islands. Some cultivated for ornament.

stilt root A **tree** root that arises from the lower part of the trunk and extends to the ground at an angle, providing support for the tree.

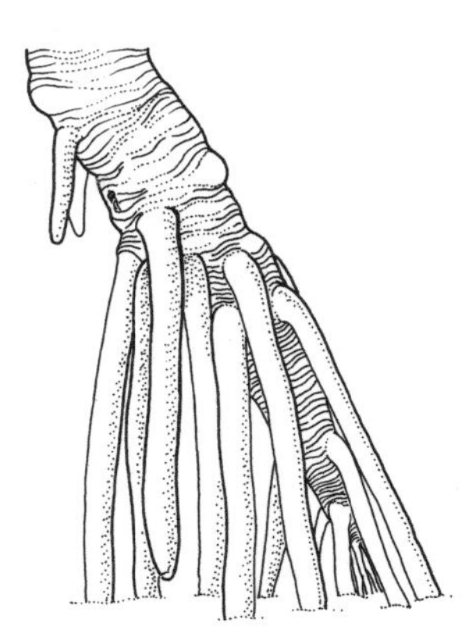

The stilt root arises from the lower part of the stem and provides extra support for the plant.

stinging nettle (*Urtica dioica*) *See Urtica*.

stingless bees A group of about 500 species of **eusocial** bees (**Apidae**) in which the sting is much reduced and cannot be used, although the bees will bite if disturbed. They store **pollen** and honey, though seldom in commercial quantities. The bees occur in the tropics and subtropics, and are active at all times of year.

stingless wasps *See* Trichogrammatidae.

S

stink badgers *See* Mephitidae.

stink bugs *See* Pentatomidae.

stinking smut *See* bunt.

Stipa (family **Poaceae**) A **monocotyledon** genus of **hermaphrodite**, mostly **perennial** grasses with tough, narrow leaves, sometimes with hairs forming a **ligule**. **Spikelets** are needle-like and borne in **panicles**. There are about 300 species occurring in temperate and tropical regions; they are characteristic of **steppe** and **prairie** grasslands.

stipe The stalk of a seaweed or large fungal **fruiting body**.

stipulate Possessing **stipules**.

stipule An outgrowth at or close to the base of a leaf **petiole**. Stipules often occur in pairs and may be leaf-like, or hard with a sharp point, and may provide a protective sheath to the young leaf, or be **adpressed** to the twig or petiole, or be **amplexicaul**.

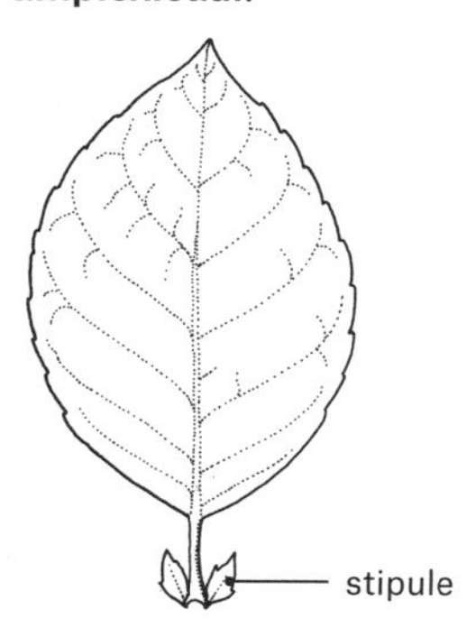

A growth, often hard, pointed, and resembling a leaf, that usually develops near the base of the leaf petiole.

stoats *See* Mustelidae.

stock 1. A plant onto which a **scion** is grafted (*see* graft). **2.** (rootstock) A root or part of a root that is used to propagate a plant. **3.** The common name for several species of *Matthiola* and *Malcolmia* (**Brassicaceae**).

stolon A horizontal **stem**.

stoma (pl. stomata) A pore in the **epidermis**, opened and closed by **guard cells**, through which a plant exchanges gases with the outside air.

stomata *See* stoma.

stomium An area of thin-walled cells in a **pollen sac** or **sporangium** that ruptures when dry to release the **pollen grains** or **spores**.

stoneworts *See* Charophyceae.

storage organ A plant organ that is modified to store energy, usually as **carbohydrate**, or water. Storage organs include **bulbs**, **corms**, **pseudobulbs**, **rhizomes**, **taproots**, **tubers**, etc.

storm A wind of 25–28 m/s. *See* appendix: Beaufort Wind Scale.

Stramenopila *See* Heterokontophyta.

stramineous Straw-coloured.

strangler A plant that depends on another plant for physical support and eventually envelops its host with **branches** or aerial roots, often killing it.

strangler fig One of several ***Ficus*** species native to tropical forests that first grows as an **epiphyte** in the crown of a host tree from seeds dropped by birds or monkeys. Roots from the seedling descend the tree and root when they reach the ground. The roots then anastomose (*see* anastomosis), finally surrounding and strangling the host tree. When the host has decomposed, the fig survives as a very large tree with a hollow centre. Banyans (*F. benghalensis*) are strangler figs that also produce **adventitious** roots from their **branches**, producing a grove of densely crowded roots.

strangleweed *See* japweed.

Strasburgeriaceae (order **Crossosomatales**) A family of **evergreen trees** with **alternate**, **obovate**, **coriaceous**, **petiolate**, **sinuate** or **dentate** leaves. Flowers **actinomorphic**, **bisexual**, hypogynous (*see* hypogyny), with 8–11 **imbricate sepals**, usually 5 imbricate **petals**, 10 **stamens**, **ovary superior** of 5 sometimes 4 or 5–7 fused **carpels**. **Inflorescence** a solitary flower. Fruit **indehiscent**. There are two genera of two species occurring in New Zealand and New Caledonia.

stratiform Forming horizontal layers (strata).

stratocumulus (Sc) A genus (*see* cloud classification) of low, white, grey, or white and grey clouds that form patches, sheets, or layers, always with dark, round masses or rolls.

stratopause The upper boundary of the **stratosphere**, its height varying with latitude and season from about 48 km over the equator to 60 km over the poles.

stratosphere The layer of the atmosphere that lies between the **tropopause** and the **stratopause**.

stratus (St) A genus (*see* cloud classification) of uniformly grey, low clouds.

strawberry (*Fragaria*) *See* diageotropism.

strawberry aphid *See Chaetosiphon fragaefolli.*

strawberry blossom weevil *See Anthonomus rubi.*

strawberry mite *See Phytonemus pallidus* ssp. *fragariae.*

strawberry rhynchites *See Neocoenorrhinus germanicus.*

strawberry seed beetle *See Harpalus rufipes.*

strawberry tortrix moth *See Acleris comariana.*

strawberry tree *See Arbutus.*

strawbreaker *See* eyespot.

straw-coloured apple moth *See Blastobasis decolorella.*

streak Any **virus** disease of plants, usually **monocotyledons**, in which yellow or necrotic streaks appear on leaves.

streak lightning A **lightning stroke** with branches from the main **lightning channel**.

streaming *See* cytoplasmic streaming.

streamline The track moving air follows.

Strelitzia (family **Strelitziaceae**) A genus of **monocotyledon**, large, **evergreen**, **perennial herbs**, some with woody stems, that have **distichous** leaves up to 2 m long and 80 cm wide, resembling those of the banana, but with a longer **petiole**. Flowers form a horizontal **inflorescence** emerging from a stout **spathe**; both flower and **bracts** are brightly coloured orange and blue. They are pollinated by sunbirds. There are five species **endemic** to South Africa. Some cultivated for ornament as bird-of-paradise flower or crane flower.

Strelitziaceae (order **Zingiberales**) A **monocotyledon** family of **perennial herbs** with large, **sympodial rhizomes**, either **acaulescent** or tree-like with **pseudostems** formed by sheathing leaf bases. Leaves **alternate** and **distichous**, **entire**. Flowers **zygomorphic**, **bisexual**, with 3 **tepals**, 6 or 5 **stamens**, **ovary inferior** of 3 fused **carpels**. **Inflorescence** terminal or lateral with tough, boat-shaped **spathes** each supporting a cincinnate (*see* cincinnum) cluster of flowers. Fruit is a woody **capsule**. There are three genera of seven species occurring in tropical South America, eastern southern Africa, and Madagascar. Some cultivated as ornamentals, e.g. ***Strelitzia***.

Streptocarpus (family **Gesneriaceae**) A genus of **annual** or **perennial herbs** and **subshrubs**. They have 1 (unifoliate) or many (plurifoliate) leaves. Plurifoliate species may have a basal rosette of leaves (the rosulate form), or 2 or 3 leaves appearing after the first leaf. Unifoliate forms have only 1 leaf, which grows continuously from the base; many unifoliate species are also **monocarpic**. Flowers **zygomorphic** and **pentamerous**, the **corolla** forming a tube with the ends flattened. There are about 155 species occurring in tropical and southern Africa and

Madagascar. Several cultivated for ornament, as Cape primrose.

Streptomyces A genus of **Actinobacteria** that are Gram-positive (*see* Gram reaction), filamentous **aerobes** that produce variously coloured filament networks resembling a fungal **mycelium** on which they bear **spores**. Most are **saprotrophs** living in soils, and many produce antibiotics.

Streptopelia decauocto (collared dove, Eurasian collared dove) A slender, grey, pinkish grey, or buff dove with a distinctive black half-collar edged in white, red legs, and a black beak. It is 300–330 mm long. At the end of the 19th century it occurred in subtropical and warm temperate Asia. It then expanded and by the end of the 20th century it occurred in the Faroe Islands, arctic Norway, the Ural Mountains, the Canary Islands, and North Africa, and central and northern China. It has also been introduced elsewhere, including North America, where it is widespread. It is gregarious and is often found near human dwellings. It feeds on seeds.

Streptophytina A subphylum that includes the **Charophyceae** and **Embryophyceae**.

striate Marked with fine furrows or ridges.

S

Strigiformes (owls) An order of birds of prey, mostly nocturnal, which vary greatly in size. They have large heads, eyes that face forward, short necks, and short, hooked beaks. Their wings are broad, rounded, and have soft tips so the birds fly silently. Their large eyes are more or less fixed, so to see to the sides an owl must turn its head. Owls have excellent hearing and the characteristic facial disc and ruff channel sound waves to the ears. Owls inhabit forests, open country, grassland, farmland, and desert, feed on small mammals, birds, reptiles, fish, and insects, and nest in hollow trees or burrows. There are two families: Tytonidae (barn owls) and Strigidae (typical owls), together comprising 27 genera and more than 200 species, found worldwide.

strigose With stiff hairs all aligned in the same direction.

striped cucumber beetle (*Acalymma vittatum*) *See* bacterial wilt.

strobilus 1. A group of **sporophylls** bearing sporangia (*see* sporangium) and arranged around an **axis. 2.** (cone) In **angiosperms**, any **cone**-shaped structure, e.g. the fruits of alder (***Alnus***) and hop (*Humulus*), not to be confused with the cone of **gymnosperms.**

stroma 1. The matrix of **chloroplasts** in which the grana (*see* granum) are embedded. **2.** A mass of fungal **mycelium** in or on which **spore**-bearing structures may develop.

Strombosiaceae (order **Santalales**) A family of mostly **evergreen trees** and **shrubs** with spiral, **alternate**, or **distichous**, **exstipulate**, **petiolate**, **simple**, **entire**, **glabrous** leaves. Flowers **bisexual**, **tetramerous** or **pentamerous**, **calyx** cup-shaped alternating with **petals**, 4–5 or 8–10 **stamens adnate** to petals, **ovary superior**. **Inflorescence** an **axillary fascicle** of **cymes** or **racemes**. Fruit is a **drupe**. There are 6 genera of 18 species with a scattered pantropical distribution.

strong breeze Wind of 11–14 m/s. *See* appendix: Beaufort Wind Scale.

Stropharia aeruginosa (verdigris agaric) A species of **agaric fungi** in which the **fruiting body** has a bell-shaped **pileus** 20–80 mm across, initially bright blue-green with white remnants of the **partial veil** around the edges and very slimy, turning more ochre as it matures. The blue or green **stipe** is 30–80 mm long, about 10 mm thick, and cylindrical. The **gills** are pale grey, turning purple-brown later. A

saprotroph, the fungus grows on rotten wood in grass, in woodland, on roadside verges, and in gardens on wood-chip mulches, and occurs throughout Europe and in parts of Asia and North America. It may be poisonous.

structural gene A **gene** that codes for an **amino acid**.

strychnine An **alkaloid** ($C_{21}H_{22}N_2O_2$) produced by *Strychnos* species (**Loganiaceae**), especially the tree *S. nux-vomica*, and stored in the seeds. It is extremely poisonous to animals.

stubby-root nematodes Nematodes (**Nematoda**) of the family Trichodoridae (*see Trichodorus*) that are **ectoparasites** feeding on plant roots, causing stunting and giving the roots a stubby appearance. They puncture holes in plant cells, feeding mainly on **meristem** cells. They infest turf, tomatoes, and some trees and occur in most temperate and subtropical regions.

Sturnus vulgaris (starling, European starling) A **passerine** bird, 200 mm long, that has glossy, speckled black plumage, pink legs, and a beak that is black in winter and yellow in summer. It lives in the lowlands wherever there are holes for nesting and vegetation for feeding. They feed on seeds, fruit, other plant material, and invertebrate and vertebrate animals. Except during the breeding season, they tend to congregate in vast flocks. Starlings occur throughout the **Palaearctic** and have been introduced to North America, so today they are found worldwide except for the New World tropics.

Sturt's desert pea (*Swainsona formosa*) *See Swainsona.*

style The **flower** structure growing upward from the **carpel** that supports the **stigma**.

Stylidiaceae (order **Asterales**) A family of **annual** and **perennial herbs**, climbers, creepers, **subshrubs**, and **cushion plants** with leaves often as a basal rosette, alternatively **alternate** with a terminal rosette, whorled, or **imbricate**, **simple**, **sessile** or **petiolate**, **exstipulate**. Flowers usually **zygomorphic** occasionally **actinomorphic**, **bisexual**, **calyx** with 5 free or 5–9 fused lobes, the tube **adnate** to the **ovary**, **corolla sympetalous** with 5 free or 4 or 5–9 fused lobes, 2 **stamens**, ovary **inferior. Inflorescence** a terminal **raceme**, **panicle**, or **cyme**, rarely **corymb** or flowers solitary. Fruit is a **capsule**. There are 3 genera of 245 species occurring scattered from southeastern Asia to New Zealand and southern South America.

Styracaceae (order **Ericales**) A family of **evergreen** or **deciduous trees** and **shrubs**, most **pubescent** or scaly, with **alternate**, **simple**, **exstipulate**, usually **entire** leaves. Flowers **actinomorphic**, usually **bisexual** or female (plants **gynodioecious**), **calyx** tubular, persistent, with 4–5 free or 5–9 fused lobes, **corolla** with 5 sometimes 4 or 5–7 **valvate** lobes, twice as many **stamens** as corolla lobes, **ovary superior**, semi-**inferior**, or inferior of 5 fused **carpels** with 3–5 **locules. Inflorescence** a terminal or **axillary raceme**, **cyme**, or **panicle**, rarely a **fascicle** or flowers solitary. Fruit is a **capsule** or **drupe**. There are 11 genera of 160 species occurring in northern warm temperate to tropical regions of both hemispheres. **Bark** from the trees, especially *Styrax benzoin* yields **resin** used medicinally (e.g. friar's balsam) and in incense.

subalpine Describes a mountainside immediately below the **tree line**.

subalpine forest Forest dominated by conifers and resembling **boreal forest**, but composed mainly of different species, that grows in **subalpine** regions in temperate latitudes with a few extensions into the northern tropics, the elevation of the forest increasing with decreasing latitude.

suberin A waxy, water-repellant substance found in the **cell walls** of endodermal cells (*see* endodermis) in vascular plants (**Tracheophyta**); in the roots this tissue forms the **Casparian strip**. It is the main ingredient of **cork**. Suberin waterproofs and protects tissues.

suberization The deposition of **suberin**.

subhymenium A layer of tissue beneath a fungal **hymeniium**.

sublimation The change of ice directly into water vapour, without passing through a liquid phase.

submergence marsh The seaward part of a **salt marsh**, extending between the level of mean high water and mean high water of neap tides.

subpolar low A belt of low **atmospheric pressure** in latitudes 60°–70° where the **polar easterlies** and **midlatitude westerlies** converge in both hemispheres.

subpolar region The area between the edge of cool temperate or cool desert vegetation and the poleward boundary of the **tundra**.

subshrub A plant that is smaller than a **shrub** and woody only at the base, with **branches** arising from the base, the upper parts of which die back at the end of each growing season.

subsidiary cell (accessory cell) One of the cells in the **epidermis** that surround the **guard cells**.

subsidence The general sinking of air, producing high surface pressure.

subsoil The lower part of the soil, from the B **soil horizon** to the **parent material**.

subsoiling The breaking up of compacted **subsoil** without inverting it, usually by dragging a chisel-like implement through it.

subspecies A **race** of a **species** that is given a Latin name. The process of naming a subspecies is somewhat arbitrary, but to qualify the plants should be geographically distinct, different from other geographic populations, and a population rather than a phenotypic variation.

substrate 1. The reactant on which an **enzyme** acts to catalyze a chemical reaction. **2.** (substratum) The surface to which an organism is attached. **3.** An underlying substance or layer.

substratum An underlying layer, or the material to which an organism is anchored and on which it grows. *See* substrate.

subtropical fronts The boundaries between the poleward edges of the **Hadley cells** and the equatorward edges of the **Ferrel cells**, at about latitude 30° in both hemispheres.

subtropical high One of several semi-permanent **anticyclones** situated over the subtropical oceans. They tend to block the eastward movement of midlatitude **cyclones**.

subtropics The regions in both hemispheres between the **tropics** (latitude 23.5°) and latitudes 35°–40°.

succession A sequential change in the plants and associated animals inhabiting an area, due to a change in the **environment** or to the properties of the organisms and the relationships among them. Classically, the term describes the **colonization** of a new site and the changes that occur before a stable **climax** is reached. *See* primary succession, secondary succession.

succinyl-coenzyme A (succinyl-CoA) An important intermediate product in the **citric acid cycle**, where it is synthesized from alpha-ketoglutaric acid and coenzyme A, then catalyzes a reaction that produces **adenosine triphosphate** (ATP).

succulent Fleshy, usually because the tissue is storing food or water.

sucker An underground shoot that arises from the root or lower part of the stem of a woody plant and forms a new plant, initially nourished by the parent.

S

sucrase (invertase) An **enzyme** that catalyzes the **hydrolysis** of **sucrose** to **fructose** and **glucose**.

sucrose (saccharose, table sugar) A disaccharide sugar ($C_{12}H_{22}O_{11}$) formed from **fructose** and **glucose** that occurs widely in plants as an energy store. It is produced commercially from sugar beet (*Beta vulgaris*) and sugar cane (*Saccharum* spp.).

Sudanese park-steppe floral region The area that includes most of the Sahel region, from the Atlantic coast of Africa to the border of Ethiopia, part of the **Palaeotropical region**. The vegetation is sparse and dominated by palms, grasses, and thorn trees (*Vachellia* spp., *see Acacia*).

sudden oak death *See Phytophthora ramorum.*

suffrutescent Describes a plant that is intermediate between a **herb** and a woody plant.

suffruticose chamaephyte A **chamaephyte** in which the parts above ground partially die back at the onset of unfavourable conditions and **buds** arise on the lower, persistent part of the stem.

sugar Any member of a large group of **carbohydrates** produced by plants. Simple sugars (monosaccharides) are joined to form a wide variety of disaccharides and **polysaccharides**.

sugar apple (*Annona squamosa*) *See Annona.*

sugar beet *See* Amaranthaceae.

sugar beet nematode *See Heterodera schachtii.*

sugar cane (*Saccharum officinarum*) *See Saccharum.*

sugar-kelp (*Saccharina latissima*) *See* sea belt.

sugar plum *See Amelanchier.*

sugar snow *See* depth hoar.

sulcal *See* sulcate.

sulcate (sulcal) With surface ridges or furrows.

sulcus A furrow or groove.

sulphur 1. (S) An element that is an essential plant **macronutrient**. It stabilizes the structure of some **amino acids** and is involved in **redox reactions** and **photosynthesis**. Deficiency causes **chlorosis** and **etiolation**. **2.** *See* sulphur fungus.

sulphur butterflies *See* Pieridae.

sulphur cycle The cyclical flow of **sulphur** from rocks, through living organisms, air, and water. Sulphur enters the cycle as sulphur dioxide (SO_2) and sulphate (SO_4) in gases erupted by volcanoes and by the **chemical weathering** of rocks. Living organisms absorb sulphate and incorporate its sulphur in **proteins**. Sulphur dioxide dissolves in water and is transported to the oceans, from where it enters the air, principally as **dimethyl sulphide**, and is carried back over land.

sulphur fungus (chicken fungus, chicken mushroom, chicken of the woods, rooster combs, sulphur polypore, sulphur shelf) The **fruiting bodies** of about 12 *Laetiporus* species, which appear as brackets, typically in tiers, with an uneven upper surface resembling suede, initially bright yellow or orange, later fading to dull white, and a yellow underside. The **spore print** is white. They occur on many **broad-leaved** trees, especially oak, throughout Europe and North America, and their presence indicates that the tree is dying or already dead. They are edible and delicious.

sulphuric horizon A subsurface **soil horizon**, at least 15 cm thick, in which sulphuric acid (H_2SO_4) forms through the oxidation of sulphides. It is extremely acid, with a **pH** of less than 3.5.

sulphur oxidation The oxidation of sulphur and sulphur compounds by soil

S

sulphur **bacteria**. *Thiobacillus denitrificans* is capable of anaerobic **respiration**, in which it oxidizes sulphur to sulphate (SO_4), which is soluble and available to plants. *Thiobacillus thiooxidans* oxidizes sulphur to sulphuric acid (H_2SO_4).

sulphur shelf *See* sulphur fungus.

sumach (*Coriara*) *See* root nodule.

sumac *See Rhus.*

sun cracks *See* desiccation cracks.

sundew *See Drosera.*

sunflies *See* Syrphidae.

sunflower (*Helianthus annuus*) *See Helianthus.*

sunrose *See Helianthemum.*

superadiabatic Describes the condition where the **environmental lapse rate** is steeper than the **dry adiabatic lapse rate**.

supercell A **convection cell** that develops in very large **cumulonimbus** clouds. Currents carrying air upward rise at an angle to the vertical, so instead of falling from high in the cloud into the rising air, thereby cooling and suppressing it, the cold precipitation falls to the side, allowing the cell to survive and continue growing.

supercilium A white mark, resembling an eyebrow, above the eye of a bird.

S

supercooling Reducing the temperature of water droplets to below freezing without triggering the formation of ice. Most large clouds contain supercooled droplets.

supergene A segment of a **chromosome** that is transmitted intact from one generation to the next, i.e. it is not affected by **crossing over**. It comprises several linked (*see* linkage) **gene** loci (*see* locus) that control a single **character** or group of interrelated traits.

superimposed drainage *See* epigenetic drainage.

superior Describes a plant **ovary** that is attached to the **receptacle** above the level at which the **calyx**, **corolla**, and **stamens** are attached, and free from them. *Compare* half-inferior, inferior.

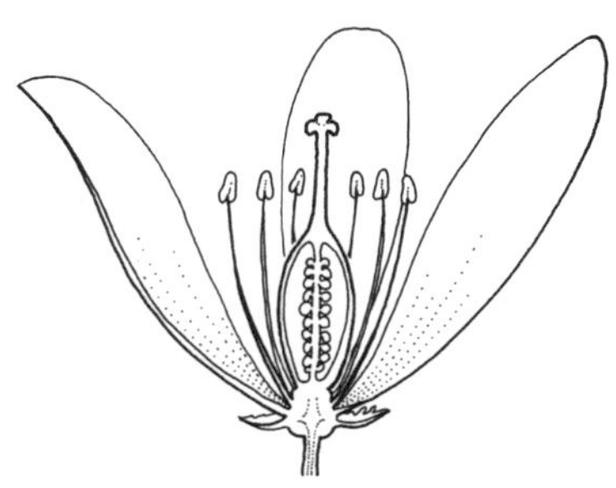

A superior ovary is inserted above the other reproductive organs.

superphosphate A slow-release phosphate **fertilizer** that is made by treating **rock phosphate** with sulphuric acid: $Ca_3(PO_4)_2 + 2H_2SO_4 \Rightarrow Ca(H_2PO_4)_2 + 2CaSO_4$; $Ca(H_2PO_4)_2$ is monocalcium phosphate, also called single superphosphate (SSP). Triple superphosphate is made by reacting rock phosphate with phosphoric acid (H_3PO_4).

supersaturation The condition of air when the **relative humidity** exceeds 100 percent without triggering **condensation**.

supertramp A species that is widely distributed, distributes its seed efficiently, and colonizes readily, but does not compete well and so tends to be excluded from **habitats** supporting a range of species.

supporting services *See* ecosystem services.

suppressor mutation A second **mutation** that masks the phenotypic (*see* phenotype) effects of an earlier mutation. It occurs at a different site in the **genome**, so it is not a **reverse mutation**.

surface boundary layer *See* friction layer.

surface inversion (ground inversion) A **temperature inversion** that commences at ground level.

surface runoff (quick flow, runoff) The flow of rainwater or melting snow across the ground surface and into streams, rivers, ponds, or lakes.

surface soil The A **soil horizon**, which lies immediately below the O horizon of largely undecomposed plant litter.

surface temperature *See* shelter temperature.

surface tension The mutual attraction of molecules at the surface of a liquid, where they experience lateral attraction and attraction from below but not from above, that allows them to resist pressure, as though there were a film across the surface.

surface-water gley soils Slowly permeable soils that are seasonally waterlogged and show prominent mottling with 40 cm of the surface. They comprise a major group in the **soil classification** devised by the Soil Survey for England and Wales.

surface wetness gauge *See* dew gauge.

Surianaceae (order **Fabales**) A family of **trees** and **shrubs** with **alternate**, **simple**, **elliptical**, or **pinnate** leaves. Flowers **actinomorphic**, **bisexual**, with 5 **sepals**, 5 free **petals** or **apetalous**, 5 or 10 **stamens**, 5 or no **staminodes**, **ovary superior** of 1, 2, or 5 free **carpels**. **Inflorescence** a terminal or **axillary panicle**. Fruit is a **drupe**, **berry**, or **nut**. There are five genera of eight species with a pantropical distribution, especially Australia.

suspension A mixture of solid particles that are dispersed through a liquid. The solid particles are usually larger than 1 μm, which means they are large enough to settle and form a sediment if the mixture remains undisturbed, and to remain in suspension the mixture must be periodically agitated.

suspensor 1. Following the asymmetrical division of the zygote in seed plants (**Spermatophyta**) and some vascular **cryptogams**, a large cell that elongates, pushing the **embryo** into the **endosperm**. **2.** During reproduction in **Mucorales**, a **hypha** that supports each **gametangium**.

Svalbard Global Seed Vault A **seed bank** located near Longyearbyen, on the Norwegian island of Spitsbergen that contains seeds that duplicate those held in other seed banks around the world as insurance against the accidental loss of those banks.

Swainsona (family **Fabaceae**) A genus of **herbs** and small **shrubs**, usually **glabrous**, with leaves **alternate**, **pinnate** with many **leaflets**, often **stipulate**. Flowers with a **calyx** tube with teeth of different lengths, **petals** in a butterfly shape, 10 **stamens**, 9 of them fused, **ovary superior**. **Inflorescence** is an **axillary raceme**. Fruit is a **pod**. There are 85 species, all but one **endemic** to Australia, the other (*S. novae-zelandiae*) endemic to New Zealand, some grown for ornament. *Swainsona formosa* (Sturt's desert pea) is the floral emblem of South Australia.

swallow *See Hirundo rustica.*

swamp An area that is ordinarily covered by water throughout the year, with a vegetation comprising **emergent** aquatic plants. In Europe, herbaceous wetlands are described as swamps; in North America these are called marshes and swamps must be forested.

swarm cell *See* swarmer.

swarmer (swarm cell) A flagellated (*see* flagellum), **haploid**, **amoeboid** cell, lacking a **cell wall**, that is produced when a **Myxogastria spore** germinates. The cell matures into a **gamete**. If the gamete encounters another of the correct **mating type** they fuse to form a **zygote**; zygotes then grow into a **plasmodium**.

S

swarm spore *See* zoospore.

sweat flies *See* Syrphidae.

sweat gland *See* eccrine gland.

sweeping The mechanism by which **raindrops** grow by merging with smaller droplets they encounter as they fall, and other small droplets are swept into the wake of the bigger drops, also merging with them.

sweet box *See Sarcococca.*

sweet chestnut *See* Fagaceae.

sweet flag *See* Acoraceae and *Acorus.*

sweet marjoram (*Origanum majorana*) *See Origanum.*

sweet pea *See Lathyrus.*

sweet pepper (*Capsicum annuum*) *See Capsicum.*

sweet potato (*Ipomoea batatas*) *See* Convolulaceae.

sweet potato chlorotic stunt A disease of sweet potato plants in which middle and lower leaves turn red or chlorotic (*see* chlorosis). It is caused by a crinivirus and transmitted by the whitefly ***Bemisia tabaci***. It occurs throughout the tropics and especially in sub-Saharan Africa.

sweetpotato whitefly *See Bemisia tabaci.*

sweetsop (*Annona squamosa*) *See Annona.*

sweet william (*Dianthus barbatus*) *See Dianthus.*

Swietenia (mahogany) *See* Meliaceae.

Swiss cheese plant (*Monstera deliciosa*) *See* Araceae.

syconium The fruit of a fig (***Ficus*** spp.), comprising an enlarged **receptacle** enclosing 50–7000 **florets**, depending on the species, on its inner surface. Syconia may be **monoecious** or **dioecious**. The florets are pollinated by specialized wasps (*Blastophaga* spp.) and the true fruit is an **achene** or **drupe.**

syllepsis The condition in which an **axillary** shoot grows from a **bud** in the same season the bud forms, and grows at the same time as the **apical meristem**, so side **branches** and leaves appear as the stem elongates. *Compare* prolepsis.

symbiont An organism that participates in a **symbiosis**.

symbiosis Any relationship in which members of different species live in close association. *See* commensalism, mutualism, parasitism.

sympetalous Having the **petals** fused, at least at the base.

Symphoricarpus (snowberry) *See* Caprifoliaceae.

Symphyta (sawflies, woodwasps) A suborder of **Hymenoptera** that lack the 'wasp waist' characteristic of ants, bees, and wasps. The large **ovipositor** is used to pierce plant tissue into which the female lays eggs. Larvae resemble **caterpillars**, but differ in possessing up to seven pairs of abdominal **prolegs**. There are about 8000 species with a worldwide distribution.

symplast The **cytosol** of a cell together with the **plasmodesmata** and interconnecting channels between the **cytoplasm** of adjacent cells. This provides a transport route for water and solutes through the cytoplasmic continuum.

Symplocaceae (order **Ericales**) A family of **evergreen** occasionally **deciduous trees** and **shrubs** with **distichous** or spirally arranged, **simple**, **dentate**, **entire**, **petiolate**, **exstipulate** leaves. Flowers **actinomorphic**, **hermaphrodite**, usually **pentamerous**, with 5 free or 3–5 fused **sepals**, 5 free or 3–11 fused **petals**, sometimes 4, usually 5–60 **stamens**, **ovary inferior** or semi-inferior of 2–5 fused **carpels** each with 1 **locule. Inflorescence axillary** sometimes terminal, a **spike**, **raceme**, **cyme**, **thyrse**, **panicle**, or flowers solitary.

Fruit is a **drupe**. There are 2 genera of 320 species occurring in tropical and subtropical regions, but not in Africa.

sympodial A form of branching in which each **branch** arises from the tip of the preceding one, giving the appearance of a single **axis**.

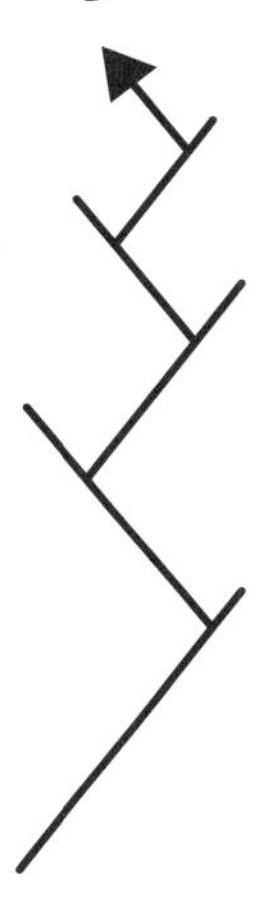

In sympodial branching, the main stem is formed from a sequence of lateral branches growing from nodes near the apex.

synandrous Having the **stamens** fused.

synangium A cluster of fused sporangia (*see* sporangium).

Synanthedon tipuliformis (currant clearwing) A small moth with a wingspan of 17–20 mm that superficially resembles a wasp, having a black body with three narrow yellow bands and clear wings. It flies by day between late May and July. Its larvae bore into the stems of ***Ribes*** species to feed and are sometimes a serious pest. The moth occurs throughout the temperate Northern Hemisphere and in Australia.

synanthrope An animal that lives close to humans because it benefits from environmental modifications made by humans, e.g. feral pigeon (***Columba livia***), house sparrow (***Passer domesticus***), and house mouse (*Mus musculus*).

syncarp A fruit consisting of two or more **carpels**.

syncarpous Having the **carpels** fused together along their edges (i.e. **concrescent**).

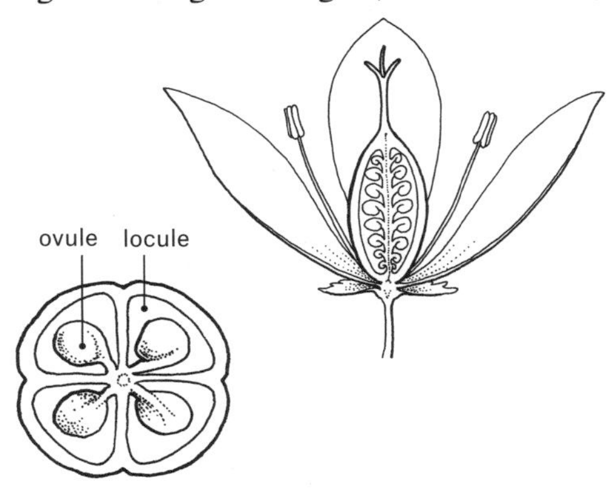

Syncarpous. The carpels are fused together at their edges, in this example containing four locules each with one ovule.

Synchytrium endobioticum A species of chytrid (*see* Chytridiomycota) **Fungi** that does not produce a **mycelium**. It forms sporangia (*see* sporangium) each containing 200–300 **zoospores** and gathered in sori (*see* sorus). These overwinter (winter sporangia) in the soil and germinate when the soil is moist and the temperature reaches about 8°C, releasing the zoospores which penetrate plant cells. The fungus is an **obligate parasite** of *Solanum* species, especially potatoes (*S. tuberosum*) and causes potato wart disease, also called potato canker and black scab. Symptoms are abnormal growth leading to the formation of warts on all underground parts except the roots. The fungus occurs throughout Europe and in parts of North Africa, southern Africa, Asia, New Zealand, and Canada, but has been eradicated in the United States.

syncytium A multinucleate cell that results from the fusion of many single cells, each with 1 nucleus. *Compare* coenocyte.

syndetocheilic Describes a **stoma**, found in some **gymnosperms**, in which the **guard**

cells and **subsidiary cells** are derived from the same mother cell. *Compare* haplocheilic.

synecology The study of entire plant and animal communities.

syneresis cracks *See* desiccation cracks.

synergism (synergy) The interaction among forces, processes, or substances all of which tend in the same direction so the overall effect is larger than the sum of the individual components.

synergy *See* synergism.

syngamy Following **fertilization**, the fusion of the nuclei of two **gametes** to form a **zygote** nucleus.

syngeneic *See* isogeneic.

syngenesious Describes an **androecium** in which the **anthers** are fused, e.g. in **Asteraceae.**

synoptic Describes a view over a wide area.

synoptic chart A geographic map marked with **isobars** and symbols so it represents atmospheric conditions over a large area at a particular time.

synpetalous Having the **petals** fused.

synsepalous Having the **sepals** fused.

syntepalous Having the **tepals** fused.

synthetic theory The modern theory of evolution that includes Darwinian theory, Mendelian inheritance (*see* Mendel's laws), and **genetics.**

synusia A plant community that is defined by its life form or height, e.g. shrub layer, rather than the species it contains.

synzoochory *See* zoochory.

Syringa (family **Oleaceae**) A genus of **shrubs** and small **trees** (lilac) with **opposite** (or occasionally in **whorls**), **simple, cordate** to **lanceolate**, occasionally **pinnate** leaves. Flowers **bisexual**, with a 4-lobed **corolla. Inflorescence** a **panicle.** Fruit is a **capsule.** There are 20–25 species occurring from southeastern Europe to eastern Asia. They are widely cultivated.

syringa *See Philadelphus.*

syrinx The vocal organ of birds.

Syrphidae (hoverflies, drone flies, flower flies, sweat flies, sunflies) A family of flies (**Diptera**) that vary greatly in size. Adults may have yellow and black bands giving them a superficial resemblance to wasps (**Batesian mimicry**), or be marked with stripes or spots, and feed mainly on **nectar** and **pollen.** The maggot-like larvae are **herbivores** feeding inside plants or on roots, **carnivores** feeding on aphids (**Aphididae**), thrips (**Thysanoptera**), and other larvae, or scavengers living in mud, stagnant water (e.g. rat-tailed maggot, the larvae of the drone fly, *Eristalis tenax*), wasp and bee nests, and decaying vegetation. There are about 6000 species found worldwide.

systemic Describes a **herbicide** or **insecticide** that spreads throughout a plant exposed to it. A systemic herbicide kills all of the plant, rather than only the exterior tissues exposed to the spray. A systemic insecticide renders the whole plant toxic to insects.

Syzgium aromaticum (clove) *See* Myrtaceae.

T

2,4,5-T (2,4,5-trichlorophenoxyacetic acid) A **herbicide** closely related to **2,4-D** and **MCPA** that is cheap to manufacture and highly effective as a defoliant of **broad-leaved** plants. It achieved notoriety when Agent Orange, a mixture of 2,4-D and 2,4,5-T, was used by U.S. forces during the Vietnam War and contained dioxin, a contaminant formed during manufacture. Although the concentration of dioxin was very small it was responsible for birth defects and possibly cancer in people exposed to it. It has been banned in the United States and Germany, but is still used elsewhere under strictly controlled conditions.

2,4,5-trichlorophenoxyacetic acid *See* 2,4,5-T.

table sugar *See* sucrose.

Taccaceae (order **Dioscoreales**) A monogeneric, monocotyledon family (*Tacca*) of perennial herbs with rhizomes and basal, dissected, petiolate leaves. Flowers actinomorphic, **bisexual**, with 12 petaloid tepals in 2 whorls, 6 stamens in 2 whorls, ovary inferior of 3 **carpels** with 1 **locule**. Inflorescence **cymose** or umbel-like on a radical scape, subtended by an **involucre** of 4 occasionally 2 or up to 12 **bracts**. Fruit is a **berry**, rarely a loculicidal **capsule**. There are 12 species with a pantropical distribution. Tubers of *T. leontopetaloides* are a source of arrowroot.

Tachinidae (parasitic flies) A family of robust-looking, bristly flies (**Diptera**) most of which are drab but with some species brightly coloured. Their maggots are cylindrical rather than being tapered and most are **parasitoids**, developing inside a living host and killing it; others are parasites that do not kill their hosts. There are more than 10,000 species and they occur worldwide.

Tachycineta bicolor (tree swallow) A species of migratory **passerine** birds, 140 mm long, that are iridescent green-blue on the head, back, wings, and tail, and white on the underside. They live in open areas near water and feed mainly on insects caught in flight, augmented by plant material. They breed throughout northern and central North America and winter in the southern United States and along the eastern coast of Central America.

tachytely A rate of evolutionary change of a group within a **taxon** that is markedly faster than the average rate for that taxon. It usually occurs through **adaptive radiation** when a **species** enters a new environment. *Compare* horotely.

tactic movement *See* taxis.

tadpole An amphibian (**Amphibia**) larva.

tagma (pl. tagmata) In segmented animals, a group of segments that form a functional unit, e.g. head, thorax, and abdomen.

tagmosis During the evolution of segmented animals, the fusion of segments to form tagmata (*see* tagma).

taiga Depending on the authority, either the whole of the **boreal forest**, or the **lichen woodland**.

taipans *See* Elapidae.

takyric horizon A surface **soil horizon** that develops in soil that is periodically flooded in an overall arid climate (*takyr* is Uzbek for **barren** land). There is a surface **crust** with a platy layer below it. When the soil is dry, **desiccation cracks** form that are at least 2 cm deep.

Talauma (family **Magnoliaceae**) A genus of **evergreen shrubs** and **trees** closely related to ***Magnolia***, but differing in having the **carpels** forming a **dehiscent** woody or cartilaginous mass separate from the **axis.** Leaves **petiolate**, **stipulate**, **obovate** to **oblong**, **coriaceous.** Flowers with 9–15 **tepals** in 3 or 4 **whorls**. There are about 50 species occurring in tropical and subtropical America and Asia.

Talinaceae (order **Caryophyllales**) A family of **herbs**, **lianas** and **shrubs**, often with **tubers** that has been recognized by the **Angiosperm Phylogeny Group** to address long-standing phylogenetic difficulties in the order. There are 2 genera of 27 species occurring in America and Africa, including Madagascar.

Talpa europaea (mole, European mole) A mammal, 120–160 mm long, with a cylindrical body covered in short, black, velvety fur, short tail, very short legs, large front claws adapted for digging, tiny but functional eyes, and no external ears. Moles live wherever the soil is not shallow, stony, waterlogged, or very acid, and they spend almost all their time in their networks of tunnels that they dig themselves, feeding mainly on earthworms and insect larvae, but also slugs, centipedes, and millipedes. They occur throughout temperate Europe.

talus (scree slope) A sloping mass of loose rock that accumulates at the foot of a hillside.

Tamaricaceae (order **Caryophyllales**) A family of **trees**, **shrubs**, and **subshrubs**, some **xeromorphic**, with **alternate**, scale-like, **exstipulate** leaves. Flowers **actinomorphic**, **bisexual**, without **bracts**, 4 or 5 free **sepals** and **petals**, 5–10 or many **stamens**, **ovary superior** of 2, 4, or 5 fused **carpels** with 1 **locule. Inflorescence** a **spike** or **raceme**, or a solitary flower. Fruit is a **capsule**. There are 5 genera of 90 species occurring throughout Eurasia and Africa. Several *Tamarix* species grown for ornament.

Tandonia budapestensis (keeled slug, Budapest slug) A slender, yellowish brown or grey **slug** with black spots, up to 50–70 mm long, with a prominent **keel** of a lighter colour extending from the tail to the **mantle**. It is active only in summer. The slug lives close to humans, in farmland, parks, gardens, and greenhouses, and is a pest of root vegetables, especially potatoes. It occurs throughout Europe.

tangle *See* oarweed.

tannins A group of compounds composed of **phenols**, **glycosides**, or hydroxy acids that bind to and precipitate **amino acids**, **proteins**, and **alkaloids**. They have a bitter, astringent taste. Many plants produce them, probably to render their tissues unpalatable. Traditionally they were extracted and used in tanning leather and making ink.

Tansley, Arthur George (1871–1955) An English ecologist and conservationist who believed botany should be approached through the study of plants in their natural surroundings, and since plants occur as communities studies should embrace the structure of those communities.

Tapesia yallundae *See* eyespot.

tapetum 1. In vascular plants (**Tracheophyta**), a layer of cells that surrounds the **sporocyte** and is involved in the formation of **pollen grains. 2.** In ferns (**Pteridophyta**), a layer of nutritive tissue in the **sporangium. 3.** A layer of cells containing crystals of zinc and a **protein** (often riboflavin, which fluoresces) in or immediately outside the retina in the eyes of many nocturnal mammals. The tapetum reflects light back through the retina, thus increasing the eye's sensitivity to dim light. If an animal turns to look into a bright light at night, when its pupils are fully dilated, light reflected from the tapetum will cause the eyes to shine ('night shine').

tapeworms *See* Platyhelminthes.

Taphrina A genus of **ascomycete fungi** in the subphylum **Taphrinomycotina** that grows as a **yeast** for part of its life cycle, then infects woody **angiosperms** and forms **hyphae** that form a layer of asci (*see* ascus) on the surface. The infection causes leaf curl, leaf blister, and catkin curl, and the formation of witch's brooms. There are about 100 species occurring throughout temperate regions.

Taphrinomycotina A subphylum of **ascomycete fungi** that are dimorphic (*see* dimorphism) plant parasites (*see Taphrina*) with both a **yeast** and filamentous stage in the life cycle, and that do not produce **fruiting bodies.** It comprises 1 order (Taphrinomycetes [Taphrinales]), with 2 families, 8 genera, and 140 species.

tapioca (*Manihot esculenta*) *See* Euphorbiaceae.

Tapisciaceae (order **Huertales**) A family of **shrubs** and small **trees** with leaves **alternate** sometimes **opposite, petiolate, imparinnate** with 3 **leaflets**, leaflets **dentate, exstipulate** or with **caducous stipules.** Flowers **actinomorphic, hermaphrodite, pentamerous, ovary superior** of 2 fused **carpels. Inflorescence** a terminal or **axillary** drooping **panicle** or **raceme.** Fruit a **berry** or **drupe.** There are two genera of five species, *Tapiscia sinensis* **endemic** to China, the others occurring in the West Indies and South America.

taproot A root that grows vertically downward.

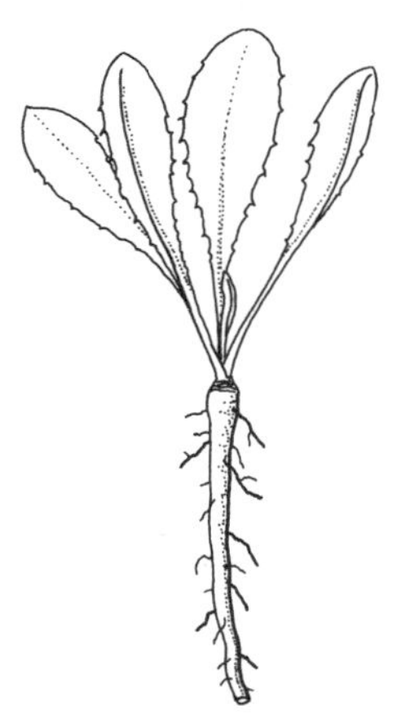

A taproot grows vertically downward.

Taraxacum (family **Asteraceae**) A genus of **biennial** or **perennial herbs** with **taproots**, many producing **latex.** Leaves in a basal rosette, **simple, entire** or lobed. Flowers in heads surrounded by **bracts** and comprising only ray **florets** that develop into spherical seed heads ('clocks'), each containing an **achene** attached to a **pappus.** There are about 34 species, but they are apomictic (*see* apomixis), so there are about 2000 microspecies. They occur throughout Eurasia and America. All are edible and some cultivated as salad plants. *Taraxacum officinale* (dandelion) is a cosmopolitan weed; *T. erythrospermum* (red-seeded dandelion) is a weed of North America.

Tardigrada (water bears, moss piglets) A **phylum** of short, plump animals, up to

1 mm long (but most smaller), with four pairs of legs bearing claws. Most live in the films of water coating mosses and **lichens**. Some species are parthenogenic, but most reproduce sexually, females laying eggs that they leave inside shed **cuticles**. Most feed on plant material or **Bacteria**; some are predators. Tardigrades tolerate extreme temperatures and pressures, as well as high levels of radiation and many toxins, and they can suspend their metabolism for several years to survive unfavourable conditions, and then revive. There are 930 species of which about 150 are marine.

tarnished bug *See Lygus rugulipennis.*

tarragon (*Artemisia dracunculus*) *See Artemisia.*

tarsal bone One of the bones of the hind foot in **tetrapods** articulating with the **metatarsals**.

Tarsonemidae (broad mites, thread-footed mites, white mites) A family of elliptical or round, translucent mites (**Arachnida**) that are less than 0.3 mm long, females being larger than males. Most feed on **Fungi** but some species are serious pests of cultivated plants. There are more than 500 species with a worldwide but mainly tropical and subtropical distribution.

tar spot *See Rhytisma.*

tarsus 1. In tetrapods, the part of the hind limb between the **tibia** and **fibula** and the **metatarsus**; the ankle. **2.** In birds, the lower long bone of the leg. **3.** In vertebrates, **connective tissue** in the eyelid. **4.** In insects, one of the small segments forming the **distal** end of the tibia.

tassel bush *See Garrya.*

tawny grisette (*Amanita fulva*) *See Amanita vaginata.*

Taxaceae (order **Pinales**) A family of **gymnosperm evergreen trees** and **shrubs** with needle-like, **alternate** or **opposite** or spirally arranged, **simple**, **linear** to linear-**lanceolate**, **decurrent** leaves. Plants **dioecious** or **monoecious**. Male **cones** solitary or clustered, **globose** to ovoid, **axillary** on year-old **branches**, shed annually, female cones of one to two **ovules** subtended by inconspicuous, **decussate bracts**, maturing in one to two years. There are 6 genera of 30 species occurring scattered throughout the Northern Hemisphere and in New Caledonia. *See Taxus.*

Taxidea taxus (American badger) A stocky, nocturnal mustelid (**Mustelidae**) with short, strong legs and a short tail, 520–875 mm long, with grey or reddish fur and distinctive facial markings comprising a white stripe from the nose over the head, a pale chin and throat, and black facial patches. Badgers are solitary. They feed on small mammals and ground-nesting birds, digging burrows to seek prey, in which they also rest. They occur throughout most of temperate North America in dry grasslands and pastures.

taxis (tactic movement) A change in the direction of movement of a **motile** organism or cell in response to an external stimulus.

taxon (pl. taxa) A group of organisms of any taxonomic rank, e.g. a species, genus, family, etc.

taxonomy The scientific classification of biological organisms.

Taxus (family **Taxaceae**) A genus of **dioecious** or **monoecious gymnosperm trees** and **shrubs** (yew) with **distichous**, **mucronate** but soft-pointed leaves. Male **cones globose**, yellowish, female cones with one **ovule**, seed maturing in one season, brown in a scarlet to orange, sweet-tasting **aril**. There are seven species occurring mainly in the northern temperate zone, also in Mexico and mountains of Malesia.

All parts extremely poisonous except for the fleshy aril (but including the seed).

TBV *See* tulip breaking virus.

tea (*Camellia sinensis, C. assamica*) *See Camellia*, Theaceae.

teak (*Tectona*) *See* Lamiaceae.

tea oil (*Camellia oleifera*) *See Camellia.*

teasel *See Dipsacus.*

tea tree *See Leptospermum.*

tebufenozide An **insecticide** that mimics **ecdysone**, causing premature and fatal moulting in insect larvae, and is used against **caterpillars**. It is highly selective and of low toxicity to mammals.

Tecophilaceae (order **Asparagales**) A **monocotyledon** family of **perennial herbs** with **corms**. Leaves spirally arranged and generally basal (**cauline** in *Walleria* spp.), **entire**, **glabrous**, **linear** to **lanceolate**-linear, occasionally **undulate**. Flowers **zygomorphic** or **actinomorphic**, **bisexual**, with 9 membranous **tepals** in 3 **whorls** fused into a tube, 6 **stamens** in 2 whorls, some **staminodes**, **ovary** semi-**inferior** or inferior of 3 **carpels** and **locules**. **Inflorescence** a **panicle** or **raceme** or flowers solitary or in small groups. Fruit is a **loculicidal capsule**. There are 7 genera of 25 species occurring in Africa, United States, and Chile. Some cultivated for ornament.

tectonic plates *See* lithosphere.

tegmen (endopleura) The inner, protective layer of a seed **testa**.

teleconnections Atmospheric events that occur in widely separated locations because they are linked to a common cause.

teleomorph The sexual reproductive stage in **Ascomycota** and **Basidiomycota**. *Compare* anamorph.

teleutosorus *See* telium.

teleutospore *See* teliospore.

teliospore (teleutospore) A thick-walled resting **spore** formed by **basidiomycete fungi** and borne in a **telium** in which a **basidium** develops.

telium (teleutosorus) Fungal tissue bearing **spores** in a plant infected with a **rust** fungus.

telocentric Describes a **chromosome** in which the **centromere** is located at one end of the chromosome. *See also* acrocentric, holocentric, metacentric.

telophase The final stage of **mitosis** and of both divisions of **meiosis** (but sometimes omitted in the first meiotic division in many vascular plants (**Tracheophyta**). During telophase the **spindle** disappears, nucleoli (*see* nucleolus) reappear, **nuclear envelopes** begin to form around the daughter **chromatids** and the **chromosomes** extend, ceasing to be visible.

temperate belt The regions between latitudes 25° and 50° in both hemispheres.

temperate deciduous forest **Forest** dominated by **broad-leaved**, **deciduous trees** that occurs widely in middle latitudes of Europe, eastern Asia, and North America, but in the Southern Hemisphere is confined to Chilean Patagonia.

temperate grassland A grassland **biome** that has probably been shaped by fire and grazing, and that occurs extensively in middle latitudes of both hemispheres where low seasonal precipitation produces a water deficit during part of most years. *See* pampas, prairie, steppe, veld.

temperate phage A **bacteriophage** that rarely causes the **lysis** of the host cell.

temperate rain forest **Forest** that occurs in temperate regions where the average annual rainfall is typically 1500–3000 mm or there is frequent **fog**. The forest contains **broad-leaved evergreen** and often coniferous trees, and abundant

epiphytes. These forests occur in the northwestern and coastal areas of the southeastern United States, southern Chile, parts of Australia and New Zealand, China, and southern Japan.

temperate zone The regions in both hemispheres between the **tropics** (latitude 23.5°) and the Arctic and Antarctic Circles (latitude 66.5°).

temperature inversion The condition in which the air temperature increases with increasing height through a layer of the atmosphere.

temporary wilting Loss of **turgor**, i.e. wilting, that a plant experiences in hot weather, when the rate of **transpiration** exceeds the rate at which the plant is able to absorb moisture from the soil. The plant will recover when the temperature falls.

tendon A band of **connective tissue** that links a muscle to a bone and is able to withstand tension.

tendril A thread-like aid to climbing formed from a modified and commonly twisting **stem**, **leaf**, or **petiole**.

tendrillate Having **tendrils**.

ten-striped potato beetle *See Leptinotarsa decemlineata.*

ten-striped spearman *See Leptinotarsa decemlineata.*

tentacle A long, flexible, slender structure rich in sensory receptors that an invertebrate animal uses to obtain information about its surroundings. Some animals also use tentacles to manipulate objects and to seize food items.

Tenthredinidae A family of black, brown, or brightly patterned sawflies **(Symphyta)**, 5–20 mm long, that lay eggs into incisions they make into leaf tissue or the **bark** of twigs. In most species the larvae feed on foliage, but some are **leaf miners**, stem borers, or **gall** makers. There are more than 6000 species distributed worldwide.

teosinte *See Zea.*

tepal A member of the perianth in **flowers** where there is no distinction between **petals** and **sepals**.

tepalostemon A tube formed by **stamen filaments** fused to **tepals**.

terbutryn A selective **herbicide** that is applied to the soil, where it is absorbed by the roots of grass and **broad-leaved** weeds, and inhibits **photosynthesis.** It is used for pre- and post-emergent weed control on a variety of crops and on weeds in static or slow-moving water. Its toxicity to animals is low.

terete Circular in cross-section.

terminal deletion *See* deletion.

ternary fission A variety of cell division that results in three **daughter cells.**

ternate **Compound** and divided into three parts.

terpenes A group of **hydrocarbon** compounds produced by many plants, especially conifers that comprise two or more **isoprene** units. Many have a strong and pleasant aroma. They are the principal ingredients in **resin** and turpentine.

terpenoids (isoprenoids) A large group of compounds similar to **terpenes** and derived from 5-carbon **isoprenes** assembled in many different ways. They comprise the largest group of natural biological compounds and are found in most organisms. Terpenoids form part of membrane-bound **steroids**, **carotenoids**, **chlorophyll**, **abscisic acid**, and many other substances, and contribute to the aromas and flavours of many plants and spices.

terracing A system of management in which sloping ground is cut into a series of steps, comprising level areas (terraces) bounded by near-vertical edges. This improves ease of **cultivation** and water retention, and reduces **erosion**.

Terrapene carolina (common box turtle, eastern box turtle) A species of **Chelonia**, approximately 110 × 80 mm in size, that have a hinged **plastron**, allowing them to close their shells almost completely, like a box. The **carapace** has a high dome and a **keel**, with variable markings. It inhabits open woodland and pasture, usually close to water, and is omnivorous. It occurs throughout the eastern United States and is popular as a pet.

terrapins *See* Chelonia.

terrestrial raw soils Soils that occur in material that formed recently and has been little altered by soil-forming processes. There is a surface layer of organic or mixed organic and mineral matter less than 5 cm thick, and **buried soil** may form a **soil horizon** below 30 cm, but otherwise there are no horizons. This is a major group in the **soil classification** devised by the Soil Survey for England and Wales.

terric horizon An **anthropedogenic horizon** that develops over a long period through the addition of mud, **compost**, or manure. It has a **base saturation** greater than 50 percent.

territory An area occupied by an animal or group of animals that will be defended against intruders.

tertiary consumer A **carnivore** that feeds on other carnivores.

tertiary structure The configuration of a **protein** molecule that develops when the **secondary structure** folds on itself, **amino acids** linking to each other by **disulphide bridges**. *See* primary structure, quaternary structure, secondary structure.

tertiary vein A leaf **vein** that branches from a **secondary vein**.

tessellated Forming a chequered or mosaic pattern.

test A protective covering or shell of certain **Protozoa** and invertebrate animals.

testa The outer protective coat of a **seed**.

Testudines *See* Chelonia.

Testudinidae (land tortoises) A family of tortoises (**Chelonia**), all of which are terrestrial. They have a **carapace** with a high dome and ranging in size from less than 120 mm to 1.3 m, and a **plastron** that is usually without a hinge. They have thick legs and scaly feet with claws. They are **herbivores**, although some species eat carrion. They inhabit many kinds of environments, from rain forests to deserts, and occur in all warmer parts of the world except Australia. There are about 11 genera and 40–50 species.

Tethys Sea The sea that formed a large inlet between the northern and southern sections of **Pangaea**.

Tetracarpaeaceae (order **Saxifragales**) A **monotypic** family (*Tetracarpaea tasmannica*) which is an erect, subalpine, **evergreen shrub** or **subshrub** with leaves **alternate**, **petiolate**, **lanceolate** or **elliptical**, **cuneate** and **entire** in the upper part, rounded and **crenate** to **serrate** in the lower part. Flowers **actinomorphic**, **bisexual**, **tetramerous**, with 4 small **sepals**, 4 **petals**, 8 **stamens**, **ovary superior** of 4 free **carpels**. **Inflorescence** a terminal **raceme**. Fruit is a group of 4 **follicles**. The family is **endemic** to Tasmania.

Tetrachondraceae (order **Lamiales**) A family of prostrate, mat-forming, **perennial herbs** either ascending from a **taproot** or with roots at the **nodes**. Leaves **opposite**, **ovate** or **elliptical** to **linear**. Flowers small, inconspicuous, **actinomorphic**, **tetramerous**, **corolla** tubular **rotate**, 4 **stamens**, **ovary superior** and 4-lobed or semi-**inferior** and **syncarpous**. Flowers borne in leaf **axils** or in short **cymes**. Fruit of 4 **nutlets** or **capsule**. There are two genera of three species occurring in Patagonia, Australia, and New Zealand.

tetrad 1. During the first **prophase** and **metaphase** of **meiosis**, a bundle comprising four **homologous chromosomes. 2.** The four **haploid** cells produced by meiosis.

tetradynamous Describes a **flower** that has four long **stamens** and two short ones.

Tetragonia tetragonioides (New Zealand spinach) *See* Aizoaceae.

Tetramelaceae (order **Cucurbitales**) A family of **dioecious trees**, often with **buttress roots**, and with wood that fluoresces. Leaves **alternate**, **petiolate**, **ovate**, **entire** or **dentate**. Flowers **actinomorphic** to slightly **zygomorphic**, 4 or 6–8 **sepals**, **corolla** when present of 6–8 **petals**, 4 or 6–8 **stamens**, **ovary inferior**, **syncarpous. Inflorescence** a terminal or **axillary panicle** or **raceme**. Fruit is a **capsule**. There are two genera of two species occurring in Indo-Malesia.

Tetrameristaceae (order **Ericales**) A family of **evergreen trees** and **shrubs** with **alternate**, spirally arranged, **coriaceous**, **simple**, **ovate**, **entire**, **exstipulate** leaves. Flowers **tetramerous** or **pentamerous**, **ovary superior** of 4–5 fused **carpels** each with 1 **locule**. **Inflorescence axillary**, **pedunculate**, an **umbel**-like **raceme**. Fruit is a **berry**. There are three genera of five species occurring in Central and South America and western Malesia.

tetramerous With **flower** parts in fours.

Tetranychidae (spider mites) A family of mites (**Arachnida**), variable in colour and most less than 1 mm long, with well-developed **pedipalps**, that feed on the contents of plant cells. They are usually found on the underside of leaves, where they often spin silk webs as protection against predators. There are about 1200 species, found worldwide. They cause considerable damage to plants. Infestations usually occur in warm conditions, when a female may lay hundreds of eggs during her two- to four-week life.

Tetranychus cinnibarinus (carmine spider mite) A **spider mite** that is oval, about 0.4 mm long, in which the females are red in summer, with two or four dark spots. They lay spherical eggs about 0.14 mm across which hatch into six-legged larvae that are just slightly larger and colorless with red **eyespots**. The larvae go through two **instars** as green, eight-legged **nymphs**. The mites feed on more than 180 species of plants, including more than 100 cultivated species. They are believed to have a **cosmopolitan distribution**. *See also Tetranychus urticae.*

Tetranychus urticae (two-spotted spider mite) A **spider mite** that differs from ***T. cinnibarinus*** only in the colour of its females in summer. These are yellow to dark green with two or four dark spots.

tetraploid Having four sets of **chromosomes** (4*n*).

tetrapyrroles A group of natural pigments composed of four **pyrrole** rings linked by **covalent bonds** or by 1-carbon bridges, and may be linear with the pyrrole rings joined side by side, or cyclic. They are involved in many processes, including **photosynthesis** and **respiration**.

tetrapod A vertebrate animal that has four limbs.

tetrarch Primary xylem that has four strands.

tetrasomy The condition of having four copies of a particular **chromosome**. *See* tetraploid.

tetro-allelic Describes a type of **polyploidy** in which four different **alleles** are present at a particular **locus**.

Texas ratsnake *See Elaphe obsoleta.*

thale cress *See Arabidopsis thaliana.*

thallophyte A plant-like organism that is not differentiated into stem, root, and leaves, i.e. algae (*see* alga), **Fungi**, and **lichens**. These were formerly classified as the Thallophyta, but the name is no longer used.

thallose Having a **thallus**, e.g. thallose liverworts (**Marchantiophyta**).

thallus A plant body that is not differentiated into stem, root, and leaves. It is composed of plates of cells or **filaments** and may consist of a single cell or have a complex, branching structure. It occurs in algae (*see* alga), **Fungi**, **lichens**, and **Marchantiophyta**.

Thamnophis sirtalis (common garter snake, eastern garter snake) A species of snakes, 460 mm–1.4 m long, that are highly variable in colour but usually have three pale, longitudinal stripes. They inhabit many different **habitats**, especially those with damp grass, and are common in towns and suburbs. They are active by day, often basking, and feed on earthworms, leeches, slugs, snails, insects, frogs, toads, and other snakes. They are beneficial in gardens, and are easily tamed. They occur throughout most of North America.

Thanatephorus *See Rhizoctonia solani.*

Theaceae (order **Ericales**) A family of usually **evergreen trees** and **shrubs** with a unicellular **indumentum** and leaves **alternate** and spirally arranged or **distichous**, **simple**, **coriaceous**, usually **dentate**, **exstipulate**. Flowers solitary, borne in the leaf **axil**, **actinomorphic**, **hermaphrodite**, 5 or 6 (occasionally more) **sepals**, 5 or rarely more **petals**, 20 or more **stamens**, **ovary superior** of 5 sometimes 3–5 or 5–10 fused **carpels**. Fruit usually a **loculicidal capsule**, occasionally a **drupe**. There are 9 genera of 195–460 species, most occurring in southeastern Asia, southeastern United States, Malesia, and tropical South America. Many ***Camellia*** species cultivated for ornament; *C. sinensis* and *C. assamica* are the source of tea.

theca A shell-like covering enclosing the single cell in some algae (*see* alga).

thermal capacity *See* heat capacity.

thermal high An area of high **atmospheric pressure** that is produced when air comes into contact with a cold surface and its density increases as its temperature falls.

thermal low (heat low) An area of low **atmospheric pressure** that is produced when air comes into contact with a warm surface and its density decreases as its temperature rises.

thermal weathering (insolation weathering, thermoclastic weathering) A type of **mechanical weathering** that is due to stresses arising from the repeated heating and cooling of a rock exposed to sunlight, if the stresses are sufficient to cause failure. The low thermal conductivity of rock causes a surface layer to expand and contract, but not the layer below it. This causes splintering, called spalling.

thermal wind A wind that is generated when the air temperature changes by a large amount over a short horizontal distance. The **jet stream**, at the top of the **polar front** separating polar and tropical air, is a thermal wind.

thermoclastic weathering *See* thermal weathering.

thermophile *See* extremophile.

Theophrastus (ca. 372–ca. 287 BCE) A Greek philospher who succeeded Aristotle as head of the Lyceum in Athens. He cultivated plants in his own botanical garden and described more than 500, classifying them as **trees**, **shrubs**, **subshrubs**, and **herbs**. He was the first to distinguish between **monocotyledons** and **dicotyledons**, recognized the difference between coniferous and flowering plants, noted that some flowers have **petals** and others do not, and observed that cultivated plants often reverted to the wild type but wild plants bred true. Two of his works survive: *De historia plantarum* (On the history of plants) and *De causis plantarum*

(On the reasons for plant growth). He is known as the 'father of botany'.

therophyte An **annual** or **ephemeral** plant that completes its life cycle rapidly when conditions permit and survives unfavourable periods as seed. It is one of the categories in the classification devised by Christen **Raunkiær**.

thiamine Vitamin B_1 ($C_{12}H_{17}ClN_4OS$), produced by plants, **Fungi**, and **Bacteria** and essential to all organisms, its phosphate derivatives being involved in many cellular processes. Thiamine pyrophosphate is a **coenzyme** that helps catalyze **sugars** and **amino acids**.

thiamethoxam *See* neonicotinoid.

thigmotaxis (stereotaxis) A change in the direction of movement of a **motile** organism or cell in response to a stimulus of touch.

Thiobacillus denitrificans *See* denitrifying bacteria.

thiocarbamate pesticide A group of compounds, some used as **fungicides** to protect seeds, seedlings, turf, vegetables, fruit, and ornamentals, and others as **herbicides**. They are of low to moderate toxicity to animals but can cause skin, eye, and respiratory irritation.

Thismiaceae (order **Dioscoreales**) A family of **perennial saprophytes** with **rhizomes**. They lack **chlorophyll**. Leaves much reduced, **alternate** or **opposite**, **distichous**, membranous, **sessile**, **simple**, **entire**. Flowers small, with **bracts**, **actinomorphic** or very **zygomorphic**, **trimerous**, **perianth** of 6 **petaloid tepals** in 2 **whorls**, 6 or 3 **stamens**, **ovary inferior** of 3 **carpels** with 1 **locule**. **Inflorescence racemose** or flowers solitary. Fruit is a **capsule**. There are 5 genera of 45 species with a widespread, mostly tropical and subtropical distribution.

Thomandersiaceae (order **Lamiales**) A **monogeneric** family (*Thomandersia*) of small **trees** and **shrubs** with **opposite**, **entire**, **elliptical**, **acuminate**, **petiolate** leaves. Flowers **zygomorphic**, **bisexual**, **calyx** 5-lobed, **campanulate**, **corolla** tubular, 4 fertile **stamens** and 1 **staminode**, **ovary superior** of 2 fused **carpels**. **Inflorescence** a terminal or **axillary raceme**. Fruit is a **capsule**. There are six species occurring in West and Central Africa.

thorax The anterior part of the body of a mammal, containing the heart and lungs and separated from the abdomen by the **diaphragm**. In insects (**Insecta**), the three segments of the body between the head and the abdomen to which legs and wings are attached.

thorn A hard, sharply pointed projection from a plant surface that may be an outgrowth from the **epidermis** or a modified plant organ.

thorn-apple *See Datura.*

thorn forest (thorn scrub, thorn woodland) Tropical vegetation found in semi-desert regions that comprises thorny **shrubs** and mostly small **trees** (*see Acacia*) that are scattered on otherwise bare ground, with few herbs or grasses.

thorn scrub *See* thorn forest.

Thornthwaite classification A **climate classification** scheme that defines the boundaries of **climate types** by temperature and precipitation, based on the concept of **potential evapotranspiration** and an index of moisture. The scheme was proposed in 1933 with a revised version in 1948 by the American climatologist Charles Warren Thornthwaite (1899–1963).

thorn tree *See Acacia.*

thorn woodland *See* thorn forest.

thread-footed mites *See* Tenthredinidae.

threadworms *See* Nematoda.

three-banded garden slug *See Lehmannia valentiana.*

three-cell model A description of the **general circulation** of the atmosphere that consists of three types of vertical cells: **Hadley cells** transport air in the tropics and **polar cells** in high latitudes, and these drive the **Ferrel cells** of middle latitudes.

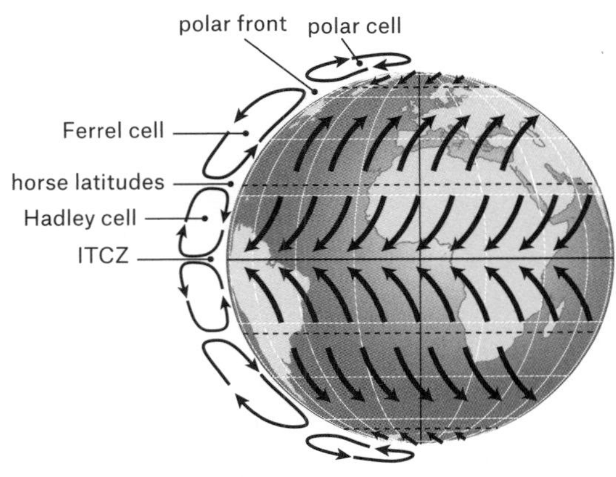

Three-cell model. Air rises at the equator and subsides in the subtropics, forming the Hadley cell, and subsides over and moves away from the poles forming the polar cell. The Hadley and polar cells drive the Ferrel cell. The polar front is at the boundary between the Ferrel and polar cells, and the horse latitudes occur near the boundary between the Hadley and Ferrel cells. The intertropical convergence zone (ITCZ) is where the winds on either side of the equator converge.

three-domain system A taxonomic system (*see* taxonomy) that divides living organisms into three major groups called **domains: Bacteria, Archaea,** and **Eukarya,** on the basis of differences in certain of their genes.

three-legged bugs *See* Reduviidae.

threshold velocity The lowest velocity at which the wind or water will dislodge a mineral particle.

thrift (*Armeria*) *See* Plumbaginaceae.

thrips *See* Thysanoptera.

throughflow *See* interflow.

thrum-eyed *See* heterostyly.

thrushes *See* Turdidae.

Thryothorus ludovicianus (Carolina wren) A species of birds, 120–140 mm long, with a rust-brown back and cinnamon underside, with a white chin and throat and black bars on the wings and tail. They spend much time on the ground, seldom flying far, and feed on insects and spiders. They prefer woodland, but inhabit a variety of **habitats** including suburban gardens. They are year-round residents of the southeastern United States.

Thunbergia alata (black-eyed Susan) *See* Acanthaecae.

Thunbergioideae *See* Acanthaceae.

thunder The sound produced by the explosive expansion of air that is rapidly heated by a **lightning stroke.**

thunderstorm A storm that delivers heavy precipitation, strong wind **gusts,** and **thunder** and **lightning.**

Thurniaceae (order **Poales**) A **monocotyledon** family of **perennial herbs** with leaves V-shaped in section, in three ranks, sometimes **serrate,** with parallel sides and sheathing base. Flowers **bisexual** with 5 **trimerous whorls,** 6 **stamens** in 3 whorls, **ovary superior. Inflorescence** a terminal **raceme.** Fruit is a **loculicidal capsule.** There are two genera of four species occurring in flooded areas of South Africa, Guyana, and Amazonia.

thylakoid A membranous sac or disc that is a principal subunit of a **granum** in a **chloroplast.**

thyme (*Thymus*) *See* Lamiaceae.

Thymelaeaceae (order **Malvales**) A family of **trees, shrubs,** and a few **lianas** and **herbs** with **alternate** or **opposite, simple, exstipulate, entire** leaves. Flowers **actinomorphic, bisexual, unisexual,** sometimes **polygamous** (plants **dioecious** sometimes **monoecious**), with 4–5 free or 3 or 6 fused **sepals** often **petaloid,** as many or twice as many **petals** as sepals, 3–5 or 8–10 (or up to 100) **stamens**, **ovary superior** of 2–5 sometimes 5–8 or up to 12 fused **carpels** and as many **locules**. **Inflorescence** a **raceme**, **capitulum**, or **fascicle**. Fruit is a **drupe**, **berry**, **achene**, or **capsule**. There are 46–50 genera of 891 species with a worldwide distribution, especially in tropical Africa and Australia.

thymine One of the **pyrimidine** bases found in **DNA**. It binds to adenine and in **RNA** it is replaced by **uracil**.

thyrse A dense **inflorescence** consisting of a central **raceme** with **cymose** lateral **branches**, e.g. lilac (***Syringa***), horse chestnut (***Aesculus hippocastanum***).

Thysanoptera (thrips) An order of slim, pale or blackish, **hemimetabolous** insects (**Insecta**), most 0.5–2.0 mm long, with short antennae (*see* antenna), short legs, and asymmetrical mouthparts. Some adults are wingless, others with very narrow wings with a fringe of long hairs. Some thrips are predators of mites and other small invertebrates, but most feed on plants, often in flowers, and are serious pests. There are more than 4500 species with a worldwide distribution. The name thrips is both singular and plural; there is no such thing as a 'thrip'.

tibia 1. In **tetrapods**, the anterior long bone of the hind limb; the shin bone. **2.** In insects, the segment of the leg between the **femur** and **tarsus**.

T

tibiotarsal joint In birds, the joint between the **tibiotarsus** and the **metatarsals**.

tibiotarsus 1. In birds (and some dinosaurs), a bone formed by the fusion of the **tibia** and **tarsal** bones.

tick-borne diseases Diseases that are transmitted by ticks to their hosts through the transfer of pathogenic bacteria, viruses, or protozoa, or by the direct release of toxins.

Ticodendraceae (order **Fagales**) A **monotypic** family (*Ticodendron incognitum*), which is a wind-pollinated, **evergreen tree** with **alternate**, **simple**, **ovate** to **elliptical**, **serrate** leaves with a short **petiole**. Flowers **actinomorphic**, **unisexual** (plants **dioecious** or **polygamodioecious**), with no **perianth**, male flowers consisting of 8–10 **stamens** subtended by **bracts**, female flowers with **superior ovary** of 2 **carpels** with 2 **locules** per carpel. Male **inflorescence** a dense **axillary** cluster of **sessile** flowers, female inflorescence axillary, a solitary terminal flower. Fruit is a **drupe**. The tree is **endemic** to Central America.

TICV *See* tomato infectious chlorosis virus.

tile drain A drain composed of short sections of ceramic or concrete pipe that are laid end to end, or a continuous length of perforated plastic piping. A series of drains are buried, aligned downslope, and spaced to allow them to carry surplus water into a ditch or stream. A tile drain will remove water to a distance on either side equal to about eight times its depth.

Tilia (family **Malvaceae**) A genus of **deciduous trees** (lime, linden, basswood) with **alternate**, usually **cordate**, **petiolate** leaves. Flowers **actinomorphic**, **hermaphrodite**, with 5 free **sepals** and **petals**, many **stamens**, and an **ovary** of 5 **carpels**. **Inflorescence** is a **cyme** with a large **bracteole** fused to the inflorescence stalk. The fruit is a **nut**. There are about 30 species occurring throughout the temperate Northern Hemisphere.

till 1. Sediment that has been deposited by glacial action, without the involvement of liquid water. *See* glacial till. **2.** To prepare soil for the growing of crops.

tiller A lateral shoot, arising at ground level, in grasses (**Poaceae**).

Tilletia *See* bunt.

Tilletiales An order of **basidiomycete smut fungi** comprising about 150 species, all of which infect grasses (**Poaceae**) except for *Erratomyces*, which infects legumes (**Fabaceae**).

till plain A smooth-surfaced plain formed on **till**.

tilth The physical condition of the upper layers of soil with regard to its suitability for growing crops.

timber line The upper boundary on a mountainside to **forest** composed of erect, tall **trees** with a closed **canopy**.

timber rattlesnake *See Crotalus horridus.*

tinsel *See* flagellum.

Tipulidae (crane flies, daddy-long-legs, leatherjacket) A family of long-legged, slender flies (**Diptera**) with long antennae (*see* antenna) that have six segments. They hold their wings outstretched when at rest. Crane flies fly weakly, are easily caught, and readily discard legs that become trapped. Eggs take 1–2 weeks to hatch, larvae go through 4 **instars**, and pupae take 12 weeks to develop into short-lived adults. The larvae (leatherjackets) live in the soil, among roots, or in swamps and marshes, some as predators, others feeding on plants. Where present in large numbers they are destructive pests. There are about 15,000 species with a worldwide distribution.

tissue A group of similar cells, in plants usually bound together by their **cell walls**, that collectively perform a particular function.

tissue culture Single cells that are grown together in a sterile medium supplied with nutrients for purposes of research or to grow exact copies (**clones**) of the original plant.

titmice *See* Paridae.

tits *See* Paridae.

toads *See* Anura, Bufonidae.

toadstool Any umbrella-shaped fungal **fruiting body** that is inedible or poisonous.

tobacco *See Nicotiana.*

tobacco mosaic virus A species of ***Tobamovirus*** that infects tobacco plants and other members of the **Solanaceae**, as well as species in nine other families, causing mottling and distortion of the leaves. It enters cells through **plasmodesmata** and is able to spread through **phloem**.

tobacco necrosis virus A **satellite RNA virus** that can infect plants only when accompanied by its helper, tobacco necrosis necrovirus. It produces lesions in leaves or necrosis of the whole plant. It infects common bean (*Phaseolus vulgaris*), tulip (*Tulipa gesneriana*), and tobacco (*Nicotiana* spp.), and is transmitted by a fungus, ***Olpidium*** *brassicae*.

tobacco rattle virus (TRV) A **virus** (*Tobravirus*) that is a pathogen of more than 400 species of plants; it was first identified in tobacco (*Nicotiana tabacum*) and its name also refers to the sound from fields of dried-out tobacco leaves. It produces a variety of symptoms including mottling, **chlorosis**, necrotic lesions, and ring spots. It is transmitted by several *Paratrichodorus* species of stubby-root nematodes (**Nematoda**) and can be present in seeds and transmitted on garden tools. It occurs in Eurasia, North America, Central America, South America, and in parts of Australia and New Zealand.

Tobamovirus A genus of **RNA viruses** that are pathogens of a number of plants including tobacco where they are known as **tobacco mosaic virus** (the name is an acronym: TOBAccoMOsaicVIRUS). They also infect other members of the **Solanaceae**, brassicas (**Brassicaceae**), cucurbits (**Cucurbitaceae**), and members of the **Malvaceae**.

T

Tobravirus *See* tobacco rattle virus.

tocopherol Vitamin E, a group of **terpene** compounds produced by many plants that function in cells as antioxidants. The richest sources are wheat germ oil, sunflower oil, and almond oil.

ToCV *See* tomato chlorosis virus.

Tofieldiaceae (order **Alismatales**) A **monocotyledon** family of **herbs** with **ensiform**, **distichous** leaves growing from the base. Flowers usually **trimerous** with 6 **tepals** and **stamens** in 2 **whorls**, **ovary** of 3 **carpels**. **Inflorescence racemose** rarely **spicate** with a 3-**bracteole calyculus**. Fruit is a **capsule**. There are 3–5 genera of 31 species occurring in northern temperate regions, southeastern United States, and northern South America.

tomato (*Solanum lycopersicum*) *See Solanum*.

tomato chlorosis virus (ToCV) A species of ***Crinivirus*** that causes **chlorosis**, leaf brittleness, necrotic patches in tomato plants; it also infects 24 other plant species. It is transmitted by several whitefly species, especially *Trialeurodes abutilonea* and ***Bemisia tabaci***. It occurs in North and Central America, South Africa, Taiwan, southern Europe, and North Africa.

tomato infectious chlorosis (TICV) A species of ***Crinivirus*** that causes **chlorosis** in tomato plants, especially in older leaves, and also infects a number of wild plants. Symptoms are identical to those caused by **tomato chlorosis virus**, but TICV differs in being transmitted only by the greenhouse whitefly (*Trialeurodes vaporariorum*). It occurs in North America, Japan, Taiwan, and parts of Europe.

Tombusviridae A family comprising 9 genera and 52 species of **RNA viruses**, including *Tombusvirus*, the tomato bushy stunt virus. The viruses occur in soil; some are transmitted by **zoospores** of **Fungi** of the order Chytridiales (*see* Chytridiomycota) but the **vectors** for others are not known.

tomentose Covered with fine hairs; woolly.

tomentum A mat of fine, woolly hairs found on some **lichen** thalli (*see* thallus) and the underside of some ***Rhododendron*** leaves.

tonnhäutschens *See* clay skins.

tonoplast The membrane that surrounds a **vacuole**.

toothed wrack *See* serrated wrack.

top dressing The application of a mixture of soil material and **fertilizer** to the surface of turf in order to accelerate the decomposition of dead grass (thatch), improve **drainage**, and increase the retention of nutrients.

topogenous mire (topogenous peat) A **bog** that forms in climates with a summer **drought** and only moderate annual rainfall, conditions that restrict wetland vegetation to low-lying areas, e.g. valley bottoms.

topogenous peat *See* topogenous mire.

topsoil The uppermost part of the soil, comprising the O and A **soil horizons**, which contains most of the organic matter.

tornadic storm A storm that produces **tornadoes** or is capable of doing so.

tornado A rapidly spinning vortex of air that extends downward from a large **cumulonimbus** cloud, appearing as a **funnel cloud** until it touches the surface. *See* Fujita tornado intensity scale.

torpor A state of hypothermia that some **endotherms** enter in order to conserve energy. During torpor the body temperature may fall to within 1°C of the ambient temperature, which may be close to, or even slightly below freezing, and

T

metabolic processes slow to as little as 5 percent of their normal rate.

Torricelliaceae (order **Apiales**) A family of **shrubs** and small **trees** with **alternate**, **simple**, **petiolate**, **entire**, **dentate** or dissected leaves. Flowers mostly **unisexual** (plants **monoecious** or **dioecious**), with 5 (or 3–5 in female flowers) **sepals**, 5 **petals** (only in male flowers), 5 **stamens**, **ovary inferior** of 3–4 **carpels** with 2, 3, or 4 **locules**. **Inflorescence** a terminal or near-terminal **panicle** or **raceme**. Fruit is a **drupe**. There are three genera of ten species occurring in Madagascar, southeastern Asia, and western Malesia.

torsion In **Gastropoda**, the twisting of the body through 180 degrees in the course of development, giving the nervous and digestive systems a U-shape and placing the anus, gills (in aquatic species), and **mantle** behind the head.

tortoises *See* Chelonia, Testudinidae.

tortoise scales *See* Coccidae.

tortrices *See* Tortricidae.

Tortricidae (tortrix moths, tortrices, bell moths, leaf-rollers, fruit moths) A large family of small moths in which adults have rough scales on the upper side of the head, a well-developed **proboscis**, and broad wings that often form a bell shape when the moths are at rest. The **caterpillars** feed on litter between leaves that they sew together with silk, or inside fruit or **galls**. Some roll leaves. Many are serious pests. There are more than 10,300 species with a worldwide distribution.

tortrix moths *See* Tortricidae.

torus Describes an object that is shaped like a doughnut.

Torymidae (chalcid wasps, chalcid seed flies) A family of long-bodied (1–8 mm) wasps (**Hymenoptera**), most of which are metallic green, yellow, or black. The female has an **ovipositor** that is much longer than her abdomen. Most are parasites of flies (**Diptera**) or **gall** wasps, but some species eat seeds. The parasites lay eggs in galls or larvae. There are 986 species with a worldwide distribution.

totara (*Podocarpus totara*) *See Podocarpus.*

Tournefort, Joseph Pitton de (1656–1708) A French botanist who devised a system for plant classification and nomenclature, describing in *Institutiones Rei Herbariae* 9000 species and grouping them into 700 genera. He travelled in the eastern Mediterranean region and in *A Voyage into the Levant* (1718) described the plants he found, including the first descriptions of azaleas and rhododendron.

Tovariaceae (order **Brassicales**) A **monogeneric** family (*Tovaria*) of strong-smelling **herbs** and **shrubs** with green **bark**. Leaves are **alternate**, **trifoliate**, **stipulate**. Flowers **actinomorphic**, **hermaphrodite**, hypogynous (*see* hypogyny), with 8 **sepals**, 8 **petals**, 8 **stamens**, **ovary superior** of 6–8 fused **carpels**. **Inflorescence** a terminal, pendulous, **raceme**. Fruit is a **berry**. There are two species occurring in tropical America.

toxaphene An **insecticide** that is a mixture of about 200 closely related compounds, most of which have been banned owing to their high toxicity. Toxaphene was used on a variety of crops, but mainly cotton. It is now banned in most countries.

trabecula A beam or bar that extends from a **cell wall** across a **lumen** or **lacuna**.

trace element *See* micronutrient.

trachea 1. The series of **tracheids** that comprise **xylem** tissue, providing mechanical support to the plant and acting as a conduit for the movement of water and mineral nutrients. **2.** In air-breathing vertebrates, the windpipe, extending from the throat to the lungs. **3.** In **Insecta**, one of the tubes passing through the **cuticle** that together comprise the respiratory system.

tracheary element A **tracheid** or **vessel element** that is involved in the transport of water.

tracheid One of the dead cells in the **trachea** that conduct fluid through the **xylem** tissue. It is long, tapered, cylindrical, and the **cell walls** contain bands of **lignin** that add strength.

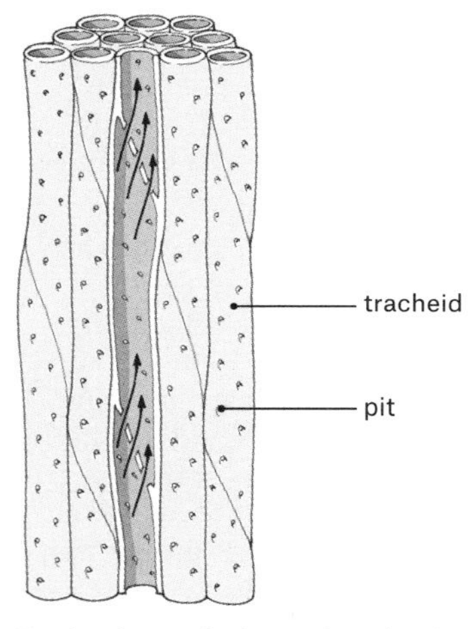

Tracheids are the long, thin, dead cells that are joined end to end to form xylem. Materials are exchanged with adjacent cells through the pits.

T

Trachemys scripta (pond slider) A species of semi-aquatic turtles (**Chelonia**), 125–290 mm long, that have an oval, somewhat flattened, keeled (*see* keel) **carapace** and a yellow **plastron**. There are three subspecies. The red-eared slider (*T. scripta elegans*) has a red or orange stripe behind each eye. The yellow-bellied slider (*T. scripta scripta*) has a yellow patch behind each eye. The Cumberland turtle (*T. scripta troostii*) resembles the red-eared slider but has narrower stripes behind the eyes and fewer and wider ones on the head, neck, and legs. Sliders inhabit slow-moving water with a muddy bottom and places on land to bask. Adults are mainly herbivorous but also eat invertebrates; juveniles are more carnivorous. Sliders occur throughout the central and southern United States, Central America, and northern South America. ↗

Tracheophyta (kingdom **Plantae**) A **phylum** comprising all the vascular plants, i.e. plants with vascular tissue (**xylem** and **phloem**) through which water and nutrients are transported. The tracheophytes include seagrasses and all terrestrial and flowering plants. ↗

Tradescantia *See* Commelinaceae.

Tragopogon (family **Asteraceae**) A genus of **biennial** or **perennial forbs** with taproots, erect stems, and **alternate**, **linear**, **entire** leaves, upper leaves short, somewhat grass-like, sheathing, lower leaves longer, **glaucous**, **recurved**. Yellow or purple flower heads with outer ray **florets** terminating in a fringe of five small teeth, inner florets shorter, the head subtended by eight leaf-like **bracts**. Fruit is an **achene** attached to a **pappus**, the seed head resembling that of a dandelion. There are more than 140 species occurring throughout Eurasia. Some cultivated for their edible roots (salsify, goatsbeard).

trait *See* character.

trama Fungal tissue that supports a **hymenium**.

tramp species Plants and animals that people have inadvertently distributed around the world, e.g. Japanese knotweed (*Fallopia japonica*), water hyacinth (*Eichhornia crassipes*), brown rat (*Rattus norvegicus*), house mouse (*Mus musculus*).

transaminase (aminotransferase) An **enzyme** that helps catalyze a **transamination** reaction.

transamination (aminotransfer) A chemical reaction, catalyzed by **transaminase**

with a **coenzyme**, in which an **amine** group (NH_2) from an **amino acid** is exchanged with the oxygen (O) group of a keto acid; the amino acid then becomes a keto acid and the keto acid becomes an amino acid. It is one of the principal ways in which organisms synthesize **non-essential amino acids**.

transcellular streaming The movement of substances in **protoplasm** through **vascular tissue** along tubular strands that cross the lumina (*see* lumen) of **sieve tubes**. It is the most important mechanism for **translocation**.

transcription The formation of a strand of **RNA** in a sequence of **nucleotides** that complements a strand of **DNA**.

transcription network *See* gene regulatory network.

transcriptome All the types of **RNA** present in a cell.

transcytosis The movement of substances within a cell by membrane-bound carriers.

transferase An **enzyme** that catalyzes reactions involving the transfer of a functional group from one molecule to another.

transfer cell A specialized **parenchyma** cell that has many protruberances in its **cell wall** caused by infoldings of the **cell membrane**, increasing its surface area. Transfer cells are most abundant in regions where nutrients are being absorbed, in **xylem** and **phloem**, and in **nectaries**. They facilitate the movement of solutes from a source, e.g. leaves, to a sink, e.g. fruits.

transfer-RNA (t-RNA) During **protein** synthesis, a group of **RNA** molecules, each 70–80 **nucleotides** long and arranged in a clover-leaf pattern, that binds to **amino acids** and transfers them to a **ribosome** to which an **m-RNA** is attached. The next **codon** in the m-RNA sequence binds to a corresponding anticodon in the t-RNA, allowing the amino acid to be attached to the end of the growing **polypeptide**.

transgenic Describes an organism that carries **genes** derived from another organism other than by reproduction.

transglycosylation A process for forming glycosidic (*see* glycoside) bonds, especially during the synthesis of **polysaccharides**.

transient polymorphism A type of **polymorphism** in which one of the alternate **alleles** at a particular **locus** is progressively displaced by another.

translation The polymerization of **amino acids** into a **polypeptide** chain, i.e. the synthesis of a **protein** molecule.

translocation **1.** The movement of dissolved substances within a plant. **2.** An alteration in the location of a **chromosome** segment, most commonly through the exchange of segments between non-**homologous** chromosomes, or by transposition. **3.** The movement of dissolved or suspended soil materials from one **soil horizon** to another.

transpiration The loss of water vapour from a plant to the outside air, mainly through leaf stomata (*see* stoma) and stem **lenticels**.

transpiration ratio The ratio of the weight of water a plant transpires (*see* transpiration) during a growing season to the weight of dry matter it produces.

transposable element (transposon) A segment of **DNA** that is able to change its location in a **genome**, either through an intermediary **RNA** (**retrotransposon**) or by direct copying from a donor to a target site.

transposition The movement of a **chromosome** segment to a different location in the same or a different chromosome without a reciprocal exchange.

T

transposon *See* transposable element.

Trapa natans (water chestnut) *See* Lythraceae.

tree A woody plant usually growing more than 10 m tall, typically with a single stem although many are multi-stemmed, supporting leaves, often on **branches**, well clear of the ground, in which, apart from shedding leaves, the aerial parts do not die back at the end of the growing season.

tree line The boundary beyond which the climate is too cold for **trees** to survive, average summer temperature remaining below 10°C. Latitudinally it is where **tundra** gives way to bare rock, snow, and ice. The altitude of the tree line on mountains varies with latitude.

tree lungwort The **foliose lichen** *Lobaria pulmonaria*, which has a broadly lobed **thallus** with an upper surface that is green when wet and greenish grey when dry, and marked with hollows and ridges, resembling a lung. In addition to the algal and fungal symbionts, *L. pulmonaria* also includes *Nostoc*, a cyanobacterium (*see* cyanobacteria). The lichen occurs widely in northern regions of the Northern Hemisphere, mainly as an **epiphyte** in **ancient woodland**.

tree moss *See Usnea.*

tree of heaven (*Ailanthus altissima*) *See* Simaroubaceae.

tree ring (annual ring, growth ring) One of a series of alternately pale and dark concentric circles that are visible in a cross-section of the stem or a **branch** of a woody plant that produces **secondary growth**. The rings are sheaths around the stem or **branch** made from **xylem** cells produced each year from the **vascular cambium**. Cells produced in spring are typically large, thin-walled, and form a pale layer, those produced in late summer and autumn are smaller, with thicker walls, and darker.

tree's dandruff *See Usnea.*

tree swallow *See Tachycineta bicolor.*

tree veld South African grassland (**veld**) with scattered trees, resembling parkland, and probably maintained by fire.

trembling aspen (*Populus tremuloides*) *See Populus.*

Tremellomycetes A class of dimorphic (*see* dimorphism) **basidiomycete fungi** in which the basidia (*see* basidium) are divided by septa (*see* septum) and the **fruiting bodies** are typically jelly-like when wet and leathery when dry. The class includes **saprotophs**, animal parasites, and **mycobionts** in **lichens**. There are 50 genera and 377 species.

Trialeurodes abutilonea (banded-wing whitefly) *See Abutilon* yellows, *Diodia* vein chlorosis, tomato chlorosis virus.

Trialeurodes vaporariorum (greenhouse whitefly) A species of whiteflies (**Aleyrodidae**) that are 1–2 mm long, pale yellow, and with four wings coated with wax. Females usually lay eggs on the underside of leaves. The newly hatched larvae (crawlers) are mobile, but later **instars** are **sessile**. Adults and all stages of larvae feed on **phloem sap** and excrete **honeydew**, which provides a substrate for fungal infection. The whiteflies also transmit viral diseases. They are active throughout the year and infest a variety of plants, including wild plants, and the crops most at risk are tomatoes, potatoes, and cucurbits. The **parasitoid** wasp ***Encarsia formosa*** is used in **biological control**. The insect occurs throughout temperate regions. *See* beet pseudo-yellows virus, potato yellow vein, tomato infectious chlorosis virus.

triallate A **thiocarbamate herbicide** that is used for the selective control of grasses. It is applied to the soil and kills the weeds prior to their emergence. It is slightly toxic to mammals.

triallelic Describes a type of **polyploidy** in which three **alleles** occur at a particular **locus.**

triarch **Primary xylem** that consists of three strands.

tribe A **taxon** between **genus** and **family** in rank that comprises genera with shared features that identify them as a group distinct from other genera. Tribe names bear the suffix *-eae*. Tribes may be grouped into subfamilies and divided into subtribes.

tricarboxylic acid cycle *See* citric acid cycle.

Trichia hispida *See Trochulus hispidus.*

trichoblast A hair-like protrusion. In root **epidermis**, a cell that becomes a **root hair.**

Trichoderma A genus of **ascomycete fungi**, which are present in most soils and form mutualistic (*see* mutualism) relationships with the roots of several plant species. They also parasitize other Fungi, controlling most pathogenic fungi. Most species produce only asexual **spores** and since they do not undertake **meiosis** different strains have different numbers of **chromosomes** and most cells have many nuclei, some with more than 100. There are at least 33 species.

Trichodorus (stubby-root nematodes) A genus of small, cigar-shaped **Nematoda** that feed on the roots of a variety of crop plants, including maize (corn), potatoes, carrots, parsnips, and peas. Their feeding inhibits root growth, causing stubby roots, and can cause the death of the plant. They also transmit viral diseases. The nematodes occur worldwide, preferring sandy soils with good **drainage.**

Trichogrammatidae (stingless wasps) A family of wasps (**Hymenoptera**) most of which are less than 1 mm long. They cannot fly strongly and are distributed by the wind. All but a few species are parasites of the eggs of other insects, especially **Lepidoptera**, making them very important agents of **biological control**. There are more than 840 species found worldwide.

trichogyne 1. Prior to **fertilization** in **red algae** and some **ascomycete** and **basidiomycete fungi**, an often hair-like protrusion from a female **gametangium** that receives the male **gamete** or nucleus. **2.** A genus in the family **Asteraceae.**

trichome 1. A hair-like outgrowth from a plant **epidermis. 2.** A chain of cells in certain **Bacteria** and **cyanobacteria.**

trichothallic growth Growth that occurs by cell division only in certain defined regions, e.g. in some species of **brown algae** (Phaeophyta).

tricolpate Describes a **pollen grain** that has three colpi (*see* colpus).

trifoliate Describes a leaf consisting of three **leaflets.**

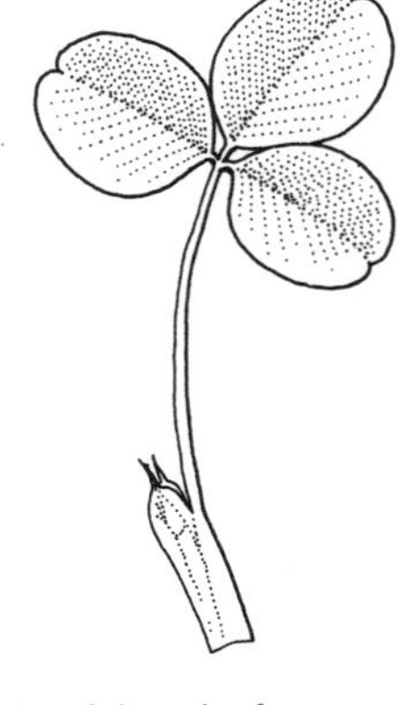

A trifoliate leaf comprises three leaflets.

Trifolium (family **Fabaceae**) A genus of **annual, biennial**, and short-lived **perennial herbs** with **trifoliate, petiolate** leaves with **stipules adnate** to the petioles. Flowers with persistent **petals** in dense, **racemose** heads. Fruit is a pod enclosed in the **calyx.** There are about 300 species with a **cosmopolitan distribution.** Many are grown as fodder crops, e.g. *T. pratense* (white clover) and *T. repens* (red clover).

trifurcate Forming three **branches.**

Trigonella (family **Fabaceae**) A genus of **annual herbs** with **trifoliate, dentate** leaves. Blue, yellow, or off-white flowers are small, resembling those of peas, solitary or in clusters, the fruit is a long, curved **pod.** There are 37 species occurring in the Mediterranean region, Asia, Macronesia,

South Africa, and Australia. *Trigonella foenum-graecum* is widely cultivated as fenugreek.

Trigoniaceae (order **Malpighiales**) A family of **evergreen trees** and **lianas** with **opposite** sometimes **distichous** or spirally arranged, **simple**, **stipulate** leaves. Flowers **zygomorphic**, **hermaphrodite**, somewhat **papilionoid**, with 5 **sepals**, 5 **petals** sometimes forming a **saccate keel**, 5–7 free or 7–12 fused **stamens**, some as **staminodes**, **ovary superior** of 3 **carpels**, **syncarpous**, with 3 **locules. Inflorescence** a terminal **raceme** or **panicle**. Fruit is a **capsule** or **samara**. There are 5 genera of 28 species occurring in Central and South America, Madagascar, and western Malesia.

trigonous Triangular in cross-section.

trilete Describes a **spore** marked by three lines radiating from a central pole, showing that the four spores produced by **meiosis** were in contact before splitting apart. *Compare* monolete.

trilocular With three **locules**.

trimerous With **flower** parts in threes.

Trimeniaceae (order **Austrobaileyales**) A family of **trees**, **shrubs**, and **lianas** with **opposite**, **petiolate**, **simple**, **ovate** to **lanceolate**, **entire** or **serrate** leaves. Flowers **actinomorphic**, **unisexual** (plants **andromonoecious**) or **bisexual**, **perianth** of 2–40 or more **sepaloid tepals**, 7–25 **stamens**, **ovary superior** of 1 **carpel** (rarely 2) with 1 **locule. Inflorescence** an **axillary** or terminal **raceme**, **cyme**, or **panicle**. Fruit is a **berry**. There are one or two genera of six species occurring from News Guinea and Australia to Fiji.

trinerved Describes a **leaf** with three main **veins**.

tripeptide *See* peptide.

triple superphosphate *See* superphosphate.

triploid Having three sets of **chromosomes** ($3n$).

triquetrous Triangular, with three sharp angles.

triradiate Y-shaped; branching three ways.

trisomy The presence of three sets of **chromosomes**, i.e. the **triploid** condition.

Tristar *See* chloronicotinyls.

tristyly The existence in a species population of three types of **flowers** distinguished by the relative heights of their **anthers** and **stigmas**. If the stigma is high, one anther is at medium height and the other low; if the stigma is at medium height, one anther is high and the other low; if the stigma is low, one anther is high and the other at medium height.

Triturus cristatus (great crested newt, northern crested newt) A species of dark, grey-brown, spotted newts (**Salamandridae**) in which males have a serrated crest along the back and a smooth crest along the upper and lower tail during the mating season. They are terrestrial but breed in water and feed on tadpoles, small froglets, other newts, worms, aquatic snails, and insect larvae. They occur throughout most of Europe, but are now uncommon and protected throughout Europe.

Triturus helveticus (palmate newt) A species of brown or olive-green newts (**Salamandridae**) in which males are about 85 mm long and females 95 mm. Males have a smooth, low crest along the back continuing into a slightly higher crest on the tail. The newts live close to fresh water and feed on small invertebrates. They occur throughout most of western Europe and are protected.

Triturus vulgaris (smooth newt, common newt) A species of pale brown or olive newts (**Salamandridae**), about 100 mm long, with orange spots. They have a laterally flattened tail to aid swimming. Males are darker than females during the breeding season. The newts emerge from hibernation in spring and prefer ponds and shallow, still water. It occurs throughout Europe and is protected.

Triuridaceae (order **Pandanales**) A **monocotyledon** family of small colourless, yellow, or purple, **perennial**, **mycotrophic**, **saprophyte herbs** that lack **chlorophyll**. Leaves very small, **alternate**, scale-like. Flowers small, **actinomorphic**, usually **unisexual** (plants **monoecious** or **dioecious**) occasionally **bisexual**, **perianth** of 3–6 approximately triangular **tepals**, 2–6 **sessile stamens**, female flowers sometimes also with **staminodes**, **ovary superior** of 10 to many free **carpels**. **Inflorescence** a terminal **bracteate raceme**. Fruit is a **follicle** or **raceme**. There are 11 genera of 50 species with a pantropical distribution.

Trochodendraceae (order **Trochdendrales**) A family of **evergreen trees** with **alternate** but in **pseudowhorls** at **branch** tips, **coriaceous**, long **petiolate**, **simple**, **elliptical** to **obovate**, **entire**, **serrate**, **exstipulate** leaves. Flowers **actinomorphic**, **unisexual** (plants **androdioecious**) or **bisexual**, **asepalous**, **apetalous**, 40–70 **stamens**, **ovary superior** of 4–11 free or 11–17 fused **carpels**. **Inflorescence** a terminal **raceme**. Fruit a ring of **connate follicles**. There are two genera of two species, with a scattered distriution in southeastern Asia.

Trochodendrales An order of **evergreen trees** comprising one family of two genera and two species. *See* Trochodendraceae.

Trochulus hispidus (hairy snail) A **snail** (formerly known as *Trichia hispida*) with a brown to cream shell 3–6 mm high and 5–11 mm wide that is covered densely with curved hairs 0.2–0.3 mm long. It inhabits humid areas in woodland and cultivated ground, often associated with nettles (***Urtica** dioica*). It occurs throughout western Europe.

Troglodytes aedon (house wren) A species of wrens, 110–130 mm long, that are brown with a grey throat and darker markings, that perch like all wrens, with the tail erect. They inhabit woodland, but also urban and suburban gardens. They feed on insects and other invertebrates and occur throughout the Americas.

Troglodytes troglodytes (wren, Eurasian wren, winter wren) A species of wrens, 90–105 mm long, that are **rufous** on the back, grey on the underside, with dark grey and brown bars. It inhabits forests and hedgerows, especially in upland areas, and feeds mainly on insects and other arthropods. It occurs throughout Eurasia.

Tropaeolaceae (order **Brassicales**) A **monogeneric** family (*Tropaeolum*) of twining, **caespitose herbs** with **alternate**, **peltate**, sometimes lobed, **petiolate**, the **petioles** sometimes twining around supports, **exstipulate** leaves. Flowers **zygomorphic**, **bisexual**, with 5 free **sepals** with 1 modified to form a long **nectar** spur, 5 free **petals**, 8 **stamens**, **ovary superior** of 3 fused **carpels** each with 3 **locules**. Flowers usually borne singly in leaf **axils**. Fruit is a **schizocarp**. There are 105 species occurring in America, especially in the Andes. *Tropaeolum majus* is nasturtium.

trophic cascade The movement of nutrients released by **secondary** or **tertiary consumers** through three or more **trophic levels** of a **food web**.

trophic fountain The movement of a large quantity of nutrients released suddenly near the base of a **food web** through three or more **trophic levels**.

trophic level A stage in the transfer of energy or nutrients through a **food web**, from **producers** to **primary**, **secondary**, and **tertiary consumers**, each level appearing as a horizontal bar in an **ecological pyramid**. Because energy is dissipated as heat

T

at each level, there are seldom more than four levels.

tropical cyclone An area of intensely low surface **atmospheric pressure** that develops between latitudes about 5° and 20° in either hemisphere when the sea surface temperature is at least 27°C and there is no wind shear at high level to remove rising air. Such a system begins as a tropical disturbance and grows into a tropical depression before becoming a tropical storm, at which point it is assigned a name. It becomes a tropical cyclone when its sustained wind speeds exceed 120 km/h. Tropical cyclones are known as hurricanes if they develop in the Atlantic or Caribbean, typhoons in the Pacific, and cyclones in the northern Indian Ocean or Bay of Bengal, but are often called hurricanes regardless of where they form.

tropical moist forest **Tropical rain forest** and **tropical seasonal forest** considered together.

tropical montane forest **Forest** that grows on tropical mountainsides. Lower montane forest consists of tall **trees**; trees in upper montane forest are shorter, typically 10–12 m tall, and form a single layer.

tropical rain forest A term coined by in 1898 by Andreas **Schimper**, who defined the vegetation as '**evergreen**, at least 30 m tall, rich in thick-stemmed lianes, and in woody as well as herbaceous **epiphytes**'. The trees are often described as forming four strata, although these can be difficult to discern, and the tallest trees, up to 60 m or more tall, stand isolated and emerge through the **canopy**.

tropical seasonal forest Tropical forest that grows in regions with a yearly dry season. It is dominated by both **evergreen** and **deciduous**, **broad-leaved trees**, and there are fewer **lianas** and other climbers than in rain forest.

tropical subalpine rain forest A **forest** of small, stunted trees with small leaves that grows immediately below the **tree line** on the highest tropical mountains.

tropic movement *See* tropism.

tropics Latitudes 23.5° N (Tropic of Cancer) and 23.5° S (Tropic of Capricorn) where the noonday Sun is directly overhead at one of the **solstices**, and the region lying between them but to either side of the narrow equatorial belt.

tropism (tropic movement) A movement or growth by a plant or plant organ toward (positive tropism) or away from (negative tropism) the source of a stimulus.

tropopause The boundary between the **troposphere** and **stratosphere**, at an average height of 16 km over the equator, 11 km in middle latitudes, and 8 km at the poles.

troposphere The lowest layer of the atmosphere, extending from the surface to the **tropopause**. It contains almost all of the atmospheric moisture and is the region in which **convection** and winds ensure that the air is thoroughly mixed.

trough A long, narrow protrusion from an area of low pressure into a region of higher pressure.

trout worm *See Dendrodrilus rubidus*.

true flies *See* Diptera.

TRV *See* tobacco rattle virus.

Tsutsusi *See Azalea*.

tube nucleus (vegetative nucleus) The large **nucleus** of the **pollen tube** cell, formed within a **pollen grain**, that moves to the tip of the tube as it grows, apparently directing its development. Once the tube has penetrated the **nucellus** and the two generative nuclei have left the tube, the tube nucleus disintegrates.

tuber A swollen root or underground stem that stores nutrients. Stem tubers (e.g. potato) often produce **buds** from which aerial stems arise the following

T

season. Root tubers produce no buds, or buds only at the point where the tuber is attached to the root.

tubercle A hemispherical projection, like a dome.

Tuberolachnus salignus (giant willow aphid, large willow aphid) A large, dark brown aphid (**Aphididae**), 5 mm long and covered in fine grey hair, with a large, conical tubercle on its back, making it very distinctive. No males have ever been recorded and they may not exist. It overwinters in both adult and juvenile form, and feeds on willow (***Salix***) trees. It has a **cosmopolitan distribution**.

tubule A small tube or cylinder.

tubulin *See* microtubule.

tufted titmouse *See Baeolophus bicolor.*

tulip *See Tulipa.*

Tulipa (family **Liliaceae**) A genus of **monocotyledon**, **perennial herbs** that grow from **bulbs**, typically with 2–6, occasionally up to 12 strap-shaped, **alternate** leaves with a waxy coating. Flowers with 3 **sepals** and 3 **petals** almost identical, 6 **stamens**, **ovary superior** of 3 **carpels**. Most with 1 flower per stem, some with several on a **scape**. Fruit is a **capsule**. There are more than 100 species occurring from southern Europe to China. Several species and **hybrid cultivars** are grown for ornament, most cultivars descended from *T. gesneriana* (garden tulip) introduced to Europe from Turkey in the sixteenth century.

tulip breaking virus (TBV) A **virus** of the **Potyviridae** that infects tulip **bulbs** and produces streaks, stripes, or other patterns of contrasting colour on petals, i.e. the virus breaks the solid colour pattern. It also weakens and eventually kills the bulb. The virus is transmitted by aphids (**Aphididae**).

tulip tree (tulip poplar) *See Liriodendron.*

tumblebugs *See* Scarabeidae.

tumbleweed *See* diaspore.

tumour (neoplasm) Tissue that results from uninhibited and unstructured growth, producing a form unlike any organ. Tumours occur in plants as well as vertebrate animals. Plant tumours include crown galls caused by ***Agrobacterium tumefaciens***, which are shapeless proliferations of cells. **Mutations** may also lead to the formation of tumours. Unlike animals, however, plants have no circulation system, so tumour cells cannot move about, and plant cells are less specialized than animal cells, so tumour cells cannot invade tissue of a different type. Consequently, plant tumours can be disfiguring but they are not fatal.

tundra A treeless plain found in high latitudes of both hemispheres, with sedges, rushes, and grasses, as well as **perennial herbs**, dwarf **trees** and **shrubs**, **lichens**, and mosses (**Bryophyta**).

tundra soil A **zonal soil** that develops in cold climates under acid conditions on ground that drains poorly, usually because of a **permafrost** layer. It is 30–60 cm deep, has a high content of organic matter, and its surface configuration is shaped by repeated freezing and thawing.

tunica In **angiosperms**, the cap of cells on the **apical meristem**, above the **corpus**.

turban fungus *See Gyromitra esculenta.*

Turbellaria A class of flatworms (**Platyhelminthes**), most of which live on the seafloor, but some of which occur in fresh water and in moist places on land in temperate and tropical regions. They are **acoelomate**, have a gut but no anus, ejecting undigested food through the mouth, and sense organs that respond to light and darkness, gravity, and chemical stimuli. They move by means of cilia (*see* cilium). Some marine species are very colourful but other species are drab. There are about 3000 species. It is uncertain whether the Turbellaria is monophyletic (*see* monophyly).

turbinate Conical, like a spinning top, and attached at the point.

Turdidae (blackbirds, bluethroats, chats, nightingales, redstarts, rubythroats, shortwings, thrushes, wheatears) A family of small to medium-sized **passerine** birds that inhabit woodland and open country, often visiting gardens, and often feed on the ground, mainly on invertebrates but they also eat fruit. There are about 50 genera and more than 300 species, found worldwide.

Turdus merula (blackbird, Eurasian blackbird) A bird 235–290 mm long with a long tail in which the male is glossy black with a yellow eye-ring and yellow or orange beak and the female is brown with a yellow or brown beak. They live in **deciduous** woodland and gardens. They feed mainly on the ground on invertebrate animals, pulling earthworms from the ground, as well as other invertebrates, seeds, and berries. They occur throughout Eurasia and North Africa and have been introduced to Australasia.

Turdus migratorius (American robin) A migratory songbird with a red breast, 230–280 mm long with a wingspan of 310–410 mm, a black or grey head, white eye arcs, and white supercilia (*see* supercilium), a white throat with black streaks and white underside and white **wing coverts**. It breeds throughout most of North America and winters in the southern United States and Mexico, and along the Pacific Coast. It feeds on berries, other fruits, and invertebrates.

Turdus philomelos (song thrush) A thrush that is 200–235 mm long with a brown back and yellow or cream underside with black spots. Song thrushes nest in undergrowth in woodland, parks, and gardens, and feed on invertebrates, berries, and fruit. It occurs throughout most of Eurasia.

Turesson, Göte Wilhelm (1892–1970) A Swedish evolutionary botanist who demonstrated that plant populations adapted to local conditions are genetically distinct from other populations. He coined the term **ecotype**.

turgor Rigidity of a plant or plant cells due to hydrostatic pressure.

turion 1. An overwintering shoot produced from a modified shoot **apex** by an aquatic plant in response to deteriorating conditions, e.g. falling temperature, decreasing day length. A turion often acts as a storage organ. **2.** A young shoot or sucker arising from a **rhizome**, e.g. an emerging shoot of ***Asparagus***.

turmeric (*Curcuma longa*) *See* Zingiberaceae.

turnip fly *See Delia radicum.*

turnip gall weevil *See Ceutorhynchus assimilis.*

turnover rate A measure of the rate at which an element moves through a **biogeochemical cycle**, calculated as the rate of flow into and out of a specified **reservoir pool**.

turnover time A measure of the rate at which an element moves through a **biogeochemical cycle**, and the reciprocal of the **turnover rate**, calculated as the amount present in a specified **reservoir pool** divided by the rate at which the element is entering and leaving the pool. The calculation reveals whether the amount of the element in the pool is increasing or decreasing and, if it is decreasing, how long it will take to empty the pool.

turtle dove *See Zenaida macroura.*

turtles *See* Chelonia.

twisted wrack (*Fucus spiralis*) *See* spiral wrack.

two-lipped door snail *See Balea biplicata.*

two-spotted spider mite *See Tetranychus urticae.*

two-winged flies *See* Diptera.

tyloses Hollow outgrowths from **parenchyma** cells in **xylem** vessels that often fill with pigmented materials, e.g. **resin**, **tannin**, and that may fall away, causing blockage. This prevents the spread of infection by **Fungi** or other **pathogens**.

tylosis The process of forming **tyloses**.

Typha (family **Typhaceae**) A **monocotyledon** genus of strong **herbs** with **rhizomes** that have erect, unbranched stems up to 2 m tall. Leaves **alternate**, **linear**, sheathing, most growing from the base. Flowers **unisexual** (plants **monoecious**), male flowers reduced to 2 **caducous stamens** and forming a **spike**, female flowers numerous, forming a spike below the males. Fruit is an **achene**. There are 10–12 species with a **cosmopolitan distribution**, mainly Northern Hemisphere, in shallow water. With many common names, e.g. bulrush, reedmace, and cattail; leaves of *T. latifolia* are used to weave baskets and mats.

Typhaceae (order **Poales**) A **monocotyledon** family of **perennial herbs** with **rhizomes** and stems up to 2 m tall. Leaves often **distichous**, sheathing to form a false stem, **linear**, sometimes with a **keel**. Flowers very small, **monoecious**, **perianth** of 1 or more inconspicuous **tepals** or many scales or bristles, 3 **stamens**, **ovary superior** of 2 or 3 **carpels**. **Inflorescence** terminal and **spike**-like with an upper male and lower female section, or **racemose**, female heads below the males. Fruit is a **drupe**, **achene**, or **follicle**. There are 2 genera of about 25 species with a more or less worldwide distribution. ***Typha*** species used to make baskets and mats.

typhoon A **tropical cyclone** that develops over the Pacific Ocean.

U

ubac Sloping ground that faces away from the equator and is permanently shaded.

ubidecarenone *See* ubiquinone.

ubiquinone (ubidecarenone, coenzyme Q) A group of oil-soluble compounds present in most eukaryote cells, especially in mitochondria (*see* mitochondrion) that act as electron carriers in cell respiration by generating adenosine triphosphate (ATP). It also acts as an antioxidant.

Ulmaceae (order Rosales) A family of trees (elm) and a few shrubs and lianas, with leaves alternate rarely opposite, simple, crenate or dentate rarely entire, stipules caducous and small. Flowers actinomorphic, unisexual (plants monoecious) sometimes bisexual, tetramerous or pentamerous, 4–5 tepals, as many or twice as many stamens, ovary superior of 2 carpels with 1 free or 2 fused locules. Inflorescence a terminal or axillary cyme sometimes in clusters, or female flowers solitary. Fruit is a samara or drupe. There are 6 genera of 35 species occurring in northern temperate regions, especially Asia, a few tropical. *Ulmus* spp. (elm) formerly grown for timber but much reduced due to Dutch elm disease.

ulna The longer and thicker bone of the forelimb in tetrapods.

ultisols Leached (*see* leaching), acid soils that have an argillic horizon with a base saturation of less than 35 percent, and are red due to a high concentration of iron oxide. They form in humid subtropical forests. Ultisols comprise an order in the soil taxonomy of the U.S. Department of Agriculture.

ultrabasic rock An igneous rock that contains no free quartz and less than 45 percent silica. It consists almost entirely of ferromagnesian minerals.

umbel An inflorescence in which all the pedicels arise at the tip of an axis. It is usually umbrella-shaped and commonly compound.

Umbel.

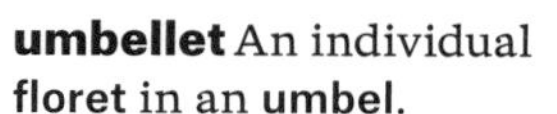

umbellet An individual floret in an umbel.

Umbelliferae *See* Apiaceae.

umbilicate 1. Of a fungus, having a small protuberance (umbo) in a central depression in the pileus. **2.** Of a lichen, having an approximately circular thallus attached by a central point to the substrate. **3.** Supported by a central stalk.

umbric horizon A dark-coloured surface soil horizon that is rich in organic matter and has a base saturation of less than 50 percent.

umbrisols Soils that have an umbric horizon. They develop mainly in mountain areas with a cool, humid climate.

Umbrisols are a reference soil group in the **World Reference Base for Soil Resources**.

unavailable water Water that is adsorbed onto soil particles, where it is held so tightly that plant roots cannot absorb it rapidly enough to meet their requirements.

unconfined aquifer *See* aquifer.

underdominance An unstable **genetic polymorphism** in which the **heterozygote** is less fit than the **homozygote**, which gradually displaces it.

underflow The flow of **groundwater** through an **alluvial** sediment along a channel that is lower than and parallel to a river channel.

underwings *See* Noctuidae.

undulate With wavy margins.

A leaf with an undulate margin.

unguligrade Descries a gait in which only the tips of the digits, covered with hoofs, touch the ground, as in cattle.

uniaxial With a single, unbranched, central **axis**.

unifoliate Having a single leaf.

unifoliolate Describes a **compound** leaf with a single **leaflet**. It can be distinguished from a **simple** leaf by having the **lamina** articulated with the **petiole** or by the presence of a **pulvinus** where the petiole is attached to the **rachis**.

unilocular Having one **locule** or chamber.

uniramous With one **branch**.

uniseriate In a single row.

unisexual flower A **flower** that possesses either **stamens** (male) or **carpels** (female) but not both.

univalent During **meiosis**, a single **chromosome** that is visible when **bivalents** are also present. The univalent has no partner with which to pair.

universal soil loss equation An equation that relates a number of factors to the rate at which soil will be lost by **erosion** at a particular site. The equation is: A = RKLSCP, where A is the rate of soil loss, R is the erosive force of a normal year's rain, K is the susceptibility of the soil to erosion based on a cultivated soil in continuous **fallow** on a 9 percent slope 22 m long, L is the length of the slope, S is the slope gradient, C is the cropping management, and P is the measures taken to minimize erosion, e.g. through contouring, **terracing**, etc.

universal veil In some **agaric fungi**, a membrane that covers the developing **fruiting body**, rupturing as the fruiting body grows larger.

unit leaf rate (ULR) The rate of **photosynthesis** per unit area of leaf. **Primary productivity** is ULR × LAI (**leaf-area index**).

univoltine Producing one generation each year.

unnatural classification A taxonomic grouping of organisms that does not reflect their degree of relationship and is therefore false.

unsaturated *See* fatty acid.

unsaturated zone *See* soil-water zone.

upland chorus frog *See Pseudacris feriarum*.

upper atmosphere All of the atmosphere above the **tropopause**.

Uppsala School of Phytosociology *See* Du Rietz, Gustaf Einar.

upregulation *See* RNA interference.

upslope fog *See* hill fog.

upwind In the direction from which the wind is blowing.

upwind effect The higher precipitation that falls on the **upwind** side of high ground.

uracil One of the **pyrimidine bases** in **RNA**. It binds to adenine and replaces **thymine** in **DNA**.

urban boundary layer The layer of air over a city that extends from the top of the **urban canopy layer** to the highest level at which the properties of the air are affected by its proximity to the surface.

urban canopy layer The air that lies below the level of the rooftops in a city.

urban canyon A city street that is lined by tall buildings on both sides.

urban climate The climate of a city, which is warmer, wetter, dustier, and less windy than the climate of the adjacent countryside.

urban dome The approximately dome-shaped body of warm air that forms beneath the **temperature inversion** that forms above an **urban heat island**.

urban heat island The area around a city where the air temperature is higher than that of the surrounding countryside, so the urban area resembles an island of warm air surrounded by cooler air.

urceolate Shaped like a flask.

urea herbicides A group of **herbicides** that inhibit **photosystem II** in **photosynthesis** and are used to control weeds in crops and in other settings. They are of low toxicity to animals.

U

urediniospore *See* urediospore.

uredinium (uredium, uredosorus) A reddish pustule that appears on a leaf of a plant infected with a **rust** fungus. It consists of a mass of fungal tissue within which **urediospores** develop.

urediospore (uredospore, urediniospore) A thin-walled **spore**, borne on a thin **pedicel** in a **uredinium**, produced in summer by a **rust** fungus, usually on a grass leaf.

uredium *See* uredinium.

uredosorus *See* uredinium.

uredospore *See* urediospore.

Urodela *See* Caudata.

uropygium The posterior of a bird to which the tail **feathers** are attached.

Urtica (family **Urticaceae**) A genus of **annual** and **perennial herbs** (nettles) and a few **shrubs** with **rhizomes**. Stems erect, ascending, or spreading, branched or unbranched, often square in section. Leaves **opposite**, **simple**, **elliptical**, **ovate**, or **peltate**, **serrate** to coarsely **dentate**, with many stinging hairs. Flowers small, green, **unisexual** (plants **monoecious** or **dioecious**), with 4 **sepals**, 4 **petals**, and 4 **stamens**, borne in **panicles** along **axillary**, **spike**-like stems. Fruit is an **achene**. There are 24–39 species with a **cosmopolitan distribution**, mainly temperate. *Urtica dioica* is the stinging nettle.

Urticaceae (order **Rosales**) A family of **annual** or **perennial herbs**, **lianas**, and **trees** with **cystoliths** on leaves and stems and many with stinging hairs (*see Urtica*). Leaves **alternate** or **opposite**, **simple**, sometimes deeply lobed, **entire** or **dentate**, usually **stipulate**. Flowers small, usually green, **actinomorphic**, female flowers sometimes **zygomorphic**, **unisexual** (plants **monoecious** or **dioecious**) rarely **bisexual**, male flowers with 3–5 free or 1–2 fused **tepals**, female flowers with 3–5 tepals or absent, as many **stamens** as tepals, **ovary superior** of 2 **carpels** only 1 of which develops. **Inflorescence** an **axillary** or terminal **cyme** sometimes clustered into a **panicle** or reduced to 1 or a few flowers. Fruit is an **achene**. There are 54 genera of 2625 species with a worldwide, mainly tropical, distribution. Some produce useful fibres, some edible.

Usnea (beard lichens, old man's beard, tree's dandruff, woman's long hair, tree moss) A genus of **fruticose lichens** that hang from tree **branches**, resembling shrubs. They reproduce both sexually and asexually, but grow very slowly, establishing themselves on sick or old trees that have lost their **canopy**, allowing more light to penetrate. They are highly sensitive to air pollution, especially to sulphur dioxide, and their presence is an indicator of unpolluted air. **Usnic acid** ($C_{18}H_{16}O_7$) obtained from *Usnea* is used as an antibiotic and antifungal agent, and the lichen has medical uses. There are probably several hundred species.

usnic acid A **secondary metabolite** found in many **lichens** that is a powerful antibiotic effective against *Staphylococcus*, *Streptococcus*, and other Gram-positive (*see* Gram reaction) **pathogens**.

Ustilaginomycetes (smut fungi) A class of **basidiomycete fungi** that are **obligate parasites** of vascular plants (**Tracheophyta**), causing smut diseases, so named for the black discoloration they produce. There are about 70 genera with more than 1400 species found worldwide.

Ustilaginomycotina A subphylum of **basidiomycete fungi** that includes the classes **Entorrhizomycetes**, **Exobasidiomycetes**, and **Ustilaginomycetes**.

Ustilago A genus of **basidiomycete fungi** all of which cause smut diseases on grasses (**Poaceae**). There are about 200 species found worldwide.

Ustilago maydis A species of **basidiomycete fungi** that parasitizes maize (sweetcorn, *Zea mays*), causing the disease corn (maize) smut, in which the flowers, stems, and leaves swell. The fungus occurs wherever maize is grown and the swollen tissue is a popular delicacy in Mexico, although some people suffer toxic effects from it.

utile (*Entandrophragma utile*) *See* Meliaceae.

utricle 1. A single-seeded, **indehiscent**, dry fruit resembling a thin-walled **bladder** produced by some members of the **Amaranthaceae**; it is a type of **achene**. **2.** In *Utricularia* species of bladderworts, modified leaves in the form of compartments lined with sensitive hairs that trap insects.

Utricularia (family **Lentibulariaceae**) A genus of carnivorous, aquatic freshwater or terrestrial plants or **epiphytes** (bladderworts), most of which produce long **stolons**, lying below the soil or water surface, from which photosynthetic leaf shoots arise and to which the **bladder** traps (**utricles**) are attached. Only the flowers rise clear of the surface. They have 2 asymmetric, lip-like **petals**, usually with the lower petal larger than the upper. Some produce closed, self-pollinating flowers, others open, insect-pollinated flowers; aquatic species often have open flowers above the surface and closed flowers submerged. There are about 220 species with a **cosmopolitan distribution**. Some cultivated for ornament.

U

V

Vaccinium (family **Ericaceae**) A genus of **shrubs**, often **evergreen**, some trailing, with **alternate**, **simple** leaves. Flowers **tetramerous** or **pentamerous** with a **campanulate** or urn-shaped **corolla** and **inferior ovary**. Fruit is a **berry**. There are about 450 species most occurring in northern temperate regions, some tropical. Several are cultivated for their edible fruits, e.g. *V. macrocarpon* and *V. oxycoccus* (American and European cranberry). *Vaccinium myrtillus* is the bilberry.

vacuole A membrane-bound **organelle** found in all plant and fungal, and some bacterial, protist, and animal cells. Vacuoles are filled with water, containing molecules and particles of other substances.

vadose zone *See* soil-water zone.

vagile Capable of moving about. *Compare* sessile.

vagility The ability of an organism to move independently.

Vahliaceae A **monogeneric** family (*Vahlia*) of **annual** to **perennial herbs** and **subshrubs** that have not been placed in an order. Leaves **opposite**, **simple**, **entire**, **ovate** to **linear**, **sessile** or nearly so. Flowers **actinomorphic**, **bisexual**, **pentamerous**, with 5 **sepals**, **petals**, and **stamens**, **ovary inferior**. **Inflorescence** an **axillary cyme**. Fruit is a **capsule**. There are eight species occurring from Africa and Madagascar to India.

valency A measure of the ability of a chemical element to form bonds with other elements. This is determined by the number of electrons in the outer shell of its atoms.

valerian *See Valeriana.*

Valeriana (family **Caprifoliaceae**) A genus of **dioecious perennial herbs** (valerian) with **opposite** or wholly **radical**, sometimes **connate** leaves. Flowers small with a **sympetalous corolla**, borne in a **cymose inflorescence** with **bracts**. There are 200 species occurring in northern and South American temperate regions and in the Andes. Some have medicinal uses.

Valerianella (family **Caprifoliaceae**) A genus of **dioecious perennial herbs** with **opposite** or wholly **radical**, sometimes **connate** leaves. Flowers small with a **sympetalous corolla**, borne in a **cymose inflorescence** with **bracts**. There are 16 species occurring in northern and South American temperate regions and in the Andes. Several are cultivated as corn salad, or lamb's lettuce.

vallecular canal (cortical canal) One of a number of large intercellular channels filled with air, located between the **vascular bundles**, that extend the full length of each **internode** in ***Equisetum*** and its **fossil** relatives.

valley bog A **mire** that develops in valley bottoms and hollows where **drainage** is poor.

valvate 1. In **aestivation**, having the **petals** or **sepals** meeting at their edges without overlapping. **2.** Opening by **valves**. **3.** Possessing **valves**.

valve 1. One of the pieces into which a **fruit** splits when ripe. **2.** A mollusc (**Mollusca**) shell. **3.** An **epithecium** or **hypothecium**. **4.** A **theca**.

Vapona *See* dichlorvos.

vapour pressure The **partial pressure** exerted by water vapour.

variegation The occurrence of patches of two or more colours on leaves or **petals**.

variety In **taxonomy**, a group within a **species** or **subspecies** that is sufficiently distinct to be recognizable but too similar to consistitute a separate **taxon**.

varve A layer of **silt** and **sand** that settles each year on the bed of a lake, especially lakes close to ice sheets. It forms two bands, one pale in colour and coarse-grained, that consists of material released into the lake when ice melts, the other darker and finer, so it settles more slowly. Each varve comprises one pale and one dark band. Varves are counted to calculate the age of glacial deposits.

vascular bundle In vascular plants (**Tracheophyta**), a strand of **vascular tissue** together with **cambium** groups of which form continuous channels for the transport of water and nutrients to all parts of the plant.

vascular cambium **Cambium** that generates **secondary xylem** on its inner side and **secondary phloem** on its outside.

vascular cryptogam A vascular plant (**Tracheophyta**) that reproduces by **spores** rather than **seeds**.

vascular tissue Plant tissue that forms channels through which water and nutrients move to all parts of the plant. *See* vascular cambium, phloem, xylem.

vascular wilt A fungal disease in which the pathogen blocks **vascular tissue**, depriving the plant of water and nutrient and thereby killing it.

Vavilov, Nikolai Ivanovich (1887–1943) A Russian plant geneticist and geographer who proposed that the greatest variation in species occurs in particular areas (**centres of diversity**), which he believed were also **centres of origin** of those species. He collected many plant specimens and built the world's first **seed bank**. He fell foul of Stalin and Trofim **Lysenko**, however, and died in prison from starvation.

vector An organism that conveys a pathogen which has infected it to another organism, thus transmitting the infection.

veering A clockwise change in the wind direction.

vegetation tension zone An area where **phytochoria** overlap.

vegetative Describes a structure, organ, or stage in the life cycle of an organism that is concerned with feeding and growth rather than reproduction.

vegetative cell Any plant cell that is not involved in the production of **gametes**.

vegetative cloning *See* vegetative reproduction.

vegetative multiplication *See* vegetative reproduction.

vegetative nucleus *See* tube nucleus.

vegetative propagation *See* vegetative reproduction.

vegetative reproduction (vegetative cloning, vegetative multiplication, vegetative propagation) **Asexual reproduction** in which new individuals develop from the roots, stems, or leaves of a parent plant.

vegetative state 1. The condition of a plant that is reproducing asexually, i.e. by **vegetative reproduction**. **2.** The non-infective stage in the life cycle of a

bacteriophage during which its **genome** multiplies, thereby controlling the synthesis by the host **bacterium** of materials needed to form more bacteriophages and their **DNA**, which are released by **lysis** of the cell.

vein 1. (nerve) A **vascular bundle** or group of vascular bundles that lie parallel and close together in a **leaf**. **2.** A blood vessel through which blood flows toward the heart.

velamen Several spongy layers of dead cells on the **epidermis** of the aerial roots of certain **epiphytes** and semi-epiphytes; the velamen absorbs moisture from water flowing over it.

veld Grassland with scattered **trees** and **shrubs** that occurs in the eastern interior of South Africa.

Velloziaceae (order **Pandanales**) A **monocotyledon** family of **xeromorphic herbs** and **shrubs** with **dichotomously** branching stems and three-ranked, **linear**, **entire** or **denticulate** leaves usually clustered at the ends of stems. Flowers **actinomorphic**, **dioecious**, **bisexual** rarely **unisexual**, with 6 **tepals** in 2 **whorls**, 6 or many **stamens**, **ovary inferior** or semi-inferior of 3 **carpels** and **locules**. **Inflorescence** terminal with a **caducous scape** or a **capitulum** with a few flowers. Fruit is a **loculicidal capsule**. There are 5 genera of 240 species occurring in South America and from Africa and Madagascar to Arabia and China.

velocity head *See* hydraulic head.

velum 1. In certain ferns (**Pteridophyta**), a flap of membranous tissue that protects the **sporangium**. **2.** The remnants of the ruptured **partial veil** in some mushrooms and toadstools.

velutinous Covered in soft, short hairs, like velvet.

velvet *See* antlers.

vena cava In **tetrapods**, the principal **veins**; the vena cava superior serves the forelimbs and head, the vena cava inferior serves all of the body behind the forelimbs. The vena cava superior is usually a pair of veins but in many mammals only the right vein persists in adults. The vena cava inferior is a single vein, the largest in the body.

venation (nervation) The arrangement of **veins** in a **leaf**.

Venezuela and Guiana floral region The area that covers the Orinoco Basin and uplands of Venezuela, part of the **Neotropical region**. There are about 100 **endemic** species.

venter The swollen base of an **archegonium** containing the **megaspore**.

ventral Nearest to the substrate, usually the underside.

Venturia inaequalis A species of **Ascomycota** that causes the disease apple scab. Infection occurs in spring, when mild temperatures and high **humidity** stimulate the release of **ascospores** that settle on host trees. Lesions appear on leaves and blossoms, and later on fruit. The fungus infects several tree species including apple, hawthorn, and rowan. ↗

Venus's car *See Lamprocapnos spectabilis*.

Verbenaceae (order **Lamiales**) A family of **annual** and **perennial herbs**, **shrubs**, **lianas**, and **trees** with **opposite** occasionally **alternate** or whorled, **entire** or **dentate**, **exstipulate** leaves, sometimes reduced to scales. Flowers with a tubular to **campanulate**, persistent **calyx** of 4–5 sometimes 2–4 fused **sepals**, **corolla** irregular often 2-lipped, 4 sometimes 2–4 or 5 **stamens**, **ovary superior** of 2 **carpels** each with 2 **locules**. **Inflorescence racemose**, sometimes with an **involucre** of **bracts**. Fruit is a **schizocarp** or **drupe**. There are 31 genera of 918 species with a pantropical to warm temperate, mainly American, distribution. Several cultivated for ornament.

verdigris agaric *See Stropharia aeruginosa.*

vermiculite A **clay mineral** with a 2:1 structure, comprising layers each with one sheet of octahedral crystals bounded by two sheets of tetrahedral crystals. It expands to up to 30 times its volume when heated, and swells and shrinks when wetted and dried. It has a high **cation exchange capacity**. Vermiculite has many industrial uses. It is also used as a medium for germinating seeds, storing **bulbs** and root vegetables, and improving soil structure.

vermiculture The growth of colonies of earthworms (**Annelida**), usually as part of a composting process.

vernalization Prolonged exposure of a plant or seeds to a low temperature in order to stimulate flowering to promote seed production.

vernation (ptyxis) The arrangement of young leaves or **bud** scales in a shoot bud. They may be rolled with the sides to the centre of the underside (involute), rolled with the sides to the centre of the upper side (revolute), rolled with one side around the other (convolute), folded with each leaf clasping those next to it inside the fold (conduplicate), pleated lengthways (plicate), or rolled lengthwise (circinate).

Veronica (family **Plantaginaceae**) A genus of **annual** and **perennial herbs** and **shrubs** with **opposite**, **simple** or lobed, **exstipulate** leaves. Flowers **bisexual**, **calyx** fused with 4 or 5 lobes, the upper lobe very small, **corolla** in a tube of 4 lobes, 2 **stamens**, **ovary superior** of 2 **carpels**. **Inflorescence** an **axillary** or terminal **raceme** or 5-lobed flowers solitary in leaf **axils**. Fruit is a **capsule**. There are about 500 species occurring in temperate regions. Some edible some cultivated for ornament (speedwell).

verrucate *See* verrucose.

verrucose (verrucate) Warty; covered in bumps.

vertebra One of the bone segments that form the vertebral column (backbone, spinal column) in vertebrates. There are five types. Cervical vertebrae form the neck; **dorsal** vertebrae form the section from the neck to the pelvis, the dorsal vertebrae attached to the ribs being known as the

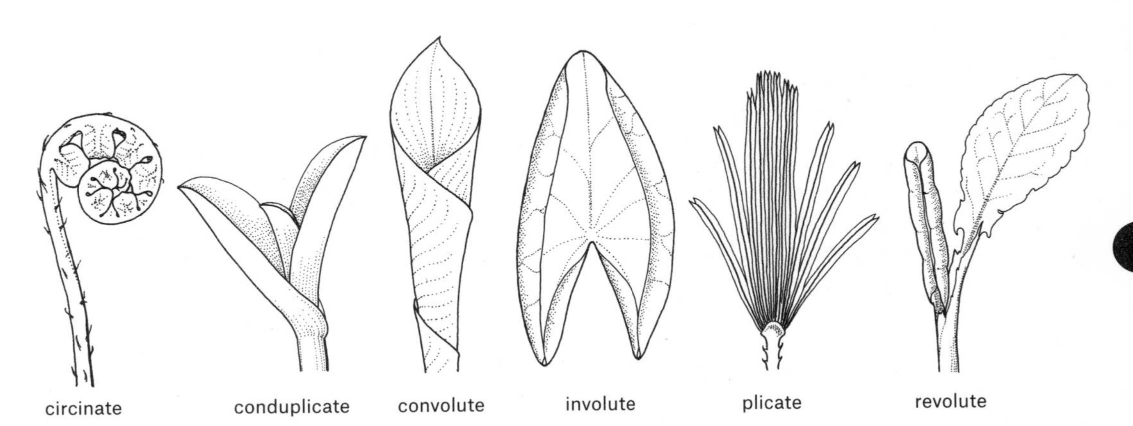

Vernation reflects the way a young leaf is folded in the bud.

thoracic vertebrae and those not attached to ribs being the lumbar vertebrae; sacral vertebrae are in the pelvic region and when several are fused together they form the sacrum; **caudal** vertebrae form the tail.

vertic horizon A subsurface **soil horizon**, at least 25 cm thick, containing **peds** with smooth, polished surfaces produced by repeated swelling and shrinking of **clay**, of which it contains at least 30 percent.

verticillate In one or more **whorls**.

Verticillium A genus of **ascomycete fungi**, some of which are pathogens of plants and other Fungi and others that are **saprotrophs**. The plant pathogens cause verticillium wilts in more than 400 **angiosperm** species, with *V. dahliae*, *V. albo-atrum*, and *V. longisporum* accounting for 300. There are 51 species with a worldwide distribution. The fungus enters from the soil and grows inside the vascular tissue. The lower leaves of infected plants turn yellow, the discoloration then spreading upward, and cut stems show brown discoloration, eventually turning black as the plant dies.

verticillium wilt *See Verticillium.*

vertisols **Mineral soils** that contain at least 30 percent swelling **clay**, e.g. **montmorillonite**; due to the repeated swelling and shrinking the soil is self-inverting (*see* self-mulching soil). Vertisols form from **parent materials** rich in suitable clay, in a climate with pronounced wet and dry periods. They are fertile, but difficult to cultivate. They comprise an order in the **soil taxonomy** devised by the U.S. Department of Agriculture.

V

very deep soil *See* effective soil depth.

very shallow soil *See* effective soil depth.

vesicle **1.** An **organelle** comprising a small sphere enclosed by a double **lipid** membrane found in **eukaryote** cells. **2.** Any structure resembling a **bladder** and containing liquid.

vesicular structure A structure that contains many small cavities, e.g. the structure of a rock such as pumice, filled with holes formed by the expansion of gases in a cooling **magma** as the rock enters a region of lower pressure as it approaches the surface.

Vespidae (wasps) A family of wasps (**Apocrita**), which includes most of the **eusocial** wasps and many solitary species. Most have abdomens banded in yellow and black and white or yellow face markings. Colonies of temperate social wasps usually last for only one season, only the **queens** surviving the winter, hibernating in sheltered places. Some species make nests of papery **cellulose**. Social wasps feed their larvae on pre-chewed insects. Solitary wasps feed their young on paralyzed prey. There are more than 4500 species, found worldwide.

vessel element An elongated cell with thickened walls and **perforation plates** at either end, found in **angiosperms**. Many vessel elements joined end to end form **xylem**.

vetch *See Vicia.*

vetchling *See Lathyrus.*

vibrissae Stiff hairs or modified **feathers** which protrude from the face and in some species from the limbs, i.e. 'whiskers'. Vibration of the vibrissae stimulates sensory nerves in the skin.

Vicia (family **Fabaceae**) A genus of **herbs** (vetch) and climbers by means of leaf **tendrils**, with **opposite**, **pinnate**, **lanceolate** or **oblong** leaves. Flowers pea-like, solitary or in clusters of up to 3, in leaf **axils**. Fruit is a **pod**. There are about 140 species, most occurring in northern temperate regions, but some in South America and tropical East Africa. Many cultivated for food or fodder, e.g. *V. faba* (broad bean), *V. sativa* (common vetch, tare).

Victoria (family **Nymphaeaceae**) A genus, named in honour of Queen Victoria, comprising two species of water-lilies renowned for the size of their flat, almost circular leaves with raised rims that lie on the water surface. Those of *V. amazonica* are up to 3 m in diameter. Flowers, up to 40 cm across, open at night and are white on their first night and pink on subsequent nights. *Victoria cruziana* has slightly smaller leaves. The plants occur on rivers and lakes in the Amazon region.

villous Covered with soft, unmatted hairs.

Vinca (family **Apocyanaceae**) A genus of **subshrubs** and **herbs** (periwinkle) with trailing stems that often strike where they touch the ground, and **opposite**, **simple**, broadly **lanceolate** to **ovate** leaves. Flowers with a tubular **corolla** of 5 **petals**. Fruit is a **follicle**. There are six species occurring in Europe, southwestern Asia, and northwestern Africa. Several cultivated for ornament and (*V. major* and *V. minor*) ground cover.

vinclozolin A **fungicide** that inhibits **spore** formation in several species of **Fungi** and is widely used on a variety of vegetable, fruit, and ornamental crops, and on turf. It is an endocrine disrupter, mimicking male **hormones**. It is slightly toxic to mammals, moderately irritating to the skin, and moderately toxic to fish, and breaks down fairly slowly and incompletely in soil.

Viola (family **Violaceae**) A genus of **perennial** and **annual herbs** and **shrubs** with **alternate**, **simple**, **cordate** or **palmate** leaves, some **acaulescent** with leaves as a basal rosette. Flowers usually solitary, **zygomorphic**, **pentamerous**, **corolla** often with spurs, **ovary superior**. Fruit is a **capsule**. There are 400–500 species with a worldwide distribution. Many are cultivated for ornament as pansies and violets.

Violaceae (order **Malpighiales**) A family of **trees**, **shrubs**, **subshrubs**, climbers, and **herbs** with usually **alternate**, **linear** to **reniform**, **entire** or **serrate**, **stipulate** leaves. Flowers **actinomorphic** or **zygomorphic**, **bisexual** or **unisexual**, hypogynous (*see* hypogyny) or **perigynous**, 5 **sepals**, 5 **petals**, 5 or 3 **stamens**, **ovary superior** of 3 fused **carpels**. **Inflorescence** usually a **thyrse**, **dichasium**, **cyme**, or **raceme** in woody species, a **fascicle** or flowers solitary in herbaceous species. Fruit is a **capsule**, rarely a **berry**, **nut**, or **follicle**. There are 23 genera of 800 species with a worldwide distribution. Many cultivated for ornament, e.g. ***Viola***, some with medicinal uses.

violent storm Wind of 29–32 m/s. *See* appendix: Beaufort Wind Scale.

violet *See Viola.*

violet aphid *See Myzus ornatus.*

violet ground beetle *See Carabus violaceus.*

Viperidae (vipers) A family of **solenoglyphous**, venomous snakes **(Serpentes)** that are stocky, most with a broad, triangular head, short tail, keeled (*see* keel) scales, and drab colour. Most have vertically elliptical pupils and are nocturnal, ambushing their prey. Many are **ovoviviparous**. They can open their mouths almost 180 degrees when striking. Viperid venom contains **enzymes** that cause blood poisoning. The venom is used to kill prey and secondarily in self-defence. The snake can vary the amount of venom it injects and when striking non-prey, e.g. a human, it may stab without injecting any venom. Bites with venom are invariably painful and potentially very serious even though they may not be fatal. There are 32 genera and 224 species, found worldwide except for Antarctica, Australia, New Zealand, Ireland, Madagascar, Hawaii, and other islands.

vipers *See* Viperidae.

virga (fallstreaks) A grey, wispy, veil-like extension to the base of a cloud, consisting of precipitation falling into dry air, where it vaporizes before reaching the ground.

Virginia buttonweed (*Diodia virginiana*) *See Diodia* vein chlorosis.

Virginia creeper *See Parthenocissus,* Vitaceae.

Virginia deer *See Odocoileus virginianus.*

Virginia striatula (rough earth snake) A brown, unpatterned, colubrid snake (**Colubridae**), up to 250 mm long, that spends much of its time below ground. It inhabits places with loose soil that do not flood. It feeds on soft-bodied invertebrates and occurs throughout the southeastern United States.

virion A single **virus** particle.

viroid A section of single-stranded **RNA**, much shorter than in a **virus** and lacking a **capsid**, that is arranged in a circle and is capable of infecting a plant. They replicate in a cell **nucleus** or **chloroplast**, and then move from cell to cell through **plasmodesmata** and sometimes through the **phloem**. They may cause disease, e.g. potato spindle tuber disease, hop stunt.

virulent phage A **bacteriophage** that causes the **lysis** of its host cell, killing it.

virus A body that consists mainly or only of an **RNA** or **DNA genome** enclosed by a **protein** envelope (capsid), and sometimes also a **lipoprotein** envelope. A virus has no metabolism of its own; it is therefore debatable whether it is alive. In order to reproduce it must invade a host and redirect the host metabolism to produce more **virions**, which are then released to invade other cells, so repeating the process.

viscotaxis A change in the direction of movement of a **motile** cell or organism in response to a change in viscosity of the surrounding medium.

visceral hump In **Mollusca**, the main part of the body containing the digestive, respiratory, and reproductive systems, and the **mantle**.

Viscum (family **Santalaceae**) A genus of woody **obligate hemiparasites** (mistletoes) of **shrubs** and **trees** with **whorls** or **opposite** pairs of **simple**, **entire**, **exstipulate** leaves that perform some **photosynthesis**. Flowers are inconspicuous, 1–3 mm in diameter, **unisexual** (plants **monoecious** rarely **dioecious**), with 3–4 sometimes 4–6 **valvate tepals**, as many **stamens** as tepals, **ovary inferior**. Flowers solitary or in **spikes**. Fruit is a **berry** containing several seeds in very sticky juice; when birds (commonly mistle thrushes, hence the common name) pick the berries they clean their bills by rubbing them on tree **branches**, thereby transferring the seeds to a new host. There are 70–100 species occurring in temperate and tropical regions of Eurasia, Africa, and Australia.

Vitaceae (order **Vitales**) A family of **lianas**, **shrubs**, small **trees**, and **herbs** with **alternate**, **simple**, often coarsely **dentate** rarely **entire** leaves. Flowers **actinomorphic**, **bisexual** rarely **unisexual** (plants **monoecious** rarely **dioecious**), 4–5 free or 3 fused **sepals** and **valvate**, often **deciduous petals**, as many **stamens** as petals, **ovary superior** of 2 **carpels** and 2 **locules**. **Inflorescence cymose** or **racemose**, opposite the leaves or terminal. Fruit is a **berry**. There are 14 genera of 850 species with a pantropical and warm temperate distribution. *Vitis vinifera* is the most widely cultivated grapevine, ***Parthenocissus*** *quinquefolia* is Virginia creeper.

Vitales An order of plants comprising 1 family (**Vitaceae**) of 14 genera and 850 species.

vitric horizon A surface or subsurface **soil horizon**, at least 30 cm thick, that is made predominantly from volcanic material, especially glass (Latin *vitrum*, glass).

Vitrinidae (glass snails) A family of **snails**, less than 20 mm long, that have a glassy, translucent shell so small that the snails resemble **slugs**; some species are unable to retract into their shells. The snails occur in cool, damp places in many parts of Europe, Africa, and North America.

Vivianaceae (order **Geraniales**) A family of **shrubs** and **herbs** with **opposite**, **simple** or deeply 3-lobed, **entire**, or sometimes **crenate**, or **dentate**, **sessile** or shortly **petiolate**, **exstipulate** leaves. Flowers **actinomorphic**, **tetramerous** or **pentamerous** rarely 8- or 10-merous, 8–10 **stamens**, **ovary superior** of 3–5 fused **carpels. Inflorescence cymose**. Fruit is a **capsule**. There are 4 genera of 18 species occurring in South America especially southern Brazil.

viviparous 1. Describes a plant that produces seeds which germinate inside the fruit while still attached to the plant. **2.** Describes a plant that reproduces vegetatively (*see* vegetative reproduction). **3.** Describes an animal that gives birth to live young.

vivipary Animal reproduction in which the **embryo** develops inside the mother's body, is nourished directly by the mother, and is born active. *Compare* ovipary, ovovivipary.

Vochysiaceae (order **Myrtales**) A family of **trees** and **lianas** with **opposite** or whorled, **simple**, **stipulate**, **petiolate** leaves. Flowers **zygomorphic**, **bisexual**, 5 **sepals**, 3 sometimes 1, rarely 5 **petals** or **apetalous**, 1 fertile **stamen**, **ovary superior** or **inferior**. **Inflorescence** terminal or **axillary racemose** or a few axillary flowers. Fruit a **loculicidal capsule**. There are 7 genera of 190 species occurring in lowland tropical America and Africa.

volatile oil *See* essential oil.

volcanism (volcanicity, vulcanicity, vulcanism) All of the processes that are involved in the movement of **magma** and the volatile substances dissolved in it from the interior of the Earth to the surface.

voles *See* Cricetidae.

volunteer plant A domesticated plant that grows where it has not been sown or planted.

volva In the fruit body (*see* fruiting body) of certain **agarics**, a cup-like sheath that surrounds the base of the **stipe**. It is a remnant of the **universal veil**.

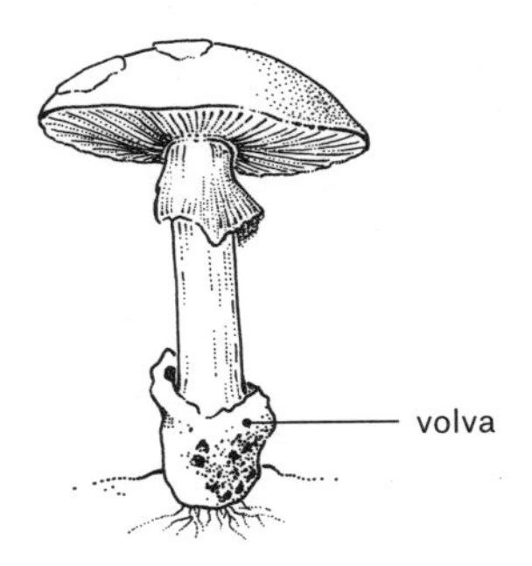

The volva is what remains of the universal veil.

vulcanicity *See* volcanism.

vulcanism *See* volcanism.

Vulpes vulpes (red fox) The largest of all species of fox, adults measuring 350–500 mm with a tail 320–530 mm. They have soft fur that is red, silver or black, or a cross between these. They inhabit a wide range of **habitats**, increasingly including urban areas and suburban gardens. They feed on rodents, rabbits and cottontails, insects and other invertebrates, fruit, and carrion, but they also attack poultry. They occur throughout most of the Northern Hemisphere.

vulpine Pertaining to foxes (***Vulpes vulpes***).

W

waggle dance *See* dance language.

wagtail dance *See* dance language.

Wahlenbergia (family **Campanulaceae**) A genus of **annual** and **perennial herbs** with **taproots** and sometimes extensive **rhizomes**. Leaves **alternate**, **opposite**, or whorled, more or less **sessile**, **denticulate**. Flowers with 5 **sepals**, **corolla campanulate** or **rotate** usually of 5 **petals**, usually 5 **stamens**, **ovary inferior**. **Inflorescence** terminal **cymose**, or flowers solitary. Fruit is a **capsule**. There are about 200 species with a **cosmopolitan distribution**, but none in North America, most occurring in the Southern Hemisphere. Many cultivated for ornament (Australian bluebell).

Wallace, Alfred Russel (1823–1913) An English naturalist who worked in the East Indies (now Indonesia). His studies of the species he found there led him to propose a boundary, now called Wallace's line, between the fauna of Asia and that of Australasia. He also developed a theory of evolution by natural selection, which he described in a letter to Charles **Darwin**. Wallace's paper on the subject was presented with that by Darwin at a meeting of the Linnean Society in London on 1 July 1858.

Wallace's line *See* Wallace, Alfred Russel.

wall cloud An extension that appears beneath a large **cumulonimbus** cloud that contains a **mesocyclone**. It indicates that the mesocyclone is expanding downward, and marks a region where warm, moist air is being drawn into the mesocyclone and moisture is condensing. The wall cloud rotates cyclonically (*see* cyclonic). Its appearance warns of the imminent possibility of a **tornado**.

Wallemiomycetes A class of **basidiomycete fungi** comprising a single genus, *Wallemia*, with three species, found worldwide. They form moulds that tolerate dry conditions and often occur in dry foods, e.g. bread and cakes.

walnut *See Juglans*, Juglandaceae.

walnut orb-weave spider *See Nuctenea umbratica*.

warm cloud A cloud in which the temperature throughout is above freezing.

warm front A **front** with air behind the front warmer than the air ahead of it.

Warming, Johannes Eugenius Bülow (1841–1924) A Danish botanist who believed plant communities should be studied in the context of their surrounding environment, thus stimulating the study of plant **ecology**.

warm sector The wedge of warm air below the crest of a **frontal wave**, bounded by the **cold** and **warm fronts**.

warm wave A sudden rise in temperature that occurs in middle latitudes, usually in summer. It often heralds wet weather.

Washington giant earthworm *See Driloleirus americanus.*

wasps *See* Apocrita, Hymenoptera, Vespidae.

water bears *See* Tardigrada.

water chestnut (*Eleocharis dulcis*) *See* Cyperaceae; (*Trapa natans*) *See* Lythraceae.

water cloud A cloud that consists entirely of water droplets.

water deficit The difference between the amount of water plants require for healthy growth and the amount delivered by precipitation where this is smaller.

water equivalent The depth of snow after this has been melted. Since snow varies greatly in its density, snowfall amounts are always reported as the water equivalent.

water hyacinth (*Eichhornia crassipes*) *See* Pontederiaceae.

water lilies *See* Cabombaceae, *Nymphaea*, Nymphaeaceae.

waterlogging The saturation of the ground with water, such that the **water table** is higher than the depth required for plant roots. Water fills all soil **pores**, producing anaerobic conditions that inhibit root **respiration**.

watermeal *See Wolffia.*

water milfoil (*Myriophyllum* spp.) *See* Haloragaceae.

water moulds *See* Oomycota.

water plantain *See* Alismataceae.

water potential The energy with which water moves from a region of high water potential to one of low water potential. *See* capillarity, osmosis, surface tension.

watershed *See* catchment, divide.

water snake *See Natrix natrix.*

water surplus The difference between the amount of water plants require for healthy growth and the amount delivered by precipitation where this is greater.

water table The upper boundary of the saturated zone of the soil (*see* groundwater). It is not sharply defined, but comprises a layer in which water is moving upward by **capillarity** into the **capillary fringe**.

watery soft rot *See Sclerotinia sclerotiorum.*

wattle 1. A bare, fleshy area of skin possessed by some birds. It is often brightly coloured and pendulous and is used in **display**. **2.** *See Acacia.*

wave clouds *See* lenticular cloud.

wave cyclone *See* wave depression.

wave depression (wave cyclone) A depression that forms at the crest of the wave where a **cold** and **warm front** meet and warm air is beginning to rise above the denser cold air.

wax flowers *See Eriostemon.*

wax scales *See* Coccidae.

waxy laccaria *See Laccaria laccata.*

weakening A decrease in a **pressure gradient**, with an associated reduction in wind speed.

weasels *See* Mustelidae.

weather The state of the atmosphere over a short period at a particular place and time.

weathering The breakdown of rocks and minerals at or near the Earth's surface by physical (**mechanical weathering**) and chemical (**chemical weathering**) processes. Despite the name, weathering is not confined to such meteorological phenomena as freezing and thawing, wind and rain.

wedge A **ridge** of high pressure where the **isobars** meet in a V-shape, like a wedge.

weed A plant that is unwanted, because it is considered unattractive or because it is growing where people do not wish it to

W

grow, or that is competing with cultivated crops.

weevils *See* Curculionidae.

Welwitschiaceae (order **Pinales**) A **monotypic** family (*Welwitschia mirabilis*), which is a **gymnosperm** with a deep taproot and a massive stem ensheathed by two very long, strap-like leaves that become dead, curled, and tattered at the tips. Flowers are small, **dioecious**, and covered by **bracts**. The plant is **endemic** to the Namibian Desert.

West African rain forest floral region The area that extends from Guinea to Cameroon and the Congo Basin, including the offshore islands, part of the **Palaeotropical region**. The **flora** is rich but poorly documented; it is the source of *Coffea liberica*.

West and Central Asiatic floral region The large area that extends from Armenia to the Tibetan Plateau and covers southern Russia, the Iranian highlands, Turkestan, and Mongolia, part of the **boreal region**. Barley, wheat, and other crop plants originated in this region and there are about 150 **endemic** species.

western flower thrips *See Frankliniella occidentalis.*

western jackdaw *See Corvus monedula.*

western rat snake *See Elaphe obsoleta.*

western scrub jay *See Aphelocoma californica.*

western tanager *See Piranga ludoviciana.*

wet-bulb temperature The temperature registered by a **wet-bulb thermometer**.

wet-bulb thermometer A thermometer that has the bulb wrapped in wet muslin, the muslin extending into a reservoir of water so it acts as a wick. Water evaporating from the wick absorbs **latent heat** from the bulb, depressing the temperature.

wetland An area defined in 1971 by the Ramsar Convention on Wetlands of International Importance as 'all areas of marsh, **fen**, peatland, or water, whether natural or artificial, permanent or temporary, with water that is static or flowing, fresh, **brackish**, or salt, including areas of marine water the depth of which at low tide does not exceed 6 metres'.

wet spell In Britain, a period of at least 15 days during which at least 1 mm of rain falls every day.

wetting front The boundary between wet and dry soil, marking the extent to which water has penetrated following rain.

wheatears *See* Turdidae.

wheel animalcules *See* Rotifera.

whiplash *See* flagellum.

whippoorwill storm *See* frog storm.

whipscorpions *See* Arachnida.

whirling psychrometer (sling psychrometer) A **psychrometer** with an attached chain and handle, allowing the operator to swing the instrument around, thereby ensuring the thermometer bulbs are exposed equally to moving air.

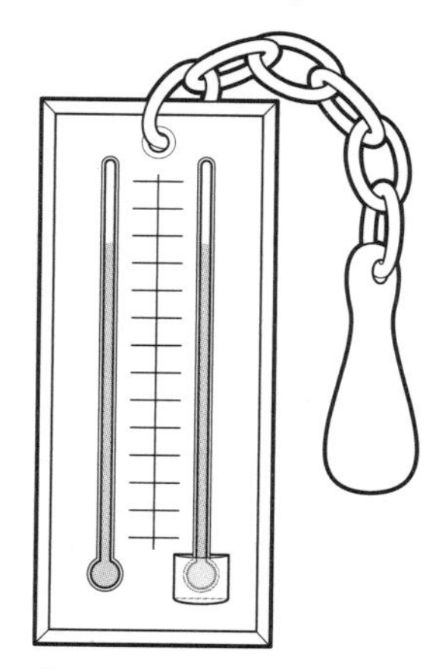

Swinging the whirling psychrometer through the air ensures that the bulb of the dry-bulb thermometer and the wick of the wet-bulb thermometer are evenly exposed to the ambient air.

whitebeam (*Sorbus aria*) *See Sorbus.*

white blister *See Albugo.*

white-breasted nuthatch *See Sitta carolinensis.*

white butterfly *See* Pieridae, *Pieris rapae.*

white cabbage butterfly *See Pieris brassicae.*

white chestnut *See Aesculus.*

white cinnamon (*Canella winterana*) *See* Canellaceae.

white clover (*Trifolium pratense*) *See Trifolium.*

white dew Dew that freezes after it has formed.

whitefly *See* Aleyrodidae.

white-lipped snail *See Cepaea hortensis.*

white mites *See* Tenthredinidae.

white mould *See Sclerotinia sclerotiorum.*

white muscardine disease *See Beauveria bassiana.*

white poplar (*Populus alba*) *See Populus.*

white potato cyst nematode *See Globodera pallida.*

white rot *See Phellinus.*

white rot of onions *See Sclerotium cepivorum.*

white rust *See Albugo candida.*

white-spotted slimy salamander *See Plethodon cylindraceus.*

whitetail *See Odocoileus virginianus.*

white-tailed deer *See Odocoileus virginianus.*

white-throated sparrow *See Zonotrichia albicollis.*

whiteworm *See Enchytraeus buchholzi.*

whorl 1. The arrangement of **organs** that all arise at the same level, encircling the **axis**. **2.** One of the coils in a **snail** shell.

wiggler *See Dendrodrilus rubidus.*

wilderness An extensive area of land that has never been occupied permanently by people or used intensively for agriculture, forestry, or mineral extraction, and that is in a natural condition or nearly so.

wild plum *See Amelanchier.*

wild service tree (*Sorbus torminalis*) *See Sorbus.*

wild type 1. The original form of an organism, as it appears in nature. **2.** The **allele** observed most frequently at a particular **gene locus**. **3.** The **phenotype** of a **species** as it occurs naturally.

Willdenow, Karl Ludwig von (1765–1812) A Prussian medical botanist who noted patterns in the distribution of plants, which he described in *Grundriß der Kräuterkunde zu Vorlesungen*, published in 1792, and in English in 1805 as *Principles of Botany.*

willow *See Salix.*

Willughby, Francis (1635–72) An English ornithologist and ichthyologist who was a student and friend of John **Ray**. Willughby was independently wealthy and when Ray lost his academic post at Cambridge University through refusing to sign the Act of Uniformity aiming to standardize the Church of England liturgy, Willughby supported him financially and collaborated with him; they planned to publish a joint work, but Willughby died before his contribution could be completed, so Ray completed the work.

wilting The loss of **turgor** that occurs when plant tissues contain insufficient water to hold them rigid. *See* permanent wilting point, temporary wilting.

wilting coefficient *See* permanent wilting point.

wilting point *See* permanent wilting point.

wind chill The sensation of additional cold people feel when exposed to the

W

wind. It is due to the removal by the wind of a layer of warm air from the surface of the body.

wind erosion The **detachment** and transport of soil particles by the wind. It removes the most fertile layer of soil, sometimes including seed and **fertilizer**, and the eroded soil pollutes the air.

wind flower *See Anemone.*

windowing Pest damage to a plant leaf that removes areas of leaf tissue from the upper or lower surface, leaving an area of leaf that is much thinner than the surrounding tissue, resembling a window.

wind rose A diagram that shows the frequency with which the wind blows from each direction at a particular location. The wind direction is noted at the same time each day and a line indicating the direction is extended by the same amount each time that direction occurs.

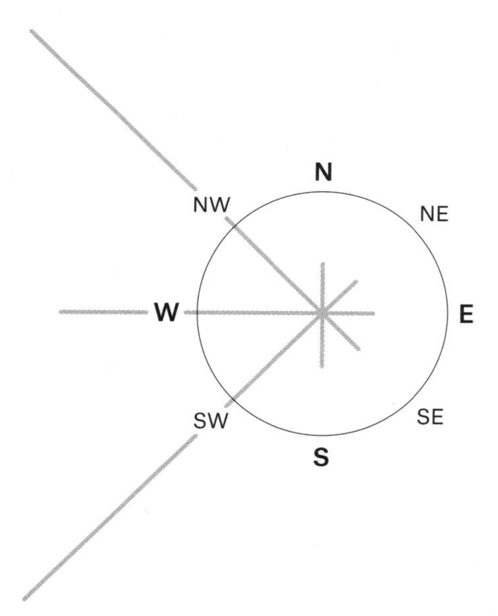

The length of the lines on a wind rose represents the frequency with which the wind blows from a particular direction.

windward Describes the side facing into the direction from which the wind is blowing.

wing covert A **feather** that covers the base of a large feather on a bird's wing.

winged kelp *See* dabberlocks.

wing nut *See Pterocarya.*

Winteraceae (order **Canellales**) A family of **evergreen trees** and **shrubs** with **alternate**, **simple**, **entire**, often **coriaceous**, **exstipulate** leaves, usually **glaucous** on the underside. Flowers **actinomorphic**, **hermaphrodite**, 2–3 sometimes 3–6 **valvate** or calyptrate (*see* calyptra) **sepals**, 2–25 or more **imbricate petals** or none, a few or many **stamens**, **ovary superior** of 1 to many free **carpels**. **Inflorescence** a terminal or **axillary cyme**. Fruit **baccate** or a **follicle**. There are 5 genera of 60–90 species occurring in tropical mountains, but absent from mainland Africa. Some cultivated for ornament.

wintergreen *See* methyl salicylate.

winter moth *See Operophtera brumata.*

winter wren *See Troglodytes troglodytes.*

wireweed *See* japweed.

wireworm *See* Elateridae.

witches' broom A dense mass of shoots growing from a single point on a **tree** or other woody plant that forms a mass resembling a sweeping broom or large bird nest. There are several possible causes including infection by insects (**Insecta**), mites (**Arachnida**), nematodes (**Nematoda**), **Fungi**, oomycetes (**Oomyceta**), and **viruses**. *See* aster yellows, *Phytoplasma*, *Taphrina.*

witch hazel (*Hamamelis virginica*) *See* Hamamelidaceae.

Wolffia (family **Araceae**) A genus of **monocotyledon** floating **herbs** (duckweed, watermeal) with a **thallus** but no roots or leaves, and a flower with 1 **stamen** and 1 pistil in a depression on the plant surface. Fruit is a **utricle**. *Wolffia arrhiza* is the smallest known **angiosperm**, approximately 1 mm in diameter. There are 9–11 species with a **cosmopolitan distribution**. Some edible and a good source of protein.

wolf's milk *See Lycogala epidendrum.*

wolverines *See* Mustelidae.

woman's long hair *See Usnea.*

wonga vine (*Pandorea pandorana*) *See Pandorea.*

wood anemone (*Anemone nemorosa*) *See Anemone.*

wood blewit *See Clitocybe.*

wood cauliflower *See Sparassis crispa.*

wood duck *See Aix sponsa.*

woodland 1. A plant community that comprises mature trees with spreading crowns that do not touch to form a closed **canopy**. Typically the tree crowns cover up to 40 percent of the total area. **2.** A wooded landscape.

woodlice *See* Isopoda, Porcellionidae.

wood mouse *See Apodemus sylvaticus.*

wood-pasture **Woodland** on which deer or farm livestock are permitted to graze.

wood pigeon *See Columba palumbus.*

wood rays *See* medullary rays.

wood sorrel (*Oxalis acetosella*) *See Oxalis.*

wood sugar *See* xylose.

wood thrush *See Hylocichia mustelina.*

woodwasps *See* Symphyta.

woolly aphid *See Eriosoma lanigerum.*

woolly apple aphid *See Eriosoma lanigerum.*

woolly conifer aphids *See* adelgids.

woolly vine scale *See Pulvinaria vitis.*

worker In **eusocial** insects, the **caste** that maintains the nest, forages for food, and tends the eggs and larvae. Bee, wasp, and ant workers are sterile females, termite workers may be male or female and a true worker caste occurs only in the family Termitidae.

World Conservation Union *See* International Union for Conservation of Nature.

World Reference Base for Soil Resources (WRB) A scheme for **soil classification** that was developed between 1961 and 1974 by the Food and Agriculture Organization (FAO) of the United Nations and the U.N. Educational, Scientific, and Cultural Organization (UNESCO), with the support of the U.N. Environment Programme (UNEP) and the International Society of Soil Science. It has been revised several times and received its present name in 1998. The WRB groups soils into 32 soil groups and 170 subunits.

worm snake *See Carphophis amoenus.*

wormwood (*Artemisia absinthum*) *See Artemisia.*

wound response Metabolic activity that responds to physical damage to plant tissues. In vascular plants (**Tracheophyta**) it is controlled by **plant hormones** and typically involves an increase in the synthesis of **callose**, the forming of more **endoplasmic reticulum**, increased **mitosis**, and finally the production of new **buds** and **roots**.

wrack Any seaweed of the family Fucaceae. They occur on rocky shores and consist of a **holdfast** and a **stipe**, parts of which form wide, flat **blades** of varying shape. *See* bladder wrack, channelled wrack, egg wrack, serrated wrack, spiral wrack.

WRB *See* World Reference Base for Soil Resources.

wren *See Troglodytes troglodytes.*

Xanthomonas A genus of **Gammaproteobacteria** that are rod-shaped, Gram negative (*see* Gram reaction), aerobic (*see* aerobe), **motile** with a single **polar flagellum**, and yellow because they contain xanthomonadin, a yellow pigment. There are 20 species, all plant pathogens causing diseases in almost 400 species of plants involving **canker**, necrosis, **blight**, and spots.

xanthophylls Yellow **accessory pigments** that are present in all leaves. They are oxygenated **carotenoids** synthesized within **plastids** and absorb light at wavelengths **chlorophyll** does not absorb.

Xanthorrhoeaceae (order **Asparagales**) A **monocotyledon** family of **perennial herbs** some with **rhizomes**, including many **xerophytes**, and **pachycaul trees** with **simple**, **linear**, sheathing leaves in tufts at the tips of **branches**. Flowers **actinomorphic**, **bisexual** or **unisexual** (plants **dioecious**), **bracteate**, with 3 **scarious sepaloid** and 3 **petaloid tepals** in 2 **whorls**, 3+3 **stamens**, **ovary superior** of 3 fused **carpels** and **locules**. **Inflorescence** a **panicle**, **spike** on a **scape**, or cluster. Fruit is a **loculicidal capsule** or single-seeded **nut**. There are 35 genera of 900 species with a widespread distribution in temperate and tropical Eurasia, Africa, Australia, and western South America.

Xanthosoma (family **Araceae**) A **monocotyledon** genus of **herbs** with starchy **corms** and leaves that are **sagittate**, **hastate**, or divided into many segments, and are up to 2 m long. **Inflorescence** is a **spadix** with **staminate** flowers at the top, sterile flowers in the middle, and **pistillate** flowers at the base, all enclosed in a **spathe** prior to opening. Some species are protogynous (*see* protogyny). On opening, the flower releases warmth and a scent that attracts pollinating beetles. The top of the inflorescence opens first, the middle and lower parts later, and the inflorescence lasts for two nights. There are 73 species occurring in tropical America and the Caribbean. Several are cultivated for their edible corms, others, known as elephant ear, for ornament.

X-chromosome The **sex chromosome** that occurs twice in the homogametic sex and once in the heterogametic sex.

xenia The effect of **pollen** on **endosperm**, e.g. using pollen from a maize (*Zea mays*) plant with yellow seeds to fertilize one with purple seeds will result in offspring with some yellow and some purple seeds. *See also* metaxenia.

xenogamy **Pollination** that involves the transfer of pollen between plants that are genetically distinct.

xeric Describes an area that is extremely dry.

xeromorphic Tolerant of **drought**.

Xeronemataceae (order **Asparagales**) A **monocotyledon**, **monogeneric** family (*Xeronema*) of **perennial herbs** with **distichous**, **ensiform**, sheathing leaves growing from the base. Flowers **actinomorphic**, **bisexual**, with 3+3 **tepals** forming a tube, 3+3 **stamens**, **ovary superior** of 3 **carpels** and **locules**. **Inflorescence** is a **spike**. Fruit is a **loculicidal capsule**. There are two species **endemic** to New Zealand.

xerophile *See* extremophile.

xerophyte A plant that grows in dry conditions and can withstand prolonged **drought**.

xerosere A **succession** that begins in an arid environment.

Ximeniaceae (order **Santalales**) A family of **evergreen** or **deciduous trees** and **shrubs** with **alternate**, **simple**, **entire**, **exstipulate**, papery to **coriaceous**, **glabrous** leaves. Flowers **bisexual**, or functionally **unisexual**, **tetramerous** or **pentamerous**, **corolla apopetalous**, usually 8 **stamens**, **ovary superior**. **Inflorescence** an **axillary umbel**. Fruit is a **drupe**. There are 4 genera of 13 species with a pantropical distribution. *Ximenia* has axillary thorns and is a root **hemiparasite**; it is cultivated for its edible fruits (hogplum).

xylan A complex **polysaccharide** made from linked **xylose** units. It is a **hemicellulose** found in plant **cell walls** and in some green algae (**Chlorophyta**).

xylary Pertaining to **xylem**.

xylem Plant tissue through which water and dissolved nutrients move from the roots to the leaves of a vascular plant (**Tracheophyta**).

xylose (wood sugar) A **monosaccharide** ($C_5H_{10}O_5$) first isolated from wood that is a precursor to **hemicellulose** and a major constituent of **xylan**.

Xyridaceae (order **Poales**) A **monocotyledon** family of **perennial** a few **annual**, **caulescent**, **monopodial herbs** with **distichous** or spirally arranged, sheathing, occasionally **ligulate** leaves. Flowers **zygomorphic**, **bisexual**, **bracteate**, with an outer **whorl** of 3 **sepals** and inner whorl of 3 **petals**, 3 **stamens**, **ovary superior** of 3 **carpels**. **Inflorescence** a lateral or terminal **scape** usually with 2 **bracts** ending in a **spike** or **panicle** of spikes. Fruit is a **loculicidal capsule**. There are 5 genera of 260 species with a pantropical and warm temperate distribution.

X

Y

yam (*Dioscorea* spp.) *See* Dioscoreaceae.

Y-chromosome The **sex chromosome** that occurs only in the heterogametic sex.

yeast Any fungus (*see* Fungi) that can exist as a single-celled organism reproducing by cellular fission or by **budding**.

yellow bell *See Allamanda.*

yellow-bellied slider *See Trachemys scripta.*

yellow dock (*Rumex crispus*) *See* Polygonaceae.

yellow potato cyst nematode *See Globodera rostochiensis.*

yellows Any plant disease that produces **chlorosis** or yellow discoloration.

yellow-soled slug *See Arion hortensis.*

yellow tail worm *See Octolasion cyaneum.*

yellow warbler *See Setophaga petechia.*

yerba maté (*Ilex paraguariensis*) *See* Aquifoliaceae, *Ilex.*

yermic horizon A surface **soil horizon** typical of deserts, which usually consists of rock fragments called desert pavement embedded in a **loam crust** with a **vesicular structure**, covered by a thin layer of **aeolian sand** or **loess**.

yew *See Taxus.*

Yponomeuta padella (small ermine moth) A white or grey moth (**Lepidoptera**) with small black dots on the forewings and a wingspan of 15–26 mm. They lay batches of eggs in August. These hatch in autumn but remain dormant until the following spring, when the **caterpillars** feed on the foliage of hawthorn and blackthorn until midsummer, when they pupate. The larvae produce characteristic webs in the **shrubs** where they feed and can defoliate patches of hedgerow.

Yponomeutidae (ermine moths) A family of mainly white moths (**Lepidoptera**) with black spots. In many species the **caterpillars** spin communal tents in trees and can cause serious **defoliation** from which the plants recover when the larvae pupate. Adults are minor pollinators, caterpillars are minor pests. There are several hundred species with a worldwide but mainly tropical distribution.

Y

Z

Zamiaceae (order **Cycadales**) A family of **evergreen**, **perennial**, **dioecious gymnosperms** that resemble palm trees or ferns with underground, or erect usually unbranched stems. Leaves are **simple**, **pinnate** with **leaflets** sometimes with **dichotomous branching**, spirally arranged, **coriaceous**, **entire**, **dentate**, or **spinose**. **Pinnules** are straight in **vernation**. Male and female **sporophylls** are borne in **cones** along the **axis**, female cones being scale-like with a thickened base. There are 9 or 10 genera of 200 species with a scattered distribution throughout the tropics and subtropics. All are poisonous.

Zantedeschia (family **Araceae**) A **monocotyledon** genus of **perennial herbs** with **rhizomes** and leaves up to 45 cm long. **Inflorescence** is a **spathe** with a central **spadix**. There are eight species occurring in southern Africa. Many cultivated for ornament, e.g. *Z. aethiopica* (arum lily).

Zea (family **Poaceae**) A **monocotyledon** genus of robust **annual** and **perennial** grasses that use the **C4 pathway** of **photosynthesis** and produce separate male and female **inflorescences**, the male as a terminal flower (tassel) and female flowers as **spikes** in lower leaf **axils**. There are five species occurring in Central America. Wild species are known as teosinte, the cultivated *Z. mays* (maize, corn) is the world's third most important cereal (after rice and wheat).

zebra spider *See Salticus scenicus.*

Zenaida macroura (mourning dove, American mourning dove, rain dove, turtle dove) A pale grey and brown dove (**Columbidae**), 225–360 mm long with a wingspan of 142–150 mm, with white tips to the tail, black beak, and red legs and feet. They inhabit woodland edges, grasslands, fields, and suburban areas, feeding on seeds, fruit, and insects. Their common name refers to their call. They occur throughout North America.

Zetaproteobacteria A class of **Proteobacteria** that consists of a single species (*Mariprofundus ferroxydans*) found in the ocean and estuaries.

Zetzellia mali (yellow predatory mite) A bright yellow mite (**Arachnida**) that lives on fruit trees, especially apples. It overwinters in crevices in and beneath **bark** as a mated female and produces up to four generations a year. It is able to survive but not reproduce on a diet of **pollen**, **sap**, and fungal **spores**, but thrives as a predator of other mites. It has a **cosmopolitan distribution**.

Zeuzera pyrina (leopard moth) A nocturnal moth (**Lepidoptera**) with a white body, pale wings with blue blotches and spots, and a wingspan of 40–60 mm, that is strongly attracted to bright lights. Its **caterpillars** are white with black spots and a black head, and grow to about 50 mm long. They bore into the stems of many species of shrubs and trees to feed on the **heartwood**, where they spend two to three years before pupating beneath the **bark**. They cause serious damage, especially to apple, pear, and plum trees. It is native to Europe but has become established in the eastern United States.

zineb A general-use **pesticide** that functions mainly as a **fungicide** against a wide range of fungal diseases of vegetable and fruit crops. It is a skin and eye irritant and moderately toxic if ingested.

Zingiberaceae (order **Zingiberales**) A family of **monocotyledon**, aromatic, **perennial herbs** with thick, branched **rhizomes**, many with **tubers** and aerial, unbranched **pseudostems**, most short but some up to 8 m tall. Leaves **distichous** or in tufts, **petiolate**, **entire**, **elliptical**, **pinnate**. Flowers last only a day and are **zygomorphic**, **bisexual**; **calyx** 3-lobed, **corolla** 3-lobed and tubular, 6 **stamens** in 2 **whorls**, **ovary inferior** of 3 **carpels** and 1 or 3 **locules**. **Inflorescence** a **thyrse**, rarely a **spike** or **raceme**, or flowers solitary. Fruit is a **capsule**. There are 46–52 genera of 1075–1340 species distributed throughout the subtropics, especially from southeastern Asia to Malesia. Rhizomes of *Zingiber officinale* are ginger, those of *Curcuma angustifolia* yield arrowroot, and those of *C. longa* yield turmeric.

Zingiberales A **monocotyledon** order of giant **herbs** with no aerial stem except when they are flowering, comprising 8 families of 92 genera and 2151 species. *See* Cannaceae, Costaceae, Heliconiaceae, Lowiaceae, Marantaceae, Musaceae, Strelitziaceae, and Zingiberaceae.

zinc (Zn) An element that is an essential plant **micronutrient** involved in the production of **auxins**, the activation of **enzymes** in **protein** synthesis, the formation of **starch**, other **carbohydrates**, and **chlorophyll**, and the rate of stalk and seed maturation. Deficiency causes stunted growth, sometimes producing a rosette leaf formation.

Zn *See* zinc.

zonal soils **Mature soils** that have developed over a long period during which conditions remained stable, and that reflect the climate and vegetation of the area more than the **parent material**. They usually have well-defined **soil horizons** and occur over large geographic areas (zones).

zone of saturation *See* phreatic zone.

zones of hardiness Regions in which specified **perennial** plants are able to survive the season when conditions are most severe, i.e. the low temperatures of winter or prolonged aridity during the dry season.

Zonotrichia albicollis (white-throated sparrow) A **passerine** bird, 260–290 mm long with a 230 mm wingspan, that occurs in two forms, one with a white stripe on the head and white **supercilium**, the other with a tan stripe and supercilium. The throat is white with a black edge, the underside near-white, the back brown with dark streaks, and there are yellow patches between the eyes and beak. The sparrow inhabits forests and feeds on seeds, fruit, and insects. It occurs in eastern Canada and the northeastern United States.

zoochory (synzoochory) Dispersal of **spores** or seeds by animals.

Zoopagales An order of **Fungi** belonging to the **Zygomycota**, most of which are parasites or predators of other fungi, **Protozoa**, **Rotifera**, and similar small organisms. Many produce haustoria (*see* haustorium). There are 22 genera with 190 species.

zooplankton *See* plankton.

zoosporangium A **sporangium** in which **zoospores** form.

zoospore (swarm spore) A **motile**, asexual **spore** that moves by means of a **flagellum**. It is produced by certain algae (*see* alga), **Bacteria**, and **Fungi** as a means of propagation.

zooxanthellae Single-celled dinoflagellates (**Pyrrophyta**) that live symbiotically (*see* symbiosis) with certain corals.

Zosteraceae (order **Alismatales**) A **monocotyledon** family mostly of **perennial** aquatic **herbs** (eel grasses) with creeping, **monopodial rhizomes**. Leaves **distichous**, **linear**, grass-like, lacking **stomata**. Flowers **unisexual** (plants **monoecious** or **dioecious**), lacking **tepals**, 2 **stamens**. **Inflorescence** a flattened **spadix** enclosed by a **spathe**, in monoecious plants the male and female flowers alternating along it. Fruit is an **achene**. There are 2 genera of 14 species with a temperate and subtropical distribution.

Zosterophyllopsida A group of plants, now extinct, which are among the earliest vascular plants (**Tracheophyta**) known. They occurred worldwide from the Late Silurian epoch and throughout the Devonian Period (422.9–359.2 million years ago). They had **circinate** stems that were smooth or covered with small spines and branched **dichotomously**, but had no leaves or roots.

Zürich-Montpellier School of Phytosociology *See* Braun-Blanquet, Josias.

zwitterion A molecule that has both positive and negative electrical charges which cancel each other, so it is neutral overall. **Amino acids** usually exist in this form at the **pH** at which they are electrically neutral; the **amino group** is positive (NH_3^+) and the carbonyl group negative (COO^-).

zygomorphic (irregular) Bilaterally symmetrical, therefore capable of being divided into equal halves in only one plane.

Zygomycetes A class of fast-growing **Fungi** belonging to the **Zygomycota** that have coenocytic (*see* coenocyte) **hyphae**. They live as **saprotrophs** or parasites and include **mould** fungi. There are 12 species found worldwide.

Zygomycota A **phylum** of **Fungi** that produce resistant, spherical **spores** with **cell walls** of chitosan, derived from the more usually **chitin**. The phylum contains about 1 percent of all fungi and includes the fast-growing **moulds** that spoil foods high in sugar. They reproduce sexually by the fusion of gametangia (*see* gametangium) to form zygospores, and asexually with sporangia (*see* sporangium). There are approximately 1060 species.

Zygophallaceae (order **Zygophyllales**) A family of **annual** and **perennial herbs** and **deciduous** and **evergreen shrubs** and **trees** with **opposite** or **alternate**, **pinnate** leaves with up to 10 pairs of **sessile**, **entire** to **pinnatisect**, **stipulate leaflets** sometimes reduced to 1 pair with or without a terminal leaflet and some with only a terminal leaflet or a basal pair of fused leaflets; **stipules** sometimes form spines. Flowers **actinomorphic**, **bisexual** or **unisexual** (plants **dioecious**), **tetramerous** or **pentamerous**, some **apetalous**, twice as many **stamens** as **petals**, **ovary syncarpous** of 5 free or 4 fused **carpels** and **locules**. **Inflorescence** a terminal or **axillary cyme** or flowers solitary or in pairs in leaf **axils**. Fruit is a **capsule**, **schizocarp**, or **drupe**. There are 22 genera of 285 species occurring in dry, warm temperate and tropical regions. *Balanites aegyptiaca* (desert date) cultivated for its edible fruit, *Guaiacum officinale* yields lignum-vitae, one of the hardest of all woods, *Larrea tridentata* is the creosote bush.

Zygophyllales An order of plants that comprises 2 families of 27 genera and 305 species. *See* Krameriaceae and Zygophyllaceae.

zygophore In some members of the **Mucorales**, specialized **hypha** bearing a **zygospore**.

zygospore A **zygote** formed by the fusion of isogamous (*see* isogamy) **gametes**. Zygospores occur in certain algae (*see* alga) and **Fungi**. *Compare* parthenospore. *See* Zygomycota.

zygote The fertilized **ovum** of a plant or animal, formed from the fusion of **gametes**.

Appendix

Beaufort Wind Scale

FORCE	SPEED (M/S)	NAME	DESCRIPTION
0	Less than 1	Calm	Air still; smoke rises vertically
1	1–2	Light air	Wind vanes and flags still; rising smoke drifts
2	2–3	Light breeze	Drifting smoke indicates wind direction
3	4–5	Gentle breeze	Leaves rustle; small twigs move; light flags stir
4	6–8	Moderate breeze	Loose leaves and pieces of paper blow about
5	9–11	Fresh breeze	Small trees in full leaf sway
6	11–14	Strong breeze	Difficult to use open umbrella
7	14–17	Near gale	Wind pushes strongly against people walking into it
8	17–21	Gale	Small twigs torn from trees
9	21–24	Severe gale	Chimneys blown down; tiles and slates torn from roofs
10	25–28	Storm	Trees broken or uprooted
11	29–32	Violent storm	Trees uprooted and thrown some distance; buildings destroyed
12	More than 33	Hurricane	Widespread devastation; buildings destroyed; many trees uprooted

1 metre per second (m/s) = 3.6 km/h

The Genetic Code

AMINO ACID	ABBREVIATION	CODONS RNA	CODONS DNA
Alanine	Ala	GCA, GCC, GCG, GCU	GCA, GCC, GCG, GCT
Arginine	Arg	AGA, AGG, CGA, CGG, CGC, CGU	CGT, CGC, CGA, CGG
Asparaginine	Asn	AAC, AAU	AGT, AGC
Aspartic acid	Asp	GAC, GAU	GAC, GAT
Cysteine	Cys	UGC, UGU	TGC, TGT
Glutamic acid	Glu	GAA, GAG	GAA, GAG
Glutamine	Gln	CAA, CAG	CAA, CAG
Glycine	Gly	CGA, GGC, GGG, GGU	GGA, GGC, GGG, GGT,
Histidine	His	CAC, CAU	CAC, CAT
Isoleucine	Ile	AUA, AUC, AUU	ACT, ACA, ATA
Leucine	Leu	CUA, CUC, CUG, CUU, UUA, UUG	TCA, TCG, CTT, CTC, CTA
Lysine	Lys	AAA, AAG	AAA, AAG
Methionine	Met	AUG	ATG*
Phenylalanine	Phe	UUC, UUU	TTC, TTT
Proline	Pro	CCA, CCC, CCG, CCU	CCA, CCC, CCG, CCT
Serine	Ser	AGC, AGU, UCA, UCC, UCG, UCU	AGC, AGT, TCT, TCC, TCA, TCG
Threonine	Thr	ACA, ACC, ACG, ACU	ACA, ACC, ACG, ACT
Tryptophan	Trp	UGG	TGG
Tyrosine	Tyr	UAC, UAU	TAC, TAT
Valine	Val	GUA, GUC, GUG, GUU	GTT, GTC, GTA, GTG
Start codon		AUG	
Stop codon		UAA, UAG, UGA	TAA, TAG, TGA

* When with a gene, otherwise ATG signals start of transcription.
A = adenine
C = cytosine
G = guanine
T = thymine (DNA only)
U = uracil (RNA only)

Further Reading

Acacia Explanation of the controversy over *Acacia* taxonomy. http://www.anbg.gov.au/cpbr/taxonomy/acacia-conserved-2004.html

Acrididae Description of the Acrididae. http://bugguide.net/node/view/155

Acris crepitans Description of the northern cricket frog. http://animaldiversity.ummz.umich.edu/accounts/Acris_crepitans/

Acris gryllus Description of the southern cricket frog. http://animaldiversity.ummz.umich.edu/accounts/Acris_gryllus/

Adanson, Michel Details of the life and work of Adanson. http://huntbot.andrew.cmu.edu/HIBD/Departments/Archives/Archives-AG/Adanson.shtml

adenosine triphosphate Explanation of ATP and the ATP–ADP reaction. http://hyperphysics.phy-astr.gsu.edu/hbase/biology/atp.html

adhesion Explanation of adhesion and cohesion. http://ga.water.usgs.gov/edu/adhesion.html

Aegithalos caudatus Description of the long-tailed tit. http://animaldiversity.ummz.umich.edu/accounts/Aegithalos_caudatus/

aflatoxin Aflatoxins, their sources, and their effects. http://www.ansci.cornell.edu/plants/toxicagents/aflatoxin/aflatoxin.html

Agaricomycetes Decription of the Agaricomycetes. http://tolweb.org/Homobasidiomycetes

Agaricomycotina Description of the Agaricomycotina. http://tolweb.org/Agaricomycotina

Agaricus campestris Description of the field mushroom. http://www.first-nature.com/fungi/agaricus-campestris.php

Agkistrodon contortix Description of the copperhead. http://animaldiversity.ummz.umich.edu/accounts/Agkistrodon_contortrix/

Agkistrodon piscivorus Description of the cottonmouth. http://animaldiversity.ummz.umich.edu/accounts/Agkistrodon_piscivorus/

Agrobacterium Use of *Agrobacterium* in genetic engineering. http://www.apsnet.org/publications/apsnetfeatures/Pages/Agrobacterium.aspx

agroforestry Explanation of agroforestry and its advantages. http://www.agroforestry.co.uk/agover.html

Aix sponsa Description of the wood duck. http://animaldiversity.ummz.umich.edu/accounts/Aix_sponsa/

Albugo candida Description of the organism causing white rust. http://www.extento.hawaii.edu/kbase/crop/Type/a_candi.htm

Allee effect Description of the Allee effect. http://www.nature.com/scitable/knowledge/library/allee-effects-19699394

Amanita muscaria Description of fly agaric. http://www.kew.org/plants-fungi/Amanita-muscaria.htm

Amanita pantherina Description of panther cap. http://www.first-nature.com/fungi/amanita-pantherina.php

Amanita phalloides Description of death cap. http://www.first-nature.com/fungi/amanita-phalloides.php

Amanita vaginata Description of the grisette. http://www.first-nature.com/fungi/amanita-vaginata.php

ambrosia fungi Description of the ambrosia fungi. http://www.ambrosiasymbiosis.org/ambrosia-fungi/who-are-the-fungi/

Ambystoma maculatum Dscription of the spotted salamander. http://animaldiversity.ummz.umich.edu/accounts/Ambystoma_maculatum/

Ambystoma opacum Description of the marbled salamander. http://animaldiversity.ummz.umich.edu/accounts/Ambystoma_opacum/

amine Introducing Amines. http://www.chemguide.co.uk/organicprops/amines/background.html

angiosperm Description, evolutionary history of, and relationships among angiosperms. http://tolweb.org/Angiosperms/20646

Angiosperm Phylogeny Group Home page of the Angiosperm Phylogeny Group. http://www.mobot.org/MOBOT/research/APweb/welcome.html

Annelida Description of annelid worms. http://animaldiversity.ummz.umich.edu/accounts/Annelida/

Anolis carolinensis Description of the green anole. http://animaldiversity.ummz.umich.edu/accounts/Anolis_carolinensis/

Anthocerotophyta Description of hornworts. http://www.ucmp.berkeley.edu/plants/anthocerotophyta.html

anthocorid bugs Description of the Anthocoridae. http://bugguide.net/node/view/33701

Anthonomus pomorum Description of this weevil and its life cycle. http://www7.inra.fr/hyppz/RAVAGEUR/6antpom.htm

anthracnose Diagnosis and treatment of anthracnose. http://www.ipm.ucdavis.edu/PMG/PESTNOTES/pn7420.html

Antirrhinum Hudson, Andrew, Joanna Critchley, and Yvette Erasmus. 'The Genus *Antirrhinum* (Snapdragon): A Flowering Plant Model for Evolution and Development.' http://cshprotocols.cshlp.org/content/2008/10/pdb.emo100.full

Anura Description of the frogs and toads. http://animaldiversity.ummz.umich.edu/accounts/Anura/

Aphelinidae Description of Aphelinidae, with illustrations. http://www.nhm.ac.uk/research-curation/research/projects/chalcidoids/aphelinidae.html

Aphis fabae Description of black bean aphid. http://rhs.org.uk/advice/profile?pid=797

Apis Description of bee species. http://www.fao.org/docrep/x0083e/x0083e02.htm

Apodemus sylvaticus Description of the field mouse. http://animaldiversity.ummz.umich.edu/accounts/Apodemus_sylvaticus/

Apterygota Description of the Apterygota. http://www.nhc.ed.ac.uk/index.php?page=24.25.298.299

Archaea Introduction to the domain Archaea. http://www.microbeworld.org/types-of-microbes/archaea

Archaeopteris Description of the tree, with an illustration. http://www.devoniantimes.org/who/pages/archaeopteris.html

Archilochus colubris Description of the ruby-throated hummingbird. http://animaldiversity.ummz.umich.edu/accounts/Archilochus_colubris/

Area of Outstanding Natural Beauty Information about AONBs. http://www.landscapesforlife.org.uk

Arion hortensis Description of *A. hortensis*. http://idtools.org/id/mollusc/factsheet.php?name=Arion%20hortensis%20group:%20Arion%20hortensis

Arionidae Description of round back slugs. http://molluscs.at/gastropoda/terrestrial.html?/gastropoda/terrestrial/arionidae.html

Ascomycota Description of the ascomycete fungi. http://tolweb.org/Ascomycota

ascorbic acid Nutritional significance of vitamin C. http://umm.edu/health/medical/altmed/supplement/vitamin-c-ascorbic-acid

asparagus beetle Description of both species. http://www1.extension.umn.edu/garden/insects/find/asparagus-beetles/

aspergillosis Information about aspergillosis. http://www.aspergillus.org.uk/

Asteroxylon Description and illustrations of *Asteroxylon*. http://www.abdn.ac.uk/rhynie/aster.htm

aster yellows Description of the disease. http://ipm.illinois.edu/diseases/rpds/903.pdf

auxin Description of auxins and explanation of their action. http://users.rcn.com/jkimball.ma.ultranet/BiologyPages/A/Auxin.html

avermectin Description of avermectin and its action. http://www.beyondpesticides.org/infoservices/pesticidefactsheets/toxic/Abamectin.php

Aves Introduction to the birds. http://www.ucmp.berkeley.edu/diapsids/birds/birdintro.html

Bacillus thuringiensis Description of *B. thuringiensis*. http://www.ext.colostate.edu/pubs/insect/05556.html

Bacteria Introduction to the Bacteria. http://www.ucmp.berkeley.edu/bacteria/bacteria.html

bacteriorhodopsin Explanation of the action of bacteriorhodopsin. http://www.rcsb.org/pdb/101/motm.do?momID=27

Baeolophus bicolor Description of the tufted titmouse. http://animaldiversity.ummz.umich.edu/accounts/Parus_bicolor/

Baragwanathia longifolia Description of *Barangwanathia longifolia*. http://www.adonline.id.au/plantevol/tour/baragwanathia-flora/

barley yellow dwarf Description of the disease and its cause. http://www.apsnet.org/edcenter/intropp/lessons/viruses/Pages/BarleyYelDwarf.aspx

Basidiomycota Description of the basidiomycete fungi. http://tolweb.org/Basidiomycota

batrachotoxins Description of the poison and it mode of action. http://www.chm.bris.ac.uk/motm/batrachotoxin/batrac.htm

Bauhin, Gaspard Biography of Bauhin. http://www.oilsandplants.com/bauhin.htm

Bennettitales An introduction to the Bennettitales. http://www.ucmp.berkeley.edu/seedplants/bennettitales.html

Bentham, George Biography of Bentham. http://adb.anu.edu.au/biography/bentham-george-2979

biological pesticide Description of biological pesticides. http://www.epa.gov/pesticides/biopesticides/whatarebiopesticides.htm

biosphere reserve Information about biosphere reserves. http://www.unesco.org/new/en/natural-sciences/environment/ecological-sciences/biosphere-reserves/

Bipalium kewense Description of the land planarians. http://entnemdept.ufl.edu/creatures/misc/land_planarians.htm

Biston betularia Explanation of the experiments on industrial melanism and the attacks on them. http://faculty.unife.it/giorgio.bertorelle/didattica_insegnamenti/biologia-evoluzionistica-1/Bistonbetularia_History.pdf

bolete Description of boletes with recipes for cooking them. http://www.mssf.org/cookbook/boletes.html

Boletus edulis Description of this edible bolete. http://www.first-nature.com/fungi/boletus-edulis.php

Bombus Identification guide to British bumblebees. http://www.nhm.ac.uk/nature-online/life/insects-spiders/identification-guides-and-keys/bumblebees/

Bombycilla cedrorum Description of the cedar waxwing. http://animaldiversity.ummz.umich.edu/accounts/Bombycilla_cedrorum/

Braconidae Description of braconids. http://tolweb.org/Braconidae

Brown, Robert Biography of Brown. http://adb.anu.edu.au/biography/brown-robert-1835

brown algae Introduction to the Phaeophyta. http://www.ucmp.berkeley.edu/chromista/phaeophyta.html

Bruchidae Description of bruchids. http://delta-intkey.com/britin/col/www/bruchida.htm

Bryophyta A detailed description of mosses. http://tolweb.org/Bryophyta

Bryopsida A detailed description of the bryopsids. http://tolweb.org/Bryopsida

buffer Explanation of biological buffers. http://scifun.chem.wisc.edu/chemweek/biobuff/biobuffers.html

Bufo fowleri Description of Fowler's toad. http://animaldiversity.ummz.umich.edu/accounts/Anaxyrus_fowleri/

Calypte anna Description of Anna's hummingbird. http://animaldiversity.ummz.umich.edu/accounts/Calypte_anna/

canalization Article explaining canalization. http://www.math.tau.ac.il/~illan/Articles/53.pdf

canker Diagnosis and treatment of canker. http://www.gardenersworld.com/how-to/problems/fruit-and-nuts/canker/378.html

Cape floral region UNESCO description of the region's plants. http://whc.unesco.org/en/list/1007

Carabidae Description of ground and tiger beetles. http://ento.psu.edu/extension/factsheets/ground-beetles

carbon cycle Description of the cycle. http://dilu.bol.ucla.edu/home.html

carbon-nitrogen ratio Explanation of the C:N ratio and its importance. http://whatcom.wsu.edu/ag/compost/fundamentals/needs_carbon_nitrogen.htm

carboxysome Description and function of carboxysomes. http://www.microbemagazine.org/index.php?option=com_content&view=article&id=1863:the-carboxysome-and-other-bacterial-microcompartments&catid=464&Itemid=826

Cardinalis cardinalis Description of the northern cardinal. http://animaldiversity.ummz.umich.edu/accounts/Cardinalis_cardinalis/

Carduelis tristis Description of the American goldfinch. http://animaldiversity.ummz.umich.edu/accounts/Carduelis_tristis/

Caribbean floral region Description of the flora. http://botany.si.edu/projects/cpd/ma/ma-carib.htm#flora

Carpodacus purpureus Description of the purple finch. http://animaldiversity.ummz.umich.edu/accounts/Carpodacus_purpureus/

carposporophyte Description of the life cycle of red algae. http://www.nilauro.com/plocamium/lh_both.htm

catena Description of a catena. http://www.educationscotland.gov.uk/images/Soil%20catenas_tcm4-481670.pdf

cation exchange capacity Explanation of cation exchange capacity. http://nmsp.cals.cornell.edu/publications/factsheets/factsheet22.pdf

Caudata Description of the salamanders and newts. http://tolweb.org/Caudata/14939

Cecidophyopsis ribis Description of the blackcurrant gall mite and its control. http://homeguides.sfgate.com/black-currant-gall-mites-25304.html

cell wall Description of plant cell walls. http://www.plantphysiol.org/content/154/2/483.full

centriole Description of centrioles and their function. http://www.bscb.org/?url=softcell/centrioles

Cepaea nemoralis Description of *C. nemoralis* and its ecology. http://animaldiversity.ummz.umich.edu/accounts/Cepaea_nemoralis/

Cervidae Description of deer. http://animaldiversity.ummz.umich.edu/accounts/Cervidae/

Chaetosiphon fragaefolli Description of the strawberry aphid. http://edis.ifas.ufl.edu/hs253

channelled wrack Description of channelled wrack. http://www.theseashore.org.uk/theseashore/SpeciesPages/Additional%20Species/Pelvetia.jpg.html

chaparral Description of the chaparral. http://kids.nceas.ucsb.edu/biomes/chaparral.html

Charadrius vociferus Description of the killdeer. http://animaldiversity.ummz.umich.edu/accounts/Charadrius_vociferus/

Charophyceae Description of charophytes. http://www.charophytes.com/cms/index.php?option=com_content&view=article&id=47&Itemid=27

check dam Description of check dams and their construction. http://www.polytechnic.edu.na/academics/schools/engine_infotech/civil/libraries/hydrology/docus/checkdam.pdf

Chelydra serpentina Description of the common snapping turtle. http://animaldiversity.ummz.umich.edu/accounts/Chelydra_serpentina/

Chilopoda Decription of the Chilopoda. http://bugguide.net/node/view/20/bgpage

chloronicotinyls Description of these insecticides. http://hyg.ipm.illinois.edu/pastpest/200320e.html

chlorophyll Description of chlorophyll. http://www.chm.bris.ac.uk/motm/chlorophyll/chlorophyll_h.htm

Chrysemys picta Description of the painted turtle. http://animaldiversity.ummz.umich.edu/accounts/Chrysemys_picta/

Chrysophyta Introduction to the Chrysophyta. http://www.ucmp.berkeley.edu/chromista/chrysophyta.html

citric acid cycle Description of the stages in the cycle. http://www.elmhurst.edu/~chm/vchembook/611citricrx.html

Citrus Botanical classification of citrus fruits. http://citruspages.free.fr/classification.html

cladode Description of cladodes. http://www.botgard.ucla.edu/html/botanytextbooks/generalbotany/typesofshoots/cladode/

Cladoxylopsida Introduction to the Cladoxylopsida. http://www.ucmp.berkeley.edu/plants/cladoxylopsida/cladoxylopsida.html

Claviceps purpurea Description of ergot. http://www.apsnet.org/edcenter/intropp/lessons/fungi/ascomycetes/Pages/Ergot.aspx

Clifford, George Detailed description of Clifford's herbarium. http://www.nhm.ac.uk/research-curation/scientific-resources/collections/botanical-collections/clifford-herbarium

Closteroviridae Description of the closteroviruses. http://www.els.net/WileyCDA/ElsArticle/refId-a0000747.html

cloud seeding Explanation of cloud seeding. http://www.weatheronline.co.uk/reports/wxfacts/Cloud-seeding.htm

Coccinellidae Description of coccinellid biology and folklore. http://entnemdept.ufl.edu/creatures/beneficial/lady_beetles.htm

coefficient of inbreeding Explanation of the coefficient and its application. http://www.genetic-genealogy.co.uk/Toc115570144.html

Colaptes auratus Description of the northern flicker. http://animaldiversity.ummz.umich.edu/accounts/Colaptes_auratus/

Coleoptera Description of the beetles and their classification. http://www.cals.ncsu.edu/course/ent425/library/compendium/coleoptera.html

Collembola Description of springtails. http://www.cals.ncsu.edu/course/ent425/library/compendium/collembola.html

Coluber constrictor Description of the black racer. http://animaldiversity.ummz.umich.edu/accounts/Coluber_constrictor/

Columba livia Description of the domestic pigeon. http://animaldiversity.ummz.umich.edu/accounts/Columba_livia/

Columbidae Description of the pigeons and doves. http://animaldiversity.ummz.umich.edu/accounts/Columbidae/

Cricetidae Description of the Cricetidae. http://animaldiversity.ummz.umich.edu/accounts/Cricetidae/

Crotalus adamanteus Description of the eastern diamondback rattlesnake. http://animaldiversity.ummz.umich.edu/accounts/Crotalus_adamanteus/

Crotalus horridus Description of the timber rattlesnake. http://animaldiversity.ummz.umich.edu/accounts/Crotalus_horridus/

Cryphonectria parasitica Description of chestnut blight. http://www.columbia.edu/itc/cerc/danoff-burg/invasion_bio/inv_spp_summ/Cryphonectria_parasitica.htm

cryptomonads Description of cryptomonads and their significance. http://tolweb.org/Cryptomonads

cucumber mosaic virus Description of the disease and its treatment. http://apps.rhs.org.uk/advicesearch/profile.aspx?pid=143

cyanobacteria Introduction to the cyanobacteria. http://www.ucmp.berkeley.edu/bacteria/cyanointro.html

Cyanocitta cristata Description of the blue jay. http://animaldiversity.ummz.umich.edu/accounts/Cyanocitta_cristata/

Cycadales Description of the Cycadales. http://www.conifers.org/zz/Cycadales.php

cytochrome Role of cytochromes in electron transport. http://hyperphysics.phy-astr.gsu.edu/hbase/organic/cytochrome.html

***Cytophaga-Flavobacterium* group** Description of these bacteria. http://microbewiki.kenyon.edu/index.php/Cytophaga_Flavobacterium

2,4-D Detailed information about 2,4-D. http://npic.orst.edu/ingred/24d.html

daminozide Technical description of daminozide. http://www.gpnmag.com/successful-use-pgr-daminozide

dance language Description and explanation of the bee dance. http://www.cals.ncsu.edu/entomology/apiculture/pdfs/1.11%20copy.pdf

DDT Introduction to DDT. http://people.chem.duke.edu/~jds/cruise_chem/pest/pest1.html

Delia radicum Description of *D. radicum*, the damage it causes, and treatment. http://apps.rhs.org.uk/advicesearch/profile.aspx?pid=646

Dendrodrilus rubidus Description of *D. rubidus* as an invasive species. http://www.issg.org/database/species/ecology.asp?si=1697&fr=1&sts=&lang=EN

Dermaptera Description of earwigs. http://www.cals.ncsu.edu/course/ent425/library/compendium/dermaptera.html

Desmognathus fuscus Description of the dusky salamander. http://animaldiversity.ummz.umich.edu/accounts/Desmognathus_fuscus/

destroying angel Description of the fungus. http://www.first-nature.com/fungi/amanita-virosa.php

Diadophis punctatus Description of the ringneck snake. http://animaldiversity.ummz.umich.edu/accounts/Diadophis_punctatus/

diatom Description of diatoms. http://tolweb.org/Diatoms/21810

Dikarya Description of the Dikarya. http://website.nbm-mnb.ca/mycologywebpages/NaturalHistoryOfFungi/DikaryaDiscussion.html

dimethyl sulphide Explanation of the formation and significance of DMS. http://www.csa.com/discoveryguides/dimethyl/overview.php

DNA Explanation of the structure and function of DNA. http://ghr.nlm.nih.gov/handbook/basics/dna

Dodoens, Rembert Biographical notes on Dodoens. http://galileo.rice.edu/Catalog/NewFiles/dodoens.html

Dothideomycetes Description of the Dothideomycetes. http://www.tolweb.org/Dothideomycetes/

double fertilization Simple explanation of double fertilization. https://www.boundless.com/biology/flowering-plants/angiosperm-life-cycle/double-fertilization-in-plants/

downy mildew Description of downy mildew and its treatment. http://apps.rhs.org.uk/advicesearch/profile.aspx?pid=683

drumlin Description of drumlins. http://www.sheffield.ac.uk/drumlins/drumlins

dry bubble disease Description of this disease of mushrooms. http://extension.psu.edu/plants/vegetable-fruit/mushrooms/fact-sheets/diseases/verticillium-dry-bubble

Dryocopus pileatus Description of the pileated woodpecker. http://animaldiversity.ummz.umich.edu/accounts/Dryocopus_pileatus/

Dumetella carolinensis Description of the gray catbird. http://animaldiversity.ummz.umich.edu/accounts/Dumetella_carolinensis/

dust mulch Description of dust mulches and their effects. http://puyallup.wsu.edu/~linda%20chalker-scott/Horticultural%20Myths_files/Myths/magazine%20pdfs/Dust%20mulches.pdf

elater Explanation of how elaters work. http://www.anbg.gov.au/bryophyte/elaters-liverworts.html

Elateridae Description of the elaterids. http://tolweb.org/Elateridae/9190

electron-transport chain Detailed explanation of the electron-transport chain. http://www.elmhurst.edu/~chm/vchembook/596electransport.html

enantiostyly Development of enantiostyly. http://www.amjbot.org/content/90/2/183.full

Encyrtidae Description of encyrtids. http://www.nhm.ac.uk/research-curation/research/projects/chalcidoids/encyrtidae.html

endoplasmic reticulum Description of the structure and function of ER. http://www.princeton.edu/~achaney/tmve/wiki100k/docs/Endoplasmic_reticulum.html

endosymbiotic theory Explanation of the theory. http://learn.genetics.utah.edu/content/begin/cells/organelles/

eocyte Explanation of the eocyte hypothesis. http://www.pnas.org/content/105/51/20049.full

Equisetum Description of *Equisetum*. http://www.bio.umass.edu/biology/conn.river/equisetm.html

Erinaceus europaeus Description of the hedgehog. http://animaldiversity.ummz.umich.edu/accounts/Erinaceus_europaeus/

Eriostemon Explanation of the change in classification. http://anpsa.org.au/e-aus.html

Erwinia amylovora Description of fire blight and its treatment. http://apps.rhs.org.uk/advicesearch/profile.aspx?pid=160

Eukarya Introduction to the Eukarya. http://www.ucmp.berkeley.edu/alllife/eukaryota.html

Eulophidae Description of the Eulophidae. http://www.nhm.ac.uk/research-curation/research/projects/chalcidoids/eulophidae1.html

Eumeces fasciatus Description of the five-lined skink. http://animaldiversity.ummz.umich.edu/accounts/Plestiodon_fasciatus/

Eurotiomycetes Description of the Eurotiomycetes. http://website.nbm-mnb.ca/mycologywebpages/NaturalHistoryOfFungi/Eurotiomycetes.html

Eurycea cirrigera Description of the southern two-lined salamander. http://animaldiversity.ummz.umich.edu/accounts/Eurycea_cirrigera/

Eurycea wilderae Description of the Blue Ridge two-lined salamander. http://animaldiversity.ummz.umich.edu/accounts/Eurycea_wilderae/

exchangeable sodium percentage Explanation of ESP. http://www.terragis.bees.unsw.edu.au/terraGIS_soil/sp_exchangeable_sodium_percentage.html

extremophile Description of extremophiles. http://serc.carleton.edu/microbelife/extreme/extremophiles.html

Fabaceae Description of the Fabaceae. http://academics.hamilton.edu/foodforthought/Our_Research_files/beans_peas.pdf

fairy ring Description of fairy rings and their cause. http://apps.rhs.org.uk/advicesearch/profile.aspx?PID=158

feather Description of the different types of feather. http://www.ornithology.com/Lectures/Feathers.html

fibre Description of different types and uses of fibres. http://waynesword.palomar.edu/traug99.htm

Filicopsida Description of the Filicopsida. http://tolweb.org/Filicopsida

Fistulina hepatica Description of the beefsteak fungus. http://www.first-nature.com/fungi/fistulina-hepatica.php

five-kingdom system Explanation of the system. http://www.ruf.rice.edu/~bioslabs/studies/invertebrates/kingdoms.html

flask fungi Description of flask fungi. http://www.anbg.gov.au/fungi/two-flask.html

florigen Explanation of florigen and its discovery. http://www.plantcell.org/content/18/8/1783.full

Forbes, Edward Obituary of Forbes summarising his life and work. http://aleph0.clarku.edu/huxley/UnColl/Gazettes/EdwForbes.html

Frankliniella occidentalis Description of western flower thrips. http://www.extento.hawaii.edu/kbase/Crop/Type/f_occide.htm

Fungi Description of the Fungi. http://tolweb.org/Fungi/2377

furbelows Description of furbelows. http://www.marlin.ac.uk/speciesfullreview.php?speciesID=4284

Garden Organic Home page of Garden Organic. http://www.gardenorganic.org.uk/

Garrulus glandarius Description of the Eurasian jay. http://animaldiversity.ummz.umich.edu/accounts/Garrulus_glandarius/

Gastrophryne carolinensis Description of the eastern narrowmouth toad. http://animaldiversity.ummz.umich.edu/accounts/Gastrophryne_carolinensis/

Gastropoda Introduction to the Gastropoda. http://www.ucmp.berkeley.edu/taxa/inverts/mollusca/gastropoda.php

GenBank GenBank home page. http://www.ncbi.nlm.nih.gov/genbank/

gene bank How gene banks help maintain biodiversity. http://web.mit.edu/12.000/www/m2015/2015/gene_banks.html

genetic code Explanation of the genetic code. http://users.rcn.com/jkimball.ma.ultranet/BiologyPages/C/Codons.html

genetic engineering Explanation of genetic engineering. http://www.csiro.au/en/Outcomes/Food-and-Agriculture/WhatIsGM.aspx

germination Movies showing germination in several plants. http://plantsinmotion.bio.indiana.edu/plantmotion/earlygrowth/germination/germ.html

Glomeromycota Description of the Glomeromycota. http://tolweb.org/Glomeromycota

glucosinolates Describes the chemistry of glucosinolates. http://www.cancer.gov/cancertopics/factsheet/diet/cruciferous-vegetables

gnetophyte Description of the gnetophytes. http://comenius.susqu.edu/biol/202/archaeplastida/viridiplantae/gymnosperms/gnetophyta/default.htm

goat tang Description of goat tang. http://www.marlin.ac.uk/speciesinformation.php?speciesID=4168

Golgi body Description of the Golgi body. http://hyperphysics.phy-astr.gsu.edu/hbase/biology/golgi.html

Golgi, Camillo Biographical note on Golgi. http://www.nobelprize.org/nobel_prizes/medicine/laureates/1906/golgi-bio.html

gradient wind Explanation of the gradient wind. http://ww2010.atmos.uiuc.edu/%28Gh%29/guides/mtr/fw/grad.rxml

granite moss Description of granite mosses. http://www.ohio.edu/plantbio/vislab/moss/mckinney.htm

grapevine leafroll Description of the disease. http://www.nysipm.cornell.edu/factsheets/grapes/diseases/grape_leafroll.pdf

green algae Description of the green algae. http://www.ucmp.berkeley.edu/greenalgae/greenalgae.html

green manure Description of green manure and its use. http://apps.rhs.org.uk/advicesearch/profile.aspx?pid=373

Gyromitra esculenta Description of the false morel. http://www.first-nature.com/fungi/gyromitra-esculenta.php

Harmonia axyridis Description of this ladybird. http://www.issg.org/database/species/ecology.asp?si=668

heat capacity Explanation of heat capacity. http://chemwiki.ucdavis.edu/Physical_Chemistry/Thermodynamics/Calorimetry/Heat_Capacity

heirloom plant Explanation of heirloom plants. http://www.sciencedaily.com/articles/h/heirloom_plant.htm

hemiparasite Explanation of hemiparasitism. http://www.ncbi.nlm.nih.gov/pmc/articles/PMC3115071/

Hemiptera Description of bugs. http://www.ucmp.berkeley.edu/arthropoda/uniramia/hemiptera.html

Hennig, Emil Hans Willi Biography of Hennig. http://www.cladistics.org/about/hennig.html

Heterobasidion annosum Description of annosum foot rot. http://www.cals.ncsu.edu/course/pp728/heterobasidion/heterobasidion_annosum.html

Heterodera schachtii Description of *H. schachtii*. http://plpnemweb.ucdavis.edu/nemaplex/taxadata/G060S7.HTM

Heterodon platirhinos Description of the eastern hognose snake. http://animaldiversity.ummz.umich.edu/accounts/Heterodon_platirhinos/

Heteroptera Description of the Heteroptera. http://www.cals.ncsu.edu/course/ent425/library/compendium/heteroptera.html

Hirundo rustica Description of the swallow. http://animaldiversity.ummz.umich.edu/accounts/Hirundo_rustica/

homology Examples of homologies in plants. http://evolution.berkeley.edu/evolibrary/article/lines_04

Hooke, Robert Biography of Hooke. http://www.ucmp.berkeley.edu/history/hooke.html

Hooker, Sir Joseph Dalton Account of Hooker's explorations. http://www.plantexplorers.com/explorers/biographies/hooker/

Hooker, Sir William Jackson Biography of Hooker. http://www.nndb.com/people/276/000102967/

Howard, Luke Brief biography of Howard. http://www.rmets.org/weather-and-climate/observing/luke-howard-and-cloud-names

Hyalopterus pruni Description of this aphid. http://influentialpoints.com/Gallery/Hyalopterus_pruni_Mealy_Plum_Aphid.htm

hybrid speciation Speciation in plants. http://evolution.berkeley.edu/evosite/evo101/VC1iSpeciationPlants.shtml

hydroid Description of hydroids. http://plantphys.info/plant_biology/copyright/moss.html

Hygrophoropsis aurantiaca Description of the false chanterelle. http://www.first-nature.com/fungi/hygrophoropsis-aurantiaca.php

Hyla chrysoscelis Description of Cope's gray tree frog. http://animaldiversity.ummz.umich.edu/accounts/Hyla_chrysoscelis/

Hyla cinerea Description of the green tree frog. http://animaldiversity.ummz.umich.edu/accounts/Hyla_cinerea/

Hyla squirella Description of the squirrel tree frog. http://srelherp.uga.edu/anurans/hylsqu.htm

Hylocichia mustelina Description of the wood thrush. http://animaldiversity.ummz.umich.edu/accounts/Hylocichla_mustelina/

Hymenoptera Description of the Hymenoptera. http://www.cals.ncsu.edu/course/ent425/library/compendium/hymenoptera.html

Ichneumonidae Description of the ichneumons with a list of the subfamilies. http://bugguide.net/node/view/150

Icterus bullockii Description of Bullock's oriole. http://animaldiversity.ummz.umich.edu/accounts/Icterus_bullockii/

Icterus galbula Description of the Baltimore oriole. http://animaldiversity.ummz.umich.edu/accounts/Icterus_galbula/

imbibition Explanation of imbibition. http://waynesword.palomar.edu/pljuly96.htm

index of abundance Explanation of population indices. http://www.afrc.uamont.edu/whited/Estimating%20Population%20Abundance%20Indices%20and%20Complete%20Counts.pdf

Insecta Description of the insects. http://animaldiversity.ummz.umich.edu/accounts/Insecta/

insecticidal oil Description of insecticidal oils and their uses. http://www.ext.colostate.edu/pubs/insect/05569.html

insecticidal soap Description of insecticidal soaps and their use. http://www.ext.colostate.edu/pubs/insect/05547.html

insectivorous plants Description of insectivorous plants. http://www.botany.org/Carnivorous_Plants/

integrated pest management Explanation of integrated pest management. http://www.epa.gov/opp00001/factsheets/ipm.htm

intermediate filament Explanation of the role of intermediate filaments. http://www.ncbi.nlm.nih.gov/books/NBK21560/

International Code of Nomenclature for Algae, Fungi, and Plants Full text of the Code. http://www.iapt-taxon.org/nomen/main.php

International Union for Conservation of Nature Home page of the IUCN. http://www.iucn.org/

iodine cycle Explanation of the cycle. http://www.uea.ac.uk/~e780/iodcycle.htm

Ixodidae Description of hard ticks and their life cycle. http://www.cvbd.org/en/tick-borne-diseases/about-ticks/developmental-cycle/life-cycles-ixodidae/

jet stream Explanation of the jet stream. http://www.metoffice.gov.uk/learning/wind/what-is-the-jet-stream

Juan Fernández floral region Description of region. http://www.oikonos.org/projects/jfic.htm

Junco hyemalis Description of the dark-eyed junco. http://animaldiversity.ummz.umich.edu/accounts/Junco_hyemalis/

kelp Description of kelp around Scottish coasts. http://www.snh.gov.uk/about-scotlands-nature/species/algae/marine-algae/kelp/

Laboulbeniomycetes Description of the Laboulbeniomycetes. http://website.nbm-mnb.ca/mycologywebpages/NaturalHistoryOfFungi/Laboulbeniomycetes.html

Lamarck, Jean-Baptiste Pierre Antoine de Monet, chevalier de Biography of Lamarck. http://www.ucmp.berkeley.edu/history/lamarck.html

Lampropeltis getula Description of the eastern kingsnake. http://animaldiversity.ummz.umich.edu/accounts/Lampropeltis_getula/

lectin Description of plant lectins. http://www.ansci.cornell.edu/plants/toxicagents/lectins.html

Leotiomycetes Description of the Leotiomycetes. http://tolweb.org/Leotiomycetes/29048

Lepidoptera Description of the Lepidoptera. http://www.cals.ncsu.edu/course/ent425/library/compendium/lepidoptera.html

Leporidae Description of the leporids. http://animaldiversity.ummz.umich.edu/accounts/Leporidae/

Leptinotarsa decemlineata Colorado beetle homepage. http://resistance.potatobeetle.org/index.html

lichen zone List of the lichen zones and explanation of their significance. http://www.air-quality.org.uk/19.php

lignin Occurrence, structure, and function of lignin. http://www.lignoworks.ca/content/what-lignin

Lilioceris lilii Description of the lily beetle. http://www.rhs.org.uk/science/plant-pests/lily-beetle

Limacidae Description of the keel-back slugs. http://molluscs.at/gastropoda/terrestrial.html?/gastropoda/terrestrial/limacidae.html

limestone forest Description of limestone forest. http://edukit.atbioversity.net/tiki-index.php?page=Limestone+forest

Linnaeus, Carolus Biography of Linnaeus and explanation of his taxonomic system. http://www.ucmp.berkeley.edu/history/linnaeus.html

Linyphiidae Description of money spiders. http://www.jorgenlissner.dk/Linyphiidae.aspx

lipid Description of lipids. http://www.lignoworks.ca/content/what-lignin

lithotroph Explanation of how lithotrophs work, with examples. http://faculty.weber.edu/mzwolinski/Lithotrophs.pdf

love dart Description of love darts and explanation of their function. http://molluscs.at/gastropoda/index.html?/gastropoda/morphology/love_dart.html

Lumbricus terrestris Description of the common earthworm. http://www.issg.org/database/species/ecology.asp?si=1555&lang=EN

Lycopodiophyta Description of the Lycopodiophyta. http://petrifiedwoodmuseum.org/solycopodiophyta.htm

Lycopsida Description of the Lycopsida. http://www.ohio.edu/plantbio/staff/rothwell/PBIO_3080-5080-12/Lab-4,%20Lycophytes.pdf

lysosome Description of lysosomes. http://hyperphysics.phy-astr.gsu.edu/hbase/biology/lysosome.html

manganese Functions, sources, deficiencies of manganese. http://www.spectrumanalytic.com/support/library/ff/Mn_Basics.htm

Marchantiophyta Detailed description of the liverworts http://comenius.susqu.edu/biol/202/archaeplastida/viridiplantae/bryophytes/hepatophyta/default.htm

meiosis Explanation of meiosis. http://www.emc.maricopa.edu/faculty/farabee/biobk/biobookmeiosis.html

Melanerpes carolinus Description of the red-bellied woodpecker. http://animaldiversity.ummz.umich.edu/accounts/Melanerpes_carolinus/

Meles meles Description of the Eurasian badger. http://animaldiversity.ummz.umich.edu/accounts/Meles_meles/

Meloidogyne hapla Description of the root-knot nematodes. http://www.apsnet.org/edcenter/intropp/lessons/Nematodes/Pages/RootknotNematode.aspx

Melospiza melodia Description of the song sparrow. http://animaldiversity.ummz.umich.edu/accounts/Melospiza_melodia/

Mendel, Gregor Johann Biography of Mendel. http://astro4.ast.vill.edu/mendel/gregor.htm

Mendel's laws Explanation of Mendel's laws of heredity. http://anthro.palomar.edu/mendel/mendel_1.htm

Mephitidae Description of skunks and stink badgers. http://animaldiversity.ummz.umich.edu/accounts/Mephitidae/

Meripilus giganteus Description of the giant polypore. http://www.first-nature.com/fungi/meripilus-giganteus.php

meristem Description of meristems. http://www2.mcdaniel.edu/Biology/botf99/tissimages/meristematic.html

Mermis nigrescens Description of this nematode and its life cycle. http://entnemdept.ufl.edu/creatures/beneficial/misc/mermis_nigrescens.htm

mesophyll Description of mesophyll, its structure, and functions. http://dwb4.unl.edu/chem/chem869p/chem869plinks/www.rrz.uni-hamburg.de/biologie/b_online/e05/05e.htm

Microchiroptera Description of the echolocating bats. http://animaldiversity.ummz.umich.edu/accounts/Chiroptera/

Micrurus fulvius Description of the eastern coral snake. http://animaldiversity.ummz.umich.edu/accounts/Micrurus_fulvius/

middle lamella Explanation of cell wall structure. http://sites.bio.indiana.edu/~hangarterlab/courses/b373/lecturenotes/cellwall/cellwall.html

Mimus polyglottos Description of the northern mockingbird. http://animaldiversity.ummz.umich.edu/accounts/Mimus_polyglottos/

mitochondrion Description of a mitochondrion. http://www.nature.com/scitable/topicpage/mitochondria-14053590

mitosis Explanation of mitosis. http://www.life.umd.edu/cbmg/faculty/wolniak/wolniakmitosis.html

Mollusca Description of the molluscs. http://www.ucmp.berkeley.edu/taxa/inverts/mollusca/mollusca.php

Morchella Description of the morel. http://eol.org/pages/19583/overview

Mustelidae Description of the mustelids. http://animaldiversity.ummz.umich.edu/accounts/Mustelidae/

mycorrhiza Description of the different types of mycorrhizae. http://www.anbg.gov.au/fungi/mycorrhiza.html

Myzus persicae Description of green peach aphid. http://entnemdept.ufl.edu/creatures/veg/aphid/green_peach_aphid.htm

Nectria galligena Describes plant diseases caused by *Nectria* fungi. http://www.missouribotanicalgarden.org/gardens-gardening/your-garden/help-for-the-home-gardener/advice-tips-resources/pests-and-problems/diseases/cankers/nectria-canker-and-dieback.aspx

Nematoda Description of nematodes. http://animaldiversity.ummz.umich.edu/accounts/Nematoda/

nematophagous fungi Description of how these fungi trap prey. http://www.biological-research.com/philip-jacobs%20BRIC/

Neuroptera Description of the Neuroptera. http://www.ento.csiro.au/education/insects/neuroptera.html

nicotinamide adenine dinucleotide Explanation of the function of NAD. http://hyperphysics.phy-astr.gsu.edu/hbase/organic/nad.html

nitrogen fixation Explanation of the process. http://www.nature.com/scitable/knowledge/library/biological-nitrogen-fixation-23570419

Nothia Description of *Nothia*. http://www.abdn.ac.uk/rhynie/nothia.htm

Notophthalmus viridescens Description of the eastern newt. http://animaldiversity.ummz.umich.edu/accounts/Notophthalmus_viridescens/

nuclear envelope Description of the structure and function of the nuclear envelope. http://hyperphysics.phy-astr.gsu.edu/hbase/biology/celnuc.html

nucleolus Description of the nucleolus. http://www.ncbi.nlm.nih.gov/books/NBK9939/

Odocoileus virginianus Description of the white-tailed deer. http://animaldiversity.ummz.umich.edu/accounts/Odocoileus_virginianus/

Oomycota Introduction to the Oomycota. http://www.ucmp.berkeley.edu/chromista/oomycota.html

Opheodrys aestivus Description of the rough green snake. http://animaldiversity.ummz.umich.edu/accounts/Opheodrys_aestivus/

organochlorine Description of organochlorines. http://www.fws.gov/pacific/ecoservices/envicon/pim/reports/contaminantinfo/contaminants.html

Orthoptera Description of the Orthoptera. http://www.cals.ncsu.edu/course/ent425/library/compendium/orthoptera.html

osmotic pressure Explanation of osmosis and osmotic pressure. http://chemwiki.ucdavis.edu/Wikitexts/Simon_Fraser_Chem1%3A_Lower/Solution_Chemistry/Osmosis_and_Osmotic_Pressure

oxidation-reduction potential Explanation of oxidation-reduction potential and its measurement. http://hyperphysics.phy-astr.gsu.edu/hbase/chemical/redoxp.html

PAMP-triggered immunity Research article describing PAMPs and PRRs. http://botanika.biologija.org/predmeti/MB-4L-domacanaloga2008-clanek.pdf

Pantherophis guttatus Description of the corn snake. http://animaldiversity.ummz.umich.edu/accounts/Pantherophis_guttatus/

parapodia Description and illustration of parapodia. http://rmbr.nus.edu.sg/polychaete/feet.html

Passer domesticus Description of the house sparrow. http://animaldiversity.ummz.umich.edu/accounts/Passer_domesticus/

Passeriformes Description of the passerine birds. http://tolweb.org/Passeriformes

Passerina cyanea Description of the indigo bunting. http://animaldiversity.ummz.umich.edu/accounts/Passerina_cyanea/

Paxillus involutus Description of the brown roll-rim. http://www.first-nature.com/fungi/paxillus-involutus.php

permanent wilting point Explanation of permanent wilting point. http://www.cprl.ars.usda.gov/pdfs/Tolk-Permanent%20Wilting%20Pt-Ency%20Water%20Sci.pdf

Peromyscus leucopus Description of the white-footed deer mouse. http://animaldiversity.ummz.umich.edu/accounts/Peromyscus_leucopus/

peroxisome Description of peroxisomes and their functions. http://www.ncbi.nlm.nih.gov/books/NBK9930/

Pezizaceae List of the most common species of cup fungi. http://www.mushroomexpert.com/cups.html

Pezizomycotina Description of the Pezizomycotina. http://tolweb.org/Pezizomycotina/29296

pH Explanation of the pH scale and how it is calculated. http://www.elmhurst.edu/~chm/vchembook/184ph.html

phenology Description of phenology projects. http://www.rbge.org.uk/science/plants-and-climate-change/phenology-projects

Pheucticus ludovicianus Description of the rose-breasted grosbeak. http://animaldiversity.ummz.umich.edu/accounts/Pheucticus_ludovicianus/

Pheucticus melanocephalus Description of the black-headed grosbeak. http://animaldiversity.ummz.umich.edu/accounts/Pheucticus_melanocephalus/

phosphorus cycle Description of the cycle. http://filebox.vt.edu/users/chagedor/biol_4684/Cycles/Pcycle.html

photomorphogenesis Demonstration of photomorphogenesis in *Arabidopsis*. http://www.botany.wisc.edu/photomorphogenesis.htm

photophosphorylation Explanation of this process. http://www.tutorvista.com/content/biology/biology-iv/photosynthesis/photophosphorylation.php

photorespiration Explanation of photorespiration. http://hyperphysics.phy-astr.gsu.edu/hbase/biology/phoc.html#c5

photosynthesis Description of photosynthesis. http://biology.clc.uc.edu/courses/bio104/photosyn.htm

phyllosphere Description of the phyllosphere and its inhabitants. http://aem.asm.org/content/69/4/1875.full

PhyloCode The current draft of the PhyloCode. http://www.ohio.edu/phylocode/

phylogenetic tree Explanation of phylogenetic trees and how to read them. http://www.nature.com/scitable/topicpage/reading-a-phylogenetic-tree-the-meaning-of-41956

phylogeny Explanation of phylogeny. http://tolweb.org/tree/learn/concepts/whatisphylogeny.html

phytochrome Description of the structure and function of phytochrome. http://www.mobot.org/jwcross/duckweed/phytochrome.htm

Phytophthora infestans Description of *P. infestans* and late blight of potato. http://www.apsnet.org/edcenter/intropp/lessons/fungi/oomycetes/Pages/LateBlight.aspx

Phytophthora ramorum Description of sudden oak death. http://apps.rhs.org.uk/advicesearch/profile.aspx?pid=329

Picoides pubescens Description of the downy woodpecker. http://animaldiversity.ummz.umich.edu/accounts/Picoides_pubescens/#physical_description

Picoides villosus Description of the hairy woodpecker. http://animaldiversity.ummz.umich.edu/accounts/Picoides_villosus/

Piptoporus betulinus Description of the birch polypore. http://www.first-nature.com/fungi/piptoporus-betulinus.php

Piranga olivacea Description of the scarlet tanager. http://animaldiversity.ummz.umich.edu/accounts/Piranga_olivacea/

plant hormone Explains plant hormones with links to individual ones. http://www.ext.colostate.edu/mg/gardennotes/145.html

Plantlife Home page of Plantlife. http://www.plantlife.org.uk/

Plasmodiophora brassicae Description of clubroot. http://www.cals.ncsu.edu/course/pp728/Plasmodiophora/Plasmodiophora.html

Plasmopora viticola Description of downy mildew of grapes. http://www.apsnet.org/edcenter/intropp/lessons/fungi/Oomycetes/Pages/DownyMildewGrape.aspx

plastid Description of plastids. http://www.plantcell.org/content/11/4/549.full

Platyhelminthes Introduction to the flatworms. http://www.ucmp.berkeley.edu/platyhelminthes/platyhelminthes.html

Plethodon cylindraceus Description of the white-spotted slimy salamander. http://animaldiversity.ummz.umich.edu/accounts/Plethodon_cylindraceus/

Pleurotus ostreatus Description of the oyster fungus. http://www.first-nature.com/fungi/pleurotus-ostreatus.php

plum pox Description of the disease and its treatment. http://www.fera.defra.gov.uk/plants/publications/documents/factsheets/plumpox.pdf

Podocarpaceae Description of the Podocarpaceae. http://www.conifers.org/po/

Polyphagotarsonemus latus Description of the mite. http://entnemdept.ufl.edu/creatures/orn/broad_mite.htm

Polyporus squamosus Description of the dryad's saddle fungus. http://www.first-nature.com/fungi/polyporus-squamosus.php

polytomy Explanation of polytomy. http://evolution.berkeley.edu/evolibrary/article/phylogenetics_03

primary pigments Description of photosynthetic pigments. http://hyperphysics.phy-astr.gsu.edu/hbase/biology/pigpho.html

prostheca Description of prosthecae and prosthecate Bacteria. http://cronodon.com/BioTech/Prosthecate_bacteria.html

protoplasm The composition and functions of protoplasm. http://www.botany.uwc.ac.za/sci_ed/grade10/cells/protoplasm.htm

Pseudacris crucifer Description of the spring peeper. http://animaldiversity.ummz.umich.edu/accounts/Pseudacris_crucifer/

Pseudomonas syringae* pv. *phaseolicola Description of halo blight of the common bean. http://www.ncbi.nlm.nih.gov/pubmed/21726364

Pterygota Description of the Pterygota and of all the orders. http://homepages.abdn.ac.uk/nathist.museum/classify/animalia/uniramia/pterygota/

Puccinia graminis Description of the fungus causing black stem rust. http://web.aces.uiuc.edu/vista/pdf_pubs/108.PDF

Pucciniomycetes Description of the Pucciniomycetes. http://tolweb.org/Pucciniomycetes/51246

Pucciniomycotina Description of the Pucciniomycotina. http://tolweb.org/Urediniomycetes

punctuated equilibrium Explanation of punctuated equilibrium. http://evolution.berkeley.edu/evosite/evo101/VIIA1bPunctuated.shtml

pyrethrins Description of pyrethrins. http://pmep.cce.cornell.edu/profiles/extoxnet/pyrethrins-ziram/pyrethrins-ext.html

pyrethroid insecticides Description of pyrethroids. http://citybugs.tamu.edu/factsheets/ipm/ent-6003/

Pyrrophyta Description of dinoflagellates. http://botany.si.edu/projects/algae/classification/DINOPHYTA.htm

Pythium Description of the root rot caused by *Pythium*. http://extension.psu.edu/pests/plant-diseases/all-fact-sheets/pythium

Quadraspidiotus perniciosus Description of San José scale. http://jenny.tfrec.wsu.edu/opm/displaySpecies.php?pn=490

Ralstonia solanacearum Description of the bacterium and bacterial wilt. http://extension.psu.edu/pests/plant-diseases/all-fact-sheets/ralstonia

Rana clamitans clamitans Description of the green frog. http://animaldiversity.ummz.umich.edu/accounts/Lithobates_clamitans_clamitans/

Rana palustris Description of the pickerel frog. http://animaldiversity.ummz.umich.edu/accounts/Lithobates_palustris/

Rana temporaria Description of the common frog. http://animaldiversity.ummz.umich.edu/accounts/Rana_temporaria/

Raunkiær, Christen Christensen Explanation of Raunkiær's life-form classification. http://courses.eeb.utoronto.ca/eeb337/B_How/307B2life_forms.html

red algae Introduction to the red algae. http://www.ucmp.berkeley.edu/protista/rhodophyta.html

Red List Description of the Red List and explanation of how it is used. http://www.iucn.org/about/work/programmes/species/our_work/the_iucn_red_list/

Reduviidae Description of reduviids. http://bugguide.net/node/view/166

resurrection plant Description of resurrection plants. http://www.howplantswork.com/2010/04/06/resurrection-plants-how-do-these-plants-come-back-to-life-after-near-total-dehydration/

Rhizoctonia solani Description of *R. solani*. http://www.cals.ncsu.edu/course/pp728/Rhizoctonia/Rhizoctonia.html

Rhododendron Classification of the genus *Rhododendron*. http://www.flounder.ca/FraserSouth/Goetsch-Eckert-Hall.asp

ribulose-1,5-biphosphate carboxylase Description of RuBisCo and its action. http://hyperphysics.phy-astr.gsu.edu/hbase/organic/rubisco.html

Ricinus Description of castor oil plant and its products. http://www.kew.org/news/kew-blogs/millennium-seed-bank/Notes-on-the-illustrious-castor-bean.htm

root nodule Explanation of the formation and operation of root nodules. http://www.plantphysiol.org/content/124/2/531.full

Rotifera Introduction to the Rotifera. http://www.ucmp.berkeley.edu/phyla/rotifera/rotifera.html

***r*-selection** Explanation of *r* and *K* selection. http://www.bio.miami.edu/tom/courses/bil160/bil160goods/16_rKselection.html

Saccharomycotina Description of the ascomycete yeasts. http://tolweb.org/Saccharomycotina/29043

Salamandridae Description of the 'true' salamanders and newts. http://tolweb.org/Salamandridae/15445

saline soil Description of saline, sodic, and saline-sodic soils. http://www.ext.colostate.edu/pubs/crops/00521.html

salt stress Explanation of salt stress and its effects. http://www.faculty.ucr.edu/~jkzhu/articles/2007/ELS%20Zhu.pdf

saponin Description of saponins. http://www.ansci.cornell.edu/plants/toxicagents/saponin.html

Sayornis phoebe Description of the eastern phoebe. http://animaldiversity.ummz.umich.edu/accounts/Sayornis_phoebe/

Scarabeidae Detailed description of chafers. http://bugguide.net/node/view/187

Sceloporus undulatus Description of the eastern fence lizard. http://animaldiversity.ummz.umich.edu/accounts/Sceloporus_undulatus/

Sciuridae Description of the squirrels. http://animaldiversity.ummz.umich.edu/accounts/Sciuridae/

Sciurus carolinensis Description of the grey squirrel. http://animaldiversity.ummz.umich.edu/accounts/Sciurus_carolinensis/

Sciurus vulgaris Description of the red squirrel. http://animaldiversity.ummz.umich.edu/accounts/Sciurus_vulgaris/

Sclerotium cepivorum Description of the fungus and the disease it causes. http://www.cals.ncsu.edu/course/pp728/sclerotium_cepivorum/Sclerotium_cepivorum.html

secondary metabolite Explanation of plant secondary metabolites with a list of the most common. http://www.biologyreference.com/Re-Se/Secondary-Metabolites-in-Plants.html#b

sedimentary rock Description of sedimentary rocks. http://hyperphysics.phy-astr.gsu.edu/hbase/geophys/sedime.html

seed fern Description of seed ferns. http://taggart.glg.msu.edu/bot335/sfern.htm

seed plants Introduction to the seed plants. http://www.ucmp.berkeley.edu/seedplants/seedplants.html

Selasphorus rufus Description of the rufous hummingbird. http://animaldiversity.ummz.umich.edu/accounts/Selasphorus_rufus/

self-incompatibility Explanation of self-incompatibility. http://users.rcn.com/jkimball.ma.ultranet/BiologyPages/S/SelfIncompatibilty.html

Sialia sialis Description of the eastern bluebird. http://animaldiversity.ummz.umich.edu/accounts/Sialia_sialis/

silicon Explanation of the role of silicon in plants. http://www2.fiu.edu/~chusb001/GiantEquisetum/Silicon_and_Plant_Health.html

Sistrurus miliarius Description of the pigmy rattlesnake. http://animaldiversity.ummz.umich.edu/accounts/Sistrurus_miliarius/

Sitta carolinensis Description of the white-breasted nuthatch. http://animaldiversity.ummz.umich.edu/accounts/Sitta_carolinensis/

soil monolith Description of how to make a soil monolith. http://forces.si.edu/soils/02_03_02.html

soil testing A list of 10 simple soil tests. http://www.organicgardening.com/learn-and-grow/10-easy-soil-tests?page=0,0

Sparassis crispa Description of the cauliflower fungus. http://www.first-nature.com/fungi/sparassis-crispa.php

spindle Description of the spindle apparatus. http://www.nature.com/scitable/definition/spindle-fibers-304

Spongospora subterranea Description of powdery scab and the organism causing it. http://www.cals.ncsu.edu/course/pp728/Spongospora/Spongospora_subterranea.htm

spore print Instructions for making a spore print. http://www.mushroomexpert.com/spore_print.html

Staphylinidae Description of rove beetles. http://entnemdept.ufl.edu/creatures/misc/beetles/rove_beetles.htm

stem Description of stems and their structure. http://www.botany.uwc.ac.za/sci_ed/grade10/anatomy/stems.htm

Streptopelia decauocto Description of the collared dove. http://www.issg.org/database/species/ecology.asp?si=1269&fr=1&sts=&lang=EN

Strigiformes Description of the owls. http://animaldiversity.ummz.umich.edu/accounts/Strigiformes/

Stropharia aeruginosa Description of the verdigris agaric. http://www.first-nature.com/fungi/stropharia-aeruginosa.php

Sturnus vulgaris Description of the starling. http://animaldiversity.ummz.umich.edu/accounts/Sturnus_vulgaris/

sulphur cycle Description of the cycle. http://www.enviroliteracy.org/article.php/1348.html

sulphur fungus Description of sulphur fungus. http://www.wvdnr.gov/wildlife/magazine/archive/06summer/sulphurshelf.pdf

sulphur oxidation Explanation of soil sulphur oxidation. http://filebox.vt.edu/users/chagedor/biol_4684/Cycles/Soxidat.html

surface runoff Explanation of runoff processes. http://ga.water.usgs.gov/edu/runoff.html

surface tension Explanation of surface tension. http://hyperphysics.phy-astr.gsu.edu/hbase/surten.html

syconium Explanation of fig pollination and description of the syconium. http://www.botgard.ucla.edu/html/botanytextbooks/economicbotany/Ficus/

2,4,5-T Description of the herbicide and the controversy surrounding it. http://www.ch.ic.ac.uk/rzepa/mim/environmental/html/245t.htm

Tachycineta bicolor Description of the tree swallow. http://animaldiversity.ummz.umich.edu/accounts/Tachycineta_bicolor/

Talpa europaea Description of the mole. http://animaldiversity.ummz.umich.edu/accounts/Talpa_europaea/

tannins Information on tannins. http://www.ansci.cornell.edu/plants/toxicagents/tannin.html

Taxidea taxus Description of the American badger. http://animaldiversity.ummz.umich.edu/accounts/Taxidea_taxus/

Terrapene carolina Description of the common box turtle. http://animaldiversity.ummz.umich.edu/accounts/Terrapene_carolina/

Thamnophis sirtalis Description of the common garter snake. http://animaldiversity.ummz.umich.edu/accounts/Thamnophis_sirtalis/

three-domain system Explanation of the system. http://www.usc.edu/org/cosee-west/Nov30_2011/Three%20domains%20of%20life.pdf

Thryothorus ludovicianus Description of the Carolina wren. http://animaldiversity.ummz.umich.edu/accounts/Thryothorus_ludovicianus/

Thysanoptera Description of thrips. http://www.cals.ncsu.edu/course/ent425/library/compendium/thysanoptera.html

tissue culture Outline of the process. http://www.flytrapcare.com/tissue-culture-basics.html

Tortricidae Description of the Tortricidae. http://www.tortricidae.com/morphology.asp

Torymidae Description of the Torymidae. http://www.nhm.ac.uk/research-curation/research/projects/chalcidoids/torymidae.html

Trachemys scripta Description of the pond slider. http://animaldiversity.ummz.umich.edu/accounts/Trachemys_scripta/

Tracheophyta An alphabetical list of all tracheophytes. http://www.sms.si.edu/irlspec/Phyl_Trache1.htm

tree lungwort Description of tree lungwort. http://www.treesforlife.org.uk/forest/species/tree_lungwort.html

Trialeurodes vaporariorum Description of the pest, damage, and control methods. http://apps.rhs.org.uk/advicesearch/profile.aspx?PID=193

triallate Description of triallate. http://pmep.cce.cornell.edu/profiles/extoxnet/pyrethrins-ziram/triallate-ext.html

Trichoderma Description of the *Trichoderma*. http://www.biocontrol.entomology.cornell.edu/pathogens/trichoderma.php

tristyly Explanation of tristyly. http://biology.duke.edu/rausher/tristyly.htm

Troglodytes aedon Description of the house wren. http://animaldiversity.ummz.umich.edu/accounts/Troglodytes_aedon/

Turbellaria Description of the Turbellaria. http://animaldiversity.ummz.umich.edu/accounts/Turbellaria/

Turdus migratorius Description of the American robin. http://animaldiversity.ummz.umich.edu/accounts/Turdus_migratorius/

unnatural classification A brief history of taxonomy explaining the difference between natural and unnatural systems. http://www.bihrmann.com/caudiciforms/div/hist2.asp

Ustilaginomycetes Description of the smut fungi. http://tolweb.org/Ustilaginomycetes

Utricularia Description of the bladders and how they work. http://botany.org/Carnivorous_Plants/Utricularia.php

valency Explanation of how to calculate valency. http://classroom.synonym.com/calculate-valency-2790.html

Vavilov, Nikolai Ivanovich Biography of Vavilov. http://russiapedia.rt.com/prominent-russians/science-and-technology/nikolay-vavilov/

Venturia inaequalis Description of the fungus causing apple scab. http://www.apsnet.org/edcenter/intropp/lessons/fungi/ascomycetes/Pages/AppleScab.aspx

vernalization Explanation of vernalization. http://www.plantcell.org/content/16/10/2553.full

Verticillium Description of verticillium wilt. http://www.eppo.int/QUARANTINE/fungi/Verticillium_alboatrum/VERTSP_ds.pdf

vinclozolin Description of vinclozolin. http://pmep.cce.cornell.edu/profiles/extoxnet/pyrethrins-ziram/vinclozolin-ext.html

virus Description of a virus. http://www.ucmp.berkeley.edu/alllife/virus.html

Vitrinidae Description of the glass snails. http://molluscs.at/gastropoda/terrestrial.html?/gastropoda/terrestrial/vitrinidae.html

Vulpes vulpes Description of the red fox. http://animaldiversity.ummz.umich.edu/accounts/Vulpes_vulpes/

water potential Explanation of water potential. http://biology.kenyon.edu/HHMI/Biol113/movemet%20of%20water%20in%20plants.htm

wind erosion Description of the causes and consequences of wind erosion. http://milford.nserl.purdue.edu/weppdocs/overview/wndersn.html

wrack Description of wracks. http://www.pznow.co.uk/marine/wrack.html

xylem Description of xylem. http://www.botany.uwc.ac.za/sci_ed/grade10/plant_tissues/xylem.htm

Zenaida macroura Description of the mourning dove. http://animaldiversity.ummz.umich.edu/accounts/Zenaida_macroura/

zineb Description of zineb. http://pmep.cce.cornell.edu/profiles/extoxnet/pyrethrins-ziram/zineb-ext.html

Zonotrichia albicollis Description of the white-throated sparrow. http://animaldiversity.ummz.umich.edu/accounts/Zonotrichia_albicollis/

Zoopagales Description of the Zoopagales. http://zygomycetes.org/index.php?id=8

Zosterophyllopsida Description and illustration of Zosterophyllopsida. http://palaeos.com/plants/lycopodiophyta/zosterophyllopsida.html

Zygomycota Description of the Zygomycota. http://tolweb.org/Zygomycota

Illustration Credits

Endpaper illustrations by Alan Bryan and Marjorie Leggitt

Alan Bryan: pages 11, 13, 14, 28 (left), 37, 40, 44, 72 (bottom), 76, 87, 99, 104, 132, 136, 137, 140, 150, 159, 175, 183, 187, 209, 210, 229, 243, 249, 265, 267 (top), 282, 285 (right), 294, 296, 341, 352, 356, 360, 365, 370, 384, 389, 390, 395, 412, 431, 440, 442, 454, 455, 457, 458, 468, 469 (right), 475 (right), 476, 482, 485 (right), 489, 505, 513, 519

Dave Carlson: pages 33, 46, 67, 78, 115, 236, 375, 393, 400, 502

Anna Eshelman: pages 27, 98, 121, 127, 138, 149, 163, 169, 181, 257, 267 (bottom), 283, 324, 347, 349, 362 (top), 394, 399, 421, 450, 464, 467, 469 (left), 472, 485 (left), 512, 528

Kate Francis: pages 28 (right), 110, 129, 141, 174, 190, 199, 211, 212, 241, 251, 258, 336, 340, 373, 404, 414, 422, 460, 461, 463, 475 (left), 526

Marjorie Leggitt: pages 39, 226, 388, 523

Mike Morgenfeld: pages 59, 72 (top), 173, 184, 224, 231, 239, 285 (left), 362 (bottom), 497

About the Author

ANDREW GRAHAM-WEALL

Michael Allaby is an enthusiastic, prolific, award-winning science writer who has written, edited, or co-authored over 100 books on environmental science. Of these, 17 were about atmospheric science. His 2-volume *Encyclopedia of Weather and Climate* and *Dangerous Weather: Hurricanes* won awards and *DK Guide to Weather* won the 2001 Junior Prize of the Aventis Prizes for Science Books. His *Plants and Plant Life* won *Booklist* Editor's Choice for 2001. He is editor of four science dictionaries for Oxford University Press. Before becoming a full-time writer in 1973 he worked in the police force, the RAF, and as an actor. See michaelallaby.com for more information.